OSCILLOSCOPE MEASURING TECHNIQUE

OSCILLOSCOPE MEASURING TECHNIQUE

PRINCIPLES AND APPLICATIONS OF MODERN CATHODE RAY OSCILLOSCOPES

J. CZECH

1965

SPRINGER-VERLAG BERLIN HEIDELBERG GMBH

Translated from the German by A. Smith Hardy, Eindhoven

This book contains xviii + 620 pages and 659 illustrations.

U.D.C. No. 621.317.755

Library of Congress Catalog Card Number: 66-18568

Original German edition:

English edition:

Originally published by N.V. Philips' Gloeilampenfabrieken, Eindhoven, The Netherlands in 1965
Softcover reprint of the hardcover 1st edition 1965

ISBN 978-3-662-38761-0 ISBN 978-3-662-39653-7 (eBook)
DOI 10.1007/978-3-662-39653-7

First published in England

PREFACE

The first edition of the German version of my book DER ELEKTRONENSTRAHL-OSZILLOGRAF was sold out within a year and stocks of the reprint were exhausted within the following ten months. The English edition THE CATHODE RAY OSCILLOSCOPE was also received with considerable interest and generally good reviews. It could, therefore, be assumed that the presentation of the subject matter was such as to satisfy the highest standards. Since the completion of the original manuscript the technique of the electron cathode ray oscilloscope has in many ways substantially advanced and the field of application of this instrument has consequently widened so that it seemed no longer justified once again to issue a reprint of this book or even consider a so-called enlarged and revised edition. Thoroughly revising parts of my first book and enlarging it with essential supplementary material – in particular with regard to modern measuring techniques – made it necessary to publish a new version in 1959 under the title OSZILLOGRAFEN-MESSTECHNIK. The great success of the English translation of the first edition also warranted the issue of an English translation of the new version. A translation of this kind and the work involved unfortunately requires a not inconsiderable amount of time. We have, however, made use of this time by making some further smaller and larger alterations to the text as well as adding important supplementary material. Thus this version likewise presents a comprehensive picture of up-to-date developments in equipment and measuring technique. This meant in detail that some chapters of Part I, again dealing with the instrument itself, were incorporated without substantial alterations as developments here were more or less completed. The new English version, however, is enlarged by a chapter on "Display Storage Tubes". Others again – in particular the chapters on "Time Base Generator" and "Deflection Amplifier" – had to be thoroughly revised and enlarged. Furthermore, in the present English version a new chapter was introduced discussing the sampling oscilloscope. Part II dealing with general measuring techniques has also been supplemented.

Part III in particular has been greatly enlarged. Examples from different fields of practical oscilloscope measuring techniques including several in which measurements of non-electrical values are discussed, are now described in detail in 18 chapters and the evaluation of the oscillograms is discussed at great length. Chapter 19 – "Investigation into TV Technique" – has been completely re-written for the English edition, with a great number of new oscillograms.

Owing to the considerably higher measuring accuracy of modern cathode ray oscilloscopes the photographic recording for quantitative evaluation gains in importance. This is discussed in detail in one paragraph of Part IV. The English version of this Part has again been somewhat enlarged by a description of the polaroid system. A further chapter of this Part deals with the large screen projection of oscillographs which is particularly useful for training and instruction purposes.

The building instructions for two oscilloscopes and a simple microscope time-base generator given in Part IV of the first edition have been omitted in this edition since the construction of home-made measuring instruments plays a minor roll in modern measuring techniques. On the one hand, the construction of individual high capacity oscilloscopes involves expenses far in excess of the costs of a factory-built instrument, and on the other hand, there are instruments of average qualities on the market at very reasonable prices. Home-made construction is thus only important in those rare cases where building is done out of sheer interest in the job for its own sake; in that case, however, the technical journals repeatedly give ample advice.

I should like to thank here all those who kindly co-operated in compiling this book or furnished material for it. I am particularly indebted to the management of Philips Industrie Elektronik GmbH., Hamburg, who put at my disposal the measuring equipment and other means necessary for recording the oscillograms. I am also grateful to Messrs. Telefunken GmbH, Ulm, and Valvo GmbH, Hamburg, for supplying me with valve data. By kind permission of Philips Gloeilampenfabrieken, Eindhoven, I was able to use numerous data from their trade information quoted in detail in the bibliography. I must not forget to thank the translator Mr. A. Smith-Hardy for the complete translation of my text and also wish to express my gratitude to Mr. Harley Carter, A.M.I.E.E., for going through the translation with such care and understanding.

Finally, I should like to express my gratitude to the publishers for the admirable production of the book and for their generous compliance with my wishes with regard to lay-out and above all for their ready co-operation.

Hamburg, August 1964

J. CZECH

TABLE OF CONTENTS

PART I

THE CATHODE RAY OSCILLOSCOPE

PART II

GENERAL MEASURING TECHNIQUE

PART III

PRACTICAL EXAMPLES

PART IV

PHOTOGRAPHIC RECORDING AND LARGE PICTURE PROJECTION OF OSCILLOGRAMS

Part I

THE CATHODE RAY OSCILLOSCOPE

CHAPTER 1

CONSTRUCTION OF CATHODE RAY OSCILLOSCOPES

The oscilloscope is an instrument which, in addition to the actual measuring device — the cathode ray tube — contains all the auxiliary devices requisite for the general run of tasks. These include the power supply unit, at least one vertical deflection amplifier (Y-axis) and the time base generator. This is often combined with a horizontal amplifier (X-axis).

Fig. 1-1 is a block diagram to illustrate the way in which the individual sections work in combination (the mode of operation and applications of the oscilloscope will be dealt with in detail in later chapters). The power supply unit delivers all the supply voltages (AC and DC) needed for working the cathode ray tube (and the auxiliary devices, as well as E.H.T. for the cathode ray tube) [1].

Apart from the component elements named, the diagram also indicates the synchronizing device, which enables the time base frequency to be locked either with the vertical deflection voltage, with a voltage taken from an external source or with the mains voltage according to choice.

In modern oscilloscopes provision is always made for triggering the time base singly, periodically, or at random as well as in a controlled manner. The time base velocity can be continuously adjusted independently of the frequency of the signal voltage. Part I, Chapter 4 deals with these features in detail. The synchronizing device is nearly always combined with the time base, both in circuitry and in practical construction. This has been brought out particularly in Fig. 1-1 in order to emphasize its many possible app'ications, which, as experience has shown, are seldom exploited to the full.

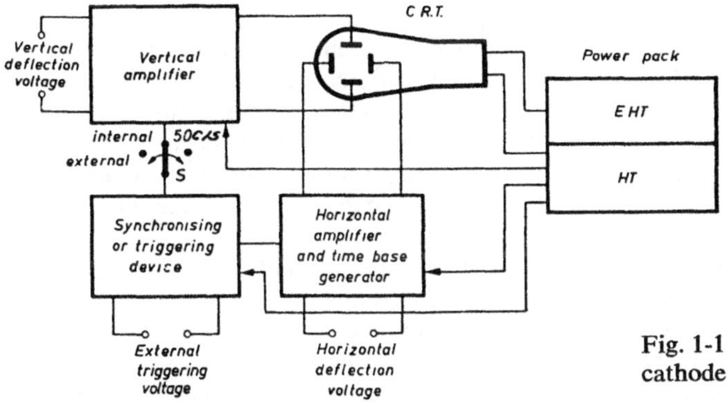

Fig. 1-1 Schematic lay-out of a cathode ray oscilloscope

Fig. 1-2 shows the exterior of an oscilloscope with its control panel. The probe at the input permits the signal voltage to be attenuated in a scale of 20 : 1 with a uniform frequency response. On this picture the viewing hood is turned upwards, so that the

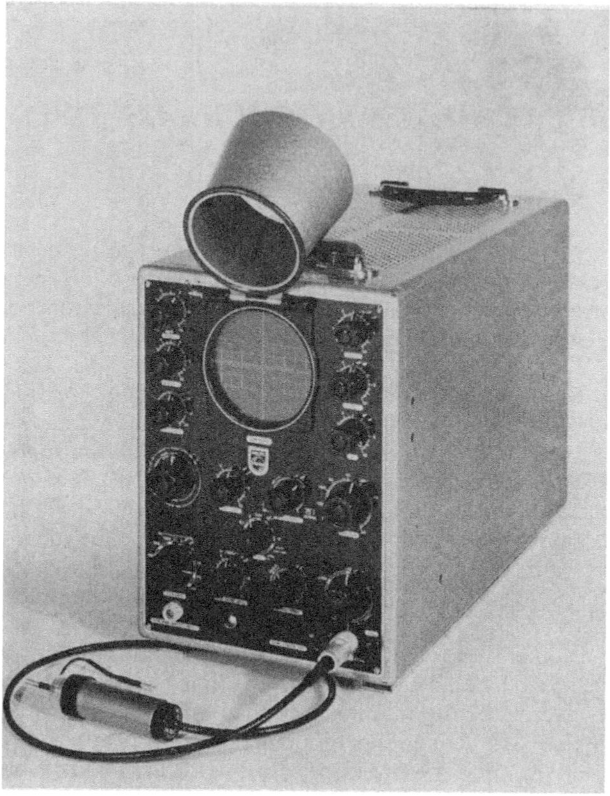

Fig. 1-2 Front view of Philips "GM 5662" oscilloscope

illuminated graticule can be seen. The heat generated in the apparatus can escape through vents liberally provided in the casing.

Fig. 1-3 gives an internal view of the structure of the well-known Philips "GM 5662" oscilloscope in which the individual components are indicated as in Fig. 1-1. In particular, the mains transformer, the two EY 51 valves for the extra high tension rectifiers and the valves and circuit components of the signal amplifier are clearly to be seen. The high upper frequency limit (14 Mc/s) of this amplifier demands short connections and a construction which ensures the lowest possible capacitances of the circuit components to the chassis.

The mains transformer, the filter chokes of the power supply unit and the potentiometer for the optimum correction of tube astigmatism can be seen. Both HF-chokes which prevent the radiation of harmonics of the time base voltage lie directly between transformer and mains connection (not visible here). This picture also shows that the mains transformer is of generous rating. This is required to keep its stray field at a minimum, so that undesirable deflection of the electron beam may be avoided. The cathode ray tube is further screened by a high grade mu-metal cylinder in order to reduce, as far as possible, interference by magnetic fields.

Increased use is being made of printed circuits in the development of cathode ray

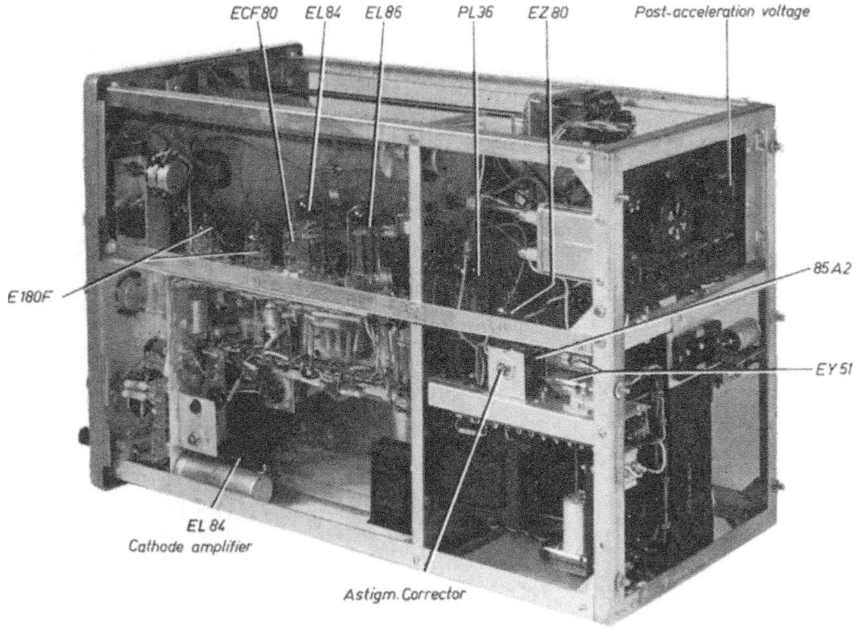

Fig. 1-3 Interior view of "GM 5662" wide-band oscilloscope showing vertical amplifier and power pack

oscilloscopes. Their advantages, particularly freedom from faults and the fact that they need but little adjustment, are of great importance here. Fig. 1-4 shows the circuit and construction of such an oscilloscope.

The individual component parts, such as the cathode ray tube, power supply unit time base unit and signal amplifier will be described in detail in the following parts of the book with the help of basic circuit diagrams.

The basic elements of a cathode ray oscilloscope described in Fig. 1-1 are, in the types of apparatus shown in Figs. 1-2, 1-3, and 1-4, assembled into complete instruments which are unalterable in their essential properties. This is the normal construction for oscilloscopes in the lower and medium price ranges. The need to limit the cost of such instruments must, of course, lead to some restriction in performance in certain fields of application, depending on the price of the particular apparatus, such as, for example small oscilloscopes for radio and television servicing, LF or HF oscilloscopes and so forth.

Higher priced instruments and particularly the most expensive, are designed for great versatility in application; hence the very considerable outlay involved. This requirement is satisfied to a very great extent by supplying the amplifiers and sometimes even the sweep time base units, in the form of replaceable plug-in units. Often it is only the preamplifier stages or the time base generator which are replaceable, the output stages, whose bandwidth must be broad enough for all applications ranges, forming a permanent part of the basic set.

Fig. 1-4 Philips "GM 5602" oscilloscope. Showing latest form of frame construction and printed circuit

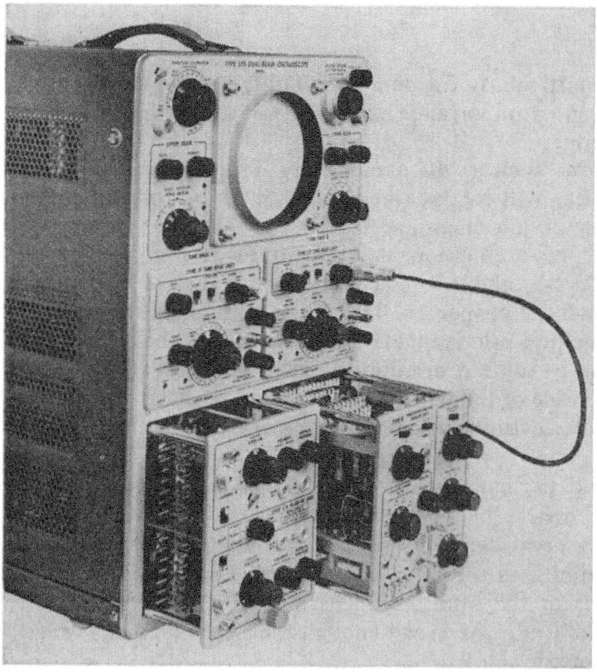

Fig. 1-5 Tetronix oscilloscope, type 555.

In such oscilloscopes, only the cathode ray tube and power supply unit and usually the output stages for X- and Y-deflection are permanent. Sometimes a sweep unit having a suitably wide choice of time base ranges is also a permanently mounted element, but in most cases it is interchangeable.

Fig. 1-5 shows oscilloscope type 555 made by the firm of Tektronix, who produce a whole range of such units which are typical oscilloscopes with plug-in units. The unit shown is the basic unit of a twin-beam oscilloscope, in which both Y-preamplifiers and both sweep units are replaceable. The power supply unit is mounted on the oscilloscope carriage as a separate unit. In the illustration the plug-in units of the amplifier are shown withdrawn for convenience. The left-hand unit is a double image-difference wideband preamplifier (dual trace preamplifier plug-in unit CA) which together with the output stage, has an upper cut-off frequency of 24 Mc/s corresponding to a rise time of 15 ns. With the single channel preamplifier K the upper cut-off frequency is 30 Mc/s, corresponding to 12 ns rise time. The right-hand plug-in unit (R) is a special unit for measuring the pulse behaviour of transistors. It contains all the voltage sources for the operating currents of the transistors, preamplifier stages, and a mercury-switch pulse generator for signal pulses with a rise time of 5 ns.

In addition to Tektronix, EMI (UK), Hewlett-Packard (Fig. 4-97), Siemens, Telequipment (UK) etc. all produce similar oscilloscopes with replaceable plug-in units.

By using units possessing the properties most desirable for the particular investigation, the outstanding performance of the highest-grade equipments, which, for instance, use high acceleration voltages to give a very bright trace, can be used for the most varied fields of application without any debasement of these optimum characteristics. They have therefore gained wide popularity. In recent years increasing numbers of medium-priced equipments have been put on the market with similar replaceable units.

CHAPTER 2

THE CATHODE RAY TUBE [1)]

2. 1 The cathode

An indirectly heated cathode is generally used as the electron source [1]. The electrons are emitted from an oxide coating on the tip of a small nickel cylinder sealed at one end. The coated tip is clearly shown in Fig. 2-1*a* as a white surface marked (*k*). The filament (*f*) or "heater" is reproduced in Fig. 2-1*b*, the cathode cylinder having been removed for this purpose. As can be seen, it is bent double. This is done to obtain maximum heating by the most economical means. To insulate it from the cathode, the filament is coated with a layer of kaolin. The filament can be heated by direct or alternating current.

Fig. 2-1*b* Heating filament of DG 10-6 C.R.T. Heater supply 6.3V/0.3A

Fig. 2-1*a* Electrode system of the Philips DG 10-6 C.R.T. showing the cathode, *k* = electron emitting layer; *g* = control grid with end partly opened; *A* = aperture for the electron beam

2. 2 Electrode arrangement - electron optics

In its simplest form a cathode ray tube might be built up as indicated in Fig. 2-2. Opposite the indirectly heated cathode is a metal disc or diaphragm with a central aperture. This is the anode, which must be maintained at a high positive voltage with respect to the cathode. The negative electrons emitted by the cathode are attracted by the positive anode. The great majority of the electrons land on the metal disc and can return to the cathode via the voltage source. Some however, accelerated by the attrac-

---

1) First described by Prof. Braun in 1897 and called "Braun tube" after him. The term cathode ray tube is now usually employed.

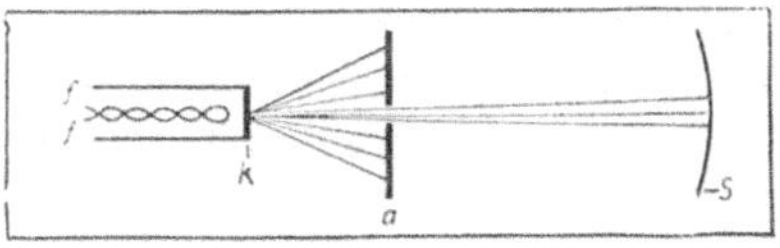

Fig. 2-2 Diagram of the electrode system of a simple C.R.T.
f = filament or heater; k = cathode; a = anode; S = fluorescent screen

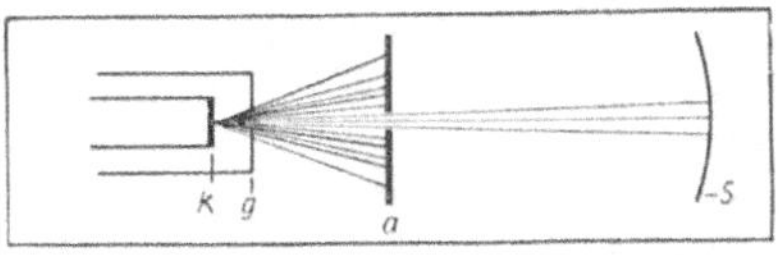

Fig. 2-3 Arrangement of electrodes of C.R.T. with control grid (Wehnelt cylinder). k = cathode; g = grid (Wehnelt cylinder) a = anode; S = luminescent screen

tion of the anode voltage, pass through the opening in the anode disc and proceed in a straight line to the end wall of the tube. If coated with a luminescent material, this part of the envelope becomes luminescent when the electron beam impinges on it.

Within certain limits, the brightness of the spot is proportional to the density of the electron beam. The colour of the luminescent glow depends upon the properties or the composition of the screen material. By analogy with radio valves, the arrangement of electrodes described corresponds to that of a diode. To control the strength of the electron beam, another electrode is added, consisting of a metal cylinder which is placed around the cathode and closed at the outer end except for one small aperture. This arrangement is shown schematically in Fig. 2-3.

This electrode, called the Wehnelt cylinder after its inventor, or simply the grid (g) by analogy with the corresponding electrode in an amplifying valve, is maintained at a negative (direct) potential with respect to the cathode. By varying this potential it is possible to control the intensity of the electrons attached by the positive anode through the opening in the Wehnelt cylinder. The voltage at which the electron beam is just suppressed, or cut off, is known as the grid cut-off voltage. This voltage is given in all tube data published by the manufacturers. In standard commercial tubes it lies between —30 and —100 V. With a grid bias of 0 V the electron beam is not influenced, and the majority of the emitted electrons then pass through the aperture in the electrode. Figs. 2-4*a*, *b* and *c* show schematically the effect on the electron beam of different grid bias voltages, and Fig. 2-5 illustrates the actual construction of the Wehnelt cylinder [2].

It is important to remember that the voltage on the Wehnelt cylinder must never be

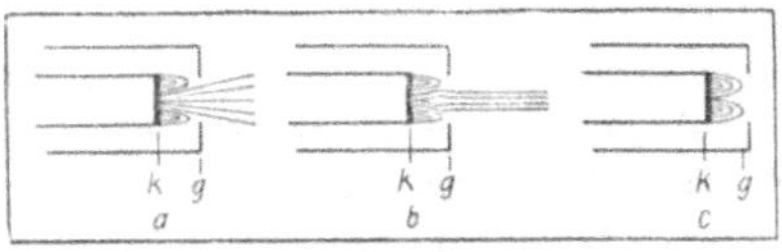

Fig. 2-4 Control of the electron current of a C.R.T. by means of the grid
a) bias *O*, maximum electron emission
b) medium bias, medium emission
c) Cut-off bias, electron emission suppressed

Fig. 2-5 Control grid of a C.R.T. g = Wehnelt cylinder (grid); A = aperture for electron beam; B = opening for assembly

positive, as otherwise grid current will flow and the cathode may be damaged by excessive emission. The electrons emitted by the cathode represent like charges; they therefore mutually repel each other and the beam tends to diverge. It is essential, however, to have as fine a beam as possible, and hence special measures are taken to concentrate the electrons into as narrow a beam as can be obtained. It is clear that the beam will be narrower if the electrons are made to travel faster, since they will then have less time to scatter before reaching the screen.

Thus, to obtain a small spot on the screen, it is desirable to make the anode potential as high as possible. Reasons of economy, however, set a practical limit to this. Only by introducing further electrodes between the anode and the grid can a sufficiently narrow beam been obtained with a reasonable anode voltage. Fig. 2-6*a* illustrates the basic arrangement and action of these electrodes. Between anode and grid a cylindrical electrode is added and receives a positive voltage with respect to the cathode, namely about one third of the anode voltage. A potential difference now exists between the anode and the additional electrode amounting to about two-thirds of the anode voltage, and a corresponding electrostatic field is set up.

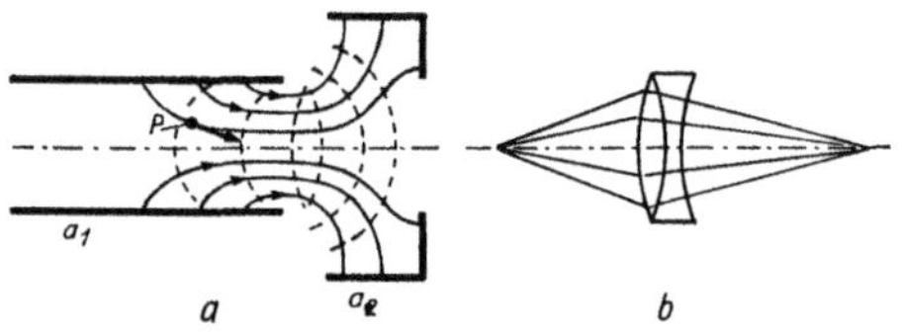

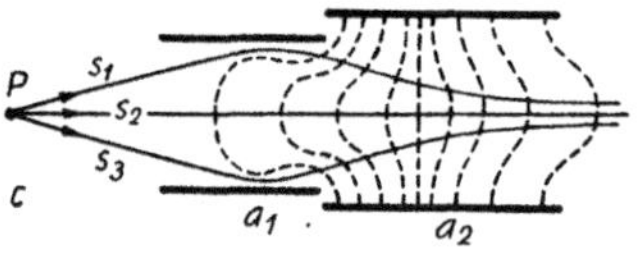

Fig. 2-6 Electron optics – light optics.
a) Voltage relationship between anode a_2 and auxiliary anode a_1 and its effect on the electron beam.
b) Analogy with *a*) behaviour of light rays in a glass lens.
c) Influence of electrostatic field on electron beam for rays S_1, S_2 and S_3, emanating from point *P*

The lines of force of this field are shown in Fig. 2-6*a*. The arrows indicate the direction from the low to the higher potential and also the direction of the force acting on the electrons. The concentrating effect of this new electrode on the electron beam can plainly be recognized. The planes that cut the lines of force vertically and have the same potential at all points, are known as "equipotential planes". The cross-sections through these planes are shown as broken lines in Fig. 2-6*a*. Comparison with Fig. 2-6*b* makes it clear that these surfaces have the same form as the boundary surface of a corresponding optical lens. Their influence on the electron beam is in fact very similar to the influence of an optical lens on rays of light, and their action is termed "electron optics". Fig. 2-6*c* shows the influence of the "electron lens" on the electron beam. Altering the voltage on the focussing electrode a_1 varies the electrostatic field between anode and this electrode, thus regulating the focussing effect on the electron beam.

In this way the spot of light on the screen can be sharply adjusted for definition, just as, for instance, an object to be photographed is focused by a camera. Where particularly sharp focussing of the spot is demanded, it is not uncommon in practice to employ not merely one "electron lens" but, as in light optics, to use a combination of such lenses involving complicated electrode systems. Fig. 2-7 shows different stages of spot defini-

Fig. 2-7 The spot on the screen with different voltages on anode a_1

tion when the voltage on the first (focusing) anode is varied in the region of sharp focus [3] [4] [5].

The spot on the screen shown here is actually the reduced image of the emitting surface of the cathode. It follows from this, of course, that it must also be possible to reproduce in this way a magnified image of the cathode.

With the aid of a suitable electron lens it is in fact possible to achieve magnifications of this sort far exceeding the capabilities of optical microscopes [6].

Most oscilloscopes are able to reproduce an enlarged image of the cathode on the screen; or of a section of it, which is determined by the apertures in the electrodes through which the beam passes. When the grid bias is set at zero, i.e. when the tube is "turned up", it requires only a low potential on a_1 to obtain distinctly on the screen a magnified image of the cathode. In this way it is possible to gain an impression of the condition of the cathode in a given tube. "C.R.T. spot photographs" of this sort are shown in Fig. 2-8*a* and *b*.

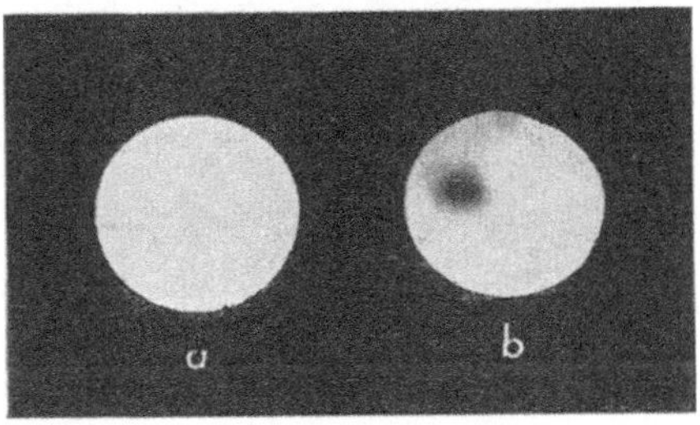

Fig. 2-8 Image of the cathode of a C.R.T. *a*) Uniformly emitting cathode. *b*) Cathode with an area of low emission

2. 3 Deflecting the beam

The electron beam can be deflected from its path by electromagnetic and also by electrostatic fields. For electromagnetic deflection, two coils are fitted opposite each other over the neck of the tube, and the current which causes deflection of the spot on the screen is passed through them. This method of deflecting has the disadvantage that, even when using coils with a large number of turns, a fairly large current and therefore relatively high power is needed for deflecting the beam. Moreover, self-induction in the coils may cause extra distortion of a possibly non-sinusoidal alternating current passing through them. Since it is precisely the behaviour of the current which is to be observed, magnetic deflection is only used in television or for special tasks such as observation of very strong currents [7].

To deflect the electron beam, and thus the spot on the screen, by electrostatic fields, a pair of deflection plates is arranged as close behind the anode a_2 as possible, in such a way that when no potential is applied between the plates the electron beam passes centrally between them. In order to deflect the beam the voltage to be examined is applied between the deflection plates. The deflection plates in effect constitute a capacitor with the vacuum of the tube as dielectric.

However, its capacitance, together with that of the leads in the tube, amounts to only a few picofarad (in practice between 1 and 3 pF), so that the alternating current flowing remains extremely small even at frequencies of some Mc/s.

The load on the voltage source is therefore a minimum. Electrostatic deflection of the beam is thus particularly suitable for electrical measurements and is almost exclusively used in such applications. It will therefore be useful to examine the electrostatic deflection in greater detail, making reference to the schematic representation in Fig. 2-9.

When, for instance, one of the deflection plates is positive, the electron beam will be attracted towards it, while it will be repelled by a negative plate. If, therefore, an alternating voltage is applied between the two plates, the electron beam will be alternately attracted and repelled by each plate in a way corresponding exactly to the waveform and polarity of the voltage.

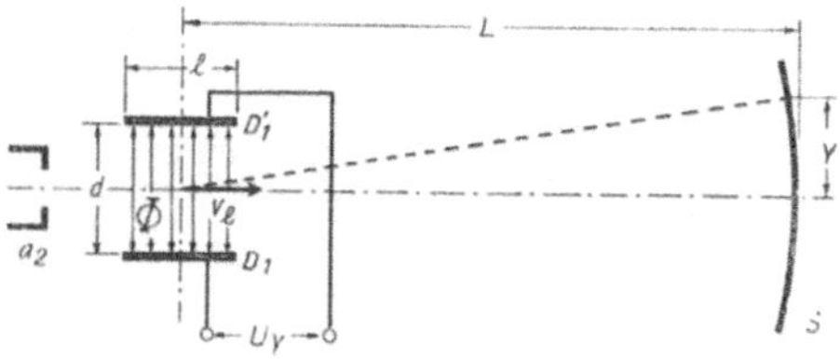

Fig. 2-9 Electrode system for electrostatic beam deflection. a_2 = anode; D_1, D'_1 = deflection plates; S = screen

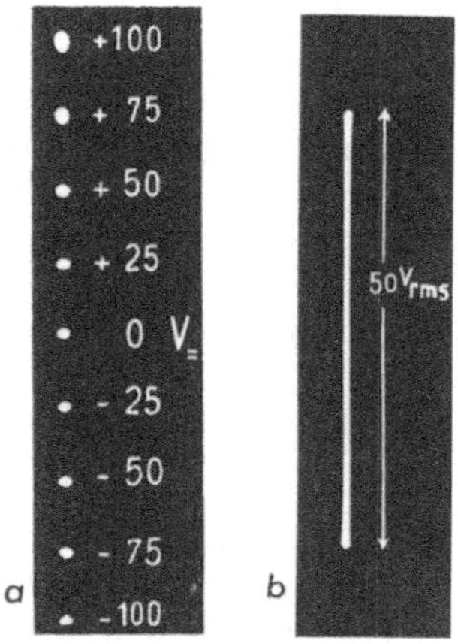

Fig. 2-10 Display on the screen with DC and AC deflection
a) DC voltage in steps of 25 V;
b) AC voltage 50 V_{rms}

Fig. 2-10*a* shows the positions of the spot when direct voltages up to +100 V or —100 V are applied to the plates in steps of 25 V, and Fig. 2-10*b* shows the trace under otherwise identical conditions with an alternating voltage of 50 V_{rms} on the plates.

As the alternating voltage moves the spot to and fro on the screen at a speed corresponding to the frequency, the individual positions of the spot can no longer be seen and now appear to the eye as a single luminous line, the length of which corresponds to the positive and negative amplitudes of the alternating voltage. The increase in brilliance and thickness at the ends of this line is due to the fact that the velocity of a sinusoidal voltage diminishes in the region of the maximum value of the voltage. Since, within certain limits, the screen becomes brighter the more electrons strike it within a given interval of time, the line traced on the screen by a sinuoidal voltage appears brighter at the ends than in de middle.

What stands out most in Fig. 2-10*a* is that the distance between each position of the spot is absolutely identical for identical increases of voltage. From this the important fact emerges that the beam deflection is in linear relationship with the deflection voltage.

2. 4 Calculating the beam deflection

The following considerations assume that care has been taken in the tube construction to ensure that a sufficiently uniform field exists between the plates within the limits required for deflection. After passing through the anode, the average longitudinal velocity of the electrons in the beam is v_l. The amount of work exerted upon the electrons is:

$$W = V_a \cdot e \tag{a}$$

(e = charge of an electron).
This work must be equal to the kinetic energy, thus:

$$W = V_a \cdot e = \frac{1}{2} \cdot m \cdot v_l^2 . \tag{b}$$

From this it follows that:

$$v_l^2 = 2 \cdot \frac{e}{m} \cdot V_a . \tag{c}$$

The electrons now arrive in the field between the deflection plates, the distance between the plates being d and potential differences V_Y.
The field strength E is:

$$E = \frac{V_Y}{d} . \tag{d}$$

The force exerted upon an individual electron is thus:

$$F = E \cdot e . \tag{e}$$

The transverse acceleration of the electron is therfore equal to:

$$a = \frac{F}{m} = E \cdot \frac{e}{m} . \tag{f}$$

If the electron remains under the influence of the transverse field for a time t_1, its transverse velocity is:

$$v_Y = a \cdot t_1 = E \cdot \frac{e}{m} \cdot t_1 . \tag{g}$$

The time t_1 is determined by the velocity with which the electron passes through the length l of the deflection plates, thus:

$$t_1 = \frac{l}{v_l} . \tag{h}$$

From this its follows that:

$$V_Y = E \cdot \frac{e}{m} \cdot \frac{l}{v_l}. \tag{i}$$

The electron therefore retains this velocity while passing through the distance L, for a time of

$$t_2 = \frac{L}{v_l}. \tag{j}$$

The electron thus covers a distance at right angles to the axis of the tube (the deflection from its original path) equal to:

$$Y = t_2 \cdot v_Y = \frac{V_Y}{d} \cdot \frac{e}{m} \cdot \frac{l}{v_l} \cdot \frac{L}{v_l}. \tag{2.1}$$

If, instead of v^2_l, the value obtained from (*c*) is inserted in the equation, the deflection on the screen can be expressed by:

$$Y = \frac{1}{2} \cdot \frac{L \cdot l}{V_a \cdot d} \cdot V_Y. \tag{2.2}$$

With this formula all values needed for deflecting the beam can be calculated [8] [9] [10]. The formula also shows that the deflection is in linear proportion to the deflection voltage. A linear reduction of the deflection occurs with increasing anode voltage. As a measure of the deflection characteristics of a tube it is usual to indicate how far the spot can be displaced by a deflection voltage of 1.0 V. Thus, the deflection sensitivity can be expressed by the equation:

$$DS = \frac{Y}{V_Y} = \frac{1}{2} \cdot \frac{l}{d} \cdot \frac{1}{V_a} \cdot L. \tag{2.3}$$

The deflection coefficient is also frequently used for describing the behaviour of a tube and is expressed by the reciprocal of the deflection sensitivity:

$$DC = \frac{V_Y}{Y} = \frac{2 \cdot d}{l \cdot L} \cdot V_a = k_= \cdot V_a. \tag{2.4}$$

In this equation $k_=$ is a tube constant depending on the value of d, l and L. The deflection coefficient indicates what direct voltage is needed to deflect the spot over a unit of length on the screen (e.g. 1 cm). It therefore increases linearly with the accelerating voltage V_a.

When an alternating voltage is applied to the deflection plates the spot follows the voltage to its peak values. If the voltage is sinuoidal (the peak-to-peak values then correspond to $2 \cdot \sqrt{2}$ times the rms values), the deflection sensitivity is equal to:

$$DS_\sim = \frac{\sqrt{2} \cdot l \cdot L}{d \cdot V_a}. \tag{2.5}$$

and the deflection coefficient is:

$$DC_{\sim} = \frac{d}{\sqrt{2} \cdot l \cdot L} \cdot V_a = k_{\sim} \cdot V_a \cdot \qquad (2.6)$$

For tube DG 10-6, to take one example, the tube data give a deflection sensitivity of 0.3 mm/$V_=$ for an anode voltage of 2000 V. From Eq. (2.3) the deflection is expressed by:

$$Y_= = DS_= \cdot V_Y . \qquad (2.7)$$

With a direct voltage of 40 V, therefore, a deflection of 12 mm is obtained.
The deflection coefficient (for 1 cm) would thus amount in this case to:
$\frac{10 \text{ (mm)}}{0.3 \text{ (mm/V)}} = 33^1/_3$ V. In other words, a deflection of 10 mm is achieved with a direct voltage of $33^1/_3$ V. With a sinusoidal alternating voltage the same deflection is achieved with a voltage whose rms value is $\frac{1}{2 \cdot \sqrt{2}} = 0.355$ times the DC value. In the example given, therefore, a deflection of 10 mm is achieved by $33^1/_3 \cdot 0.355 = 11.8\ V_{rms}$. In pulse technique (television, radar, electronic control, etc.) non-sinuoidal voltages are very frequently encountered.

Voltages such as these can only be indicated by the oscillogram and expressed as the peak-to-peak (V_{pp}) values. The peak voltage is obtained by comparison with a sinusoidal voltage causing the same beam deflection. The rms value of this voltage multiplied by $2\sqrt{2} \approx 2.83$ gives the corresponding V_{pp} value of the voltage being investigated.

2. 5 Two-dimensional deflection of the beam

Although cathode ray tubes with one pair of plates can be put to quite a considerable number of uses, the present wide range of applications first became possible by using tubes with two pairs of deflection plates. As shown in Fig. 2-11, the pairs of plates are mounted one behind the other in such a way that the deflection due to one pair is at right angles to the deflection due to the other pair.

Thus deflecting the electron beam in two mutually perpendicular directions corresponds to the customary scientific method of representing the dependence of one quantity upon the values of another by means of rectangular coordinates. [2])

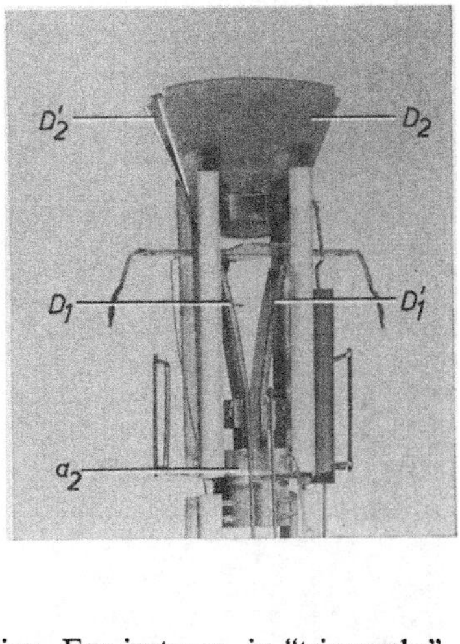

Fig. 2-11 Anode and deflection plates of DG 10-78 tube for double electrostatic deflection

[2]) There are, of course, other possible methods of deflection. For instance, in "triographs", (used for investigation of heart muscle contractions) the electron beam is deflected by three pairs of plates, each displaced at 120° with respect to the others [11].

In this method, the values of the quantity whose dependence is to be illustrated are shown in the vertical direction, along the X-axis ("ordinate"), while the values of the reference quantity are shown in the horizontal direction, i.e. along the X-axis ("abscissa"). To use the terms of analysis, y is represented as a function of x, thus:

$$y = f(x). \tag{2-8}$$

If the voltage to be examined is applied to the pair of plates for vertical deflection, and if the voltage upon which the dependence is to be shown is applied to the plates for horizontal deflection, the spot will be moved on the screen in a manner corresponding to the influence of both voltages.

In this way it is possible to display the interdependence of two variables. The visible path traced on the screen represents the "locus diagram" of the resultant or the vector sum of the two voltages. In addition to actual electrical voltages, practically any other measurable phenomena can be converted into proportional electrical voltages by means of a suitable transducer (condenser microphone, photocell, mechanical vibration pick-up, strain gauge, etc.) and displayed on the cathode ray tube.

It is usual in circuit diagrams to represent the cathode ray tube by the deflection plates only, as shown in Fig. 2.12. The remaining electrodes for controlling brilliance, focus etc. are not essential parts for the measurement itself.

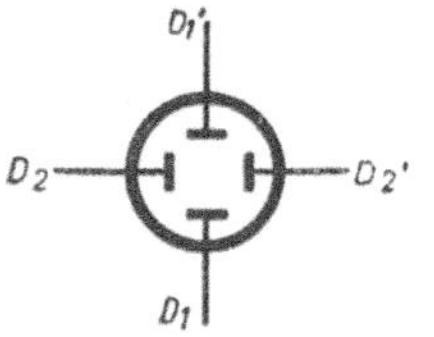

Fig. 2-12 Symbolic representation of a C.R.T.
D_1 D_1' = Y-plates
D_2 D_2' = X-plates

The photograph reproduced in Fig. 2-13*a* gives an example of the positions of the spot when direct voltages are applied in uniform steps of 20 V to the pairs of plates singly or simultaneously. By reversing the polarity of the voltage source the spot can be moved upwards as well as downwards.

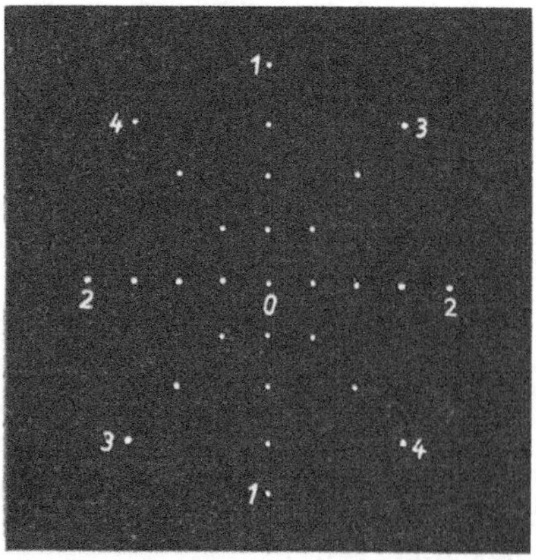

Fig. 2-13*a* Spot deflection with DC voltage on the plates. 1) DC voltage on Y-plates only. 2) DC voltage on X-plates only. 3) and 4) DC voltage on X- and Y-plates simultaneously but by changed polarity

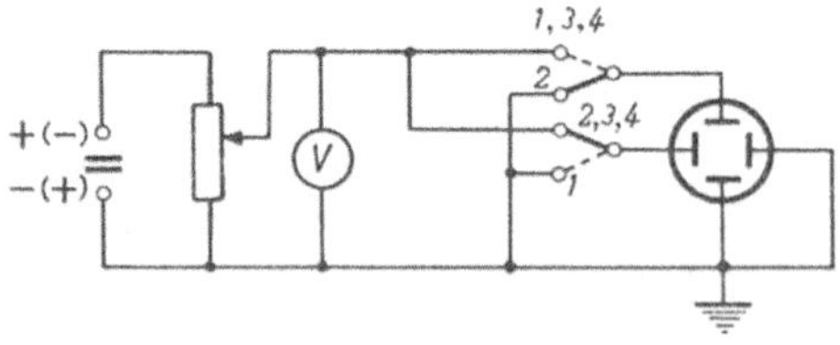

Fig. 2-13*b* Circuit for 2-13*a*. The switch positions correspond to the relevant rows of spots on the screen

The spots of row 1 lie in a line representing the Y-axis. Row 2 represents the X-axis, while row 3 and 4 represent the position of the spot when the voltages on the Y-plates and on the X-plates are changed simultaneously with differing polarities. Row 3 is situated in quadrants I and III, and row 4 in quadrants II and IV of this system of coordinates. For row 4 the connections of a pair of deflection plates must be reversed. If a voltage of 80 V is applied to both pairs of plates, the spot is deflected beyond the useful area of the screen.

It is particularly noticeable in this photograph that the deflections along the Y-axis for each of the voltage stages are greater than along the X-axis. Thus, the line of spots, when an equal voltage is applied to both pairs of plates, is not, as might be expected, at an angle of 45° but steeper. This is due to the fact that both pairs of plates do not act upon the beam at the same point, since, as can be seen from Figs. 2-11 and 2-16, they are arranged one behind the other. The distance L in Eq. (2.1) or Eq. (2.2) is therefore not the same for both axes of deflection, with the result that there are different values of deflection sensitivity.

Fig. 2-14 shows two patterns on the screen produced by two different alternating voltages with the same frequency applied to both pairs of plates.

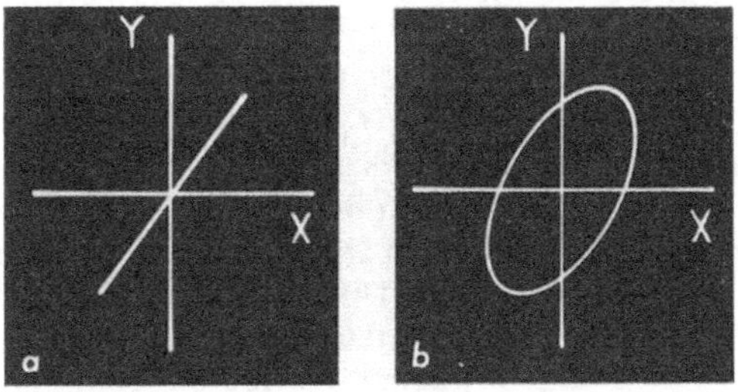

Fig. 2-14 Alternating voltages applied to the two pairs of deflection plates. *a*) Both voltages reach maximum and minimum simultaneously; they are in phase. *b*) The voltages do not reach maximum and minimum simultaneously; a phase shift exists

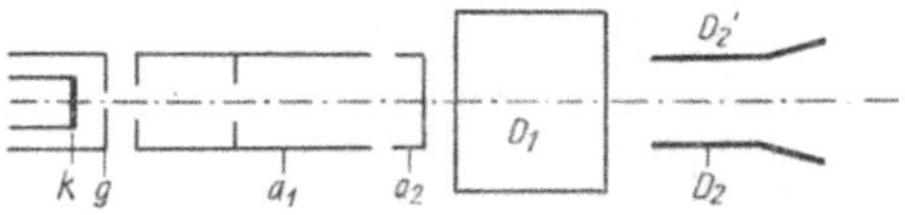

Fig. 2-15 Complete electrode system of a C.R.T. k = cathode; g = grid; a_1 = auxiliary anode; a_2 = anode; D_1 = Y-plate; D_2, D_2' = X-plates

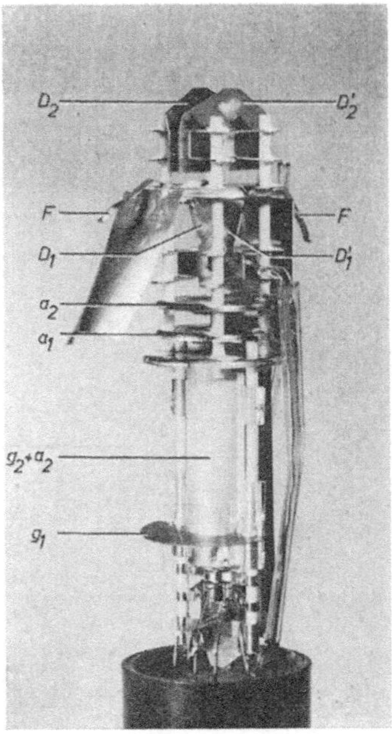

Fig 2-16 Electrode system of the Philips DG 10-74 cathode ray tube

A highly significant feature of this method of displaying two interdependent quantities is the ability represent the behaviour of a given phenomenon within a given interval of time; that is to say, a quantity can be shown as a function of time ($y = f(t)$. For this purpose it is customary to apply the voltage to be measured to the vertical deflection (Y)-plates, while the spot is made to move along the horizontal axis uniformly with time — linear with time — by applying an appropriate voltage. For this reason the Y-plates are sometimes known as the measuring plates and the X-plates as the time plates. Fig. 2-15 shows schematically the arrangement of electrodes in a tube with two pairs of deflection plates. Fig. 2-16 shows the assembly of electrodes in tube DG 10-74.

2. 6 Connecting the deflection plates

There must be no considerable voltage difference between the anode and the deflection system, as otherwise the velocity of the electrons would be increased or decreased according to the polarity of this voltage. As the voltage to be measured is usually connected in some way with earth potential, the anode is earthed in cathode ray tubes, and not the cathode as is customary in the case of amplifier valves. The cathode is therefore negative with respect to earth.

For sharp and undistorted patterns on the screen it is also necessary to consider the operation of the plates with respect to earth. Fig. 2-17 for example, shows a simplified arrangement of the first pair of plates in their circuit relationship with the anode. Plate D_1 is connected to earth, i.e. with the anode. If a potential difference exists between the deflection plates, there will be a voltage drop between them — a condition which always exists in practice. This means that with this circuit the potential at the mid point of the deflection system differs from that of the anode. It fluctuates between zero and half of the voltage V_p on the "hot" plate. This results in additional acceleration of the beam, following the fluctuations of the deflection voltage.

The field is therefore distorted as shown in Fig. 2-17, being distributed from plate D_l' to plate D_1 and to the anode a_2. As the accelerating field does not originate from symmetrical electrodes, undesirable distortion of the spot takes place; however, much depends upon whether the beam passes nearer to the earthed plate or to the voltage carrying plate.

If it passes nearer to the earthed deflection plate its velocity is not affected. The sensitivity in the X-direction vertical to the horizontal plane remains unchanged.

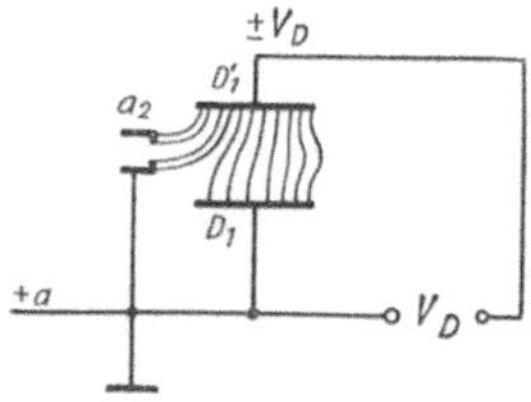

Fig. 2-17 Unbalanced circuit for deflection plates

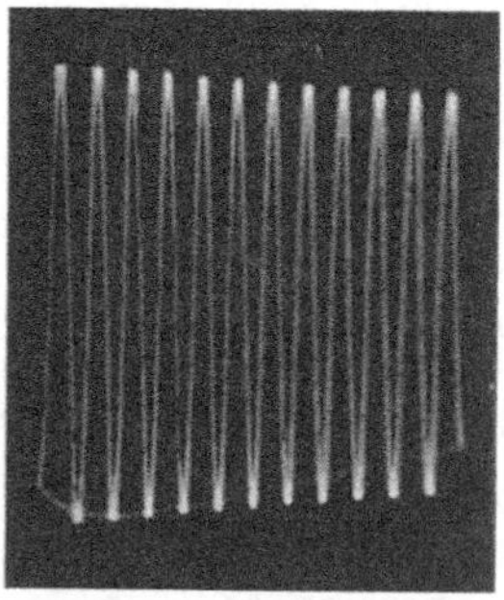

Fig. 2-18 Oscillogram of asymmetrical voltage on balanced X-plates

If it passes nearer to the "hot" plate (which is positive), it will be accelerated correspondingly. According to Eq. (2.2), however, this results in a reduction of the deflection sensitivity, as the accelerating voltage is now $V_a + V_D$.

With alternating deflection voltages on both pairs of plates, the spot should describe a rectangular pattern on the screen. Owing to the voltage asymmetry on the respective plates, the amounts of deflection in the vertical direction vary and the pattern described by the spot becomes trapezoidal, a phenomenon known as trapezium distortion and represented in Fig. 2-18.

These unwanted phenomena (partial defocusing — see the series of spots in Fig. 2-10*a* — and trapezium distortion) can be wholly avoided in practice by using a deflection system balanced with respect to earth. As shown by Fig. 2-19, the lines of the field are then symmetrically distributed between the plates, the average field remaining constant. The voltages on the plates now fluctuate symmetrically with respect to the last accelerating electrode by $\pm \frac{V_D}{2}$, so that there is no noticeable deterioration of the spot nor trapezium distortion. The deflection system can be balanced by voltage dividers or by centre-tapped transformers as shown in Fig. 2-20*a* and *b*. As a rule, however, amplifier valves in push-pull are used for this purpose. This method is discussed in detail in Part I, Ch. 5.

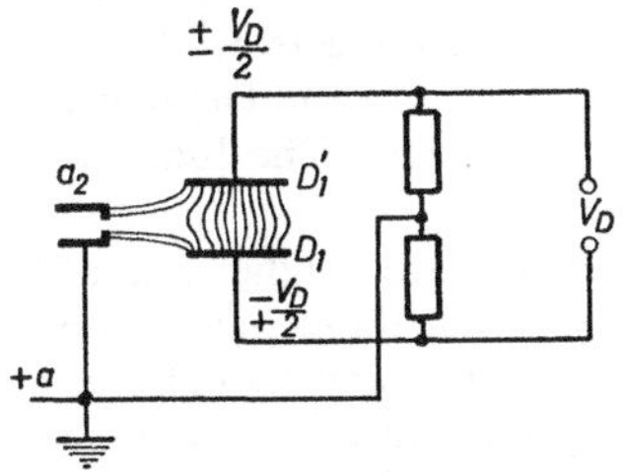

Fig. 2-19 Balanced circuit for deflection plates

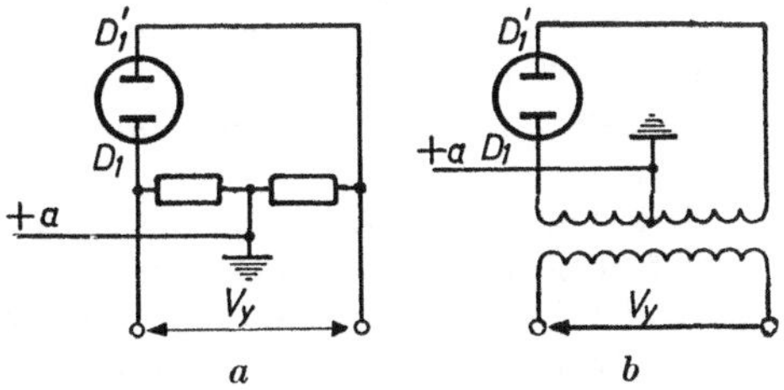

Fig. 2-20 Circuits for balancing the signal voltage.
a) Balancing by means of a resistive voltage divider
b) Balancing by means of a transformer

For reasons of economy, other means of compensating for these defects were sought by improving the deflection plates and by the use of auxiliary electrodes [12] [13] [14] [15] [16].

This is satisfactory as regards the pair of plates nearer to the screen. The pair of plates nearer to the anode, however, should always have as balanced a deflection voltage as possible applied to them, unless only small deflections are required or general concessions can be made for the geometry of the pattern. A solution of this sort is indicated in Fig. 2-21*a* and *b*. Two small wire hooks have been soldered as auxiliary electrodes *–d–* to the non-earthed plate for horizontal deflection. If the voltage on this plate increases in the positive direction, the deflection sensitivity will normally decrease. But this auxiliary electrode also attracts the beam, so that the deflection in this direction is greater. The fact remains, however, that where high demands are made on pattern geometry, the voltage balance for both pairs of plates must be produced and symmetrical electrodes used for deflection.

Fig. 2-21 Compensation for trapezium distortion
a) Non-earthed *X*-plate alters the velocity of the electron beam
b) Compensating for loss of sensitivity by means of two auxiliary electrodes *d*

2. 7 Load due to deflection plates

As already mentioned, the deflection plates together with their leads in the tube possess a certain amount of capacitance, small though it may be. With an alternating voltage on the plates, a current increasing with rising frequency will flow through this capacitance.

Because of the low value of this capacitance, however, this current need be taken into account only for high frequency measurements.

There is, however, still another way in which the deflection plates represent a certain load upon the voltage source. When the electrons strike the screen their kinetic energy is partly converted into light and partly serves to release new electrons from the material of the screen. These are known as secondary electrons. (Secondary emission is brought about intentionally. Details are given in Part I, Ch. 2.9 "The luminescent screen".)

The screen secondary electrons constantly strive to move towards higher potentials, which means that they will move to the deflection plates when these have a higher potential than the anode at a given instant. Other electrons (stray electrons) also attach themselves to the deflection plates, so that a compensating current can arise between each of a pair of plates, resulting in a further load upon the voltage source. In order to avoid these effects, the inside of the glass envelope of the C.R.T. is coated with a colloidal graphite preparation known as "aquadag".

By means of two light springs (see spring marked *"F"* in Fig. 2-16) this coating is connected with the anode and thus to earth. The electrons are thereby conducted to earth, and additional load on the voltage source is avoided. Rough treatment during transport may have caused damage to the contact between the graphite coating and the springs, or even have broken the springs. Defect of this sort can be recognized immediately by severe distortion of the pattern if the screen is touched during operation. It can also happen in such cases that the pattern on the screen becomes erratic. The charge on the screen is no longer able to leak away, and after reaching a certain potential caused by the accumulation of electrons it breaks down at irregular intervals determined by the remaining possibilities of escape. It is also possible in this way for "islands" to form, that is to say, areas where no luminescence occurs. Similar defects are more likely to be found in tubes working at excessively low anode voltages. The potential due to the accumulated electrons on the screen is then higher than the accelerating voltage between cathode and anode, either over the whole screen or at certain parts of the screen.

For various reasons the alternating voltage for the vertical deflection plates must be applied via coupling capacitors. This is primarily necessary in order to isolate the plates from the anode direct voltage of an AC amplifier. At the same time the deflection plates must always be directly connected to earth via leak resistors, as otherwise they can obtain random charges from stray electrons and thus cause the beam to wander. The leak resistors should be rated between 1 – 10 MΩ to keep the load on the voltage source to a minimum. It is therefore possible, as far as the connection of the signal voltage to the deflection plates is concerned, to speak with a considerable amount of truth of an electrostatic display.

Usually, however, this signal voltage reaches the deflection plates via an amplifier. The load on the source of the voltage is determined by the input impedance of this amplifier. It usually consists of a capacitance of 15 – 60 pF in parallel with a resistor of from 0.5 to 2 MΩ.

In cases where this impedance is still too low, a voltage divider probe or cathode follower probe, according to the task, is connected between the amplifier input and the source of the signal voltage (Part I, Ch. 5-29 "Probes").

2. 8 Influence of electron transit time on deflection sensitivity

The displacement of the spot on the screen corresponds to the instantaneous value of the alternating deflection voltage only if the electron transit time within the pair of plates concerned is sufficiently short compared with the duration of one cycle. Otherwise the deflection sensitivity will suffer. The frequency f, at which the deflection sensitivity is reduced by a portion p for a plate length l and an anode voltage V_a can be found from the equation:

$$f\,[\text{Mc/s}] = 46 \cdot \frac{\sqrt{p \cdot V_a\,[V]}}{l\,[\text{cm}]}\,. \tag{2.8}$$

In tube DG 10-6, for example, $l = 2$ cm. If, with an anode voltage of 1000 V, a deflection error of 2% ($p = 0.02$) is permissible, the limiting frequency can be calculated as: $f = 46 \cdot \frac{\sqrt{0.02 \times 1000}}{2} = 103$ Mc/s approx.

In the case of cathode ray tubes having sufficiently short pairs of deflection plates, and especially when working with even higher accelerating voltages, the equation shows that the deflection error remains within reasonable limits of accuracy on the screen up to and exceeding 100 Mc/s [17].

The reduction in dynamic deflection sensitivity at high frequencies, due to the no longer negligible transit time within the pair of plates, opposes the intended increase in deflection sensitivity obtainable by using as long deflection plates as possible. A solution to this problem has been found by breaking up the deflection system into a number of pairs of subdivided plates [18]. These pairs of short plates are connected across small inductance. Together with the capacitances of the pairs of subdivided plates they form a delay line (see also Ch. 5.23 "Delay lines"). Starting from the first pair of plates, the voltage arrives at successive pairs of plates with a certain delay each time. By correspondingly rating the individual links it is possible to ensure that as the electron beam passes through the deflection system the voltage surge reaches each individual pair of plates at the same speed, corresponding to the acceleration speed.

This method is of particular value for cathode ray tubes in which the beam voltage *before* deflection is kept relatively low for the purpose of obtaining satisfactory deflection sensitivity, the greater part of the beam acceleration taking place after deflection (see Ch. 2.11 "Cathode ray tube with post-acceleration helical electrode"). This has made it possible to develop cathode ray tubes of high deflection sensitivity when the total acceleration voltage is high, even at frequencies of up to 100 Mc/s and more. Fig. 2-22 shows the output of the distributed delay line amplifier, and the connection of the deflection plate system in the Type 580 Tektronix oscilloscope, in which the C.R.T. is constructed in this way.

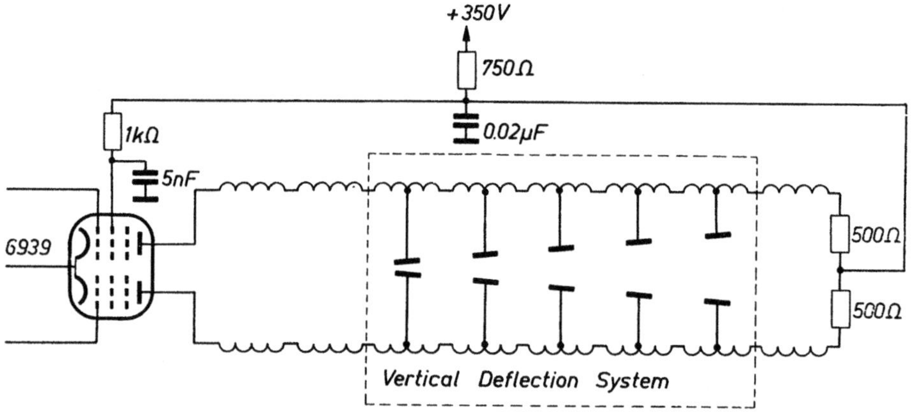

Fig. 2-22 Deflection system with delay line coupling in the Tektronix type 580 oscilloscope (travelling-wave tube)

The inductances for the signal delay between the pairs of reduced plates are placed in this instance outside the cathode ray tube. The upper cut-off frequency for Y-deflection in this oscilloscope is quoted as 100 Mc/s corresponding to a rise time of 3.5 ns. The deflection coefficient with two-stage delay line amplifier and output stage is 100 mV/cm. An upper cut-off frequency as high as 1000 Mc/s has been obtained with the tube of the Tektronix Oscilloscope 519.

In another suggested method for beam deflection with voltage rise times in the nanoseconds range, the beam is deflected by a helical electrode (travelling-wave deflection helices) [19] [20]. The pattern on the screen is unusually small and has to be enlarged for interpretation. The spot diameter is given as 2.5 μm (0.001″) in the publication quoted. By this means readily interpretable pictures of processes are obtained up to a cut-off frequency of about 45 Mc/s (see also bibliography [3] of Part II, Ch. 6).
Where the time difference between the voltages on the two pairs of plates is to be determined (e.g. in phase measurements by ellipse), the transit time of the beam between both pairs of plates must be taken into consideration. At high frequencies an error occurs which has the effect of increasing the apparent-phase angle. For a phase error amounting to an angle ψ, the limiting frequency f is found from the equation:

$$f\ [\text{c/s}] = \frac{\Psi \sqrt{V_a}}{s_E} \cdot 1.65 \cdot 10^5. \tag{2.9}$$

Taking the phase angle error of 2°, which is easily visible in an elliptical pattern, a limiting frequency $f = 4.2$ Mc/s is obtained when $V_a = 1000$ V, and the average distance between the plates is $s_E = 2.5$ cm (DG 10-6). This transit time error can be important in phase measurements in radio engineering and most especially in television and short-wave technique ($f > 3$ Mc/s).

To overcome this effect, Hollman suggested dividing the relevant pair of plates into two pairs of divided plates and placing them before and after the pair of deflection plates in the other coordinate direction [17].

In this way both halves of the deflection system can bring about equal but opposed phase shifts which cancel each other out. The deflection effect of this system thus remains free from phase errors even at high frequencies.

2. 9 The luminescent screen

As already stated, certain chemical substances used for the screen in cathode ray tubes become luminescent when bombarded by electrons. The brightness and colour of the light thus emitted depend upon the characteristics of the substance used. The light is produced by conversion of a part of the kinetic energy of the electrons into light energy. The electrons receive their kinetic energy while traversing the voltage drop between the cathode and the final accelerator electrode. The power inherent in the beam current can accordingly be calculated from the product:

$$N = I_S \cdot V_a \text{ [3]}, \qquad (2.10)$$

and expressed in watts or mW. The amount of light emitted is not directly proportional to the power of the beam but is also dependent upon the velocity of the beam, and thus upon V_a. For a given product $I_s \cdot V_a$, therefore, more light is obtained with a high anode voltage and small current than with a lower voltage and correspondingly higher current (Fig. 33-3). An upper limit is set to the brightness by the fact that beam currents exceeding a certain value cause "spot burns" resulting in a noticeable loss of efficiency in that part of the screen. It is obvious that this danger is greatest when the area under bombardment is small, i.e. a spot. It should be a rule, therefore, always to keep the beam moving in some way. But even as a trace, burning can be caused if the screen is subjected to bombardment for a prolonged period, as can often be observed on oscilloscopes which have been in use for a considerable time. It is interesting to note that the danger of burning is apparently greater at lower anode voltages than at higher ones. Whereas it is difficult to produce a screen with a satisfactory life for anode voltages under 500 V, screens of tubes working with anode voltages of several kV are comparatively free from the phenomenon of burning. No complete explanation of this apparent paradox has so far been offered, as the process of energy conversion in the luminescent screen is not yet sufficiently understood [21] [22] [23] [24] [25] [26] [27].

It is suggested that with low accelerating voltages, only the upper layer of screen atoms takes part in the conversion of energy. At higher voltages the electrons of the beam penetrate deeper. Therefore deeper layers of the screen will also emit light, and thus the specific electrical load on the screen material will not be so great. The screen is continuously supplied by the electron beam with a negative charge, so that its potential would progressively increase. This charge could have an undesirable effect upon the spot deflection and might even prevent further electrons reaching the screen. However, by choosing a suitable luminescent material, the electrons penetrating the

---

[3]) The beam current I_s must not be confused with the anode current to I_{a2}. It is generally appreciably smaller than the anode current and is usually not easily measured.

screen (primary electrons) can be made to release new electrons, known as secondary electrons. The majority of these secondary electrons possess enough energy actually to leave the screen.

They then move towards the positive anode, or towards the graphite layer on the inside of the tube, which is also connected to the anode. In this way they are able to return to the electron source, i.e. to the power supply unit for the C.R.T. anode voltage.

A condition of balance results when the number of secondary electrons emitted from the screen equals the number of primary electrons striking the screen. The luminescent layer in general purpose tubes then assumes a constant potential, which experiments have shown to be about 100 V below the accelerating voltage. The maximum possible luminous intensity is thus directly related to the value of the upper limit of potential for a given material. This is known as the "sticking potential". Every screen has its "sticking potential" beyond which an increase in luminous intensity is no longer possible, even with higher accelerating voltages (unless metal-backed screens are used). The sticking potential is about 8 kV for standard screen materials. In the case of metal-backed screens accelerating voltages of > 3 kV result in brighter and better contrasted spots than can be obtained without this feature [28].

A great variety of screen materials is available for producing a luminous trace in practically any desired colour [29] [30]. As, in most cases, oscillograms are observed visually, it is desirable to use a screen whose spectral energy maximum corresponds as far as possible with the spectral maximum sensitivity of the eye. The maximum sensitivity of the average human eye is about 550 nm [4]), so that screens emitting a greenish-yellow light are the most suitable. With such screens satisfactory brightness of image can be achieved with a smaller beam current than would be necessary for a screen with a less suitable colour. This means, of course, that the spot is smallest and the picture "sharpest" when a greenish-yellow screen is used.

The curves in Figs. 2-23 *a*, *b* and *c* show the spectral energy distribution of various Philips screens. The letter prefixes are those used by the Philips and Telefunken companies to indicate the types of screen. The current indication employed in the U.S.A. for the same types of screen is added in brackets in the text. In Fig. 2-23*a*, which is for the screen *N* (*P*2), the sensitivity curves of the human eye, adapted both to brightness and to darkness, have been added for purposes of comparison. It can be seen that this type of screen as well as the *G* (*P* 1) and the *H*-screen (*P* 31) (fig. 2-23*b* and fig. 2-23*c*) are particularly suitable for visual observation. It should be noted that the spectral energy distribution changes according to the degree of excitation. The two curves in Fig. 2-23*b* for the *H*-screen show that, when the brightness is greater, the blue component of the emitted light, which is of particular importance for photographic recording, is more than doubled.

Fig. 2-23*c* shows the spectral energy distribution of the blue-fluorescent *B*-screen (*P* 11) which is very suitable for photographic recording of the *G*-screen (*P* 1) and of the long-persistence double-layer *P*-screen (*P* 7).

Production of light when the screen is struck by the electron beam requires a certain amount of time, as does the decay of this light. While the excitation time is short enough to be generally neglected, persistence lasts relatively longer. Screens with particularly long persistence have been deliberately produced for observing non-recurrent or repeating, slowly decaying processess.

[4]) This depends on the adaptation of the eye. The maximum sensitivity of the light-adapted eye is around 554 nm and is about 513 nm for the eye that is completely adapted to the dark.

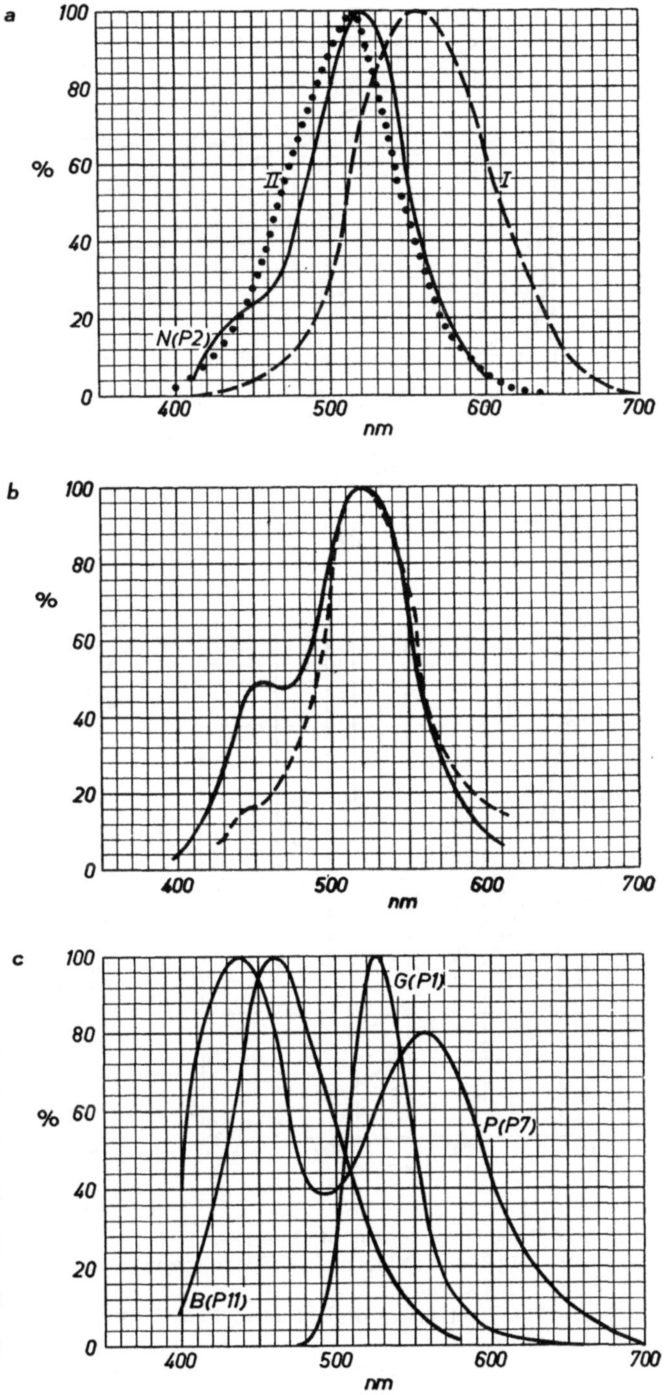

Fig. 2-23 Spectral energy distribution of fluorescent screens
a) *N* (P2) screen and eye sensitivity
I eye adapted to bright light
II eye adapted to dull light
b) *H* (P31) screen at various intensities
c) *G* (P1)-, *P* (P7)- and *B* (P11) screens

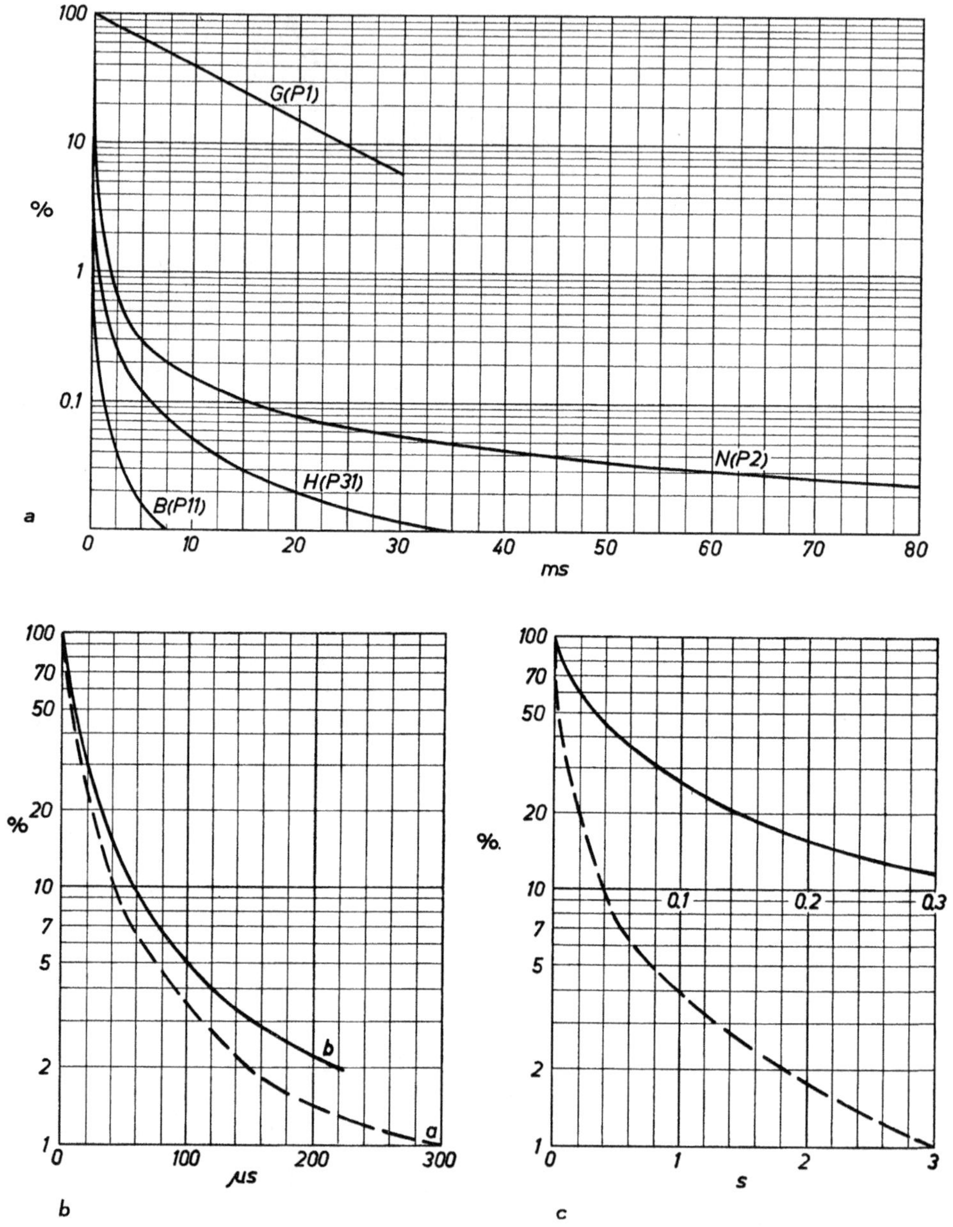

Fig. 2-24 Time characteristic of afterglow

a) *G* (P1)-, *N* (P2)-, *B* (P11)- and *H* (P31)- screen after medium excitation

b) and *c)* ***P*** (P7) screen

b) purple-blue light share: straight curve $V = 3$ kV, $I = 1{,}7\ \mu$A; dotted curve $V = 3$ kV, $I = 17\ \mu$A

c) yellowish-green light share: straight and dotted curves for different long intervals

The persistence curves of a number of different screens are shown in Figs. 2-24*a*, *b* and *c*. When studying these curves, it should always be noted that the duration of afterglow is very dependent on the intensity of the excitation. It is therefore only possible to come to conclusions by analogy. The behaviour of the screen under any given conditions can be estimated only on the basis of experience under widely differing conditions of practical use. The blue-fluorescent *B*-screen (*P* 11) is one of the short-persistent screens, its light decaying after only about 10 ms to less than 1 % of its initial value. The *G*-screen (*P* 1) differs fundamentally from the others in its persistence behaviour. After 24 ms it still emits at 10% of its initial intensity. It is therefore unsuitable for moving film recording, as the persistence causes additional exposure and blurring effects would occur behind the oscillogram (see also Fig. 32-18). It is usually preferable therefore, to use the *H*-screen (*P* 31), the spectral energy distribution of which corresponds closely to that of the human eye, but its persistence is short. The persistence curve of the *H*-screen lies roughly between the curves of the *B*- and *N*-screens. The *N*-screen (*P* 2) has a great variety of uses. Its spectral energy distribution corresponds very closely to the sensitivity curve of the human eye (Fig. 2-23*a*), and thus results in a clearly visible trace.

In addition, its persistence enables non-recurrent phenomena or repeating processes having a moderate repetition frequency to be observed. Its light decays to below the 1% value so quickly, however, that it can very readily be used for moving film recording. When deciding whether to use the *H*- (*P* 31) or *N*- (*P* 2) screen, the screen efficiency under the prevailing working conditions should be taken into account. Whereas the *N*-screen does not achieve a high degree of efficiency at acceleration voltages below about 2.5 kV, the efficiency in the case of the *H*-screen is very satisfactory even at low acceleration voltages.

The *P*-screen (*P* 7) was developed for observing particularly slow repeating processes or slow non-recurrent voltage waveforms. It consists of two layers. The layer which is struck by the electron beam emits a blue light, that is to say it fluoresces. The second layer, which faces the observer, is excited by the blue light and emits a yellowish light, which is of long persistence, i.e. it phosphoresces. How long the persistence of the various oscilloscope tubes can be observed by the eye depends mainly on the brightness of the surroundings [31]. In a darkened room the oscillogram can be observed for one or more minutes on the *N*- and *P*-screens.

An impression of the photographic effect of the afterglow is given by the oscillogram in Fig. 2-25. For this photograph, the spot was moved from left to right at a speed of approx. 1 m/s (on the screen) by the time base generator of a standard oscilloscope. Photographs were taken of the afterglow from 4 tubes with screens of different persistence using in each case an exposure of $^1/_{100}$ s. Whereas practically only the spot path could be retained on the *B*-screen, the other types of screen still show afterglow

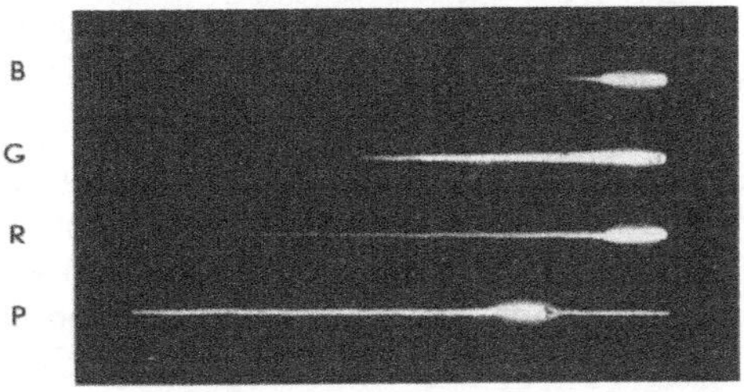

Fig. 2-25 Photographic recording of afterglow persistance on four different screens

traces varying both in brightness and length. The *G*-screen, as one would expect from the curve in Fig. 2-24*a*, shows short but initially fairly strong afterglow. The *R*-screen shows a sharp initial drop and then a longer duration of afterglow. The *R*-screen is a yellow-persistent single-layer screen, but has been little used in recent times. The afterglow effects on the *P*-screen overlapped during the individual periodic movements of the spot, so that the whole length of the trace was recorded, falling off a little during the exposure and thereafter. Of course, these oscillograms can give only an approximate idea of the persistence of the various screens. When the photographs were taken, the most favourable conditions for each type of screen could not be selected, as otherwise there would have been no basis of comparison. Particularly with regard to the long-persistence screens *N* and *P*, the difference between the luminous intensity upon excitation and that of the afterglow is so great (as shown by the curves in Fig. 2-24) that it can hardly be overcome by an ordinary photographic film.

To avoid glare effects on the eyes during the observation of slow-speed phenomena or single transients on long-persistence screens, advantage can be taken of the fact that the light emitted as a direct result of excitation (fluorescence) has a different colour from the light given off during afterglow (phosphorescence). If, for instance, a yellow-orange filter is placed over the *P*-screen, much of the bluish-green light emitted during excitation will be absorbed, and the yellow afterglow trace will stand out more clearly. Filters of this sort, using the colour of the trace of a particular tube, can be applied with advantage to other screens also. They prevent the ambient light from whitening the whole screen. The ambient light has to penetrate the filter twice (there and back) before it reaches the eye of the observer. For example, if a green filter is used in front of a *H*-tube, a bright green oscillogram appears upon the dark green screen. Thus the contrast of an image on the screen is considerably heigthened. (In television tubes a grey glass screen is used for similar reasons.) Care must be taken, however, that the transmission characteristic of the filter corresponds as closely as possible to the spectral energy distribution of the screen, as otherwise loss of light will result.

Photographic recording of oscillograms is possible with all standard types of film. The overall sensitivity of the film as well as its spectral sensitivity in relation to the spectral energy distribution of the screen must naturally be taken into account. For yellow-green screens, orthochromatic high-sensitivity film is especially suitable, e.g. Agfa "Isopan Record" and Gevaert "Scopix G". High-contrast developers must be employed (no soft tones). For general purposes, standard 21 DIN (80 ASA) Pan film is recommended, as this produces a finer grained photograph than the high- sensitivity films mentioned above. Blue *B*-screens are useful when, for instance, a short afterglow is needed for moving-film recordings, or when a large number of photographs are to be made on inexpensive recording paper. Low colour-sensitized material has the highest sensitivity in the blue-violet part of the spectrum, so that it gives the best results when a blue-fluorescent screen is used [32] [33] [34] (see also Part IV, Ch. 32).

2. 10 Post-acceleration

The aim in tube designing has always been to achieve the brightest possible spot on the screen. This means, of course, that the energy of the beam must be as high as possible. A certain increase in brilliance can of course be obtained by increasing the beam current, but beyond a certain point this results in beam scattering, so that the spot size increases. The only alternative is to increase the anode voltage. If this is stepped up, the brightness of the spot does indeed increase quite considerably, because the higher anode voltage not only boosts the beam power but also improves beam concentration and screen efficiency. Unfortunately, however, with high anode voltages the deflection sensitivity deteriorates. This entails very high voltages for beam deflection and thus a relatively high outlay for deflection amplifiers or time base unit. Intensive research has nevertheless shown a way of increasing the spot brightness without causing any appreciable deterioration in deflection sensitivity [35] [36] [37] [38].

The method is to effect the main acceleration of the beam after deflection by using a further electrode (*a*). In this way a reduction in deflection sensitivity due to the increased "pre"-acceleration voltage can be largely avoided. If a voltage positive with respect to earth is applied to this electrode, the electron beam will enter the new accelerating field *after* deflection but *before* it reaches the screen, and thus the beam velocity and the brilliance of the spot will be correspondingly increased. Tubes employing this device are known as "post-acceleration tubes".

A certain longitudinal attractive force exerted by this electrode cannot be altogether avoided, so that the deflection sensitivity still suffers a slight reduction. Nevertheless it remains considerably higher than it would be if the anode voltage on a_2 alone were to be raised to the same value (see also Part 1, Ch. 2.11 "C.R.T. with post-acceleration helical electrode").

A cahode ray tube of this type, DG 10-74, is illustrated in Fig. 2-26*a*, which shows the connection for the post-acceleration voltage and the cone-shaped graphite-layer electrode for a_3.

With a potential of 1 kV on a_2 in the DG 10-74, the deflection sensitivity declines by only 20% for a post-acceleration potential of 1 kV on a_3, and by $33^1/_3$% for 2 kV on a_3. A distinct increase in luminous intensity and spot sharpness is noticeable if only the ordinary direct voltage of 250 – 400 V (for the supply of the deflection amplifier and time base unit) is applied to a_3. In this case, deflection sensitivity declines by about 5% only.

a)

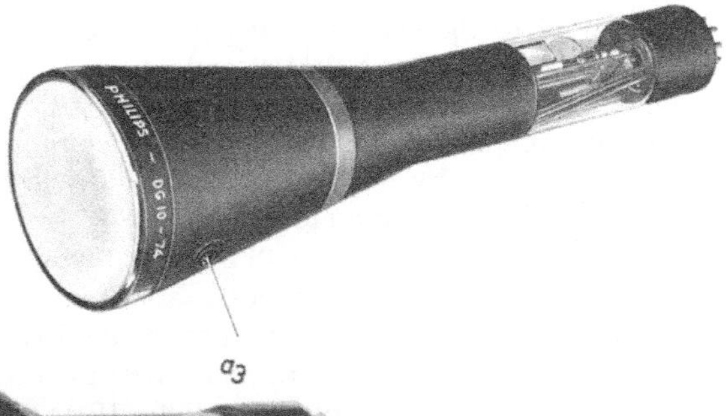

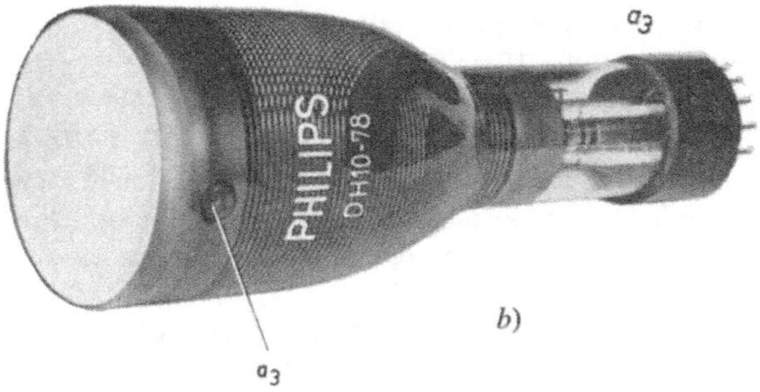

b)

Fig. 2-26 External view of 10 cm cathode ray tubes
a) Philips DG 10-74 tube with post-acceleration electrodes (a_3) in the form of a truncated cone coating;
b) Philips DH 10-78 tube with a helical post-acceleration electrode

The increased brightness of the image on the screen due to post-acceleration makes it possible, on ordinary cathode ray tubes, to photograph movements of the spot at speeds of up to about 1000 km/s. By means of special tubes, recording speeds of up to 10,000 km/s and above have been achieved (Part I, Ch. 2.14 "Multi-ray tubes; high-performance tubes"). It also enables the oscillogram to be projected on screens of up 2 m^2 ([38] [39] and Part IV, Ch. 33).

2. 11 Cathode ray tubes with a post-acceleration helical electrode

The development of oscilloscopes for observing high-frequency phenomena and short time pulses of low repetition frequency has been facilitated by the design of various new types of cathode ray tubes.

In the arrangement of the post-acceleration electrode which was normal until recently, i.e. in the form of a large-area conductive layer on the inner wall of the envelope (Fig. 2-26*a*), this layer, together with that part of the wall coating connected to the last acceleration electrode (in front of the deflection system) forms an electron lens. If the potential difference between these electrodes is too great, the refractive power of this lens becomes so high that the image produced by the effect of the deflection has excessive barrel distortion. With such tubes it is recommended that the total post-acceleration voltage should not exceed twice the voltage difference between cathode and the last acceleration electrode before the deflection system.

The latest cathode ray tubes, however, exploit the fact that the undesirable refractive power of the post-acceleration lens can be kept very low by generating the post-acceleration field not as a sudden increase in potential, but by making it rise gradually over a certain distance in the direction of the tube axis.

Various methods of achieving this have been suggested from time to time. In the latest development, as suggested by Schwarz, a resistor helical [43] is mounted on the inner wall of the tube bulb behind the deflection system. One end of this resistance helical is connected to the last acceleration electrode before the deflection system, and the other end to the source of the post-acceleration voltage. Depending on the resistance of this helix (50 to 300 MΩ), only a small current flows through it. The rise in potential of the post-acceleration field can be so adjusted by choosing a suitable pitch of the helix, bulb shape and the location of the beginning and end of the helix, that the image distortion caused by the post-acceleration field is negligibly small, even up to voltage ratios of 1:4, and even op to1:6.

Hence it is possible to produce cathode ray tubes with high deflection sensitivity, which, in spite of grat image brightness, satisfy increasing precision demands. Such tubes are produced by Tektronix, Telefunken, Philips, etc.

Where as Tektronix only supply tubes in their own oscilloscopes, Telefunken offers a 7 cm tube (DG 7-18), a 10 cm tube (DG 10-18) and several 13 cm tubes (DG 13-18, DG 13-38 and DG 13-58) separately. Philips offers a 7 cm tube (DH 7-78), a 10 cm tube (DH 10-78) and four 13 cm tubes (DH 13-10, DH 13-78, DH 13-76 and DH 13-79) with post-acceleration helical electrode. The deflection coefficient of the 7 cm tube DH 7-78 is only about 4.8 V/cm for a total acceleration voltage of 1.6 kV. For the maximum permissible total acceleration voltage of 4 kV it rises to about 12 V/cm.

In the case of the Telefunken tube DG 13-38, with a voltage ratio of 1:5 and a total accelerating voltage of 6 kV, the deflection coefficient of the *Y*-plates is only about 3.5 V/cm.

An exterior view of the DH 10-78 tube is shown in Fig. 2-26*b*. There is no doubt that such cathode ray tubes will be developed still further. The Philips tube DH 13-10 (5 CLP 31) has several staggered helical post-acceleration systems, by means of which a deflection coefficient for the *Y*-plates of 1.85 V/cm is obtained with a total accelerating voltage of 10 kV. With an accelerating voltage of 15 kV it rises to only about 2.7 V/cm. It will be clearly realized that such low deflection coefficients offer considerable advantages for the rating of amplifiers. A limit would be set to further reduction of the deflection coefficients by the resulting increase in sensitivity to stray electromagnetic fields.

Dimensions, base connections and a schematic arrangement of the electrodes for tube DH 10-78 are given in Fig. 2-27. Characteristic curves of I_{a2} and I_s (screen current) as functions of the grid voltage are reproduced in Fig. 2-28 with respect to a DG 10-6 tube, and are also valid for the DH 10-74 tube.

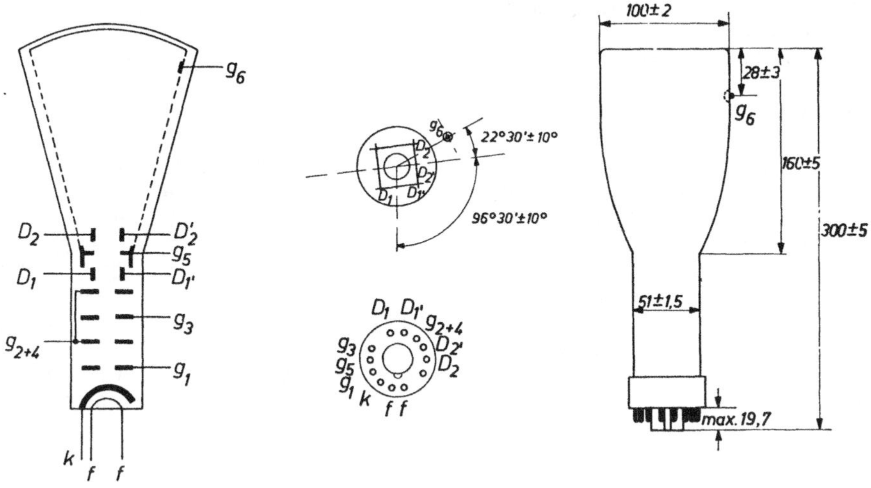

Fig. 2-27 Dimensions and base connections of the Philips DG 10-78 C.R.T.

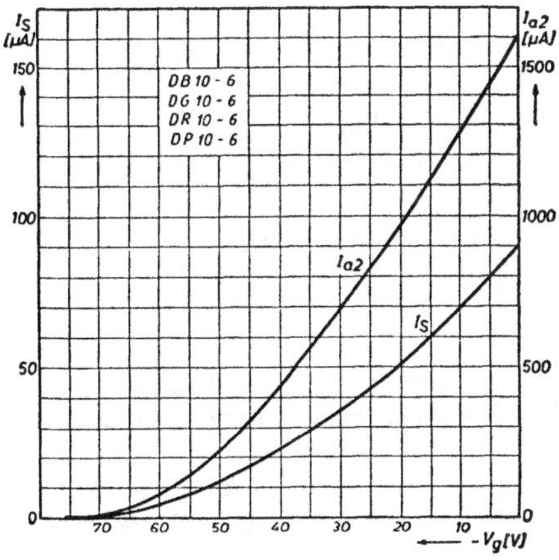

Fig. 2-28 Current to anode a_2 and screen current I_s (I_{a2}) of tubes DG 10-6 and DG 10-74 as a function of the voltage on grid g_1

2.12 Data

Operating and limiting data as well as the main capacitances are given for cathode ray tubes as they are for other thermionic valves. The following data apply to tube DH 10—78:

TABLE 2-1 OPERATING AND LIMITING DATA

Capacitances

Grid- other electrodes [5]	C_{g1}	= 5.0 pF
Cathode- other electrodes	C_k	= 3.4 pF
First pair of plates- all other electrodes	$C_{D1}, C_{D'1}$	= 1.7 pF
Second pair of plates- all other electrodes	$C_{D2}, C_{D'2}$	= 2.1 pF
Deflection plate D_1- all other electrodes	C_{D1}	= 3.5 pF
Deflection plate D_1'- all other electrodes	$C_{D1'}$	= 3.5 pF
Deflection plate D_2- all other electrodes	C_{D2}	= 4.5 pF
Deflection plate C_{D2}'- all other electrodes	$C_{D2'}$	= 4.5 pF

Operating data

Beam voltage on screen	V_{g6} =	2000	4000	4000	V
Voltage across electrode g_5	V_{g5} =	2000	2000	1000	V
Voltage across electrodes g_{2+4}	V_{g2+4} =	2000	2000	1000	V
Focusing voltage	V_{g3} =	400 . . . 700	400 . . . 700	200 . . . 350	V [6]
Control grid voltage	$-V_{g1}(I_1{=}0)$ =	45 . . . 75	45 . . . 75	22.5 . . . 37.5	V
Deflection coefficient of first pair of plates	DC_Y =	14.4 . . . 17.8	16.6 . . . 20.0	9.8 . . . 11.9	V/cm
Deflection coefficient of second pair of plates	DC_X =	34.4 . . . 43.4	43.4 . . . 55.6	30.3 . . . 38.5	V/cm
Line width		0.45	0.35	0.45	mm [7]

Limiting data

Total acceleration voltage	V_{g6}	=	max. 8000 V min. 1500 V
Voltage across electrode V_{g5}	V_{g5}	=	max. 2200 V
Voltage across electrodes g_2 and g_4	V_{g2+4}	=	max. 2100 V min. 1000 V
Voltage ratio	V_{g6}/V_{g2+4}	=	max. 4, min. 1
Focusing voltage	V_{g3}	=	max. 1500 V
Control grid voltage	$-V_{g1}$	=	max. 200 V
	$+V_{g1}$	=	max. 0 V

Further data are given on the data sheet of this tube, obtainable from the manufacturers on request.

---

[5]) All electrodes of oscilloscope tubes except the cathode have now been designated by g. This method of designation, as distinct from the rest of the book, has been used for the data of the DH 10-78 tube.

[6]) I_{g3} = min. —30 μA, max. + 15 μA.

[7]) In the case of I_1 = 0.5 μA, measured on a circle with a diameter of 50 mm.

2.13 Cathode ray tubes with particularly high deflection sensitivity of the *Y*-plates for wide-band oscilloscopes

The highest signal frequency which can be sufficiently amplified and therefore satisfactorily displayed on a C.R.T.-screen is determined to a large extent the g_m/C ratio of the amplification stages, as is discussed in Ch. 5 in detail, but particularly in Ch. 5.14 "Loss of gain at the upper frequency limit".
These properties can be characterized by the

$$\text{electron tube figure of merit} = \frac{g_m}{2\,\pi \cdot C_R}. \tag{2.11}$$

This is the frequency at which the amplification becomes unity. In that case, only the total capacitance C_R resulting from the valve and circuit capacitances is effective as the anode impedance. (The sum of input and output hot capacitance is to be taken as the valve capacitance, and 5 pF as the circuit capacitance.) As the product of amplification and bandwidth in a single amplification stage is known to be constant, an impression can be gained of the amplification obtainable from a given tube over a required frequency range. For a wide frequency range correspondingly more amplifier stages are needed. Even though every type of valve has a limit beyond which increase in the number of amplifying stages brings about no increase in the product of amplification and bandwidth, yet, using suitable tubes (e.g. E 180 F, $g_m/C = 1.17$; $\frac{g_m}{2\,\pi \cdot C_R} = 138$ Mc/s) wide-band amplifiers with frequency ranges of up to about 60 Mc/s can now be successfully built.

There is, however, the added requirement that the output stage of the amplifier has to supply a voltage output of sufficient amplitude even at these high frequencies, to enable the spot on the oscilloscope screen to be adequately deflected.

Since the upper cut-off frequency is given by the equation $R_a = 1/\ (2 \cdot \pi \cdot f \cdot C_a)$, it is essential that the capacitive load imposed by the C.R.T.-deflection plates and their connecting leads shall be as small as possible.

Only in this way it is possible to select the value of the anode resistance R_a of the amplifier such that the required amplitude of the output voltage is achieved without distortion at a certain permissible alternating current. Moreover, it is particularly important in the case of a wide-band oscilloscope tube that it should have as great a deflection sensitivity as possible (if necessary, at the expense of other qualities). However, it must be borne in mind that wide-band and pulse oscilloscopes require the screen to be as bright as possible, so that even short pulses with a low repetition frequency can still be readily observed. This, however, requires high acceleration voltages, which, of course, reduces the deflection sensitivity.

This problem can be solved only by making concessions on other tube characteristics. It is fortunate, however, that in practice the oscillogram is usually contained within a rectangular area, the image height in the case of 10 cm tubes being generally between 35 and 40 mm. Consequently, cathode ray tubes have been produced in which the plates in the vicinity of the cathode are brought closer together than was formerly customary. As a result, the field strength and, according to Eq. (g), the deflection effects are increased at a certain value of the deflection voltage. Of course, only a limited part of the image height can be displayed. If the deflection becomes greater, the electron beam strikes the corresponding deflection plate, and the image disappears from the screen.

Tube type DG 10-54 is an example of such a tube. The electrode system of this tube, including this pair of deflection plates, is shown in Fig. 2-11. Maximum deflection by the *Y*-plates is 55 mm. It is admitted that these deflection plates are somewhat long, in fact as long as is permissible, bearing in mind the transit time effects at high frequencies, see Eq. (2.8). Hence the electron beam remains in the deflection field longer on its way from the cathode to the screen than is usual in the system employed in conventional tubes (Fig. 2-16, DG 10-74). With a total accelerating voltage of 2000 V, the *Y*-plate sensitivity of the DG 10-54 tube amounts to 0.55 mm/V, while for the DG 10-74 tube at the same accelerating voltage it is only 0.36 mm/V. With this tube, however, the deflection can extend over the whole fluorescent screen surface.

In the DG 10-54 tube, which is also obtainable with a *B*- or *P*-screen, the leads to the *Y*-plates are brought through the tube base. Thus the mutual capacitance and that of the plates to the other electrodes, which, in fact, load the output stages of the signal amplifier capacitively, are not much lower than those of the DG 10-74 tube, as can be seen from Table 2-2. In this connection it should be pointed out that only slight electrostatic coupling is desirable between the pairs of plates, as otherwise, particularly in the case of capacity unbalance, electrostatic "cross-talk" from one pair to the other can occur [40] [41].

TABLE 2-2 CAPACITANCES OF THE DEFLECTION PLATES OF A STANDARD OSCILLOSCOPE TUBE COMPARED WITH THOSE OF SPECIAL TUBES WITH A DIMINISHED VIEWING AREA FOR WIDE-BAND AMPLIFIERS

	DG 10-74	DG 10-54	DG 13-54
$C_{D1,D'1}$ — rest earthed . . . approx.	2.8	2.2	1.5 pF
$C_{D2,D'2}$ — rest earthed . . . approx.	2.8	3.3	2.0 pF
C_{D1} — rest and D'_1 earthed	5.8	4.0	2.8 pF
$C_{D'1}$ — rest and D_1 earthed	5.8	4.7	2.8 pF
C_{D2} — rest and D'_2 earthed	7.6	5.9	3.0 pF
$C_{D'2}$ — rest and D_2 earthed	7.6	5.8	3.0 pF

The DG 13-54 tube, in which all these considerations have been taken into account, is particularly suitable for use with wide-band amplifiers. The connections for the deflection plates are in the tube-neck, in the immediate vicinity of the deflection plates, as they are also and for the same reason, in the DBM 13-34 double beam tube (Fig. 2-30). (This tube will be described in a later section.) The result is not only lower lead capacitance, but also lower lead-in inductances. The deflection plates are similar to those of the DG 10-54 tube (Fig. 2-11). With a total accelerating voltage of 2000 V the *Y*-plate sensitivity is approximately 1.0 mm/V; at 4000 V it is still about 0.8 mm/V. Maximum deflection through the plates cannot exceed 65 mm, while for the plates nearest the screen (for the time base) it is limited by a circle 105 mm in diameter. The tube has a flat screen and gives a remarkably sharply defined image [42].

This tube is therefore very suitable for use in oscilloscopes with an upper frequency limit of the signal amplifier up to 30 Mc/s and gives quite good picture quality. Further increase in deflection sensitivity seems feasible if the whole cathode ray gun assembly is made as compact as possible and is located close to the cathode. This would make it possible to have a greater distance between the screen and the centre of the plate [*L* in Fig. 2-9 and Eq. (2.2)]. As the electrons would then have time to spread out in transit through this tube, however, the concentration of the electron beam would be less satisfactory (see also last two paragraphs in Ch. 2.11 "Cathode ray tubes with post-acceleration helical electrode"). This indicates once more what limits are set for the increase of the upper frequency range in the use of normal wide-band amplifiers. This matter will be discussed in more detail in Part I, Ch. 5, and it will be shown how, with the help of delay-line couplings, in which the tube and circuit capacitances are also used as circuit components, extremely high frequency ranges can be amplified (see also Ch. 2.8 "Influence of electron transit time on deflection sensitivity").

2. 14 Multi-beam tubes; high-performance tubes

It is often necessary to show two or more processes on the luminescent screen at the same time [8]).

The technique of potential surge measurements makes the additional demand that the oscillogram of extremely rapid non-recurrent phenomena should be satisfactorily recorded. Both requirements are met in the DBM 13-34 Telefunken tube. The properties of this tube as a characteristic type will be described later in more detail.

2.14.1 MULTI-BEAM TUBES

The electron beams required to indicate simultaneously the various changes of state can be generated by means of several independent beam-generating systems, or by splitting up the single electron beam emerging from one cathode [44] [45]. For the time base, a pair of time base plates is usually provided for each ray system. It is, of course, also possible to deflect a single electron beam linearly with time in the X-direction [46] and to split it up into several rays later. The advantage of this for both systems is that deflection in the time direction is absolutely synchronous. The relatively high capacitance of the measuring plates, a certain limitation of the screen surface available for each system and the fact that it is impossible to modulate both rays separately and thus to key in various time marks, are disadvantages. Tubes having several independent beam gun assemblies avoid these disadvantages.

2.14.2 HIGH-PERFORMANCE TUBES

By using sufficiently high acceleration voltages, the spot intensity has been increased to such an extent that it is possible to record photographically writing speeds of 1000 km/s and over. It is therefore also possible to use cathode ray oscilloscopes with permanently sealed tubes for many tasks formerly requiring high-performance cathode ray oscilloscopes with continuously pumped tubes and with the electron beam directly focused on to the photographic material [47] [48] [49] [50] [51] [52]. In modern high-

---

[8]) The possibility of showing two or more changes of state on a single beam oscilloscope tube by means of an amplifier changeover electronic switch is dealt with in greater detail in Part II Ch. 9.

performance tubes also both post-acceleration and grid voltage can be increased for short periods in order to increase beam intensity during periods of maximum writing speed. Care must of course be taken that the deflection sensitivity is not altered, at least not to any serious extent [53] [54].

If there are two deflection systems in such a high-performance tube, it is possible,

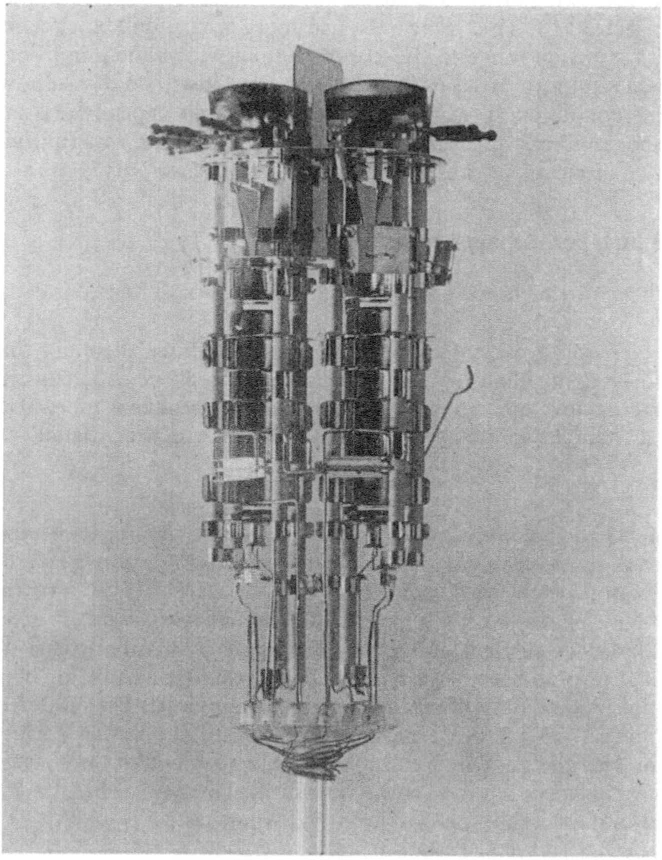

Fig. 2-29 Electrode structure of the Telefunken DBM 13-34 double beam tube

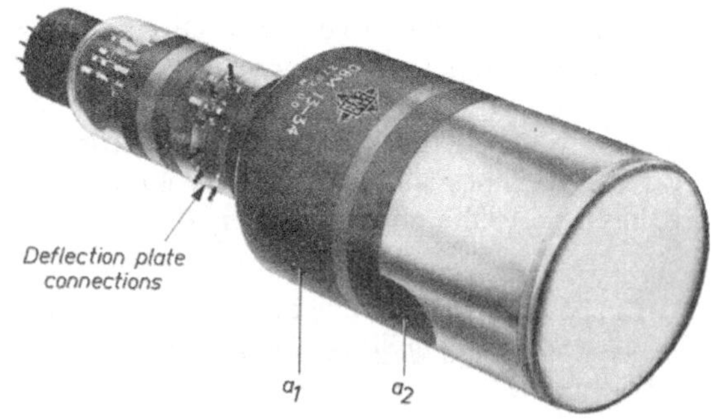

Fig. 2-30 External view of the Telefunken DBM 13-34 tube

for instance, to show the general picture of a transient voltage by means of one system and the front of the wave (over a greater period of time) with the other system. Hence it is especially desirable to use a two-beam tube for this purpose.

Fig. 2-29 shows the internal construction of such a tube with the two systems of electrodes, and Fig. 2-30 shows an external view. The terminals of both pairs of measuring plates are so situated on the tube-neck that the terminal connections are as short as possible and the capacitances of these electrodes remain low. The anode a_1 terminal and that of the post-acceleration electrode a_2 can also be seen on the bulb of the tube. The total accelerating voltage can be a maximum of 24 kV.

A total working voltage of, say, 22 kV is best subdivided in such a way that there are 11 kV between cathode and anode a_1, and a further 11 kV between this electrode and the post-acceleration anode a_2. The maximum voltage between any point in the system and the chassis is then only 11 kV, negative at the cathode and positive at anode a_3. Thus the cost of the power supply unit is relatively low, in spite of the high total accelerating voltage required to ensure a high spot intensity. It has proved possible to increase the oscillogram luminosity still further by means of a layer of aluminium on the surface of the luminescent screen facing the cathode. This layer of metal prevents a part of the emitted light from being radiated back into the tube bulb and thus lost. It is, in fact, reflected by the thin metal layer, so that it is added to the light emerging forwards. Metal backing the luminescent screen in this way is a further advantage, namely that for photographic recording of the oscillogram the shutter can be opened long before the occurrence of the awaited phenomenon without the photographic material becoming pre-fogged by the direct, reddish light of the cathode. This is particularly useful in view of the uses to which this tube is put.

As the transient voltages to be measured are usually earthed at one side, the deflection systems have been so designed that symmetrical or asymmetrical voltages can be connected to the measuring plates at will without the risk of additional marginal lack of definition in asymmetrical working.

The following are recommended operating data:

TABLE 2-3 OPERATING DATA OF THE DBM 13-34 DOUBLE BEAM TUBE

Heater voltage	V_f	= 6.3 V (indirect)
Heater current (per system)	I_f	= 0.3 A
Anode voltage	V_{a1}	= 8 kV
Total accelerating voltage (cathode-anode a_2)	V_{tot}	= 16 kV
Focusing voltage	V_{g3}	= 1.8 to 2.5 kV
Screen grid voltage	V_{g2}	= 3.0 kV
Grid cut-off voltage	V_{gc}	= —140 to —60 V
Deflection sensitivity of the pair of plates nearest the cathode	DS_{pc}	= 0.1 mm/V
Deflection sensitivity of the pair of plates nearest the screen	DS_{ps}	= 0.085 mm/V

Capacitances (per system)

Mutual capacitance of pair of plates nearest the screen (rest earthed)	1.6 pF
Mutual capacitance of pair of plates nearest the cathode	1.3 pF
Pair of plates nearest the screen to all	4.9 pF
Pair of plates nearest the cathode to all	5.0 pF

More detailed data on these and other oscilloscope tubes are to be found in the various published tube data sheets [55] or in the catalogues of the manufacturing firms.

2. 15 Special luminescent screens

2.15.1 EXTREMELY SHORT PERSISTENCE SCREEN

In the recording of oscillograms on photographic material which is moved continuously, thus providing the time base a high persistence is generally undesirable. After the actual recording it causes so-called "blurring", even in the case of the "*B*-screen" speed of the recording material is greater than 1 m/s, particularly with intense exposure. For such cases the *Z*-screen with its extremely short persistence has been developed. As the decay time curve of this screen shows in Fig. 2-31, the intensity of the light emitted from the screen decreases to 10% of its original value after 2.8μs [9]).

In general the blue-fluorescent *B*-type or green-fluorescent *H*-(P31), *N*-(P2) type tubes will prove satisfactory, particularly when there is no over-exposure. The intensity of the trace quickly falls below the level which can be recorded photographically.

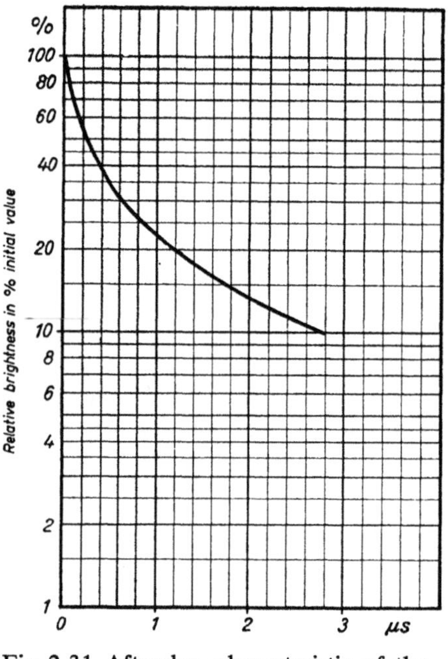

Fig. 2-31 Afterglow characteristic of the very short-persistence *Z* screen (Telefunken)

2.15.2 DARK-TRACE TUBES

In this connection the luminescent screen of the so-called dark-trace tubes deserves mention. The screen consists of a sheet of mica (in the case of the MS 17-21 Lorenz tube, for instance, a usable surface of about 8 × 12 cm) with a layer of potassium chloride on the inner side [56] [57]. With an accelerating voltage of about 10 kV an electron beam incident on this layer discolours it, giving rise to a dark blue-violet trace. In the case of the other luminescent screens described previously, the fluorescent trace of the beam path can be the more readily perceived, the lower the ambient illumination. In contrast to this, with the dark-trace tube a certain amount of illumination of the tube screen is necessary to enable the spot trace to be followed. Very weak traces can only be perceived if the screen illumination is so chosen as to offer the best contrast sensitivity to the eye of the observer. (Too great brightness abnormally accelerates the erasure of the picture in any case).

It is a peculiarity of these dark-trace screens that the trace of the spot path remains visible for from a few seconds to a maximum of several days according to the intensity of the excitation. The picture can be erased by heating or illumination; the screen then assumes its original whitish-blue colour. The mica sheet has on its outer side a trans-

[9]) In all the persistence curves shown here (see also Figs. 2-23, 2-24*a* and *b*) the starting point was an excitation light of medium intensity. Strong excitation produces a relatively faster afterglow and slight excitation a relatively longer one.

parent semiconductor layer connected to the two terminal contacts on the envelope of the tube by silver conducting strips. When it is desired to erase a picture, a voltage is connected to these electrodes; the current flowing heats the semiconductor layer and hence the mica sheet together with the screen, so that the spot trace disappears. But the coloured trace on the screen disappears after a certain time in any case even due to the light required for observation. For example, its contrast after excitation by 10 kV at room temperature without illumination drops to between 12% and approximately 7% after about an hour, but falls to zero after about 10s, when illuminated from a distance of 75 cm with a 50 W incandescent lamp.

After extinction, a certain recuperation time is required before a new trace can be written on the same place again. This time is however dependent to a great extent on the demands made by the observer as regards the contrast qualities of the new trace. For normal working conditions, in which an observation time of a few minutes should generally suffice, a recuperation time of about 15 sec is sufficient.

In order that a non-recurrent phenomenon may still produce a recognizable oscillogram, the electron beam must produce the necessary charge density on the screen. Up to a certain limit the charge density and thus the contrast can be controlled by the beam current. This limit is determined by the quality of the electron optics and the emissivity of the cathode.

With anode voltages of 7 kV to 10 kV and a beam current of 200 μA and for normal requirements for non-recurrent phenomena, a recording velocity of 100 to 400 m/s can be employed. As has been reported by Dietrich [58] as the result observations under working conditions, this depends on the required contrast threshold.

With dark trace tubes with electrostatic focusing and electrostatic deflection the obtainable writing speeds are per force considerably lower still since a large part of the electron beam has to be marked out in order to achieve sufficient marginal sharpness. This is not the only limitation of maximum writing speed restricting the field of application of dark trace tubes. The evaporation of potassium chloride when the screen is heated for erasing the picture cannot be completely avoided. This sooner or later leads to a poisoning of the cathode. Furthermore, a dark trace tube of this kind is highly sensitive towards overloading. The risk of poisoning is therefore greater than in normal luminescent screens. Moreover, the manufacturing costs of a dark trace tube are substantially higher than those of an oscilloscope tube with a normal luminescent screen. As already mentioned, the contrast between the spot tracing and the rest of the screen area is relatively small. For these reasons dark trace tubes have so far only been used to a limited degree for special purposes [59] [60] [61].

A further possibility of retaining an oscillogram directly on the tube for a longer period, so that it can be studied, is given by the oscillograph viewing storage tube described in the following section.

2. 16 Viewing storage tubes

In all regular electronic charge-storage tubes the information is reproduced as a charge image on the surface of a non-conductor, the storage layer. The secondary emission effect is used for this. There are many types of such charge-storage tubes. Input and output of information can be done by electric signals or by means of a visual picture. The period between information input and output can usually be chosen freely within wide limits. The well-known TV-camera tubes such as Iconoscope, Orthicon and

Vidicon are among these. In this connection we shall only deal with those types of storage tube which are designed for visible reproduction of oscillograms. Even with this restriction the processes taking place in storage tubes which with their detailed function determine very different effects, can only be discussed in principle here. More detailed information is given in special publications, in particular the book by Knoll and Kazan, which discusses the function of storage tubes in general and in detail [62] as well as [63], [64], [65], [66], [67], [68], [69].

The secondary emission is of fundamental importance in the functioning of storage tubes; this then above all will have to be dealt in with detail in so far as it is applied in storage tubes.

2.16.1 SECONDARY EMISSION

Fig. 2-32 indicates the most essential electrodes for the electron movement and for the secondary emission in this type of storage tube. From an electron beam gun — here indicated as an electrode system consisting only of cathode and anode — an electron beam is emitted which strikes the storage layer (the target) T consisting as a rule of insulating material. (An auxiliary electrode P (back-plate electrode) is here shown as a

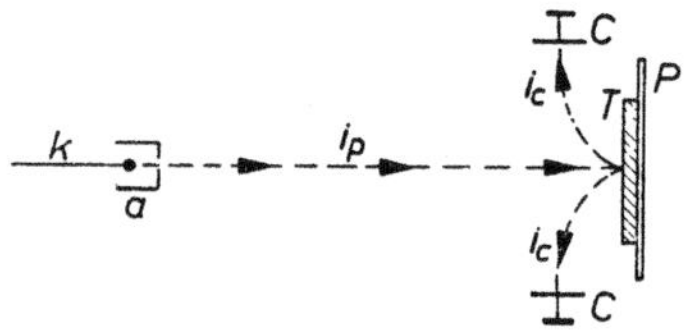

Fig. 2-32 Arrangement of electrons and electronic moving in a (direct) viewing storage tube
k = cathode; a = anode; C = collector; i_p = primary electrons; i_c = collector current; T = storage surface; P = backplate electrode

carrier of the storage layer. In the actual design of storage tubes this is a fine-meshed grid.) As soon as the electrons of such a beam in a high vacuum strike insulated elements (conductor or non-conductor) at a certain speed, they produce secondary electrons. The secondary electrons thus emitted are collected at a usually cylindrical electrode, the collector. The ratio of secondary electrons to primary electrons is termed secondary emission rate ($\delta_e = i_s/i_{pr}$). It is dependent on the acceleration energy of the primary electrons. Fig. 2-33 shows the dependence of the secondary emission rate on the volt velocity V_{pr} (electron volt). With low and high values of primary electron energy we obtain two voltage values V_{cr1} and V_{cr2} at which the secondary emission rate becomes One. They are thus known as the "cross-over voltage points" of the secondary emission. This means that there are as many secondary electrons being emitted as there are primary electrons bombarding. These cross-over point voltages are dependent on the material. The lower cross-over voltage V_{cr1} is usually below 100 V, while the upper voltage V_{cr2} has the order of magnitude of several kV (it can amount up to 20 . . . 25 kV). This curve generally has a maximum between the two cross-over voltage points. (For specific

Fig. 2-33 Dependence of the secondary-emission of the primary electrons energy

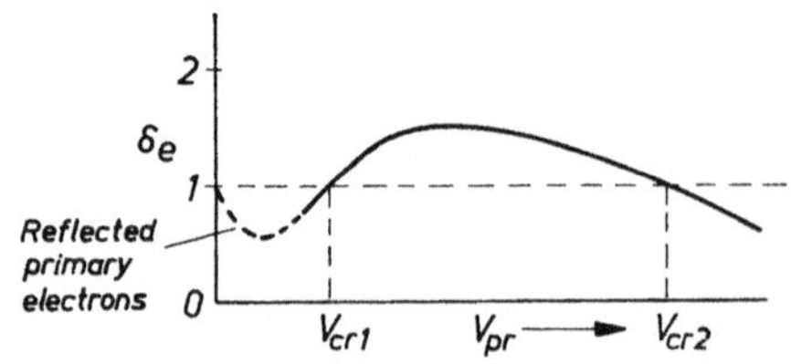

pure metals and carbon in the form of soot -Aquadag, for example- the peak value of this curve, however, is lower than one. There are no cross-over voltages with these materials. They are used if secondary emission is to be prevented in certain parts of electron tubes which are bombarded by electrons.)

With primary energies below the curve maximum the secondary emission rate δ_e rises with the beam voltage V_{pr}, as expected, because the energy of the beam increases. With primary energies above the maximum, however, δ_e decreases with increased beam voltage, as the secondary electrons are then produced increasingly in successively deeper layers of the storage screen material and as a result are prevented from being emitted even more so and are consequently absorbed. With very low primary beam energies a large portion of the primary electrons is again reflected when colliding with atoms of the storage area. These reflected electrons are now collected together with the primary electrons of the beam at the collector, and we obtain the dash-line at the beginning of the curve, which at zero volt beam voltage approaches the value 1 for δ_e. The course of this curve is thus not determined by secondary electrons only. Reflected primary electrons contribute to the collector current also in other sections of this curve. It seems permissible, however, in the following discussion on the functioning of storage tubes to study at one and the same time all the electrons arriving at the collector electrode comprehensively, the secondary as well as the reflected primary electrons.

A storage element bombarded by electrons tends to stabilize itself to one of three voltages corresponding to the radiation voltage of these electrons. With voltages below the first cross-over voltage V_{cr1} more primary electrons are bombarding than secondary electrons are being emitted. This produces an increasingly negative charge. The primary electrons arriving are slowed down to a growing extent, and the secondary emission rate δ_e is thus further reduced so that the negative charge-up develops even faster. This goes on until no more primary electrons can bombard. By ignoring the contact potential at the cathode and the output energy of the electrons, the voltage of the storage element then becomes equal to the cathode voltage. This process is known as cathode potential stabilization. With an acceleration potential equal to V_{cr1} primary electron current and secondary emission are equal. This condition, however, is unstable since a slight alteration of the storage element potential immediately results in a stabilization to the cathode potential (as described) or to the anode potential (as yet to be discussed). For acceleration potentials between the first and second cross-over voltage more secondary electrons are released than primary electrons appear. If the output energy of these electrons is sufficient to reach the field of the collector electrode, then they travel towards the latter and the potential of the relevant storage element runs into the positive. If, however, this potential increases, it intercepts at the same time the emission of further secondary electrons, so that the secondary emission rate is reduced. The potential of the storage element now stabilizes itself to a potential at which the actual current also becomes zero. These processes are shown more precisely in Fig. 2-34. The plotted curve of this illustration corresponds to the course of the effective

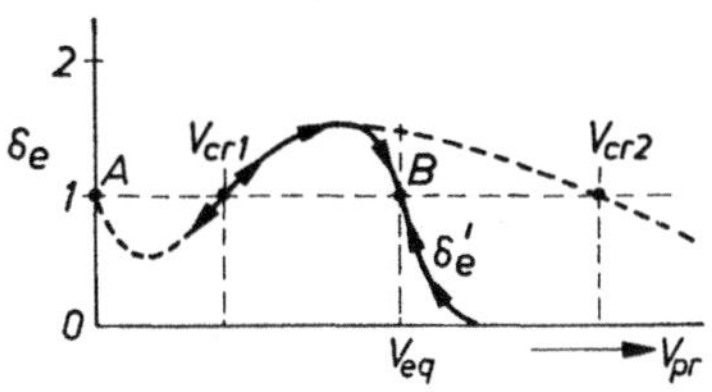

Fig. 2-34 Waveform of the effective secondary-emission ratio with equilibrium points *A* and *B*

secondary emission coefficient δ'_e under the conditions of operation described. The potential of the storage element stabilizes itself at point B on the curve at the potential value indicated by V_{eq}. (The cathode potential stabilization occurs at point A of this curve.) This method of operation is of general importance for the customary viewing storage tubes. The equilibrium is stable since a potential of the storage element increasing via V_{cq} causes a reduction of the secondary emission rate through which the potential returns again to the value V_{eq}. If the potential of the storage element is greater than V_{cr2} (the second cross-over voltage) we then have in similar fashion a stable equilibrium condition with a voltage equal to V_{cr2}.

By selecting the energy of the bombarding electrons we are capable of controlling the individual elements of the storage layer by charging them up in different degrees depending on whether they are bombarded with very slow or faster electrons. Owing to this difference in intensity of charging the storage elements, a charge diagram is produced on the storage layer. In the viewing storage tubes as described below this diagram gives us the stored display of oscillograms.

2.16.2 CONSTRUCTION OF A VIEWING STORAGE TUBE

In Fig. 2-35 following the usual description of storage tubes we reproduce a generally valid diagram of the fundamental structure of this type of tube. It contains — like normal oscillograph tubes — an electron writing gun system (the electrodes marked with the index letter w) as well as a deflector plate system for electrostatic beam deflection in both coordinate directions (D_1, D'_1 and D_2, D'_2). The electron beam of this system -the writing beam- writes its trace the usual way though not directly on to the luminescent screen, but on the storage layer (ss). A second electron beam gun (without deflector electrodes, index letter f) supplies a wide current of slow electrons -the flooding or viewing beam- spraying the entire storage area. (At times two flooding beam systems are also employed). A collimation system consisting of electrodes g_3 and g_4 distributes the current of flooding beam electrons evenly across the entire storage area and directs their beam

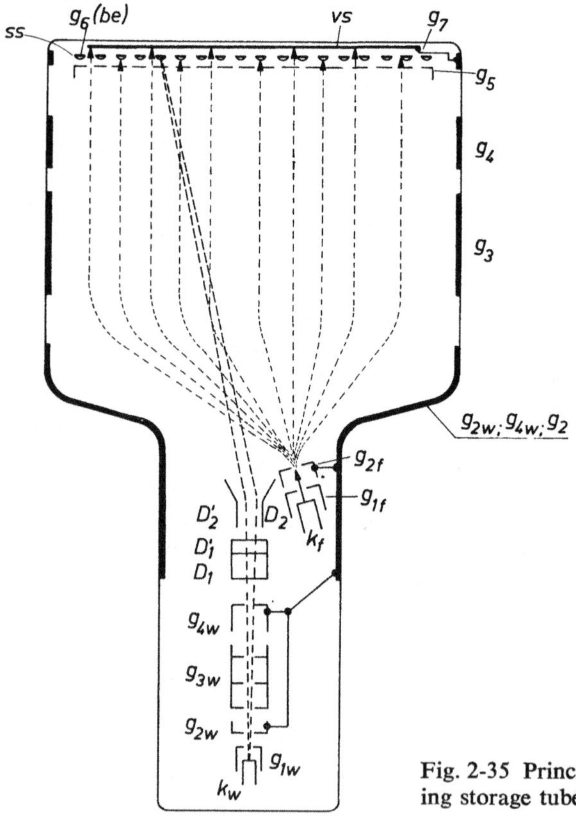

Fig. 2-35 Principal construction of a (direct) viewing storage tube for oscillographs

perpendicularly upon it. We then have the collector electrode (g_5), the storage layer (*ss*) mounted on the auxiliary electrode g_6, the backing electrode (*be*) -a very fine mesh-, the metal-backed luminescent screen (*vs*) and the post-acceleration electrode g_7. (We have given no special indices to the electrodes of the tube bulb -g_2 ... g_7- as they influence the writing beam as well as the flooding beam.) The operating voltages at the electrodes with the Mullard tube ME 1251 for example, are:

WRITING BEAM SYSTEM:

V_{kw} = — 3 kV (against V_{kf})
V_{g1w} = — 40 ... — 80 V
V_{g2w} = 3.2 kV
V_{g3w} = + 450 ... 1050 V

DEFLECTION COEFFICIENTS:

$d_1 = d_2 = 39$ V/cm

FLOODING BEAM SYSTEM:

V_{kf} = 0 V
V_{g1f} = — 50 ... — 200 V
V_{g2f} = +200 V
V_{g3f} = + 80 ... 120 V (adjust as per instruction)
V_{g4f} = + 250 V
V_{g5f} = + 85 ... 200 V (adjust as per instruction)
V_{g6f} = 0 V
V_{g7f} = + 3 kV

FURTHER DATA:

Writing speed : > 1.3 km/s (0.13 cm/μs)
Definition : 20 lines/cm
Erasing speed : 50 ... 200 ms
Max. reading time : ca. 5—15 min

As the potential of the viewing beam lies below the first cross-over voltage, the storage area becomes increasingly negative through radiation with viewing beam electrons alone, which prevents any further bombardment of electrons. The writing of the information begins when the charges produced through the writing beam become negative to a sufficently low degree (equilibrium voltage in point *B* of Fig. 2-34). Every single storage element then acts together with the other electrodes as a small electron lens and thus can be regarded as functioning as a control grid for the flooding beam electron current. Flooding beam electrons can now penetrate at these points of the charge diagram and can be attracted by the intensifier electrode, which is at a high potential, and the metal backing of the luminescent screen (g_1) connected with it. The luminescent screen is excited at those points and gives a true reproduction of the charge diagram produced by the beam trace on the storage plate. This diagram is very bright on account of the high total accelerating potential.

2.16.3 THE WRITING

At the beginning of the writing the potential of the grid net electrode g_6 (*be*) is around zero volt (or only a few volts negative depending on the tube construction), and because of the capacitive coupling with the storage area its potential also equals this voltage. The flooding beam electrons striking with low energy quickly charge the storage area to the potential of the flooding beam cathode. It then no longer accepts any electrons. All viewing beam electrons still bombarding it are attracted by the strong field of the intensifier electrode g_7 at the luminescent screen. It lights up equally brigthly across its entire area. In order to be able store an information, the storage area must first of all be made negative until the passage of further electrons is blocked and the luminescence of the screen is completely suppressed. This erasing can be achieved by a positive pulse applied to the auxiliary electrode whose amplitude must equal the corresponding cut-off voltage. Owing to the capacitive coupling the potential of the storage area then also rises temporarily up to this value; through the continuing flow of further flooding beam electrons, however, it soon returns to the potential of the flooding beam cathode. If the voltage at the auxiliary electrode now drops again at the end of the erasing pulse, then the potential of the storage area is carried along by the same amount and thus becomes correspondingly more negative. The storage area thus blocks the further flow of slow flooding beam electrons, and the luminescent screen becomes dark. If the sharply focused writing beam now writes the progress of a process upon the storage layer, then a positive charge is applied to all bombarded elements. (The generation of this charge takes less than 1 μs.) The storage layer thus becomes open to the flooding beam electrons at these points. As the current of flooding beam electrons flows in a parallel formation, we receive a true picture on the screen of the beam path on the storage area.

How long a picture thus produced can be viewed sufficiently well depends on the quality of insulation of the storage area and even more so on the ionization effects. The losses through insulation fields of the storage area can be kept very low by choosing suitable material (quartz or magnesium-fluoride). One would expect therefore that the charge diagram could thus be held for several hundred hours. Unfortunately, the gas residue in the tube space shortens this period considerably if the screen picture is to be presented continuously during this time. These residual gas molecules are bombarded by flooding beam electrons causing positive ions which travel towards the negatively charged storage area. This reduces the negative charge of this area continuously, as a result of which the entire screen background becomes slowly brighter and brighter. As a rule this storage period (max. retention time) amounts to about 60 –90 seconds, nowadays up to 10 min. It depends chiefly on the residual gas pressure but also on the construction of the tube. There are various methods of prolonging the storage time. As a rule it is usual to reduce the flooding beam current in this case. This produces fewer ions and the storage time is consequently prolonged. This, however, reduces the light capacity in proportion. Sometimes it is recommended to prolong the picture display time by not releasing the flooding beam continuously but only by tracing it on with pulses.

In practice we have to fix the most favourable compromise in this kind of operation between brightness and max. retention time.

If, after the information has been written upon the storage layer with the writing beam wich is then completely suppressed, the flooding beam current is greatly reduced or completely intercepted, then only a few ions can be formed and the storage time can then become very long (one week or longer). In that case, however, the oscillogram is invisible during that period.

If the flooding beam is released again afterwards, the oscillogram can be viewed for a limited period. In this sort of operation we have to distinguish between two periods: the actual storage time for the charge diagram on the storage area (max. retention time) and the display or reading time. As these data are not always made absolutely clear or distinct in the trade literature, it is advisable to pay special attention to them.

Considerable improvements in the storage properties of such tubes were achieved by inserting a suppressor grid in front of the collector electrode (g_5) which intercepts the ions on their way to it [70], [71]. This grid has not been specially drawn for Fig. 2-35. It is connected in that tube with the electrode g_4.

According to the process described, the storage layer can alternate between two charge conditions (bistable) –namely between the potential of the flooding beam cathode and that of the collector electrode, for instance– so that it is impossible to reproduce half-tones in this way in the oscillograms. Designs of storage tubes, however, have also been constructed in which the charge level of the storage area can be adjusted to intermediate values by intensity control of the writing beam. These tube types, however, are designed less for oscillography than for radar where the reproduction of image half-tones is of the utmost importance (Hughes "Tonatron" type 7033) [72].

2.16.4 ERASING THE SCREEN PICTURE

To do this the charge image has to be erased from the storage area. As we already explained at the beginning in the section "Writing the screen picture", the stored charge and thus also the luminescent screen picture can be erased by a short positive pulse upon the auxiliary grid (g_6, *be*) which carries the storage layer. It can also be slowly erased by a succession of very short pulses. With other constructions (Mullard ME 1251) erasing by negative pulses upon the collector electrode (g_5) is recommended. The erasing process can be influenced by means of a specific course of these pulses (rectangular or sawtooth shape). More detailed advice is given in the information charts of the individual tube manufacturers. The erasing times lie somewhere between 50 and 500 ms.

With viewing storage tubes the afterglow effect of the luminescent screen has little influence on the display time of the oscillogram. This is solely determined by the residual gas pressure, the constructional data and the working conditions chosen. As already mentioned, we have to distinguish here between maximum retention time and display time (reading time).

2.16.5 TYPES OF VIEWING STORAGE TUBES

Typical examples of such tubes for use in oscilloscopes are: the RCA tube 6866, the Hughes "Memotron" H-1038, the E 702 A tube of English Electric and the Mullard tube ME 1251. An external view of the latter is given in Fig. 2-36.

It is understandable that, owing to the complicated structure of these tubes and the high precision necessary, their price level is somewhat higher than that of oscillograph tubes of conventional design. It is to be expected, however, that in future the prices of tubes and thus also of storage oscillographs will be brought down to a minimum by reducing manufacturing costs.

Fig. 2-36 Viewing storage tube Mullard ME 1251

An interesting solution of a viewing storage tube is found by Messrs. Tektronix in their oscillograph 564. With this tube it is possible to have a choice of using the luminescent screen entirely as storage screen, purely in the conventional manner, or one half of the screen (divided vertically) for storage, or as a normal oscillograph screen. The max. retention time whit this instrument is given as one hour, and the max. writing speed as 25 cm/ms.

The firm of Hughes also supply the so-called "Memo-Corder" type 106, and the "Multitracer" model 5, as well as the wide-band oscillograph "Memoscope" type 105 (0 — 10 Mc/s, 50 mV/cm) equipped with a viewing storage tube. The Memo-Corder can be used with conventional oscilloscopes providing an additional possibility of storing the observed voltage diagram at the same time.

With the Multitracer we are able to re-trace up to 10 voltage diagrams of different processes which are displayed one line above the other.

The highest writing speed with instruments of the highest capacity is about 1 cm/μs to 2.5 cm/μs (1″/μs). Although the main field of application of the oscilloscope viewing storage tube is the display of single processes, these tubes can nevertheless be used with great advantage for investigating repetitive processes. If, for example, the processes are fast, so that the single writ does not yield a visible picture, then a satisfactory charge diagram can be produced by repeating the trace several times (integrated writing) sufficient to give a clear reproduction.

2.16.6 PRECAUTIONS IN THE USE OF VIEWING STORAGE TUBES

Due to unfavourable working conditions it may happen that the storage area potential does not reach an equilibrium at $\delta_e = 1$, but with an effective secondary emission rate $\delta'_e > 1$ steadily grows in a positive direction towards not permissible values. This kind of "run-away" charge-up can be caused by:

1. Excessive writing beam current.
2. Too low deflection speed and thus too long retention time of the writing beam on the individual surface elements of the storage layer.
3. Writing when the flooding beam is suppressed or too weak.

Such an effect manifests itself in the following ways:

1. Erasing is no longer possible.
2. Brilliant flashing at the edges of the luminescent screen as a result of flash-over between the individual storage layer elements.
3. Dark islands on the screen due to destruction of corresponding areas of the storage area or the auxiliary electrode.

Whenever these effects occur the luminescent screen voltage must be switched off at once, the potential of the auxiliary electrode must be cut off from its supply and must be connected for a few seconds to the collector electrode. When the tube is put into operation, the viewing beam current should therefore first of all always be set at its nominal value before the writing beam is released. At the end of the operation the writing beam current again should first be reduced before the flooding beam is switched off. When operating oscilloscopes with storage tubes, the relevant paragraphs of the instructions should be specially observed in order to avoid damage of this kind.

A detailed description of an oscilloscope with a storage tube is given by Cawkell and Reeves [73].

CHAPTER 3

POWER SUPPLY UNIT

3. 1 Construction

Like all other electronic measuring apparatus, cathode ray oscilloscopes are mostly designed for AC mains operation. A suitable rated mains transformer supplies all the necessary voltages on its secondary side. The tubes are generally heated by AC, which is also supplied from suitable secondary windings. The heater windings for E.H.T. rectifier valves and cathode ray tubes must be provided separately and be suitably insulated, as they are at a very high potential with respect to the chassis and to other windings.

The direct voltages required for the anode currents of amplifier tubes and the time base unit are produced by full-wave rectification of the AC mains current. Fairly large currents are required, in particular for vertical deflection amplifiers with a high upper cut-off frequency. These currents form the greater part of the load on the whole power supply unit.

3. 2 Anode voltage for the cathode ray tube; smoothing

The high anode voltage for the cathode ray tube can be obtained in the traditional way by simple half-wave rectification of a correspondingly high alternating mains voltage, if necessary using a voltage doubler or tripler etc.

Occasionally, the alternating E.H.T. to be rectified is produced by special medium frequency oscillators (40 kc/s to approximately 300 kc/s). On account of the high frequency, smoothing is then relatively simple. Electronic stabilization of the direct current so produced can also be achieved by regulating the heterodyne oscillator stage (Fig. 3-10).

For television large-picture projection especially, the E.H.T. is also produced by rectifying the decaying alternating voltage of an oscillatory circuit excited by voltage pulses [1].

The specially developed Philips tubes 1875, 1876 and 1877 are suitable for rectification. The EY 51 and EY 86 tubes are suitable for lower currents. The PL 81 power pentode, connected as a diode, has also been used for this purpose.

The basic circuit of such an E.H.T. rectifier is shown in Fig. 3-1. The voltage required for the auxiliary anode a_1 is obtained by a variable voltage divider, and by this means the focus of the spot is adjusted. The negative grid voltage required for the adjustment of luminosity is in this circuit taken from potentiometer R_2, which is in series with the filter resistors R_1 and R_2. Resistor R_4 and capacitor C_3 are used for additional smoothing and decoupling. Table 3-1 gives a summary of the ratings of the circuit elements required for the operation of some types of Philips cathode ray tubes.

As the current consumption of the anode seldom exceeds 250 to 500 μA in practice, a total standing current in the voltage divider of from 1 to 2 mA is sufficient. For this reason, only a resistor is used as the smoothing element. It may even be possible to

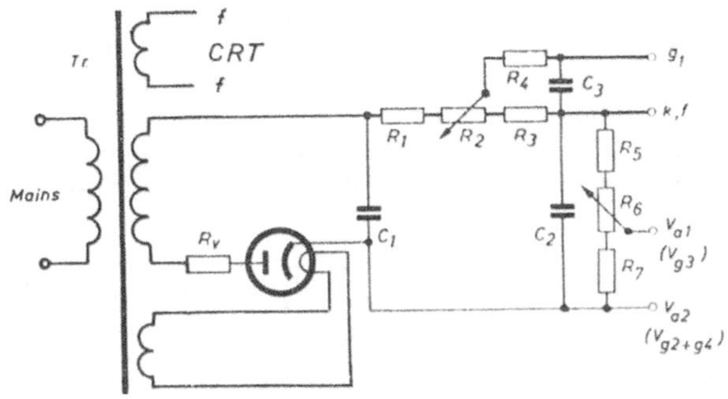

Fig. 3-1 Conventional circuit of an E.H.T. power pack

obtain satisfactory smoothing by means of a sufficiently large charging capacitor. For this purpose, two or three electrolytic capacitors rated for the highest possible operational voltages may be connected in series. Two capacitors, each of 8 μF for 550/660 V working, for instance, with a resulting capacitance of 4 μF, yield satisfactory smoothing at 1000 V. It is remarkable that anode voltage fluctuations caused by ripple manifest themselves more as a cyclical variation of deflection sensitivity than as a reduction of spot intensity. This can be explained by the fact that the auxiliary anode voltage is obtained by dividing the anode voltage. As the ratio of the anode voltage to the auxiliary anode voltage determines the spot intensity, it requires considerable anode voltage fluctuations to cause noticeable changes in spot intensity. In the oscillograms of Fig. 3-2 this is indicated in more detail by comparing the image of the anode voltage waveform with the brightness of the spot, and of the zero line and the image of a sinusoidal 300 c/s alternating voltage during three cycles of ripple voltage. In this case only charging capacitors of 0.1 μF or 0.4 μF were used for smoothing.

In order that the deflection sensitivity fluctuations caused by ripple in the accelerating voltage should not be noticeable, they must be smaller than half the spot diameter. As the spot diameter in good cathode ray tubes is half a millimetre, the change of the beam deflection as the result of ripple should be less than 0.25 mm. At a maximum deflection of 75 mm, this corresponds to a permissible ripple of 0.3%. If the anode voltage is 1 kV, for instance, this means a variation of only 3.0 V. All these consider-

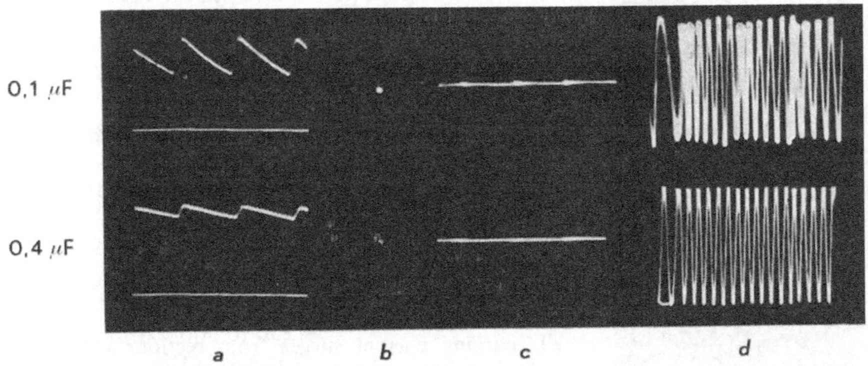

Fig. 3-2 Influence of anode voltage ripple on spot focus and deflection sensitivity

ations refer to the beam acceleration voltage before deflection. They only apply to the post-acceleration voltage in so far as the deflection sensitivity is affected by them.

The exact rating of rectifier and smoothing networks has already been dealt with in such detail elsewhere [2] [3] [4] [5], that it is unnecessary to consider the matter further here. The few hints which follow are given only for general guidance. Thus, for instance, the following rule of thumb equation can be used to calculate the ripple voltage:

$$\Delta V\,[V_{pp}] = 12 \cdot \frac{I_= \,[\mathrm{mA}]}{C\,[\mu\mathrm{F}]}\,. \tag{3.1}$$

In words this means, that for 1 mA of current and 1 μF capacitance the ripple voltage is 12 V_{pp}. (For full-wave rectification ΔV is 4.2 V_{pp} under the same conditions.) From Eq. (3.1), for a specified permissible ripple ΔV, the capacitance C is obtained from

$$C\,[\mu\mathrm{F}] = 12 \cdot \frac{I_= \,[\mathrm{mA}]}{\Delta V\,[V_{pp}]}\,. \tag{3.2}$$

The process of rectification can be examined in somewhat more detail by reference to the oscillograms in Fig. 3.3. In Fig. 3-3*a* are shown two cycles of alternating voltage across the secondary winding of the transformer, and in Fig. 3-3*b* the half cycles of this voltage, which are passed by the rectifier valve (without smoothing capacitor). The oscillogram in Fig. 3-3*c* then shows the direct voltage on a filter capacitor of 0.1 μF for a current of 1.5 mA. In order to choose the rectifier valve it is important to understand the way in which the current passes through to the charging capacitor. The voltage across the charging capacitor rises during charging, as is shown in Fig. 3-3*c*, almost up to the peak voltage of the rectified wave. After the decay of this voltage pulse, the capacitor discharges until the next voltage pulse makes further charging possible. This means, however, that during the negative half cycle current passes through the rectifier valve

a

b

c

d

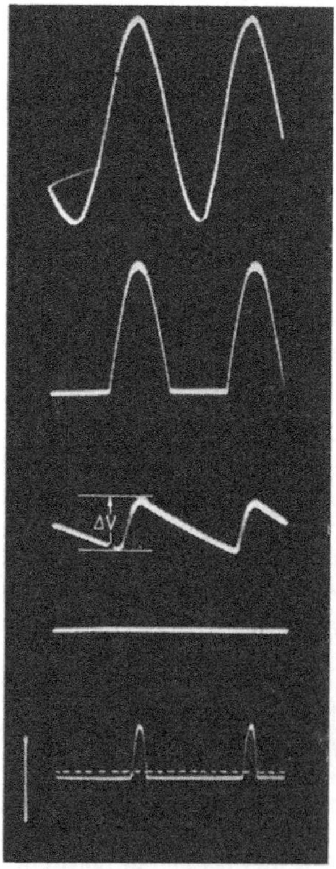

Fig. 3-3 E.H.T. voltages and currents in half-wave rectification

a) AC voltage V_{Tr} of secondary windings of transformer (approx. 850 V_{rms})

b) Rectified half-cycles of AC voltage (capacitance C not present)

c) DC voltage $V_=$, when $C = 0.1\ \mu$F and current drain is 1.5 μA; ΔV = ripple voltage

d) Charging current surges from rectifier to C. The broken line corresponds to the mean value of the constant current drain $I_= = 1.5$mA only during the short time in which the voltage across the capacitor is lower than that of the secondary winding of the transformer in this half cycle.

As the charging of the capacitor must now be correspondingly heavy, this current pulse may at times attain considerable peak values. In Fig. 3-3*d* the oscillogram curve of the current is shown for the example mentioned. The trace of a line made by an alternating current was used as a reference before this oscillogram was taken, and was found to correspond to a current of 30 mA_{pp}. It will be seen that the peak value of the charging current pulses in Fig. 3-3*d* is 18 mA.

Particularly when large capacitors are used, these current pulses must sometimes be limited, by means of a series resistor as indicated in Fig. 3-1, to the magnitude permissible for the tube in question, for instance, the EY 51. Charging, of course, then takes longer, and the voltage on the capacitor becomes even less capable of attaining the peak voltage.

The following are some further hints on rating the smoothing filter for the extra high tension supply. If the first filter capacitor is chosen at 0.5 μF, then according to Eq. (3.1) taking a current of 1.5 mA, a ripple will occur across this capacitor, its value being:

$$\Delta V = 12 \cdot \frac{1.5}{0.5} = 36 \text{ V}.$$

As a ripple voltage of only 3.0 V is permissible, the smoothing must be increased $\frac{36}{3.0} = 12$ times. To this end, another resistor and a further capacitor as shown in Fig. 3-4*a* must be added after the first capacitor. As a result of the ripple voltage across C_1, an alternating current flows through the resistor R and the capacitor C_2. It is desirable that the impedance of the capacitor C_2 should be as small as possible compared with the resistance R. In other words, the ripple voltage must, as far as possible, occur wholly across the resistor. In the example quoted, the ripple across C_2 should at most be in the ratio of 1:12 to the ripple across C_1. This means, therefore, that the ratio of the total impedance of R and C_2 to the reactance of the capacitor C_2 should be as 13:1.

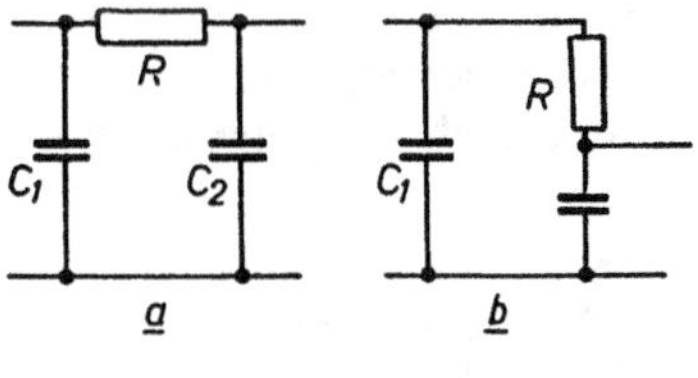

Fig. 3-4 Resistance-capacitance filter

The reactance is:

$$X_c = \frac{1}{\omega \cdot C}. \tag{3.3}$$

Here, $\omega = 2\,\pi f$, in which f is the frequency of the ripple voltage, and C is the capacitance in farads. The impedance of this circuit is:

$$Z = \sqrt{R^2 + \left(\frac{1}{\omega \cdot C_2}\right)^2}. \tag{3.4}$$

The ratio of X_c to Z, i.e. the attenuation of the ripple at C_1, thus becomes:

$$\frac{X_c}{Z} = \frac{\frac{1}{\omega \cdot C_2}}{\sqrt{R^2 + \left(\frac{1}{\omega \cdot C_2}\right)^2}}. \tag{3.5}$$

From Eq. (3.5), equations are obtained which allow the values for R ad C_2 to be calculated directly. If the attenuation ratio of $\frac{X_c}{Z}$ is taken to be s, then the following equation is obtained for the filter resistance R:

$$R = \frac{1}{\omega \cdot C_2} \cdot \sqrt{\frac{1}{s^2} - 1}. \tag{3.6}$$

When R is given, the capacitance C_2 is:

$$C_2 = \frac{1}{\omega \cdot R} \cdot \sqrt{\frac{1}{s^2} - 1}. \tag{3.7}$$

In the instance quoted, the intention was to obtain an attenuation factor of at least 13 times. If a capacitance of 1.0 μF is used for C_2, then R must be not less than 40 kΩ for the desired smoothing. At an anode current of 1.5 mA a direct voltage drop of 60 V occurs across the resistor.

The transformer voltage must therefore be chosen correspondingly higher. This fall in voltage is not entirely unavailable for the operation of the cathode ray tube. It can still be used as a grid bias, as has been demonstrated in Fig. 3-1. In practice, in such a case the resistance will be rounded off to 50 kΩ, and the voltage drop of 75 V will be accepted to ensure avoidance of "hum". In circuit diagram 3-1 the filter elements have been rated with this in mind (Table 3-1).

In order to prevent random variations of the screen image due to fluctuation of the anode E.H.T. caused by mains voltage surges, it has been recent practice to stabilize this voltage electronically in addition, by a method similar to that described in the following section dealing with high tension supply (Figs. 3-9 and 3-10).

As an example of a voltage multiplier circuit for generating the anode voltage for

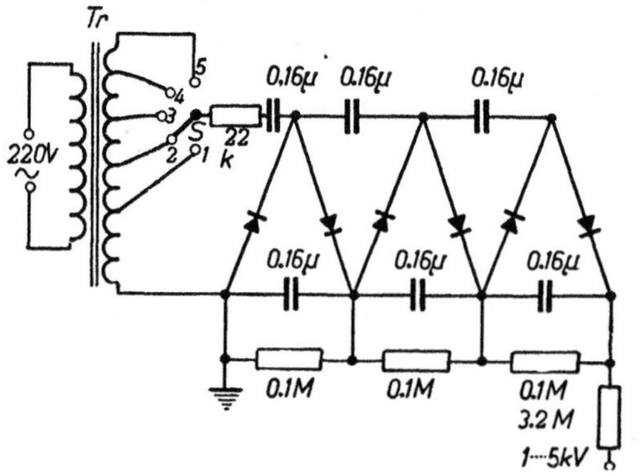

Fig. 3-5 Circuit of the Philips "GM 4188" 5kV post-acceleration unit

the cathode ray tube, Fig. 3-5 shows the circuit of the Philips "GM 4188" post-acceleration unit.

This unit makes it possible to supply a number of types of Philips oscilloscopes with additional, higher post-acceleration voltage. Here a sextuplicator circuit with rectifier cells is fed from the secondary side of the mains transformer at an alternating voltage which ensures that direct voltages of 1, 2, 3, 4 or 5 kV, as desired, are available at the output socket. The whole rectifier circuit is contained in an oil-filled sheet-metal housing because of the high voltages employed.

The maximum current which may be taken from this apparatus is 100 μA. The ripple is less than 40 V (0.8%) with a load of 50 μA and an output voltage of 5 kV. (The post-acceleration voltage has considerably less influence on the deflection sensitivity than the voltage for "pre"-acceleration [V_{a2}]. For the post-acceleration voltage a considerably greater ripple was thus permissible than for the anode voltage across anode a_2).

3. 3 High tension unit and electronic stabilization

3.3.1 HIGH TENSION UNIT IN GENERAL

When large amounts of current are required, it is sometimes necessary to connect two rectifier valves in parallel, as in Fig. 3-6. Alternatively, semiconductor rectifier elements (silicon rectifiers) are often used. As the amplifier and the time base unit are usually both fed from the high tension power unit, often using extremely low frequencies (a few c/s and less), but both require to be decoupled, it is of the greatest importance that the internal resistance of the power supply unit should also be sufficiently low at these low frequencies. The normal filter elements only fulfil this condition under certain conditions, so that a satisfactory solution can be found only by using an electronically stabilized power supply.

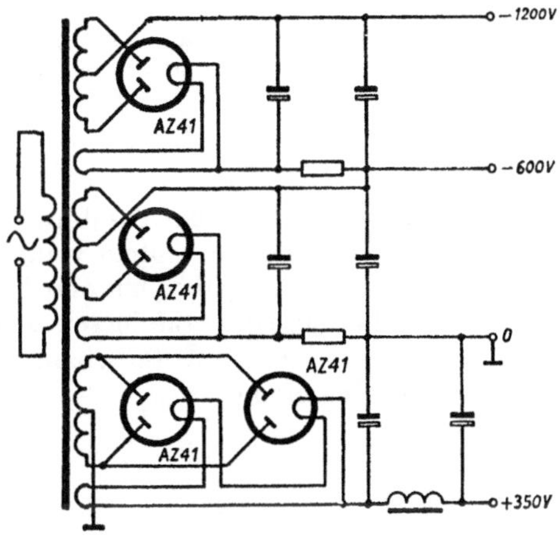

Fig. 3-6 Power supply unit of a large oscilloscope (Philips "GM 5653") H.T. section: two AZ 41 rectifiers in parallel; E.H.T. section: two full-wave AZ 41 rectifiers delivering 600 V= connected in cascade

3.3.2 ELECTRONIC STABILIZING

The simplest circuit for the electronic stabilization of direct voltage under on-load conditions is shown in Fig. 3-7. A triode of internal resistance R_i is connected between the voltage source V_i and the load resistor R_B. A constant voltage, for example, from a dry battery or a gas discharge stabilizing tube [6] [7], is applied between the grid and the negative pole of the input and output voltages. The output voltage V_{out} then adjusts itself to a somewhat higher value than the reference voltage V_g, to such an extent that

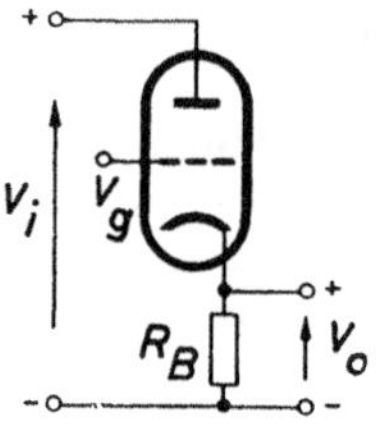

Fig. 3-7 Simplest form of electronic stabilization without amplification of control potential

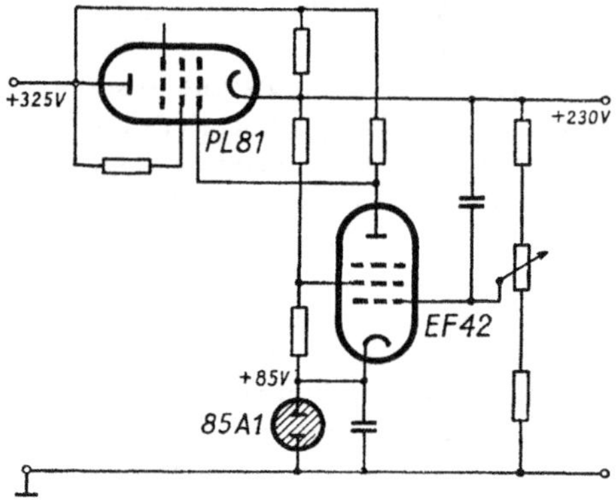

Fig. 3-8 Electronic stabilization of H.T. section with amplification of control potential

the voltage between the control grid of this tube and the cathode is exactly equal to the voltage at which the outgoing current is transmitted. (The cathode voltage of the tube therefore "follows" the grid voltage. This circuit is therefore identical with that of the cathode follower.) If the voltage V_i is increased, while the current taken remains unchanged, then the output voltage V_{out} also increases slightly by just enough to enable the increase in grid voltage V_g to compensate the effect of the increase of V_i.

But since even a slight change in grid voltage can cancel the influence of a relatively large input voltage change, the output voltage change ΔV_{out} is only a fraction of the input voltage change ΔV_i. The proportion of this change α is a measure for the stabilizing effect of this circuit, and is therefore

$$\alpha_{(\Delta I=0)} = \frac{\Delta V_{out}}{\Delta V_{in}}. \tag{3.8}$$

If, on the other hand, the current drawn by the load increases, while the input voltage remains constant, then the output voltage decreases slightly. Because of this, V_g also becomes slightly less negative, but only by the amount necessary to increase the current through the tube to equal the increased load current. The ratio of the change of the output voltage $-\Delta V_{out}-$ to the change of current ΔI is the internal resistance R_i of this stabilizing circuit.

Hence:

$$R_{i(\Delta V_i=0)} = \frac{-\Delta V_{out}}{\Delta I}. \tag{3.9}$$

As can be seen from the derivation [8] [9] [10], the measure α of the stabilizing effect in this circuit is:

$$\alpha = \frac{1}{\mu+1} \approx \frac{1}{\mu} = D, \tag{3.10}$$

where μ = amplification factor and D = penetration factor of the valves.
The internal resistance R_i is obtained from:

$$R_i = \frac{\mu}{\mu + 1} \cdot \frac{1}{g_m} \approx \frac{1}{g_m}, \tag{3.11}$$

where g_m = the mutual conductance of the valve.

This simple circuit is very often used to stabilize the voltage in circuits drawing a small current, such as those supplying negative grid voltages and screengrid voltages.

The reference voltage for V_g is taken from the stabilized voltage of the power supply by means of further amplifier valves in a way which will be described later in more detail (circuit diagram 3-8).

In this connection it should be pointed out that in similar circuits this control valve can serve simultaneously as a rectifier. This can be particularly useful if the voltage to be applied to the load is to be adjusted over a particularly wide range [11] [12]. However, the regulation is then reduced considerably. An increase in the stabilizing effect of the circuit shown in Fig. 3-7 is obtained if the change of output voltage which serves to control the valve in the circuit, is amplified by an additional valve. Fig. 3-8 shows a circuit such as is used for electronic stabilization of E.H.T. in the "GM 5653" oscilloscope, and is typical of modern practice. Between the smoothing filter of the mains rectifier unit and the load there now appears the internal resistance of the PL 81 valve connected as a triode. The grid potential of this valve and hence also its internal resistance, are controlled by the voltage across the load, a fraction of which is taken via a voltage divider and amplified by a high-slope EF 42 pentode. If, for example, the output voltage of the power supply were to drop for any reason, the voltage across the load would also drop. A slight change in load voltage is sufficient to lower the grid voltage of the PL 81 valve as amplified by the EF 42 valve, and the internal resistance of the PL 81 falls. Because of this, the voltage drop across this valve decreases, so that the voltage across the load remains substantially constant. If, for example, the mains voltage changes by 10%, the load voltage changes only by a few pro mille. The constant grid voltage for the control valve, which may be considered as the reference voltage, is supplied by the 85 A1 stabilizer valve, which has a particularly high degree of stability. The operating voltage changing of this valve is not more than 1 V over a period of 5000 hours service. In practice the change is only a few tenths of a volt. This method of electronic stabilization ensures that the voltage actually applied to the load is independent of the current taken, and that this current in turn is to a large extent independent of the mains voltage. It also means, that the internal resistance of the power supply is very small (10 to 25 Ω).

The stabilizing effect of this circuit is expressed by the equation:

$$a = \frac{1}{G_c \cdot \mu + 1} \approx \frac{1}{G_c \cdot \mu}, \tag{3.12}$$

and the internal resistance of the whole circuit is expressed by:

$$R_{i\ \mathrm{tot}} = \frac{\mu + R_i \cdot g_m}{g_m (G_c \cdot \mu + 1)} \approx \frac{1}{G_c \cdot g_m} \approx \frac{R_i}{G_c \cdot \mu}, \tag{3.13}$$

where G_{st} = the gain obtained by means of the control valve. In Eq. 3.13 the influence of the load resistance R_i of the load is now also taken into account.

By employing suitable circuits it is possible to modify the standard characteristics in order to obtain any desired degree of stabilization. It is also possible –at least within certain limits– to obtain "overcompensation" such that the output voltage increases with rising load. By using additional valves and circuit elements, the stabilizing factor α of the circuit can be reduced to a very low value [13].

The reduction of the internal resistance holds good for all frequencies from zero upwards to the range in which filter capacitance at the output functions as a short-circuit. A certain loss of voltage must, of course, be accepted (in this instance about 95 V). Naturally, this electronically controlled voltage feature, which involves some extra manufacturing cost, is used only where it is absolutely essential. The output stages of the signal amplifier with its considerable current consumption can usually be fed directly with the voltage from the smoothing filter, and only the screen grid voltages of the valves, which determine the anode current of pentodes, need perhaps be kept constant by electronic stabilization (Fig. 5-53).

3. 4 Electronic stabilization of the E.H.T. supply

Even when the E.H.T. supply is electronically stabilized, it may happen that, when there are mains voltage fluctuations, the image on the screen increases or decreases in size on account of the fluctuation of the E.H.T. applied to the cathode ray tube. This effect can be overcome only by stabilizing the E.H.T. electronically. [10])

As an example of this, Fig. 3-9 shows the circuit of the power supply unit for the cathode E.H.T. of — 1200 V in the Philips "GM 5654" oscilloscope.

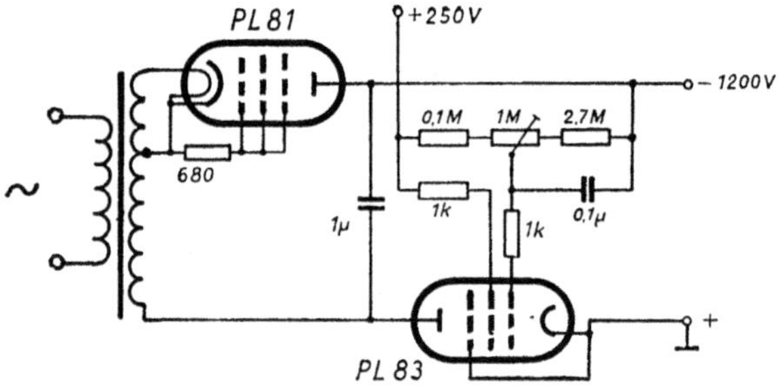

Fig. 3-9 Electronic stabilization of E.H.T. section (Philips "GM 5654")

A PL81 pentode, connected as a diode, serves here as a rectifier valve. The direct current taken from the 1 μF smoothing capacitor must flow through the PL 83 pentode connected in the positive line of this circuit. The second grid of this valve is at an electronically stabilized voltage of +250 V as described in Fig. 3-8. The control grid is fed by a voltage taken from a voltage divider connected between this +250 V stabilized supply and the output E.H.T of —1200 V. The voltage divider is so adjusted that the operating point of the valve at the rated mains voltage and normal load current lies within the desired working range of I_a/V_{g1} characteristic curve. As the voltage of this voltage divider is stabilized electronically at +250 V, every change in the negative output E.H.T. (whether due to mains voltage surges or to the cathode ray tube) causes

---

[10]) Stabilization with "cold cathode voltage stabilizer tubes" has the disadvantage that the voltage source has to supply a roughly 50% higher voltage. Moreover, with the valves usually used for the purpose, voltage rises due to sudden changes in gas pressure or to temperature influences are likely to occur.

variations of the load current drain (for example as the result of intensity adjustment a proportional change in the (negative) voltage on the first grid of the PL 83 valve. Thus the impedance of this valve changes in such a way that it inhibits all voltage fluctuations across the E.H.T. output. Although the control voltage for the PL 83 control tube is not specially amplified by an additional valve, as is the case with the stabilizing circuit of the E.H.T. supply (Fig. 3-8), nevertheless, the high mutual conductance of this valve ensures adequate stabilization of the E.H.T. voltage. In this way, too, there is a very considerable decrease in the residual ripple voltage from the smoothing capacitor. Thus, a stable oscillogram with a well defined spot trace uninfluenced by chance mains voltage surges is obtained.

For generating high direct voltages in which ripple must be very small (e.g. for colour T.V.), an "electronic" filter which exploits the Miller effect can be used, as has been shown by V. Wouk [14], to reduce the residual ripple to about $5 \cdot 10^{-5}$ of the voltage value.

3. 5 Stabilized E.H.T. supply with a medium-frequency oscillator

In power supply units in which the E.H.T. is obtained by rectifying a transformed 50 c/s alternating voltage, relatively expensive filter capacitors are required for smoothing, because of the low frequency. For this reason, the alternating E.H.T. is often generated in a special medium-frequency oscillator (30 kc/s to 300 kc/s). Very much smaller, and therefore cheaper, smoothing capacitors can then be used. Moreover, it is relatively easy to adopt electronic stabilization of the E.H.T. output in such a circuit. An example of this is the circuit of the E.H.T. unit in the Siemens oscilloscope "Oscillar II" shown in Fig. 3-10. The EL 90 valve in the Hartley oscillator delivers the alternating voltage at about 40 kc/s across transformer secondary windings for generating the direct E.H.T. as well as the heater supply for two EY 51 tubes from further secondary windings. The anode of the upper EY 51 valve in the circuit diagram is connected to the alternating voltage, and this valve delivers a positive voltage at 2 kV on the smoothing capacitor. This voltage is used as the post-acceleration voltage. The other EY 51 tube is connected to the alternating voltage at its cathode, so that a

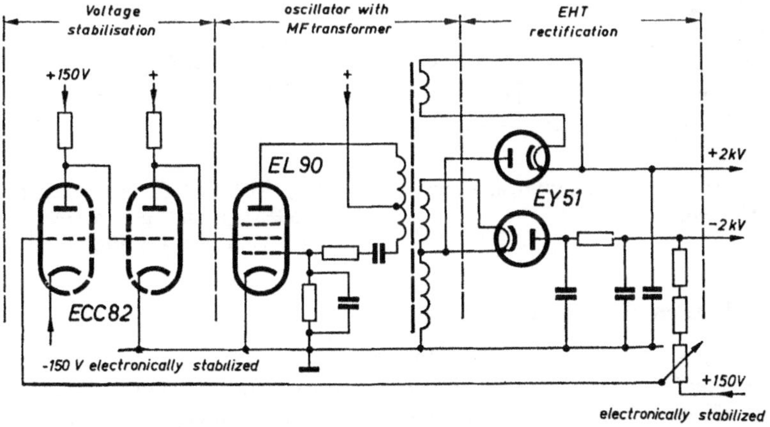

Fig. 3-10 Generating E.H.T. with medium frequency oscillator and electronic stabilization (Siemens "Oscillar II")

negative voltage of about 2 kV appears across the smoothing capacitor. Since the ripple of this voltage must be extremely small, a *CRC*-filter is provided. In addition, this voltage is stabilized electronically. For this purpose the operating point of the oscillator valve and hence its mutual conductance and the amplitude of the alternating voltage generated are controlled via a two-stage direct voltage amplifier by means of a voltage from a voltage divider, as described in the preceding section, situated between an electronically stabilized voltage of +150V and the output voltage of —2 kV. If, for any reason, this voltage goes more negative, it results in a higher grid bias at the control grid of the EL 90 oscillator valve, so that the output voltage immediately readjusts itself to the pre-determined value. The load characteristic of such a circuit results in an almost constant low internal resistance up to a certain current limit. If this limit is exceeded, however, the impedance increases rapidly and the output voltage drops. Such a circuit is therefore relatively a proof against short-circuiting [15] [16] [17] [18].

3. 6 Simplifications of the circuit in the E.H.T. section

The simplifications shown in Figs. 3-11 and 3-12 have become generally accepted as a means of reducing the cost of the power supply unit. As can be seen in Fig. 3-11, a part of the E.H.T. winding can be dispensed with if a half of the H.T. winding (*b*) is connected in series with the E.H.T. winding (*c*). The small current drawn by the E.H.T. section does not noticeably disturb the symmetry of the H.T. section. In another arrangement, shown in Fig. 3-12, the negative E.H.T. for the cathode, as well as the equally high positive voltage for the post-acceleration electrode of the C.R.T., can be obtained from one E.H.T. winding by using two reciprocally opposed rectifier valves. The load currents in this case flow through each rectifier valve in turn in both half-cycles.

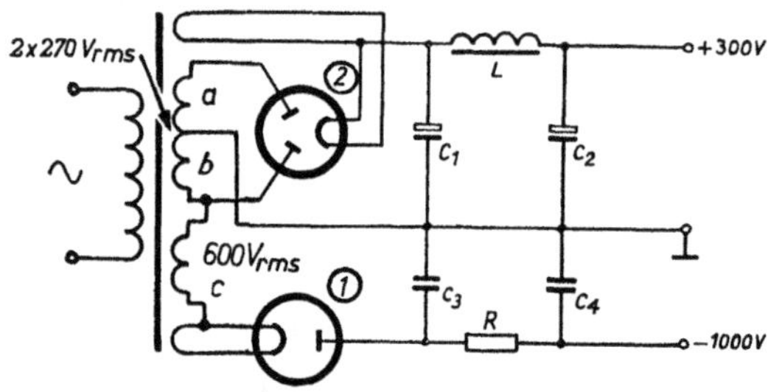

Fig. 3-11 E.H.T. power supply using one half of H.T. winding

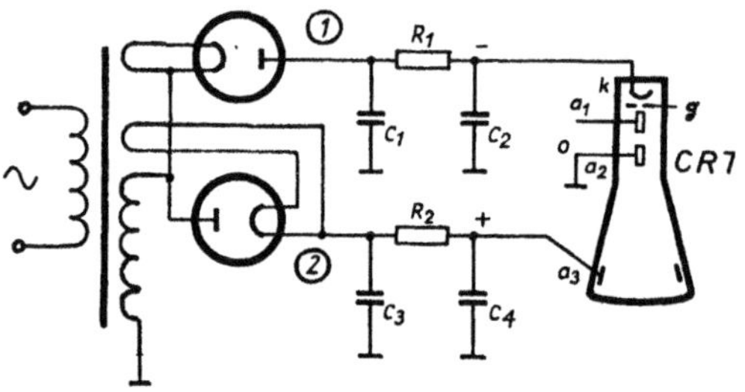

Fig. 3-12 E.H.T. power supply for cathode (–) and post-acceleration anode (+) from one winding

3. 7 Stabilization of the heater voltage

In oscilloscopes with AC amplifiers, it is generally unnecessary to maintain a constant heater voltage. Owing to the thermal inertia of the cathode, sudden voltage fluctuations have not effect, while slow emission changes almost always fall below the lowest transmission frequency, so that they too can cause no perceptible change in spot deflection.

The case is different with DC amplifiers. Then, every change in the operating point of the input valves represents a corresponding change of the output operating point, multiplied by gain of the amplifier. It is therefore advisable, first of all to operate oscilloscopes having DC amplifiers via an alternating voltage stabilizer [19] [20]. Only alternating voltage stabilizers which do not produce too great a distortion (i.e. less than 3%) of the alternating current should be used, otherwise the voltage appearing at the smoothing capacitors of the H.T. section, which is determined by the maximum amplitude of the rectified voltage, is very dependent on the phase relationship of the harmonics. In unfavourable circumstances this voltage can be so low that, when operating at under-voltage, the electronic stabilization is no longer effective.

When the gain of the signal amplifiers is not too great, stabilization of the heater supply by an iron-hydrogen barretter is sufficient. When the demands made are high, the heater current of all series-connected valves must be kept constant electronically in the manner described previously. A transistorized, electronically regulated power supply is often used for supplying heater current to filaments connected in parallel.

As will be discussed in detail in Ch. 5.27 "DC voltage amplifiers", there are various ways in which the emission of the most important valves in an amplifier can be maintained at a constant level electronically. This serves to level out continuous as well as transient (flicker effect) changes in emission.

3.8 Zero position adjustment; astigmatism control

In order to adjust the "no signal" position of the luminous spot precisely, or to move it in the event of asymmetrical images appearing on the screen, it is necessary to supply the deflection plates with direct voltages which can be suitably adjusted. These voltages must be variable over a certain range between positive and negative values. A suitable voltage is obtainable by connecting a potentiometer between two points of sufficiently high positive and negative voltage in the E.H.T. supply network, e.g. in Fig. 3-6, to "—600 V" and "+ 350 V".

If one of the deflection plates is earthed, the luminous spot can, of course, only be shifted by a direct voltage potential applied to the other -"hot"- plate, as indicated in the circuit in Fig. 3-13*a*. In the case of such an asymmetrical operation, a certain amount of unbalance of the oscillogram (trapezoidal distortion and astigmatism), as already described has to be accepted, particularly in the case of considerable image shift.

Completely undistorted, sharply defined images are obtained in symmetrical operation with equally symmetrical direct voltage shift, as is shown in the circuit in

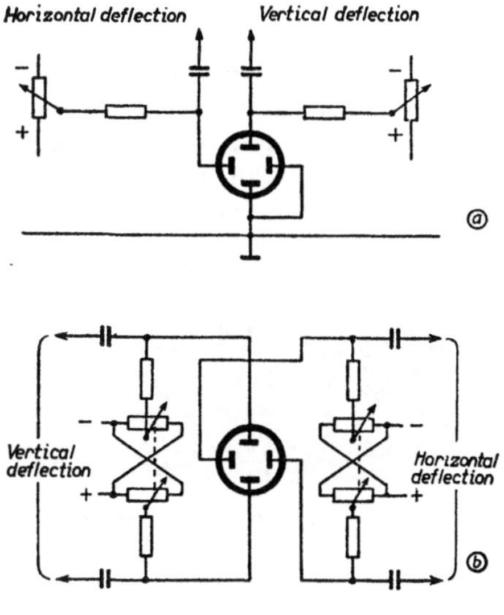

Fig. 3-13 Supply for positioning the spot.
a) for unbalanced deflection plates
b) for balanced deflection plates

Fig. 3-13*b*. For each pair of plates a tandem potentiometer is now needed, in which the two resistance paths are cross-connected. Every increase in voltage in the positive sense at one plate is inevitably accompanied by a corresponding negative one at the other. It is very important here, that the mean potential between the pair of plates remains constantly equal to that at anode a_2. The mid-position of the potentiometer must therefore actually correspond to V_{a2}.
If this is not the case, for instance if, when the spot is in the central position, both plates have a voltage of say +15 V, an intolerable astigmatism occurs, that is to say the spot is no longer circular but longish. In addition, the form of the image is dependent on the position of the spot. In the design of power supply units, very careful attention must be paid to this too. In order to be able to find the electrical centre point subsequently with the utmost precision, special adjustments are also provided to permit either the mean potential of the pair of deflection plates or the potential of anode a_2 to be correspondingly shifted; the influence on spot definition is the same in either case.

In oscilloscopes with direct voltage amplifiers, the potential of anode a_2 must be set at the same value as the mean direct anode voltage of the output valves, since here the deflection plates are directly connected to the output electrodes. As shifting of the output voltage of these valves by changes taking place in their operating points cannot be entirely avoided, it is essential that such oscilloscopes should be provided with this adjusting device. Such a device for readjusting the average plate potential (astigmatism control) makes it possible to obtain a good luminous spot even in the case of asymmetrical working of balanced deflection plates [11]). As, however, the shifting of the plate potential must always correspond to the mean value of the deflection voltage on these plates in order to obtain a well-defined screen image, this second focus adjustment is to some extent dependent on the amplitude of this voltage. The shift voltage must therefore (at least in the case of considerable changes in deflection voltage) also be correspondingly readjusted.

3.9 Screening the cathode ray tube

As the electron beam in a C.R.T. can also be influenced by magnetic fields, the stray field of the mains transformer can cause unwanted deflection of the spot unless special precautions are taken [21]. In practice, the effect of the stray field is that, instead of a spot, the electron beam produces a more or less inclined line or loop (Fig. 3-14a). If beam deflection is carried out at the same time, Then, instead of the usual straight line, an irregular line or band appears on the screen (Fig. 3-14*b* and *c*). By trial and error, the position of the mains transformer at which the minimum amount of irregularity takes place can be determined.
However, the tube must also be magnetically screened to get rid of any residual magnetic interference. This is best done by means of a mu-metal cylinder. This screening

Fig. 3-14 Effect on the screen produced by magnetic interference

11) Philips "GM 5660" oscilloscope.

TABLE 3-1

MAGNITUDE OF THE CIRCUIT COMPONENTS FOR VARIOUS PHILIPS C.R.T.'S IN AN E.H.T. UNIT OF THE TYPE SHOWN IN FIG. 3-1

C.R.T.	V_{Tr} [V_{rms}]	Valve	Capacitors [μF]			Resistors [MΩ]								Shunt current roughly [mA]	Electrode voltages [kV]	
			C_1	C_2	C_3	R_v	R_1	R_2	R_3	R_4	R_5	R_6	R_7		V_{a2} ($V_{g2} + V_{g4}$)	V_{a3} (V_{g6})
DG 7-5 DG 7-6	850	1876	0.5	0.5	0.5	—	0.1	0.05	—	0.5	0.2	0.1	0.5	1.0	0.8	—
DG 7-31 DG 7-32	640 1100	EY 51 or EY 86	0.25	0.1	0.5	0.02	0.05	0.05	0.05	0.5	0.005	0.1	0.4	1.0 1.6	0.5 0.8	—
DG 7-36	1470	EY 51 or EY 86	0.25	0.1	0.5	0.02	0.05	0.05	0.04	0.5	0.27	0.20	1.2	1.0	1.5	—
DG 10-6 [12]	850	1876	0.5	0.5	0.5	—	0.05	0.05	—	0.5	0.15	0.10	0.42	1.5	1.0	1.0
DG 10-74 [13]	1500	1875 + EY 51	0.5	0.5	0.5	—	0.05	0.10	—	0.5	0.40	0.30	1.25	1.0	2.0	2.0
DG 10-78 [13]	950	2 × EY 51 or EY 86	0.25	0.1	0.5	0.02	0.05	0.05	0.018	0.5	0.25	0.20	0.82	0.8	1.0	2.0
DG 13-2 [12]	1500	1875	0.5	0.5	0.5	—	0.05	0.10	—	0.5	0.4	0.3	1.25	1.0	2.0	2.0

[12] $V_{a2} = V_{a3}$

[13] Generation of post-acceleration voltage with an EY 51 valve, as in the circuit of Fig. 3-12 (series resistance 20 kΩ; one 0.1 μF charging capacitor is sufficient with a parallel discharging resistor of 5 to 10 MΩ).

must be carefully freed of remanent magnetism, as otherwise a permanent deflection of the luminous spot from the zero position and a certain amount of astigmatism will occur. It is easy to recognize the presence of remanent magnetism by the eccentric parallel movement of the spot on the screen when the shield is turned about the tube. For a single-layer magnetic shield with a highly permeable material, the effect of the shielding can be calculated approximately from Eq. (3.14).

$$\frac{\text{Magnetic field strength without screening}}{\text{Magnetic field strength with screening}} = 0.22 \cdot \mu \cdot \left[1 - \left(1 - \frac{d}{r_o}\right)^3\right] \quad (3.14)$$

In this equation, μ = permeability of the screen material, d = thickness of the screen and r_o = radius of the surface to be screened. For particularly high demands, several shield layers have to be used. For detailed calculations see the special bibliography [22].

3. 10 Example of a power supply unit for a high-performance oscilloscope

The target cost in designing a power supply unit for oscilloscopes depends to a large extent upon the required performance and other requirements, such as stability of the individual voltages. In Fig. 3-15, the circuit of the power supply unit for the Philips "GM 5666" oscilloscope with DC amplifier is given as an example.

Two HF chokes –L_{403} and L_{404}– are connected in the mains leads to the primary winding of the transformer T_1 in order to prevent the radiation of the sawtooth alternating voltage and its high-frequency harmonics. The numbers 1 tot 9 indicate connecting points whereby, via a rotary switch, the transformer can be adjusted for operation at any of six different mains voltages from 110 to 245 V. This switching is obtained by varying the connections of windings S_1, S_1' and S_1''. A thermal cut-out VL_1 and two fuses VL_2 and VL_3 suitably protect the transformer against damage resulting from a long period of medium overloading or a short period of severe overloading.

The transformer has ten secondary windings. The various AC heater currents are delivered by windings S_3, S_6 and S_9. The signal lamp La_3 is also supplied from winding S_3 and the illumination lamps La_1 and La_2 for the floodlit scale from winding S_6. The illumination of the scale can be adjusted by variable resistor R_6. The alternating voltage delivered by winding S_2 is rectified by the rectifier valve 14 (GZ 34) for the H.T. section, which has to supply the largest direct current. A *CLC*-filter chain, consisting of C_{409}, L_{402} and C_{411}, smoothes out the ripple to a great extent. Choke L_{402}, the series element of this chain, is tuned to 100 c/s by the capacitor C_{408}, so that the ripple voltage is almost entirely eliminated. In those parts of the oscilloscope where high voltages are required and electronic stabilization can be dispensed with (output stages of the amplifier), a smoothed direct voltage of +270 V is thus available at the output of this filter. In addition, a voltage is taken via a control valve 15 and amplified by the pentode system of an ECF 80 (valve 12) to give an electronically stabilized output at + 170V. The input stages as well as the preamplifier stage and phase-inverter stage of the time base unit are fed from this supply. For the other valves of the time base unit, this voltage is decoupled by the *RC*-link R_{273} and C_{225}. The reference voltage for the electronic stabilization is obtained from an 85 A 2 (valve 16) stabilizing valve. This valve receives its supply from a rectifier network consisting of the winding S_2, valve 13 (EZ 80 as half-wave rectifier) and the *CLC*-filter C_{405}, L_{401} and C_{406}, which produces

Components power supply unit.

$R_{6} = 100\ \Omega$
$R_{167} = 100\ \Omega$
$R_{273} = 47\ \Omega$
$R_{315} = 680\ \Omega$
$R_{316} = 1.2\ k\Omega$
$R_{324} = 330\ \Omega$
$R_{401} = 5.6\ k\Omega$
$R_{404} = 1\ k\Omega$
$R_{405} = 180\ \Omega$
$R_{406} = 150\ k\Omega$
$R_{407} = 100\ k\Omega$
$R_{408} = 1.5\ M\Omega$
$R_{409} = 1.2\ M\Omega$
$R_{411} = 1\ M\Omega$
$R_{412} = 100\ k\Omega$
$R_{413} = 180\ \Omega$
$R_{414} = 180\ \Omega$
$R_{415} = 6.2\ k\Omega$
$R_{416} = 150\ k\Omega$
$R_{417} = 500\ k\Omega$
$R_{418} = 1\ M\Omega$
$R_{419} = 4.7\ M\Omega$
$R_{420} = 18\ k\Omega$
$R_{421} = 90\ \Omega$
$R_{423} = 90\ \Omega$
$R_{422} = 4.7\ M\Omega$
$R_{424} = 1\ M\Omega$
$R_{425} = 180\ \Omega$
$R_{426} = 180\ \Omega$
$R_{427} = 4.7\ M\Omega$
$R_{429} = 100\ k\Omega$
$R_{431} = 22\ k\Omega$
$R_{432} = 180\ \Omega$
$R_{433} = 180\ \Omega$
$R_{434} = 15\ k\Omega$
$R_{435} = 10\ M\Omega$
$R_{436} = 120\ k\Omega$
$R_{437} = 100\ k\Omega$
$R_{438} = 220\ k\Omega$
R_{439}, $R_{440} = 100\ \Omega$
$R_{446} = 2.2\ M\Omega$
R_{447}, $R_{448} = 100\ \Omega$
$R_{451} = 1.8\ k\Omega$

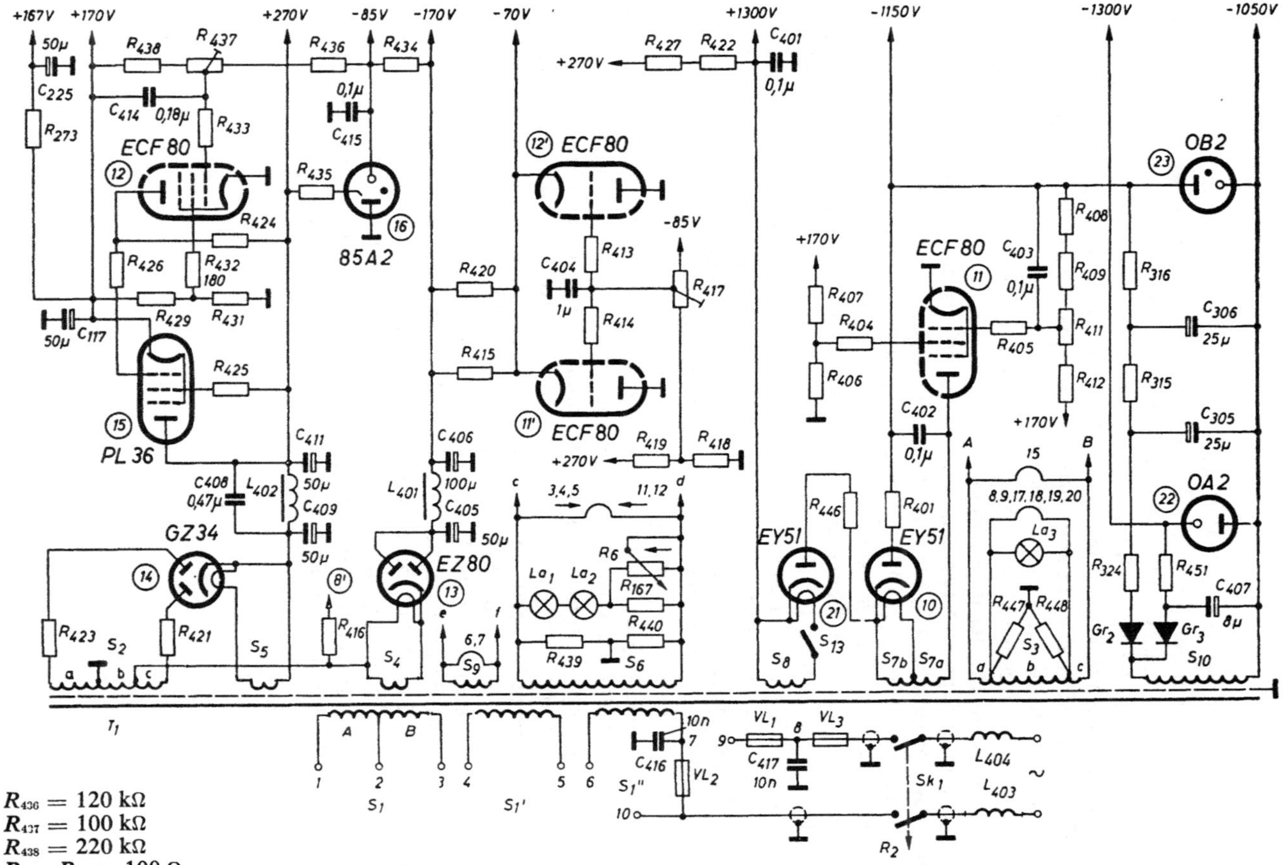

Fig. 3-15 Complete power supply unit of a high-performance oscilloscope (Philips "GM 5666")

a negative output voltage of —170 V. This voltage is required mainly as the cathode voltage for the DC amplifier. The two valves 11′ and 12′ (triode systems of two ECF 80 valves), connected as cathode followers, are provided for those stages for which a negative voltage must be taken from a low impedance source (Fig. 3-7). The grid voltage of these valves is taken from a voltage divider connected between the stable voltage of —85 V (85 A 2) and the unregulated voltage of +270 V. The grid voltage, and hence the output voltage at the cathodes of these valves, is thus to a certain extent dependent on this unregulated voltage. If the nominal voltage of +270 V rises, a corresponding rise in the negative output voltage of —70 V is also obtained, so that obviously a certain degree of compensation of the effects due to the rise of the unregulated positive voltage is brought about by the negative output voltage.

The negative E.H.T. of —1150 V for the cathode of the cathode ray tube is obtained from winding S_{7a}, via rectifier valve 10 (EY 51), electronic stabilization in the way already described (Fig. 3-9) being provided by means of valve 11 (pentode section of an ECF 80). The second E.H.T. rectifier valve 21 (EY 51) delivers an E.H.T. of +1300 V to C_{401} (after S_{13} is closed) for the post-acceleration of the cathode ray.

If this circuit is interrupted by switching off S_{13}, then the capacitor C_{401} is not charged by the rectifying valve 21, but from the +270 V voltage point via the resistors R_{422} and R_{427}. In this way a choice can be made between post-acceleration voltages of + 270 V or +1300 V according to the image intensity required. The rectifier circuit formed by winding S_{10} and the two germanium diodes D_2 and D_3 (OA 85) supplies additional voltage points of —1050 V and —1300 V, i.e. potentials of +100V and —150 V with respect to the —1150 V terminal. These voltages are required for the brightening circuit of the time base unit. This involves considerable cost, as a direct voltage coupling between the anode of the bistable multivibrator (+100 . . . . +150 V) and the control grid of the oscilloscope tube (about —1100 V) is needed. A more detailed description of the "brightening" control circuit is to be found in the section dealing with the time base unit of this oscilloscope (Fig. 4-64).

Examination of the circuit of this power unit makes it clear that meeting the greatly increased demands made on oscilloscope performance results in considerably higher costs of the power supply unit. In the Philips "GM 5662" wide-band oscilloscope, the power supply unit is the same as the one described here, apart from some non-essential details.

CHAPTER 4

TIME BASE UNIT

4.1 The display of a variable quantity

The most important property of the cathode ray oscilloscope, and upon which most of its practical applications are based, is the ability of the instrument to display in a uniquely clear and accurate manner the changing values of a magnitude over a period of time. The examination of such records often provides much valuable information: for example, it may give a clue to the cause of the discrepancy if the shape of the trace differs from what had been expected (distortion).

In order to explain this mode of operation of the cathode ray oscilloscope it will be advantageous first to consider the way in which physicists and engineers commonly represent changes in the value of a given quantity as a function of time.

Starting from the datum or zero point, the instantaneous values of the quantity under examination are plotted vertically, the successive values being moved progressively to the right, i.e. along a horizontal "time" axis, suitable scales being chosen for the vertical and horizontal axes.

Positive values of the quantity examined are plotted above the zero line, and negative values below it. The individual instantaneous values, which appear to be displaced to the right by amounts representing the time which has elapsed, are then joined up to form a continuous curve or graph which depicts the value of the quantity as a function of time.

As a simple example, Fig. 4-1 shows a voltage which changes sinusoidally as a function of time. This method is so generally used, that it is commonly referred to as the "image of the process". Other methods of display are however also possible. In the discussion of practical measuring technique this matter will be dealt with in more detail.

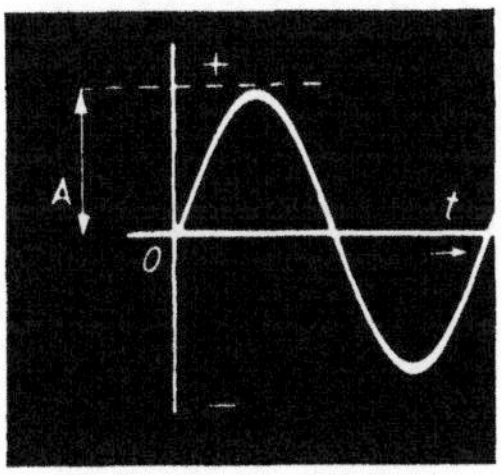

Fig. 4-1 Representation of a quantity which varies with time. A = magnitude of the quantity to be measured, t = time axis moving progressively to the right

It should be noted at this point that the representation of the mutual dependence of one magnitude or of several magnitudes on one or several other magnitudes, but not on time, can also be achieved by means of the cathode ray oscilloscope.

4. 2 Time base deflection in general

In order to obtain with the cathode ray oscilloscope an image of the variation of a given magnitude with time, it must be possible to observe the changes of value over a sufficiently long period. This means that the deflections of the luminous spot must be so presented that the individual instantaneous values of the signal magnitudes can be perceived "simultaneously" by the observer over the selected period of time. For this purpose, the Y-plates are supplied with a voltage corresponding to the magnitude to be measured, so that the luminous spot is moved vertically in accordance with the instantaneous values of this voltage.

At the same time a voltage which increases at a uniform rate during each time period is applied to the X-plates, so that the luminous spot is also moved to the right at a constant rate dependent on time. Thus the vertical deflections caused by the signal voltage do not produce a vertical line, but proceed progressively side by side to the right. Hence, in a manner completely analogous to the mode of display described earlier, the luminous spot on the screen moves through a curve which depicts the time-dependent changes of the phenomenon under observation.

Fig. 4-2 explains the formation of the picture on the fluorescent screen resulting from simultaneous vertical and horizontal deflections. It represents a time base cycle divided into 12 equal sections. The time durations of the signal voltage is shown over rather more than one cycle on the right of the actual screen trace. Below the screen image the changes in voltage at the X-plates are indicated by the voltage which increases uniformly with time, and this time period also starting from zero, is indicated as proceeding downwards. If, from the individual points of the voltage on the X-plates, verticals are drawn and are produced to meet the screen image, and horizontals are drawn and produced to the relevant instantaneous values of the signal magnitude, the

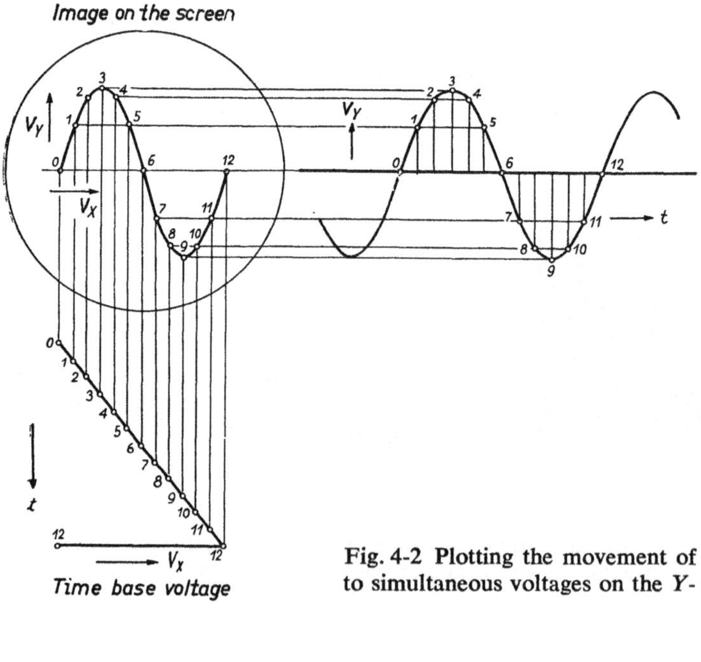

Fig. 4-2 Plotting the movement of the spot due to simultaneous voltages on the Y- and X-plates

points of intersection represent the positions of the luminous spot at the corresponding instants of time.

To simplify the explanation, it has been assumed above that the time base cycle is equal to exactly one cycle of the signal magnitude, and also that the signal and the time base voltage both commence simultaneously at zero. In this way the picture of a single cycle of signal voltage variation is displayed on the luminescent screen.

With such a non-recurrent or "single stroke" time deflection, only one cycle of the curve of the signal voltage will be recorded. Only in the case of relatively slow-decaying phenomena, and then only by employing either a tube giving long afterglow or by means of photography, it is possible to obtain a useful image. It was not until the idea was hit upon of repeating the time deflection cyclically that a satisfactory way of depicting cyclic phenomena was found. If matters are so arranged that the time base always covers exactly one or several complete cycles of the signal (synchronization), or if the sawtooth voltage of the individual time base cycles of equal duration are initiated in a controlled manner at a specified value of the amplitude of the signal voltage (triggering), then the trace of the luminous spot always occupies the same postion on the screen; the afterglow of the tube screen, and also the persistance of human vision causes the observer to perceive a static image. [14])

4. 3 Generating the time base voltage

For the time base deflection a voltage is required which increases linearly with time to a pre-set and adjustable value and then falls as rapidly as possible to zero. This process must be continuously repeatable at regular and adjustable intervals. In Fig. 4-3 three cycles of the desired pattern of this voltage are shown. A "sawtooth" waveform is the result (hence index S in V_S), so-called because the voltage always falls to zero on reaching its peak value. Among the many available circuits suitable for the production of a voltage with the required waveform is the discharge-tube time base circuit shown in Fig. 4-4 [1]. In this circuit capacitance C is charged from a direct voltage source V_b via resistor R. The glow discharge tube G_l is connected in parallel with the capacitance.

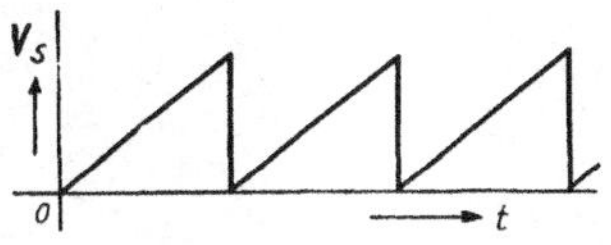

Fig. 4-3 The waveform of the time-deflection plate voltage

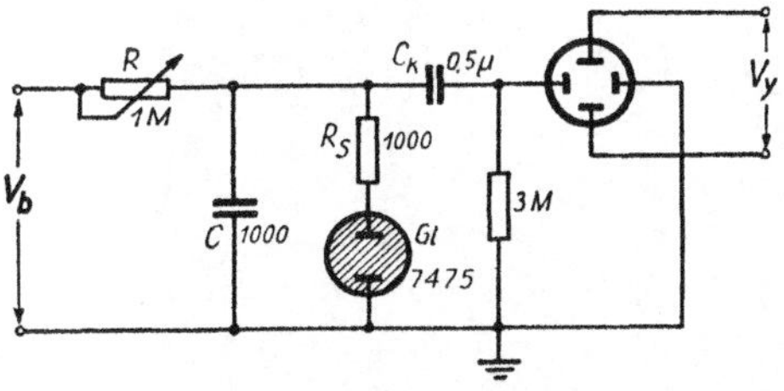

Fig. 4-4 Discharge-tube relaxation oscillator for time base voltage. R_s resistor for limiting discharge current

[14]) If the required time base frequency is lower than about 30 c/s, the picture begins to flicker more noticeably as the frequency decreases. The eye is thus more and more able to follow the path of the luminous spot. That the scanning rate required for a steady oscillogram must be higher than the minimum frame frequency in a cinema projector (minimum 16 frames per second) can be explained by the fact that the image on the luminescent screen, just as in television, does not occur as a whole, but is produced by a moving luminous spot.

When the voltage at the capacitor reaches the ignition point of the tube, the capacitor discharges through the tube until the voltage drops below the extinguishing voltage. At that instant the flow of current through the tube ceases and the charging of the capacitor re-commences. The voltage variation thus appearing across the capacitor is as shown in the oscillogram in Fig. 4-5_a, and it is seen that it is similar to the sawtooth waveform required for the time base (Fig. 4-3). The curve of the capacitor current is shown in Fig. 4-5*b*.

An essential condition for the production of such oscillations is that the discharge time is of such short duration that no considerable re-charging can take place via resistor *R* during the discharge time; in other words, the "decay-time" of the discharge must be short compared with the *RC* time constants of the charge circuit [2] [3].

The magnitude of ΔV, i.e. the voltage variation (Fig. 4-5*a*) is equal to the difference between the ignition and extinguishing voltages of the tube employed. The approximate frequency of this voltage change can be calculated from the equation:

$$f = \frac{I_c}{\Delta V \cdot C} \,. \tag{4.1}$$

For a particular tube giving a fixed voltage ΔV therefore, the frequency is governed by the average charge current I_c and by the value of the charge capacitor *C* (the timing capacitor). The current, and hence the frequency, can be adjusted by changing the

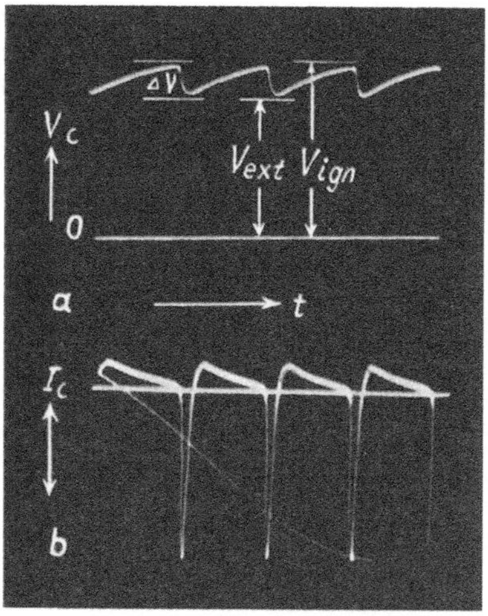

Fig. 4-5 Waveform of current and voltage across charging capacitor *C* in Fig. 4-4. *a*) voltage waveform, V_{ign} = ignition voltage, V_{ext} = extinction voltage; *b*) waveform of the capacitor current: charge above zero line, discharge below zero line

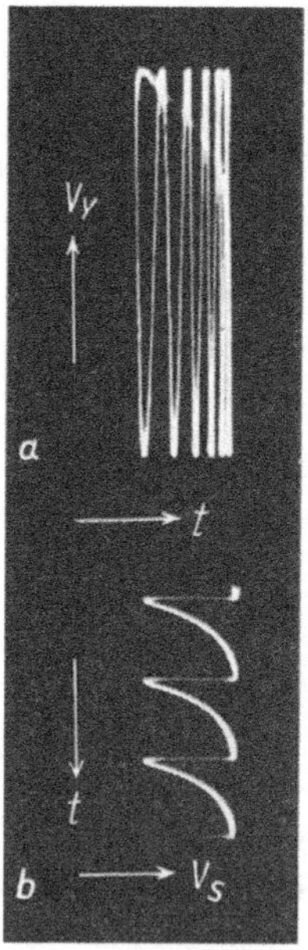

Fig. 4-6 *a*) Pattern on screen produced by a time base voltage as in circuit in Fig. 4-4; *b*) waveform of the *X*-plate voltage

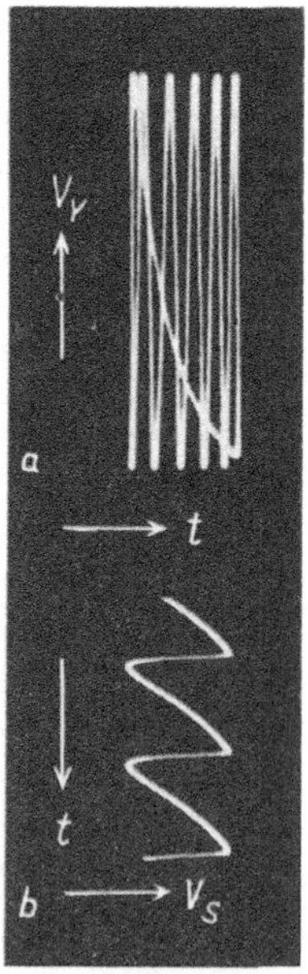

Fig. 4-7 *a*) Screen pattern as in Fig. 4-6*a*, but with higher supply voltage V_b. *b*) Waveform of the *X*-plate voltage

resistance R. In order to prevent the direct voltage V_b from reaching the deflection plates (which would shift the luminous spot sideways), a capacitor C_k must be interposed between the glow discharge tube and the deflection plate. However, in order to maintain the direct voltage at this plate at the predetermined value, this plate must be earthed via a resistor from 1 to 10 MΩ. The other deflection plate is, of course, directly connected to earth. The deflection when using the discharge-tube time base circuit is therefore "asymmetrical". Using a type 7475 Philips tube, which has an ignition voltage of 110 V and an extinguishing voltage of 85 V, the difference ΔV is 25 V. This voltage difference, the value of which is mainly determined by the type of gas-filling, is relatively low and is usually insufficient to deflect the luminous spot horizontally across the whole width of the screen.

Fig. 4-6*a* shows an oscillogram produced when a time base voltage from such a discharge tube circuit is applied to the X-plates and a sinusoidal alternating voltage of suitable frequency is simultaneously applied to the Y-plates. Apart from the insufficient image width, it is noticeable in Fig. 4-6*a* that the signal voltage curve is increasingly compressed towards the right. This can be quite simply explained from Fig. 4-6*b* which shows the waveform of the time base voltage. The voltage does not rise in the ideal manner, i.e. uniformly with time, but increases more and more slowly towards the end of the charging period. This is because the instantaneous value of the charging current is determined by the difference $V_b - V_c$ at each particular instant. Because the voltage on the capacitor increases, this difference becomes progressively less, so that the charge current progressively decreases (Fig. 4-5*b*). The charging current at any instant is given by the equation:

$$i_c = \frac{V_b - V_c}{R}\,. \tag{4.2}$$

This accounts for the characteristic waveform as shown in Figs. 4-5*a* and 4-6*b*. The instantaneous voltage across the capacitor is given by the equation:

$$v_c = V_b \cdot \left(1 - \mathrm{e}^{-\frac{t}{R \cdot C}}\right). \tag{4.3}$$

(in which v_c is the capacitor voltage after time t; V_b is the battery voltage and e is the base of the natural logarithms e = 2.718).

From Eq. (4.3) it is possible to show that the voltage waveform is less curved if a high value of V_b can be chosen. In that case the difference $V_b - v_c$ even towards the end of the charge is not much less than at the beginning, so that the charge current remains more constant and the voltage increase on the capacitor is more uniform.

This improvement is clearly visible in Figs. 4-7*a* and *b*. For Fig. 4-6 the direct voltage was V_b = 150 V, while for the oscillograms of Fig. 4-7 it was 250 V [15]). Fig. 4-8 shows that this gives a sufficient amplitude and degree of linearity for modest requirements. The waveform of two cycles can be quite clearly followed.

From the examples illustrated, however, it is quite clear that a sufficiently high value of the sawtooth voltage is not likely to be attained in this circuit. Although it would be possible to amplify this voltage as is done in other time base circuits, the discharge tube circuit has further drawbacks, so that, while serving a useful purpose

[15]) These pictures were made with a DG 9-3 C.R.T. with a voltage of about 950 V on anode a_2.

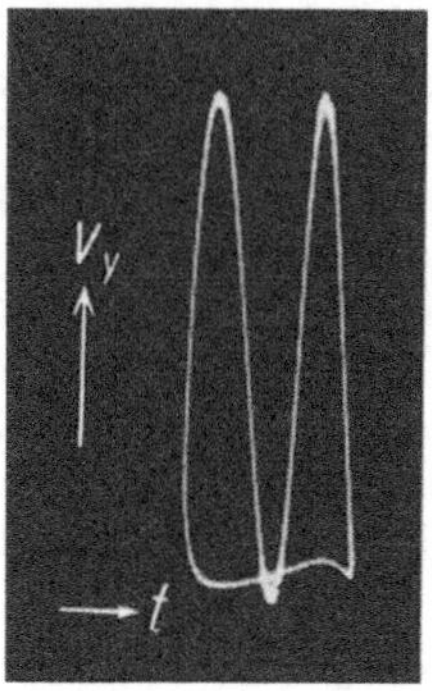

Fig. 4-8 Oscillogram of two cycles of an alternating voltage with time base deflection as in Fig. 4-7*b*

during the early stages of development of cathode ray oscilloscopes, its use is almost entirely discontinued. It has been described here merely as an example of possible circuitry for producing sawtooth voltages and because it furnishes some insight into the behaviour of gas discharge tubes, a matter which is also of interest.

4. 4 Required time base amplitude

When designing a time base generator, it is first necessary to decide on the voltage amplitude required. As time base voltages are non-sinusoidal, they cannot be measured by normal voltmeters calibrated in *rms* values. In view of the fact that the total horizontal deflection path of the luminous spot corresponds to the peak values of the deflection voltage [16]), and since, also, these peak values must be taken into account in rating the voltage source for the time base voltage, it is always necessary to specify and design on the basis of V_{pp} the "peak to peak" voltage (here the peak to peak voltage V_{pp} is, by analogy, identical with the direct voltage $V_{=}$).

The deflection voltage required is determined by the deflection sensitivity of the cathode ray tube under the prevailing operating conditions.

According to Eq. (2.3) the deflection sensitivity is determined by the relationship $DC = X/V_X$, where X is the deflection and V_X is the voltage on the X-deflection plates. The voltage required for a given deflection (direct voltage $V_{=}$ or peak voltage V_{pp}) is derived thus:

$$V_X = \frac{X}{DS_{=}} \, .$$

For the DG 10-6 tube, for instance, when $V_{a2} = V_{a3} = 2$ kV, and for an 80 mm horizontal deflection, $V_{pp} = 80/0.30 = 267$ V_{pp}.

4. 5 Time base circuits using a thyratron

Satisfactory time base voltages can be obtained by using a gas discharge valve which has a control grid and usually an indirectly heated cathode, in addition to the anode. Such valves, which are sometimes used at the present time for generating low-frequency time base voltages, are known as thyratrons.

Fig. 4-9 shows a valve of this type with its three electrodes: an indirectly heated

---

16) See Part I, Ch. 2, Fig. 2-10*b* and Chapter 2.4 "Calculating the beam deflection".

cathode, a control electrode (grid) and an anode. After evacuation, the interior of the glass envelope is filled with an inert gas (neon, argon, helium or hydrogen) at low pressure.

In the unignited state, the thyratron behaves very much like a high-vacuum valve. From the cathode surface electrons are emitted, their behaviour being determined by the potential difference between grid and cathode. The control voltage V_{contr}, which can be imagined to be in the plane of the grid, consists of the actual grid voltage and that part of the anode voltage which penetrates beyond the grid. The extent of the influence of the anode voltage depends to a large degree upon the geometrical dimensions of the grid (number and thickness of the windings), the distance from the anode, etc. It is denoted by a factor known as the penetration factor, and is designed by D.

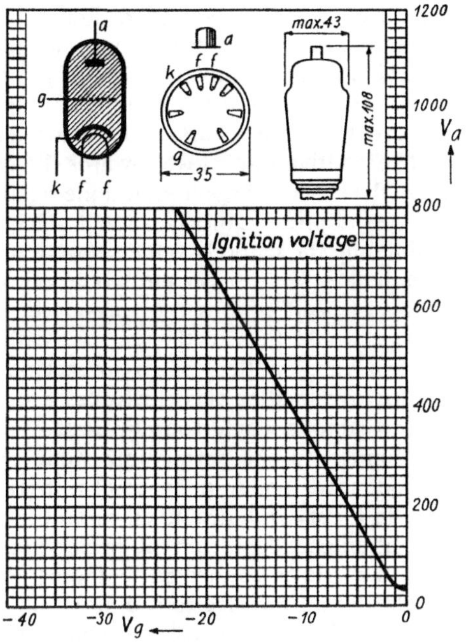

Fig. 4-9 Thyratron ignition characteristic. Arrangement of electrodes of the EC 50, dimensions and base connections

The amount of the penetrating anode voltage is $D \cdot V_a$, so that the control voltage is given by the expression:

$$V_{\text{contr}} = V_g + V_a \cdot D. \tag{4.5}$$

Conversely, it can also be shown how much greater is the influence on the anode current of a change of grid voltage than that of an equal change of the anode voltage. This number is by analogy called the amplification factor in high-vacuum valves and is designated by μ. The amplification factor is the reciprocal of the penetration factor, thus as:

$$\mu = \frac{1}{D}\,. \tag{4.6}$$

When the control voltage is less than zero, i.e. negative, only few of the electrons emitted from the cathode can pass through the grid. The remainder are repulsed and form a cloud of electrons in the vicinity of the cathode, and known as the space charge. If, however, say by increasing the anode voltage, the grid voltage becomes positive with respect to the cathode, the electrons flow in increasing numbers through the grid towards the anode. So long as the (now positive) control voltage remains below a certain value – the ionization voltage of the gas filling – it can also be assumed that in a thyratron the electrons drift to the anode as a small but measurable current. In these circumstances, as in the case of the high-vacuum valve, the anode current is mainly limited by the space charge, which determines the grid-voltage/anode-current characteristic of the valve. However, should the control voltage exceed the ionization voltage, the effect of the gas filling becomes noticeable. The electrons, now moving at considerable velocity, ionize atoms of gas setting up a chain reaction whereby new ions and electrons are constantly formed. The gas filling "ignites", that is to say a spontaneous arc discharge occurs, and a heavy electronic current flows from cathode to anode.

The positive ions move towards the cathode, where most of them capture electrons from the space charge and become neutral gas atoms once more. In this way the space charge disappears quickly, so that unless special precautions are taken, all the electrons emitted by the cathode – the saturation current – would be able to flow to the anode. This heavy electron current would not only overload the cathode and anode, but the excess of positive ions thus produced, no longer able to be neutralized by the space charge, would flow to the negative cathode, mainly to those points from which most electrons are emitted. Their impact on he cathode surface would release heat, thus further overloading this electrode.

The cathode would soon be destroyed by the heat released if the anode current was not limited to the permissible value by inserting a resistor in either the anode or cathode lead. The impedance of a thyratron is, at least in the ionized state, extremely low (in theory it is negative), so that it is very suitable for use as a discharge valve in relaxation oscillators for the production of time base voltages.

If the grid is connected to the cathode, that is to say if the grid voltage is zero, practically all the electrons emitted by the cathode can be attracted through the grid by the positive anode voltage. Only a small ignition voltage is then required for ionization. Taking the helium-filled EC 50 valve as an example, only 35 V are required for ignition. At this potential the discharge –the "arc"– is only just maintained in the tube. This value is called the arc potential and is denoted by V_{arc}.

If the voltage on the grid is not zero but has a negative value, a higher anode potential is required for ionization. The relationship between grid voltage and the ignition is shown by the characteristic curve reproduced in Fig. 4-9. It indicates what the anode voltage must be to ionize the gas at a given negative grid voltage. It is clear that for ionization to take place with this tube above 40 V, the anode potential must always exceed the negative grid potential by a certain constant factor. This factor is called "ignition factor" and is denoted by μ by analogy with the amplification factor in the case of high vacuum valves. Thus:

$$V_a = -\mu \cdot V_g. \tag{4.7}$$

In the EC 50 valve this factor is 35. This means, for instance, that, with a grid bias of —11.4 V, ionization occurs at an anode potential of 400 V. If a thyratron of the type shown in Fig. 4-4 is used to discharge the sweep capacitor, as is shown in Fig. 4-10, it is possible, by controlling the grid voltage, to adjust the potential to which the capacitor is to be charged.

The voltage V_s across the capacitor (Fig. 4-11), which is the voltage for the time base, varies between the anode potential V_a at the point of ionization and the arc potential V_{arc}, i.e.:

$$V_s = V_a - V_{\text{arc}}. \tag{4.8}$$

The capacitor potential required for a certain deflection voltage is therefore:

$$V_c = V_a = V_s + V_{\text{arc}}. \tag{4.9}$$

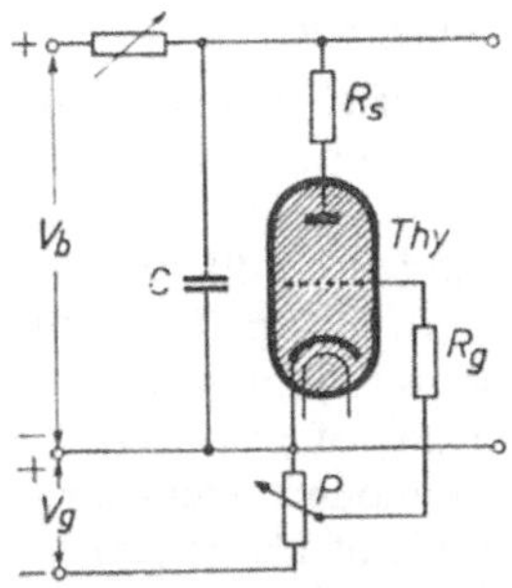

Fig. 4-10 Thyratron relaxation oscillator for generating time deflection voltage

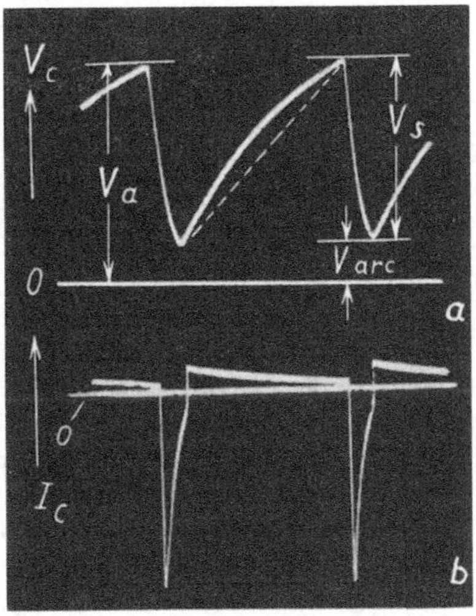

Fig. 4-11 Voltage and current waveforms across capacitor C as in Fig. 4-10

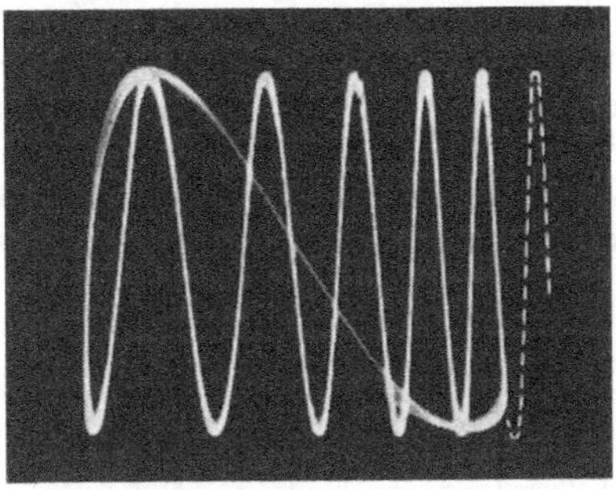

Fig. 4-12 Oscillogram of sinusoidal input signal with time base as shown in Fig. 4-10

The oscillogram in Fig. 4-11 shows the waveform of the capacitor voltage using the circuit shown in Fig. 4-10. When this is compared with Fig. 4-5, the great improvement is apparent at once. The residual direct voltage V_{arc} is considerably smaller than V_{ext} in Fig. 4-5*a*, and V_s is several times greater than V in Fig. 4-5.

The expression (4.9) should, however, be equal to (4.7), thus:

$$-\mu \cdot V_g = + V_{\text{arc}}. \tag{4.10}$$

From this is obtained the required grid voltage for a specific value of sawtooth voltage:

$$V_g = \frac{-V_s + V_{\text{arc}}}{\mu}. \tag{4.11}$$

As a deflection voltage $V_{pp} = 267$ is required for the DG 10-6 valve, according to Eq. (4.4), the thyratron EC 50 will require a negative grid voltage of $V_g = \frac{-267 + 35}{35} = -8.63$ V. This value can also be read from the characteristic curve in Fig. 4-9.

From the oscillograms in Fig. 4-11*a* and *b* the output of a thyratron time base circuit can be compared with that of a circuit employing a gas discharge tube (Fig. 4-6 and 4-7*a*). With a thyratron adequate deflection voltage amplitude can be obtained for general purposes. It is, however, noticeable in Figs. 4-11*a* and 4-12, that the deflection voltage waveform is still not linear with time.

4. 6 Linearizing the sawtooth voltage by means of a pentode

From Fig. 4-11*a* it is clear that in order to ensure linearity it is necessary to prevent a too rapid initial voltage rise; the charge must increase uniformly in each unit of time. This means that the charge resistor should have the property of allowing a constant and adjustable current to pass, even with a falling potential difference. This requirement is met, at least up to a certain lower voltage limit, by a pentode. Fig. 4-13*a* gives the characteristic curves of the EF 80 valve, and shows the dependence of the anode current upon the anode voltage, with various voltages of screen grid as parameter.

From these curves it is evident that the change in anode current for a given screen voltage is negligible at anode voltages down to a limit of about 100 V (especially at lower screen grid voltages). The impedance of a pentode is known to be high. It can be made higher still by the introduction of negative feedback. This is often effected by means of a non-bypassed cathode resistor (Figs. 4-29 and 4-40).

Fine adjustment of the time base frequency by the anode current is now best effected by controlling the screen grid voltage. Using a potentiometer with a linear characteristic, the curve which results is practically linear, in contrast to that obtained by controlling the voltage of the first grid. The curves in Fig. 4-13*b* show the dependence of the anode current of the EF 80 valve on the screen grid voltage for various voltages on the control grid. It can be seen that the anode current is in linear relationship with the screen grid voltage. A corresponding circuit for generating a time base voltage is shown in Fig.4-14.

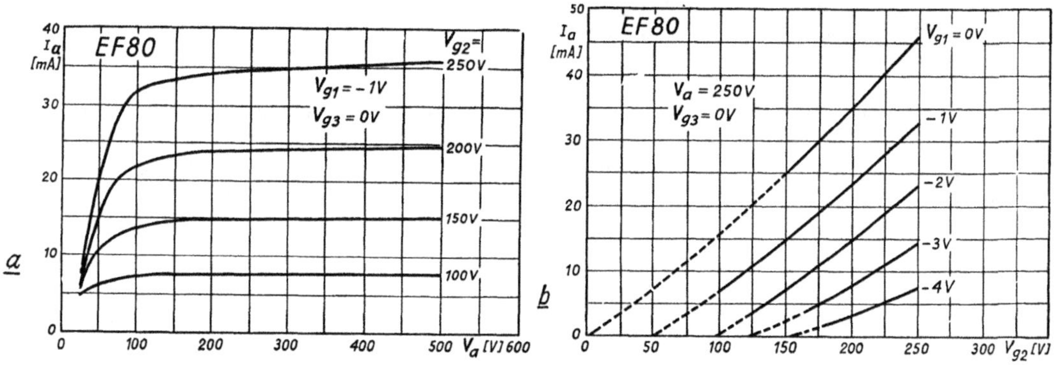

Fig. 4-13 Characteristics of EF 80 pentode.

a) Anode current vs. anode voltage *b*) anode current vs. screen grid voltage

The pentode which maintains the charging current constant is now connected to the negative lead so that it may receive the supply voltage for the charging valve in the required polarity from voltage source V_b. The operation of the circuit itself is not altered by this in any way [17]).

Fig. 4-15*a* shows the waveform of the voltage occurring across the charging capa-

[17]) Such variations are quite common in the time base circuits. There are, however, circuits in which charging takes place rapidly and linear discharge occurs through a pentode or a cathode follower.

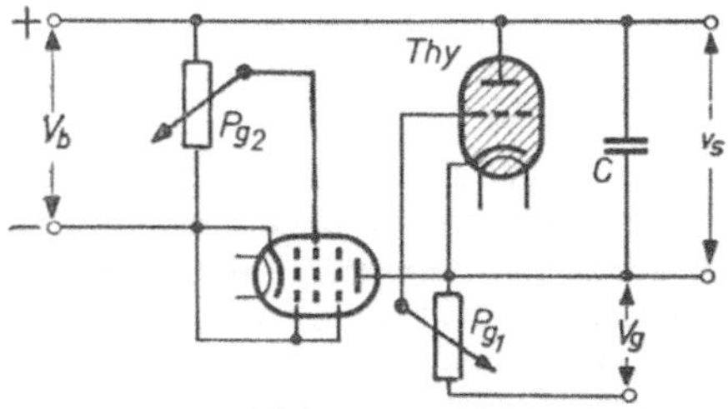

Fig. 4-14 Time base generator with a pentode as a charging-current controlling element, *Thy* = Thyratron
P_{g1} = potentiometer for controlling ignition potential of thyratron (amplitude);
P_{g2} = potentiometer for controlling charging current by varying voltage on screen grid of pentode (frequency control)

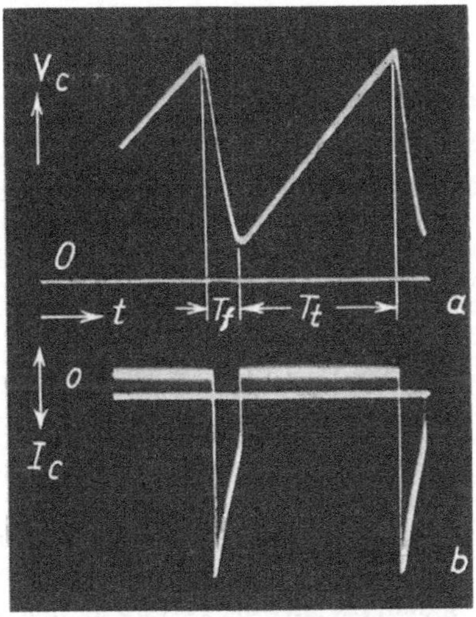

Fig. 4-15 Waveform of current and voltage across charging capacitor *C* in circuit as in Fig. 4-14.
a) Capacitor voltage; *b*) capacitor current

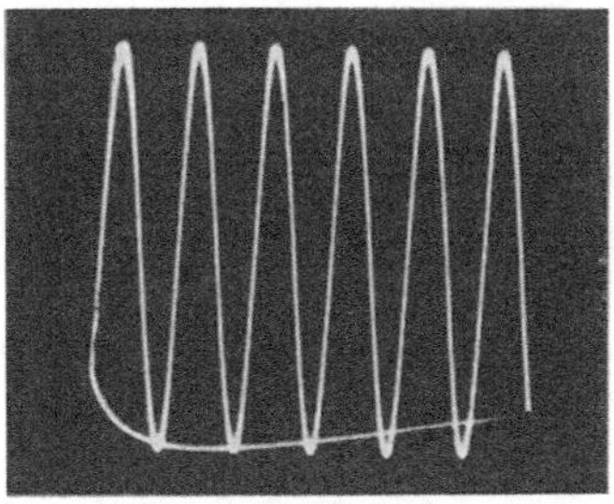

Fig. 4-16 Oscillogram of sinusoidal input signal with time base as in circuit in Fig. 4-14

citor, while Fig. 4-15*b* shows the capacitor current in this circuit. The charging current (above the zero line) is now constant and the voltage rise linear. Fig. 4-16 shows that the individual cycles of a sinusoidal signal voltage are evenly spaced.

4. 7 Flyback time

It has been assumed, so far, that the flyback of the spot to its starting position, and therefore the discharge of the capacitor in the time base generator, takes place in a negligibly short space of time. As can be seen in Fig. 4-15*a*, this is not the case in practice. The time required for discharge depends on the values of the resistors in the discharge circuit and on the capacitance of the charge capacitor (the *RC* time constant). The charge, and thus the forward trace of the spot, takes place during time T_h, and the discharge –the flyback of the spot– during time T_r. The total duration of the time base cycle is therefore the sum of these two times:

$$T_X = T_h + T_r. \tag{4.12}$$

The number of time base cycles in one second –the time base frequency– is thus the reciprocal of the duration of one time base cycle:

$$f_X = \frac{1}{T_X}\,. \tag{4.13}$$

In order to be able to study the waveform of the signal applied to the Y-plates, at least one complete cycle must be displayed. The time base frequency is therefore always chosen equal to or an aliquot part (e.g. $^1/_2$, $^1/_3$, $^1/_4$ . . .) of the signal frequency. If the time base frequency equals the signal frequency, one cycle appears on the screen; if the time base frequency is half the signal frequency, two cycles appear, and so on [18]). The total number N_t of the cycles of the signal voltage displayed on the screen during the forward trace and flyback is thus the ratio of the signal frequency f_M to the time base frequency f_X, so that:

$$N_t = \frac{f_M}{f_X} \tag{4.14}$$

With a cyclic signal, the time taken for the flyback can be read off immediately. For instance, in Fig. 4-12 the flyback accounts for approximately $^3/_4$ of a cycle of the signal, the missing part being indicated by dotted lines. In this example, in which there is a total of 6 cycles, the flyback time is therefore: $\frac{^3/_4}{6} = 12^1/_2\ \%$ approximately of one time base cycle [19]).

In general, therefore, the ratio of the flyback cycles N_r to the total number of visible cycles N_t of the signal, shows the part of the total duration of the time base cycle attribuable to flyback. This ratio will be designated Z_r. Thus:

$$Z_r = \frac{N_r}{N_t}\,. \tag{4.15}$$

From Figs. 4-11*a* and 4-15*a* it can be seen that the discharge, just as the charge, takes place according to an exponential function, but in reverse, the charge increasing exponentially and the discharge decreasing exponentially. The aim is to keep flyback time to a minimum, as it interrupts the pattern. The flyback time in this circuit is mainly determined by the resistor limiting the discharge current through the thyratron. This current drops, during discharge from the maximum permissible peak value $I_{a\ max}$, to zero. To a first approximation it can therefore be assumed that the mean discharge current is $^1/_2\, I_{a\ max}$.

The period of discharge (or flyback time) T_r is arrived at in a similar way to Eq. (4.1):

$$T_r = 2 \cdot \frac{V_X \cdot C}{I_{a\,\max}} = 2 \cdot R_p \cdot C. \tag{4.16}$$

---

[18]) These conditions are valid only for free-running (recurrent) operation of synchronized sweep circuits, which are mainly considered here. Triggering of time base voltages is governed by quite different conditions, which will be dealt with later (see Part II, Ch. 4.18 "Time base generator for self-oscillating and triggered operation").

[19]) For Figs. 4-11, 4-12, 4-15 and 4-16 the flyback time was intentionally made long in order to show clearly the phenomena under discussion. In practice it is roughly between 1 and 20%. Only in the highest frequency ranges does it occasionally rise to 40%.

According to Eq. (4.1) the time T_h required for the trace, i.e. the time required for charging, is:

$$T_h = \frac{V_X \cdot C}{I_c}, \tag{4.17}$$

where I_c = average charging current.
The ratio T_r/T_h, which corresponds approximately to Z_r for the flyback portion, is therefore:

$$\frac{T_r}{T_h} = \frac{2 \cdot I_c}{I_{a\,max}}. \tag{4.18}$$

Thus, if the flyback period is to be small, the ratio of the charging current to the discharging current should be as low as possible. However, in order to obtain a given time base frequency with a given value of capacitor C, we are compelled, according to Eq. (4.1), to choose a correspondingly high value for I_c. At higher frequencies therefore, a relatively higher flyback time will have to be accepted. Moreover, at higher time base frequencies, a certain inertia of the gasfilling becomes increasingly noticeable. This is because a certain time is necessary for the gasfilling to become ionized and thus to permit the discharge of C, so that under certain circumstances the flyback time may be several times longer than that calculated according to Eq. (4.16). It is quite conceivable that the complete ionization of the gasfilling will be accelerated if, immediately after ignition, the grid is driven sharply negative, because this will result in the free positive ions being neutralized more rapidly.

4.8 Improved time base generator circuit with thyratron

Fig. 4-17 shows a circuit which incorporates improvements on that shown in Fig. 4-14. The charging element in this case is also a pentode, but is here connected between grid and cathode of the thyratron.

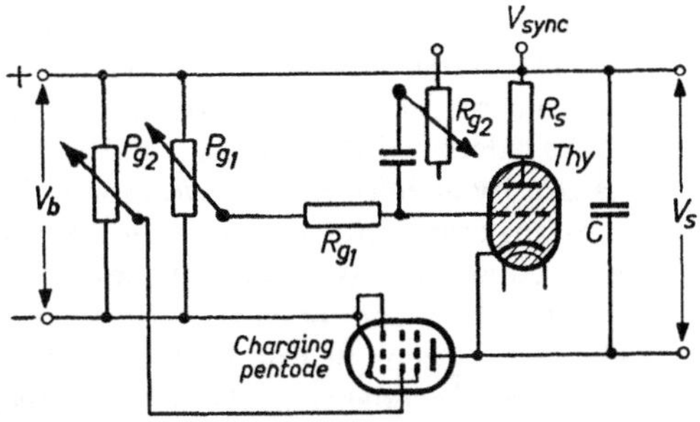

Fig. 4-17 Improved time base generator. The charging element is connected between grid and cathode of the thyratron

The time base voltage is controlled by the potentiometer P_{g1}, by which a correspondingly high positive inverse voltage is applied to the grid of the thyratron.

When timing capacitor C is discharged, a high potential difference exists between cathode and anode of the charging valve, and the grid of the thyratron is negative with respect to the cathode. As C charges up, the negative potential of the cathode goes more positive, so that the difference in potential between grid and cathode of the thyratron decreases until ionization takes place at a potential corresponding to the ignition characteristic. Fig. 4-19 reproduces oscillograms of the resultant waveforms of the grid and anode voltages of the thyratron during one time base cycle with reference to the ignition characteristic. The reduction in grid voltage necessarily corresponds to an increase in anode voltage.

With this circuit it is possible, even at higher time base frequencies, to obtain flyback times approaching those calculated from Eq. (4.16).

For high frequencies, small values of timing capacitance C must be used. This, however, increases the risk that interfering voltages may reach the unearthed connection of the timing capacitor, which is connected to one of the X-plates [20]).

For this reason a special heater winding on the mains transformer must be provided for the discharge valve, and screened from the other windings. Its centre tap must be connected to the cathode to avoid stray voltages between cathode and heater. As the capacitance of this winding is effectively in parallel with the timing capacitor, it must be kept as small as possible by means of suitable insulation.

4. 9 Maximum time base frequency and thyratron load

It is obvious that high time base frequencies are needed in order to be able to examine high frequency waveforms. The technical data published by the manufacturers concerning thyratrons refer only to the individual valves (gasfilling, electrodes, dimensions etc.) and should be taken to mean that the frequency indicated can be achieved with the valve in question in a fairly "normal" circuit.

The maximum frequency is limited not only by the maximum permissible average anode current of the thyratron, by the timing capacitance and by the capacitance in parallel with it, but also, and for another reason, by the permissible load of the thyratron. The power drawn from it, if the frequency is not too high, is found from the product: arc voltage × recharging current (the latter being also the mean current through the valve). Thus:

$$N_{th} = V_{\text{arc}} \cdot I_{a\,\max} \,. \tag{4.19}$$

In high vacuum valves the power taken causes heating of the anode (anode dissipation). In thyratrons, however, part of the current is represented by the positive ions imigrating to the cathode, where they give up their energy. It is therefore necessary to take into consideration the maximum permissible cathode load. Thus, from Eq. (4.19) it can be calculated that for the EC 50: $N_{EC\,50} = 35 \times 0.01 = 0.35$ W.

As already mentioned, however, some time elapses before the gasfilling forms an arc. This means that during this time the voltage at the anode is higher than the arc vo'tage at low frequencies. As a result, the load on the thyratrons is increased, so that at higher frequencies overloading can easily occur.

The oscillograms in Fig. 4-18*a* and *b* show these conditions in more detail. They indicate the dependence of the anode voltage V_a of an EC 50 thyratron on the voltage V_c at the timing capacitor for two cases with different capacitance values. The direction of the time cycle is shown in Fig. 4-18*a* by arrows. The heavy line which slopes upwards to the right occurs during charge (when the valve is not ignited there is the same voltage both at the anode and at the timing capacitor), while the thin line which falls back to the left shows that the anode voltage during the discharge depends on the voltage

[20]) The power supply unit must be so rated that it represents an AC short-circuit over the entire frequency range. It is then immaterial whether the positive or the negative pole is earthed.

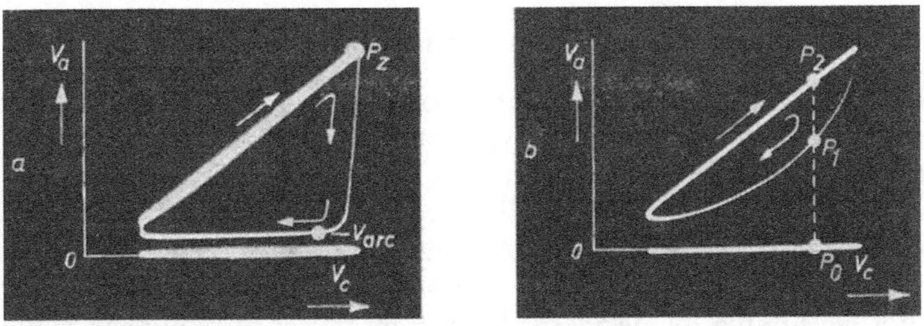

Fig. 4-18 Dependence of operating voltage of a thyratron on voltage across charging capacitance at various values of charging capacitor
a) Charging capacitor 0.1 μF; *b*) Charging capacitor 1000 pF

on the capacitor. Hence the whole image corresponds in time to the duration of a sawtooth voltage cycle. For the oscillogram in Fig. 4-18*a* the capacitance was 0.1 μF and the frequency of the sawtooth voltage about 220 c/s. It can be seen that the anode voltage –towards the end of one cycle– drops rapidly to arc voltage V_{arc} at P_{arc}. For the oscillogram in Fig. 4-18*b* a capacitor of 1000 pF was used. The discharge time constant is now so short that the ionization time of the gas, i.e. the time in which the discharge is built up, can no longer be regarded as negligible compared with the discharge, in a proportion which changes from instant to instant, into a part across the protective resistor R_s and the remaining part between the anode and cathode of the thyratron. This voltage division can be seen clearly in the oscillogram. In Fig. 4-18*b*, for instance, the distance $\overline{P_0 P_1}$ corresponds to the voltage across the thyratron and the distance $\overline{P_1 P_2}$ to the voltage across the protective resistor, taken at a moment selected at random. With still smaller capacitances the discharge constant is so small that the anode voltage remains almost equal to the capacitor voltage during discharge. Its average value is therefore roughly:

$$\frac{V_{a.\,\max} - V_{\text{arc}}}{2}.$$

As then the mean discharge current corresponds to the mean charging current, the load of the thyratron is

$$N_{th} = I_{a\,m} \cdot \frac{V_{a,\,\max} - V_{\text{arc}}}{2} \tag{4.20}$$

With an EC 50 valve, a charging current of 10 mA and an anode voltage of 500 V give $N_{EC50} = 10 \cdot 10^{-3} \cdot \dfrac{500 - 35}{2} = 2.3$ W. This is many times the normally permissible load of about 1/3 W. The valve may therefore be driven under such conditions only for a very short time, otherwise its working life would be very much shortened.

The protective resistor R_p must be at least large enough to ensure that during the voltage drop from the value V_c to V_{arc} the maximum permissible current $I_{a,p}$ for the thyratron is never exceeded. This resistor must therefore be equal to or greater than

$$R_s \geqq \frac{V_c - V_{\text{arc}}}{I_{a,p}}. \tag{4.21}$$

With the EC 50 the protective resistor must therefore be equal to $R_s = \frac{500-33}{0.75}$ $= 620\ \Omega$ for the discharge of a capacitor potential of 500 V to the arc voltage of 35 V at the maximum permissible current peak of 750 mA.

Determination of the load rating of this resistor should not be on the assumption that, when a capacitor voltage V_c is discharged, the average anode current $I_{a,m}$ flows, for that would give a load $N_{Rs} = R_s \cdot I_{a,}{}^2{}_m$ and a power of only $N_{Rs} \approx 0.06$ W would be obtained from the EC 50. A resistor with such a rating would soon be destroyed, since it would not be suitable for discharge by current pulses of high peak value. The actual power in the example quoted was found from Eq. (4.20) to be 2.3 W. Resistor R_s must be rated accordingly. Thyratron time base circuits are mainly used for low frequencies; they make it possible to obtain relatively short flyback times. Usually, particularly for higher frequencies, high-vacuum valve sweep circuits are used. These will be dealt with in detail later. The following observations on synchronization apply, of course, to time base circuits in general.

4. 10 Synchronization

As has already been said in the introduction stationary pictures of cyclic phenomena can be obtained by adjusting the time base frequency to be equal to, or an integral fraction of the input signal frequency. This is difficult to accomplish by a single adjustment, especially at higher frequencies, so that the images on the screen are seen to drift either to the right or to the left during each time base cycle. It is possible, however, to lock the sawtooth voltage with the signal being investigated. This is known as synchronization. For example, if a part of the alternating signal voltage is fed to the grid of the thyratron, it will be superimposed on the fixed grid bias (Figs. 4-17 and 4-19). As long as the grid voltage is well below the ignition point, this additional voltage has no influence whatever. But as, due to the increasing amplitude of the signal, the operating point of the valve approaches the ignition potential (Fig. 4-19), as the result of increase in the anode voltage and the drop in the grid potential, the alternating signal at the grid can cause earlier ionization than would otherwise be the case. The duration of one time base cycle T_{X1} will now be shorter; without synchronization it would be T_{X2}. In other words, the time base frequency is slightly increased. Thus, if in Fig. 4-19 the time base frequency is a little less than 1/6 of the input signal fre-

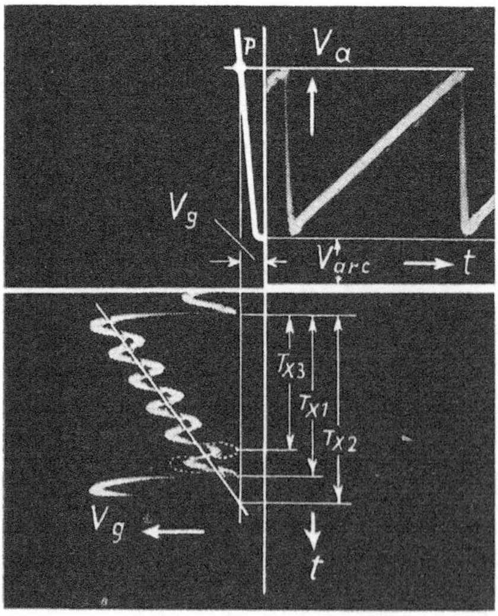

Fig. 4-19 Waveform of anode voltage and grid voltage with reference to ignition characteristic of the thyratron in circuit Fig. 4-17, showing signal superimposed on grid voltage for synchronization

quency, it is progressively controlled by the sync signal in such a way that ionization will occur at the peak of the sixth cycle of the signal. The pattern then jumps into step and remains stationary on the screen for six cycles (half a cycle here occupied by the flyback). In this way it is possible to obtain stationary waveforms on the screen up to very high frequencies. In thyratrons, sinc signals of only a fraction of a volt are sufficient for synchronization. If the sync signal is too large, ionization might occur at the fifth cycle of the input signal, as shown by the dotted cycle in Fig. 4-19. The result would then be, that the capacitor would discharge at a low anode potential, so that the time base voltage and hence the width of the image in the oscillogram would be reduced by one sixth (T_{X3}). In order to obtain the optimum value of sync voltage at will, variable resistors are connected in the lead to the grid of the thyratron.

Broadly speaking, the electrodes of the thyratron must be considered as being practically short-circuited once ionization has set in and a considerable current could flow in the sync circuit. In order to avoid this, a resistor and a capacitor should be connected in one of the sync leads. As a general guide, resistor R_{g2} should be at least 300 Ω · V_X. For a 500 V_{pp} time base voltage it must not be lower than 150 k Ω. Usually the synchronization voltage is obtained from the input signal as amplified by the vertical amplifier, or from an external source, or from the mains. To avoid mutual interference between time base generator and input signal, special sync amplifiers are used when demands are particularly high. In order to use the oscilloscope correctly it is most important to understand clearly the correct conditions for synchronization. The various possibilities available will be considered in detail in a latter section devoted to the practical uses of the oscilloscope.

4. 11 Further circuits for linearizing the sawtooth sweep

When describing the thyratron time base generator, the circuits in Figs. 4-14 and 4-17 showed how a pentode could be used to obtain a linear increase of voltage across the timing capacitor.

Circuits by means of which an exponentially rising voltage, resulting from the charging of a capacitor through an ohmic resistor, can be subsequently linearized, have obtained considerable importance [4]. They make it possible, first to generate a sawtooth voltage in the simplest way, and then to linearize it before application to the X-plates. The easiest method of doing this is offered by taking advantage of the curvature of the grid voltage/anode current characteristic of an amplifier valve. By suitably choosing the valve characteristic and operating point, it is possible, as is indicated in Fig. 4-20, to obtain a sawtooth time base voltage of good linearity in the anode circuit, this voltage, of course, being inversely proportional to the increase in anode current, since the anode current rises exponentially with increase of grid potential. The drawbacks of this method are that the operating point must be well maintained and the amplitude of the grid voltage must not vary to any large extent.

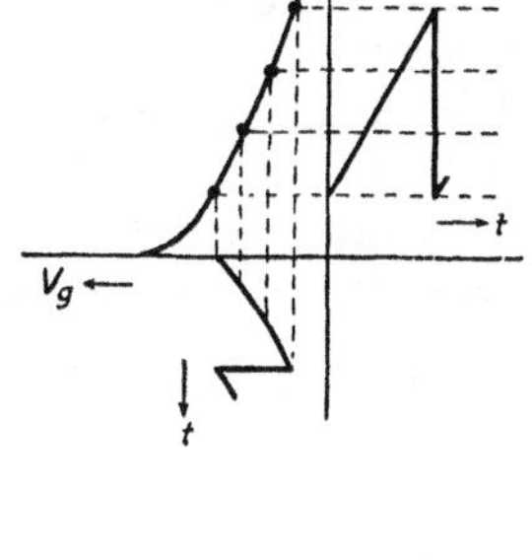

Fig. 4-20 Linearising the sawtooth voltage by means of the curvature of the grid voltage/anode current characteristic of an amplifying valve

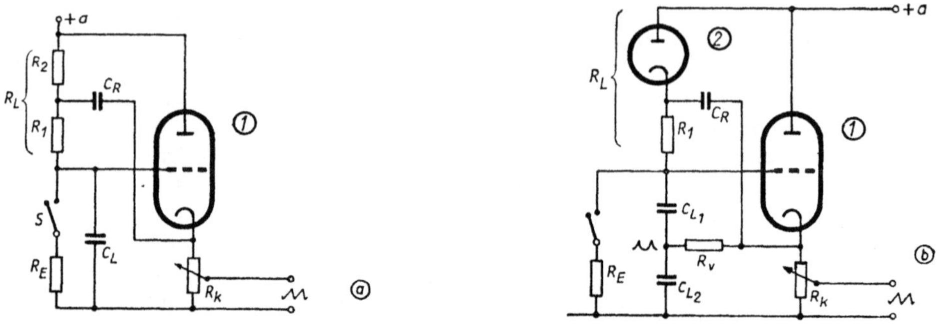

Fig. 4-21 Linearising the sawtooth voltage by means of a cathode follower stage with feedback (bootstrap circuit). *a*) Basic circuit; *b*) improved circuit using diode and integration network in the charging circuit

Fig. 4-21*a* shows another possible way of linearizing the charging characteristic of a capacitor. The charging resistor R_L now consists of two resistors, R_1 and R_2, connected in series. The voltage of the timing capacitor C_L is applied to the grid of a valve, 1, connected as a cathode-follower pre-amplifier stage [21]). The voltage across the cathode resistor increases with the voltage at the timing capacitor and is in phase with it. This voltage is now fed back via capacitor C_R to the junction of R_1 and R_2. While the capacitor is charging up through the resistor, the potential difference between capacitor C_L and positive terminal $+a$ is normally falling, so that the charging current becomes progressively smaller and the potential at the timing capacitor increases more and more slowly. In this circuit, however, an increasing current flows from the cathode of the valve (where the potential increases) via C_R to R_2. This means that the voltage at the junction between the resistors R_1 and R_2 can be kept practically constant during charging, and that the voltage at R_1 and also the charging current remain constant. Thus a considerable voltage rise linear with time is obtained across timing capacitor C_L. Discharge takes place in the usual way, for example, by means of a thyratron as already described, or in time base circuits employing high-vacuum valves, in a manner which will be discussed later. The discharge circuit is indicated diagrammatically by the switch *S* and the discharge resistor R_E. The circuit shown in Fig. 4-21*a* and *b*, which is known in America as the "bootstrap-circuit", offers two very important additional advantages resulting from the use of a cathode follower. The input impedance of a cathode follower is high, but the output impedance is low. For reasons which will be discussed in detail later, good linearity of the sawtooth voltage requires that the load of the timing capacitor should be kept as low as possible, since this capacitor must be regarded as the source of the sawtooth time base voltage i.e. it must have a high ohmic value. On the other hand, the output impedance must be low in order, amongst other things, to minimize the unfavourable influence of circuit capacitances at high frequencies. Both requirements are largely met by this circuit. The circuit can be extended as indicated in Fig. 4-21*b*. Here, a diode takes the place of R_2, so that in the direction from the junction of R_1, C_R and the cathode of valve 2 to the positive terminal of the voltage source $+ a$ the impedance is infinitely high. In consequence, only the current from the cathode resistor R_k via C_R will serve to increase

[21]) Details in Part I, Ch. 5.

the potential at this point (the diode cannot, like R_2 in Fig. 4-21*a*, load the feedback voltage) and thus also charge the capacitances C_{L1} and C_{L2}.
Moreover, the timing capacitor is made up of two elements, C_{L1} and C_{L2}. Between the junction point of these two capacitors and the cathode of the cathode follower a resistor is connected, which, with C_{L2}, forms an integrating network. The sawtooth signal from the cathode produces a parabolic upward peaking waveform at this point. The potential at the grid of the cathode follower now consists of the sawtooth voltage across C_{L1} and the parabolic voltage across C_{L2}, the peaks of which coincide with the peak of the sawtooth voltage. This enables a high degree of linearity to be obtained. Since the way the circuit components function depends upon the frequency at which they have to work, provision must be made for switching over to higher frequency ranges. This entails higher costs, which is why the circuit shown in Fig. 4-21*b* is used more in television and radar sets than in oscilloscopes [5]. Both changes in the circuits described in Fig. 4-21*a* can be used separately as, for example, in the Philips "GM 5660" oscilloscope.

To obtain the lowest possible output impedance, a tube of high mutual conductance is needed as the cathode follower in these circuits, for which reason a steep slope pentode, such as types EF 42, EF 80 etc., connected either as a triode or as a pentode is used. When used as a pentode, the screen grid must be decoupled to the cathode via a capacitor [22]).

A triode can also be used to linearize the rate of charge. By means of a relatively high cathode resistor, a heavy negative current feedback is introduced to keep the charging current constant (see also Fig. 3-7).

These widely used and characteristic examples are only a few of the many circuits, either possible or in practical use, for obtaining greater linearity of the sawtooth waveform. They are dealt with in detail in the literature mentioned under [2] and [3].

Although thyratrons can also be used in the way indicated in Fig. 4-17 to obtain time base frequencies of up to a maximum of 150 kc/s, most oscilloscopes use time base generators incorporating high-vacuum valves. By this means, time base frequencies of 250 kc/s, 500 kc/s, or even 1 Mc/s can be obtained, but naturally the cost of such equipment increases with the frequency. In some cases, these circuits work in a way similar to that of the thyratron circuit (cyclic charging and discharging of a capacitor), but in others they work on other principles, such as the "multivibrator circuit". The amplitude of the sawtooth voltage supplied by this circuit is usually too low for direct application to the horizontal deflection plates, so that further amplification is necessary. *DC* amplifiers are used almost exclusively, at least for high performance oscilloscopes, as by this means the image can be easily extended horizontally by adjusting the amplifier gain.

4. 12 Multivibrator circuit

If the amplified voltage output of each stage of a two-stage *RC*-coupled amplifier is fed back to the input of the other stage via a coupling capacitor, a circuit is obtained as shown in Fig. 4-22, which, redrawn as in Fig. 4-22*b*, is known as the multivibrator developed by Abraham and Bloch. As drawn, the circuit is not stable; the voltage across both valves swings alternately from one extreme to the other at a frequency

---

[22]) If, in such a circuit, a coupling resistor is inserted in the anode lead, then two mass-symmetrical sawtooth voltages can be taken from cathode and anode.

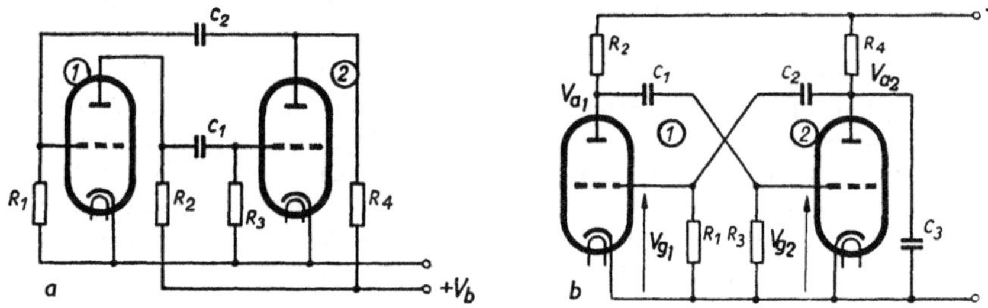

Fig. 4-22 Multivibrator. *a*) Output circuit of two-stage amplifier with capacitive feedback; *b*) conventional multivibrator circuit

determined by the time constants of the coupling components. If, for any reason, the anode voltage V_{a1} of valve 1 rises, the grid voltage V_{g2} of valve 2 goes less negative. This causes the anode voltage V_{a2} of valve 2 goes less negative, and consequently the grid voltage V_{g1} of valve 1 is driven more negative via coupling capacitor C_2. As a result the anode voltage V_{a1} rises still further, and so on, until the grid voltage V_{g1} is driven so negative that the anode current of valve 1 is cut off and V_{a1} becomes equal to V_b. V_2 is now fully conducting and no further increase in current is possible, V_{a1} being at a maximum and V_{a2} at a minimum. This state of equilibrium (the anode voltages remain at these values) lasts until, at a moment determined by the circuit time constant, the voltage V_{g1} begins to go less negative. This causes the anode current in valve 1 to rise, and the same process repeats itself rapidly in reverse; the voltages "flop" to the other extremes.

It is desirable for the trailing edge of the sawtooth time base voltage to be as steep as possible; that is, the flyback time must be short. For this reason, assymetrical rectangular pulses are generated for time base circuits in which one pulse width is made very short. If this asymmetrical rectangular pulse is used to charge a capacitor (C_3 in Fig. 4-22 and C in Fig. 4-23), the rectangular voltage is intregrated. The linear voltage rise occurs during the long portion and the drop occurs during the short portion of the cycle.

A great number of similar circuits have been developed which cannot be dealt with in detail within the scope of this book. Double triodes such as the ECC 40 or ECC 83 are generally used in these circuits, although pentodes can also be employed with advantage. The screen grids are then cross-connected and act as anodes, the output being taken from the anodes proper, so that changes in the load do not affect the oscillatory circuit. Occasionally the grid resistors are connected across a positive voltage instead of to the chassis. By this means the grid capacitor is charged until the peak of the exponentially rising charging curve is reached. This voltage rise is substantially linear, so that a good linear sawtooth voltage is available on the grid. Since in the circuit described the anode voltage is back-coupled, it is called an anode-coupled multivibrator.

Another widely used circuit is the cathode-coupled multivibrator. Here the feedback takes place via the cathode resistor, which is common to both valves. A circuit of this type is shown in Fig. 4-23, which gives the operating conditions for the ECC 40. The frequency is determined by the value of C_g and R_{g1}. Fine adjustment is effected by varying R, though this causes some fluctuation in the output voltage. For higher

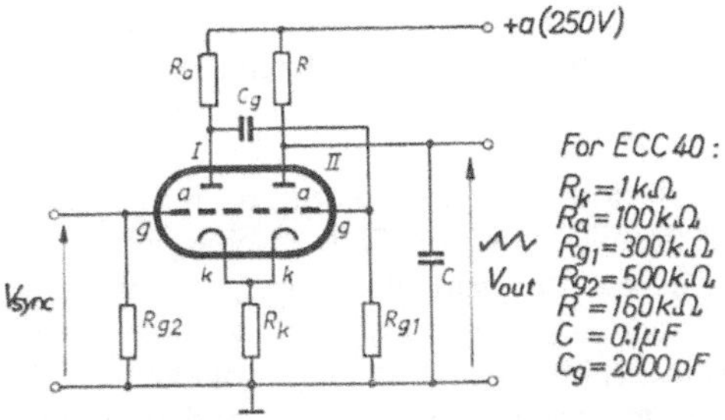

Fig. 4-23 Practical circuit of cathode-coupled multivibrator

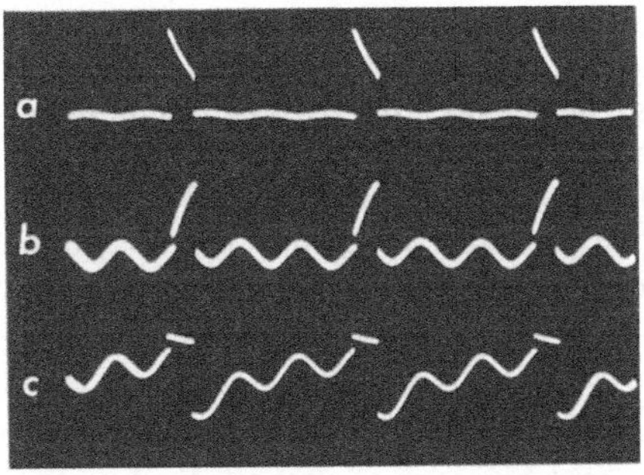

Fig. 4-25 Oscillograms as in Fig. 4-24 with synchronization using sinusoidal alternating voltage ($M : f_x = 3$)
a) Common cathode; *b*) anode I; *c*) grid I

a
100V_{pp}
b
130V_{pp}
c
5.0V_{pp}
d
30V_{pp}

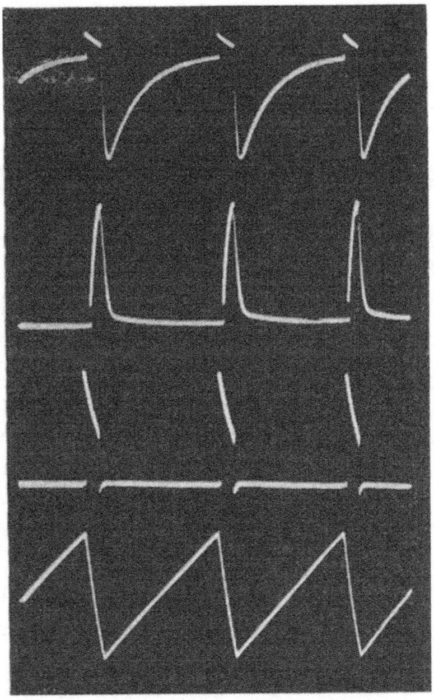

Fig. 4-24 Oscillograms of voltage waveforms in a cathode-coupled multivibrator as in Fig. 4-23. *a*) Grid I; *b*) anode I; *c*) voltage on common cathode; *d*) sawtooth voltage on anode II

frequencies, both C and C_g must be given different values; C_g should be 1/20 to 1/30 of the value of timing capacitor C, and for a sawtooth time base frequency of 50 c/s, C should be approximately 0.5 μF. At frequencies up to 1 kc/s the flyback time is about 5%, and at frequencies from 10 to 20 kc/s, about 20%. For both valve systems (I and II) the anode current is 2.6 mA. For synchronization, the sync voltage, which must be at least 1 V, can be applied to the grid of valve system 1.

The oscillograms in Fig. 4-24 show the waveforms obtained with this circuit. At the circuit component ratings specified in Fig. 4-23, the frequency of the sawtooth voltage was about 350 c/s. The waveforms are reproduced at similar amplitudes for the purpose of easy comparison, but the actual individual peak-to-peak values are given at the side. In this example, the amplitude of the sawtooth voltage was about 30 V_{pp}, so that further amplification is necessary for adequate spot deflection. The oscillograms in Fig. 4-25 *a-c* show the waveform at the cathode and anode of valve system 1 with a sinusoidal synchronizing voltage of 1050 c/s (three cycles on the screen of the oscilloscope).

A circuit of the cathode-coupled multivibrator type in wich the sawtooth voltage is produced by charging a capacitor across the grid of a multivibrator valve at a high positive voltage, is used in the Philips GM 5655/03 small oscilloscope. Both valve systems of the hexode-triode ECH 81 are used. Basically, this circuit corresponds to one in which a simple double-triode is used. In contrast to the triode system, the sync voltage can be fed to the hexode section, in which case there is no interfering reaction. The circuit provides sawtooth voltages with frequencies of from 5 c/s to 30 kc/s in 8 variable ranges (Fig. 4-26).

The operation of this circuit can best be followed by first assuming that system I is blocked, system II is under current, and timing capacitor C_L is discharged. During the time that C_L is charging, the voltage across the grid of I goes gradually more positive substantially linearly with time, since R_3 is connected to a high positive voltage, and charging continues until current begins to flow through this system. The voltage across the common cathode resistor R_{77} then begins to rise, and in addition the grid of system II is fed with a negative voltage pulse from the anode of system I, the current of which has risen sharply. System II is thus blocked, its anode voltage is suddenly increased and the timing capacitor C_L can discharge again. This means, however, that the first grid of system I goes negative because of the coupling via R_{77}, and its anode current is rapidly interrupted, thus producing the flyback. Thereupon, the charging, up of capacitor C_L can recommence, as already described. Between the anode of system I and the grid of system II there is an additional coupling via an *RC*-parallel circuit. This direct connection accelerates the flip-flop, although the flyback is prolonged somewhat by this time constant. To ensure that this coupling is not too great for the low frequency ranges, the grid voltage of system II is taken

Components of the sawtooth-generator

R_1 = 50 kΩ/lin.
R_3 = 5 M /lin.
R_{69} = 1 MΩ
R_{70} = 1 MΩ
R_{71} = 560 kΩ
R_{72} = 3.9 MΩ
R_{73} = 47 kΩ
R_{74} = 54.5 MΩ
R_{75} = 0.47 MΩ
R_{76} = 100 Ω
R_{77} = 8.2 kΩ
R_{78} = 82 kΩ
R_{79} = 10 MΩ
R_{80} = 0.33 MΩ
R_{81} = 2.7 MΩ
R_{82} = Nr. 1000 A/680

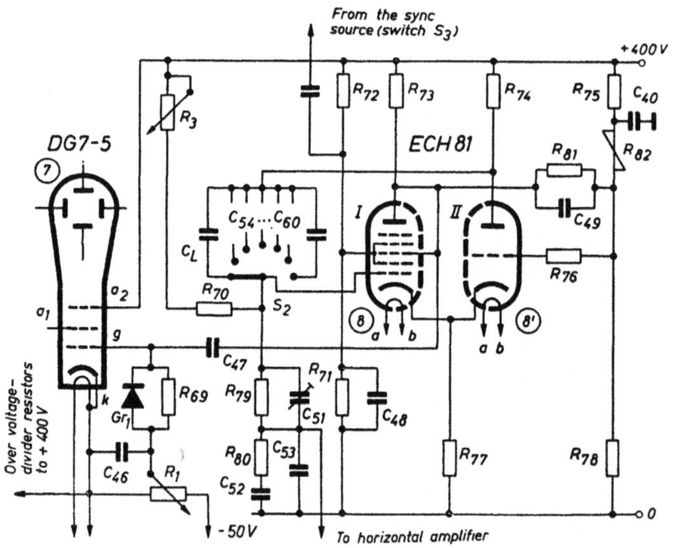

Capacitors

C_{40} = 10 nF
C_{46} = 0.47 μF
C_{47} = 22 nF
C_{48} = 390 pF
C_{49} = 12 pF
C_{51} = max. 6 pF

C_{52} = 0.22 μF
C_{53} = 560 pF
C_{54} = 82 nF

C_{55} = 56 nF
C_{56} = 15 nF
C_{57} = 4.7 nF

C_{58} = 1.2 nF
C_{59} = 270 pF
C_{60} = 82 pF

Fig. 4-26 Cathode-coupled multivibrator as sawtooth voltage generator (Philips "GM 5655/03" oscilloscope)

from a voltage divider containing a voltage-dependent resistor (R_{82} VDR resistor). Frequency adjustment is carried out, as usual, in stages, by first selecting the values of the capacitors $C_{54} \ldots C_{60}$, fine adjustment being provided by the variable resistor R_3. The frequency range is from 5 c/s to 30 kc/s.

Oscillograms of the voltages at all the electrodes of the ECH 81 valve are shown in Figs. 4-27 and 4-28 to demonstrate the operation of this circuit. The "GM 5656" oscilloscope with DC amplifier was used for this investigation, and since the oscillograms also show the zero line, it was possible not only to reproduce the waveform of the alternating voltage but also to give some impression of the voltage values, including the direct voltage level.

To keep the load on the circuit during these recordings down to the minimum, a "GM 5654" oscilloscope voltage divider probe (input impedance 10 MΩ) was modified to give a ratio of 200 : 1. The intermediate value adjuster (R_8) of the "GM 5656"

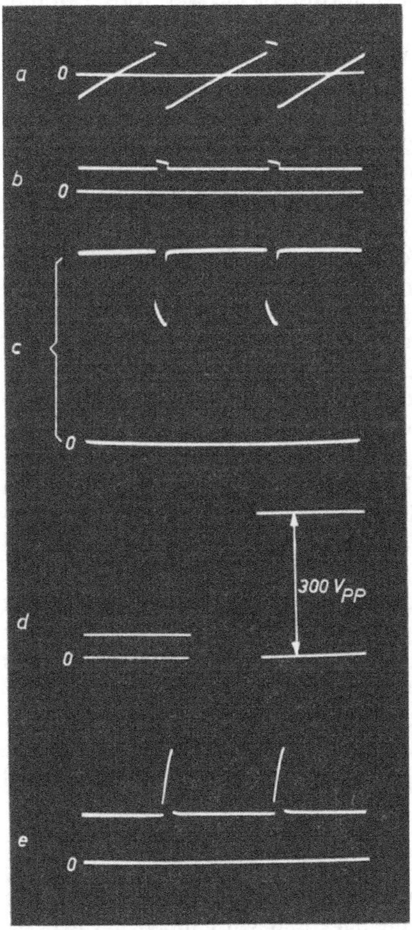

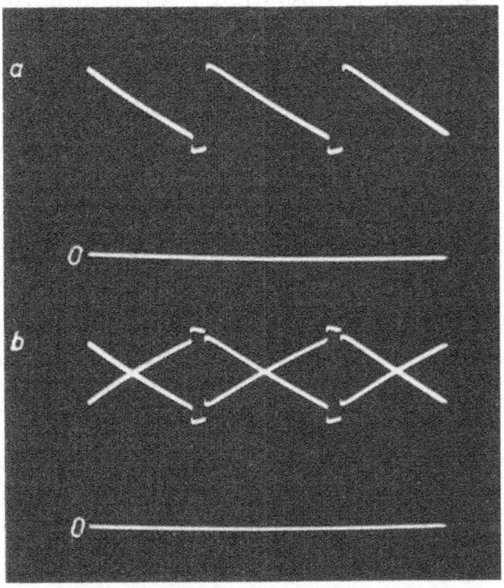

Fig. 4-28 Amplified sawtooth voltages across anodes of the horizontal amplifier.
a) Anode voltage of the valve directly controlled by the multivibrator.
b) balanced anode voltages of the two output valves by cathode-coupling control

Fig. 4-27 Voltages across the electrodes of ECH 81 valve in circuit as in Fig. 4-26. *a*) Control grid system I; *b*) voltage between cathodes and chassis; *c*) anode voltage of system I; *d*) vertical deflection by means of 300 $V_{=}$, left: voltage to grids 2 and 4 of heptode (approx. 48 V); *e*) anode voltage of system II oscilloscope was, however, moved to its extreme right-hand position, and the divider was frequency-compensated in the usual way (Fig. 5-43 *b* and Part II, Ch. 25).

These conditions were maintained for all the oscillograms, so that their voltage amplitudes could be directly compared. The oscillogram on the right in Fig. 4-27*d* is intended for amplitude calibration. It shows the zero line and its image deflected at an input voltage of 300 $V_=$. The circuit itself was working asynchronously in all these oscillograms.

Oscillogram *a* shows the sawtooth voltage appearing at the grid of valve system I together with the zero line, as already described.

The impression is given at first that the grid voltage goes strongly positive from time to time, but this is not the case. It should be borne in mind that the oscillogram in this case corresponds to the voltage between grid and chassis and not to that between grid and cathode. The voltage between cathode and chassis is shown in oscillogram *b*. According to this, the cathodes are constantly about 50 V positive, so that the sawtooth voltage oscillations actually appear only in the negative voltage range between grid and cathode. Neither overloading of the grid, nor distortion of the waveform can thus take place.

The waveform at the anode of system I is shown in oscillogram *c*. When the valve is cut off, the full feed voltage of 400 V appears at the anode. During the discharge of the capacitor the anode potential drops to about 270 V. This voltage change is also fed via capacitor C_{47} to the intensity modulation grid of the oscilloscope tube. By this means, the brightness of the trace is increased during the forward travel of the spot and reduced during the flyback. A germanium diode (G_{r1}) is connected in parallel with the leak resistor (R_{69}) to maintain the intensity level constant.

As the left-hand oscillogram in *d* shows, there is a constant direct voltage of only about 48 $V_=$ at grids 2 and 4 of the hexode system. The right-hand oscillogram which, like that on the left, covers only a short part of the complete cycle, the zero line and a line corresponding to 300 $V_=$, gives the amplitude, which in the original image was about 3.5 cm. The deflection coefficient in this set-up was thus about 86 V_{pp}/cm.

Oscillogram *e* shows the waveform at the anode of valve system II. When the valve is cut off, the voltage rises rapidly from a level of about 100 $V_=$ to about 240 $V_=$.

The sawtooth voltage is fed in ratio of 30 : 1, via a 10 MΩ compensated voltage divider, to the amplifier valve of the horizontal amplifier which is provided with a push-pull output stage just like the vertical amplifier (Figs. 5-49 and 5-55). As result, a negative going sawtooth voltage appears at the anode of the first output valve of the horizontal deflection amplifier, as can be seen from the oscillogram in Fig. 4-28*a*. By cathode coupling, the other output valve is so driven by the anodes of both valves of the output stage, as indicated in the oscillogram of Fig. 4-28*b*. These sawtooth voltages are fed to the horizontal deflection plates via coupling capacitors. At this adjustment these oscillograms show that the voltage oscillated between about +210 V and +350 V. The amplitude of the time base voltage was therefore $2 \times 140 = 280\ V_{pp}$. The image width in these signals was adjusted to about 4.5 cm, corresponding to a tube deflection coefficient of 62 V_{pp}/cm under the prevailing operating conditions (total acceleration voltage roughly 850 $V_=$).

4. 13 Triple pentode circuit

A time base circuit can also be designed using three pentodes. This circuit, which was brought out in 1936 [6], was used with slight changes and with modern valves in

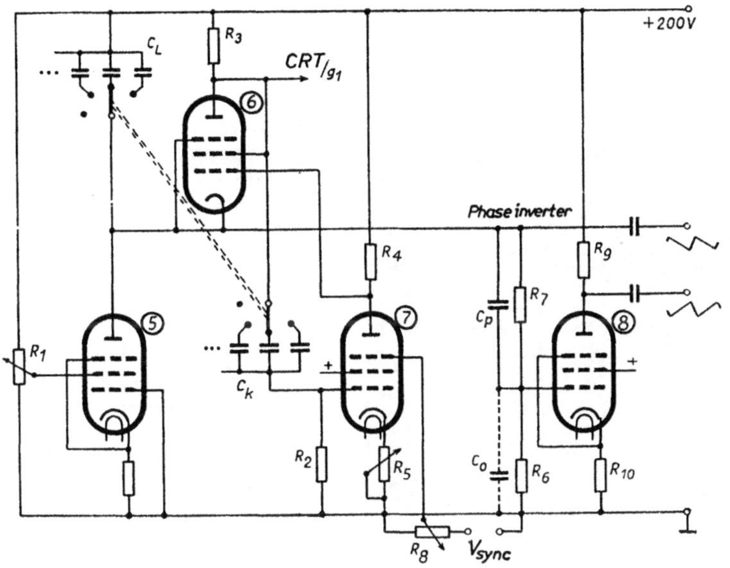

Fig. 4-29 Triple-pentode circuit (Philips "GM 5653" oscilloscope)

Philips GM 5653 and GM 5654 oscilloscopes for a frequency range of 5 c/s to 500 kc/s. Fig. 4.29 shows the basic circuit diagram.

The circuit works as follows: If, to begin with, capacitor C_L is being charged from a DC voltage source and at the same time anode current is flowing in valve 7 of such value that the voltage drop across R_4 is greater than the voltage across timing capacitor C_L, then there will be a negative bias on the grid of valve 6; the valve will be cut off, and no anode current can flow. C_L can be charged linearly with time trough valve 5. This causes the anode potential of valve 5 to fall, and with it the potential of the interconnected cathode of valve 6. As soon as the potential difference between grid and cathodes of valve 6 has become so small that anode current begins to flow, a voltage drop occurs across the anode resistor R_3 of valve 6 and is transmitted to the grid of valve 7 via capacitor C_k. The anode current of valve 7 falls and the grid of valve 6 goes more positive, so that the anode current of valve 6 increases rapidly. As a result, C_L is rapidly discharged and the cycle repeats.

This sequence of events is shown in more detail in the oscillograms in Fig. 4-30. Fig. 4-30*a* shows three cycles of the voltage waveform across C_L. The oscillogram in Fig. 4-30*b* shows the waveform of the discharge current indirectly, as a change of the voltage at the anode of valve 6. Fig. 4-30*c* gives the voltage waveform at grid 1 of control valve 7, and Fig. 4-30*d* shows the voltage waveform at the anode of this valve.

Coarse frequency adjustment is obtained by selecting various values of timing capacitance C_{11} and coupling capacitance C_k, fine adjustment by controlling the charging current of valve 5 by means of varying its screen grid voltage via R_1. The amplitude of the sawtooth voltage is adjusted by varying the cathode resistor R_5, thus varying the anode current of the control valve.

Synchronization is effected in this circuit via the third grid of valve 7. Potentiometer R_8 is connected in series for adjusting the synchronizing voltage.

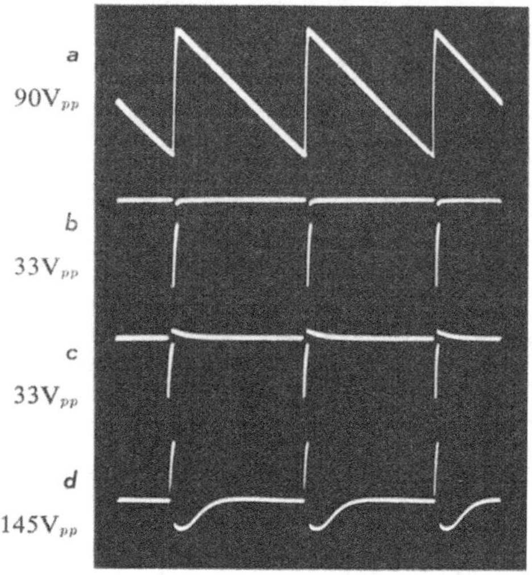

Fig. 4-30 Oscillograms for circuit in Fig. 4-29
a) Anode of charging valve 5 or timing capacitor;
b) Anode of discharge valve 6, inversely proportional to the discharge current;
c) grid 1 of control valve 7;
d) anode of control valve 7

As the time base deflection for the DG 10-6 cathode ray tube used in the "GM 5653" oscilloscope must be symmetrical, a phase inverter stage, valve 8, is used. A portion of the sawtooth voltage reaches the grid of this valve via the frequency-compensated voltage divider consisting of R_6 and R_7 with C_p and C_0. This voltage appears amplified at the anode, with its phase shifted 180°. The frequency range obtained extends from 5 c/s to 500 kc/s in ten steps. As the discharge valve in this circuit (Fig. 4-29) is connected as a triode, the description "triple pentode" may not seem quite appropriate. It became familiar, however, under this designation when the third valve was also connected as a pentode, and the name has been retained.

4. 14 Blocking oscillator circuits

In the circuits described so far, the flip-flop action was effected by means of the anode current of high-vacuum valves. There are a large number of sawtooth circuits in use, however, which employ the grid current for rhythmically charging a capacitor. In such circuits the current must, of course, be kept within permissible limits. A suitable circuit is shown in Fig. 4-31*a*, which incorporates a triode oscillator, the feedback coupling *M* being so tight that if it goes beyond the oscillation point the heavy grid current drives the voltage across capacitor *C* in the grid circuit so negative that oscillation is

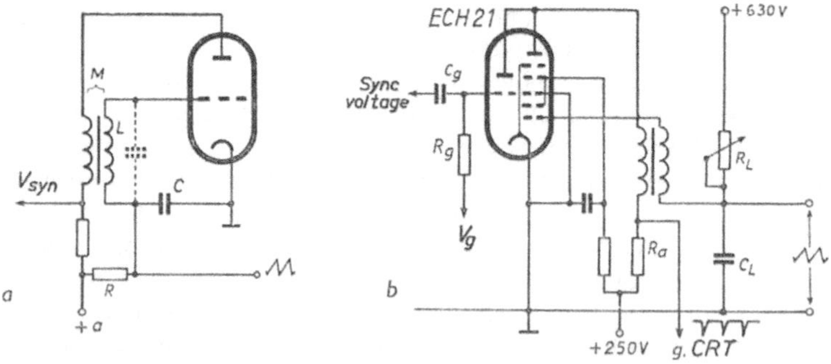

Fig. 4-31 Blocking oscillator.
a) Basic circuit with triode
b) Circuit with heptode-triode (Philips "GM 5655" oscilloscope)

cut off. The valve is now operating at a point where the curve is not steep enough to maintain the conditions for oscillation in spite of tight coupling. Capacitor *C* can now discharge through the leak resistor. The grid thus becomes less negative. As soon as the operating point moves to a region of the valve characteristic at which oscillation is possible, the cycle starts again. Every radio amateur is familiar with this process in a regenerative detector; when the feedback is increased beyond the oscillation point, the receiver begins to howl, the tone (i.e. frequency) depending on the time constant of the grid circuit.

If, as usual in radio practice, the grid resistor *R* were connected to the cathode, the capacitor would only be able to discharge in a non-linear manner because of the relatively low voltage across it. If, however, this resistor is connected to a point of high positive potential, as shown in the circuit in Fig. 4-31*a,* the voltage drop across *R* during charging will remain constant. Since the difference between the potential across the charged and across the discharged capacitor *C* is small compared with the voltage drop across resistor *R*, a good sawtooth waveform linear with time is obtained across capacitor C.

The Philips "GM 5655" oscilloscope makes use of this circuit (Fig. 4-31*b).* Here the heptode section of a heptode-triode ECH 21 [23]), connected as a pentode, acts as the oscillator. The triode section serves to amplify the synchronizing voltage; thus a smaller synchronizing input voltage is required than in the simple circuit, diagram 4-31*a,* where the synchronizing voltage is fed direct to the anode of the oscillator valve. A number of oscillograms recorded on the "GM 5655" oscilloscope are reproduced in Fig. 4-32 to illustrate the way in which the circuit functions.

Oscillogram *a* shows the voltage on the grid. During the time when the capacitor is charging negatively, it can be seen that the circuit oscillates at a high frequency for a few cycles. The sawtooth voltage at the timing capacitor C_L can be seen at *b;* the capacitor now short-circuits the high frequency voltage, so that only the pure sawtooth voltage, increasing linearly with time, appears at that point. At the anode of the valve short HF pulses occur during the high frequency oscillations, and these are shown in Fig. 4-32*c*. LF rectified pulses, as shown at *d,* can therefore be taken from the anode resistor to which the circuit capacitances are in parallel. These pulses are fed to the grid of the cathode ray tube, where they suppress the flyback of the spot, thus making it invisible. The time base frequency range in this particular circuit is 15 c/s to 25 kc/s.

Fig. 4-33 is another example of a Philips circuit using the ECC 40 valve. Valve system I serves as the oscillator, while system II connected in a bootstrap circuit, as previously shown in Fig. 4-21*a,* provides linearization of the sawtooth waveform.

In this circuit the timing capacitor is not charged by anode voltage. The charging results from the combination of the positive voltage across cathode resistor R_k and the negative charge of capacitor *C* to the peak voltage during the blocking process. With the component values specified this voltage has a mean value of 40 + 48 = 88 V.

Feedback coils particularly suitable for this circuit are normal LF filters for about 470 kc/s. Parallel capacitances must obviously be removed as well as all unnecessary connections, and care must be taken that the polarity of the coils is correct. This circuit permits high frequency oscillations of between about 4 to 5 Mc/s. Coarse frequency adjustment is obtained by switching different values of *C* and fine adjustments by

---

[23]) First model of oscilloscope "GM 5655". In the "GM 5655/03" a cathode-coupled multivibrator employing the ECH 81 valve is used (Fig. 4-26).

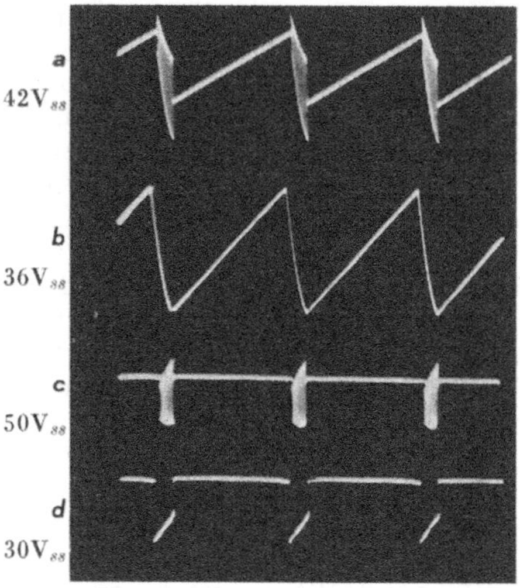

Fig. 4-32 Oscillograms of voltages in circuit of Fig. 4-30*b*.
a) Grid 1 of heptode; *b*) sawtooth voltage across charging capacitor C_L: *c*) HF pulses on anode; *d*) pulses across anode resistor R_a

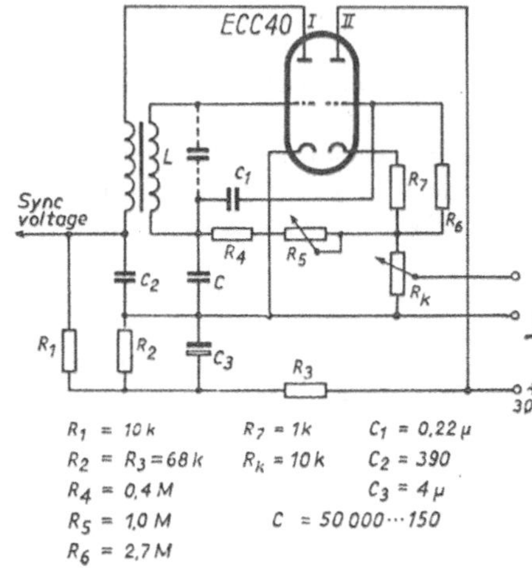

Fig. 4-33 Blocking oscillator and cathode follower feedback circuit ("Bootstrap" circuit) for linearizing using Philips ECC 40 double triode

varying R_5. The resistor R_4 is required to ensure the desired oscillating conditions under all circumstances.

The synchronizing voltage is here again superimposed directly upon the anode voltage. If required, a special valve can be provided to amplify the sync voltage as for example in the circuit shown in Fig. 4-31*b*. All the data for the circuit components are given in Fig. 4-33. If capacitance *C* can be switched in seven steps from 50.000 to 150 pF, the sawtooth frequency range will extend from 20 c/s to 20 kc/s. The output voltage is about 45 V_{pp} for a supply voltage of 300 V. Flyback times are between 2.5 and 8%, and non-linearity is between 8.8% and 5.4%.

In another type of blocking oscillator the feedback coupling is made even tighter than in the circuit described, and the grid circuit is so severely damped that high frequency oscillation can not occur. The charging of the capacitor must then take place practically within one HF cycle. Time base oscillators [24]) of this type are often used in television circuits [7].

In all such super-regenerative oscillators particular care must be taken to provide good HF decoupling of the time base generator from the power supply. Otherwise oscillator radiation can lead to severe interference with local broadcast receivers.

Among the numerous circuits which generate a sawtooth voltage linear with time by means of a single valve, usually a pentode, special significance attaches to the so-called transitron circuit.

---

24) This actually is the genuine "blocking" oscillator. The circuit so far discussed is better described as a "squegging" oscillator.

4. 15 Transitron-Miller circuit

The operation of this circuit is based on a combination of two circuit principles.

The sawtooth voltage is generated by a circuit known as a "transitron" [8], while the application of the "Miller" effect permits the use of considerably smaller capacitances [9].

The transitron circuit described by Brunetti [10] makes use of the fact that at particular values of voltage on the second and third grids of a pentode, the screen-current/screen-voltage characteristic curve has a very steep negative slope. The voltage on the first grid must therefore be slightly positive (between 0 and 0.25 V); the voltages on the second and third grid should be roughly equal and between V +40 and +80 V. The voltage on the anode is less critical. In the publication quoted [8], curves are shown for $V_{g2} = V_{g3}$ +60 V and V_a = 50V and 200 V; these curves differ very little from each other.

The transitron characteristic is very similar to the dynatron characteristic of a tetrode. In the dynatron, however, the negative slope region of the anode-potential/anode-current characteristic is due to the effect of secondary emission from the anode. As this does not occur uniformly and is neither desired (on the contrary) nor controllable, the makers of thermionic valves generally reject the dynatron circuit. With a transitron, on the other hand, use is made of the current distribution between the third and the second grids. Since this current distribution depends only on the geometric form of the grids and the voltages applied to them, there is no fundamental objection to its use. In this circuit, as always, care must be taken that the limiting data for the valves are not exceeded under any possible working conditions. The basic transitron circuit is shown in Fig. 4-34. The voltage for the third grid appears automatically across the capacitor as a result of oscillation. If, to commence with timing capacitance C_L is assumed to be discharged, then, as the charge builds up via R_L, the voltage rises until anode current begins to flow. Due to the anode current, which is controlled by the second and third grids, C_L will now discharge rapidly. This produces a sawtooth voltage at the anode and an asymmetrical rectangular voltage on the screen grid. With the operating data given in the circuit diagram, for instance, a sawtooth voltage of about 30 V_{pp} can be drawn from the anode.

In principle, the timing capacitor with its resistor could be connected in the grid circuit. The discharge would then take the form of grid current, and the voltage variation across the timing capacitor would appear amplified at the anode. Thus only a small variation in grid voltage is sufficient to produce greatly improved linearity and

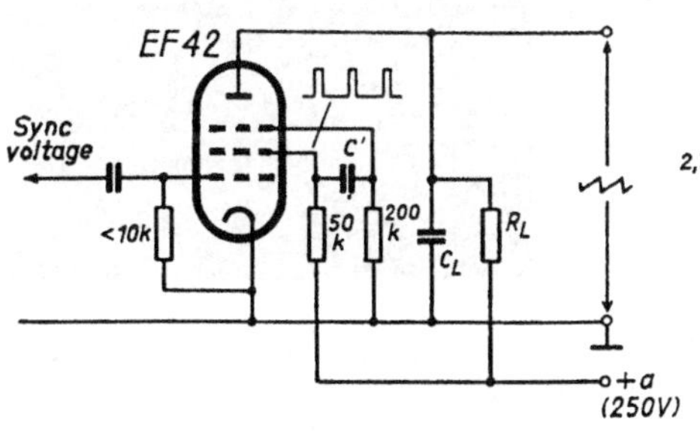

Fig. 4-34 Transitron circuit

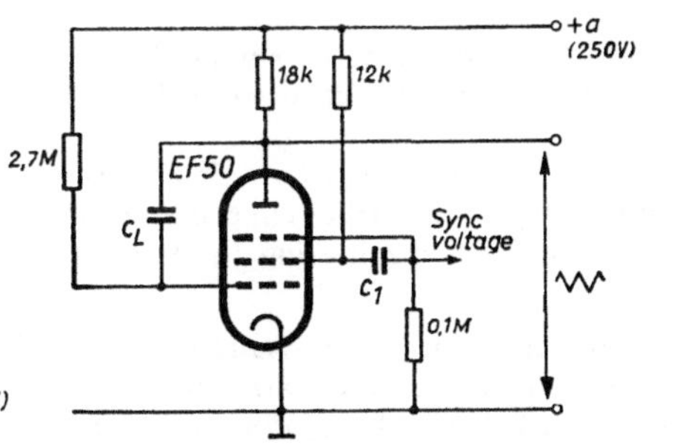

Fig. 4-35 Transitron-Miller sawtooth generator

enables a still larger sawtooth voltage to be taken off at the anode, which must then be connected to the direct voltage source via a coupling resistor. The sawtooth output suffices to deflect the beam in the cathode ray tube without subsequent amplification.

If the timing capacitor is not connected between grid and cathode, however, but between grid and anode, then for a rise in grid voltage of, say, 1 V, there will be a voltage drop of G volts at the anode, G being the gain of this stage. A given voltage variation between grid and cathode results in a $(1+G)$ time greater variation of the voltage across the capacitor C_L.

For a given charging current to produce a given voltage variation across the charging capacitor, only $1/(1+G)$ times the capacitance is now needed, compared with that necessary were it connected between grid and cathode. This effect became known as the Miller effect at the time when the influence of the grid-anode capacitance of a triode was being investigated. By itself, it does not influence the linearity of the voltage rise, as is sometimes assumed, but merely makes it possible to use very much smaller capacitances to obtain a satisfactory voltage amplitude [25]). It is because of the amplifying effect, so that the capacitor C_L need only be charged up to a small fraction of the source potential (here +250 V), that the voltage drop at the anode is substantially linear.

A basic circuit of this kind, as first shown by Cocking [11], is reproduced in Fig. 4-35, together with the component values specified by Philips for the EF 50 valve. For a time base frequency of 60 c/s, $C_L = 0.1\ \mu$F and $C_1 = 4500$ pF. The following currents were measured at a supply voltage of 250 V:

$I_a = 5.5$ mA $\quad I_{g2} = 2.4$ mA

$I_{g1} = 90\ \mu$A $\quad I_{g2} = 4\ \mu$A

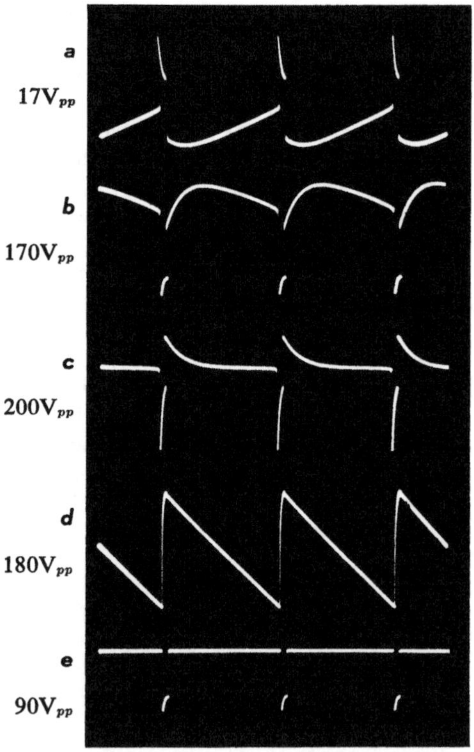

Fig. 4-36 Oscillograms of voltage in the Transitron-Miller circuit using an EF 50 valve
a) Grid 1; *b)* Grid 2; *c)* Grid 3; *d)* sawtooth voltage at anode; *e)* clipped voltage from grid 2 for suppressing the flyback

[25]) This apparent "amplification" of capacitance is made use of when time bases with a duration in the order of one minute have to be generated. The charging capacitor is then connected between the grid of the first stage of a cascade amplifier and the anode of the last stage. In this way the same effect can be obtained with a small capacitance as with one several thousand times larger between grid and cathode.

The output voltage was 130 V_{pp}, the flyback time 2% and the deviation from linearity —0.2%.

The Philips EF 42 steep-slope pentode is particularly suitable for this circuit. The oscillograms in Fig. 4-36 show the voltage waveforms obtained with an EF 50 used with the "FTO 2" oscilloscope (see bibliography [5] to Part 2, Ch. 7).

The voltage for blanking the return trace (or brightening the forward trace) is taken from the second grid. To ensure uniform brightness of the screen image during the forward trace, the voltage is "clipped" by a biased EA 50 diode. The waveform thus obtained is shown in the oscillogram in Fig. 4-36*e*. It can be seen that the positive part of the cycle is quite flat, resulting in an even intensity of brilliance on the screen.

A review of sawtooth generators of particular interest to television engineers, with detailed descriptions of the way the circuits function, is given in Chapter V of "Television" by F. Kerkhof and H. W. Werner [12].

4. 16 Dependence of amplitude on frequency control

Many of the circuits once familiar, particularly those in which post-amplification was not used (e.g. Fig. 4-29), have the drawback that controlling the amplitude of the deflection voltage influences the control of the frequency and vice-versa. With the "Transitron-Miller" circuit, in which the oscillator itself works as a "Miller" amplifier, the output is taken from the relatively low-ohmic anode resistor. As this can be connected without noticeable disadvantage as a potentiometer, the amplitude can be varied by this circuit without influencing the frequency.

The same applies also to all circuits using a cathode follower in the output for extra linearity (Fig. 4-21*a* and *b*). Here too, the normally low-ohmic cathode resistor in the output circuit can be used to control the amplitude without the frequency or the waveform being affected.

It is, of course, clear that in all circuits where the sawtooth voltage is subsequently amplified by a wide-band amplifier (e.g. in the "GM 5655/03" oscilloscope in Fig. 4-26, "GM 5662" in Fig. 4-69 and "GM 5602" in Fig. 4-50), control of the amplitude of the voltage on the time plates can be effected by the normal amplification control without influencing the frequency.

In this connection, it should be noted that another circuit [13] has been developed, in which the time base frequency is automatically controlled in such a way that, when it is so adjusted that a predetermined number of cycles of the input signal is displayed on the screen, the adjustment remains constant even when the frequency of the input signal changes considerably (up to 1:10).

The change in frequency can be read from a milliammeter which shows the average charging current. In this circuit the time base voltage amplitude also remains constant at the adjusted value.
No practical application of this circuit in industrially produced apparatus has so far been described.

4. 17 Triggered circuits

4.17.1 FREE-RUNNING TIME BASE CIRCUITS AND TRIGGERED CIRCUITS

In all time base generators described hitherto, sawtooth voltages were usually generated in a self-oscillating circuit. Stationary screen images are only obtained by synchronization, in which the spot flyback is triggered by a peak in the alternating current of the signal. It is thus inevitable that the number of time base cycles must be an integral fraction of the signal frequency, so that the images appearing on the luminescent screen can either be several complete cycles or at least one complete cycle of the signal under observation. It is possible, by the exercise of some skill, to make other images approximately stationary, but such oscillograms can hardly be considered as satisfactory, as a perfectly stationary image cannot be ensured. Since the synchronization of a self-exciting sawtooth voltage generator depends basically on cotrolling the value to which the sawtooth voltage rises (Fig. 4-19), it is clear that with synchronization the frequency rises above what it would be in the absence of a sync signal, i.e. the duration of the cycle becomes shorter. Moreover, the sawtooth amplitude decreases.

Time base circuits for triggering are so designed that, at a given value of the control voltage, which is usually obtained directly from the vertical signal voltage, one complete sawtooth cycle is displayed. In well-designed circuits for triggering, no voltage pulses are allowed to interrupt the sawtooth voltage or to influence the amplitude. This makes it possible to select the deflection speed independently of the signal frequency merely by varying the time constants of the circuit. Provision is made to render the time base infinitely variable, without having to display a certain (integral) number of cycles of the signal. Operation is thus so much simpler, that time base generators in the more recently developed oscilloscopes are designed for triggered operation, and are in general so used. This triggering of the time base circuit is especially important when as many details as possible must be clearly displayed within a single cycle of the signal. The interval of time to be displayed on the oscilloscope screen is then made shorter than a signal cycle. Before circuit details of time base generators are dealt with more specifically in the following sections, the oscillograms in Figs. 4-37, 4-38 and 4.39 will be used to study this type of time deflection thoroughly.

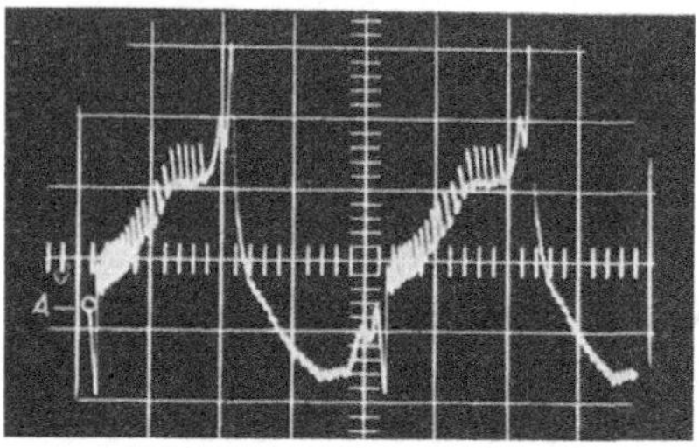

Fig. 4-37 Two cycles of AC mains voltage with strong harmonics

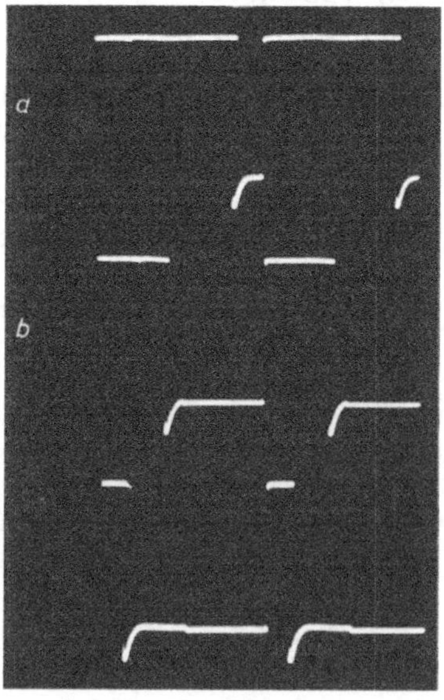

Fig. 4-39 Unblanking voltages for the oscillograms in Fig. 4-38

4.17.2 TRIGGERING

When triggering the sweep voltage, the circuit is normally first adjusted to the stand-by condition. When the signal, or other voltage which serves as a trigger for the time

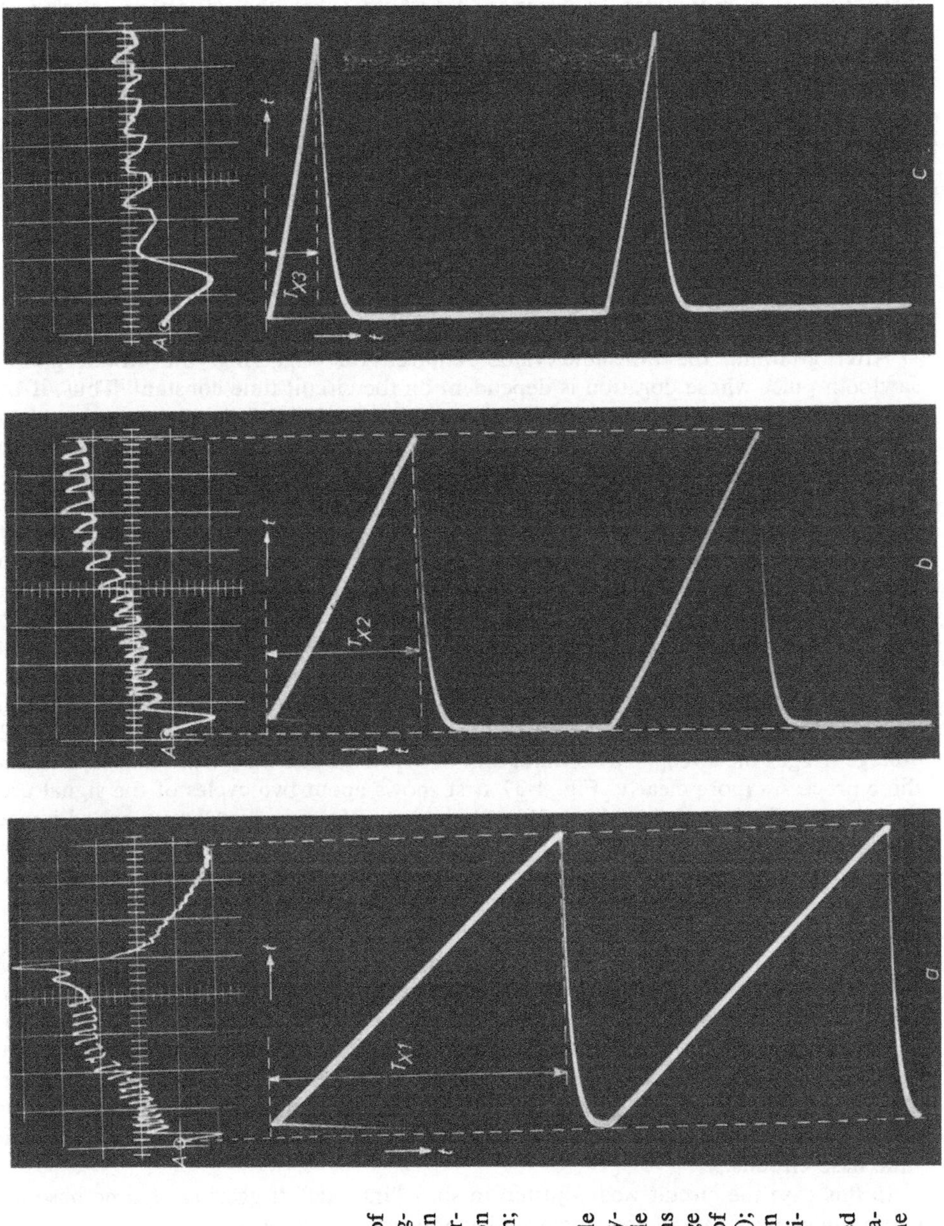

Fig. 4-38 Mode of operation of a triggered time base when deflection time varies. Above, trace on oscilloscope screen; below, deflecting sawtooth voltage.
a) Time base cycle includes (with flyback) even one cycle (20 ms) of mains alternating voltage (observed period of time about 17.6 ms);
b) higher deflection speed (period of time about 9 ms);
c) deflection speed still further increased (period of time about 3.5 ms)

deflection, rises from zero to a certain threshold value, the deflection process sets in and proceeds at a speed dependent on the set time constants. The repetion frequency of the sawtooth voltage is determined by the ratio of the duration of one signal cycle to the adjusted duration of the time deflection. If this is less than one cycle, the repetition frequency of the sawtooth voltage will be equal to the signal frequency (Fig. 4-38*a*). If the adjusted time base duration is greater than one cycle, however, but less then two cycles, the repetion frequency of the sawtooth voltage sequence will become equal to half the signal frequency. The sweep duration, and thus the signal time displayed, can then be varied within the limits:

$$\frac{1}{f_m} = T_m \quad \text{and} \quad 2 \cdot \frac{1}{f_m} = 2\,T_m\,.$$

After attaining the threshold value required for triggering, the circuit gives a sawtooth pulse whose duration is dependent on the circuit time constant. Thus, if this time is less than one, two or more complete signal cycles, depending upon the adjustment, the voltage of the time base generator remains at zero for the time until the next successive voltage threshold. If the adjustment of the time base deflection duration is altered to exceed the limit of whole cycles of the signal, then the repetition frequency of the time base generator falls abruptly in such a way that the duration of the sawtooth cycle –including flyback and stand-by time– is always equal to the total duration of the next higher whole number of signal cycles. For example, if a signal of about 50 c/s is under observation, at adjustments for observing intervals of time between one and two cycles, the repetition frequency of the time base generator reverts abruptly to 25 c/s. If the time base duration is still further reduced, the repetition frequency naturally remains at 50 c/s; the sawtooth, however, becomes even shorter and the stand-by time longer. Only occasionally in the 20 ms time interval, can the thershold voltage trigger the sweep. The oscillogram in Fig. 4-37 and 4.38 will serve to illustrate these processes more clearly. Fig. 4-37 first shows about two cycles of the signal used in this example, in order to show the complete waveform before the expanded images in Fig. 4-38 are dealt with. This voltage was obtained by very loose capacitive coupling to the discharge path of a fluorescent TL-lamp, a connecting cable, parallel to the lamp and a few cm distant from it, being led to the input of a high-pass filter, so that the harmonics of the signal are seen in an exaggerated form on the screen.

The oscillograms in Fig. 4-38 show in each case the triggered operation of the "GM 5666" oscilloscope with three different values of time base duration, and include one oscillogram of the time base voltage used to produce this image.

In addition to a fundamental oscillation with a frequency of 50 c/s, numerous harmonics are superimposed, due mainly to processes taking place in the gas discharge. This waveform therefore seemed particularly suitable for indicating the possibilities for employing reduction of the time coefficient as could only be obtained by triggered time base circuits.

In this case the circuit was adjusted to start "internal" triggering of time base at a certain negative value of the signal under observation (point *A*).

By varying the forward trace duration of the individual sawtooth pulses by means of the fine adjustment, increasingly smaller time coefficients are obtained, as can be seen in the pictures of the signals in oscillograms, *a, b,* and *c* of Fig. 4.38. The time base repetition frequency was 50 c/s in each of the three oscillograms in Fig. 4-38. The time coefficient of *a* was so adjusted that exactly one cycle of the signal is fully covered.

The missing portion of the waveform to make up a full cycle is accounted for by the flyback. Thus a time base is obtained which corresponds to the sawtooth rise, that is to say, to the time T_{X1} (about 2.3 ms/cm). In *b* and *c* the rise speed of the sawtooth was reduced to produce deflections of 1.1 ms/cm and 0.4 ms/cm respectively. The oscillograms, always beginning at point A are thus increasingly expanded to correspond to the smaller time coefficient, so that the details of the waveform become progressively more easily recognized.

In the case of such time base circuits, it is necessary so to adjust the basic level of the voltage at the grid of the cathode ray tube that the spot remains invisible in the absence of a brightening voltage. By additional brightness modulation by means of a rectangular voltage which is positive during the forward trace of the spot, during the linear rise of the sawtooth voltage it is possible to brighten the screen image only during the actual time base deflection. Without this brightness modulation, not only would the spot flyback be visible, but also a disturbing bright spot would occur, especially during the stand-by time, i.e. during that part of the sawtooth cycle of the beginning of the time sweep, when the voltage is zero. Oscillograms *a, b* and *c* of Fig. 4-39 show the unblanking voltages corresponding to the oscillograms of Fig. 4-38.

In triggered time base generators, as will be seen from the descriptions of such circuits in the following section, there are always points from which a suitable rectangular voltage for intensity control of the forward stroke of the spot can be taken.

In high-performance oscilloscopes the waveforms of these intensity pulses must meet stringent requirements. On the one hand, it is necessary that the rise and fall of voltage should be so abrupt that, at the required deflection speeds, the beginning and end of the image are sharply limited and not merely brightened gradually. Otherwise important portions of the image, particularly at the commencement of the trace, might be unrecognizable. On the other hand, in the case of oscilloscopes used for the observation of very slow changes of state, the intensity, even at very low repetition frequencies (1/3 to 1/10 c/s, i.e. 3 to 10 s deflection duration) must remain constant during the whole duration of the time sweep. This means that *RC*-coupling cannot be used, but a DC-coupling or special coupling devices must be employed. Examples of DC-couplings are shown in circuit diagrams 4-63, 4-64 and 4-65 in more detail.

At a given deflection frequency, the brightness of the oscillogram decreases as the time coefficient becomes less, and particularly as the deflection duration becomes shorter compared with the duration of the signal cycle. This is readily understandable if one considers that the time during which the path of the spot is visible becomes a progressively smaller fraction of the total observation time. It is therefore important in triggering so to adjust the sweep that the stand-by remains as short as possible, to avoid corresponding reduction of the brightness of screen image compared with that in synchronous operation. A minimum stand-by time is required, so that the next deflection cycle can start at precisely the same voltage value as the triggering voltage. This requires that the timing capacitor of the time base generator must be sufficiently discharged (in theory complete discharge takes an infinitely long time), otherwise the commencing points of the individual sweep cycles would occur at different positions on the screen, and the over-all impression made by the oscillogram would be one of "jittering". For stable triggering, especially at high sweep frequencies, it is therefore necessary, according to requirements, to use a rather complex, and therefore expensive circuit [26]). If triggering is not absolutely necessary for any reason it is best, in the

[26]) To ensure a satisfactory start of the next sweep, the more elaborate oscilloscope use special hold-off circuits. At high sweep frequencies the waiting times are often longer than the

case of high sweep frequencies, especially in simpler generators, to go over to self-exciting synchronized operation.

The smallest useful time coefficient is also limited by the fact that, beyond a certain time coefficient, which in time base deflection lasts only for a fraction of the duration of one signal cycle, the screen luminosity does not suffice for observation or photographic recording. It is therefore of paramount importance in oscilloscopes whose time base units can be triggered (particularly pulse oscilloscopes), to use the highest acceleration voltages possible. It is possible, however, to observe low-luminosity oscillograms succesfully with the help of an eyepiece tube and a rubber mask to shield the eyes from ambient light (Fig. 6-1 in Part II, Ch. 6).

Triggering is not limited to releasing the sweep for observing cyclic signals. On the contrary, it also enables non-cyclic and non-recurrent phenomena to be observed. In all these cases the onset of the sweep is determined by a certain –and preferably adjustable– signal value ("trigger level"). If this level is reached, a complete sweep cycle is performed, after which the circuit remains in the stand-by condition until the next voltage threshold occurs. In non-cyclic phenomena (light flashes from scintillation counters or the like) a screen image is obtained, the individual sweep cycles of which always begin at the same place and run at the desired deflection speed. If the periodicity of the signal cycles is altered, this is manifest only by corresponding fluctuations in the luminosity of the oscillogram.

The observation of a non-recurrent phenomenon is thus very simple. The triggering threshold value of the signal automatically sets the sweep in operation, and the unblanking simultaneously displays the moving spot on the screen. If the signals are not too rapid, the trace can be observed on a persistent screen. In the case of very rapid signals, and whenever the oscillogram is required for exact interpretation, photographic recording is essential.

The problems involved in triggering time base units have been dealt with in many publications, of which a selection is mentioned in the biblography [14] [15] [16] [17] [18] [19].

4. 18 Time base generator for self-oscillating and triggered operation

4.18.1 FUNDAMENTAL CIRCUIT

In this section the time base generator of the Philips "GM 5602" wide-band oscilloscope, the circuit of which is typical of this kind of time base generator, is described. It takes the form of a printed circuit (see Fig. 1-4) and is also used without any fundamental changes in the large Philips "GM 5603" wide-band oscilloscope. In triggered operation with the "GM 5602" oscilloscope, the coefficients can be selected in 15 ranges from 0.2 μs/cm to 10 ms/cm (expanded 5 times from 40 ns/cm). In the "GM 5603" oscilloscope the time base coefficients can be similarly adjusted from 0.2 μs/cm to s/cm in 21 steps (also 5 times expanded to 40 ns/cm).

For explanatory purposes a very much simplified fundamental circuit in Fig. 4-40 is first considered, the method of operation being explained by reference to the oscillograms in Figs. 4-41, 4-42 and 4-43, which were recorded with a Philips "GM 5601" HF oscilloscope with a DC amplifier. In the course of discussing further details,

sweep duration, often a multiple of this time and indeed often result in a corresponding loss of brilliance. The designers are therefore anxious to curtail the hold-off time as far as is consistent with maintaining stable operation.

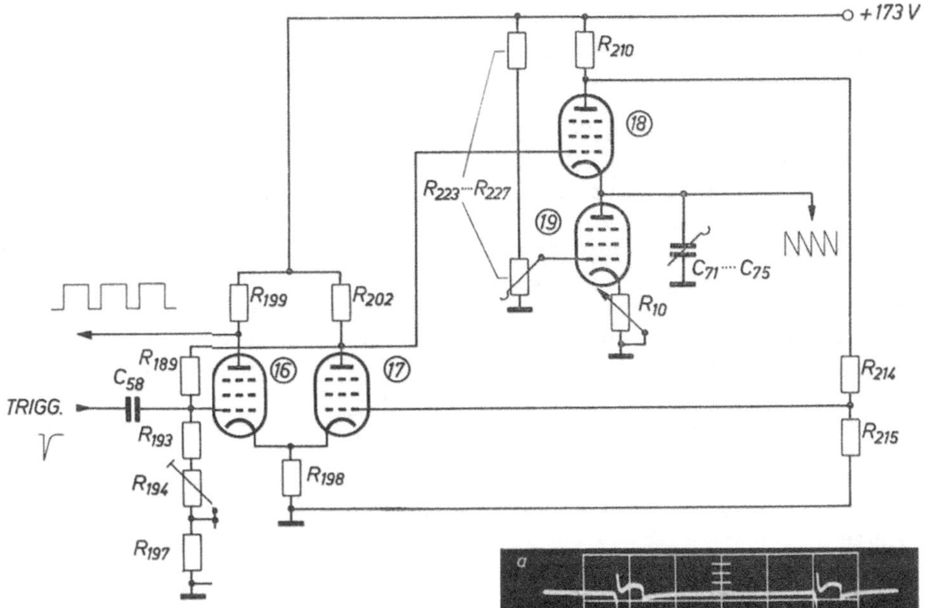

Fig. 4-40 Basic circuit for a time base unit for recurrent sweep and triggered operation (combination of the circuit elements of Fig. 4-50)

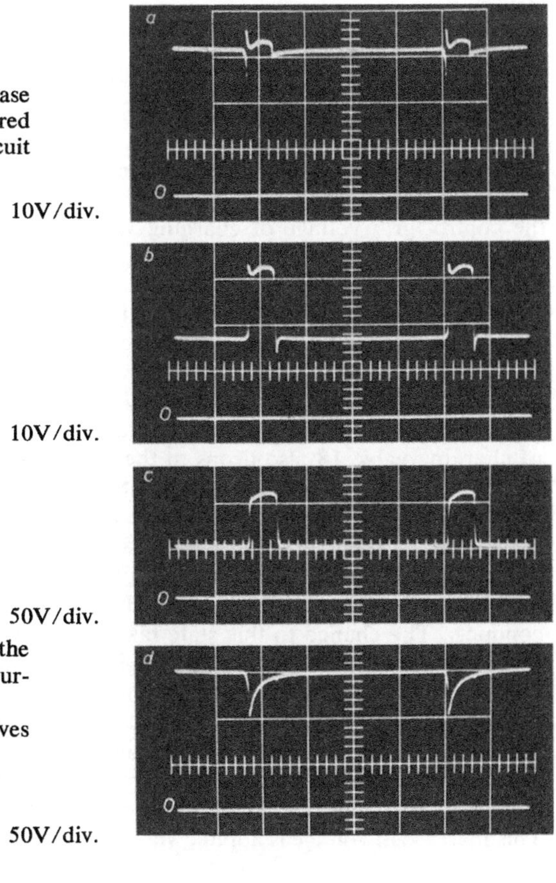

Fig. 4-41 Voltage waveforms in the Schmitt-trigger valves (16,17) in recurrent sweep operation

a) Common cathode connection valves 16 and 17

b) 1st grid valve 16

c) anode valve 17

d) 1st grid valve 17

a correspondingly expanded circuit will then be dealt with (Fig. 4-47), and finally the complete circuit diagram (Fig. 4-50), the steps required for calibrating the time scales (Fig. 4-51), and the trigger pulse shaper circuit (Fig. 4-53). The principle of the triple pentode circuit will be used for that part of the circuit which generates the time deflection sawtooth voltage (Fig. 4-29).

System 18 is the charging valve, and 19 is the discharge valve. The control of this circuit does not take place by means of a single control valve, as in the simple triple pentode circuit of Fig. 4-29 (valve 7), but by means of a multivibrator, a so-called "Schmitt-trigger". It is formed by the two valve systems 16 and 17. Such a circuit, which is described separately and in detail in the chapter on "The Schmitt-trigger circuit" on pages 145 ff, comprises a cathode-coupled multivibrator, in which the anode of the one valve system is directly connected to the grid of the other valve. A Schmitt-trigger has two stable states. In the one, the left-hand system of the valve is conducting 16, and the right-hand system 17 is blocked. In the other state the left-hand system is blocked and the right-hand system is conducting. The reversal of the circuit from the one state to the other can be triggered by a negative-going voltage applied to the control grid of the valve under current or by a positively going voltage applied to the grid of the blocked valve. The operation of this circuit in self-oscillation will first be considered.

4.18.2 SELF-OSCILLATING OPERATION

In the starting position, switch S_4 is closed. The resistors R_{202}, R_{189}, R_{194} and R_{197} form a voltage divider between the positive supply voltage and the chassis, thus serving also to determine the grid potential of valve 16. Closing switch S_4 causes the grid voltage of valve 16 to be lower than that of valve 17, so that the latter is under current and valve 16 is blocked. In these circumstances the anode voltage of valve 17, and thus the control grid voltage of charging valve 18, are both low. If at the same time the cathode voltage of valve 18 has been set sufficiently high, then the valve is blocked (see the oscillograms in Fig. 4.41).

It should be assumed that the timing capacitor selected from $C_{71} \ldots C_{75}$ in accordance with the desired time scale has previously been charged by charging valve 18. Now that charging valve 18 is blocked, however, the capacitor can be discharged via discharging valve 19 at a constant current, commencing from point A linearly with time, as can be seen from the oscillogram in Fig. 4-42. As a result, the cathode voltage of charging valve 18 also drops to the same extent. At a certain value of this voltage the potential difference between control grid and cathode of valve 18 falls to less than the cut-off voltage, and current again begins to flow. But a voltage drop now occurs across anode resistor R_{210}, and the anode voltage is reduced. This causes the grid voltage of valve 17 of the Schmitt-trigger, which has hitherto been conducting, to drop via divider R_{214}, R_{215}. The valve is thus blocked and the other valve (16) now conducts. The change to this state takes place with extreme rapidity, and the control grid voltage of charging valve 18 rises just as fast. Valve 17 is, of course, now blocked, and its anode voltage is high, thus permitting the selected capacitor ($C_{71} \ldots C_{75}$) to be rapidly charged via valve 18. The cathode voltage of this valve thus rises, and on reaching a certain level causes the flow of current to cease, as is clearly seen in Fig. 4-42, and the charging of the capacitor is brought to stop. But this makes the anode voltage of the charging valve (18) rise once more to the value of the supply voltage. This increase in voltage is applied via R_{214} to the control grid of the right-hand Schmitt-

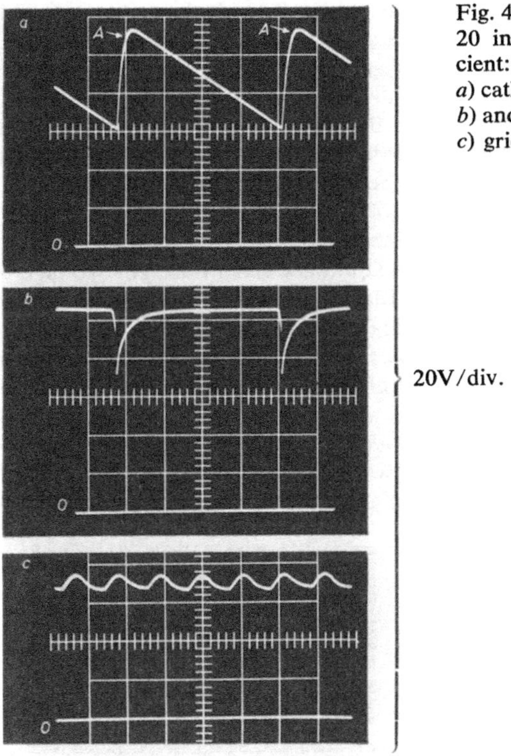

Fig. 4-42 Voltage waveforms at valves 18 and 20 in recurrent sweep operation. Time coefficient: 10 μs/cm
a) cathode valve 18
b) anode valve 18
c) grid valve 20

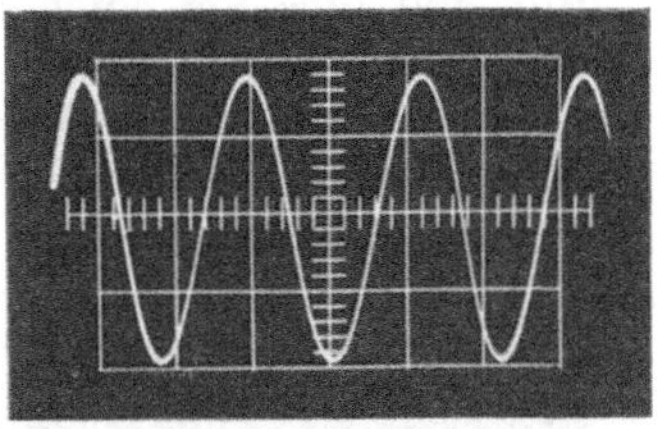

Fig. 4-43 Oscillogram of the signal voltage during the recording of the oscillograms in Figs. 4-41 and 4-42. Frequency about 90 kc/s

trigger valve (17), which now becomes conducting, so that the Schmitt-trigger flops to the other state in which valve 16 is blocked once more. Now the timing capacitor (C_{71} . . . C_{75}) can discharge linearly with time, and a new sawtooth voltage cycle can begin. This sawtooth voltage is taken from the cathode of charging valve 18 (or from the anode of discharging valve 19). The slope of the curve of the voltage drop (voltage speed) is determined by the value of the capacitor (C_{71} . . . C_{75}) and by the discharge current which is adjusted by valve 19. By selecting different timing capacitors and by changing the discharge current, for instance by changing the value of the cathode resistor of valve 19, the voltage speed and the deflection speed by means of the sawtooth voltage, that is, the time coefficient (s/cm), can be adjusted. In this way the sawtooth voltage generator is self-oscillating. The Schmitt-trigger is set to such a sensitivity that a voltage change occurring at the anode of charging valve 18 suffices to switch the circuit to either of the two operating states. As the time coefficient must be capable of fine adjustment for sync purposes, as will be described later, the calibration does not apply for self-oscillating working. The oscillogram in Fig. 4-43 is the pattern shown on the "GM 5602" oscilloscope when recording the voltages in the self-oscillating time base voltage generator for Figs. 4.41 and 4-42. The sinusoidal voltage has a frequency of about 90 kc/s [20] [21] [22] [23] [24].

4.18.3 TRIGGERED OPERATION

For triggered operation switch S_4 is opened. This increases the voltage at the control grid of valve 16, which now draws current, and valve 17 is blocked. As the anode voltage of valve 17 is now high, charging valve 18 also remains conductive. The charge of the capacitor ($C_{71} \ldots C_{75}$) now flows through discharging valve 19, but is continually replenished via the charging valve. A voltage level, determined by the ratio of the internal resistances of both the charging and the discharge valve, thus appears across the capacitor. As the charging valve has a very low internal resistance and the discharge valve has a very high one, this level is approximately the same as the supply voltage. This state is maintained until the control grid of the left-hand valve of the Schmitt-trigger is blocked by a sufficiently large negative-going trigger voltage, preferably by a negative pulse. How such a pulse can be obtained from the signal voltage or from a voltage source intended for controlling the time base unit, will be described separately later in this section under the heading "Trigger-pulse shaper". Now the left-hand valve (17) takes over the current. This causes the potential at its anode to drop and thus the potential at the grid of discharge valve 18, so that the latter is blocked.

The current flowing through discharge valve 19 can therefore discharge the capacitor ($C_{71} \ldots C_{75}$) linearly with time, as shown in the oscillogram *a* of Fig. 4-44. As soon as the cathode of valve 18 once more reaches the potential at which the

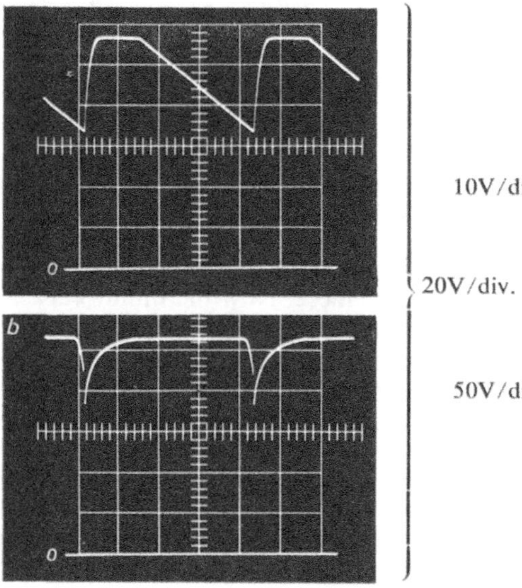

Fig. 4-44 Voltage waveform at valve 18 in triggered operation Time coefficient: 10 μs/cm
a) cathode; *b*) anode

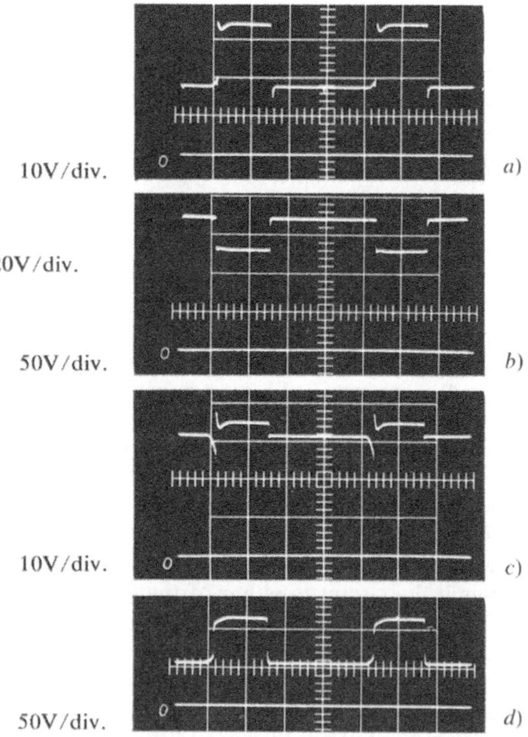

Fig. 4-45 Voltage waveforms at valves 16 and 17 in triggered operation. Time coefficient: 10 μs/cm. *a*) 1st grid valve 16. *b*) anode valve 16. *c*) common cathode connection of 16 and 17. *d*) anode valve 17.

valve becomes conducting, its anode voltage drops. and the Schmitt-trigger again flops into the state in which valve 16 is conducting and valve 17 is blocked. The anode current now flowing in valve 17 accelerates the charging process, since now the control grid of charging valve 18 becomes more positive. The oscillograms of Fig. 4-45 show the voltage waveform in the Schmitt-trigger. At the end of the charge the cathode of the charging valve clearly becomes more positive, the current flowing through the valve is reduced, and its anode voltage rises. The Schmitt-trigger is, however, so set by means of the potentiometer R_{194}, that this increase of voltage at the control grid of the right-hand Schmitt-trigger valve (17) is insufficient to cause the circuits flop. In the absence of external stimulus, the circuit remains in the same state, that is to say valve 16 is conducting and valve 17 is blocked. Charging valve 18 and discharge valve 19 are now conducting at the one and same time. The voltage across timing capacitor ($C_{71} \ldots C_{75}$) once more assumes a constant level, as was described at the outset. A new voltage sawtooth, and thus a new time base deflection process, cannot take place until a new negative voltage pulse reaching the control grid of valve 16 of the Schmitt-trigger causes this to reverse. The response sensitivity of the Schmitt-trigger is, moreover, so adjusted that it is insensitive to further trigger pulses during the discharge period, and cannot respond again until the timing capacitor is fully charged. Special switching arrangements still to be discussed (see Section 4.18.6 "Hold-off circuit for preventing picture jitter in triggered operation") ensure that the charging is practically completed before a new sawtooth begins. In triggered operation of the time base circuit, pauses must occur in the voltage waveform between one sawtooth and the next (cf. oscillogram *a* in Fig. 4-42) as distinct from self-oscillating operation, as shown in oscillogram *a* of Fig. 4-44. The voltage at the anode of charging valve 18, and thus the waveform of the control voltage on the first grid of valve 17 of the Schmitt-trigger, is shown for the sawtooth in Fig. 4-44*a* in oscillogram *b*. The oscillogram of an alternating sinusoidal voltage of about 70 kc/s displayed in the oscillogram of Fig. 4-46 was adjusted on the "GM 5602" oscilloscope for these recordings, and also for the series of oscillograms of Fig. 4-54 and 4-55 describing the method of operation of the trigger-pulse shaper.

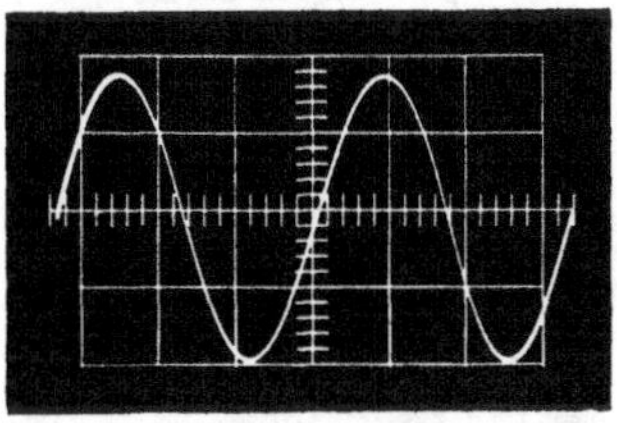

Fig. 4-46 Signal voltage waveform on the "GM 5602" oscilloscope during the recording of the oscillograms for Figs. 4-44 and 4-45. Time coefficient: 4 μs/cm

4.18.4 MEASURES FOR ENSURING THE LINEARITY OF THE DEFLECTION SAWTOOTH

4.18.4.1 Non-linearity due to delayed start of discharge

As has been described, the recharging of the timing capacitor commences immediately the cathode voltage of the charging valve has fallen, as a result of discharging, and current begins to flow once more. The drop in anode voltage so caused is intended to reverse the Schmitt-trigger as rapidly as possible. But it must be borne in mind that at the commencement of charging by valve 18 so small a current flows during the first 0.1 to 0.2 μs that the voltage drop at its anode does not suffice to make valve 17 conducting and thus to reverse the Schmitt-trigger. This means that during this short time, a further if only small amount of charge also occurs. This would make the voltage sawtooth curved and not linear at its lower end.

At lower deflection speeds this non-linear portion of the voltage curve would be of no consequence, as it is very short in proportion to the total duration of the sawtooth. At high deflection speeds (small time coefficients) however, it could be sufficient to cause serious errors in quantitative observation of the image. To avoid this, the basic circuit is modified by including two additional valve systems (20 and 20′), as can be seen in the more detailed circuit shown in Fig. 4-47. The function of these valves is to stop the discharge in good time and to permit the recharge to commence even before charging valve 18 itself begins to draw current. The anode of valve system 20 is therefore connected to that of the carging valve, so that the anode resistor R_{210} of valve 18 also carries the anode current of system 20. Valve system 20 is controlled by two voltages. The sawtooth voltage of discharging valve 19 reaches its cathode via system 20′ connected as a cathode-follower, while a direct voltage is applied to its control grid and is so adjusted by the potentiometer R_{249} that, at the end of the voltage waveform of the sawtooth voltage across is cathode, current begins to flow through the valve before it flows through charging valve 18.

The voltage across resistor R_{210} caused by this current now reverses the Schmitt-trigger and starts the renewed charging via valve 18. It is thus triggered by real voltage control before the small current passing through valve 18 at the beginning of the discharge can cause non-linearity.

The pulses on the anode of the Schmitt-trigger valve 17, which control the opening and closing of charging valve 18, have a certain rise and fall time due to the time constants of the circuit. Certain delays could occur in this way, as a result of which the charging valve could not be changed over to the blocked condition from the con-

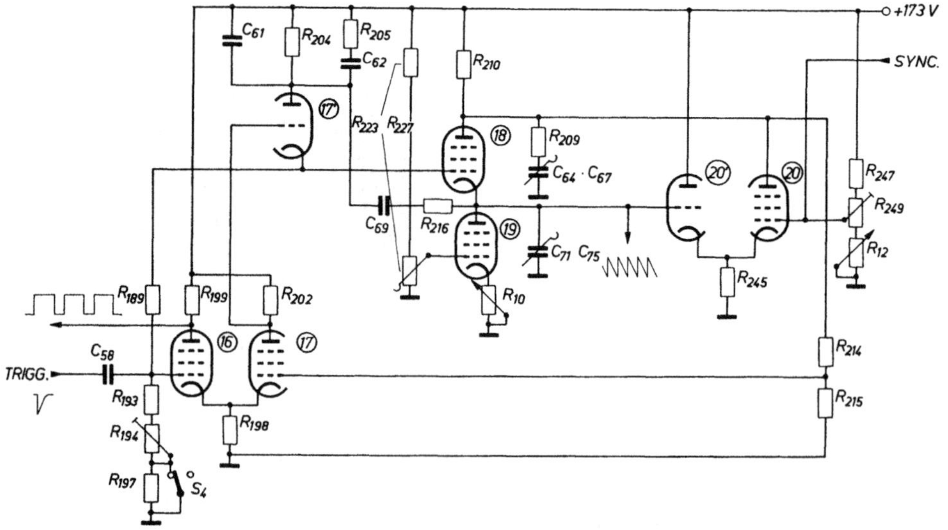

Fig. 4-47 Circuit of Fig. 4-40, extended (Arrangement of circuit elements in Fig. 4-50)

ducting state and vice versa. The cathode-follower 17′ (Fig. 4-47) is therefore connected between the anode of valve 17 and the control grid of charging valve 18. Since a cathode-follower has a high input impedance (reduced dynamic input capacitance) and a low output impedance, the transmission of the pulses is considerably accelerated.

4.18.4.2 Non-linearity at the commencement of the sawtooth due to the grid-anode capacitance of the charging valve

The grid-cathode capacitance of charging valve 18 together with that of the timing capacitor ($C_{71} \ldots C_{75}$), forms a capacitive voltage divider, so that a part of the pulse current flows from the cathode of valve 17′ to this capacitor. If small, as in the case of a high deflection speed –low time coefficient– this pulse would also cause non-linearity. This effect is compensated to a large extent by the fact that a similar current pulse is fed in anti-phase to the cathode of the charging valve from the anode of the cathode-follower 17′, in the feed line of which the resistor R_{204} is also connected. By means of a corrector network (C_{61}, C_{62} + R_{205}, C_{69} and R_{216}) this pulse is given the required waveform and most favourable phase relationship. The oscillogram in Fig. 4-48 shows the waveform of this pulse during the recording of the oscillograms of Figs. 4-44, 4-45 and 4-46.

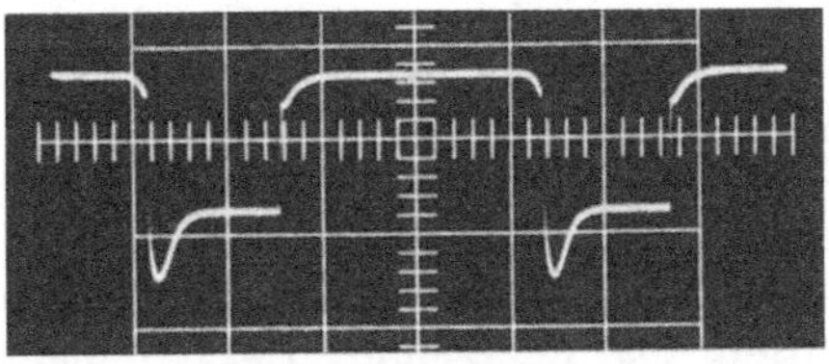

Fig. 4-48 Anode voltage waveform at valve 17. Time coefficient: 10 μs/cm

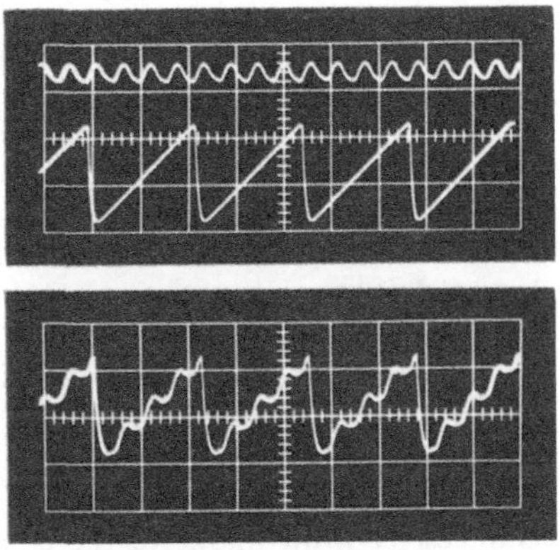

a

b

Fig. 4-49 Voltage waveforms at valve 20 for triggering the flyback. Time coefficient: 20 μs/cm
a) Above –sync voltage at 1st grid below– sawtooth voltage at cathode (shown by difference amplifier in reversed phase)
b) waveform of the sum of the instantaneous values of the two voltages of *a*)

4.18.5 SYNCHRONIZING

For Synchronization in self-oscillating operation flyback control by the synchronizing voltage is applied to the control grid of auxiliary valve 20 as well as the direct voltage from the divider resistor R_{249}. In this way the cathode of valve 20 is driven negatively by the sawtooth voltage and also controlled by the sync voltage at the grid. The waveforms of these voltages can be seen in the oscillograms of Fig. 4-49. As a negative-going voltage at the cathode has the same influence on the coltrol of the valve as a positive-going voltage at its grid, the phase of the sawtooth voltage is reversed by means of a difference amplifier at the input of the oscilloscope amplifier (Philips "GM 5603" oscilloscope). In oscillograms *a* of Fig. 4-49 both voltages are first of all shown singly, and in oscillogram *b* the voltage curve, corresponding to the difference of their instantaneous values and hence to the control effect on valve system 20, is to be seen.

The instant at which valve 20 becomes conducting and initiates the charge through valve 18 and via the Schmitt-trigger, flyback in the time deflection is no longer determined solely by the direct voltage potential and the sawtooth voltage across its control grid, but also by the voltage peaks superimposed on the direct grid voltage of the synchronizing voltage. This is clearly seen in oscillogram *b* of Fig. 4-49.

4.18.6 HOLD-OFF CIRCUIT FOR THE PREVENTION OF PICTURE JITTERING IN TRIGGERED OPERATION

When triggering the time base circuit by means of regular or irregular pulses, each deflection cycle must always begin at the same point on the screen of the oscilloscope tube. If not, the individual pictures appear irregularly beside one another. This is known as picture jittering or picture tremor.

As the sawtooth voltage drives the X-deflection plates via a DC amplifier, the point of commencement of the picture will be determined by the starting level of the voltage on the timing capacitor (C_{71} ... C_{75}). This voltage level should therefore always be precisely the same each time the X-deflection stroke commences. This condition is only fulfilled if the timing capacitor is always fully charged after each sawtooth has run its course. This means that the voltage rise across the anode of the charging valve 18 at the conclusion of each charge after the sawtooth stroke must be slowed down in such a way as to ensure that the Schmitt-trigger remains insensitive long enough to trigger new pulses ("hold-off circuit"). This is achieved by interposing a capacitor and a resistor in series between the anode of valve 18 and chassis (R_{209} and C_{64} ... C_{67} in Fig. 4-47). For every value of capacitor (C_{71} ... C_{75}) a correspondingly "delaying capacitor" must be used. Selection is automatic when switching to the various time coefficient ranges by means of switch $S_{8,I}$.

In the the case of the smallest capacitor of the time base circuit the stray capacitances from the anode of the charging valve to chassis suffice to ensure a sufficient "hold-off time".

4.18.7 COMPLETE CIRCUITS OF THE TIME BASE GENERATOR

Apart from certain resistors and potentiometers, the complete circuit shown in Fig. 4-50 is in the form of a printed circuit on a panel (unit *H*). The multiple valve, Type PCF 80, is used in all stages.

4.18.7.1 Cathode-follower

Comparing this complete circuit with the basic circuit shown in fig. 4-40, it will be seen that five cathode-follower stages are employed (16′, 17′ 18′ 19′ and 20′). In Section 4.18.4, "Measures for ensuring the linearity of the deflection sawtooth", the circuit of Fig. 4-47 was used to indicate that the cathode-follower 17′ was connected in the network for the reversal of valves 16 and 18 by means of voltage surges at the anode of the Schmitt-trigger 17. In a similar manner, for accelerating

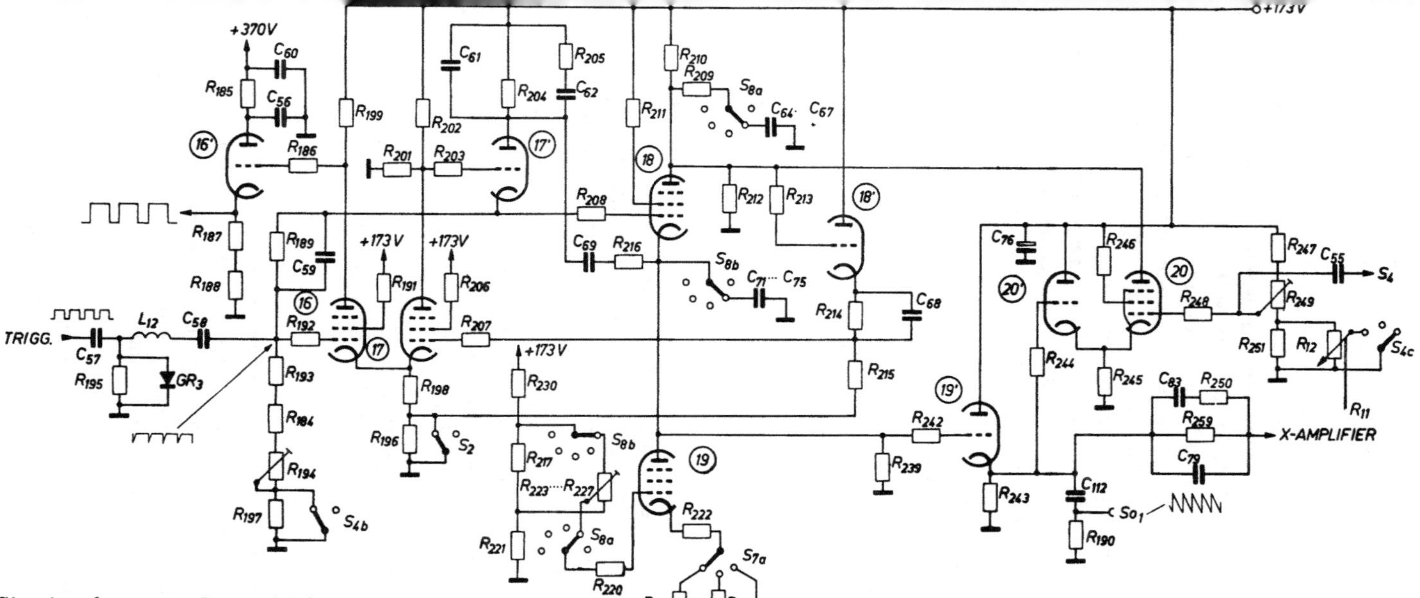

Circuits elements

$R_{10} = 50\ k\Omega$
$R_{12} = 2\ M\Omega$
$R_{184} = 140\ \Omega$
$R_{185} = 6.8\ k\Omega$
$R_{186} = 120\ \Omega$
$R_{187} = 9.1\ k\Omega$
$R_{188} = 4.7\ k\Omega$
$R_{189} = 15\ k\Omega$
$R_{190} = 10\ M\Omega$
$R_{191} = 120\ \Omega$
$R_{192} = 120\ \Omega$
$R_{193} = 4.7\ k\Omega$
$R_{194} = 1\ k\Omega$
$R_{195} = 10\ k\Omega$
$R_{196} = 8.2\ k\Omega$
$R_{197} = 390\ \Omega$

$R_{198} = 2.7\ k\Omega$
$R_{199} = 3.9\ k\Omega$
$R_{201} = 22\ k\Omega$
$R_{202} = 10\ k\Omega$
$R_{203} = 120\ \Omega$
$R_{204} = 6.8\ k\Omega$
$R_{205} = 4.6\ k\Omega$
$R_{206} = 120\ \Omega$
$R_{207} = 120\ \Omega$
$R_{208} = 4.7\ k\Omega$
$R_{209} = 6.8\ k\Omega$
$R_{210} = 0.1\ M\Omega$
$R_{211} = 120\ \Omega$
$R_{212} = 0.18\ M\Omega$
$R_{213} = 120\ \Omega$
$R_{214} = 18\ k\Omega$
$R_{215} = 6.8\ k\Omega$
$R_{216} = 3.3\ k\Omega$

$R_{217} = 6.8\ k\Omega$
$R_{220} = 120\ \Omega$
$R_{221} = 12\ k\Omega$
$R_{222} = 4.7\ k\Omega$
$R_{223} = 20\ k\Omega$
$R_{224} = 20\ k\Omega$
$R_{225} = 20\ k\Omega$
$R_{226} = 20\ k\Omega$
$R_{227} = 20\ k\Omega$

$R_{229} = 5\ k\Omega$
$R_{230} = 0.56\ M\Omega$
$R_{232} = 18\ k\Omega$
$R_{233} = 2\ k\Omega$
$R_{234} = 3.9\ k\Omega$
$R_{235} = 0.1\ M\Omega$
$R_{236} = 27\ k\Omega$
$R_{237} = 10\ k\Omega$
$R_{238} = 0.1\ M\Omega$

$R_{239} = 10\ M\Omega$
$R_{242} = 120\ \Omega$
$R_{243} = 24\ k\Omega$
$R_{244} = 120\ \Omega$
$R_{245} = 39\ k\Omega$
$R_{246} = 120\ \Omega$
$R_{247} = 0.33\ M\Omega$
$R_{248} = 120\ \Omega$
$R_{249} = 0.1\ M\Omega$
$R_{250} = 22\ k\Omega$
$R_{251} = 0.33\ M\Omega$
$R_{259} = 0.1\ M\Omega$

Capacitors

$C_{55} = 0.2\ \mu F$
$C_{56} = 22\ nF$
$C_{57} = 27\ pF$

$C_{58} = 15\ pF$
$C_{59} = 27\ pF$
$C_{60} = 22\ nF$
$C_{61} = 82\ pF$
$C_{62} = 68\ pF$
$C_{64} = 27\ pF$
$C_{65} = 560\ pF$
$C_{66} = 5.6\ nF$
$C_{67} = 56\ nF$
$C_{68} = 10\ pF$
$C_{69} = 15\ pF$
$C_{71} = 1\ \mu F$
$C_{72} = 0.1\ \mu F$
$C_{73} = 10\ nF$
$C_{74} = 1\ nF$
$C_{75} = 47\ pF$
$C_{76} = 8\ \mu F$
$C_{79} = 10\ nF$

$C_{83} = 22\ nF$
$C_{112} = 0.1\ \mu F$

Valves

16 + 16′)
17 + 17′)
18 + 18′)
19 + 18′)
19 + 19′)
20 + 20′)
each 1 × PCF 80

Semiconductor diodes

GR 3 = OA 81

Fig. 4-50 Complete circuit diagram of the time base unit in the Philips "GM 5602" oscilloscope the control valve 17 via the voltage divider (R_{214}, R_{215}), cathode-follower 18′ is interposed between the anode of charging valve 18 and the above-mentioned circuit elements. Were it not for these cathode-followers, the voltage pulses would have to charge directly the capacitive load consisting of circuit capacitance and the input capacitance of the succeeding valves. As the output resistance represented by the anode resistors is relatively high, the resulting time constant would be impermissibly high. By the interposition of these cathode-followers the voltage sources need to drive only the high input impedance of the cathode followers, while the cathode-followers, even with a low output resistance, charge the connected circuit capacitance and the input capacitance of the following valve stage at only slightly reduced voltage speed. This leads to a considerably shorter response time of the time base generator than is the case without cathode-followers. A short response time is an absolute necessity, however, if the complete picture of the triggering voltage change (rise or drop) is to be displayed. As has been described in Ch. 5.23 on "Delay lines", this can be achieved by connecting time-delay circuit elements in the Y-amplification circuit. As the upper cut-off frequency of these circuit elements is in inverse proportion to the delay time obtainable, a high upper cut-off frequency at an acceptable input can only be achieved if the required delay time is not too long. But as it must obviously be longer than the response time of the time base generator, it becomes essential for the latter to be as short as possible. By interposing the cathode-follower stages 17′ and 18′ it is only a little more than 0.2 μs, while the delay time in the line to the Y-amplifier (at a sufficiently high upper cut-off frequency) is greater than 0.3 μs (see also Ch. 5.24 on "Measuring the response time of the time base unit and the delay time in the Y-channel", together with the oscillograms in Figs. 5-49 and 5-50).

The cathode-followers 16′, 19′, and 20′ are mainly intended to provide only a small load on the voltage source. Cathode-follower 16′ thus makes possible correct load matching at the anode of the Schmitt-trigger valve 16. This cathode-follower serves for the unblanking of the forward trace of the pattern during the time deflection.

The cathode-follower 19′ is driven by the sawtooth voltage at the junction of the cathode of charging valve 18, the anode of the discharge valve 19 and the timing capacitor (C_{71} . . . C_{75}). It imposes practically no load on this point, while at its cathode the sawtooth voltage is available at low output resistance.

The voltage at this point is used for three further purposes:

1. to drive the X-amplifier,
2. to control the flyback for synchronizing stage 20′ and,
3. to make available the sawtooth voltage at socket "1" for use outside the oscilloscope.

4.18.7.2 Compensated voltage dividers

All voltage dividers which have to transmit pulse voltages or sawtooth voltages must be compensated with capacitances so that voltage waveforms over a wide band can be transmitted without distortion and frequency discrimination (see Part III, Ch. 25 "Adjusting high impedance wideband voltage dividers by means of square pulses or symmetrical rectangular pulses"). In the circuit shown, the capacitors C_{59}, C_{68} and C_{79} are used for this purpose.

4.18.7.3 Brightness control of the image during the forward stroke

The voltage at the anode of the Schmitt-trigger valve 16 is high during the forward stroke of the time deflection, but is low, however, during the flyback and at rest (in triggered operation). Its waveform is thus suitable for brightening the image. It is taken via cathode-follower 16′ and controls the oscilloscope tube grid by direct coupling via a high-resistance compensated divider. The circuit of this divider is basically similar to the circuit described in Fig. 4-63. The last link of the divider circuit is in this case, in contrast to Fig. 4-63, not a triode, but the pentode system of an ECF 80 with heavy feed back. In this way a great difference between the resistance and the differential impedance is obtained, and good control by means of the unblancking voltage results. The brightness intensification can be switched off if the *X*-amplifier is used without time base voltage.

4.18.8 ADJUSTING THE TIME COEFFICIENTS

As described in Section 4.18.2 "Self-oscillating operation", the voltage velocity of the sawtooth voltage and hence the time coefficient (s/cm) of the oscillogram is determined by the size of the timing capacitor and the discharge current for which valve 19 is adjusted. In addition, the voltage speed can be raised by increasing the gain of the *X*-amplifier (picture expansion). The fifteen time-coefficient stages of the "GM 5602" oscilloscope are set in the following way:

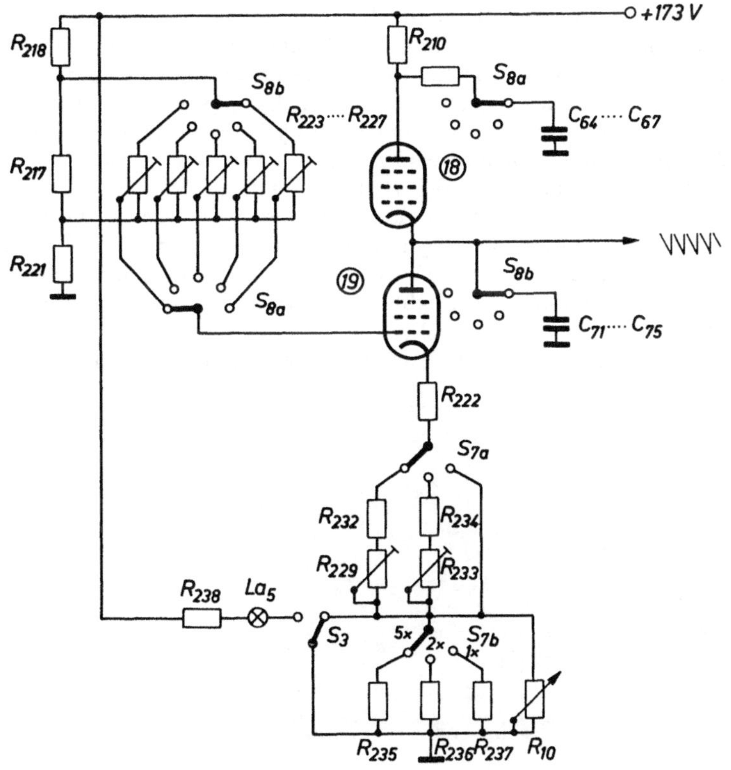

Fig. 4-51 Circuit elements for calibrating the time coefficient stages

4.18.8.1 Selecting the timing capacitors by means of switch S_8

It would be inadvisable to set these ranges only by direct selection of capacitors (C_{71} ... C_{75}). Instead, the grid voltage of the discharge valve (see Fig. 5-51) after being switched by switch $S_{8,b}$, is adjusted each time by means of potentiometer (R_{223} ... R_{227}) in such a way that the discharge current through valve 19 gives the desired time coefficient.

4.18.8.2 Adjustment of the discharging current

Three resistors of different values in the cathode lead of discharge valve 19 can be selected by means of switch S_7 for this purpose. By this means various amounts of feed-back can be obtained, thus varying the internal resistance of the discharge valve to provide three further ranges of time coefficients with factors 1, 2 and 5. At settings "2x" and "5x" adjustment is carried out by the potentiometers R_{229} and R_{233}.

Fig. 4-52 shows a part of the "GM 5602" oscilloscope control panel with the various possible adjustments for X- and Y-deflection. The double knob in the centre of the bottom row (S_7 and S_8) is used to select the time coefficient as already described. The large knob (S_8) selects the capacitors, while the smaller one (S_7), marked on the apparatus by a red arrow and pointing to red adjusting marks, is used to select the resistances in the cathode lead of the discharge valve. To read off the adjustment, the data given on the front panel have to be multiplied together. From the setting shown in the figure the coefficient is $0.2\ \mu s/cm \times 1 = 0.2\ \mu s/cm$. The large left-hand knob (S_6) is for the expansion of the picture by increasing the time coefficients. In order to calculate the finally adjusted time coefficient in the case of expansion, the result of the adjustment by knobs S_7 and S_8 must be divided by the adjustment for the picture expansion (S_6). When S_6 is switched to "5x", the smallest possible time coefficient $0.2 \times 1{:}5 = 0.04\ \mu s$ (= 40ns) is obtained. Time coefficients can, however, be continuously varied by means of potentiometer R_{10}. In this case the calibration no longer applies, as the error of such small potentiometers is very considerable. Switch S_3 is coupled with potentiometer R_{10} and controls a signal lamp (La_5) which lights as a warning that the potentiometer has been switched on.

As the time coefficient in self-oscillating operation usually has to be adjusted

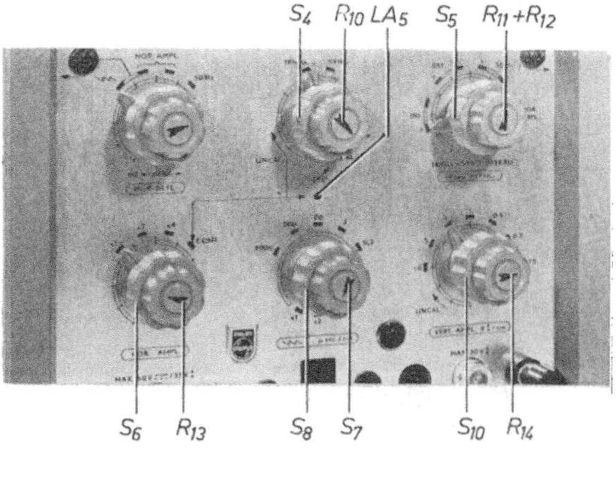

Fig. 4-52 Arrangement of the control knobs for adjusting the time coefficients and for vertical deflection

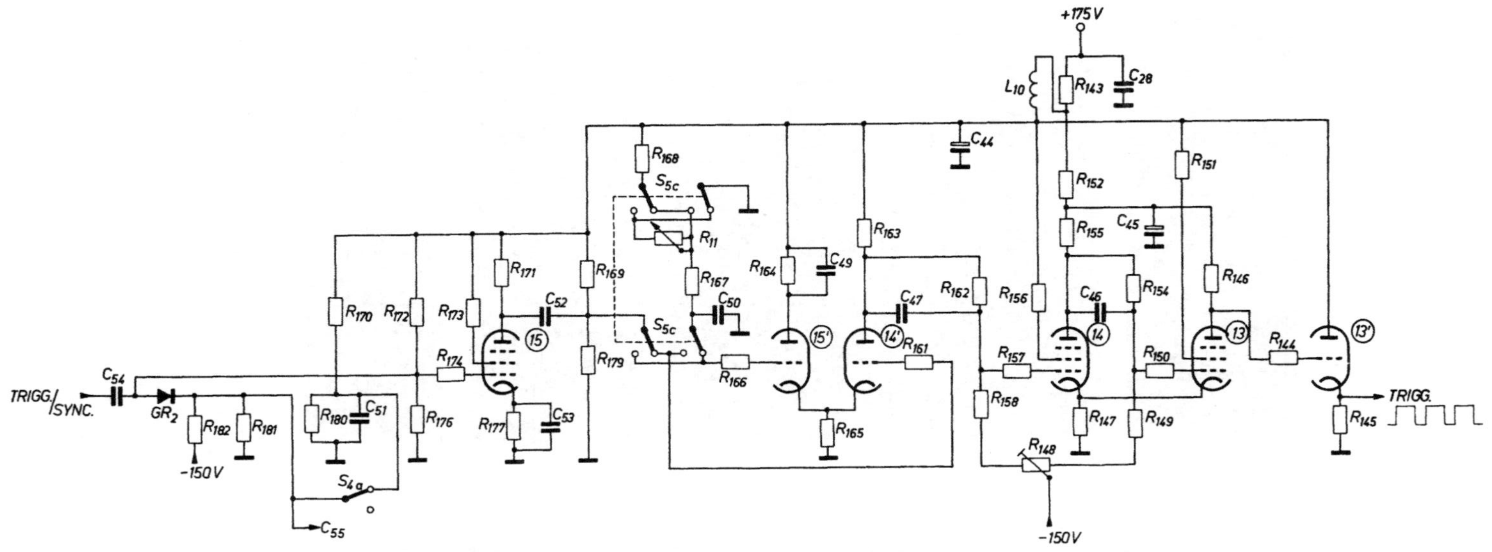

Resistances

$R_{11} = 0.1\ \mathrm{M\Omega}$
$R_{143} = 82\ \Omega$
$R_{144} = 120\ \Omega$
$R_{145} = 15\ \mathrm{k\Omega}$
$R_{146} = 1.8\ \mathrm{k\Omega}$
$R_{147} = 4.3\ \mathrm{k\Omega}$
$R_{148} = 0.1\ \mathrm{M\Omega}$
$R_{149} = 0.39\ \mathrm{M\Omega}$
$R_{150} = 120\ \Omega$
$R_{151} = 120\ \Omega$
$R_{152} = 3.9\ \mathrm{k\Omega}$
$R_{154} = 0.22\ \mathrm{M\Omega}$
$R_{155} = 1.8\ \mathrm{k\Omega}$
$R_{156} = 120\ \Omega$
$R_{157} = 120\ \Omega$
$R_{158} = 0.39\ \mathrm{M\Omega}$
$R_{161} = 120\ \Omega$
$R_{162} = 0.27\ \mathrm{M\Omega}$
$R_{163} = 4.7\ \mathrm{k\Omega}$
$R_{164} = 4.7\ \mathrm{k\Omega}$
$R_{165} = 8.2\ \mathrm{k\Omega}$
$R_{166} = 120\ \Omega$
$R_{167} = 0.39\ \mathrm{M\Omega}$
$R_{168} = 68\ \mathrm{k\Omega}$
$R_{169} = 1.5\ \mathrm{M\Omega}$
$R_{170} = 0.56\ \mathrm{M\Omega}$
$R_{171} = 4.7\ \mathrm{k\Omega}$
$R_{172} = 10\ \mathrm{M\Omega}$
$R_{173} = 120\ \Omega$
$R_{174} = 120\ \Omega$
$R_{176} = 0.47\ \mathrm{M\Omega}$
$R_{177} = 680\ \Omega$
$R_{179} = 0.56\ \mathrm{M\Omega}$
$R_{180} = 0.1\ \mathrm{M\Omega}$
$R_{181} = 0.47\ \mathrm{M\Omega}$
$R_{182} = 1.8\ \mathrm{M\Omega}$

Valves:

13 + 13′, 14 + 14′
and 15 + 15′
each PCF 80

Semiconductors
diodes

GR 2 = OA 202

Capacitors
$C_{28} = 0.22\ \mu\mathrm{F}$
$\left.\begin{matrix} C_{44} \\ C_{45} \end{matrix}\right\} = 2 \times 16\ \mu\mathrm{F}$
$C_{46} = 39\ \mathrm{pF}$
$C_{47} = 27\ \mathrm{pF}$
$C_{49} = 27\ \mathrm{pF}$
$C_{50} = 0.1\ \mu\mathrm{F}$
$C_{51} = 0.22\ \mu\mathrm{F}$
$C_{52} = 0.1\ \mu\mathrm{F}$
$C_{53} = 100\ \mathrm{pF}$
$C_{54} = 0.1\ \mu\mathrm{F}$

Fig. 4-53 Circuit of the trigger-pulse shaper

to the frequency of the phenomenon under investigation and R_{10} must be operated for synchronization, the calibration of the time coefficients does not then apply. R_{13} can also be used to obtain continuously variable picture magnification. Again the calibration does not apply, and in this case also the small warning lamp La_5

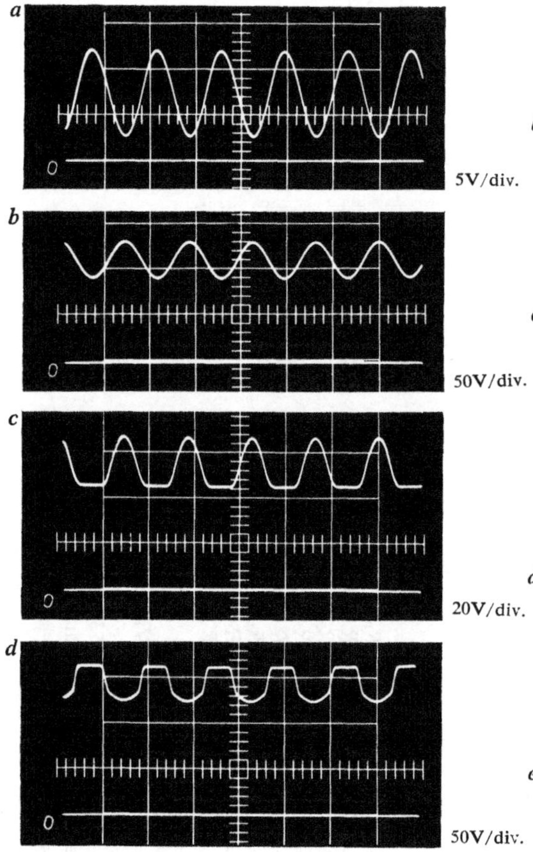

Fig. 4-54 Voltages in the pre-amplifier stage (15), the mixer stage (14′ and 15″) and the trigger-pulse shaper. Time coefficient: 10 μs/cm (Circuit in Fig. 4-53). *a*) 1st grid valve 15. *b*) anode valve 15. *c*) common cathode connection of valves 14′ and 15′. *d*) anode 14′.

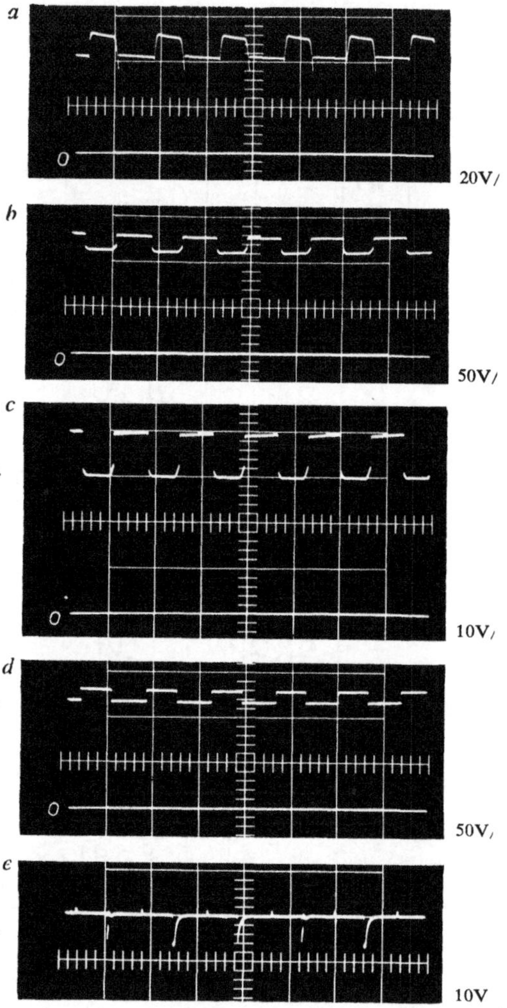

Fig. 4-55 Voltages in Schmitt-trigger (13 and 14) and output cathode-follower (13′) of the trigger-pulse shaper and across diode GR 3. Time coefficient: 10 μs/cm (circuit in Fig. 4-50). *a*) common cathode connection of valves 13 and 14. *b*) anode valve 14. *c*) 1st grid valve 13. *d*) cathode valve 13′. *e*) differentiated pulses across diode GR 3.

lights up. The other knobs, for S_{10} and R_{14}, also visible in Fig. 4-52, are used to adjust the vertical deflection of the Y-amplifier.

4.18.9 TRIGGER-PULSE SHAPER

As already stated, short control pulses are required to trigger the time base unit. As such pulses can be generated from signal or control voltages of any desired waveform, all high-performance oscilloscopes are provided with a special circuit arrangement for this purpose, the so-called trigger-pulse shaper. Fig. 4-53 shows the circuit of the trigger-pulse shaper in the "GM 5602" oscilloscope as an example. In essence it consists of the following stages:

pre-amplifier (15),
push-pull mixer stage (14′ and 15′),
Schmitt-trigger (13 and 14), and the,
output cathode-follower (13′).

The voltage, selected by means of the step switch S_5 in the position shown in Fig. 4.53 and after amplification by stage 15 is led to the control grid of mixing stage 14′. A direct voltage which can be varied by potentiometer R_{11}, is simultaneously applied to the control grid of valve 15′, thus permitting the anode current of valve 15′ to be adjusted. The direct anode current of system 14′ is also partly determined by R_{165} via the cathode coupling. The amplified alternating voltage thus occurs at the anode of system 14′ superimposed on a direct voltage which can be varied by means of R_{11}. If a waveform as shown in Fig. 4-46 is applied to the oscilloscope and is triggered by this voltage, voltage waveforms such as are shown in the oscillograms of Fig. 4-54 occur on these valves.

The anode of valve 14′ of the mixing stage is connected directly to the control grid of the left-hand Schmitt-trigger valve (14). Each time the voltage exceeds or drops below the response level of the Schmitt-trigger, the latter reverses to one of its stable states.

If, as in the present case, a repeating voltage is applied to the input of the trigger-pulse shaper, a square voltage appears at the anode of the Schmitt-trigger stage 13. This voltage is passed via the cathode-follower 13′ to the sawtooth voltage generator. As is later explained more fully by means of the oscillograms of Figs. 4-56 and 4-57, the width of the square pulses can be adjusted by varying the direct voltage level, and the trigger phase can thus be changed. This square voltage is differentiated at the input of this circuit by a CR-network (C_{57} — R_{195}), thus causing positive and negative voltage peaks to occur at the beginning and end of the square pulses respectively. The positive peaks are clipped by the diode $GR3$, so that only the negative pulses, which are intended to trigger the sawtooth voltage generator, appear at the grid of the Schmitt-trigger valve 16. The waveform of these voltages occurring in the type of operation used for the oscillograms of Fig. 5-54 is shown in Fig. 4-55.

In order to be able to trigger the time base unit as required with either the positive or the negative going slope of the control voltage, it is possible to use switch S_5 to change the polarity of the leads to the grids of valves 14′ and 15′. At the same time the connections on potentiometer R_{11} are reversed. This ensures that, when switch S_5 is used in either of its positions, the change in direct voltage level at the anode of 14′ will take place in the same direction.

4.18.9.1 Switching to self-oscillating operation

If switch S_4 (Fig. 4-52) is turned from position "TRIGG" to position "SYNC" (Fig.

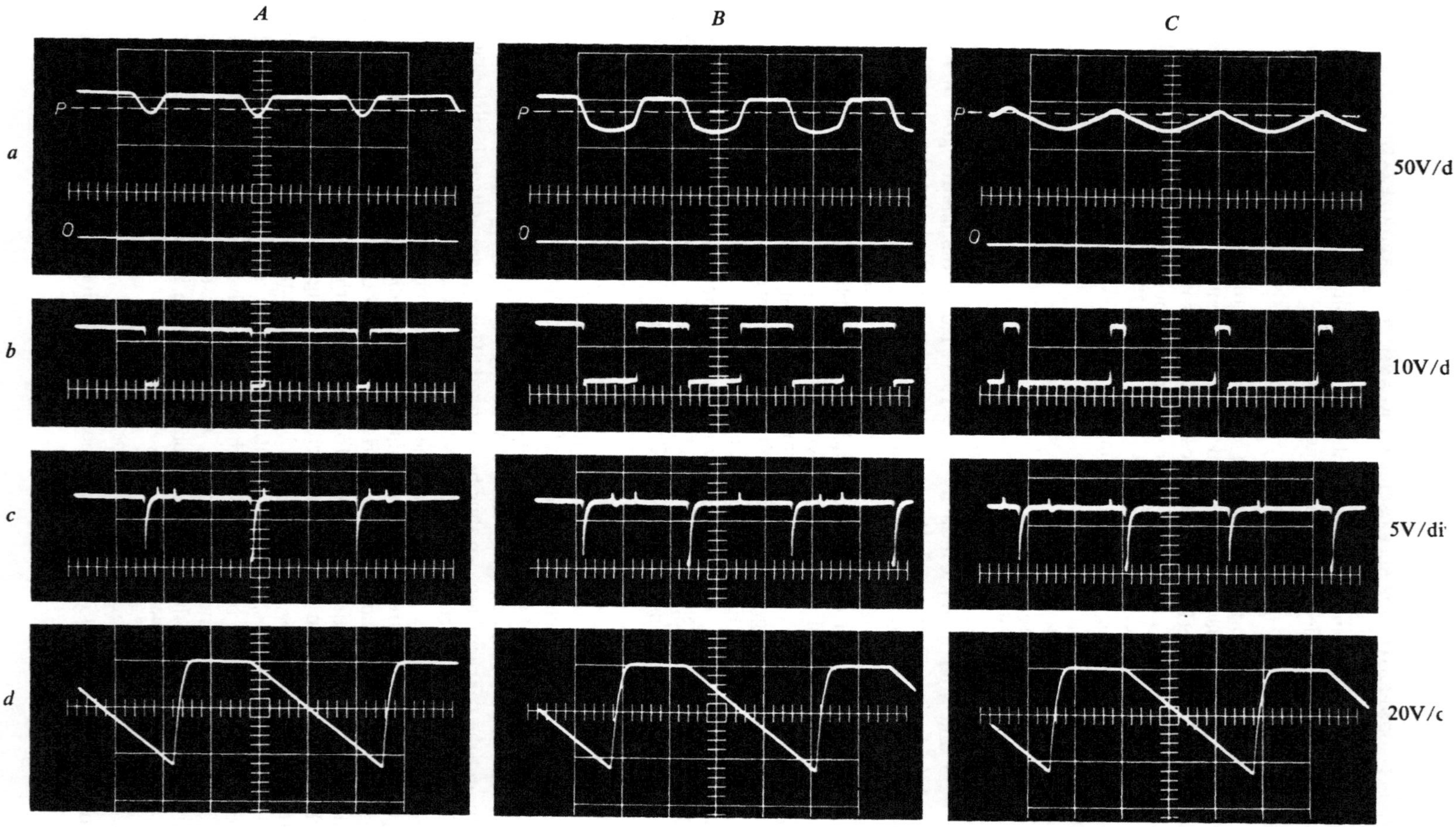

Fig. 4-56 Varying the sweep phase by adjusting the trigger level. Time coefficient: 10 μs/cm

Adjustments:
A) Maximum trigger level

Voltages:
a) anode valve 14′

d) cathode valve 18 in sawtooth voltage generator (circuit in

4-52 and 4-53), the circuit oscillates independently as already described. At the same time resistor R_{180} in the trigger-pulse shaper, which is in parallel with the divider resistor R_{181}, is switched off. The pre-amplifier valve 15 now receives such a high bias via the diode *GR*2 that it is cut off. The whole trigger-pulse shaper is thus switched off. The sync voltage for releasing the flyback now reaches the control grid of valve 20 via capacitor C_{55} (circuit in Fig. 4-50). By varying the voltage division by means of R_{12} in the circuit of Fig. 4-50 (together with R_{11} in Fig. 4-53) both of which are simultaneously adjustable, the most favourable value for the sync intensity can be chosen.

4.18.9.2 Method of operation and possible applications of the trigger level adjustment

Adjustment of the direct voltage level at the anode of valve 14′ in the mixing stage by means of potentiometer R_{11} causes distortion of the waveform of the amplified input voltage at this point. The oscillograms *a* in Fig. 4-56 show the waveforms of this voltage at the extreme adjustments (*A* and *C*) of the direct voltage level and at a more or less average setting (*B*). The three oscillograms in series *b* show the output voltage variations of the Schmitt-trigger valve 13′. Their waveform corresponds to the anode voltage variations of the Schmitt-trigger valve 13. By comparison with the oscillograms in series *a* it can be readily seen that the length of the generated square pulses is dependent on the set direct voltage level at the anode of system 14′.

The response level at the anode of valve 14 is indicated in the three oscillograms of series *a* by a dotted line. The waveform of the voltage as differentiated by C_{57} and R_{195} and rectified by *GR*3 (circuit Fig. 4-50) can be seen in the oscillograms of series *c*. As these pulses correspond in each case to the negative going voltage change of the oscillograms in series *b* (anode voltage of system 13 or output voltage of cathode-follower 13′), the trigger pulses are in each case slightly shifted in phase when the pulse width is changed. Because these pulses, as can be seen from the oscillograms of series *d*, trigger the voltage sawtooth, a phase shift can be applied for the commencement of the time deflection. This makes it possible to select the voltage level at which the time deflection should begin. The oscillograms of Fig. 4-57 show the result on a signal voltage of sinusoidal waveform. As in the case of the oscillograms *A*, *B* and *C* of Fig. 4-56, the possible limit adjustments and an approximate average position are here also recorded. According to whether triggering takes place with an increasing or decreasing voltage it is possible to choose the settings for the oscillograms of either frame *a* or frame *b*. It is thus possible,

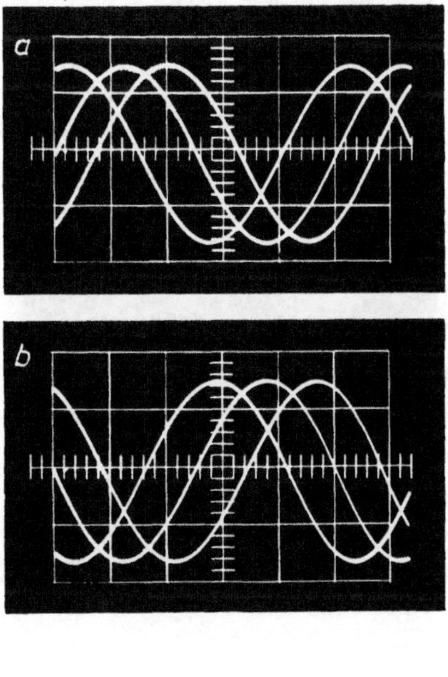

Fig. 4-57 Signal voltage patterns at various settings of the trigger phase, corresponding to the oscillograms in Fig. 4-56
Time coefficient: 4 μs/cm
a) triggering with rising voltage
b) triggering with falling voltage

merely by adjusting the triggering level, to determine the approximate point in the complete cycle of the phenomenon under investigation at which the image shall commence, so that it can be adjusted to correspond with any particularly interesting part of the waveform. By triggering with a small time coefficient and, when necessary, by magnifying the picture by means of increased X-amplification, a very close and detailed examination of the waveform can be conducted.

4. 19 Time base unit of the "GM 5650" oscilloscope

In this oscilloscope, which is extremely efficient in spite of its small dimensions (Fig. 4-58), the principle of the triple pentode circuit is also used. The time coefficient is, as usual, adjustable in steps between 15 ms/cm and 0.5 μs/cm by changing the values of the timing capacitances, and fine adjustment by varying the charging current. The triple pentode circuit is modified to the extent that the anode of the discharge valve (5′) –as can be seen in circuit diagram 4-59– is not coupled via a capacitor to the grid of the control valve (7) –a triode in this case– but is directly connected via a resistor (R_{68}). This enables the discharge of the timing capacitors (C_{27} . . . C_{35}) to be continued until the control valve ceases to pass current. At the same time repetitive operation can be obtained if the control valve (7) is set at its normal operating point. In addition, the control unit for the sawtooth generator is modified

Fig. 4-58 Philips "GM 5650" small HF oscilloscope with DC amplifier and triggered time base. View of time base section

Components of the time base unit

resistors

$R_1 =$ 0.1 MΩ/lin.	$R_{52} =$ 10 kΩ	$R_{61} =$ 39 kΩ	$R_{71} =$ 100 Ω
$R_2 =$ 50 kΩ/lin.	$R_{53} =$ 39 kΩ	$R_{62} =$ 100 Ω	$R_{72} =$ 0.68 MΩ
$R_1 =$ 0.3 MΩ/lin.	$R_{54} =$ 100 Ω	$R_{63} =$ 56 kΩ	$R_{73} =$ 100 Ω
$R_4 =$ 50 kΩ/lin.	$R_{55} =$ 10 MΩ	$R_{64} =$ 0.1 MΩ	$R_{74} =$ 3.9 kΩ
$R_{39} =$ 4.7 MΩ	$R_{56} =$ 15 kΩ	$R_{65} =$ 22 kΩ	$R_{75} =$ 0.12 MΩ
$R_{41} =$ 18 kΩ	$R_{57} =$ 0.47 MΩ	$R_{66} =$ 0.15 MΩ	$R_{76} =$ 10 kΩ
$R_{42} =$ 18 kΩ	$R_{58} =$ 100 Ω	$R_{67} =$ 100 Ω	
$R_{43} =$ 4.7 MΩ	$R_{59} =$ 0.33 MΩ	$R_{68} =$ 0.56 MΩ	
$R_{44} =$ 68 kΩ	$R_{60} =$ 1 MΩ	$R_{69} =$ 39 kΩ	
$R_{45} =$ 82 kΩ			
$R_{46} =$ 100 Ω			
$R_{47} =$ 1 MΩ			
$R_{48} =$ 15 kΩ			
$R_{49} =$ 0.22 MΩ			

capacitors

$C_{16} =$ 0.1 µF	$C_{17} =$ 0.47 µF	$C_{27}(+C_{50}) =$ 0.49 µF
	$C_{18} =$ 0.22 µF	$C_{28} =$ 0.21 µF
	$C_{19} =$ 0.22 µF	$C_{23} =$ 27 nF
	$C_{21} =$ 0.47 µF	$C_{29} =$ 6.8 nF
	$C_{22} =$ 0.22 µF	$C_{31} =$ 6.8 nF
	$C_{23} =$ 22 pF	$C_{32} =$ 2.2 nF
	$C_{21} =$ 0.1 µF	$C_{33} =$ 630 pF
	$C_{25} =$ 10 nF	$C_{34} =$ 220 pF
	$C_{26} =$ 4.7 pF	$C_{35} =$ 68 pF
		$C_{36} =$ 22 pF
		$C_{37} =$ 0.22 µF
		$C_{38} =$ 47 pF

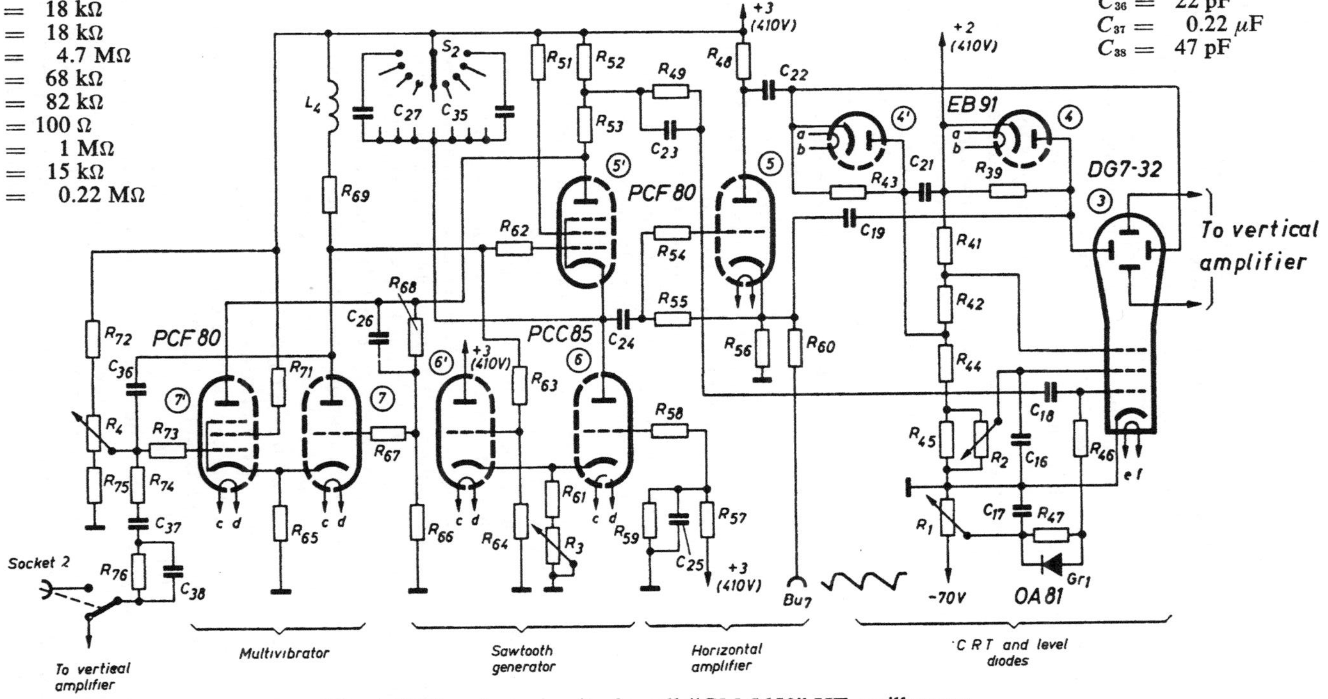

Fig. 4-59 Time base circuit of small "GM 5650" HF oscilloscope

by including an auxiliary valve (7′), which, in conjunction with the control valve (7), can function as a bistable multivibrator (flip-flop), since they are coupled by a common cathode resistor. If the auxiliary valve system (7′) is so set that both 7′ and the control system (7) are conductive, the circuit works repetitively. If the grid voltage of the auxiliary valve system is considerably increased in a positive direction by the variable resistor (R_4), this valve alone becomes conductive and the control valve is cut off. Both systems then work as a so-called flip-flop circuit, enabling the circuit to be triggered. The mode of operation of this circuit, starting with self-oscillating operation, can be followed by reference to the oscillograms made during triggered operation (Figs. 4-60 and 4-61), as follows:

Assuming first that the control system (7) is conductive, the voltage at its anode and thus also at the control grid of the discharge valve (5′) are low, so that valve (5′) is cut off. Furthermore assuming that the selected capacitor (C_{27} . . . C_{35}) is discharged, so that the cathode voltage of the discharge valve is high, and is therefore even more firmly blocked, the capacitor of the sawtooth generator can be charged via the charging valve (6) (oscillogram 4-61a) until the voltage across the cathode of the discharge valve (5′) has dropped so far that the valve becomes conductive. The voltage at the control grid of the control valve is thus decreased (oscillogram 4-60c), so that its anode voltage rises and, as a result of the increase of voltage across the control grid of the discharge valve, accelerated discharge takes place. At the end of the discharge the grid voltage of the control valve (7) rises again since discharge current through resistors R_{52} and R_{53} drops; its anode voltage therefore decreases, so that the discharge valve system is cut off once more. The auxiliary system (7′) is constantly conductive and serves as a sync amplifier.

The voltage at the control grid of (7′) influences the control grid (7) via the cathode coupling and thus the flyback. Basically, therefore, synchronization takes place in the usual way.

If, however, the control grid voltage of the auxiliary valve system (7′) is adjusted to a high positive value by means of R_4, control system (7) can be cut off. This is the adjustment at which both valve systems operate as a bistable multivibrator. In this way a triggered sawtooth voltage can be produced, since, if a negative pulse reaches the auxiliary valve system (7′), this circuit reverts abruptly. The anode voltage of control system (7′) then drops because this valve has become conductive, and the discharge valve (5′) is cut off, so that the charging of the timing capacitor can commence. The auxiliary system (7′) is meantime more or less blocked, until the forward trace has been completed. The extent of this blocking depends to a certain extent on the form of the trigger signal. A stronger positive component could open the auxiliary valve system (7′) and set the flyback off prematurely. This can be prevented, however, by adjusting the control grid of this valve to such a positive potential that triggering can only be done by the peaks of the control voltage.

If the adjustment is such that this valve system remains correctly cut off, the cathode of the discharge valve at the end of the forward trace reaches the voltage level of its grid, so that the discharge now commences. Thus the control grid of the control system receives a negative pulse which cuts this system off, and the multivibrator circuit reverts once more to the condition in which the auxiliary valve system again draws current, which can only be interrupted by a negative trigger pulse. Until then the circuit remains in the hold-off condition. For continuous adjustment of deflection speeds, the charging current is varied by changing the operating point of the charging valve by means of (R_3).

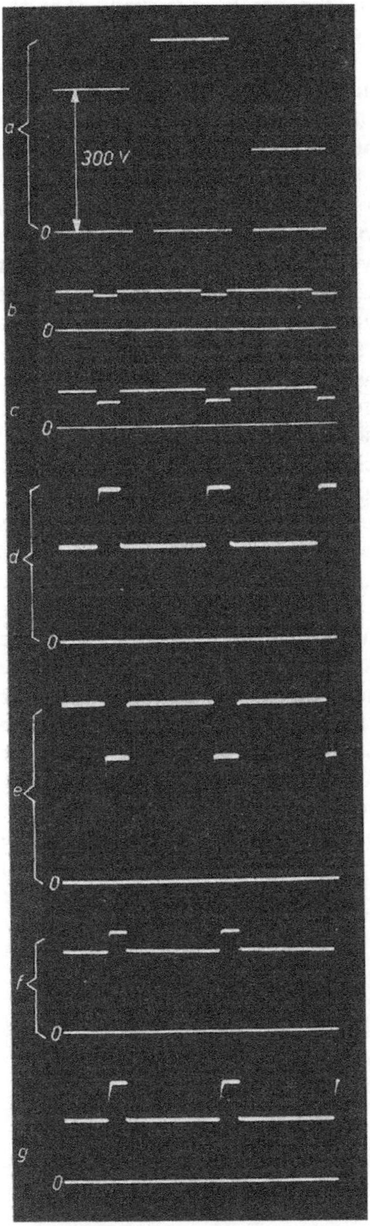

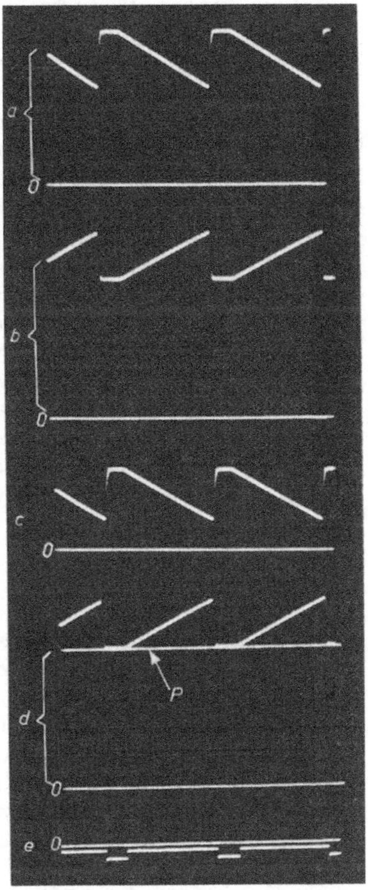

Fig. 4-61 Voltages in sawtooth generator, balancing stages and across control grid of C.R.T.
a) Anode of charging valve (6) and cathode of the discharge valve (5′); *b*) anode of balancing stage; *c*) *d*) cathode of left-hand level diode (4′) in triggered operation and level (P) with time base voltage off; *e*) Control grid of C.R.T.

Fig. 4-60 Voltages at multivibrator, charging valve and cathode follower of circuit in Fig. 4-59, for triggered operation. *a*) left: calibrating with $300V_=$; centre supply voltage +3 (approx. $410V_=$); control grid of charging valve (6); *b*) cathode of control stage (7) and auxiliary valve (7′); *c*) grid of control stage (7); *d*) anode of control stage (7); *e*) anode of auxiliary valve (7′); *f*) cathode of charging valve (6) and cathode follower (6′); *g*) cathode follower grid (6′)

If no further steps were taken, this circuit would have the drawback that the charging valve might draw current on the termination of the flyback even before the commencement of a new trigger pulse. This current would flow through the discharge valve and causes a drop in the voltage across its anode resistors (R_{52} and R_{53}). This variable voltage -dependent on the adjustment of the charging current- would then influence the operating point of the multivibrator at the grid of the control valve (7) in a corresponding manner. In order to prevent this, the system of the charging valve (6) is connected to a cathode-follower (6′) by cathode coupling. The control system (7) is not conductive during stand-by time, so that its anode voltage is high and the voltage at the control grids of the discharge valve (5′) and of the cathode follower (6′) are likewise high. The cathode follower is therefore made conductive, so that the cathode voltage of the combination of valves (6) and (6′) rises, thus cutting off the current through the charging system (6) during the stand-by time. Continuous adjustment of the deflection speed has thus no longer any influence on the operation point of the multivibrator. The sawtooth voltage generated in this circuit — triggered or recurrent — is led to a pre-amplifier stage (5). Resistors of equal value are included in the anode and cathode circuits, so that symmetrical sawtooth voltages (oscillograms 4-61*b* and *c*) can be taken between these electrodes to chassis for application to the *X*-plates. The gain in this stage is somewhat less than × 2. The coupling to the *X*-plates must be via *CR*-networks, as the average potential of these voltages is not equal. This might, when triggering with irregular pulses, cause the oscillogram to oscillate to and fro in a horizontal direction. To prevent this, level diodes (4 and 4′) are connected in parallel to the leak resistors, to ensure that the voltage level of the sawtooth voltage is maintained at its final value (oscillogram 4-61d). For the same reasons it is necessary for the operating point of the amplifier valve system (5) to be maintained constant. Starting from the stand-by condition of the time base unit, the grid potential is maintained at all times negative. By means of a high grid resistance (R_{55} = 10 MΩ) the operating point is determined by grid current. The grid/cathode circuit then functions in the same way as a level diode.

A germanium diode (*GR*1) is also connected in parallel with the leak resistor (R_{47}) of the cathode ray tube grid (3). This maintains the intensity level, so that even when triggering is carried out with irregular pulses, no fluctuations of brightness can occur (oscillogram 4-61*e*).

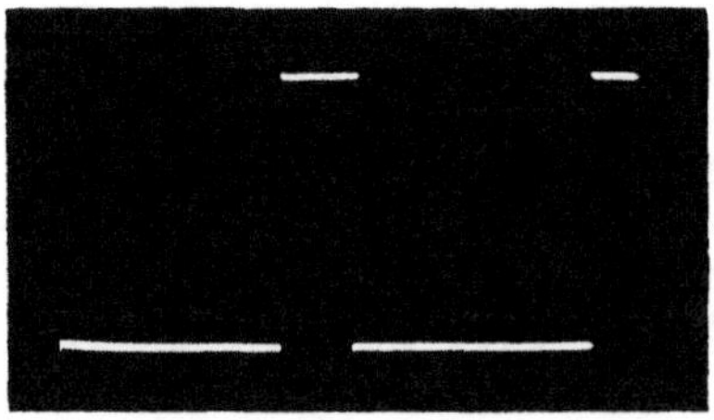

Fig. 4-62 Screen image of "GM 5650" oscilloscope during the photographing of oscillograms for Figs. 4-61 and 4-62. Pulse width 0.5 ms; pulse repetition frequency 500 c/s

Summarizing, this description shows that it is quite possible to provide a small oscilloscope with a properly functioning time base circuit for triggering, comparable in performance with that of larger equipment. The screen image obtained on the "GM 5650" oscilloscope in making oscillograms is shown in Fig. 4-62. It shows two cycles of a series of 0.5 ms pulses with a frequency of 500 c/s. Triggering was internal.

4. 20 Blanking the return stroke

As can be seen from some of the oscillograms reproduced hitherto the returning spot also leaves a trace on the screen. This visible flyback would be particularly disturbing at high frequencies, as then the flyback time is relatively longer. Because of this, circuits have been developed in which a negative voltage pulse is fed to the grid of the cathode ray tube during the discharge of the timing capacitor of the sweep voltage generator. The cathode ray is suppressed during the flyback and the flyback trace is thus rendered invisible.

A suitable voltage is available at certain points in almost all time base circuits. Mention has been made of this in the individual oscillograms given, for example in Figs. 4-26, 4-29, 4-31 and 4-35, and also in the oscillograms 4.27*b*, 4.30*b*, 4-32*d*, and 4-36*e*.

In time base circuits which provide only for self-oscillating operation, it is sufficient to blank the flyback. This can be done very simply by means of a voltage developed across a resistor, for instance, by the discharge current of the flip-flop capacitor in the sawtooth-voltage generator.

In time base circuits which can be triggered, on the other hand, the forward stroke must always be intensified, as the stand-by period during which the spot must also remain invisible must always be taken into consideration. In forward stroke intensification, the screen image is basically only visible in the forward stroke during the time base period, irrespective of whether the time base is repeating, regularly triggered (with stand-by time), irregularly triggered, or is non-recurrent. A satisfactory method of applying the brightening voltage to the time base for all these types of operation is not so easily achieved as the blanking of the flyback in a recurrent time base circuit. Moreover, a fairly complex circuit is required if the brightening is to function satisfactorily when the time base period is of considerable duration (3 to 10 s).

In the following section a number of circuits for such brightening are described, commencing at zero frequency, i.e., with direct coupling.

4. 21 Unblanking circuits in oscilloscopes with triggered time base units

Triggering of individual sawtooth pulses is almost always achieved in triggered time base units by means of a bistable multivibrator (flip-flop) or a Schmitt-trigger, which, in interaction with the sawtooth voltage generator, has two stable voltage conditions. While in one condition the forward stroke takes place, and the other corresponds to the flyback and stand-by time. Thus, there is always available in such circuits a voltage which changes suddenly from one to another of two extreme values. If the time base frequencies are not too low, a voltage suitably polarized can be fed to the intensity control grid of the C.R.T. via a *CR*-coupling link. This unblanks the forward stroke in the desired way and flyback and spot are blanked during the stand-by time.

To ensure that the intensity of the screen image should not fluctuate in the event of irregular triggering or changed pulse ratios (pulse sequence/pulse duration), circuits can be employed to maintain the intensity level as for instance in the "GM 5650" oscilloscope, circuit diagram 4-59, oscillogram in Fig. 4-61e.

Only those circuits which have direct coupling without coupling capacitors between the brightening voltage of square waveform (multivibrator) and the control electrode of the cathode ray tube (Wehnelt-cylinder) function satisfactorily. However, since the square-wave voltage occurs usually at the electrodes (anode or second grid) of the multivibrator valves, and varies between +150 and +300 V, while the cathode of the oscilloscope tube, and hence also the control grid, are at potentials in the range of —500 to —2000 V, direct connection between these two voltages in one circuit is impossible. Three practical examples given in the circuits of Figs. 4-63, 4-64 and 4-65 show how this problem has been solved.

The circuit (with DC-deflection amplifier) used in the "GM 5656" oscilloscope is shown in Fig. 4-63. In the time base circuit multivibrator the anode voltage rises to +350 V during the forward trace and falls to +150 V during the flyback and stand-by time. The connection of this voltage to the C.R.T.-grid can be made only via a high resistance voltage divider, since the cathode and control grid of the C.R.T. are both at potentials of about —750 V. The resistance of the voltage divider must be sufficiently high to ensure that the standing current through it is within acceptable limits. However, the permissible maximum resistance between grid and cathode of the C.R.T. must not be too high, since the attenuation of the square-wave voltage at the grid of the C.R.T. would then be so great that it would no longer suffice for satisfactory unblanking. At the negative end of this voltage divider, therefore, instead of a simple resistor, a triode with heavy negative feed back is connected — an arrangement which has a high impedance [Part I. Ch. 5, Eq. (5.84)]. The negative feedback is achieved by means of a relatively high cathode resistor (R_{37} = 120 kΩ), so that the waveform produced by the multivibrator brings about large voltage changes at the anode of this valve and thus at the C.R.T.-control grid. The anode current of this auxiliary valve (3) is set at about 0.5 mA. In order that the voltage divider may be independent of frequency throughout its whole operating range, and also to compensate the capacitance between grid and cathode of the C.R.T., capacitors (C_8, C_{13} and C_{14}) are connected in parallel to the divider resistors (R_{32}, R_{67} and R_{68}).

A switch socket (11) is provided for the connection of external voltages for the brilliance control. If a plug is inserted, the junction of the two divider resistors (R_{67} and R_{68}) is earthed via capacitor (C_{10}). This prevents the external brilliance control voltage from influencing the time base unit and triggering or synchronizing circuits, and disconnects the internal forward stroke brightening circuit. In order to keep the grid voltage of the C.R.T. independent of the current taken by the tube, which would otherwise tend to counterbalance the effect of the brightening circuit, the C.R.T. cathode is linked to that of a triode connected as a "constant current valve" (Fig. 3-4). The tendency of such a valve is to maintain its anode current constant once it has been set by the bias (R_{11}). With a constant load resistance (R_{36}) its cathode voltage and also the cathode voltage of the C.R.T. remains at a fixed value independent of changes in the anode current of the C.R.T. This value, and thus the basic intensity of the screen image, is selected by means of the anode current of the auxiliary valve system (3) and its grid voltage adjuster (R_1). In such a circuit constancy of the initial brightness set by R_{11} depends on the constancy of the voltage sources and of the resistors forming the voltage divider chain.

Another good solution to this problem is the unblanking circuit used in the Philips "GM 5662" oscilloscope. Its essential details are shown in Fig. 4-64. The square wave voltage of the multivibrator in the time base unit controls the grid voltage of the C.R.T. indirectly via the cathode follower (22) and a compensated voltage divider (R_{297}, R_{298}, R_{305}, R_{306}, R_{420} and C_{230}, C_{302}), first switching another bistable multivibrator (11 and 11'), which in turn applies intensity control to the C.R.T. The cathodes of the valves of this second multivibrator are now, however, roughly at the same voltage as the cathode of the C.R.T., thus making it possible to connect the brilliance

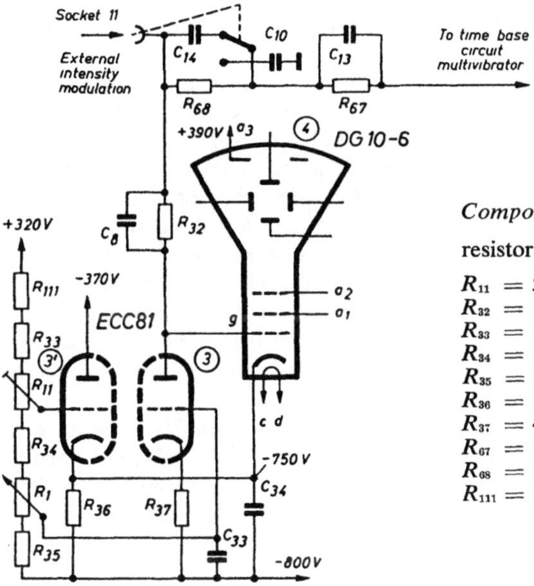

Fig. 4-63 Unblanking network in Philips "GM 5656"

Components of brilliance control

resistors	capacitors
R_{11} = 50 kΩ/lin.	C_8 = 33 pF
R_{32} = 1.5 MΩ	C_{13} = 330 pF
R_{33} = 1.0 MΩ	C_{14} = 220 pF
R_{34} = 0.18 MΩ	C_{33} = 220 pF
R_{35} = 33 kΩ	C_{34} = 220 pF
R_{36} = 0.12 MΩ	
R_{37} = 47 kΩ	
R_{67} = 0.27 MΩ	
R_{68} = 0.39 MΩ	
R_{111} = 0.82 MΩ	

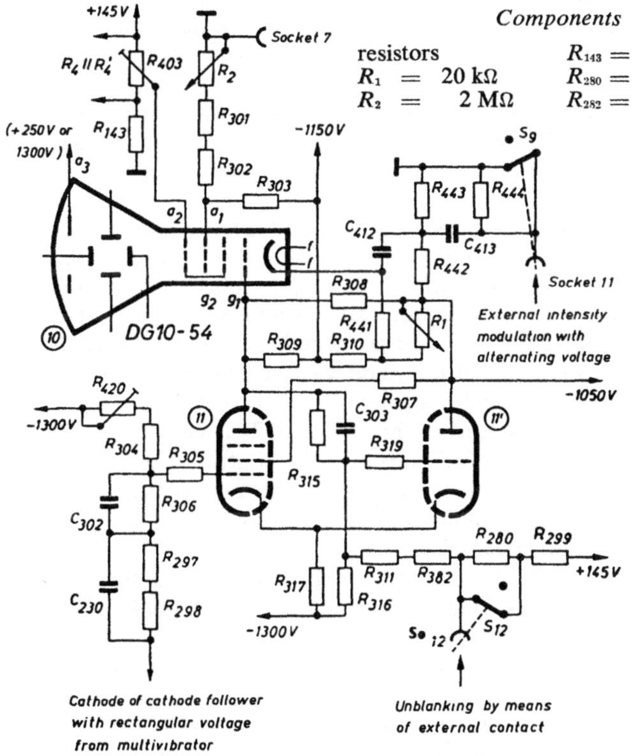

Components of brilliance control

resistors

R_1 = 20 kΩ	R_{143} = 18 kΩ	R_{297} = 1.5 MΩ
R_2 = 2 MΩ	R_{280} = 270 kΩ	R_{298} = 680 kΩ
	R_{282} = 1,5 MΩ	R_{299} = 680 kΩ
		R_{301} = 1 MΩ
		R_{302} = 1.2 MΩ
		R_{303} = 1.8 MΩ
		R_{304} = 390 kΩ
		R_{305} = 180 Ω
		R_{306} = 2.2 MΩ
		R_{307} = 180 Ω
		R_{308} = 6.8 kΩ
		R_{309} = 12 kΩ
		R_{310} = 47 kΩ
		R_{311} = 2.2 MΩ
		R_{315} = 1 MΩ
		R_{316} = 330 kΩ
		R_{317} = 39 kΩ
		R_{319} = 180 Ω
		R_{403} = 50 kΩ
		R_{420} = 200 kΩ
		R_{441} = 4.7 kΩ
		R_{442} = 6.8 MΩ
		R_{443} = 6.8 MΩ
		R_{444} = 1.8 MΩ

capacitors

C_{230} = 5.6 pF
C_{302} = 10 nF
C_{303} = 3.3 pF
C_{412} = 0.33 μF
C_{413} = 0.33 μF

Fig. 4-64 Unblanking circuit in Philips "GM 5662" HF wide-band oscilloscope control grid directly to the anode of the corresponding valve system of this second multivibrator.

If the control grid of the left-hand multivibrator valve (11) receives a positive pulse from the multivibrator of the time base unit at the commencement of the flyback, this valve system becomes conductive and its anode voltage drops. This also causes the voltage at the control electrode of the C.R.T. to go more negative with respect to chassis, and the electron beam is suppressed until the triggering of a new forward stroke via the voltage divider makes the grid voltage of the left-hand multivibrator valve go so negative that the value is cut off and anode current can no longer flow. This reduces the voltage drop across the cathode resistor (R_{317}) common to both systems, so that the right-hand valve (11′) can draw current. The anode voltage of the left-hand valve (11) is then increased and the screen image brightened. The desired critical voltage conditions at which the circuit reverts to blanking can be selected by adjusting the grid bias of the left-hand valve by means of R_{420}.

A small current flows through the variable resistor (R_1) which is in series with resistor (R_{310}) between the potential points — 1150 V and —1050 V and permits adjustment of the basic brightness by varying the cathode bias of the C.R.T. The screen can also be periodically brightened by means of an alternating voltage to provide time marks, or switched from dark to bright by an external contact.

External intensity control operates in the following manner. If alternating voltage or cyclic intensity control, which may be sinusoidal or of suitable waveform for producing marker pulses, is applied via a plug inserted in switching socket (11), the switch (S_9), which hitherto has short-circuited the input resistor (R_{444}) to prevent interfering voltages reaching the circuit, is opened. A current flows from the external voltage source via the input resistor, and the alternating voltage across this resistor is applied to the cathode of the C.R.T. It controls the intensity of the screen image, the negative-going part of each cycle producing a bright trace and the positive-going position a dark trace.

If a plug is introduced into the lower switch socket (12), the short-circuit is removed from resistor (R_{280}). This resistor is part of the divider connected between the +145 V terminal and the —1300 V terminal, its tapping point being connected to the control grid of the right-hand multivibrator valve. Removal of the short-circuit causes the control grid potential of this valve to go negative, so that the right-hand multivibrator valve (11′) is cut off and the left-hand valve (11) draws current. This causes the anode potential of the left-hand system to drop, so that the screen image is blanked. If, however, the connecting lead to this socket (12) is earthed, e.g., via a relay or by means of a switch operated by a camshaft, the control grid of the right-hand valve obtains practically the same potential at that instant as before without operating the switch in the socket. The multivibrator therefore reverts to its former condition and the screen image is intensified.

The Tektronix company uses quite a different circuit in its apparatus. Fig. 4-65 shows the part of type "515" which is used for the time base unit unblanking. The extra high tension in these units is always supplied by the rectification of an alternating voltage of medium frequency — about 60 kc/s — produced by a special auxiliary oscillator (Fig. 3-10). The cathode of the C.R.T. received a voltage of –1675 V. The bias for the R.C.T.-grid is not obtained via a voltage divider in the conventional manner but from a special rectifier valve ("5642") which is completely isolated from the rest of the power supply and is adjusted according to the cut-off

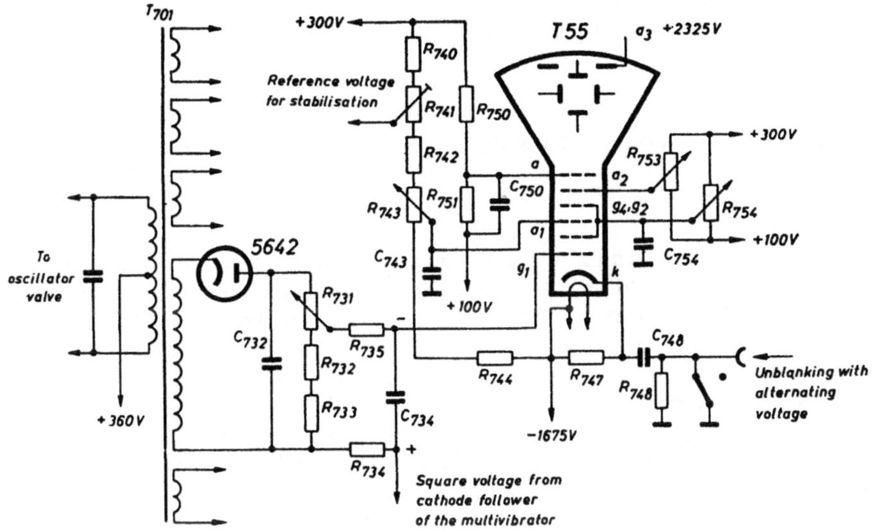

Components of the unblanking circuit

Resistors			Capacitors
$R_{731} =$ 1 MΩ	$R_{740} =$ 2.2 MΩ	$R_{748} =$ 1 MΩ	$C_{732} =$ 15 nF
$R_{732} =$ 6.8 MΩ	$R_{741} =$ 2 MΩ	$R_{750} =$ 100 kΩ	$C_{734} =$ 15 nF
$R_{733} =$ 6.8 MΩ	$R_{742} =$ 3.3 MΩ	$R_{751} =$ 120 kΩ	$C_{743} =$ 1 nF
$R_{734} =$ 100 kΩ	$R_{743} =$ 2 MΩ	$R_{753} =$ 100 kΩ	$C_{748} =$ 15 nF
$R_{735} =$ 1 MΩ	$R_{744} =$ 1 MΩ	$R_{754} =$ 100 kΩ	$C_{750} =$ 10 nF
	$R_{747} =$ 27 kΩ		$C_{754} =$ 10 nF

Fig. 4-65 Unblanking circuit in Tektronix oscilloscope Type "515"

voltage required. The positive pole of this bias supply is also connected to the square-wave voltage developed by the multivibrator in the time base unit via a cathode follower in a way similar to that shown in the circuit diagram of the Philips "GM 5602" oscilloscope (Fig. 4-50). The potential of the complete bias unit is thus controlled in accordance with the waveform of the multivibrator voltage, and unblanking of the forward stroke and blanking of the flyback and stand-by time is thus obtained.

The voltage of the auxiliary rectifier is isolated from the chassis, and opposes the cathode voltage of the C.R.T. No current therefore, flows from it to the cathode. Provision for unblanking control by means of external voltage sources in this instrument is limited to the use of an alternating voltage for controlling the cathode potential, although a "switching" control would be possible.

4.22 Load on the time base generator and linearity

When, as in some circuits, the time base voltage is not taken from an amplifier stage (cathode follower), but direct from the timing capacitor, non-linearity can occur if the external load is too great. The additional non-linearity of a capacitor charged by a current I_L to voltage V_X with a total load resistance R_B, is given to a first approximation by the expression:

$$\text{non-linearity due to load} = \frac{V_X}{R_B \cdot I_L}, \tag{4.22}$$

where R_B is the total shunt resistance. If this ratio is not to be more than 10% (i.e. the fraction 1/10), the load resistance R_B must be:

$$R_B = 10 \cdot \frac{V_X}{I_L}.$$

For $V_X = 400\ V_{pp}$ and $I_L = 2.0$ mA,

$$R_B = 10 \cdot \frac{400}{2 \cdot 10^{-3}} = 2\ M\Omega.$$

This would therefore mean a leak resistance of about 4 MΩ, while the impedance of the charging valve should be 4 MΩ. Usually, however, the impedance of the charging valve is not so high, but it is relatively simple to obtain such values [24] by negative feedback. Thus, in order to obtain satisfactory linearity at lower frequencies it is advisable that the value over the control range of the charging current should not be too small. Even if the time base frequency is used to control a second unit, e.g., a wobbulator for representing resonance curves, care must be taken that the load impedance of the unit remains sufficiently high, usually at least 5 MΩ. It is also essential that the insulation resistance of the timing capacitor is so high that it can be neglected as a component of the total load (time constant/product of insulation resistance and capacitance $>$ 1000 MΩ/μF). Otherwise this would cause much worse non-linearity than that resulting from the influences of the normal load. For this reason only first-class capacitors should be used in time base generators.

4. 23 Permissible ripple of the anode voltage in the time base generator

This supply voltage must have a minimum hum content. Fig. 4-66 shows an oscillogram of two cycles of a sinusoidal voltage with a frequency $\gg$ 50 c/s, the time base voltage amplitude being 350 V_{pp} with a ripple of about 6.5%. It is clear from this that the ripple amplitude of the anode voltage should be less than 0.5%. There must therefore be very efficient smoothing. This is also very important in order to avoid undesired triggering of the time deflection or synchronization with the mains frequency.

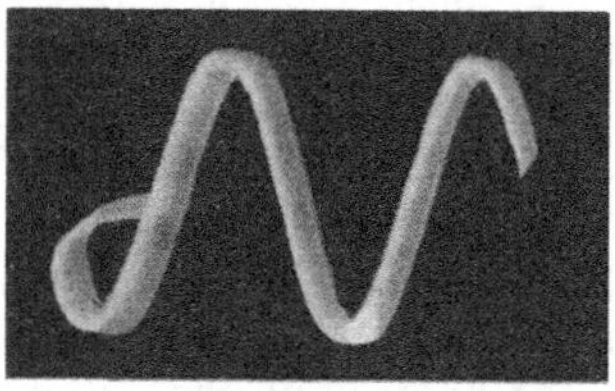

Fig. 4-66 Oscillogram of two cycles of an alternating input signal when the sweep voltage is distorted by ripple

4. 24 Screening the time base voltage sources

The necessity of carefully screening all ciruit components carrying time base voltages becomes obvious when it is considered that this voltage is often as much as several hundred volts and is of high harmonic content because of the required waveform (steep voltage drop). It is not only when using the higher frequencies but also in low frequency ranges (a few hundred c/s) that the harmonics falling within the reception bands of radio receivers are considerable; without satisfactory screening serious interference with local radio reception may result.

It is, moreover, necessary to isolate the time base voltage and its harmonics from the Y-plates and particularly from the input of the vertical deflection amplifier.

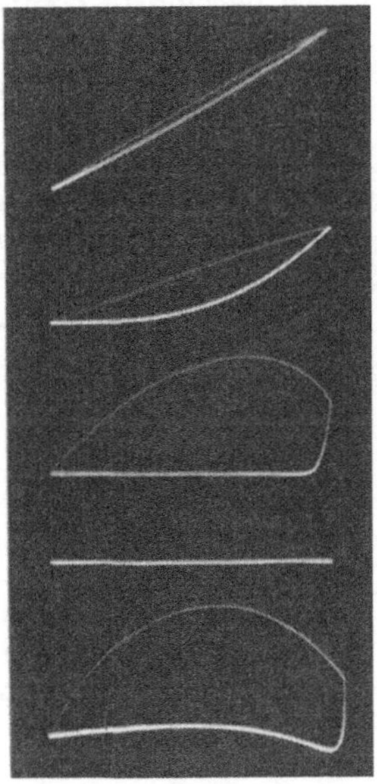

Fig. 4-67 Examples of distortion of the sweep voltage due to interference voltage on the Y-plates.
a) Sweep voltage with simultaneous interference from fundamental and harmonics
b) Fundamental of interference voltage attenuated with respect to harmonics (capacitive coupling)
c) Interference from harmonics only (coupling via small capacitance)
d) Sweep voltage without interference
e) Y-amplifier transmits interfering voltage in wrong phase

Otherwise, even in the absence of a signal, vertical deflection would occur. As this deflection occurs simultaneously with the time base deflection, the time base would become twisted or curved. If harmonics alone are responsible for spurious vertical deflection then a peak appears at one end of the time base.

As experience shows that these deflection often occur in oscilloscopes, typical effects are illustrated in Fig. 4-67 and should be of help in recognizing the cause of the faults.

For reasons stated above, thorough screening of all leads and circuit components carrying time base voltages is essential. Care should be taken, however, that the additional wiring capacitance is kept to a minimum, so that the higher frequencies are not affected. Usually it is also advisable to insert a good RF choke in the power supply line.

4. 25 Characteristics of time base

The electrical efficiency of a time base generator is sometimes indicated only by stating the frequency ranges, but this is not a sufficient indication of performance. The maximum width of deflection for a given frequency should also be known. Moreover, in triggering time base generators in particular, the flyback time must also be taken into consideration since in practice only the duration of the (unblanked) forward stroke T_X determines the possible coefficient for a set time base. The faster the rise of the deflection voltage V_X, the shorter will be the time T_X required for a single cycle. The time base generator permitting the display of the most rapidly changing phenomena will be the one with the greatest voltage rise per unit time, i.e. the highest voltage velocity.
This value is obtained from the relationship:

$$\text{velocity of sweep voltage} = v_{VX} = \frac{V_X}{T_X}. \qquad (4.23)$$

If $1/f$ is substituted for T_X in the case of repeating operation of a time base generator, ignoring the flyback time, the sweep voltage velocity is found to be the product of the time base frequency and voltage amplitude

$$v_{VX} = V_X \cdot f_X. \qquad (4.24)$$

To be able to compare the efficiencies of sweep units, all the available sweep velocities must be compared. If, for instance, a maximum frequency of 40,000 c/s at 300 V_{pp} is given for repeating operation, this indicates a maximum voltage velocity of 300 V × 40,000 c/s = 12 · 10⁶ V/ or 12 V/μs. To compare the highest possible time base velocities on the screens of different oscilloscopes, the deflection sensitivity for the *X*-direction must be taken into account. The sweep velocity shows only the increase in voltage per unit time. The distance covered on the screen along the *X*-axis is obtained by multiplying the sweep voltage velocity by the deflection sensitivity of the pair of plates in question, as follows:

$$\text{deflection velocity } v_X = \frac{V_X}{T_X} \cdot DS_= \,. \tag{4.25}$$

Or, for repeating-synchronized operation, neglecting the flyback time

$$v_X = f_X \cdot V_X \cdot DS_= \,. \tag{4.26}$$

The time base frequency and the length of trace (denoted by *X*) are usually known. As the image width is given by $X = V_X \cdot DS_=$, the time base velocity on the screen is obtained from the product:

$$v_X = f_X \cdot X. \tag{4.27}$$

Taking a DG 7-6 tube as an example, when $V_X = 300\ V_{pp}$, $f_X = 40{,}000$ c/s and $DS_= = 0.26$ mm/V, the time base velocity is $40{,}000 \times 300 \times 0.26 = 3.12 \times 10^6$ mm/s or 3.12 mm/μs and $X = 300 \times 0.26 = 78$ mm.

Although the signal frequency in cyclic processes is quite often known, at least approximately, the range of frequency adjustment of the time base should be quoted, but the deflection velocity or its reciprocal the time coefficient, should also be insisted upon. In repeating operation at the higher frequency ranges, where the flyback time is already relatively long, and more particularly, where deflection is triggered, the forward stroke speed alone is decisive. In such cases, only Eq. (4.25) in which T_X forward stroke duration can be used.

The shortest duration of a changing phenomenon which can be recognized on an oscilloscope depends not only on the maximum deflection speed of the spot but also on the diameter of the spot. It may be assumed that the smallest distinguishable time difference –time resolution– is given, just as when evaluating the vertical deflection, by the distance between the centres of two adjacent spots, equal to the diameter of a single spot, so that the minimum discrimination on the screen

$$\Delta X_{\min} = 2 \cdot d_s. \tag{4.28}$$

To illustrate these relationships Fig. 4-68a shows a part of an oscillogram of a sinusoidal voltage the frequency of which was chosen such that the spaces between the successive peaks at the upper or lower ends of the waveform were each equal to the diameter of the spot.

In Fig. 4-68*b*, the part enclosed by broken lines in *a* is reproduced magnified about 2¹/₂ times, and *c* shows a part of the oscillogram with the input frequency doubled. The end peaks of the sinusoidal curves now touch each other and are indistinguishable from one another [28]).

[28]) This setting might also be used to define the maximum time resolution.

The smallest time interval of the signal to be measured which can be distinctly recognized — $T_{M\min}$ — is seen from Eqs. (4.23) and 4.26) to be

$$T_{M\ \min} = 2 \cdot \frac{T_X \cdot d_s}{V_X \cdot DS_=} = 2 \cdot \frac{d_s}{v_X}. \quad (4.29)$$

In the example quoted, in repeating operating and neglecting the flyback time $T_X = \frac{1}{f_X}$, and with a spot diameter of 1.0 mm [29]), $T_{H\min}$ is:

$$T_{M\ \min} = \frac{2 \times 1.0}{3.12 \times 10^6}$$

$$= 0.64 \times 10^{-6} \text{ s or } 0.64\ \mu\text{s}.$$

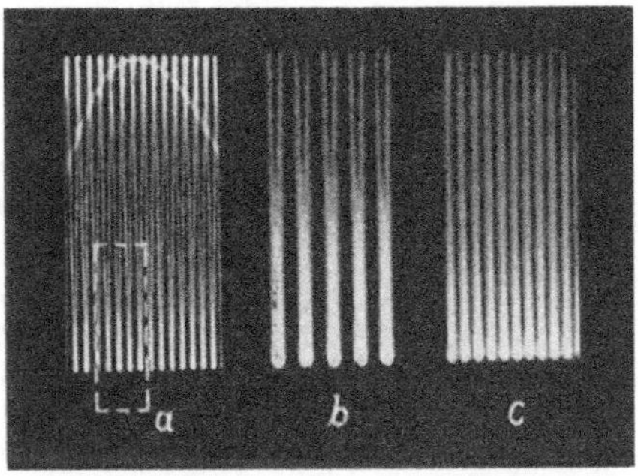

Fig. 4-68 Examples of oscillograms for determining boundary value of maximum possible time resolution

The highest signal frequency which has a still distinguishable waveform (Fig. 4-68) is clearly the reciprocal of $T_{M\min}$:

$$f_{M\ \max} = \frac{1}{2} \cdot \frac{v_X}{d_s}. \quad (4.30)$$

Referring to the same example once more, we get:

$$f_{M\ \max} = \frac{3.12 \times 10^6}{2 \times 1.0} = 1.56 \text{ Mc/s.}$$

The coefficient *TC* has been introduced as a basic unit for the measurement of the representation of changes of state dependent on time. It shows what period of time at the prevailing adjustment corresponds to the unit of length (usually one centimeter) on the screen in the direction of the time base deflection. The time coefficient is the reciprocal of the time deflection velocity $\left(\frac{1}{v_X}\right)$:

$$TC = \frac{V_X \cdot DS_=}{v_X \cdot X} = \frac{T_X}{X}. \quad (4.31)$$

It is therefore seen to be also the quotient of the deflection duration and the width of image. According to deflection velocity, it is given in s/cm, ms/cm, μs/cm, or, for particularly small time scales, [30]) in ps/cm.

For the reasons given above the adjustable scales of oscilloscopes with triggered time base generators are now quoted more frequently as the time coefficient than in terms of deflection frequency.

---

[29]) At the turning point.

[30]) The shorter the duration of the process under observation, the smaller must be the time coefficient.

4. 26 Rating the coupling components for the time base voltages

For application to the deflection plates and within the oscillator circuit it is often necessary that the time base voltage is coupled via *CR*-networks so that the direct component is blocked. When selecting values for these coupling components it should be remembered that the sawtooth voltage consists of a sinusoidal fundamental with a large number of harmonics having specific phase relationships. The coupling components should therefore be so rated that the fundamental and harmonics are passed without attenuation, and, what is even more difficult, should be transmitted without phase distortion. These problems are dealt with in detail in Part I, Ch. 5.

4. 27 Scale down of time coefficient by amplifying deflection voltage

At a given frequency of the sawtooth voltage the time resolution can be increased — in other words the time coefficient can be reduced — if the sawtooth amplitude is increased by means of amplification. By this means the voltage velocities and the deflection speeds are magnified in a manner corresponding to the deflection coefficient of the deflection plates (Part I, Ch. 4.25 "Characteristics of time base"). The amplifier must, of course, be conservatively rated so that it introduces no distortion of the waveform. As can be seen, however, in Part I, Ch. 5 (see p. 219), if very large voltage outputs are required, an input voltage which increases very rapidly is required. This imposes the condition that not only must the fundamental of the sawtooth voltage be amplified without distortion, but the numerous harmonics must also be faithfully transmitted and, what is particularly important, without phase shift with regard to the fundamental.

It is, however, quite sufficient if the amplification is such that only the part of the time base waveform visible on the screen is sufficiently linear with time to meet the demands. If, beyond this region, the waveform becomes distorted by over-driving, there will be no ill effects, so long as these parts of the waveform have no influence on the image visible on the screen. This, however, is not the case with alternating voltage amplifiers. If over-driving occurs, additional rectified voltages are produced which, depending on the amount of over-driving, vary the control grid working-point. If the deflection voltage were increased the image would be shifted and other disturbing effects could occur. The use of DC-coupled amplifiers renders it easy to avoid these difficulties, as is discussed in more detail in Part I, Ch. 5.37 on DC voltage amplifiers.

Thus, without excessive cost, the sawtooth voltages can be amplified from five to tenfold and more, without producing undesirable shifts in the operating point. The function of such an amplifier is illustrated by the following practical example.

Fig. 4-69 shows the *X*-amplifier of the Philips "GM 5662" H.F. wideband oscilloscope. The input stage is cathode-follower (21′), so that, even when used with a horizontal amplifier, a particularly low input load is imposed on the voltage source. The grid of one valve (24) of the symmetrical wideband stage (f_{cl} = 800 kc/s) is controlled by the cathode of valve (21′). The other output valve (25) is also controlled in push-pull by cathode coupling via a high cathode resistor (R_{219}). The horizontal deflection plates of the oscilloscope tube are directly connected to the anodes of the output stage valves. As the vertical deflection plates are located between these plates and the first accelerating electrode (a_2), the fact that the average voltage at these electrodes is about +145 V is immaterial. The input to the amplifier may be, according to requirements, the sawtooth voltage of the time base generator, an external

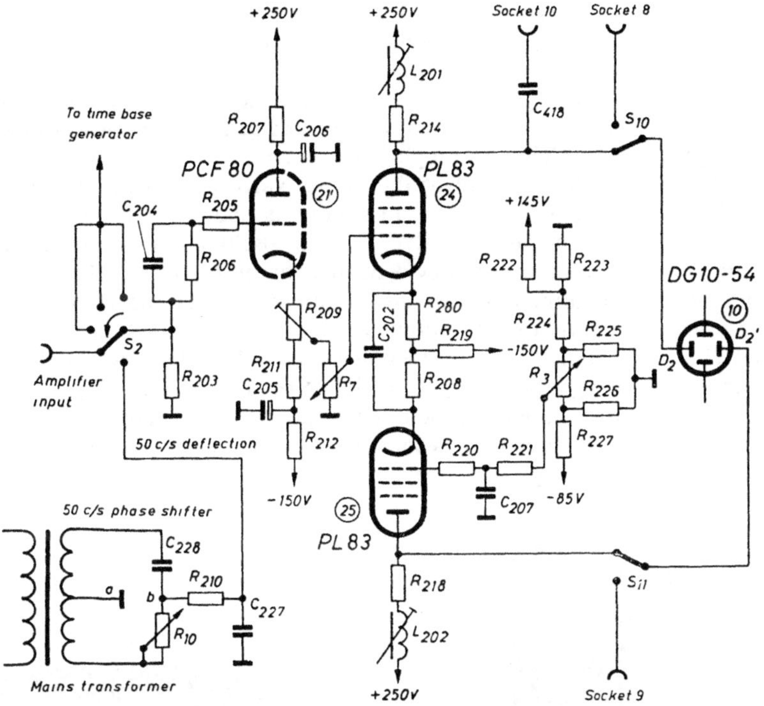

Fig. 4-69 *Y*-amplifier of Philips "GM 5662" HF wide-band oscilloscope with phase shifter for 50 c/s voltage

voltage, or a 50 c/s sine voltage, the phase of which is adjustable by means of (R_{10} and C_{228}).

In order to avoid excessive grid current at positive values of the grid voltage, and also in order to protect the grid/cathode circuit of the input valve from the effects of accidental application of excessively high voltages, a high resistance (R_{203} = 470 kΩ is included in the grid circuit of valve 21′.

As soon as grid current begins to flow, the effective grid voltage decreases compared with the input voltage due to the drop in this series resistor, thus maintaining the grid voltage within permissible limits. As usual, a capacitor is connected in parallel with this series resistor, so that the latter may not, together with the input capacitance of the valve, form a low pass filter and reduce the upper cut-off frequency; wideband transmission is thus ensured. The cathode resistors of the input valve are connected to a —150 V point in the power supply. This makes it possible to connect the potentiometer (R_7), which adjusts the voltage derived from the input valve and applied to the output stage, at a point in the cathode resistor chain which can be precisely chosen by using the adjuster (R_{209}) and which has no direct voltage difference to chassis. Thus, no current flows along this path and, in spite of the fact that direct coupling between the two valves is employed, no shift of the operating point will occur even if amplification undergoes a change through the potentiometer (R_7). The cathode current flows only to the voltage source of —150 V, which, of course, must be suitably rated for this current.

The symmetry of the operating point of the output stages is determined by a potentiometer (R_3) which adjusts the grid voltage of the output valve (25) via the cathode coupling (R_{219}). (For more details see explanation of such a circuit in Fig. 5-59.)

The oscillograms in Fig. 4-70 show the operation of the amplifier in normal working. At *a* the output voltage of the sawtooth generator is once more seen (see also oscillogram 4-44*a*). This, in amplified form, reaches the voltage divider potentiometer (R_7) and finally the grid of the output valve (24) in the upper part of the circuit. The voltage at the anode of this valve is shown in *b* and the voltage at the anode of the lower output valve (25) in oscillogram *c*. At *d* both traces are shown, superimposed, and it is evident that the push-pull waveforms are symmetrical. Finally, oscillogram *e* shows the output voltage at the cathode end of the coupling resistor common to both output valves, i.e. at the junction with the resistors R_{208} and R_{280}. At about "zero" it fluctuates to a relatively very small extent. This voltage change is enough, however, to modulate the lower valve (25).

If the voltage between input stage and the output valves is increased by the adjusting potentiometer (R_7), the latter may become overdriven, and waveforms like those seen in the oscillograms of Fig. 4-71 are obtained. The sawtooth voltage over certain

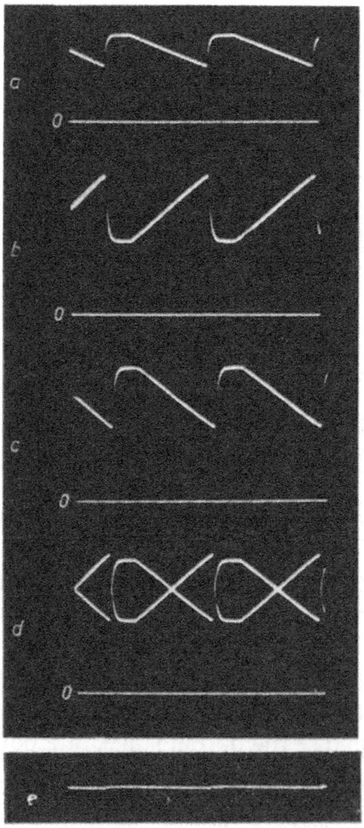

Fig. 4-70 Voltages in X-amplifier circuit of Fig. 4-69 during normal operation driven by sawtooth voltages

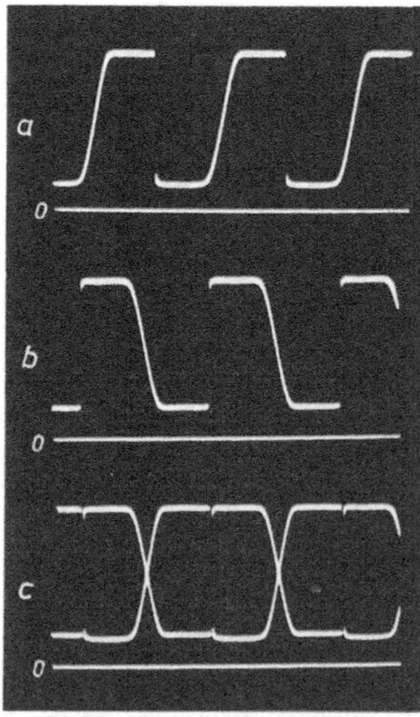

Fig. 4-71 Output voltages of horizontal amplifier as in Fig. 4-69, but with sawtooth voltages magnified fivefold

limiting values is cut off at both ends of the stroke. The total voltage change has become roughly twice as great as in the oscillograms in Fig. 4-71. Since the image at these voltages was already about 8 cm wide, the deflection resulting from the

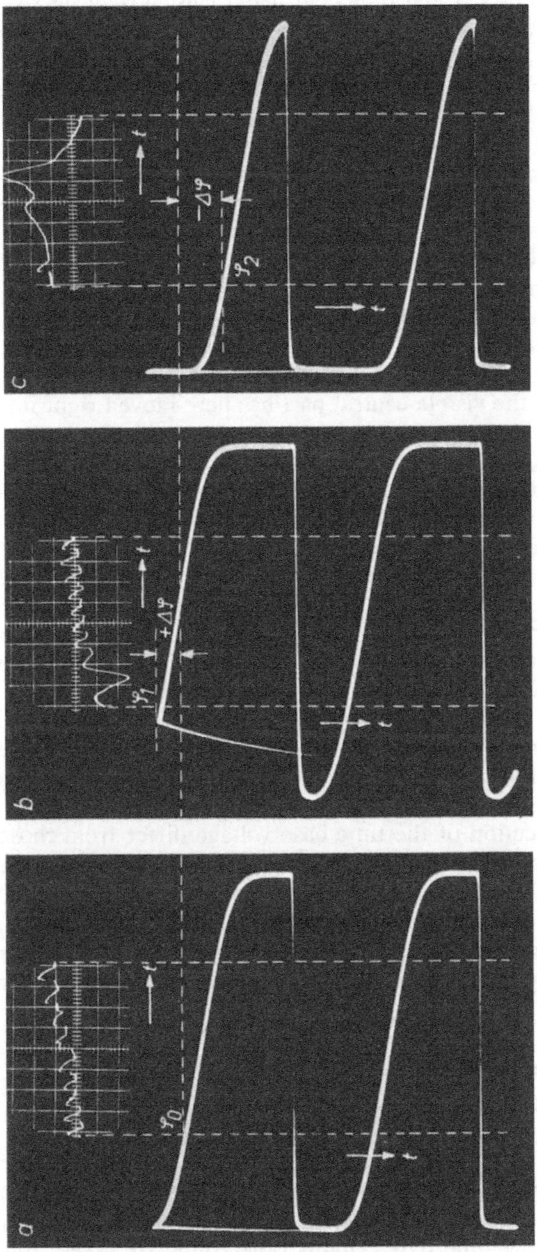

Fig. 4-72 Reduction of time coefficients by magnifying sawtooth voltage
a) Operating points of the X-amplifier adjusted roughly in balance. Centre part of oscillogram is expanded
b) Operating points so shifted that left-hand part of originally unexpanded oscillogram appears expanded on screen
c) Operating points shifted in the reverse direction from that for *b*). The right-hand part of the oscillogram appears expanded on screen

amplified sawtooth voltages will be considerably greater than the oscilloscope screen, the maximum diameter of which is less than 10 cm. The increase in steepness of the slope of the sawtooth voltage curve indicates a correspondingly increased voltage velocity and thus a magnified time resolution or a correspondingly decreased time coefficient. In the "GM 5662" oscilloscope the time coefficient can be reduced fivefold in this way. In conjunction with the time resolution due to the maximum possible time base frequency a minimum time coefficient of about 70 ns/cm can be obtained.

The opportunities offered for time resolution by magnifying the time base voltage are shown in more detail by the oscillograms in Figs. 4-72 *a, b* and *c*. Before the sawtooth voltage was amplified, an image was adjusted to an oscillogram width of about 8 cm, as has already been seen in Fig. 4-38*a*. For the oscillograms of Fig. 4-72, the time base voltage was amplified to the highest possible value. The illustrations show, at the top, the screen image on the Philips "GM 5666" oscilloscope, and below, the amplified sawtooth voltage at different adjustments of output stage symmetry. In *a*) the originally normal screen image in the centre was merely amplified fivefold. The centre part of the image now appears expanded about five times, i.e., at a time coefficient reduced fivefold. The oscillogram, which was normally adjusted at first, was rather more strongly expanded than it was in Fig. 4-38*a*, with the result that a part of the oscillogram to the left of the visible central part has here moved right into the centre.

The part of the time base voltage which determines the visible portion of the oscillogram is indicated by the vertical broken reference lines. It can be seen that the output stages of the horizontal amplifier have been so conservatively rated that the time resolution of the visible portion of the image is sufficiently linear with time.

If the symmetry of the output stage is shifted in such a way that either the left-hand part of the sawtooth voltage (*b*) or the right-hand part (*c*) falls within the region which determines which part of the section of the oscillogram is visible on the screen, it is possible to select various time intervals during the voltage change initiated at the time of triggering within the time selected for the duration of deflection (forward stroke). As the time base starts in a certain voltage phase –which, as will later be described, can also be selected according to requirements–, it is also possible to change the phase of the portion under observation with reference to the total waveform, as can be seen in the oscillograms in Fig. 4-72.

Oscillograms obtained by magnification of the time base voltage differ from those obtained with phase-shifted triggering, in that they always extend to the edge of the screen, while in the case of phase-shifted triggering (without voltage magnification) the oscillogram can be limited in width to the visible screen surface. It is, of course, possible to use both processes together, and adjust highly time-magnified screen images to the desired phase to show small portions of a cycle of the change of state under observation.

4.28 Phase-delayed triggering of the time base

4.28.1 NECESSITY FOR PHASE-DELAYED TRIGGERING

The oscillograms in Fig. 4-38 (a to c) showed how, in triggered deflection, the time coefficient can be chosen independently of the frequency of the signal voltage. The expansion of the voltage trace takes place automatically at the triggering point *A*, at which the signal voltage triggers the time base. Unless other steps are taken, this determines the start of the oscillogram. In the oscillograms of Fig. 4-38, for instance, the progressively decreasing time coefficient permits a progressively smaller portion of the original waveform, in fact, only that part which immediately follows the beginning of the time expansion, to be seen. This would, however, severely limit observation. It is, however, very desirable to be able to observe any time portion, which can be selected at will, of a repetitive cycle of changes of state, the duration of which should also be adjustable by altering the deflection velocity. In order to achieve this, it is necessary to produce an adjustable time-delay (phase-delay) between the instant at which the signal voltage reaches the value necessary for triggering the time base deflection and the actual commencement of the forward stroke.

If the time base deflection can be triggered by a sinusoidal voltage, or if the signal voltage is fully synchronized with a sinusoidal voltage (e.g., alternating voltages of mains frequency), such phase delay can be simply obtained by means of a phase shift network.

4.28.2 DELAYED TRIGGERING OF TIME BASE DEFLECTION BY MEANS OF A PHASE SHIFTER FOR SINUSOIDAL VOLTAGES

A phase shifting device for the alternating mains voltage has been built into the Philips "GM 5602", "5603", "GM 5662", "GM5666" oscilloscopes to permit the examination of mains-synchronous voltage changes by phase-delayed triggering of the time base. Fig. 4-69 shows this simple device as part of the circuit of the horizontal amplifier of the "GM 5662" oscilloscope. A 50 c/s alternating voltage can be taken from the phase-shift network (R_{10} and C_{228}) at points *a* and *b*, the phase of which can be adjusted over almost 180° by a variable resistor (R_{10}). To preserve the sinusoidal waveform as closely as possible –particularly if the voltage is to be used for the horizontal deflection itself– an additional *RC*-filter (R_{210}, C_{227}) is connected to the output of this voltage. By means of the choice between positive or negative-going control voltages of the time base triggering pulse, by means of switch $S_{5,c}$ in Fig. 4-53, it is possible to choose the voltage phase for triggering the sweep at any angle up to almost 360°.

The oscillograms reproduced in Fig. 4-73 make it possible to study mains frequency waveforms still more closely. The sweep is carried out in (*a*) directly by the –negative going– signal voltage as in Fig. 4-38. The time coefficient in Fig. 4-73 was, in this case, similar to that in Fig. 4-38*c*. The result is a correspondingly expanded image of the voltage waveform at the beginning of the oscillogram of Fig. 4-38*a*.

If, however, the phase of the voltage controlling the triggering is altered, the sawtooth voltage commences, shown in the example in Fig. 4-73*b*, at the phase angle φ_1, i.e. with a longer delay. There is therefore a corresponding shift of the oscillogram, so that a different part of the voltage waveform appears on the screen.

With an even greater phase shift –e.g., of φ_2– the greatly expanded image of the waveform at the voltage peak is obtained as in *c*. This simple phase shifter makes it possible to display on the oscilloscope screen portions at any time resolution required over almost the complete cycle.

These features have been incorporated in almost the same form in the AEG HT-"0 15" oscilloscope. The phase setter is mounted on the front panel of this oscilloscope. This input is isolated from the potential of the voltage source to earth by an

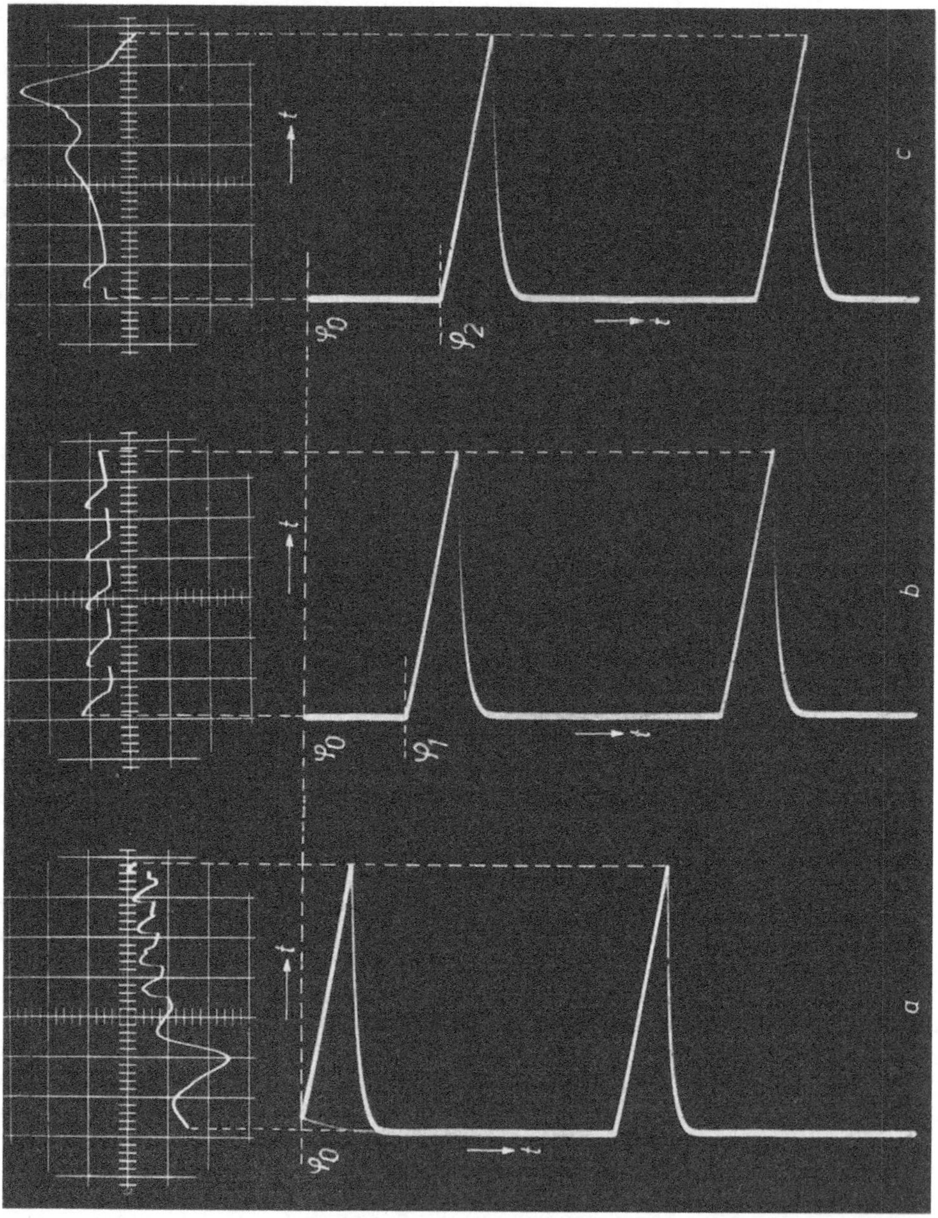

Fig. 4-73 Effect of phase-shifted triggering on screen image. *a*) Immediate release of sawtooth pulses. *b*) Sawtooth pulses released with delay of angle φ_1. *c*) Sawtooth pulses with delay of angle φ_2 ($\varphi_2 > \varphi_1$).

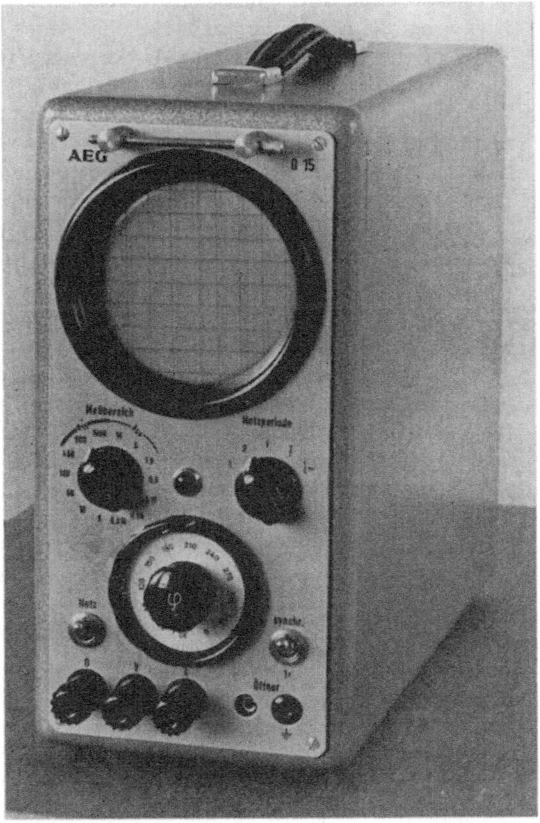

Fig. 4-74 AEG power current "015" oscilloscope in which the trigger phase can be shifted 360°

instrument transformer (Fig. 4-74). It is possible by this means to alter the triggering phase through almost 360° with an accuracy of about 3°. Moreover, it is possible to use the change-over switch to select the time period in stages for 3, 2, 1, $^1/_2$ or $^1/_6$ cycles of the signal frequency. The unit is therefore adequate to meet all the normal requirements of power engineering. For a vertical deflection of 4 cm, seven voltage measuring ranges and seven current measuring ranges from 18 mA_{pp} to 18 A_{pp} are provided. The total current taken for the voltage ranges is 6 mA_{pp}, for the current ranges the voltage drop is 3 V_{pp} (about 1 V_{rms}). As has been described in connection with the oscillograms in Fig. 4-73, it is also possible to examine details of waveforms very carefully by adjusting the time deflection to fractions of a cycle. This is of particular value, for instance, in the adjustment and checking of controlled rectifier circuits. This oscilloscope was in fact designed for this and similar tasks. Similarly, phase-delayed triggering of the time base can be achieved using sinusoidal alternating voltages of other frequencies than mains frequency, with the aid of a simple phase shifter. However, this is possible only when the signal voltage is in phase-locked synchronization with this sinusoidal voltage. Normally, these conditions will not everywhere exist, but they will, however, occur when signal pulses are triggered by a sinusoidal voltage which is also used to trigger the time base. If the phase difference is adjusted to somewhat less than a complete cycle (less than 360°), it is easy to display the leading edge of the pulses from their commencement or even slightly earlier, i.e. a portion of the preceding time base cycle. A more detailed description of this process is to be found in Part II, Ch. 13, "Representation of the rising flank of pulse-shaped voltages with oscilloscopes without delaying elements in the vertical amplifier".

4.28.3 DELAYED TRIGGERING OF THE TIME BASE BY ADJUSTING THE TRIGGER LEVEL

Another type of delayed time base triggering is that in which the instant of triggering is that at which the signal voltage reaches a predetermined value. In this system the triggering delay can be made to commence at any instant between the minimum value of the operating voltage of the sweep time base circuit and the peak value of the signal voltage. If the signal voltage has a large sinusoidal fundamental component, this means, for example, that the delay is adjustable within approximately the first 160° of the fundamental cycle. The ability to choose the amplitude level at which the time base can be set in motion is particularly advantageous when investigating voltage waveforms in a number of definite stages, as, for instance, in television engineering.

The voltage of the multivibrator in the time base unit which determines the operating mode of the sawtooth generator in such an oscilloscope, is derived from a second bistable multivibrator (Ch. 29 "The Schmitt-trigger circuit"). The operating threshold in this multivibrator is adjustable, the multivibrator changing over at the selected amplitude of the triggering voltage (amplitude discriminator).

A voltage of rectangular waveform is again obtained across the anodes of this pair of valves, the commencement of which is determined by the operating voltage level set. At the anode of the valve previously cut off, for example, the voltage drops when the valve again becomes conductive at the instant of reversal, and increases once more when the circuit reverts to the stand-by state. By means of differentiation with a CR-network a negative pulse is obtained from this voltage at the instant of reversal, and this serves to trigger the time base circuit. Hence this instant can be preset at a particular value of the signal voltage amplitude, e.g. by "internal" triggering (Fig. 4-53).

4.28.4 PHASE-DELAYED TRIGGERING IN ARBITRARILY CHOSEN TIME RANGES

In order to carry out this process satisfactorily, it is necessary that the time interval between the instant at which the triggering voltage reaches the threshold value and that at which the actual triggering of the time base occurs should be adjustable as desired. This means, for example, that in order to observe the waveform of a single television picture line, the time base triggering delay of about 20 ms with respect to the vertical sync pulse, should be adjustable up to a maximum of 40 ms of the pulse sequence if both field images are to be scanned. If the time coefficient is adjusted in such a way, only the waveform of one line, which corresponds to a time of 64 μs is displayed, it means that the time coefficient has been reduced by a factor of 1/300 as compared with the intervals occurring in a pulse sequence of 20 ms. In order that the successive sweeps may always occur at the same place in the voltage waveform to give a stable display, the time intervals of the triggering pulses must remain equally constant.

All the known devices used for this purpose make use of valve (or transistor) circuits. Two types of circuits can be singled out in which a multivibrator stage again is the most important element.

In one of these, a multivibrator is *directly* triggered by the controlling voltage. The rectangular output voltage of this multivibrator is differentiated by a *CR*-network. This causes positive peaks when the voltage flips in a positive direction and negative peaks when the voltage flops in the opposite direction. The negative pulses, which correspond in time to the right-hand edge of the rectangular pulses, serve in the way already described for the triggerable sweep circuit to trigger the time base. By changing the multivibrator time constants, i.e. the ratio of the positive to the negative parts of the voltage waveform, the symmetry of the rectangular voltage can be altered. Thus the time interval of the trailing edge is moved away from the leading edge of the rectangular pulses, and the time difference between starting the multivibrator and the starting of the triggering pulse of the time base circuit is also altered. It is therefore possible, by altering the width of the rectangular pulses, to delay the triggering of the time base to times which can be adjusted. The stability of oscillograms so obtained is dependent of course on the constancy of the width of the rectangular pulses that can be maintained. This sets the limit for the smallest usable time coefficient by means of which a portion of the pulse sequence under observation can be studied under sufficiently stable conditions. With normal input, a decrease to about 1/10,000 of the time taken by the triggering pulses to follow one another will represent the absolute minimum.

This process will be dealt with in greater detail in the description of the "time base expansion unit" on page 149. In view of its use in television engineering only at frame frequency, it is designed to operate at the lighting mains frequency of 50 or 60 c/s. The process can, however, be used analogously for other ranges of repetition frequencies which can either be selected in stages or continuously adjustable.

In the most common circuits of this type, an auxiliary sawtooth generator is used to determine the time delay. In addition to the adjustment of the surge duration of these sawtooth voltages, the time section within which the delay occurs is adjustable.

A circuit in which the delay time can be adjusted by altering the cut-off voltage of a diode is shown in the block wiring diagram in fig. 4-75.

The cathode of this diode (valve 1) is connected to a point on the voltage divider (R, R_1 and R_2) which is connected between a positive voltage and chassis. The output of the auxiliary sawtooth voltage generator which, as is usual, must be synchronized or triggered (time t_0) is connected to the anode of the diode. Depending on the positive voltage at the cathode, current can flow from the sawtooth generator through the diode only if the output voltage of the auxiliary sawtooth generator becomes greater than the cathode voltage. As the rise time of the sawtooth is selected by the time constant of this generator, it is thus possible, by adjusting the cathode voltage of the diode, to determine the point in time (t_1) at which the diode becomes conductive. The remaining part of the rising sawtooth curve appears at the cathode, as is indicated in the voltage indications in the diagram. This voltage is greatly amplified in two stages in the following amplifier, so that the sawtooth is made steeper and is clipped at the top. At its output the voltage closely approximates to a rectangular voltage. This voltage is differentiated, so that at the time (t_1), determined by the cathode voltage of the diode, a sharp positive peak occurs, and at the end of the sawtooth a negative peak. A bistable multivibrator can be started, in the way already described, by the positive pulse, and the

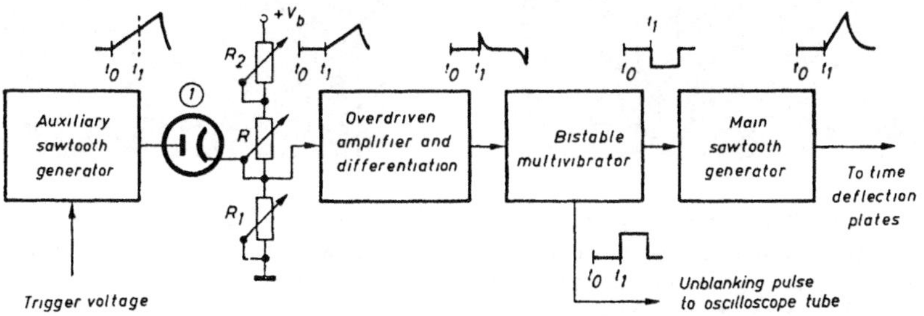

Fig. 4-75 Delayed triggering of time base by an auxiliary sawtooth generator and a diode with adjustable bias

multivibrator in turn controls the main time base generator. The time coefficient can be selected here too in the usual way. The total duration of the cycle of this sawtooth voltage must naturally be shorter than, or at most equal in duration to the cycle of the auxiliary sawtooth voltage. In the same way, an unblanking pulse can be taken from the multivibrator of this sawtooth voltage generator and used to intensify the forward stroke on the oscilloscope screen.

Such circuits are also used in radar engineering [25].

Instead of a biased diode, a circuit of this type can also employ a cathode-coupled multivibrator in which the operating level, as already described, is changed by adjusting the direct grid-voltage. Likewise, the instant of the flip-flop reversal can be made dependent on the duration of the rise of the auxiliary sawtooth voltage.

In the Tektronix oscilloscope types "535" and "545", in which there is a similar delayed sweep, it is possible to connect the auxiliary sawtooth and the main sawtooth generator alternately to the horizontal deflection plates. In this case it is advisable first to connect the auxiliary sawtooth generator to the *X*-plates, so that the time base deflection can be adjusted on the oscilloscope screen, from which a part can be selected for closer study. As by this means the rectangular unblanking voltage is added to that of the auxiliary sawtooth generator and the unblanking voltage also from the main time base generator, that part of the oscillogram corresponding to the time portion adjusted by the main sawtooth generator is brighter than the rest of the oscillogram. If the trigger phase is shifted by adjusting the operating level in the multivibrator provided for the purpose, this bright part of the oscillogram moves to a corresponding extent over the screen image, as can be seen in the illustrations in Figs. 4-76*a* and 4-77*a*. In these oscillograms a series of sync pulses from a television signal generator, triggered by the frame pulse, is shown. Between the start of the oscillogram at the first

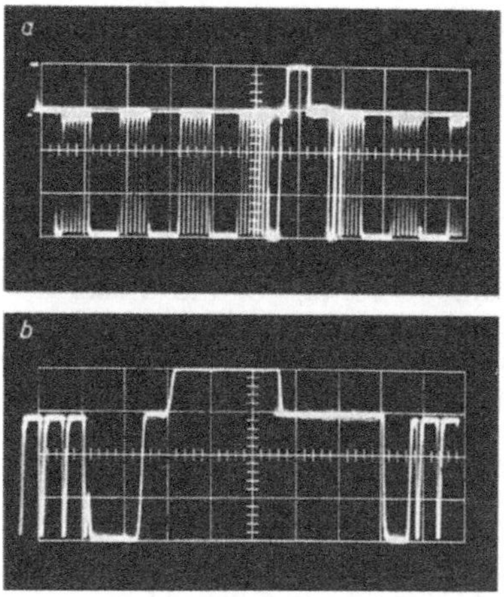

Fig. 4-76 Intensifying the brightness of a part of an oscillogram to be expanded. *a*) Composite oscillogram made with auxiliary sawtooth generator with brightened section; *b*) 5.4 times expanded oscillogram corresponding to brightened section in *a*)

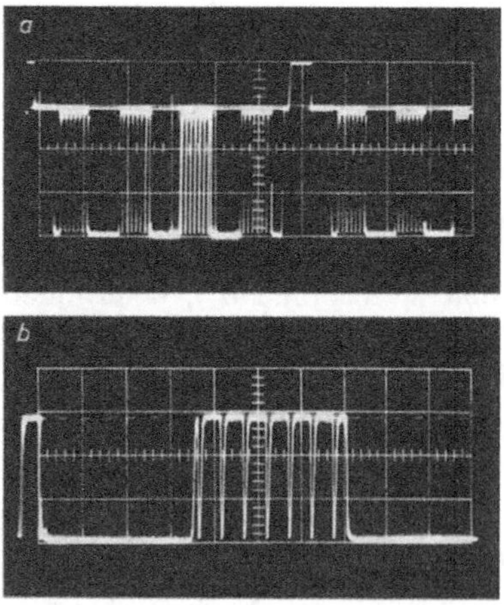

Fig. 4-77 Similar oscillograms like Fig. 4-76 but phase shifted. *a*) Composite oscillogram corresponding to brightened section of *a*)

sync pulse and the following sync pulse, four groups of voltage rises, making vertical lines in the test signal, can be seen. In the oscillogram in Fig. 4-76*a* the sync pulse itself and the voltage waveforms immediately preceding and following it are intensified. If the connection is now made to the main sawtooth generator, an oscillogram as in Fig. 4-76*b* is obtained showing the part of the original oscillogram that was shown intensified in Fig. 4-76*a*, but now expanded over the whole screen. For Fig. 4-77, the third group of the voltage rises was selected (*a*) and shown in expanded form (*b*).

For certain investigations it might be desirable for the purpose of comparison to study the remaining part of the oscillogram simultaneously with the expanded portion. This means that the time base would require different deflection speeds. Farlay [26] has described such a time base circuit. The circuit is shown in its essential details in Fig. 4-78.

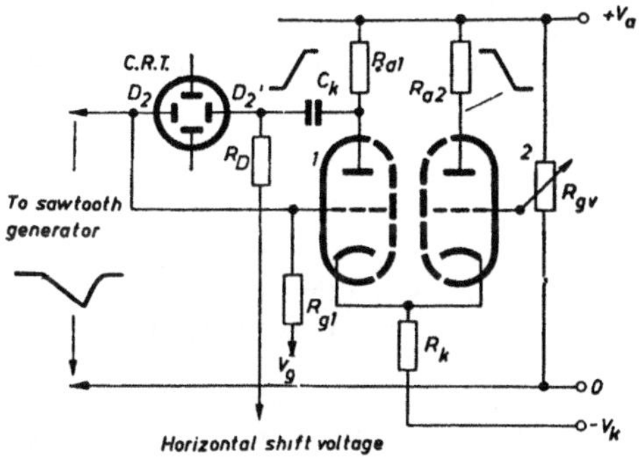

Fig. 4-78 Circuit for generating time base voltage with parts at various voltage speeds

The normal waveform of the time base voltage is supplied to the left-hand deflection plate (D_2) by a sawtooth generator of one of the usual types (e.g., a Transitron-Miller integrator). The right-hand plate (D'_2) obtains its voltage via a symmetrical DC amplifier stage. For this purpose the sawtooth voltage is fed to the grid of the left-hand valve (valve 1), the grid bias of which is so adjusted that this valve system is conductive. The other valve system (valve 2) is cut off because its grid potential has been made more negative that that of valve (1). When a negative-going input sawtooth reaches the grid of valve (1), its anode current begins to fall. The right-hand valve commences to draw current and the anode voltage of the left-hand valve, which was hitherto low, rises to a new level until the valve is cut off, as indicated in the diagram. This voltage rise –which is an amplified version of part of the sawtooth, apart from any distortion caused by the valve characteristic– is fed to the right-hand deflection plate. In this circuit the voltage on the plates is only completely symmetrical when the point at which the sawtooth voltage is cut off corresponds to half the maximum amplitude. It is more or less asymmetrical at other values. The effects of the deflection voltages at the deflection plates are that the spot is moved horizontally by the input sawtooth at a low speed and also by this selected part of the sawtooth voltage at an increased deflection speed across part of the screen.

The oscillograms in Fig. 4-79 and 4-80 serve to illustrate this in a circuit set up temporarily for the purpose. In the first place, Fig. 4-79 shows the waveform of the input sawtooth voltage (here positive going), and in *b*) the waveform of the selected

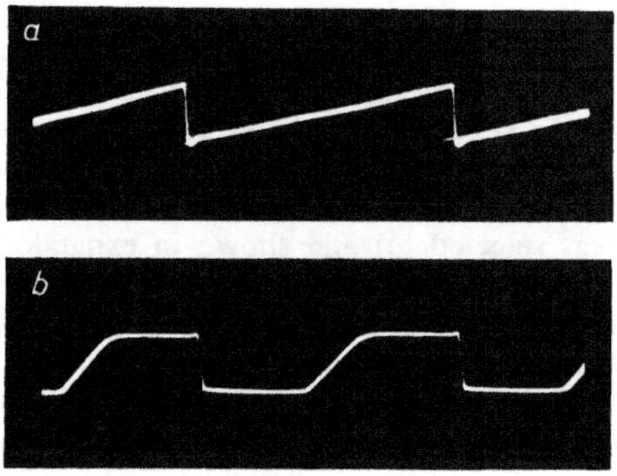

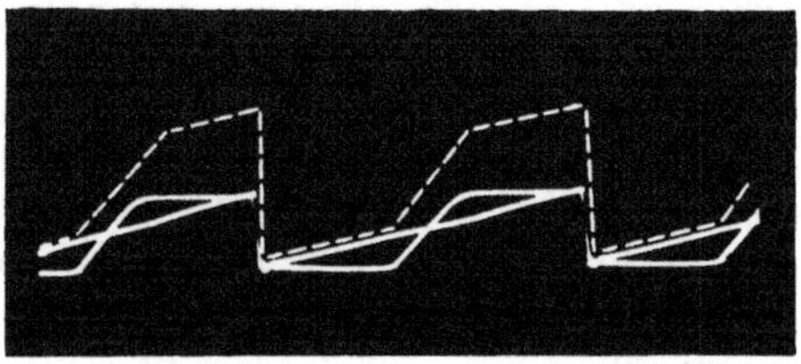

Fig. 4-80 Addition of voltages of fig. 4-79 to form deflection voltage with parts at different speeds

Fig. 4-79 (left) Oscillograms made with a circuit as in Fig. 4-78.
a) Input sawtooth voltage; b) clipped sawtooth

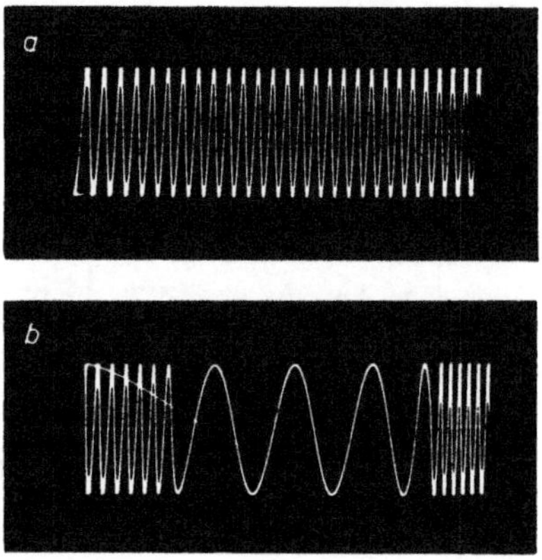

Fig. 4-81 Effect of a time base voltage with parts of different speed on the oscillogram of a sinusoidal voltage
a) Oscillogram with uniform time base speed. *b*) Oscilogram with a time base corresponding to voltage sum in Fig. 4-80

part of this voltage is seen. (This voltage was photographed in anti-phase in order to demonstrate the sum of the deflection effects more clearly.)

Both oscillograms are reproduced again in Fig. 4-80 and the sum of both voltage curves is shown in chain line. The sum curve corresponds to the deflection effect on the cathode ray in the tube. The performance which can be achieved with this circuit by way of time expansion of the screen image can be seen in the two oscillograms of a sinusoidal alternating voltage reproduced in Fig. 4-81. The normal oscillogram is seen at (*a*), while (*b*) shows the effect obtained from a time base circuit as described above. The observer gains the same impression as if a cylindrical lens was placed over the oscillogram at this particular point.

As an example of the practical use of this circuit, Fig. 4-82 shows at (*a*) the oscillogram of a series of television line pulses and modulation for five vertical bars derived from a simple television signal generator (Philips "GM 2891"). Illustration (*b*) shows the centre portion of oscillogram a) greatly expanded by the sync pulse. By adjusting the grid voltage of the right-hand valve (valve 2), the postion of the expanded portion within the total screen image can be selected. Thus, in oscillogram (*c*), the rectangular voltage rises for the vertical bars commencing at the right near the first sync pulse, have been expanded.

The firm of Du-Mont uses similar circuits, but in a form developed to suit particular requirements, in their "323-A/329-A" and "326-A" oscilloscopes under the description "notch-sweep".

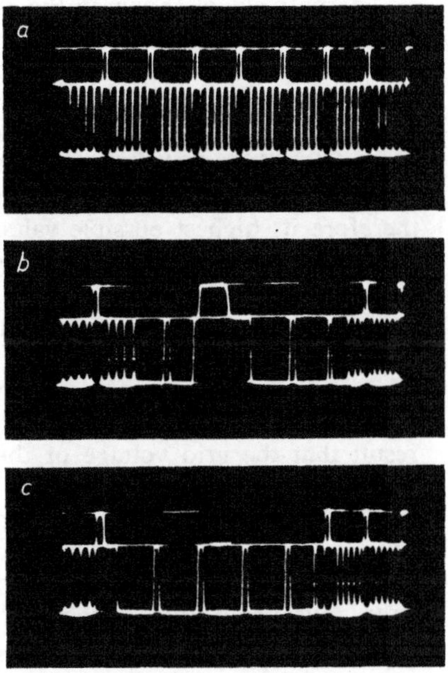

Fig. 4-82 Examination of a T.V. signal-pulse sequence by means of a time base voltage with parts at different speeds
a) Series of line sync. pulses represented in traditional way and a rectangular modulation for the test picture
b) Voltage waveform considerably expanded near the sync pulse by increased speed
c) Voltage waveform of rectangular modulation voltage (preceding sync pulse) considerably expanded

4. 29 The Schmitt-trigger circuit

It is an advantage in all time base circuits if the synchronization of triggering can be performed by a sharp voltage pulse. This is necessary, particularly if only the wave-form of the signal voltage itself is regular (e.g. a sinusoidal voltage), but a considerable time resolution is required in the case of triggered operation.

Its is therefore necessary, to ensure stable pictures (without "jitter"), that the commencement of the time base deflection should be as clearly defined as possible and its timing exact. For such purposes, circuits have been developed which make it possible to "form" a sequence of pulses of the same frequency out of practically any waveform; they are thus known as "pulse formers".

Two-valve flip-flop circuits are generally used for this. The circuit developed by Schmitt in 1938 is no doubt the best known [30]. Its mode of operation is shown in more detail in Fig. 4-83*a*. As can be seen, it employs a cathode-coupled multivibrator, including a pair of valves (usually one double-system valve), the anode of one of the valves (valve 1) being connected to the grid of the other (valve 2) by DC coupling. This coupling can be obtained by connecting the input valve grid to the reference earth (chassis) directly via its resistor (R_3). In order to prevent the voltage on the grid of the second valve from going positive with respect to the cathode, the lower end of the coupling voltage divider (R_5 and R_7) is connected to a point of a higher negative potential.

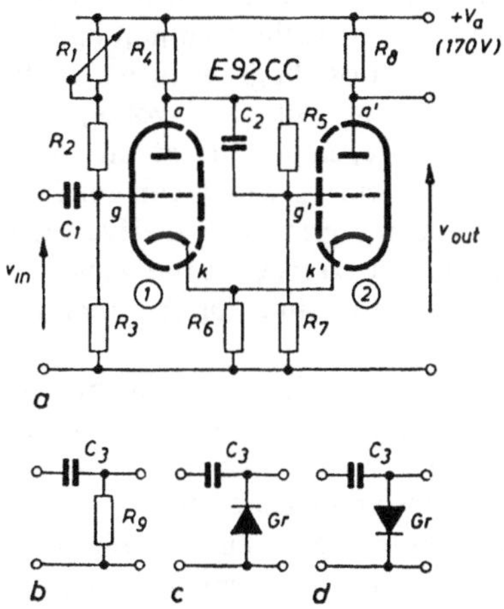

Fig. 4-83 Example of the Schmitt trigger circuit
a) Basic circuit; *b)* differentiation network with resistor; *c)* and *d)* differentiation networks with diodes connected with different polarities

Alternatively, as shown in Fig. 4-83*a*, the coupling voltage divider can be connected to chassis; the bias potential on the grid of the first valve must then be raised by means of a voltage divider so "high" with respect to the positive voltage, that the first valve (valve 1) is adjusted to the correct operating point (Ch. 5 "Deflection Amplifiers". Section 5.27.2 on "Coupling the individual stages of a DC voltage amplifier"). This potential will be such that the first valve, in the absence of input voltage, will be cut off. The potential at its anode is then equal to that of the anode voltage source; it has therefore its highest possible value. Thus the grid of the second valve is at so low a negative potential that a correspondingly high anode current flows. This is one stable state of this circuit.

If, however, due to an input voltage (V_{in}), the potential at the grid of the first valve rises above that of the grid of the second, then the first valve (valve 1) also begins to draw current. This causes its anode voltage, and also the grid voltage of the second valve, to drop. The voltage across the cathode resistor common to both systems falls, with the result that the grid voltage of the first valve now goes even more positive and its anode current rises still further. The voltage at the grid of the second valve then quickly goes so negative that the anode current of the valve is cut off and only the first valve is under current. This is the other stable state of this circuit, which is maintained until, as a result of a change of the input voltage, the potential of the grid of the first valve falls below that at the grid of the second valve, that is to say it goes more negative than the potential on the grid of the second valve. At this moment the second valve again begins to draw current; this, however, reduces the current through the first valve. This is partly due to the fact that its grid voltage decreases because of the drop in the input voltage, and partly because the voltage across the cathode resistor rises due to the current of the second valve, so that the grid goes more negative with respect to the cathode. This leads once again to the initial stable state.

The sequence of these processes is shown in the oscillograms in Figs. 4-84 and 4-85 at two different adjustments of the grid voltage of the first valve. Fig. 4-84 shows the waveform at the grid of the first valve (Δv_g) with the input voltage– in this case sinusoidal with harmonics– and the voltage change at the grid of the second valve ($\Delta v_g'$). The voltage change at the anode of the first valve is to be seen in oscillogram (*b*). While current is flowing, the waveform of the anode voltage corresponds to, but is in anti-phase with, that of the grid voltage; this valve then functions in the manner of a normal amplifier valve. The waveform of the voltage across the common cathode resistor is seen in (*c*). This also contains the portion of an oscillogram in which the time base zero line has been deflected by a direct voltage of 150 V for voltage calibration. Oscillogram (*d*) shows a rectangular voltage pulse, appearing at the anode of the second valve. Its duration is determined by the time during which current flows through the first valve. By adjusting the voltage level at the grid of this valve (or at the grid of the second valve) it is possible — depending, of course, on the waveform of the input voltage– to define very accurately two points in time for the commencement and end of this pulse.

The oscillograms in Fig. 4-85 show the same voltages as in Fig. 4-84; in this case the potential at the grid of the first valve was made much more negative. Now the duration of current flow of the first valve is only about 1/3 of that in Fig. 4-84, and thus the pulse width at the anode of the second valve is also much shorter.

To control the exact instant of switching, i.e. the beginning or end of this pulse, it is advantageous to differentiate the pulse by means of a *CR*-network (C_3 and R_9

ig. 4-84 and Fig. -85. Voltages in ıe Schmitt trigger rcuit shown in Fig. -83 at two different oltage levels at the rid of first valve. ig. 4-84 (Left) Voltge level at the grid f first valve low

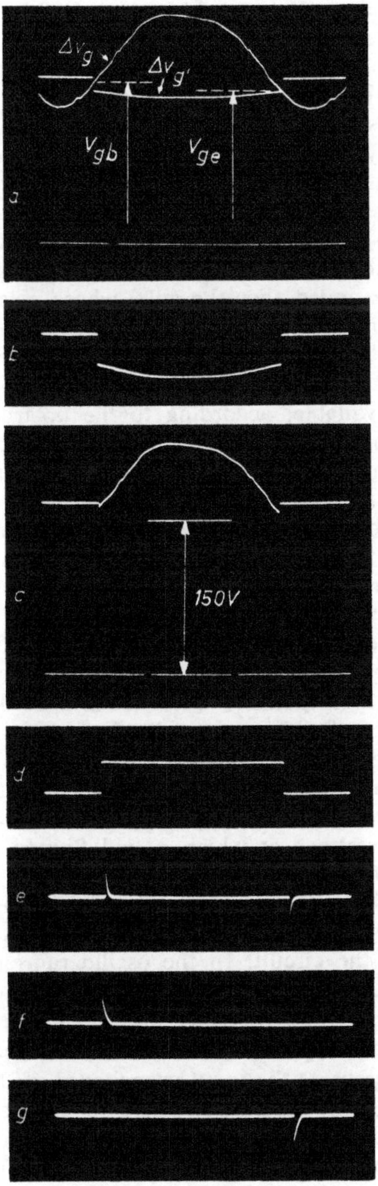

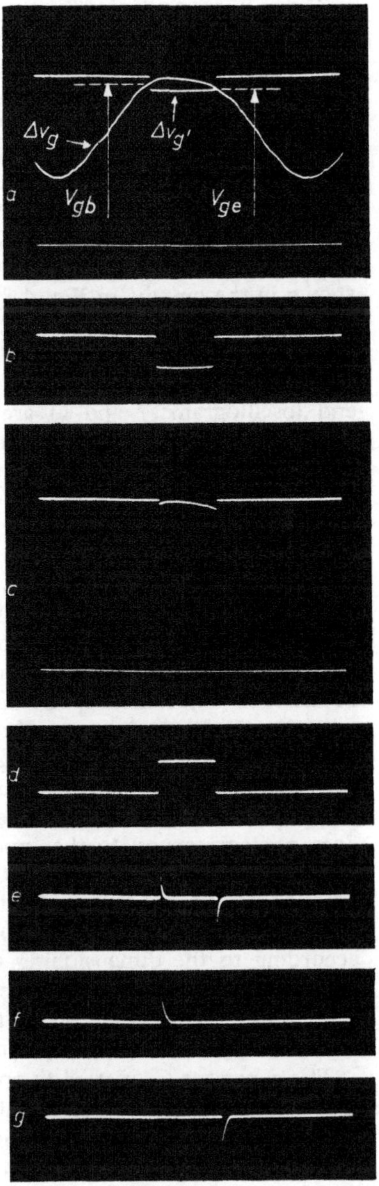

Fig. 4-85 (right). Voltage level across grid of first valve very negative. *a*) Voltages on grid *g* and *g'*. *b*) Voltage change $\Delta\ V_a$ on anode of first valve. *c*) Cathode voltage waveform ($\Delta\ V_{k\ k'}$) and deflection by means of calibrating voltage $V_{cal\ pp} = 150$ V. *d*) Voltage change on anode of second valve ($\Delta\ V_a'$). *e*) Differentiated output voltage clipped by diode. *g*) Positive voltage peaks clipped in Fig. 4-83*b*). At the beginning of the pulse, as can be seen in both oscillograms in (*e*), a correspondingly sharp positive peak is obtained and likewise a sharp negative peak at the end of the pulse. (These pictures were made with an input voltage of low frequency. The time constant of the differentiation network could not therefore be too short, as otherwise these pulses would be very short compared with the width of the rectangular pulse and thus very hard to photograph. Usually in such circuits, R_9 is of the order of several kΩ. With such values, pulses of short duration are obtained with which a time base circuit can be triggered at equally precisely defined times.)

In most cases only one of the two pulses for triggering the following circuit is suitable. If the resistor of the differentiation circuit is replaced by a germanium diode, as shown in the circuits in Fig. 4-83 *c* and *d*, then the pulse whose polarity corresponds with the low-resistance (forward) direction of the diode is cut off. During the blocking time, the blocking resistance of the germanium diode and the capacitor form the differentiation network, so that the output pulse either at the beginning or the end (oscillograms *f* and *g*) are available, according to the polarity of the diode. It is also quite clear, from these oscillograms, how the phase relation of the output control pulse can be adjusted with respect to that of the input voltage by adjusting the voltage level of the grid of the first valve. (Ch. 28 "Phase-delayed triggering of the time base", Section 4.28.4 on "Phase-delayed triggering in arbitrarily chosen time ranges".)

The oscillograms in Fig. 4-86 are also intended to give an impression of the time shifts of the output pulses to which the adjustments of the level of the oscillogram series in Figs. 4-84 and 4-85 correspond. Both pulse oscillograms (for positive and negative pulses) are shown one above the other for this purpose. This circuit can also be used, by adjustment of the operating point of the first valve, to determine, in the case of pulse-shaped input voltages of different amplitudes, which pulses exceed a certain level (radiation measuring technique). Such a circuit is called a "discriminator" [3] [27] [28]. In this application in particular, it will be seen that the voltage level to which the circuit reverts when the input voltage flops back to its original state, does not correspond to the input voltage at which it had flipped. It is somewhat greater when it flips than when it flops. This difference can be seen in the oscillograms (*a*) of Figs. 4-84 and 4-85, and is indicated by the voltage V_{gb} (beginning) and V_{ge} (end). This backlash of the circuit amounts from several tenths of a volt to about 20 V according to the dimensioning of the circuit. In the oscillograms referred to it is about 6.5 V. If a smaller cathode resistor (R_6) is employed, the backlash of the circuit decreases. This, however, alters the normal characteristics of the circuits, so that the circuit values must be based on a compromise between the various requirements.

The capacitor C_2 is used to compensate the input capacitance of the second valve. It is usual to choose it somewhat greater than is necessary for perfect compensation. By using a capacitor which is slightly too large, peaks are obtained at the grid of the valve during flip-flop, and these shorten the response time of the circuit, which can be of advantage in the case of high pulse repetition frequency. These peaks can, of course, also occur in the anode voltage. As they must be very short, however, they are only visible in oscillograms having a short switching time. In these oscillograms the pulse duration on the anode of the second valve was too long in relation to these peaks, so that they are not shown.

The possibility of using this circuit to choose the instant of response of a triggered time base circuit depending on the amplitude of the signal voltage has already been

Fig. 4-86 Time shift of output pulses by both level adjustments of grid of first valve
a) Negative peaks clipped (beginning of rectangular pulses).
b) Positive peaks clipped (end of rectangular voltages)

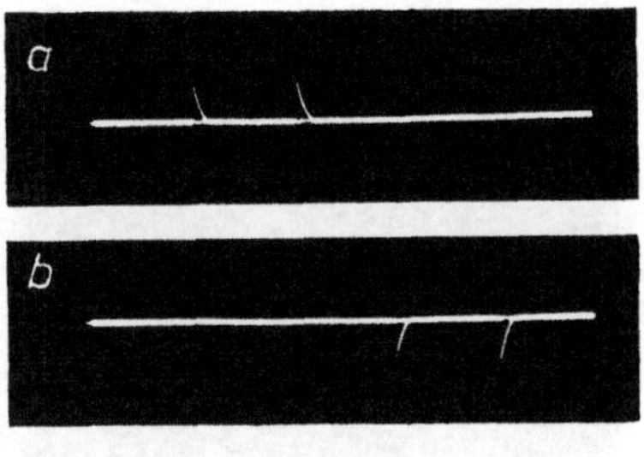

indicated on p. 166, "Delayed triggering of the sweep by adjusting the trigger level" (Fig. 4-56).

A pulse former stage of this sort is generally connected between the phase selecting and amplifier stage (separator stage) of the trigger voltage and the multivibrator triggering the sawtooth generator.

4. 30 Time base expansion unit

The arrangement of the Philips "GM 4584" equipment, especially designed for the expanded display of television signals, is shown in block diagram form in Fig. 4-87.

The sawtooth voltage for the time base is generated by an EF 80 pentode, connected as a Transitron-Miller oscillator (I). It is applied in anti-phase to symmetrical deflection plates after passing through a phase inverter formed by a section of an ECC 81 double triode.

The circuit components of the sawtooth oscillator are so rated that the transitron section can generate a frequency approximately equal to the mains frequency (50 or 60 c/s). Low values of the "Miller" capacitor (C_L in Fig. 4-35) make it possible to obtain at the same time a sawtooth cycle which, unlike that in usual time base oscillators, is definitely shorter than would correspond to the total duration of one cycle of the signal under observation (50 c/s — 20 ms, 60 c/s — 16.7 ms).

The duration of the time base can be adjusted by switching to different values of capacitance (C_L in Fig. 4-35) and can be continuously controlled by the charging resistor (2.7 MΩ in Fig. 4-35).

The oscillograms in Fig. 4-88 show, each at the starting position of the fine fre-

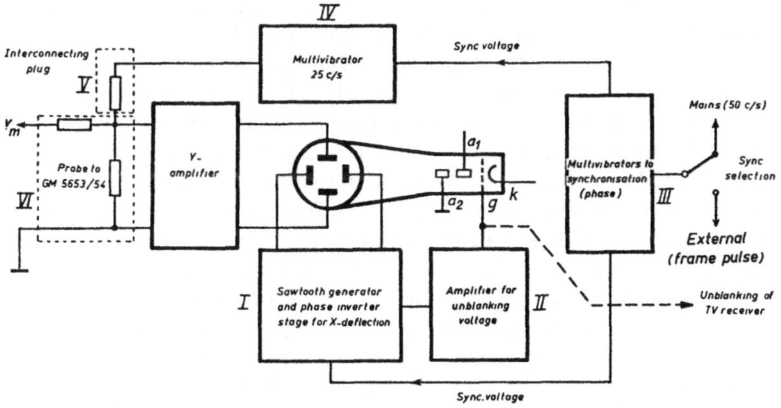

Fig. 4-87 Block diagram of Philips "GM 4584" time base expansion unit

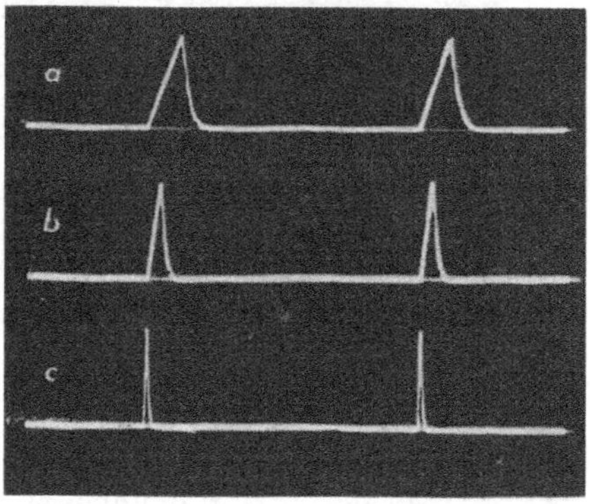

Fig. 4-88 Waveform of deflection voltages of the "GM 4584" time base expansion unit for various settings of sweep velocity

quency control (longest sawtooth pulse), the voltage waveforms during two cycles (40 ms) of the appropriate range.

The portion of the voltage rising linearly with time corresponds in a) to about $2^1/_4$ ms, in b) to about 1 ms, and in c) to about 0,4 ms. These voltages are applied to the X-plates, the normal time base generator of course being switched off. (Triggerable time base units can also be triggered by phase adjustment of the sawtooth voltage of the "GM 4584", a differentiating network being interposed.) A voltage derived from this waveform is amplified via II (second triode of the ECC 81) and used for unblanking the spot during time base deflection. In order to be able to select any desired portion of the input signal, the phase changing range for synchronizing the time base should exceed 360°. This is achieved in the following way: a multivibrator –(III) ECL 80– is locked in phase with the reference frequency (the mains or frame pulse from the source under investigation). The trailing edge of the rectangular voltage from the multivibrator is used, after differentiation, to trigger the sawtooth voltage generator. Thus, by changing the pulse width, the phase of the synchronized pulse can be shifted. Since with only one multivibrator the maximum phase change is less than 360°, two such multivibrators are connected in cascade in this apparatus.

The sync pulses of the first trigger the second, and it is the latter which locks in with the sawtooth voltage generator. In this way the time base can be adjusted in phase with respect to the reference voltage over more than 400°.

In order to be able to represent separately the waveform of the composite television pulse of both fields, a third multivibrator –(IV) ECL 80– generating a symmetrical rectangular pulse with precisely half the frequency (25 c/s or 30 c/s) of the input signal is provided. This multivibrator, too, is locked in with the first multivibrator in III. If the output volage of (IV) is fed to the input of the vertical deflection amplifier, two time base lines are obtained one above the other, similar to the result obtained when using a dual-trace amplifier. The distance between them is determined by the output voltage of the multivibrator and by the gain to which the Y-amplifier is adjusted.

In these oscilloscopes a plug together with a probe forms a resistive T-network. In this way, as indicated in Fig. 4-87, the voltage to be measured $-V_m-$ can be added to the rectangular voltage of the 25 c/s (or 30 c/s) multivibrator without any considerable risk of reaction. The pulse sequences of both fields appear on the oscilloscope screen one above the other. Fig. 4-89a shows a double oscillogram of this sort taken with the smallest time base resolution of the "GM 4584" unit. The phase of the time base voltage was adjusted in such a way as to make the image of the sync pulse combination visible above the frame blanking pulse. The adjustment ranges for the time coefficients for the indivdual stages are roughly as follows:

Stage	Time coefficients
I	30 μs/cm . . . 100 μs/cm
II	12 μs/cm . . . 45 μs/cm
III	7 μs/cm . . . 16 μs/cm

The normal oscillogram of a 50 c/s waveform is equivalent to a cycle time of 20 ms, so that at a picture width of 8 cm the oscillogram corresponds to a time base velocity of $8/20 = 0.4$ cm/ms. From this it follows that the "GM 4584" time base expansion unit is capable of expanding a normal oscillogram by more than 300 times. Of course, the brilliance of the screen image will be correspondingly reduced. For this reason the cathode ray tube is usually operated at rather high acceleration voltages.

The unblanking pulse for the linear part of the sawtooth voltage has a further very useful application. When investigating a composite television pulse, using the time base expansion unit, the pulse sequences during "horizontal" deflections –the "line signals"– appear on the oscilloscope screen according to the time base expansion employed. If the unblanking pulse for the oscilloscope is fed simultaneously to the grid of the cathode ray tube of the television receiver, a part of the image corres-

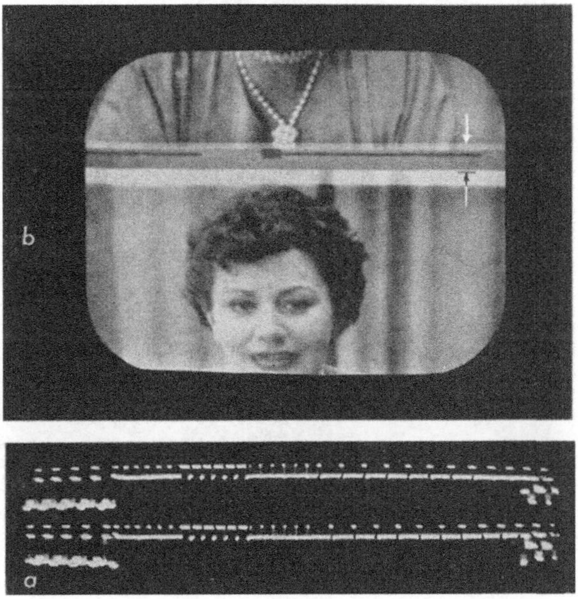

Fig. 4-89 Expanded oscillogram of composite television pulse. *a*) Vertical blanking and synchronizing pulse *b*) corresponding brightened lines on television screen

ponding to the oscillogram under observation will be brightened [29] [30] [31] [32] [33]. It is thus possible to identify clearly the lines in the image corresponding to the oscillogram observed. Fig. 4-89b shows the oscillogram corresponding to the television picture in Fig. 4-89a. As this oscillogram embraces the normally invisible frame blanking, the frame frequency was shifted in such a way that the blanking gap was included in the picture. The first line pulse is identical with the top line in the unblanked strip. As the amplitude of the deflection voltage of the "GM 4584" time base expansion unit is greater than is required for deflection over the whole width of the screen, more lines appear intensity-modulated in the received picture than line pulses seen in the oscillogram. The white arrows in the received picture indicate the lines actually contained in the oscillogram. Further similar pictures are reproduced in Part III, Ch. 19.

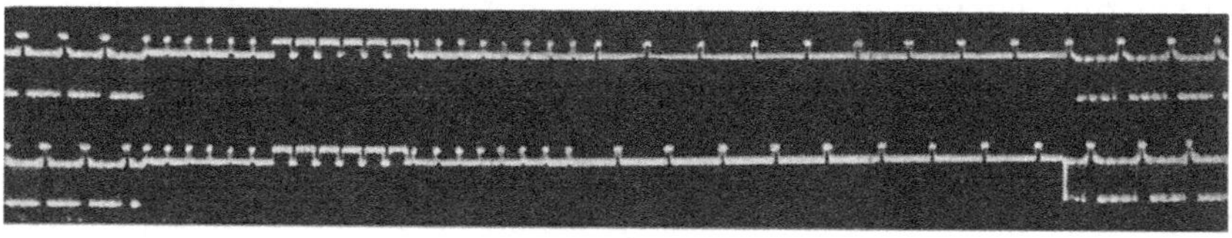

Fig. 4-90 Expanded oscillogram of composite television pulse from Philips "GM 2657" large signal generator. (Three pictures joined together)

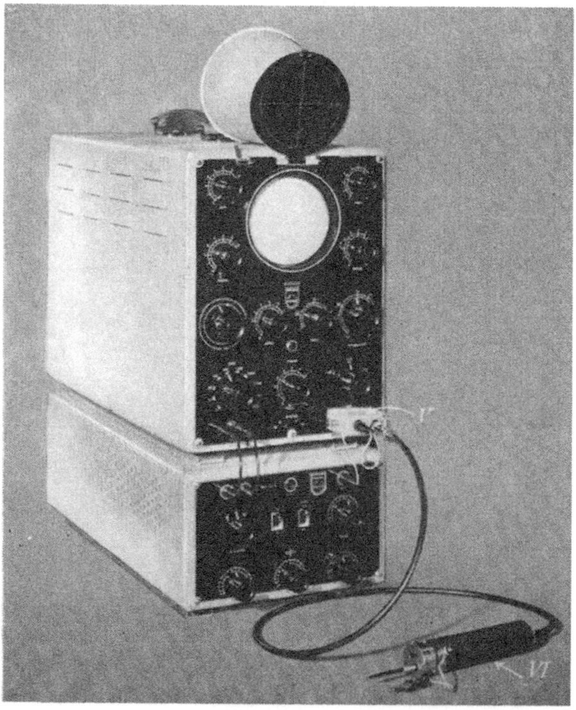

Fig. 4-91 "GM 4584" time base expansion unit with "GM 4585" oscilloscope

In conclusion, Fig. 4-90 reproduces the combination of the signal and the pulse during the frame blanking from the "GM 2657" television signal generator. The complete picture was obtained, however, by joining together three oscillograms with low time coefficient and corresponding phase angles. It can be clearly seen that the frame synchronizing pulse from this signal generator is accompanied by pre- and post-equalizing pulses in accordance with CCIR standards, and in general satisfies the conditions laid down. Mains synchronization cannot be used for low time coefficients, even in a mains-coupled television system, as phase changes in the mains between signal generator and receiver would cause horizontal jitter. In such cases external synchronization with the frame pulse is essential, taken from either the amplitude separator, or from one side of the frame deflection output transformer in the receiver, or from the test picture generator. In repeater synchronizing signal generators with quartz crystal control of the sync pulses –completely in accordance with the standards– external synchronization must always be applied.

Fig. 4-91 shows an external view of the "GM 4584" time base expansion unit together with the "GM 5653" oscilloscope set up for working. The application of this apparatus is, of course, not confined to television. All other voltage changes with repetition frequencies from 50 to 60 c/s can be observed expanded along the time axis e.g. for relay testing, waveforms of electronic control feed circuits etc.

4. 31 Special time deflection process

A time base deflection can also be obtained from sinusoidal voltages. Use is then made of the fact that the rate of change of the sinusoidal voltage in the vicinity of the zero cross-over is substantially constant (in other words, it is linear with time). In most cases, however, it is only the rising portion of the sinusoidal voltage which is used; the descending portion (flyback in the pattern) is blanked out on the control grid of the cathode ray tube by a voltage shifted in phase by 90° with respect to the deflection voltage. As the pattern is only visible during a portion of the horizontal deflection (the sinusoidal voltage is only linear near the zero cross-over), the oscillograms obtained will be time-expanded 3 to 10 times. This method of operation is possible with every normal oscilloscope, in which, for example, a sinusoidal voltage of 50 c/s can be applied to the X-amplifier.

As the pattern in the horizontal direction can be enlarged at least threefold by increasing the amplification, a display which is approximately linear with time is obtained for the middle portion. (The flyback can be suppressed by means of a voltage shifted in phase by 90°). When observing signals which are synchronous with the mains frequency, it is an advantage if the phase of this time deflection voltage can be varied relatively simply (see Fig. 4-69 and Ch. 4.28 on "Phase-delayed triggering of the time base").

As it is very difficult to generate sufficiently large sawtooth voltages with frequencies of 10 Mc/s and above, sinusoidal voltages have been used in a similar way for time deflection at these high deflection speeds. In a unit described by DAMMERS, the use of sinusoidal voltages with frequencies of 60 kc/s to 6 Mc/s made it possible to obtain time coefficients corresponding to sawtooth frequencies of 0.9 Mc/s to 90 Mc/s [34].

Because it was also very difficult to synchronize very high frequency sawtooth voltages with any degree of stability, investigation was limited to the observation of voltage waveforms of the highest frequency by means of photographs of non-recurrent time deflection. With such arrangements it was possible to take photographs of phenomena with frequencies of several thousand Mc/s [35]. This is the only way of working with non-recurrent phenomena. WHITEWAY has described such a high-duty apparatus [36].

4. 32 Sampling oscilloscope (Stroboscope oscilloscope)

Apart from the difficulty, previously dealt with, of generating sufficiently rapid and large horizontal time deflection voltages to display voltage waveforms of very high frequency with a sufficiently small time coefficient, there is usually a limit, in the order of a few hundred Mc/s, to the frequency at which such voltage can be efficiently amplified and also at which effective deflection of the electron beam by conventional means can be maintained. Even when using distributed amplifiers, the upper cut-off frequency is limited to about 100 Mc/s, because a voltage having at least a certain minimum amplitude is required for the deflection of the electron beam. This amplitude can be obtained –at a fairly economical outlay– only by means of a given higher value of the characteristic impedance of the delay line. A higher characteristic impedance, however, again inevitably limits the upper cut-off frequency of the amplifier section (section "Distributed amplifier"). In this way, at a correspondingly increased cost and by use of a cathode ray tube with subdivided Y-deflection plates (travelling-wave deflector system), a 1000 Mc/s oscilloscope has been evolved (Tektronix "519"), but the picture amplitude is limited to 2 cm, the smallest time coefficient is 2 ns/cm, and the Y-deflection coefficient (without amplifier) is 10 V/cm.

As has already been described in Ch. 2.8 "Influence of electron transit time on deflection sensitivity", errors in deflection occur in the case of high-frequency deflection voltages when the transit time of the electron beam within the deflection plates is no longer negligibly small compared with the duration of the cycle, these errors increasing rapidly as the frequency increases. However, GERMERSHAUSEN, GOLDBERG, MCDONALD, PIERCE, M. VON ARDENNE and others have described special oscilloscope tubes with a very small, sharply defined luminescent spot and specially formed deflection electrodes, which make it possible to obtain interpretable deflections at 1000 Mc/s and more [36] (see also bibliography [19] and [20] of Section 2 as well as [3] of Part II, Ch. 7). All the above equipments, however, on account of their special properties and relatively high costs, are suitable only for special laboratory applications. Moreover, there are still a number of disadvantages which remain to be overcome, such as very small picture dimensions, poor brightness when the repetition frequency is low, danger of exposure to X-rays and a still insufficiently small time coefficient.

The only solution of these difficulties would be to transform the high-frequency signal voltage to a low-frequency at which investigation can be readily carried out by conventional means. This is the method followed in the sampling oscilloscope [37]. This process not only differs from that in conventional oscilloscopes in so far as a different kind of time deflection is concerned, but also by the method in which the Y-deflection is carried out –proportional to the signal amplitude– since in this equipment an additional way was found to rich very small time coefficients (0.1 ns/cm and less). It is for this reason that the whole process will be dealt with in this section.

4.32.1 FUNDAMENTAL METHOD OF OPERATION OF A SAMPLING OSCILLOSCOPE

In a large number of applications, particularly in the investigation of phenomena of extremely short duration, the cathode-ray oscilloscope is used to display the waveform of the voltage under investigation as a function of time. The sampling oscilloscope which is intended for such use, differs from conventional oscilloscopes in that the voltage waveform is not directly displayed as the continuous path of the luminous spot, but at every repetition of the process only one point of the curve path is displayed, corresponding to a given instantaneous value of the signal. One very important difference from the conventional oscilloscope is that the voltage under investigation itself must not be amplified, but instead it is scanned relatively slowly point by point at a great number of spots by means of extremely short pulses. These pulses, which are amplitude-modulated by the vo!tage under investigation, are later spread out and amplified.

To ensure that the pulse edge of the voltage waveform is completely displayed from the outset, the method normally used with HF oscilloscopes can be employed, that is to say the signal voltage directly triggers the time deflection process, although a delay line is included in the lead to the sampling device and subsequent Y-deflection. Most sampling oscilloscopes, however, give a pre-pulse, with which the individual cycles of the signal process can be triggered at time t_v before the sampling process. This method is used whenever possible as a delay line always involves some limitation of the frequency range and is, moreover, only available at low characteristic impedance (50 Ω). In contrast to this, the direct connection to the sampling probe gives a considerably higher impedance and a correspondingly smaller load on the signal voltage source.

As the sketch in Fig. 4-92 indicates, the trigger pulse, after the commencement of the time deflection, initiates the process at time t_0, and the sample after a given interval of time t_v triggers (see also Fig. 4-93a and b and the accompanying text). Amplitude A of this sample pulse corresponds to the average value of the sampled voltage over the duration of the sample pulse. The luminescent spot is displayed in the corresponding position on the screen of the cathode ray oscilloscope and indicates this amplitude, as will be explained in more detail. When the signal voltage starts once more, the same sampling takes place, but a short time later. Corresponding to this, the time deflection is not continuous, but it takes place in little steps. Accordingly, the deflection voltage has the shape of a staircase. Each step corresponds to a pulse sample, so that the luminous spot on the screen also moves forward a little each time in the X-direction as well. Once the whole voltage curve has been sampled the whole process is repeated cyclically. The time coefficient of the oscillogram determines the selected voltage speed of a sawtooth voltage, which is only used in a coincidence circuit for this purpose but not for

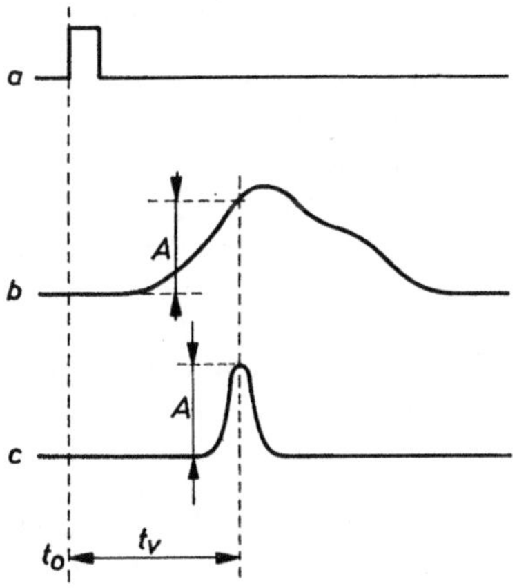

Fig. 4-92 Basic method of operation with pulse sampling
a) External trigger pulse
b) Voltage form under study
c) Sampling pulse

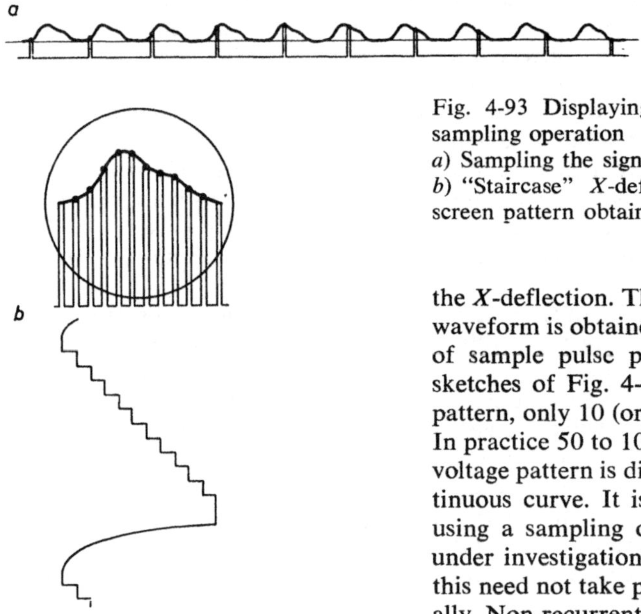

Fig. 4-93 Displaying the screen pattern in pulse sampling operation
a) Sampling the signal voltage
b) "Staircase" *X*-deflection and structure of the screen pattern obtained

the *X*-deflection. Thus the pattern of the voltage waveform is obtained on the screen from a series of sample pulse patterns, as indicated in the sketches of Fig. 4-93. In order to simplify the pattern, only 10 (or 12) sample points are taken. In practice 50 to 1000 steps are used, so that the voltage pattern is displayed as a more or less continuous curve. It is, of course, a condition for using a sampling oscilloscope that the process under investigation occurs repeatedly, although this need not take place regularly, that is, cyclically. Non-recurrent processes cannot be displayed, however, for then only one single point in the voltage waveform would be obtained.

The sampling oscilloscope can therefore be regarded as the electrical equivalent of the principle of the stroboscope for the investigation of mechanical rotational and oscillating movements. An oscilloscope of this sort is therefore occassionally described as a "stroboscopic oscilloscope". When, in normal optical stroboscopic observation, the succession of flashes of light always strikes the rotating object, for instance, at the same phase of movement, the object appears to be stationary. But if the repetition frequency of the light pulses is somewhat slowed down or slightly accelerated at every rotation, that is to say, if there is a difference between the speed of rotation and the frequency of the flash, the observer gains the impression that the movement of the object is taking place at a considerably reduced speed. The same principle is the basis of the method of operation of sampling oscilloscope. In both processes the solution is dependent on the shortness of the flash of light or on the duration of the sampling pulse. Moreover, both stroboscopic observation and sampling oscilloscopy require storage effect. In stroboscopic observation this is achieved by the inertia of the eye; in the case of the sampling oscilloscope it is obtained by means of integration, special storage devices and a pulse stretcher.

The use of the principle of sampling for observing electrical processes is not new and was in fact used more than a hundred years ago. An interesting survey of the history of this technique is to be found in a work by REEVERS [38]. By means of the process mentioned in that work, the limitations in frequency of the measuring devices available at that time were successfully overcome.

The decisive advantage of the pulse-sampling method over the conventional oscilloscope consists mainly in that the actual high-frequency and short-pulse technique is limited to the sampling probe and the generation and supply to this probe of a sampling pulse which is as short as possible. All else can take the form of conventional techniques without any special difficulties being encountered.

Fig. 4-94 shows the circuit of a sampling stage of the oscilloscope as described by MCQUEEN in 1952 [39]. Such stages are usually made in the form of a probe which can be applied directly to the signal voltage source. The oscilloscope described in the publication quoted earlier had the quite respectable bandwidth of 300 Mc/s, corresponding to a rise time of 1.2 ns. In the circuit shown in Fig. 4-94 the valve is normally cut off by a correspondingly high grid bias. The voltage under investigation is applied to its control grid, either directly (connection "1") or attenuated by a capacitive divider (connection "2"). The short, negative voltage pulses of the pulse generator (width about 1 ns) are fed to the cathode of the valve. By this means it is conducting for a short period. Anode current pulses occur, whose amplitude is determined by these of the generator and of the signal voltage across the control grid at the instants of the individual pulses. There is thus a sequence of pulses, whose amplitude variations correspond to the variation of the signal voltage at the individual sampling instants. To ensure that the variation of the amplitude of the change in signal voltage in this combined process is as linear as possible at the instant of sampling, the modulation must take place in the linear part of the valve characteristic. The modulation depth is usually less than 25% of the average pulse amplitude. The sampling valve therefore functions in this case as a mixer stage. In the sketches in Fig. 4-94 this has been taken into account. On the other hand the diode sampling circuit, which will be described later, works as a switch which, at the instant of sampling, switches the signal voltage to a capacitance for a short time.

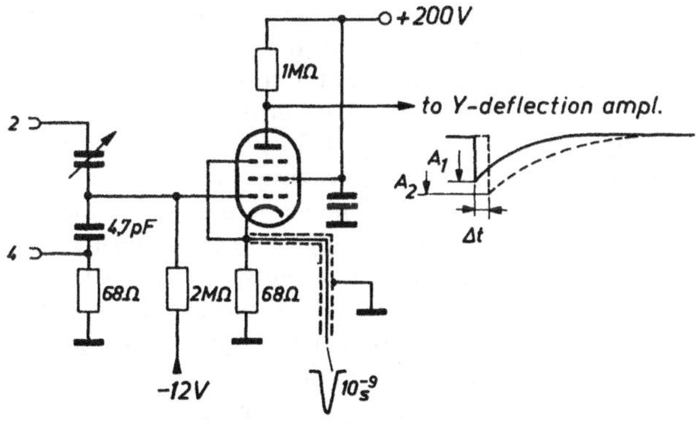

Fig. 4-94 Sampling-mix circuit with electron tube (according to J. G. McQueen)

These pulses are spread (integrated) by the capacitance across the anode and are further amplified in the subsequent amplifier. As the circuit diagram of such an arrangement in Fig. 4-95 shows, the amplifier is followed by another storage circuit and a pulse expander which lengthens the voltage pulse still more, after which the pulse is fed to the *Y*-amplifier. (The height of the individual spread pulses continues to be proportional to the sampled signal voltage amplitude, see A_1 and A_2 in Fig. 4-94). Brightening of the pattern on the screen takes place at the same time, and this lasts until shortly before the next scanning process. By this means the picture brightness remains constantly good, independent of the repetition frequency of the process under study. For the same reason, normal cathode ray tubes with average acceleration voltages

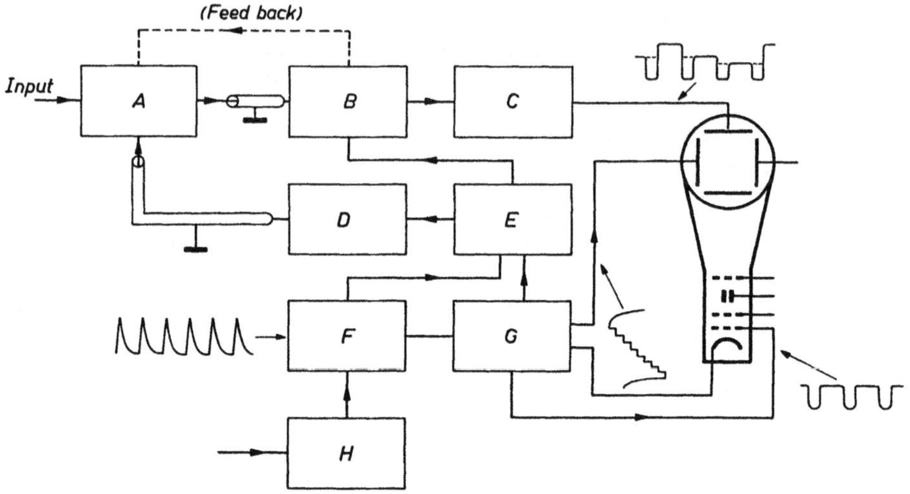

Fig. 4-95 Circuit diagram of a pulse sampling oscilloscope (possibilities of additions indicated in dotted lines or in brackets)

(about 4 kV) can be used to obtain a screen pattern of satisfactory brightness. To ensure that the time interval between successive sampling pulses is slightly increased every time, the pulses are released through a circuit in which the "staircase" voltage for the X-deflection is added to and compared with the voltage of the sawtooth voltage generator which determines the time coefficient. This can be done by a Schmitt-trigger for instance, to both of whose grids these voltages are connected. In the sketch of the voltage waveform of Fig. 4-96 it can be seen in *e*) that in this way the triggering instant for releasing the sampling pulse is reached in every single sawtooth at a given amplitude difference and hence with a given increase of delay time.

The sawtooth passes through one cycle during each step, its duration corresponding to the interval of time (Δt) between two successive samplings. The ratio of the sampling duration of the whole voltage waveform to its repeating time determines the number of picture points which go to make up the oscillogram (Figs. 4-93 and 4-99).

Hitherto, only one of many possible variants of this process has been described. A circuit diagram showing the fundamental construction of such a sampling oscilloscope can be seen in Fig. 4-95. Some details will be dealt with more fully later. Many developments and improvements have been introduced and described. Thus, for instance, two voltage waveforms which are independent of one another can be displayed on one picture by alternately switching between two sampling equipments. In the apparatus treated in the publication by MCQUEEN use is made of such double picture display. For this purpose two complete sampling equipments must be available, their outputs being switched alternately to the Y-plates of the cathode ray tube by means of an electronic switch.

A number of accessory units (Tektronix plug-in *N*) have been brought out or described, making it possible to make use of the sampling technique with conventional HF oscilloscopes [40] [41] [42]. Whereas the oscilloscope described by JANSSEN and

Fig. 4-96 Pulse diagram of a sampling oscilloscope
a) Signal voltage cycles; *b*) trigger pulses; *c*) sampling pulses; *d*) sampled voltage pulses; *e*) comparison of sawtooth voltage and step voltage for triggering the sampling pulses; *f*) voltage waveform on the second storage circuit; continuous lines without feedback, dotted lines with feedback; *g*) unblanking pulses

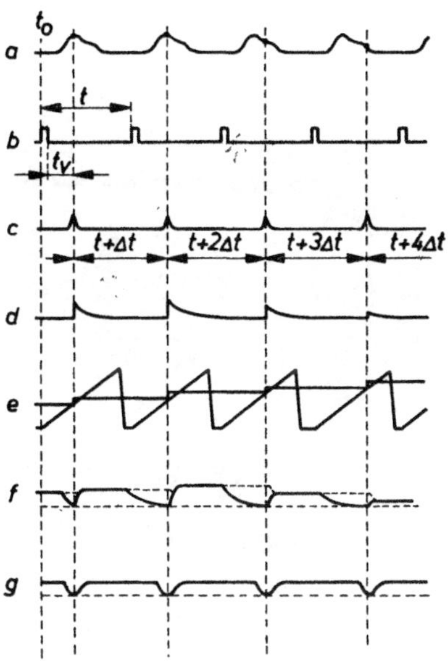

MICHELS [37] was intended for higher repetition frequencies (about 100 kc/s), the apparatus described by McQueen is equipped for repetition frequencies of 100 to 4000 c/s. A unit described by FARBER [43] operates on repetition frequencies up to 30 Mc/s. A sampling oscilloscope with a rise time of 0.4 ns corresponding to a cut-off frequency of about 900 Mc/s has been produced by the firm of LUMATRON and described by BUSHOR [44]. Microwave diodes are used in this sampling circuit. In addition, transistors are very widely used in all such apparatus.

The sampling oscilloscope, type 185 B/187 B, produced by the firm of HEWLETT-PACKARD, should be included in the class of high-duty apparatus of this sort [45] [46] [47]. Its performance is characterized by a rise time of 0.45 ns and by the upper cut-off frequency, which is given as 1000 Mc/s. As a number of particularly interesting features are included in this apparatus, indicating ways in which the maximum capacity of a sampling oscilloscope can be exploited, it will now be dealt with in more detail.

4.32.2 HEWLETT-PACKARD OSCILLOSCOPE, TYPE 185 B/187 B

In this apparatus, an external view of which is illustrated in Fig. 4-97, the special suitability of avalanche transistors and of switch-diodes for rapid switching operations has been exploited to the full. (SUGARMAN had, earlier, in 1957, described the use of microwave diodes for the sampling circuit [48]). In Fig. 4-97 the two-channel unit type 187 B is used for displaying a dual trace, for which purpose two sampling probes are also employed. The two channels can be used either as independent channels for displaying the dual traces, or as a difference amplifier (Section 5.27 "DC voltage amplifiers"). The probes each contain a four-diode sampling circuit with a cathode-follower, the fundamental circuit of which can be seen in Fig. 4-98. In the apparatus itself this is followed by a switchable voltage divider in the cathode lead of the cathode-follower, by means of which switching can be carried out in five stages with deflection coefficients of 10 mV/cm to 200 mV/cm. By means of a fine adjuster the deflection coefficient can be reduced to 3 mV/cm. For still greater deflection coefficients and hence for higher input voltages, an external voltage divider should be used. The input impedance of the probe is 100 kΩ, 3 pF.

The sampling pulse generator (Fig. 4-95) supplies a symmetrical gate pulse, about 1/3 ns in width, to the four-diode gate circuit in the probe, from a blocking-oscillator

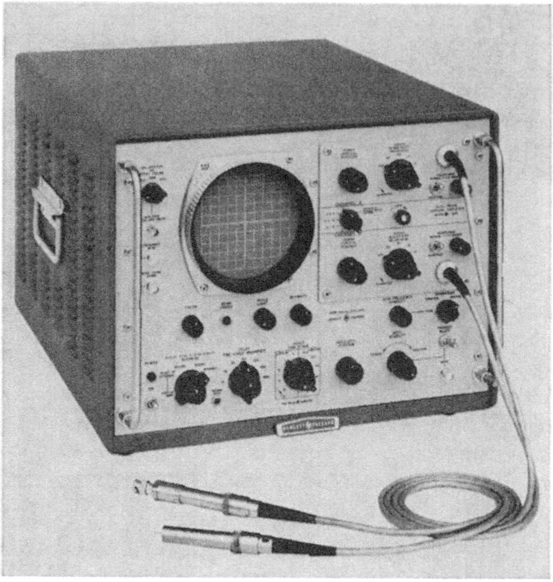

Fig. 4-97 Hewlett-Packard sampling oscilloscope, type 185 B/187 B with two sampling probes

Fig. 4-98 Four-diode sampling circuit in the probe of the Hewlett-Packard oscilloscope 185 B/187 B

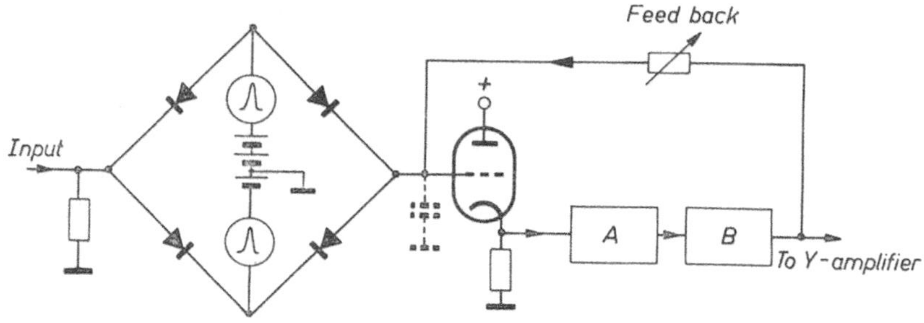

with avalanche transistors. There is a pin in the immediate vicinity of this gate circuit for contact with the signal voltage source. The gate circuit is blocked by a gate voltage of several volts and made as symmetrical as possible with respect to the sampling pulse voltage source. This ensures a minimum of feedback into the input source from the sampling bridge and also achieves a minimum of pedestal generation and a maximum dynamic range in each of the gating circuits. Because of the extremely short gating pulses, the diode circuit is open for only a short time in each case, and therefore operates as a switch. Hence the input capacitor of the cathode-follower is connected for a short time to the signal voltage and charged by current pulses the amplitude of which is proportional to the amplitude of the signal voltage averaged over the duration of the pulse during the sampling time. These charges are stored here in accordance with the time constant of the cathode-follower input. The cathode-follower of the sampling stage is followed by an amplifier with a pulse stretcher and another storage circuit in which pulses of standard width are forced. A special feature of this apparatus is that this storage circuit has adjustable positive feedback to the grid of the cathode-follower of the input circuit (shown in dotted lines in Fig. 4-95).

The loop gain in the feedback circuit is usually adjusted as accurately as possible to unity. This feedback prevents the charge stored on the grid capacitor of the cathode-follower from leaking away too quickly. The effect of a large time constant is obtained, although only a small charge is required for the relatively small input capacitor. By means of the feedback from the second store via a large time constant, on the other hand, the input voltage of the cathode-follower remains at a practically constant level until the next scanning. This has very considerable advantages. If the voltage on the sampling probe were constantly to fall back to the zero level as a result of small time constants, then the voltage source would have to supply an amount of charge at each new sampling, which would correspond to the full potential from zero to the relative instantaneous value (shown in Fig. 4-96 *d* and *f* in the form of an extended curve). If, however, the input voltage is maintained at the level of the preceding scanning by means of the feedback from the second store –after amplification – then the signal voltage source has to supply at the next sampling only an amount of charge corresponding to the potential difference. This waveform is completed in the voltage diagram shown in Fig. 4-96 *f* by a dotted line. With a sufficient number of constant voltage waveforms and with 1000 samplings per picture (100/cm), this potential difference is relatively small, so that the result is a considerably smaller load on the signal voltage source than would be the case without feedback. Since only small voltage changes now occur across the grid of the cathode-follower, considerably less noise is experienced into the bargain. There are also the following additional advantages: 1) The value of the sampled voltage is to a considerable extent independent of the constancy of the sampling pulses and the pre-amplifier. 2) The diode characteristics have a smaller influence on the sampling pulses. Since in this way only changes in the signal voltage make reversal of the charge necessary, this circuit can sample even extremely large signal amplitudes capable of rising to the level of the cut-off voltage of the diodes (about $\pm$ 2 V). The dynamic range which can be covered is thus about 1000 : 1, while the noise level is only about 2 mV. Self-excitation of this system is prevented by synchronized blocking of the feedback loop.

In applications in which the signal voltage fluctuates irregularly about an average waveform, or where it includes a large noise component, or even if its starting time fluctuates a great deal (jittering), it is possible, by reducing the loop gain and correspondingly adjusting the sampling, to average several charges for the display of one point of the curve patterns. By this means the pictures obtained are more stable, and the noise is reduced by about 2/3. The noise level estimated at 2 mV (in normal operation) then sinks to below 1 mV; jitter, assumed to be about 0.1 ns, is now reduced to 0.05 ns. The "normal-smoothed" switch is used for selecting this type of operation.

A particular advantage of the pulse-sampling system is its considerable greater ability for overdriving as compared to amplifiers in conventional oscilloscope. This is due to the fact that there are relatively long pauses between the individual samplings during which the amplifiers can return to their normal conditions of operation. In this way its is possible to expand patterns of signal voltages in the *Y*-direction as well, and thus to observe clearly small changes in voltage taking place in the course of complex voltage waveforms. While, in the case of conventional wideband oscilloscopes, the picture height must generally remain limited to 4 or 6 cm (see Ch. 2.13 on "Cathode ray tubes with particularly high deflection sensitivity of the *Y*-plates for wide-band oscilloscope"), the relatively wide pulses of low frequency with which the vertical deflection is carried out in sampling oscilloscopes can readily be amplified to voltage amplitudes for pattern heights of more than 10 cm. The possibilities for overdriving which have been described permit also this vertical stretching of the pattern.

In the technique used for measuring short pulses it has been found to be of very special advantage if two voltage patterns can be displayed at one and the same time, and this can be done without particular difficulty with the sampling oscilloscope. All that is required is two independent sampling channels operating alternately on the main Y-amplifier (in Fig. 4-95 only one of the amplifiers is shown). In order to obtain the best possible synchronism in alternate switching, both sampling probes are controlled by a common sampling pulse generator. If the sampling probes and their connecting cables are made as nearly alike as possible, both mechanically and electrically, the time errors between the two samples do not exceed 0.1 ns. By means of additional balancing it is possible to reduce the time difference still further –to less than 0.05 ns.

If an external trigger signal is used, it must be connected to one of the input sockets on the front panel of the oscilloscope, terminated by 50 Ω. Its amplitude in this unit must be sufficient for pulse durations greater than 20 ns, i.e. at least + 50 mV or —50 mV. For pulses of a duration of 1 ns an amplitude of +0.5 V or —0.5 V is required. If this load is too high for the triggering voltage source, an input voltage divider with an input resistance of 1000 Ω can be connected. The trigger signal is first fed to a buffer to prevent feed-back to the trigger voltage-source. Otherwise instability could occur, especially at high repetition frequencies. Then follows a transistorized blocking oscillator, operating a hold-off circuit which does not respond to trigger signals until all circuits in the time base unit have returned to their starting condition. The highest repetition frequency permitted by this discrimination circuit is 100 kc/s. Synchronization can still be carried out, however, with multiples of this frequency, up to 1000 Mc/s. This is rendered possible by making available only a certain fraction of the trigger voltage cycles across a frequency divider via a hold-off device in the circuit for triggering. It is also possible to allow the blocking oscillator to be self-oscillating. This is convenient for calibration purposes. The blocking oscillator then triggers a transistorized pulse generator, which in turn trips an oscillating circuit (a normal LC-resonant circuit for 50 Mc/s, or a tank circuit for 500 Mc/s). These voltages are available at sockets on the left of the oscilloscope front panel; the maximum error in frequency is 1%. A transistorized multivibrator also supplies a square-wave voltage which can be used for amplitude calibration; the limit of error of the calibration voltage amplitude is then 3%. The blocking oscillator of the trigger-pulse shaper generates a square pulse the duration of which is greater than that of the longest sawtooth cycle corresponding to the highest range of the time coefficient. The sawtooth voltage is both triggered and terminated by this voltage. The actual time base circuit ensures that each single sampling takes place a short and constant time after the preceding one. This is done by a voltage comparison circuit, or "coincidence circuit", in which the sawtooth voltage and the staircase voltage for the X-deflection are compared. As can be seen from sketches c and e of the diagram of the voltage waveform in Fig. 4-96, this circuit responds later in every sawtooth voltage cycle by a time difference Δt. The sampling pulse generator produces the pulse at this instant. If the staircase voltage has then reached its highest value, the circuit elements are discharged and the process thereupon commences anew. The staircase voltage is produced by charging a capacitor with pulses via a diode circuit. By changing the values of this capacitor the staircase steps, and hence the number of points on the patterns, can be altered. A large capacitor results in short intervals, but requires a longer scanning time for the whole pattern.

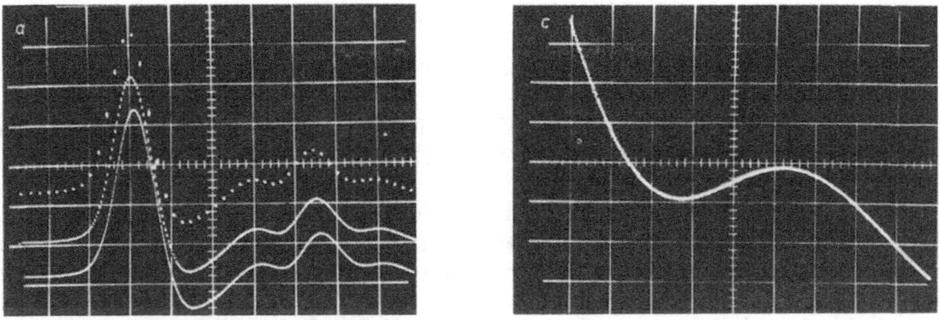

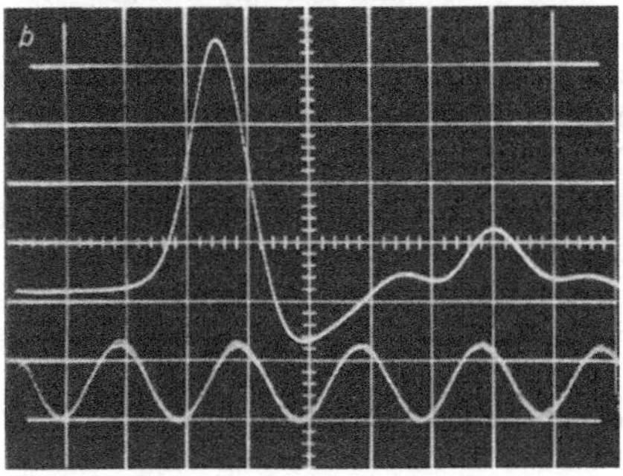

Fig. 4-99 Oscillograms recorded with the Hewlett-Packard sampling oscilloscope type 185 B/187 B

a) pulse displayed by 50, 200 and 1000 picture points. time coefficient: 1 ns/cm

b) double oscillogram. Pulse of *a*) with 1000 picture points and time calibration voltage 500 Mc/s, time coefficient: 1 ns/cm

c) decay at the end of the pulse in *b*), time coefficient: 0.2 ns/cm

According to requirements the pattern can be composed of 50, 200 or 1000 points, as indicated in the oscillograms in Fig. 4.99*a*. For signal voltages of low repetition frequencies it is generally necessary to be content with fewer pattern points, as otherwise the pattern repetition frequency will be too low.

If the pattern consists of 50 sampling points and the signal frequency of 50 c/s, for instance, a pattern is obtained only once every second. Under such conditions the pattern on the screen can only be satisfactorily observed if a long-persistence screen, the *P*- (*P*7) screen is used. For exact interpretation, especially when more picture points are required and hence a slower pattern repetition frequency results, photographic recording is essential.

Another feature of this oscilloscope is that a variable direct voltage can be led to this point of the circuit by means of a potentiometer on the front plate of the oscilloscope. It is then possible to scan the voltage waveform of the screen pattern by hand. For this method of operation connections at the back of the apparatus are provided for a *X*-*Y*-recorder (potentiometer recorder); available for *X*-direction 17 V at 1000 Ω and *Y*-direction ± 1 V at 2000 Ω. It is thus possible by turning the knob of the DC potentiometer, to record every phenomenon displayed on the oscilloscope, even the most rapid, on a slowly recording apparatus. Various special features of the conventional oscilloscope, such as picture magnification in the horizontal direction and time-delayed triggering, can also be obtained in sampling oscilloscopes. As these processes involve only the time-base circuit, and have no direct influence on the structure of the pattern on the screen, the sampling oscilloscope has the advantage over conventional ones in that the normal brightness of the pattern remains unchanged even when magnified.

In the Hewlett-Packard oscilloscope 185 B here described there are four basic time coefficients ranging from 10 ns/cm to 100 ns/cm and a 7-stage multiplier having ratios between 1× and 10×. The smallest calibrated time coefficient range is 1 ns/cm. By expansion it is possible to obtain up to 0.1 ns/cm. By means of the adjuster for time delay (time-delayed deflection), it is possible, by using a smaller time coefficient in a way similar to that in conventional oscilloscopes, to observe any required section of a pattern displayed without any reduction in spot brightness. The oscillograms in Fig. 4-99 give a number of examples of the results obtainable with such an sampling oscilloscope.

As has already been mentioned in the general description of the sampling oscilloscope, there is also an external control pulse available with this apparatus, which makes it possible to trigger the signal process in a controlled manner as distinct from delayed triggering of the commencement of time deflection. This pulse is also available at a socket on the front panel. It has an amplitude of —1.5 V and a rise time of 4 ns; the source resistance amounts to 50 Ω. If the time base circuit is self-oscillating, the frequency is about 100 kc/s; if separately controlled, it always corresponds to the sampling frequency.

For direct triggering by the signal voltage, the required delay for the *Y*-deflection must be obtained by connecting a 50 Ω coaxial delay cable between the signal voltage source and the sampling probe, the time delay then being 120 ns.

The above description of the Hewlett-Packard instrument, type 187A/187B, is intended as a typical example of such apparatus.

Similar units have been put on the market by Lumatron (type 112) and by Tektronix, as has already been mentioned. The last-named firm supplies plug-in units suitable both for the large basic equipments for plug-in assembly (530 and 540 series) and for the medium 560 series, which also operate on the sampling principle (plug-ins *N*, as "3 S 76" and "3 T 77").

The special sampling oscilloscope type 661 is capable of meeting the most stringest demands. The two-channel plug-in unit 4 S1 for vertical deflection gives a rise time of 0.35 ns (= 350 ps) with deflection coefficients of 2 mV/cm to 200 mV/cm using the scanning device, corresponding to a cut-off frequency of 1000 Mc/s. The time base plug-in unit 4 T1 makes available time coefficients in 16 calibrated stages from 1 ns/cm to 100 ps/cm. When expanded 100-fold, counting speeds of 0.1 ns/cm (= 100 ps/cm) can be attained.

No doubt many further types of oscilloscopes on the sampling principle will be developed from time to time. The particular advantages offered by this type of apparatus should also provide a demand for low-price equipment of limited performance.

4.33 Calibrating the time scale

In order to interpret an oscillogram accurately, it is also necessary to know the time coefficient exactly. In so far as the photographic recording of non-recurrent or cyclical waveforms triggered by means of time base circuits is concerned, the calibration of the screen image time coefficient is relatively simple. Either before or after the waveform to be studied is photographed an oscillogram of an alternating voltage from a suitable generator, the frequency of which is known with the required accuracy, is taken.

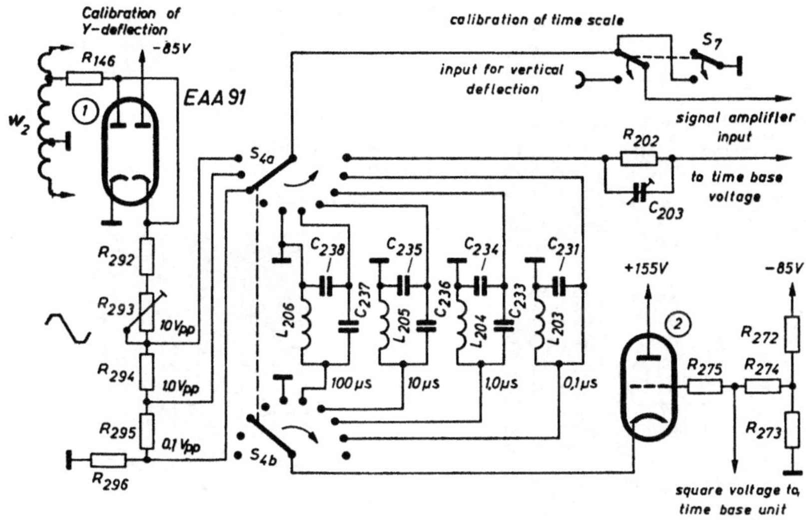

Fig. 4-100 Circuit for calibrating time scale and vertical amplification in the "GM 5662" HF wide-band oscilloscope

This method was used in several of the oscillograms reproduced in this book to illustrate the practical examples in Part III (Figs. 15-2 and 15-3 and also Figs. 28-8, 29-1, 29-3, 29-5, 29-6, 29-7 and 29-8).

In the case of oscilloscopes in which the determination of the time coefficient is of special importance, suitable "calibrating" oscillators are incorporated in the equipment. For example the Philips "GM 5662" oscilloscope includes the so-called "start-stop" oscillator shown in the diagram reproduced in Fig. 4-100. Four resonant circuits, which can be selected by means of switch (S_4), are included in the cathode lead-in of a cathode-follower stage (2). The grid of the cathode-follower is driven by the square-wave voltage from the sweep unit (oscillogram 4-45*b*). During the part of each square-wave cycle in which the grid voltage is positive, current flows through the cathode-follower tube. Its output resistance is then low, so that the oscillating circuit switched in the cathode lead is heavily damped, and no oscillations can be generated. When the grid voltage of the cathode follower corresponds to the negative portion of the square wave, the cathode follower is blocked, thus removing the damping on the oscillating circuit. At the same time the circuit is pulsed by the voltage rise so that it begins to oscillate at its fundamental frequency. These oscillations decay slowly at a rate depending on the circuit damping. By comparing the pattern of the decaying oscillation with the oscillogram to be interpreted, the time coefficient of the latter is obtained. What is notable in this case is that, even should the time deflection happen to be non-linear, the interpretation remains accurate since it applies to both patterns. As an example of this type of calibration of the time coefficient Fig. 4-101 shows the oscillograms of the time-expanded rising flank of a rectangular pulse and of the 10 Mc/s voltage (TC = 70 ns/cm). In *a* the trace of the calibration voltage is shown below the actual oscillogram, but if the calibration voltage is written into the oscillogram, as in *b*, it is then possible to enlarge the oscillogram under investigation. This is advantageous, since the vertical deflection in wideband oscilloscopes is usually

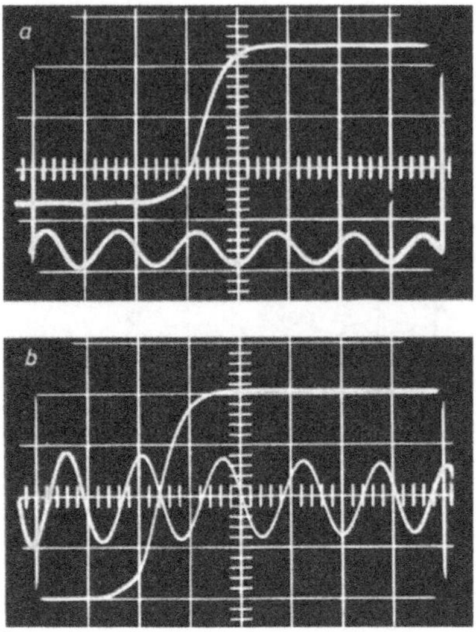

Fig. 4-101 Calibrating the time scale by double recording of decaying voltage of an excited resonant circuit (frequency 10 Mc/s corresponding to 0.1 μs).

a) Recording the calibration voltage below the image recording;

b) Calibration oscillogram photographed into screen image

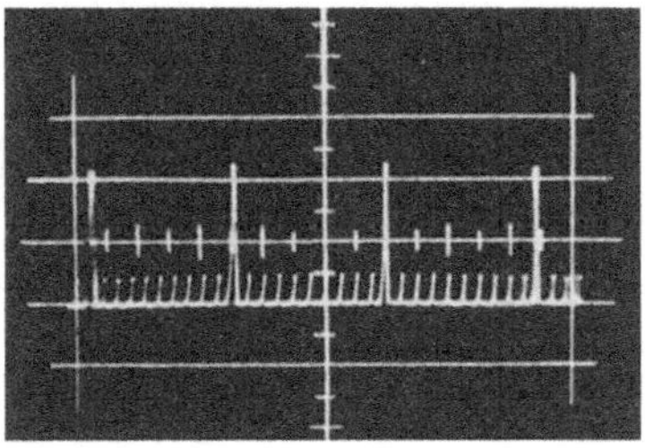

Fig. 4-102 Calibrating by means of a "timing comb"

limited. Special pulse voltage generators are used instead of sine voltage generators, and usually incorporate special circuits to ensure that the amplitude of every tenth pulse is greater than that of the others. Fig. 4-102 shows such an oscillogram. The calibration of the time axis, with its pulse decades, is very easy to follow and interpret [49]. When greater accuracy is required, the time marker indicators are controlled by quartz crystal oscillators. These oscillators are generally supplied as additional units; their highest frequency error is less than 0.03%.

The calibration of the time coefficient by means of intensity control of the oscillogram by an alternating voltage of known frequency is also widely used (Part II, Ch. 10). For this, the intensity of the oscillogram is modulated from bright to dark by changes of the polarity of the calibrating voltage. This intensity control voltage must also be generated by start-stop oscillators, to ensure that the start of the alternating voltage cycles is always in phase with the sweep and that stationary marks are obtained even in cyclic processes.

The oscillograms in Fig. 4-103 *a* and *b* are examples of greatly time-expanded

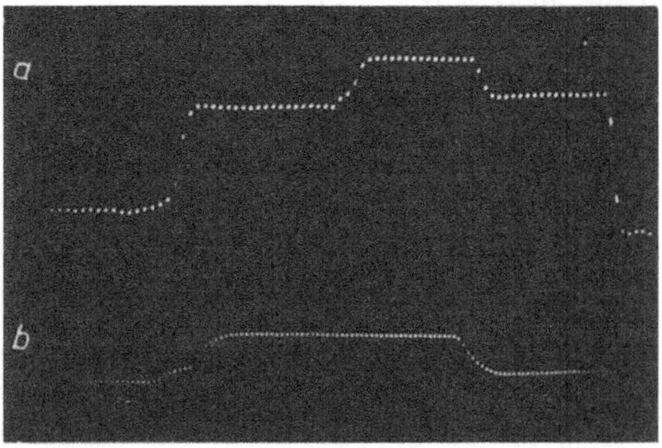

Fig. 4-103 Calibrating time scale by intensity modulation

a) TV signal voltage intensity modulated at about 3.2 Mc/s ($^1/_2$% of the duration of the line deflection).

b) Intensity modulation at 10 Mc/s

oscillograms of a sequence of television pulses, the line synchronizing waveform of which it was particularly desired to obtain in *a* clear form. These oscillograms were made with the aid of brilliance control. The duration of the part shown in *a* is about 25 μs, and in *b* only about 12 μs. In a the time mark intervals were exactly 1/2% of the total duration of a line deflection. (64 μs), which corresponds to a frequency of 3.2 Mc/s. For the even more expanded picture in *b*, which only shows the actual sync pulse, the frequency of the intensity control voltage was 10 Mc/s, the distances between consecutive points corresponding to 0.1 μs. Although in *a* only two intensity modulations occur in the rear flank of the waveform –the intervening parts of the waveform remain invisible–, in general, time marks with the smallest interval of 0.1 μs (10 Mc/s) are adequate. This type of time coefficient calibration could not, however, satisfy the requirements of the short time coefficients possible with modern wideband oscilloscopes. The shortest time coefficient in the Philips "GM 5602" HF wideband oscilloscope, for example, is 0.04 μs/cm. The rise time of the signal amplifier in this apparatus is less than 0.025 μs. In displaying the pulse of a high-performance pulse generator whose rise time is about 0.04 μs, an oscillogram is obtained with a rise time of barely 0.05 μs. With 10 Mc/s time marking (0.1 μs), the gap between the intensified marks would be twice as great as the rise time to be measured.

Since it is very difficult to obtain satisfactory brilliance modulation with frequencies higher than 10 M c/s, it was decided, for instance, in the "GM 5662" oscilloscope, to calibrate the time coefficient by switching from the signal voltage to the calibration voltage. In this way the calibration takes the form of a second trace, without interrupting the oscillogram of the waveform under observation (Fig. 4-101 *a* and *b*).

The switch is so designed that it has minimum capacitance, and therefore the input capacitance of the circuit is only slightly increased. The switch is operated by pressing a button, shown in on the right, lower corner of Fig. 1-2, so that the waveform being investigated and the image of the calibrating voltage are visible alternately.

For most new types of oscilloscope a given time coefficient accuracy is guaranteed under specified conditions (Fig. 1.4). Calibration of the time coefficient is therefore not provided for. This requires a great care in design and manufacture, and particularly close tolerances of the circuit elements, to ensure that the time base velocity it not merely within these limits but is maintained at its absolute value. The linearity of the time base must also be such that the additional deflection errors caused thereby do not exceed these limits.

STEINBERG [50] made an interesting suggestion for the calibration of the time axis, which should be mentioned at this point. An electronic switch is used to blank the signal voltage in every second time base cycle, and simultaneously the horizontal line thus obtained is intensity-modulated by means of a time marker generator. Fig. 4-104 shows this circuit. An uninterrupted trace of the waveform being studied can thus be obtained, together with a "time scale", which can be "applied" to the oscillogram by vertical shifting. Fig. 4-105 gives an impression of the screen image thus obtained. If the voltage required for this vertical shift were also made adjustable and was also calibrated, then not only the calibration of the "time coefficient", but also the measurement of the screen image in the vertical direction, would be made possible.

If a second time marker generator is available, the precision of the time calibration can be increased tenfold by the "Nonius principle". This second time marker generator is used to intensify the oscillogram with a frequency differing from that of the first marks by 10%. The speeds are then read by comparing the lines of dots and by

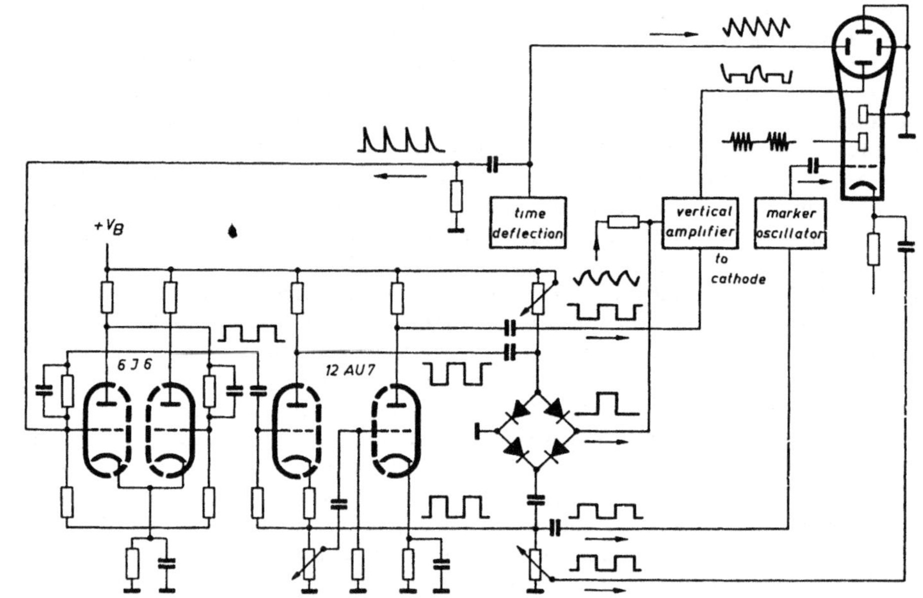

Fig. 4-104 Simultaneous representation of an intensity-modulated time base and of the uninterrupted oscillogram by electronic alternate switching

counting the dots within the time interval which is of interest. Further details are to be found in the publication referred to. The direct addition of an alternating voltage, preferably of low amplitude, to the signal voltage of known frequency can be used for the calibration of the time base (Figs. 28-2*e* and 28-6). This process is of particular advantage if it is desired to mark the duration of a certain portion of the oscillogram, for example, to measure the duration of such operations as closing or opening a contact (see Figs. 29-9, 29-10 and 29-11). A sinusoidal calibration voltage then becomes superimposed on this time portion of the oscillogram of the change of state, so that this process can be used only if no important changes of state are expected during this time (for instance, during a horizontal part of the trace). Otherwise pulses with short peaks are best used for such calibrations. It is generally sufficient if these time markers are introduced at intervalt of 1 cm, or at greater intervals if the time base linearity is good. Their regular occurrence makes them easily distinguishable from the wave-form of the phenomenon under observation. In this way even very well defined features of the oscillogram, e.g. the "dead center" in reciprocating steam engines or internal combustion engines, can be identified. The required voltage pulse is obtained by magnetic induction or photo-electrically.

GREGSON has described a process for calibrating the photographing of non-recurrent phenomena, in which a stepped voltage with precisely defined time stages instead

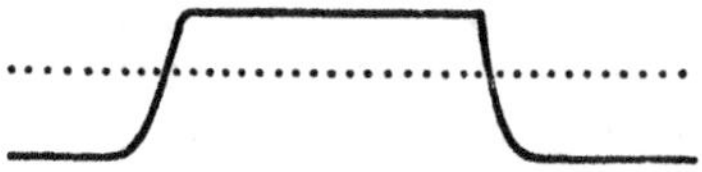

Fig. 4-105 "Time calibrating standard" made with circuit in Fig. 4-104

of the continuous time base deflection of the oscilloscope is applied [51] after recording the signal oscillogram. The vertical deflection plates receive a sinusoidal voltage simultaneously, so that vertical lines occur above the time stages of the previously photographed screen image. To ensure that only these lines stand out, they are intensified by a suitable pulse voltage. This process is only suitable for slow time bases (greater than about 1 ms/cm), such as are required, for instance, in geological investigations, as here the time stages can be relatively sharply defined by simple means.

As has already been said, it is becoming more and more common even in the medium and lower price instruments, to specify and guarantee the accuracy of the time coefficient. As this demands unusually close tolerances in the linearity of the intermediate value adjusters for the associated potentiometers, time coefficient adjustment is usually limited to adjustment in stages. The ratio of the individual stages to one another is, preferably, 1, 2, 5, 10, 20 etc., so that there is sufficient overlap.

CHAPTER 5

DEFLECTION AMPLIFIERS

5.1 General

To obtain readily interpretable beam deflection, signal voltages of from 3 to 200 V_{pp} are required for application to the deflection plates of a cathode ray tube, according to the application for which it is intended and to the design of the tube. It is necessary therefore when examining signals of lower voltage to employ an amplifier. The types of application for which an oscilloscope can be used satisfactorily are limited principally by the characteristics of this deflection amplifier.

A voltage amplifier is suitable for use as deflection amplifier for oscilloscopes if it satisfies the two following conditions:

1) the required amplification must be obtained without any change in the original waveform being perceptible on the screen;
2) the "noise" at the input of the amplifier must be negligibly small in comparison with the voltage under observation. Mains hum, ringing and valve noise are the commonest sources.

The first condition means that the frequency response characteristic of the deflection amplifier must be suitable for the application and that neither "linear" nor "non-linear" distortions should occur under acceptable operating conditions.

The second is automatically fulfilled in modern valves as regards mains hum and ringing [1] [2]. Care should be taken that amplifier gain is not excessive, for then noise would make the image of the signal voltage waveform unintelligible.

5.2 Frequency range

Modern cathode ray tubes are capable of permitting undistorted observation of direct voltage changes and of alternating phenomena of frequencies up to over 100 Mc/s without further aids and, in special versions with so-called delay line deflection systems, up to frequencies of about 1200 Mc/s [3]. If it is desired to exploit the potentialities of the cathode ray tube to the full by using an amplifier in addition, the latter would have to be capable of uniformly amplifying both direct voltages and alternating voltages up to at least 100 Mc/s. The condition would have to be made that every voltage with a frequency between 0 and 100 Mc/s must pass through the amplifier at a constant speed (constant "transit time").

These demands are physically difficult to satisfy and, particularly in the case of high gains, involve very considerable cost which in many cases is both prohibitive and unnecessary. For this reason a smaller frequency range is adopted which covers all normal requirements. The gain over the selected frequency range must not vary more than 2% however, and moreover, as has already been mentioned, the transit time of sinusoidal voltages must be practically constant over the frequency range. Otherwise considerable phase distortion takes place, which can completely disarrange the display of a given waveform (Fig. 5-13 and Fig. 5-15).

Normally at least one deflection amplifier is provided for the Y-plates of the oscilloscope. There is usually also an amplifier, possibly with a narrower frequency range for the X-deflection.

For investigation in the field of technical frequencies and voice frequencies a signal amplifier with a frequency range from about 5 c/s . . . 20 kc/s is adequate. In order to be able to reproduce the higher harmonics of the voltage under investigation, an upper frequency limit of 40 to 50 kc/s for the voice frequency would, however, be desirable.

For high frequency investigations, deflection amplifiers are required with frequency ranges up to at least between 1 and 3 Mc/s for television and pulse technique with ranges up from 5 to 50 Mc/s and for particularly stringent demands up to 100 Mc/s. Because a deflection amplifier with a very wide frequency range involves a disproportionately increasing cost, high frequency voltages are often examined by means of a simple oscilloscope in conjunction with probes incorporating a demodulating diode or a crystal detector, as will be explained in detail later.

If the RF voltage under investigation is modulated with a low frequency signal, the signal appears across the probe and can be displayed on the oscilloscope screen. If the depth of modulation is known, it is possible in this way to obtain an indirect idea of the magnitude (but not, of course, of the waveform) of the carrier frequency voltage. For physiological investigations, measurement of mechanical vibrations and the like, the lower frequency limit must be very low.

Carrier frequencies are also used to indicate magnitudes with a DC component. Here a carrier frequency is first modulated with the phenomenon and amplified to the required extent. The carrier frequency is then demodulated and the demodulation voltage is fed direct, or via an amplifier, to the Y-deflection plates [31]) [4] [5].

It is often necessary to amplify non-sinusoidal voltages faithfully, as for example, rectangular pulses or rectifier output waveforms and the like.

Mathematically, the required transmission range of the deflection amplifier for a given pulse form could be calculated with precision by Fourier analysis, but for the present discussion it will suffice to accept frequency limits obtained both mathematically and experimentally and found to be satisfactory in practice [32]). For this purpose the changes in rectangular pulses occurring when the frequency range has been wrongly chosen will be examined. These pulses can be easily converted to other curve forms, from which it can be seen whether the upper or lower frequency limits have been correctly chosen, or whether characteristic distortions have occurred in the deflection amplifier. If the lower band limit is too high, the "roof" and the "foot" of the pulse –that is, portions of the waveform that should be seen on the screen as horizontal– are seen to be tilted at an angle to the horizontal datum line (Fig. 5-1c). For this tilt to be negligible, the lower cut-off frequency f_{cl} [33]) of the amplifier must be less than

$$f_{cl} = \frac{1}{230 \cdot T_d}, \tag{5.1}$$

---

[31]) Philips "PR 9300" (formerly "GM 5536") and "PT 1200" strain gauge measuring equipments (see Part III, Ch. 31).

[32]) The derivation of the equations given here will be dealt with in detail in later parts of this book (see also Eqs. (11.13) and (11.15) in Part II, Ch. 11, and Eq. (5.34).

[33]) For the definition of "cut-off frequency" see Part I, Ch. 5.6 Amplification with thermionic valves generally – gain curve.

where T_d is the width of the pulse roof (pulse duration). If T_d is expressed in seconds, f_{cl} is obtained in c/s. If the demands in this respect are very high, DC amplifiers, which are completely free from this error, must be used.

If the upper band limit is too low, the slope of the pulse flanks is decreased, they become rounded or "worn down" (Fig. 5-1*d*).

In order that these "distortions" should be negligible, the upper cut-off frequency must correspond to

$$f_{cu} = \frac{57.5}{T_d}, \tag{5.2}$$

where T_d is again the pulse duration. If it is expressed in μs, f_{cu} is obtained in Mc/s.

Both Eqs. (5.1) and (5.2) represent very stringent demands. Some concessions have to be made to the upper band limit in particular, on the score of cost. If the factor 57.5 is substituted in Eq. (5.1) in place of the factor 230, then the distortion of a rectangular pulse produced by an amplifier of this rating becomes just recognizable. This range satisfies fairly high demands (oscillograms for $^1/_2{}^\circ$ and 2° can be seen in Fig. 11-34).

If these two conditions are fulfilled and the transit time for sinusoidal voltages of all frequencies in the range are constant, then pulses of all waveforms are faithfully transmitted [34]). In all other cases distortion must be excepted. This means that the screen image deviates more or less from the actual waveform of the input signal. Such changes

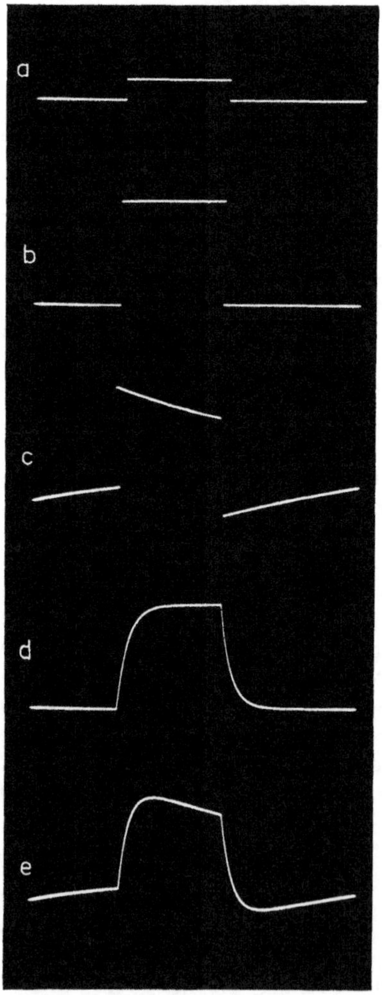

Fig. 5-1 Deformation of square pulses when bandwidth is insufficient

a) Unamplified pulse with $T_d = 1.85$ ms

b) Faithful reproduction of the pulse with five-fold amplification

c) Lower frequency limit too high

$$\left(f_{cl} = 1000 \text{ c/s}; \ \frac{1}{T_d \cdot f_{cl}} = 5.4\right)$$

d) Upper frequency limit too low ($f_{cu} = 2000$ c/s; $f_{cu} \cdot T_d = 3.7$)

e) Lower frequency limit too high (*c*) and upper frequency limit too low (*d*)

[34]) This requirement is met satisfactorily by *RC* amplifiers with normal *L*-compensation (page 199)

are known as "linear" distortions. Sinusoidal voltages are of course reproduced as sinusoidal voltages of the same frequency (slightly altered in amplitude perhaps); all other types of voltage changes appear distorted in the screen image.

The oscillograms in Fig. 5-1*a-d* illustrate this in respect of a rectangular pulse having a duration (= pulse width) $T_d = 1.85$ ms. The original waveform (*a*) is amplified some fivefold, but not distorted by a suitable amplifier (*b*). An amplifier whose lower cut-off frequency is too high (about 100 c/s) shows the roof of the curve slanting (*c*). If the upper band limit is not high enough (cut-off frequency about 2000 c/s), the leading edges are "worn down" (*d*). If the lower band limit is too high and the upper band limit too low for the transmitted pulse, not only do the leading edge flanks become distorted in this way, but the roof of the curve and the transition to the zero line are slanted (*e*) (see also Ch. 4.26 "Rating the coupling components for the time base voltages").

Something similar is also true of voltage curves consisting of a fundamental and one or more harmonics with definite phase relationships. These waveforms too, would be unacceptably distorted (oscillograms of Figs. 5-13 and 5-15*a-g*).

5.3 Non-linear distortion

Every deflection amplifier for oscilloscopes contains a number of amplifying components [35]) which can introduce non-linear distortion under unfavourable circumstances. These distortions differ from the linear distortion discussed in the preceding section, in that, for instance, non-sinusoidal output voltages are produced from sinusoidal input voltages.

Such distortions may occur if the amplifier valve (or transistor) is operated on the curved portions of the characteristic (Ch. 5.6 "Amplification with thermionic valves generally – gain curve").

The oscillograms of Fig. 5-2 *a*, *b* and *c* are photographic recordings of oscillograms of valve characteristics with the control grid alternating voltage and the anode current waveforms occuring at various operating points. (Part III, Ch. 17 "Recording the characteristics of crystal diodes, transistors and electronic valves").

The distortions illustrated can be avoided by

1) choosing suitable types of valves (transistors) the characteristic curve of which is reasonably linear over a range of grid voltages conformable with the required input control voltage, of the required output voltage ($V_a = I_a \cdot R_a$),
2) choice of the suitable operating point,
3) negative feedback.

It should be noted that a noticeable grid current begins to flow as soon as the grid is driven positive. This grid current would require a certain *control power* supplied by a driver-stage (power-stage). Deflection amplifiers should therefore be so operated that grid current does not occur; grid current exciter stages are then not required [36]). As an example of distortion due to grid current, Fig. 5-2*c* shows oscillograms of an amplifier stage with the grid driven in the region in which grid current occurs.

The distortion occurs because the grid current must flow through the input impedance of the valve, thus causing a drop in voltage, as in indicated in the simplified

---

[35]) Mostly electron tubes, but recently also transistors are more and more used [6].

[36]) For driving phase inverter stages with small gain (< 2) and for output stages of wide-band amplifiers, driver stages of a special type are required for other reasons.

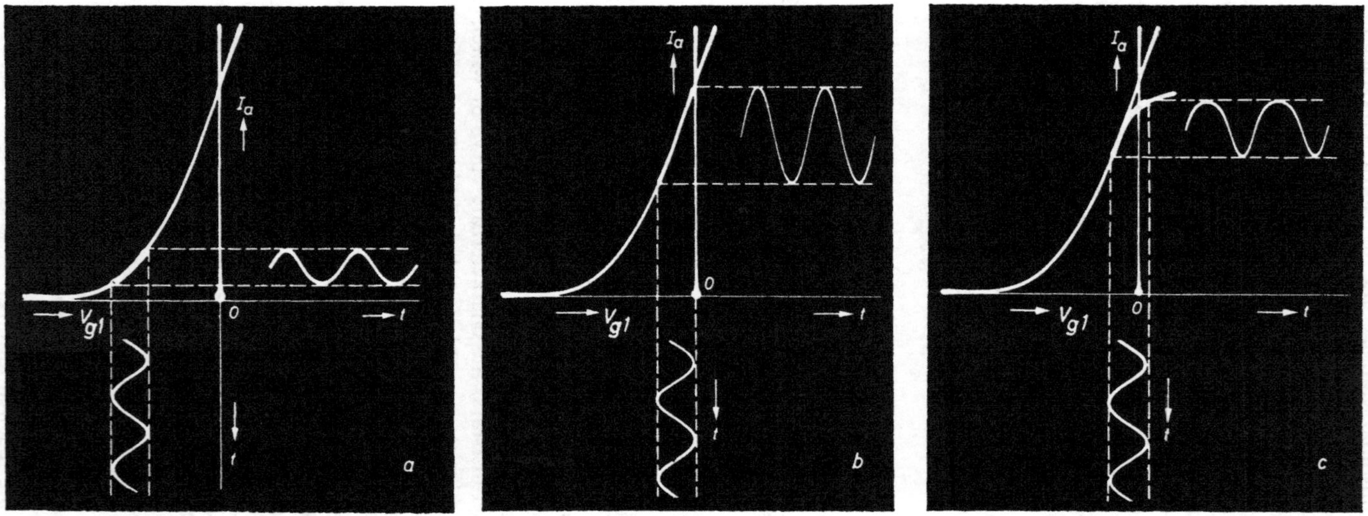

Fig. 5-2 Selection of correct operating point on the valve characteristic
a) Grid voltage too negative, distortion due to curvature of characteristic
b) Correct operating point for greatest possible operating range in linear portion of characteristic
c) Grid voltage too low; severe distortion occurs in grid current zone

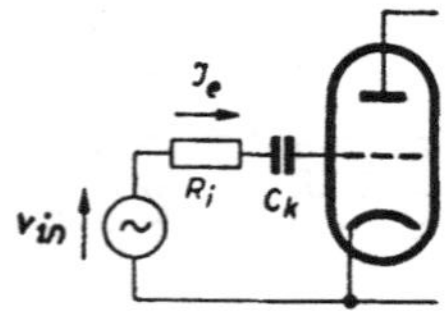

Fig. 5-3 Equivalent circuit for indicating the distortion by means of grid current (oscillogram 5 - 2 *c*)

wiring diagram (5-3). In this example, R_i was $= 50$ kΩ. A linear characteristic would correspond to $R_i = 0$. Moreover, the grid current charges the coupling capacitor; the operating point of the valve is thus shifted towards negative and returns to the original position only after a time depending on the time constant of the *RC*-circuit components. This under certain circumstances can lead to further distortion.

5. 4 Noise

Every ohmic resistor, even if not connected to a battery or other source of energy, emits very small alternating voltages, the so-called *noise* [7] [8] [9] [10]. In most practical work this noise is undetectable as is it relatively low. If sufficiently amplified, however, it can be measured and recorded on an oscillogram. Fig. 5-4a shows the oscillogram of the noise voltage of a layer resistor of 1 MΩ after about a 10^6-fold amplification over a bandwidth of 60 kc/s.

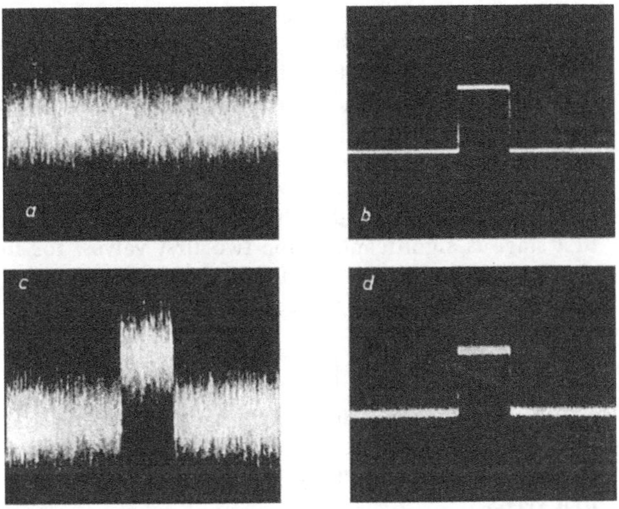

Fig. 5-4 Influence of noise voltage on interpretation of the oscillogram
a) Noise voltage produced by a resistance of 1 MΩ at about 10^6-fold amplification and $B = 60$ kc/s
b) Square wave pulse with amplitude roughly equal to noise voltage of *a*)
c) Signal and noise corresponding to *a*) of the same amplitude
d) Signal voltage of *b*) with signal-to-noise ratio 10:1

The noise voltage has an average amplitude v, calculated according to the equation

$$v^2 = 4k \cdot T \cdot R \cdot B\,[V^2]. \tag{5.3}$$

Here k = Boltzman's constant; T = the absolute temperature (at 20 °C = 293 ° abs, $k \cdot T = 0.4 \times 10^{-20}$ [Ws]) R = resistance in Ω; and B = bandwidth of the amplifier in c/s. At a bandwidth of 50 kc/s a 1 MΩ resistor thus gives an average noise voltage of about 0.03mV = 30 μV [37]).

Moreover, every electronic valve and every transistor produces noise. It is immaterial how a valve is used; it always produces a characteristic noise voltage, just as if a "noisy" resistor were connected across its grid circuit. This hypothetic resistance is known as the *equivalent noise resistance* R_{eq} of the valve.

The equivalent noise resistance corresponds only to the sources of the *shot noise,*

---

[37]) The noise voltage is changed by capacitances connected in parallel [11].

system noise and *grid noise* caused by the corpuscular structure of the stream of electrons. It does not include the so-called *flicker effect*, which plays an important role, particularly in DC amplifiers and AC amplifiers with a lower cut-off frequency (less than 1 c/s [12]).

If, in addition, a variable noise resistor of value R_{eq} at a frequency f is assumed, the flicker effect can also be included. The total noise resistance then amounts to:

$$R_{eq\ \mathrm{tot}} = R_{eq} + \frac{a}{f} \,. \tag{5.4}$$

Here a is a valve constant; according to GILLESPIE it is approximately 10^7 [11]

$$\left(\frac{a}{f} = R_{eqf}\right).$$

It is in the nature of these noise sources that if two such sources are connected in cascade, their noise voltages cannot be simply added algebraically, $v_1 + v_2$, but the total noise voltage is obtained as a geometrical sum: $\overline{v^2_{\mathrm{tot}}} = \overline{v^2_1} + \overline{v^2_2}$.
In general, the following expression is valid for any number of noise sources

$$\overline{v_{\mathrm{tot}}} = \sqrt{\overline{v^2_1} + \overline{v^2_2} + \ldots} \,. \tag{5.5}$$

It is thus understandable that in multi-stage amplifiers, actually only the noise of the first stage is significant. If the two first valves, for instance, emit noise of the same average amplitude, there occurs in the anode circuit of the second valve, assuming the same gain, G, for each stage, a noise

$$\overline{v}_{\mathrm{tot}} = \sqrt{(G^2 \cdot \overline{v}_1)^2 + (G \cdot \overline{v}_2)^2} = G \cdot \overline{v} \cdot \sqrt{G^2 + 1} \,. \tag{5.6}$$

In most cases the 1 in the right-hand expression may be ignored compared with G^2, so that, at the amplifier output, the noise v of the first valve seems to be amplified by the factor G^2. It is then fair to say that the noise is derived almost entirely from the first valve.

Carefully designed constructed amplifiers produce a noise voltage at the output which, when divided by the gain, gives an average amplitude v as shown in Table 5-1.

TABLE 5-1. NOISE WITH VOICE FREQUENCY AND WIDE-BAND AMPLIFIERS

Type of amplifier	Bandwidth	Average amplitude of noise voltage
a) voice frequency-LF-amplifier	100 c/s to 14 kc/s	$\overline{v} = 0.1\ \mu V$
b) AC wide-band amplifier	100 c/s to 10 Mc/s	$\overline{v} = 15\ \mu V$

If a signal voltage pulse is to be at all recognizable, it must be at least as large as the average amplitude v of the noise voltage. If, however, the voltage waveform is to be clearly recognized, then the signal voltage must be considerably greater than v. The oscillograms b to d in Fig. 5-4 clearly indicate this.

5. 5 Summary of the requirements for a deflection amplifier

In most practical applications — except in certain special cases —, the signal/noise ratio is so great, that the amplifier noise is relatively unnoticeable. The following are then the main demands made on a deflection amplifier for oscilloscopes:

1) sufficiently high gain,
2) satisfactorily uniform amplification within a given frequency range,
3) no obvious shift between the phase relationships of the different frequencies of the signal range (constant transit time),
4) sufficiently high output voltage within this frequency range without noticeable amplitude distortion,
5) adjustability of gain within suitable limits,
6) the output waveform must be identical with that of the input voltage.

The simultaneous fulfilment of all these conditions, which, from the technical point of view are often conflicting, demands very close study of the relevant factors. The costs involved, which will be discussed at length later, rise disproportionately when the upper frequency limit is increased, and make it essential to find a compromise between technical demands and acceptable costs.

5. 6 Amplification with thermionic valves generally-gain curve

The process of amplification will be discussed with reference to Fig. 5-5, which shows the basic circuit of a thermionic valve with its anode load resistor R_a. The gain G is equal to the ratio between the output and the input voltage, thus:

$$|G| = \left| \frac{v_{\text{out}}}{v_{\text{in}}} \right| \text{ [38]} . \qquad (5.7)$$

To calculate the voltage gain obtained with this circuit, it is customary to look upon the valve as a voltage source with an EMF $= \mu \cdot V_{\text{in}}$, which drives the anode alternating current I through the internal resistance R_i in series with the external load resistor R_a. (μ = amplification factor of the valve). However, since in the following considerations it will be shown that a number of impedances exist which are effectively in parallel with the anode load resistor, it will be more appropriate to regard the valve as a current source, the short-circuit current of which is:

$$I = V_{\text{in}} \cdot g_m \qquad (5.8)$$

(g_m = mutual conductance of the valve).

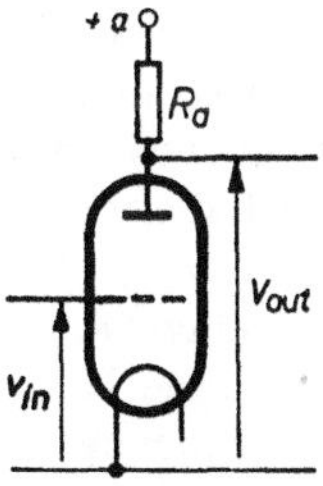

Fig. 5-5 Basic circuit of electronic valve as amplifier

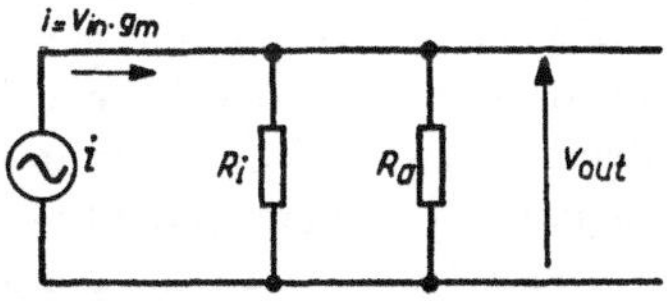

Fig. 5-6 Equivalent ciruit of valve with internal resistor R_i and external resistor R_a

[38]) In order to distinguish them from the operating direct voltages, the alternating voltages are expressed by italic letters.

The result of the analysis is not affected by using this representation, the corresponding equivalent circuit of which is shown in Fig. 5-6.

For the current source, the total load resistance R_t is equal to that of R_i and R_a in parallel [12]:

$$R_t = \frac{R_i \cdot R_a}{R_i + R_a} \,. \tag{5.9}$$

The alternating voltage at the anode is equal to:

$$-V_{\text{out}} = I \cdot \frac{R_i \cdot R_a}{R_i + R_a} \,. \tag{5.10}$$

Taking $I = V_{\text{in}} \cdot g_m$ as in Eq. (5.8), then:

$$-V_{\text{out}} = V_{\text{in}} \cdot g_m \cdot \frac{R_i \cdot R_a}{R_i + R_a} \,. \tag{5.11}$$

From this the gain $\left| \frac{V_{\text{out}}}{V_{\text{in}}} \right| = G$ is given by:

$$G = g_m \cdot \frac{R_i \cdot R_a}{R_i + R_a} \,. \tag{5.12}$$

For pentodes R_i is always greater than R_a.

In wide-band amplifiers, moreover, the anode resistance must be particularly low in order to achieve a high upper frequency limit, so that R_i should be greater than R_a for triodes also.

If, however, R_a is so small compared with R_i that it can be neglected in the denominator of Eq. (5.12) the gain is then simply

$$|G| = g_m \cdot R_a \,. \tag{5.13}$$

5.6.1. Representation of frequency response of an amplifier

It is well known, that in the frequency ranges in question (less than 50 Mc/s) amplification takes place in the valves themselves without power loss or time delay (phase shift). The values g_m, D and R_i of the valves do not introduce phase distortion. As the following observations will show, any reduction of gain at the lower and upper frequency end of the range is determined solely by the characteristics of the coupling elements between the valves, including the valve capacitances. Thus, in considering the reduction in relative gain below the mean value, it is sufficient to limit ourselves to these factors. It will be seen that the gain at the lower and upper frequency limits decreases according to definite functions. It must now be ascertained what decrease in gain is to be considered as permissible.

Taking as a starting point the fact that, by reducing the output voltage of an amplifier by 30% (corresponding to 3 dB or 0.33N), the decrease in volume is just imperceptible, it is customary in electro-acoustics to specify at what frequency the gain drops to $1/\sqrt{2} = 0.707$ times the mid-frequency gain (the "3 dB" or "half power frequency").

The curves given in Fig. 5-7 represent the loss in gain at the lower and upper ends

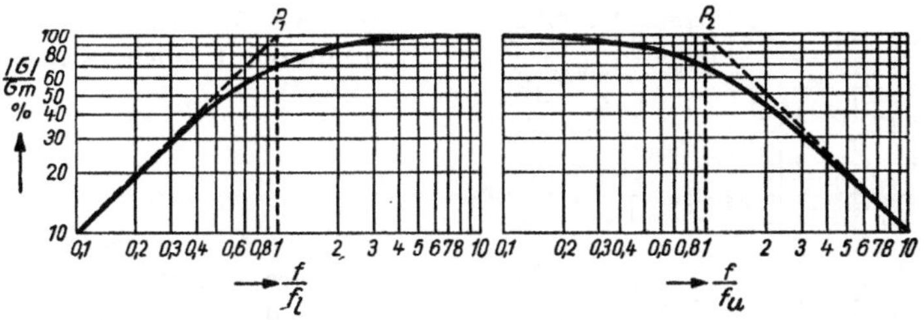

Fig. 5-7 Drop in gain at extremities of frequency band

of the frequency range respectively. The frequency at which gain drops to 0.707 at the lower end is denoted by f_{cl} (lower cut-off frequency) and at the upper end by f_{cu} (upper cut-off frequency). At each end the relative gain is indicated as a fraction of the cut-off frequency from 1/10 to 10. In accordance with usage in electro-accoustics, the ordinate scale is logarithmic.

When determining the acceptable cut-off frequencies, the stipulation is that the fall-off in gain at either end of the range must be only just perceptible. This stipulation in the case of oscilloscopy means that the limit must be set by a drop of about 1 to 3%.

For assessing the performance of amplifiers in oscilloscopes it is therefore advisible to represent the gain waveform with a linear scale for the ordinates, unless, of course, the gain for certain purposes is in actual fact logarithmically proportional to the input signal; this is only the case in oscilloscopes designed for special applications.

5. 7 Loss of gain at the lower frequency limit

The basis for the following considerations is the circuit given in Fig. 5-8, representing two stages of a conventional resistance-coupled amplifier. The components determining the frequency limits are indicated by the customary symbols. In addition to these, C_{out} and C_{in} represent respectively the output and input capacitances of the valves, C_s is the stray wiring capacitance and C_z the additional capacitance due to anode reactance. In general the symbols refer to the electrodes in whose circuit they are included, thus: C_g is the grid capacitor, C_k the cathode capacitor, etc.

For the lower frequency range, the capacitances (C_{out}, C_s etc.) shunted across the path of the amplified signal can be ignored. They offer such a high impedance at these frequencies, that, particularly as compared with the anode resistor R_a, they have no practical influence on the gain. (It is assumed in the first place that C_k in the cathode

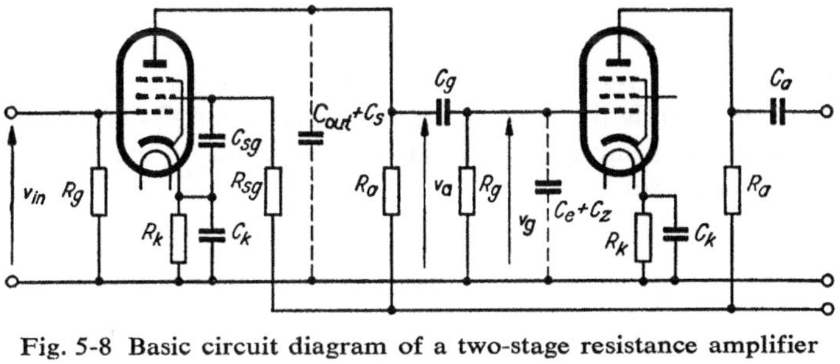

Fig. 5-8 Basic circuit diagram of a two-stage resistance amplifier

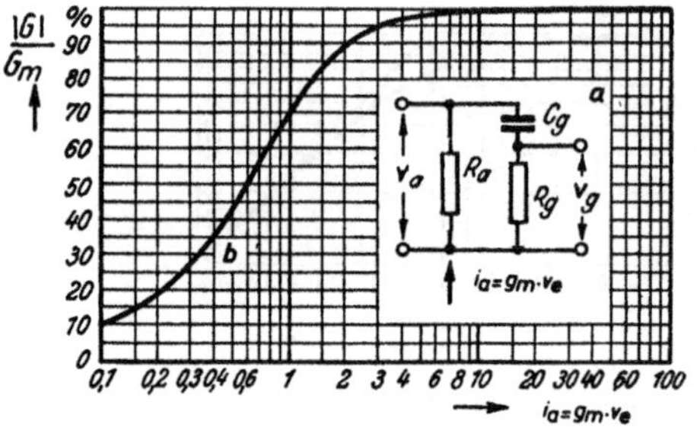

Fig. 5-9 Relative gain obtained (*b*) by means of a frequency-dependent coupling network (*a*)

lead and C_{sg} for decoupling the screen grid are infinitely large. Their influence will be dealt with separately at a later stage. The resulting equivalent is given in Fig. 5-9*a*. According to Eq. (5.8) an alternating current $I = V_{\text{in}} \cdot g_m$ flows through the anode load resistor, so that according to Eq. (5.11) there will be a voltage across this resistor of:

$$V_a = V_{\text{in}} \cdot g_m \cdot R_a \,. \tag{5.14}$$

The grid of the following valve receives this amplified voltage via the grid coupling capacitor C_g. The alternating voltage produced across the anode resistor results in a current which flows through the C_g and R_g in series. Since, however, the capacitive impedance X_c is equal to $1/\omega C_g$ and thus increases with decreasing frequency over the low frequency part of the range, the output voltage V_g of this coupling element will be smaller than V_a and will decrease with the frequency. The series connection of C_g and R_g acts as a frequency-dependent voltage divider. The drop in output voltage V_g with decreasing frequency –the "frequency response"– is shown by the curve in Fig. 5-9*b*. It indicates, in effect, the voltage ratio $V_g : V_a$ for the given frequencies. These voltages are related to the circuit impedance, thus:

$$V_g : V_a = R_g : (R_g + X_c) \tag{5.15}$$

As a capacitor produces a phase shift between the current and voltage, the voltages or the values of resistance and capacitive impedance must be brought into a *vectorial* relationship when calculating or otherwise treating these ratios (see following paragraph and Fig. 5-11).

The ratio of the voltages $V_g : V_a$ is the "transfer function" of this coupling network. Its value (ξ) may be derived from the resistance/impedance relationship:

$$|\xi| = \frac{V_g}{V_a} = \frac{R_g}{\sqrt{R_g^2 + \dfrac{1}{\omega^2 \cdot C_g^2}}} \,. \tag{5.16}$$

Since the curve in Fig. 5-9*b* includes the values of the gain in the vicinity of the lower end of the acceptable frequency range, the actual gain G in the medium frequency range can be deduced directly. This ratio is also equal to the transfer function of the coupling network, thus:

$$|\xi| = \frac{|G|}{G_m} = \frac{V_g}{V_a} \,. \tag{5.17}$$

The relative gain is therefore:

$$\left|\frac{G}{G_m}\right| = \frac{1}{\sqrt{1 + \dfrac{1}{\omega \cdot C_g^2 \cdot R_g^2}}}\,. \tag{5.18}$$

(The expression to the right of the equal sign has been simplified with respect to Eq. (5.16) by dividing both numerator and denominator of the equation by R_g). The gain G at a particular frequency is thus the product of the average gain and the transfer function for the frequency in question.

$$|G| = G_m \cdot |\xi| \tag{5.19}$$

If

$$\omega_{cl}^2 \cdot C_g^2 \cdot R_g^2 = 1,$$

then:

$$\frac{|G|}{G_m} = \frac{1}{\sqrt{2}} = 0.707\,.$$

Accordingly:

$$\omega_{cl} = 2 \cdot \pi \cdot f_{cl} = \frac{1}{C_g \cdot R_g}, \tag{5.20}$$

and thus:

$$f_{cl} = \frac{1}{2 \cdot \pi \cdot C_g \cdot R_g}\,. \tag{5.21}$$

Therefore, $f_{cl} = \dfrac{\omega_{cl}}{2\,\pi}$ is the lower cut-off frequency of the amplifier. In order to be able to make the widest possible use of the curve in Fig. 5-9*b*, the scale for the abscissa was not taken as simply the frequency, but as the product $\omega \cdot R_g \cdot C_g$. If f $(2\,\pi\,f = \omega)$ is the frequency under consideration and f_{cl} $(2\,\pi\,f_{cl} = \omega_{cl})$ is the lower frequency limit, then since $\dfrac{\omega \cdot C_g \cdot R_g}{\omega_{cl} \cdot C_g \cdot R_g} = \dfrac{f}{f_{cl}}$, instead of the product $\omega \cdot C_g \cdot R_g$ the ratio $\dfrac{f}{f_{cl}}$ can be taken as the scale for the abscissa (see also Figs. 5-7 and 5.31).

The product $\omega \cdot R_g \cdot C_g$ (for the upper cut-off frequency $\omega \cdot R_p \cdot C_p$, or the ratio $\dfrac{f}{f_l}$ for the upper cut-off frequency $\dfrac{f}{f_{cu}}$) is identical with the concept of the *standardized frequency* –Ω– of the network theory [13].

In this way the required *RC* product can be determined for a given frequency and for the permissible drop in gain. On the other hand, this curve enables the transfer function and thus the frequency response, to be ascertained for other frequencies, once the *RC* product, and with it the frequency limit, have been fixed.

If, for example, the drop in gain is not to exceed 5%, then it can be read from the curve that the value of $\omega \cdot C_g \cdot R_g$ should be 3. For a resistor $R_g = 2\text{M}\Omega$ and for a frequency $f = 30\text{c/s}$, $C_g = \dfrac{3}{\omega \cdot R_g} = \dfrac{3}{2\,\pi \cdot 30 \cdot 2 \cdot 10^6} = 8000 \cdot 10^{-12}$ F or 8000 pF. At one third of the frequency, i.e. f = 10 c/s $(\omega \cdot R_g \cdot C_g = 1)$, the attenuation is seen to be 0.707, or the cut-off frequency (the 0.707 frequency or the 3 dB frequency). Signals of three times the frequency, i.e. 90 c/s $(\omega \cdot R_g \cdot C_g = 9)$ are practically unattenuated.

These considerations, however, are valid for *one* coupling network only. There are always several such networks in a deflection amplifier.

The total transfer function $|\xi_t|$ of the amplifier, indicating the drop in gain for a given frequency, is the product of the transfer functions of each stage

$$|\xi_t| = |\xi_1| \cdot |\xi_2| \cdot |\xi_3| \ldots\ldots |\xi_n| \tag{5.22}$$

5. 8 Phase shift at the lower frequency limit

In ordinary electro-acoustical amplifiers possible shift of phase relationships are not usually important, unless special problems have to be solved, such as the rating of phase inversion stages and feedback circuits. In deflection amplifiers for oscilloscopes, on the other hand, high demands are made on the ability of an amplifier to reproduce these phase relationships faithfully. One of the most important applications of the oscilloscope is the measurement of phase shifts. Fig. 5-10*a* inset in the characteristic curve shows the vectorial relationship of the impedances and thus of the voltages in the series-circuit of R_g and C_g as shown in Fig. 5-9*a*.

The vector of the capacitive reactance of C_g or the voltage V_c, is the vertical side of a right angle triangle whose base represents the resistance R_g or the voltage V_g and is expressed symbolically by including the factor *-j*.

The effective impedance $|Z|$ resulting from the combination of R_g and $\frac{1}{j \cdot \omega \cdot C_g}$ is represented geometrically by the hypothenuse of the right angle triangle whose short sides correspond to the vectors of R_g and $\frac{1}{-j \cdot \omega \cdot C}$.

For different values of $\frac{1}{\omega \cdot C_g}$ and R_g, the vertices of the right angle triangles formed by both vectors lie on a semi-circle, the diameter of which is the vector of V_a (Fig. 5-11). The circumference of this semi-circle can be graduated with points corresponding to various values of $\omega \cdot C_g \cdot R_g$ and can thus be used as a scale. The angle formed by the short sides R_g or V_g and the resulting vector $|R|$ or V_a in the reference triangle of Fig. 5-10*a* is known as the phase angle, phase shift, phase rotation or phase distortion.

A phase shift between two alternating voltages means that they reach their respective amplitude values at different times. This time difference can be designated in

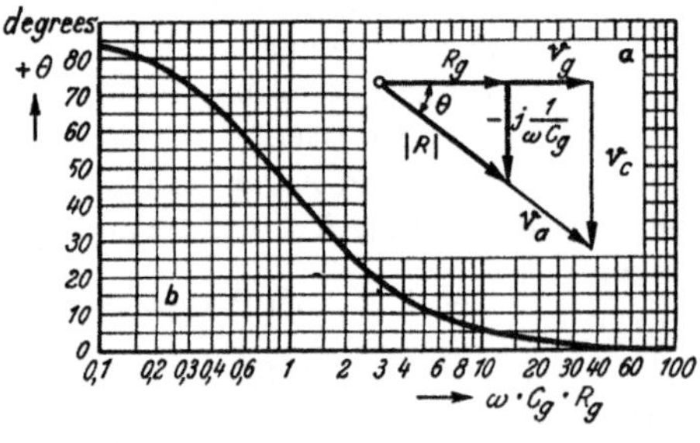

Fig. 5-10 Phase shift (*b*) by means of a *CR*-network as in circuit of Fig. 5-9*a* for various values of $\omega \cdot R_g \cdot C_g$ and vector diagram (*a*) of voltages and impedances

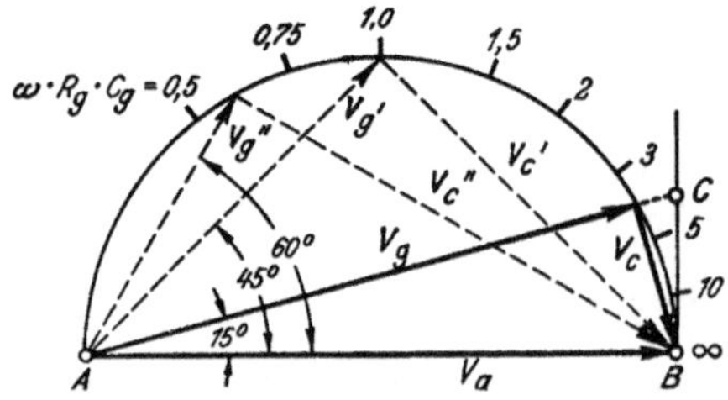

Fig. 5-11 Vector diagrams of alternating voltages in a coupling network as in equivalent circuit Fig. 5-9*a*

degrees as the angle between the voltage vectors or as a fraction of the total periodic time T. It is sometimes more convenient to indicate the phase angle by the length of the arc of radius $= 1$ corresponding to this angle. As one complete rotation of the vector –360°– is equal to the duration T of one complete cycle, corresponding to the arc 2π, an angle of 360° can be indicated as 360°, or T or 2π. The phase angles can thus also be expressed as fractions of π, where $\pi = T/2$.

In the oscillograms reproduced in Figs. 5-12 and 5-13 the phase shift is best determined as the distance between the two "zero" points, i.e. points at which the two voltages being compared cut the time axis (Figs. 5-12 and 5-13). This distance represents the time difference of the voltage vectors corresponding to the input and output voltage of a coupling element; it can be positive (leading) or negative (lagging). This "phase delay time" τ, is calculated from the equation

$$\tau = \frac{\varphi}{\omega}. \tag{5.23}$$

The angle φ must be inserted in this equation in radians. Thus, $360° = 2\pi = 6.28$. $\varphi_{\text{arc}} = \frac{6.28}{360} \cdot \varphi° = 0.0175\,\varphi°$. According to Eq. (5.23) a phase angle of 45° at 50 c/s in

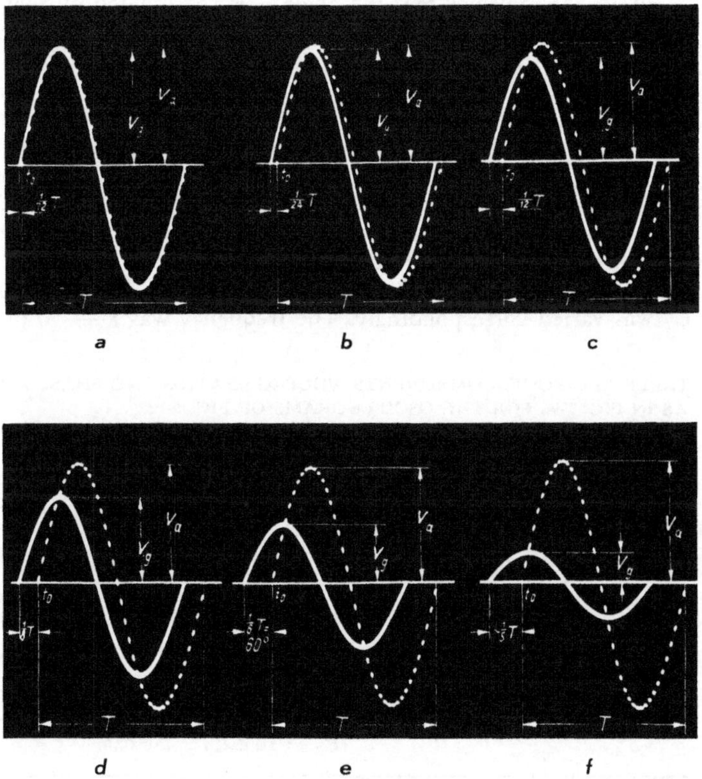

Fig. 5-12 Oscillograms of the voltages at the input and output of the CR-network of circuit as in Fig. 5-9a for various values of $\omega \cdot R_g \cdot C_g$

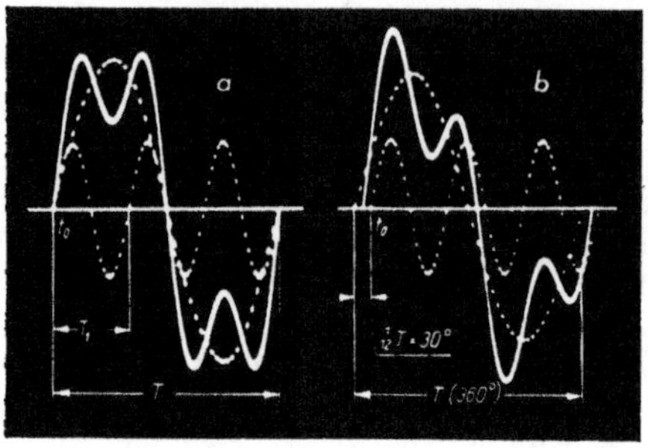

Fig. 5-13 Distortion of a complex alternating voltage by a *CR*-network
a) Sum voltage curve (full) and its components (dotted) of the original signal (fundamental and harmonic in phase)
b) Distorted curves of *a*) due to phase shift in a *CR*-network

Fig. 5-12*d*, for example, corresponds to a phase delay time of $\tau = \frac{45 \times 0.0175}{314} = 2.5$ ms. The phase angle φ can be calculated trigonometrically as

$$\cot \varphi = \omega \cdot R_g \cdot C_g, \tag{5.24}$$

and the angle φ itself is:

$$\varphi = \text{arc} \cot \omega \cdot R_g \cdot C_g. \tag{5.25}$$

Eq. (5.25) indicates that the angle φ is represented by that chord length (arc) for which the cotangential ratio is equal to $\omega \cdot R_g \cdot C_g$. (In Fig. 5-11 this ratio is the ratio $AB : BC$ for the angle 15°). Thus, for a certain value $\omega \cdot R_g \cdot C_g$, the angle φ can be read off directly from a table of trigonometrical functions. In fig. 5-10*b*, moreover, the curve (which basically corresponds to the contangential function) indicates the trend of the phase angle at the lower frequency limit for different values of $\omega \cdot R_g \cdot C_g$. The oscillograms in Fig. 5-12 illustrate the relationships under discussion. In each case, for different values of $\omega \cdot R_g \cdot C_g$, the actual oscillogram of a complete cycle of a sinusoidal voltage V_a is shown in dotted line and the corresponding cycle of the grid voltage V_g in full line. The values used for the recordings in Fig. 5-12 are set out in Table 5-2. In all cases the grid resistance was 16 kΩ while the capacitance C_g was varied correspondingly. The frequency was $f = 50$ c/s ($\omega = 2\pi f = 314$).

TABLE 5-2 CIRCUIT COMPONENTS, VOLTAGE RATIOS AND PHASE ANGLES OF A *CR* NETWORK AS IN FIG. 5-9*a* FOR THE OSCILLOGRAMS OF FIG. 5-12

Magnitude	*a*	*b*	*c*	*d*	*e*	*f*	Remarks
$\omega \cdot R_g \cdot C_g$	11.4	3.75	1.73	1.00	0.58	0.27	$= \cot \varphi$
$R_g \cdot C_g$	36.8	12.0	5.60	3.20	1.86	0.87	$\cdot 10^{-3}$s
C_g	2.3	0.75	0.35	0.20	0.116	0.054	μF
$\left\lvert \frac{V_g}{V_a} \right\rvert$	1.00	0.97	0.87	0.707	0.50	0.25	$= \cos \varphi$
φ	5	15	30	45	60	75	degrees

$R_g = 16$ kΩ; $f = 50$ c/s

The phase differences are expressed as fraction of T. From these oscillograms not only the phase shift but also the voltage attenuation between V_g and V_a can be read directly. A law governs the relationship between voltage decline and phase shift through a coupling network. It is:

$$\left|\frac{V_g}{V_a}\right| = \cos \varphi. \tag{5.26}$$

The transfer function is therefore: $\xi = \cos \varphi$, and thus:

$$V_g = V_a \cdot \cos \varphi. \tag{5.27}$$

If the vectors of the voltage across the resistor and that of the capacitor are equal (in Fig. 5-11 V'_g and V'_c), then the angle between V_a and V'_g is 45°. But cos 45° is 0.707, so that this relationship corresponds to the acceptable cut-off frequency.

A number of coupling networks included in one amplifier result in a phase shift which is equal to the sum of the phase shifts produced by each individual network. Thus:

$$\varphi_{\text{tot}} = \varphi_1 + \varphi_2 + \varphi_3 \ldots \text{etc.} \tag{5.28}$$

5.9 The influence of phase shifts on an oscillogram of a complex waveform

When all the spot positions in the oscillogram of a complex waveform (e.g. the full curve in Fig. 5-13*a*) are displaced uniformly in one direction along the time axis, the actual waveform is unchanged.

The sum-waveform in this oscillogram is the resultant of the two waveforms shown in broken lines. Fig. 5-14 shows the waveform of the harmonic of Fig. 5-13*a* and the same voltage displaced in phase by 180° $\left(\frac{T_1}{2}\right)$ (dotted). Fig. 5-12*e* shows a phase shift of the fundamental (apart from the difference in amplitude) of 60° $\left(\frac{T}{6}\right)$. A comparison of these two oscillograms confirms that to preserve the same phase relationships, the harmonic having three times the frequency of the fundamental would have to experience a phase shift approximately three times as great as that of the fundamental. (According to Fig. 5-10*b*, at certain parts of the curve there is an almost linear increase in φ with decreasing values of $\cot \varphi = \omega \cdot R_g \cdot C_g$).

If, therefore, the pattern on the screen is to correspond faithfully to the input waveform, the coupling networks must either cause no phase at all or the stipulation must be made that:

$$\varphi = k \cdot f, \tag{5.29}$$

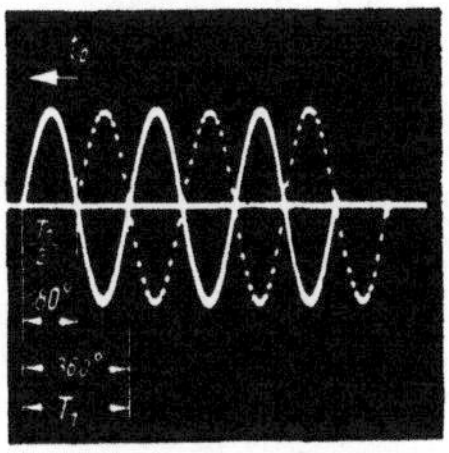

Fig. 5-14 Harmonic of Fig. 5-13 as in the original signal (dotted) and shifted in phase by 180° (full) for comparison with the oscillogram of Fig. 5-12*e*

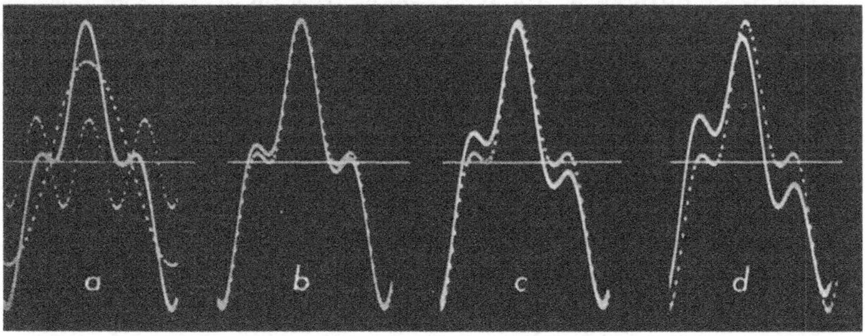

where k is a constant. This, however, can only be approximated to at the upper end of the frequency range (Figs. 5-24*a* and *b*). At the lower end the reverse is the case.

In Fig. 5-13*b*, for instance, voltage curves corresponding to Fig. 5-13*a* are reproduced under conditions in which the coupling network causes a phase shift of 30° $= T/12$ at the fundamental frequency. For the third harmonic the phase shift is only about one third that of the fundamental frequency, i.e. 10° or 1/36 of T_1.

The time period (T_1) of the third harmonic is, however, only 1/3 of T (10 mm), so that the displacement of the harmonic only 1/108 of T (in the original oscillogram about 1/3 mm). It is, therefore, scarcely perceptible. On the other hand, the displacement of the fundamental for which T on the screen amounted to about 30 mm, was $T/12 = 2^1/_2$ mm. The sum of these two curves, now displaced with respect to each other (full curve in Fig. 5-13*b*), is a very distorted image of the original waveform.

It follows from this that an essential requirement of deflection amplifiers is that phase shifts should be kept as low as possible to prevent noticeable changes in the voltage waveform appearing on the screen. Whereas with a simple sinusoidal voltage a phase shift of 5° is only just perceptible, phase shifts of the fundamental of more than 2° cause distinct distortion of the waveform of composite voltages. However, as a voltage reduction of only 1% produced by a coupling network corresponds to a phase shift of 8° (Figs. 5-9*b* and 5-10*b*) it is evident that very high demands must be made on the transfer function of the coupling elements.

Fig. 5-15 shows the distortion of a voltage waveform composed of the same frequencies, but in which the harmonic is not in phase with the fundamental as in Fig. 5-13, but is shifted in phase by 180°. The component voltages are shown dotted in *a*, and the original resultant voltage is shown as a full curve. The oscillograms *b* to *g* show the waveforms produced when, due to coupling networks, the fundamental

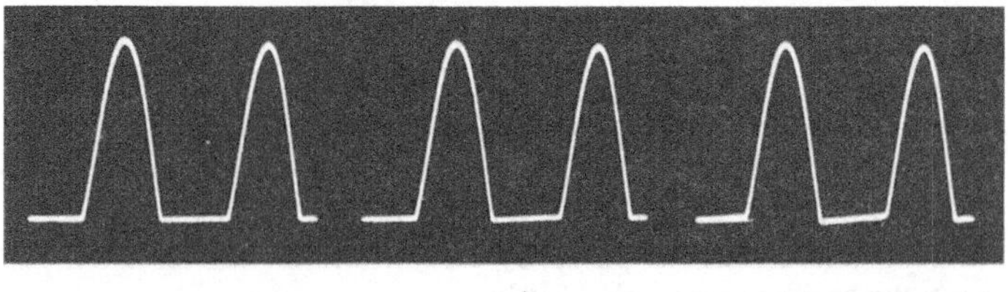

a b c

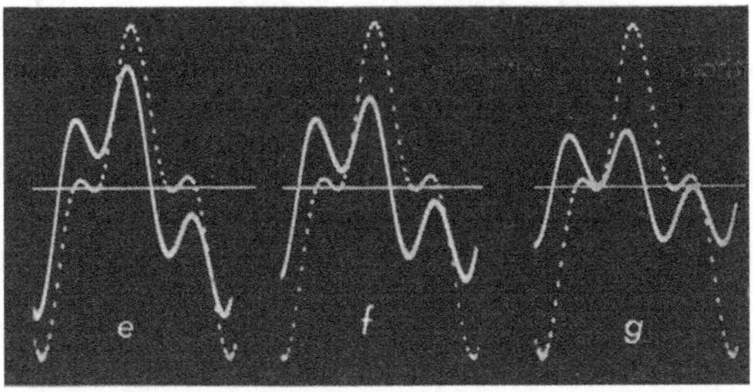

Fig. 5-15 Oscillograms showing the distortion of a complex alternating voltage due to a *CR*-network for various phase shifts of the fundamental

suffers a phase shift of 5°, 15°, 30°, 45°, 60°, or 75° respectively. For the sake of comparison the waveform of the original signal is shown dotted in each case.

5. 10 The quantitative design of coupling networks for alternating voltages having a direct current component

The highest demands on the amplifier are made when dealing with voltage waveforms which during certain parts of each cycle have a constant value, i.e. a direct component which will appear as a horizontal line on the oscillogram.

A typical example is the waveform of an alternating voltage after half-wave rectification. To ascertain the minimum values of the coupling elements necessary in this case, the *RC* product can be taken (not $\omega \cdot R \cdot C$).

This *RC* product — the time constant — must be at least large enough to prevent the drop in the output voltage of the coupling network due to the discharging of the coupling capacitor from becoming noticeable on the screen. Otherwise a tilt or even a band occurs in the line of these time portions, as can be seen from Fig. 5-16 *e* to *f*. Here are shown six oscillograms of two cycles of an alternating voltage after half-way rectification, passed through differently rated coupling networks. The voltage drop at the output of the network is particularly conspicuous at *e* and *f*. This drop corresponds to the exponential equation:

$$v_2 = v_1 \cdot \exp - \left(\frac{T}{2\,R_g \cdot C_g}\right), \tag{5.30}$$

where T = the cycle time period of the rectified alternating voltage.

Assuming that the periods of conduction and cut-off during rectification are equal,

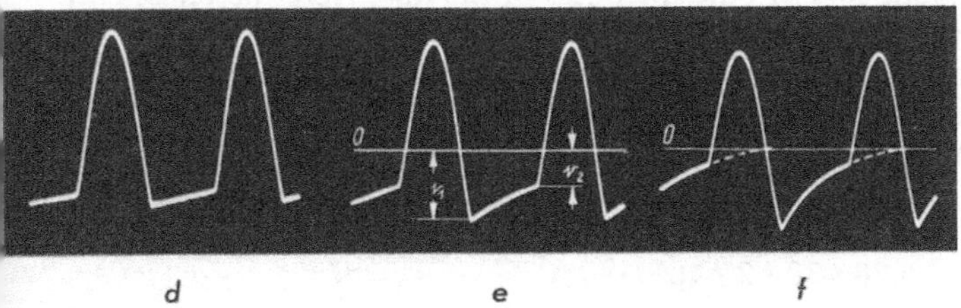

Fig. 5-16 Distortion of the waveform alternating voltage after halfwave rectification, due to various phase-shifting coupling networks

the required product $R_g \cdot C_g$ can be calculated from Eq. (5.30) for a given voltage decay to $s = \frac{v_2}{v_1}$ according to the equation:

$$R_g \cdot C_g = -\frac{1}{2 \cdot f \cdot \ln s} \tag{5.31}$$

Multiplying both sides of the equation by ω $(= 2 \cdot \pi \cdot f)$ we obtain (e.g. Eq. 5.24):

$$\omega \cdot R_g \cdot C_g = -\frac{\pi}{\ln s} = \cot \varphi.$$

Thus, for a voltage reduction to $s = 0.9$, we get:

$$\omega \cdot R_g \cdot C_g = -\frac{3.14}{\ln 0.9} = 30; \text{ arc cot } 30 = 2^\circ.$$

Fig. 5-16 serves to illustrate the process without further calculation (Table 5-3 contains the relevant data). For oscillogram a in Fig. 5-16, the voltage was applied directly. For oscillograms *b* to *f*, resistance R_g was kept at 16 kΩ, and C_g varied.

TABLE 5-3 CIRCUIT ELEMENTS, VOLTAGE RELATIONSHIPS AND PHASE ANGLES OF A *CR* NETWORK AS SHOWN IN FIG. 5-9*a* FOR THE OSCILLOGRAMS OF FIG. 5-16.

Magnitude	a	b	c	d	e	f	Remarks
$\omega \cdot R_g \cdot C_g$	—	115	28.6	11.4	3.73	1.73	$= \cot \varphi$
$R_g \cdot C_g$	—	57.5	14.3	5.7	1.87	0.87	$\times 10^{-3}$ s
C_g	—	570	143	57	18.6	7.65	10^3 pF
$\left\lvert \frac{V_g}{V_a} \right\rvert$	1.00	1.00	0.999	0.996	0.966	0.866	$= \cos \varphi$
φ	0	1/2	2	5	15	30	0

$R_g = 16$ kΩ; $f = 2000$ c/s

On closer examination of these oscillograms and following them from left to right along the time axis, Fig. 5-16*b* already shows a slight tendency for the horizontal line to tilt by a small amount about equal to the thickness of the line. Here the phase shift was only $1/2^\circ$, corresponding according to Eq. (5.24) to a value of $\omega \cdot R_g \cdot C_g = 115$.

For a coupling network, which, for example, at 30 c/s must cause a phase shift of no more than $1/2^\circ$, the coupling capacitance required with a grid resistor, $R_g = 2\text{M}\Omega$, can be calculated according to Eq. (5.25) as:

$$C_g = \frac{115}{6.28 \cdot 30 \cdot 2 \cdot 10^6} \cdot 10^6 \approx 0.3\ \mu\text{F}.$$

This value is much larger than that obtained on p. 181 for a voltage reduction of 5% (8000 pF). Thus, for amplification producing no phase direction, $\omega \cdot R_g \cdot C_g$ should be $\geqq 30$ ($\varphi = 2\%$) and, if possible, greater than 100.

If, in Eq. (5.32), time T_d is substituted, during which time a horizontal straight line ("pulse roof") is required to be shown on the oscillogram

$$\omega = 2 \cdot \pi \cdot \frac{1}{2\,T_d} = \frac{\pi}{T_d},$$

and if, according to Eq. (5.20), $\frac{1}{R_g \cdot C_g}$ is substituted by $2 \cdot \pi \cdot f_{cl}$, then, for $\varphi = {}^1/_2{}^\circ$, the equation given earlier in this section for satisfactory reproduction of a pulse [Eq. (5.1)] gives:

$$f_{cl} = \frac{1}{230 \cdot T_d}. \tag{5.34}$$

This example shows clearly how the characteristic pulse distortion can occur at relatively low values of the cut-off frequency at the low frequency end of the frequency range.

5. 11 Limination of the lower frequency limit

The lower acceptable frequency limit which can be achieved by suitable rating of the coupling components is subject to various limitations. As definite maximum ratings are laid down for the grid resistors of the amplifier valves by the valve manufacturer, the *RC* product required can only be obtained by increasing the value of the coupling capacitor. This, on the other hand, must have a high insulation resistance, so that no displacement of the working point of the following valves will be caused by the anode voltage of the preceding valves. But the higher the capacitance, the more difficult is it to ensure high insulation resistance.

The larger dimensions of coupling capacitors also increase the unwanted capacitance to chassis, thus imposing limitations on the frequency at the upper end of the frequency range (see Ch. 5.14 "Loss of gain at the upper frequency limit").

Moreover, a large time constant is synonymous with slow transit time. Thus, on the occurence of a voltage surge, due for example to applying the input signal or owing to a sudden change in supply voltage, a certain time elapses before the amplifier returns to its normal operating condition. As can be deduced, the *RC* product –the time constant– corresponds to the time in which the voltage generated on an *RC*-network falls to $1/e$ (37%) of the peak value. If $RC = 1$ s (e.g., $R = 2$ MΩ, $C = 0.5$ μF;), the voltage will take 1 second to fall to this value, whereas if $RC = 0.1$, only one tenth of a second is necessary.

Sensitivity to voltage surges can be largely reduced, for instance, by connecting the whole amplifier in push-pull and stabilizing the supply voltages. Thus, in the vertical amplifier of one oscilloscope, in spite of a time constant of $RC = 4$ s ($R_g = 2$ MΩ, $C_g = 2$ μF; giving a lower limit frequency for the complete amplifier $f_{cl} = 1/10$ c/s) [14], it was possible to obtain satisfactory operating conditions.

When making particularly high demands in this respect, DC amplifiers are necessary, as these are free from this effect.

5. 12 Influence of the cathode and screen grid capacitors

To obtain the negative bias for the control grid, a resistor is usually included in the cathode lead through which the cathode current (anode current and, in the case of a pentode, possibly also the auxiliary grid current) flows. The resultant voltage drop is applied in the conventional way to the grid resistor (R_g in Fig. 5-8). Since the AC component of the anode current also flows through the cathode resistor, an alternating voltage of a frequency of the amplified signal will reach the grid unless counter-measures are taken. This voltage will be 180° out of phase with the input voltage (except in so far as additional phase shift may have occurred), so that the effect would be to decrease the voltage on the control grid and thus to reduce the amplification of the valve (negative feedback — see also Ch. 5.18 "Feedback").

To prevent this, the cathode resistor is shunted by a capacitor whose capacitance must be so large that it represents a considerably smaller impedance (greater conductance $= \omega \cdot C$) to this frequency than the cathode resistor.

In this case the product $\omega \cdot R_k \cdot C_k$ must have a specific value. It should be noted here that the alternating voltage occurring across the cathode resistor and short-circuited by the capacitor is dependent on the valve data, particularly on the mutual conductance g_m [15].

For the sake of brevity, some values in the following equations have been combined. Thus, $x = \omega \cdot R_k \cdot C_k$ and $a = g_m \cdot \dfrac{R_k}{R_a}$ or $g_{mk} \cdot R_k$.

The transfer function of the cathode network is given by:

$$\frac{|G|}{G_m} = \sqrt{\frac{1 + x^2}{(1 + a)^2 + x^2}}\,. \tag{5.35}$$

The phase shift caused by these circuit components is:

$$\varphi = \text{arc tan}\,\frac{a \cdot x}{1 + a + x^2}\,. \tag{5.36}$$

For the parameter a, the product $g_m \cdot R_k$ must be used if the screen grid alternating current also flows through the cathode resistor. (The decoupling capacitor for the screen grid is connected to chassis). If, however, the screen grid, as in Fig. 5-8, is connected directly to cathode, then, $a = g_m \cdot \dfrac{R_k}{R_a}$ is applicable.

The cathode mutual conductance is denoted by g_{mk}. It denotes the change of the total cathode current for a given change in grid voltage, thus:

$$g_{mk} = \frac{\Delta I_k}{\Delta V_g} \tag{5.37}$$

In general, this value is not given by the manufacturers of multiple-grid valves, but for most calculations it should suffice in the case of pentodes if the equation

$$g_{mk} = g_m \cdot \frac{I_a + I_{sg}}{I_a} \tag{5.38}$$

is applied.

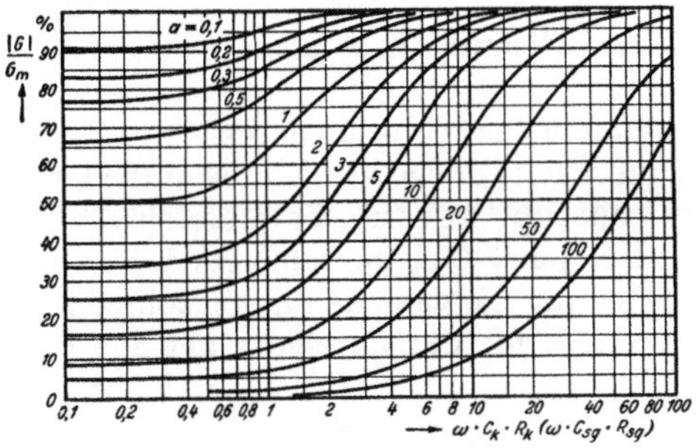

Fig. 5-17 Changes of relative gain at lower frequency limit due to circuit components in the cathode or screen-grid lead. The parameter a should be selected in accordance with the valve data

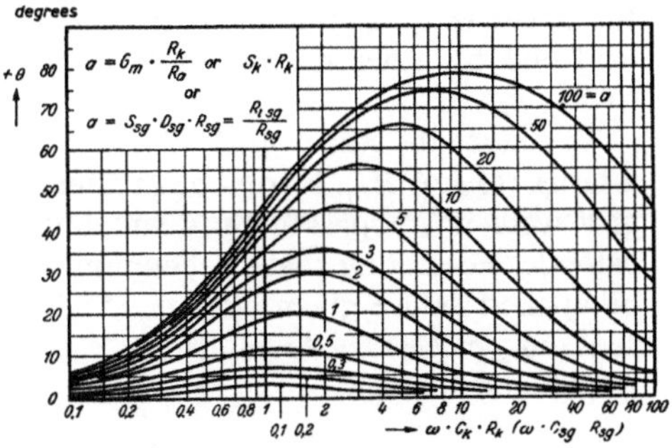

Fig. 5-18 Phase shift of amplified voltage at the lower frequency limit due to circuit components in the cathode or screen-grid lead

The curves calculated for these relationships from Eqs. (5.35) and (5.36) are given in Figs. 5-17 and 5-18 for different values of a.

If Fig. 5-17 is compared with Fig. 5-9, it is evident that for small values of $\omega \cdot C_k \cdot R_k$, the gain does not fall to zero. Even for very low values it does not fall below a certain minimum value.

It is clear, likewise from Fig. 5-18, that the phase shift at diminishing values of $\omega \cdot C_k \cdot R_k$ reaches a certain maximum value and then begins to fall again. The conductance value $\omega \cdot C_k$ of the capacitor C_k shunted across the cathode resistor then becomes so small that its contribution to the phase-shift is also reduced.

If, e.g. in pentode EF 40, $R_k = 500\ \Omega$, $R_a = 50\ \mathrm{k}\Omega$, and $g_m = 100$, then $a = 100 \cdot \dfrac{500}{50{,}000} = 1$. If the gain is not to drop below 5%, then, according to Fig. 5-17, $\omega \cdot R_k \cdot C_k$ should equal 6. Therefore, for $\omega = 30 \cdot 2 \cdot \pi \approx 200$, C_k must be: $\dfrac{6}{\omega \cdot R_k} \approx 60\ \mu\mathrm{F}$. The phase shift in this case ($f = 30$ c/s) would, according to Fig. 5-18, be about 9°.

A difficulty arises in the rating of deflection amplifiers when the cathode resistor is very small, as is often the case with steep-slope valves. Taking valve EF 80 as an example, the following conditions may occur: $R_k = 150\ \Omega$, $R_a = 3000\ \Omega$, $G_m = 20$. Accordingly, $a = 20 \cdot \dfrac{150}{3000} = 1.0$. If it is required that, when $f = 30$ c/s, $\varphi \geqq 1$, it is seen from Fig. 5-18, that $\omega \cdot C_k \cdot R_k$ must be at least $= 50$, so that $C_k = \dfrac{50}{\omega \cdot R_k} = 1700\ \mu\mathrm{F}$. Although it would be possible to produce a capacitor of this size for the low working voltage given ($V_g = -2\mathrm{V}$), it seems advisable, nevertheless, to choose some other way for achieving the desired performance. If, for example, the cathode capacitor is omitted, the gain falls to about 51% of the maximum value obtainable. At the same time, however, negative current feedback occurs, increasing the range over which amplitude distortion does not occur, in other words increasing the upper frequency limit. It is also possible to earth the cathode and to provide the grid bias by other means. In this way, full amplification is obtained corresponding to $R_a \cdot g_m$.

To reduce the screen grid voltage to the required value it is often necessary to include a series resistor R_{sg}. Here too, an alternating feedback voltage would arise if the screen grid were not bypassed by a suitably rated capacitor to earth, or preferably to the cathode. For rating these decoupling elements R_{sg} and C_{sg}, the same conditions may be applied as are given by Eqs. (5.35 and 5.36) for the cathode circuit components. It is only necessary to insert corresponding values for the variable x and the parameter a, namely:

$$x = \omega \cdot C_{sg} \cdot R_{sg} \text{ and } a = g_{m.sg} \cdot D_{sg} \cdot R_{sg} = \frac{R_{sg}}{R_{isg}}.$$

The symbols $g_{m.sg}$, D_{sg} and R_{isg} refer to the data of the screen grid. These data are not as a rule included in the standard valve data, but they will be supplied by the valve manufacturer. Some specimens of valve EF 50 were measured, and R_{isg} was found to be 40 kΩ, Therefore if $R_{sg} = 120$ kΩ, then $a = \frac{R_{sg}}{R_{isg}} = 3$. Where the phase shift not exceed 2° at $f = 30$ c/s, then, according to the curve corresponding to $a = 3$ in Fig. 5-18, $\omega \cdot C_{sg} \cdot R_{sg} = 80$. This gives

$$C = \frac{80}{\omega \cdot R_{sg}} = \frac{80}{6.28 \times 30 \times 120{,}000} \times 10^6 = 35\ \mu\text{F}.$$

5. 13 Improvement of amplifier characteristics at the lower end of the frequency range

For uniform amplification of the lower frequencies and especially for minimum phase distortion, it can be seen from the foregoing that very expensive components are required. It is therefore desirable to find a method of using more inexpensive components to compensate the undesirable loss of gain and the phase shift.

A circuit commonly used for the purpose is shown in Fig. 5-19*a*. An *RC* combination $C_v R_v$ is connected in series with the actual anode lead resistor R_a.

The resultant equivalent circuit is shown in Fig. 5-19*b*. The resistor R_a can be neglected provided it is at least ten times larger than the value of $\frac{1}{\omega \cdot C_v}$ for the given frequency range. Above a certain frequency the capacitor C_v can be regarded as a

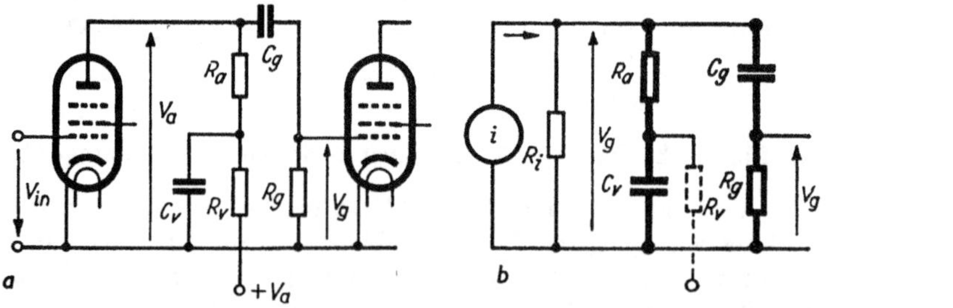

Fig. 5-19 *a*) Circuit for improving the frequency response at the lower frequency limit, by means of an *RC*-network in the anode circuit
b) Equivalent circuit to *a*)

complete short-circuit, so that it has no influence on the gain. At lower frequencies, however, the total impedance of the anode circuit (R_a and C_v in series — $R_g > R_a$) rises, and, with a constant alternating current in the anode circuit ($R_i > R_a$) the alternating voltage at the anode, in other words, the gain, rises also. In this way, the drop in amplified voltage across R_g at low frequencies can be compensated by suitably rating C_v and R_v.

Furthermore, with reduction of the frequency, capacitor C_v in the anode circuit also produces increasing phase shift of the amplified voltage. Referring once more to the equivalent circuit in Fig. 5-19*b*, it can be seen that this phase shift is in opposition to that caused by the grid network C_gR_g. An *RC*-network in the anode circuit therefore makes it possible to achieve a large measure of compensation both for the frequency response and for the phase shift due to the other circuit components. To remove the effects caused by the grid network C_gR_g, the following condition must be satisfied:

$$C_g \cdot R_g = C_v \cdot R_a \,. \tag{5.39}$$

As can be seen from the conclusions drawn from Ch. 5.10 "The quantitative design of coupling networks for alternating voltages having a direct component" and from the discussions in Part II, Ch. 11.8 "The distortion of a square wave by phase shift", the ratio of the output to the input voltage $\frac{V_2}{V_1}$ of a *RC*-network is obtained as a first approximation — if the duration of the first section of the pulse is small in proportion to the time constant of the coupling elements — from the equation

$$\frac{V_2}{V_1} \approx 1 - \frac{T_d}{R_g \cdot C_g} \,. \tag{5.40}$$

Here T_d is the time during which the amplitude is constant (a pulse roof) and is shown as a horizontal line. This means that for the reproduction of a square-wave voltage of 50 c/s ($T_d = 10$ ms) with a roof slope of 2% ($\frac{V_2}{V_1} = 0.98$), the product $R \cdot C$ must $= 0.5$ (s). With a resistor $R = 1$ MΩ this would require a coupling capacitor of 0.5 μF.

However, the reproduction of rectangular pulses can be considerably improved by choosing a suitable value for resistor R_v. It is then possible to reduce the coupling capacitor C_g without debasing the performance as regards pulse reproduction. The curves in Fig. 5-20 show the voltage ratio $\frac{V_2}{V_1}$ as a function of the ratio of the pulse

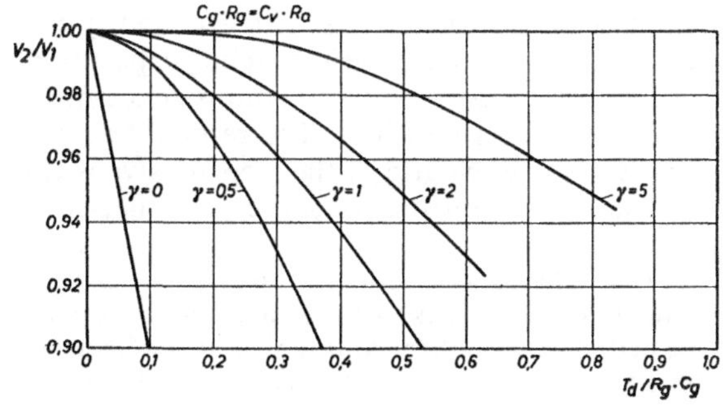

Fig. 5-20 Dependence of frequency response of a circuit as in Fig. 5-19 at the lower frequency limit on the ratio of pulse duration to time constant, with different ratios of R_g/R_a as parameter

duration (T_d) to the time constant of the circuit components ($R_g \cdot C_g$) across the grid circuit of the valves with the ratio R_v/R_a as parameter [16] [17].

From these curves it is seen that no compensation is necessary in the given example if $R \cdot C = 0.5$ ($T_d/R_g \cdot C_g = 0.02$ for $\gamma = 0$); a ratio $T_d/R_g \cdot C_g = 0.3$ suffices when $\gamma = 2$. This means, that when $R = 1$ MΩ, the coupling capacitor need only be 0.15 μF. This method of improving the characteristics of the deflection amplifier at the lower band limit is of particular value in the design of multi-stage amplifiers, as it effects considerable economies in cost.

In multi-stage amplifiers it would be impractical te compensate each stage individually. It is preferable to obtain the closest possible approximation to the desired amplifier performance by first so rating those circuit elements which are decisive at low frequencies, in the most economical way. Thereafter any further correction of the frequency and phase shift response which may be required is carried out in a single stage (in multi-stage amplifiers in every second or even fifth stage).

Thus, according to Eq. (5-39)

$$\frac{1}{C_{g1} \cdot R_{g1}} + \frac{1}{C_{g2} \cdot R_{g2}} + \frac{1}{C_{g3} \cdot R_{g3}} + \ldots\ldots = \frac{1}{C_v \cdot R_a} \,. \tag{5.41}$$

For this purpose it is of advantage to choose the stage which draws the lowest anode current (the direct voltage drop at R_v is thus a minimum). It should be pointed out here that the influence of the cathode or screen grid capacitor (providing these are not too large) can be similarly compensated within certain limits. The following conditions are valid in this case:

$$\frac{R_v}{R_k} = \frac{C_k}{C_v} = g_m \cdot R_a \tag{5.42}$$

$$\frac{R_v}{R_{g2}} = \frac{C_{g2}}{C_v} = \frac{g_{m.g2}}{\mu_1} \cdot R_a \tag{5.43}$$

$g_{m.g2}$ = screen grid mutual conductance, and μ_1 = gain when the screen is considered as an anode. If, however, larger values of capacitors are required in the stages in question, the mutual conductances g_m and $g_{m.g2}$ are no longer constant, so that a limit is set to the degree of compensation attainable.

5. 14 Loss of gain at the upper frequency limit

The basis for these observations is again the circuit of the two-stage *RC*-coupled amplifier given in Fig. 5-8. The decoupling capacitor C_{sg} for the screen grid and C_k for the cathode, and the grid coupling capacitor C_g can be considered as short-circuit [39] as far as the upper frequency limit is concerned, and are replaced in the AC equivalent circuit by direct junctions. The resultant equivalent circuit between the anode of the first valve and the grid of the second valve is given in Fig. 5-21.

The shunted resistances and capacitances can be combined into one resistance and

[39]) It should not be overlooked that the standard types of capacitors always have a certain amount of series-resistance and inductance. It is therefore essential, particularly in the case of deflection amplifiers with a high upper frequency limit, to shunt these capacitances with small, low-loss (ceramic) HF capacitors of 1000 . . . 10,000 pF.

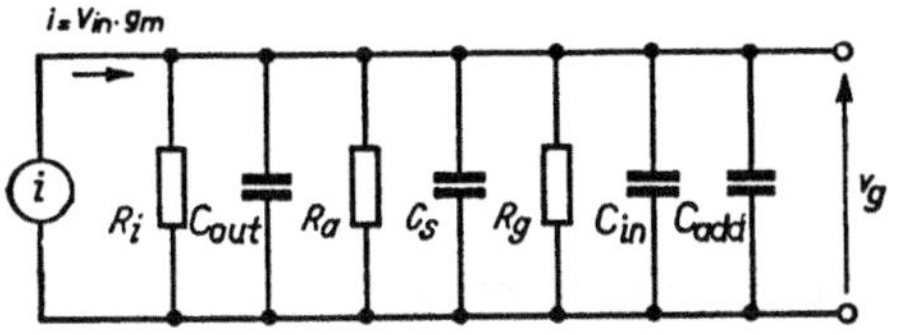

Fig. 5-21 Equivalent circuit of a resistance-coupled amplifier at high frequencies

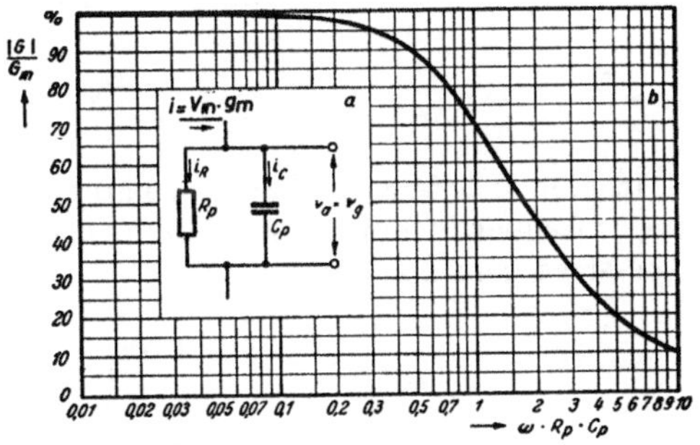

Fig. 5-22 *a*) Changes of the relative gain at upper frequency limit
b) a simplified equivalent circuit for Fig. 5-21

one capicitance respectively, producing the simplified equivalent circuit shown in Fig. 5-22*a*. In this case:

$$\frac{1}{R_p} = \frac{1}{R_i} + \frac{1}{R_a} + \frac{1}{R_g} \tag{5.44}$$

and

$$C_p = C_{\text{out}} + C_s + C_{\text{in}} + C_{\text{add}}\,^{40)}\,. \tag{5.45}$$

In wide-band amplifiers R_a is less than R_i and R_g is greater than R_a, so that, particularly when pentodes are used, R_p is for practical purposes equal to R_a. In this case the amplified alternating voltage V_g which reaches the grid of the second valve, is equal of alternating anode voltage V_a. This, according to Eq. (5.14), is equal to the product of anode alternating current and the anode load resistance. The latter is shunted, however, by the capacitance C_p, so that the resultant anode impedance decreases correspondingly and with it the gain. The impedance of the anode circuit is given by the equation:

$$|Z_a| = R_p \cdot \sqrt{\frac{1}{1 + \omega^2 \cdot C_p^2 \cdot R_p^2}}\,. \tag{5.46}$$

At frequencies at which the reactance of C_p has no influence the average gain is:

$$G_m = g_m \cdot R_p\,. \tag{5.47}$$

At higher frequencies the gain is reduced, according to Eq. (5.46), by the factor:

$$\sqrt{\frac{1}{1 + \omega^2 \cdot C_p^2 \cdot R_p^2}}\,. \tag{5.48}$$

[40]) The additional capacitance C_{add} occurs because of anode reaction ("Miller effect") in the second valve. It amounts to $C_{\text{add}} = C_{ga} \cdot (1 + G)$ in which C_{ga} is the grid-anode capacitance and G the gain. In pentodes, C_{ga} and hence C_{add} can generally be neglected.

The relative gain is therefore:

$$\frac{|G|}{G_m} = \sqrt{\frac{1}{1 + \omega^2 \cdot C_p^2 \cdot R_p^2}} \,. \tag{5.49}$$

The curve in Fig. 5-22*b* shows the trend of the relative gain for different values of $\omega \cdot C_p \cdot R_p$ between 0.01 and 10. The upper cut-off frequency is then that at which

$$R_p = \frac{1}{\omega \cdot C_p} \,. \tag{5.50}$$

In practice, the capacitances comprising C_p are always given so that for a desired upper frequency limit the maximum permissible value of R_p is also fixed by Eq. (5.50). The stage gain is then:

$$G_m = \frac{g_m}{C_p} \cdot \frac{1}{\omega_{cu}} \,. \tag{5.51}$$

From Eq. (5.51) it appears that the gain is greater than unity only when $\omega \cdot C_p < G_m$. For ω_{cu} it falls to $\frac{1}{\sqrt{2}} = 0.707$ of the maximum value. As the capacitance C_p is mainly determined by the sum of the input and output capacitances of the valves used in these stages, wide-band amplifiers should have valves having a large g_m/C_R ratio ($C_R = C_{\text{in}} + C_{\text{out}} + C_s$). The stray wiring capacitance C_s cannot be reduced below a certain minimum value, so that among a number of valves with the same g_m/C_R ratio, the valve with the greatest mutual conductance is generally the most advantageous.

EF 42, EF 80 or long-life valves E 83 F and E 180 F are particularly suitable for this purpose. The g_m/C_R ratios of these valves are: 0.52, 0.47, 0.54 and 1.17 respectively. For output stages, suitable valves are EL 41 with g_m/C_R ratio of 0.43, the PL 83 with 0.45, or the long-life E 80 L valve with a g_m/C_R ratio of 0.39. From Eq. (5.51) it can be seen that for a single amplifier stage the product of gain and upper frequency for a given g_m/C_R ratio is constant. As has already been expressed in Eq. (2.11), the relationship $g_m/(2 \cdot \pi \cdot C_R)$ can also be used as a measure for the bandwidth of a valve stage (C_R is here the sum of the input and output capacitances of the valve in question plus 5 pF for stray wiring capacitance). It indicates the frequency at which the gain is 1; it is called "the valve figure of merit". If g_m is expressed in mA/V and the capacitance C_R in pF, then the frequency is obtained in Gc/s (1Gc/s = 1000 Mc/s).

For valve E 180 F, for instance, the figure of merit is about 137.5 Mc/s. Thus, with a tenfold gain the upper frequency limit would be 14 Mc/s. In the case of the EF 42, $C_{\text{in}} = 9.4$ pF and $C_{\text{out}} = 4.3$ pF. Assuming a stray wiring capacitance of 5 pF, $C_R = 78.7$ pF. If the upper cut-off frequency is to be 3 Mc/s, then, according to (5.50), $R_p = \frac{1}{6.28 \times 3 \times 10^6 \times 18.7 \times 10^{-12}} \approx 2.8\ \text{k}\Omega$. For $g_m = 9.0$ mA/V the mid-frequency gain G_m of this stage is, according to (5.13), twentyfive-fold. At a cut-off frequency of 3Mc/s it drops to 17.8, at 9 Mc/s to 8, i.e. about a third of the average gain.

It should not be overlooked, however, that if, in this example, an upper cut-off frequency of 9 Mc/s were specified, about three times the signal voltage on the grid will be required, to produce the same alternating anode voltage. As, on the other hand, for other reasons, the operating conditions must be so selected that the maximum undistorted alternating anode voltage obtainable at medium frequencies will provide just sufficient beam deflection on the screen of the cathode ray tube, there might then be over-driving, with consequent amplitude distortion. These conditions will be further discussed in the Ch. 5.19 and Part II. Ch. 7.6 on "Output voltage requirements" and "Influence of amplifier on linearity of display". Whereas the lower frequency limit can, as already described, be reduced quite considerably by a corresponding reduction in the number of circuit elements, the upper frequency limit is determined by wiring capacitance and valve data. There is, therefore, a "natural limit" which is mainly set by the characteristics of the valves available and by the lowest obtainable stray wiring capacitance [18].

5. 15 Phase shift at the upper frequency limit

At the upper frequency limit a phase shift of the amplified voltage occurs due to the capacitance C_p, and increases with increasing frequency. The total alternating anode current $I = g_m \cdot V_{in}$ divides into the current I_R through resistance R_p and I_C through capacitance C_p (Fig. 5-22*a*).

With an increasing value of the product $\omega \cdot R_p \cdot C_p$, the proportion of the current which flows through C_p (i_C) rises, and thus the phase of the resultant current will be increasingly determined by the capacitive vector.

The voltage thus lags increasingly behind the current, and the phase angle becomes greater in the negative direction. The composition of these currents is to be seen in diagram (*a*) inset in the frame of the curve Fig. 5-23*b*. The phase angle φ is given by the relationship between the vectors i_R and i_C. As i_R is inversely proportional to the resistance R_p and i_C to the impedance $X_C = \dfrac{1}{\omega \cdot C_p}$,

$$\tan \varphi = -\frac{R_p}{\dfrac{1}{\omega \cdot C_p}}, \tag{5.52}$$

and from this the phase angle is:

$$\varphi = -\operatorname{arc\,tan}\,(\omega \cdot R_p \cdot C_p)\,. \tag{5.53}$$

In the diagram of Fig. 5-23*a*, the vectors for $\varphi = 30°$ (i_C and i_R) and 45° (dotted i_C and i'_R), have been drawn in as examples. For the upper

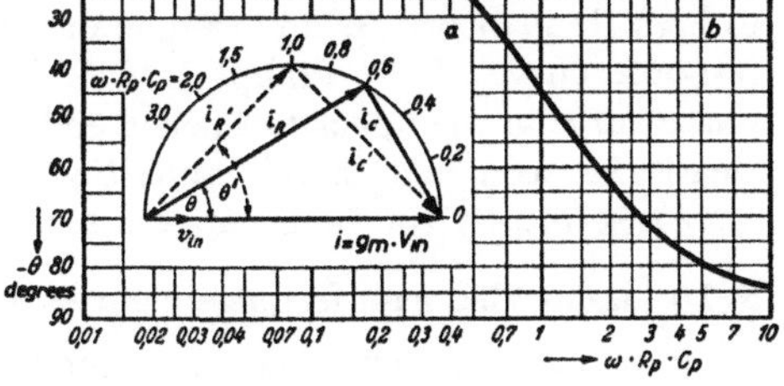

Fig. 5-23 Phase shift of the output voltage at the upper frequency limit

cut-off frequency ($R_p = \frac{1}{\omega \cdot C_p}$), $\varphi = -45°$. The curve in Fig. 5-23*b* shows the relation between the phase angle and $\omega \cdot R_p \cdot C_p$ over the range from 0.01 . . . 10. The drop in gain at the upper frequency limit at a particular phase angle is obtained from the diagram in Fig. 5-23*a*, by the relationship of i and i_R, and is

$$\frac{|G|}{G_m} = \cos \varphi. \tag{5.54}$$

A reduction of phase shift generally presupposes an increase in the upper frequency limit.

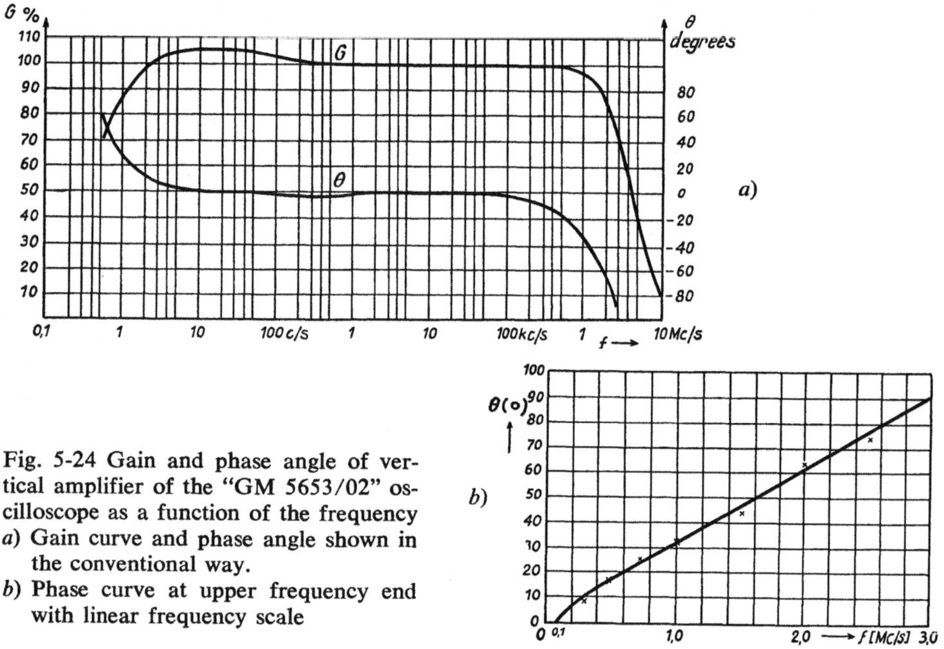

Fig. 5-24 Gain and phase angle of vertical amplifier of the "GM 5653/02" oscilloscope as a function of the frequency
a) Gain curve and phase angle shown in the conventional way.
b) Phase curve at upper frequency end with linear frequency scale

But even with the most complex and therefore costly compensating networks all phase distortion cannot be prevented above a certain frequency range. One solution of the problem has been to rate the circuit elements such that a frequency-linear increase in the phase shift of the input voltage is obtained. This fulfils the conditions of Eq. (5.29), so that the same phase delay occurs at all frequencies. It guarantees true phase reproduction even of voltages rich in harmonics (square waves etc.) up to high frequencies. As an example of this, Fig. 5-24*a* shows the frequency and phase response curve of the Philips GM 5653/02 oscilloscope. Fig. 5-24*b* reproduces the phase curve at the upper frequency end, this time with a linear frequency scale. It can clearly be seen that in this amplifier the phase shift at the upper frequency limit increases linearly (in the negative direction) with the frequency.

5.16 Improving gain frequency curve at the upper frequency limit by *L*-resonance

Various measures have been suggested for improving the linearity of the gain frequency curve at the upper frequency limit. The feature common to all is that the unwanted capacitances C_{out}, C_s and C_{in} etc. are combined with inductances to form a circuit of networks.

In the simplest circuit of this sort, an inductance is connected in series with the anode resistor. Together with the capacitance C_p (Fig. 5-25*a*) it forms a parallel-resonant circuit damped by R_a. Both parallel and series of wiring inductance and capacitance are possible. By using the distributed capacitances —C_{out} and C_s on the one side, and C_{in} and C_{add} on the other— filter couplings can also be formed. As parallel resonance involves very simple circuitry and as the phase shift of the output voltage is more linear compared with other types of coupling, this is the circuit most generally employed, and is the only one discussed here [19]. Owing to the inductance, the anode load rises with increasing frequency and the gain increases in accordance. In this way, the "natural" fall-off of the gain curve at higher frequency ranges can be compensated. Fig. 5-25*a* represents the simplified wiring diagram of an anode circuit of this sort.

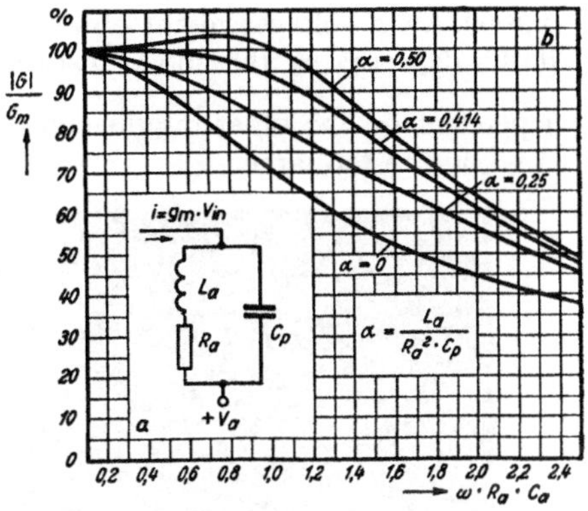

Fig. 5-25 Relative gain at upper frequency limit when using anode resonance

Denoting

$$\frac{L_a}{R_a^2 \cdot C_p} = \alpha\ ^{41)} \text{ and } \omega \cdot R_a \cdot C_p = \beta\,,$$

then the relative gain is given by the equation:

$$\frac{|G|}{G_m} = \frac{|Z_a|}{R_a} = \sqrt{\frac{1 + \alpha^2 \cdot \beta^2}{1 + (1\text{-}2\,\alpha) \cdot \beta^2 + \alpha^2\,\beta^4}}\,. \tag{5.55}$$

In Fig. 5-25*b*, the relative gain for four different values of *a* (0, 0.25, 0.414, 0.5) has been calculated according to Eq. (5.55), and is represented by four curves. The curve for the value of $\alpha = 0$ is identical with the curve in Fig. 5-22*b*. In Fig. 5-25*b*, however, in contrast to Fig. 5-22*b*, the abscissa scale of $\omega \cdot R_a \cdot C_a$ ($C_a \approx C_p$) is linear in order to show as clearly as possible the behaviour of the gain curve in the region of the frequency limit. It now remains to be seen how far the improvement of the amplification response can be taken without introducing attendant disadvantages when using it with the oscilloscope. If, for example, in Eq. (5.55) the third summand in the denominator ($a^2 \cdot \beta^4$) is neglected (which is possible if the frequency is not too high), then the relative gain would be independent of the frequency if α is given such

[41]) The factor α is identical with the square of the quality factor $Q = \frac{1}{R}\sqrt{\frac{L}{C}}$ used in other publications in this connection. Thus $\alpha = Q^2 = \frac{L}{R^2 \cdot C}$.

a value that $\alpha^2 = 1—2\alpha$. This would correspond to $\alpha = \sqrt{2}—1 = 0.414$. If, therefore, the inductance is rated $L_a = 0.414 \cdot R^2{}_a \cdot C_p$, it is to be expected that the gain will be independent of frequency over a relatively wide range.

On the other hand, the inductance cannot be selected as high as could be wished, since otherwise the damping of the anode circuit would become too small. With oscilloscope deflection amplifiers, it must be taken into consideration that the signal to be measured is not continuous, but may rise and fall in sudden jumps (rectangular pulses). Such voltages might excite these types of anode circuits into self-oscillation if the damping by the anode resistor R_a does not reach a definite minimum value dependent on the other circuit constants. In other words, the factor $\alpha = \dfrac{L}{R_a{}^2 \cdot C_p}$ must not be too large.

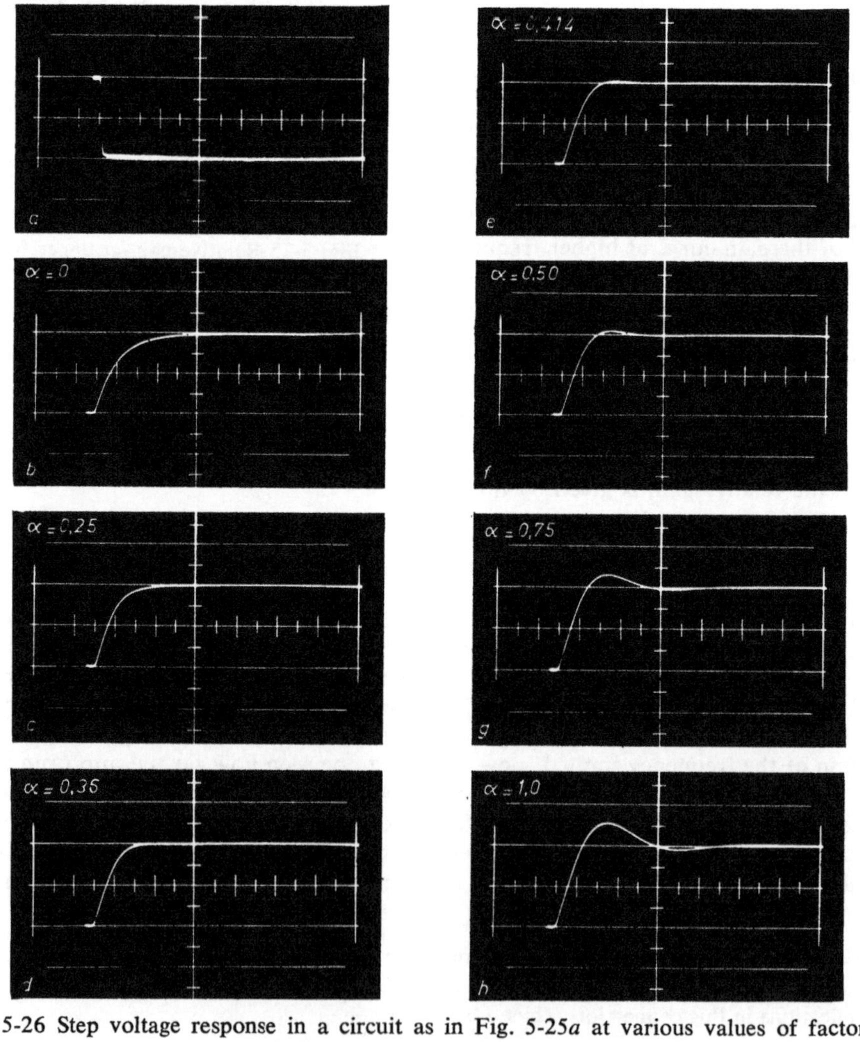

Fig. 5-26 Step voltage response in a circuit as in Fig. 5-25*a* at various values of factor α

The oscillograms shown in Fig. 5-26 show the effect of the connection of an inductance in series with the anode resistance to give seven values of a and the opportunities this offers of improving frequency linearity in the case of signal voltages which increase by sudden jumps. For this purpose, a rectangular pulse of about 25 μs duration and negative polarity was connected to the grid of a E 83 F amplifier valve from the Philips "GM 2314" pulse generator. This signal is seen in the oscillogram in Fig. 5-26*a*.

This pulse was externally triggered by the control pulse generator of the "GM 5660" oscilloscope at a repetition frequency of about 2500 c/s (for further details see Part II, Ch. 13). This oscilloscope was used to obtain the oscillograms. As the control pulse emitted by this oscilloscope is delayed by about $^1/_4$ μs following the triggering of its own time base deflection, the rectangular signal pulse is likewise delayed, so that not only is its leading-edge always represented, but also a part of the zero line preceding it. Pulse width and time coefficient were so choosen that the transient voltage rise occurring at the greatest value of the inductance corresponding to $a = 1.0$ is most pronounced, but the trailing edge of the pulse (right) remains invisible.

The oscillograms *b-h* in Fig. 5.26 show, on the same time scale of 2.5 μs/cm, as the oscillogram in Fig. 5-26*a*, the shape of the anode voltage, when the circuit elements in the anode lead (see also Fig. 5-25*a*) are so rated that the factor a $\left(= \dfrac{L}{R_a^2 \cdot C_p}\right)$ has the values of 0,0.25, 0.36, 0.414, 0.50, 0.75 and 1.0 respectively. So that the rise times of this pulse and of the oscilloscope amplifier (about 0.06 μs = 60 ns and 0.04 μs = 40 ns) could be neglected as compared with the rise time of the circuit elements in the anode circuit of the amplifier valve with an anode resistance of $R_a = 2.0$ kΩ, the value of the capacitance C_p was chosen at 625 pF, which is thirty times greater than is usual for wide-band amplifiers.

Thus, according to Eq. (5.57), a rise time of approx. 2.9 μs is obtained for the anode circuit. In the oscillogram of Fig. 5-26*b* for $a = 0$ ($L_a = 0$) a voltage rise from 10% to 90% of the highest value, occupied about 11.5mm. With a time coefficient of 2.5 μs/cm, this agrees with the calculated rise of 2.9 μs (5.57).

The other oscillograms show how the shape of the leading edge of pulses can be influenced by the addition of inductances of various values. As the inductance value rises, the rounding of the voltage rise curve becomes increasingly less. In the case of inductance values corresponding to values greater than 0.414, there occurs in addition a constantly increasing amount of overshoot. Although it was once usual to recommend $a = 0.5$, demands on the quality of pulse reproduction by oscilloscope amplifiers have since risen so sharply that $a = 0.414$ is the highest permissible value. Oscilloscopes with good pulse reproduction now have the circuit elements in the anode lead of the amplifier so rated, that $\dfrac{L_a}{R_a^2 \cdot C_p} = a$ does not exceed 0.30 or 0.36 at the most.

The influence of the amount of resonance on the rise time T_r and the overshoot O can also be calculated, as Kerkhof and Werner have shown. Fig. 5-28 in the next section will show curves demonstrating this, and the relationship between $T_r/R_a \cdot C_p$ (drop in rise time) and the factor a. This shows that there is little point in increasing a beyond 0.3 and that overshoot O —starting at zero for $a = 0.3$ increases practically linearly with a. If several amplifier stages are connected in cascade, both rise time and phase duration increase and, if the overshoot per stage is greater than 5%, the overshoot increases likewise.

For a system in which $a = 0.25$, the upper cut-off frequency of the amplifier stage is increased 1.41 times; when $a = 0.414$ it is 1.72 times. By using the previously mentioned filter couplings, the upper cut-off frequency can, in fact, be doubled. The adjusting of such stages is, however, much more difficult than in the case of a single inductance described here. The shape of the phase characteristic is also less favourable, and as this method only increases the upper cut-off frequency from 1.72 times to double, the simple tuned inductance arrangement is retained in most cases.

A very considerable increase of the upper frequency limit (up to several hundred Mc/s) of course at a corresponding increase of cost is possible by connecting several valves by means of delay lines. In this arrangement the valve capacitances are used as the shunt capacitances of the delay line, so that their harmful influence is, practically speaking, cancelled out. This type of amplifier circuit, which –together with cathode follower final stages– have found considerable favour for wide-band output stages as well, will be dealt with in more detail in Part I, Ch. 5 "Transmission line amplifier".

5. 17 Unit step response of an amplifier

As was stated in the preceding section, a rapid rise of voltage in the amplifier can, depending on the coupling elements, cause distortion of the voltage curve. Oscilloscopes are often used in pulse technique for investigating voltages with a steeply rising or falling curve, e.g. in television, so that it is usually important to know how a deflection amplifier behaves when called upon to reproduce rapid changes in signal input. These properties of an amplifier are represented by the step function response curve, which indicates the behaviour of the circuit elements in question (or of the whole amplifier) with respect to time, when the input voltage rises to a certain value (A) in an infinitely short time. The assumption that the rise time is infinitely short is made only for the purpose of the following analysis, as in practice sometime however short, is always required. (Rectangular pulses with a rise time of 10^{-9}s are, however, met with in practice). Fig. 5-27*a* shows an ideal voltage rise of the kind under discussion. The rise time actually required (T_r) is defined as the time taken by the voltage across the element under consideration to rise from 10% to 90% of its final value. Sketch 5-27*b* illustrates this. The rise time can be estimated by assuming that the *RC* network in question is to be charged by a voltage rising in an infinitely short time. This charge takes place according to the function:

$$v_t = V_o \cdot (1 - e^{-\frac{t}{R \cdot C}}) \,. \tag{5.56}$$

For a charging time from 10% to 90% of the output voltage, T_a is found to be:

$$T_a = 2.2 \cdot R \cdot C \,. \tag{5.57}$$

For an anode resistance of 2000 Ω and a capacitance of 25 pF, we obtain a rise time of $T_a = 2.2 \times 2 \times 10^3 \times 25 \times 10^{-12} = 110$ ns. It can be deduced (11.12) that there is a relationship between the rise time T_r and the bandwidth B [42]) of an *RC*-

[42]) The bandwidth can be practically equal to the upper cut-off frequency, since the lower cut-off frequency in oscilloscope amplifiers is always < 50 c/s [see also Eq. (5.2)].

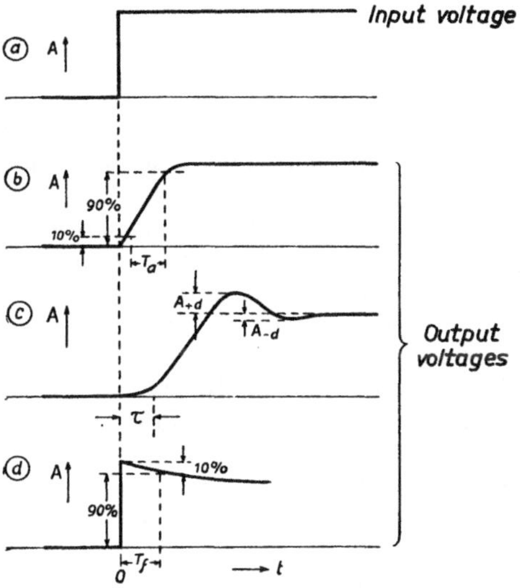

Fig. 5-27 (*a*) Unit step voltage response (*b*, *c* and *d*) output step voltage response curves

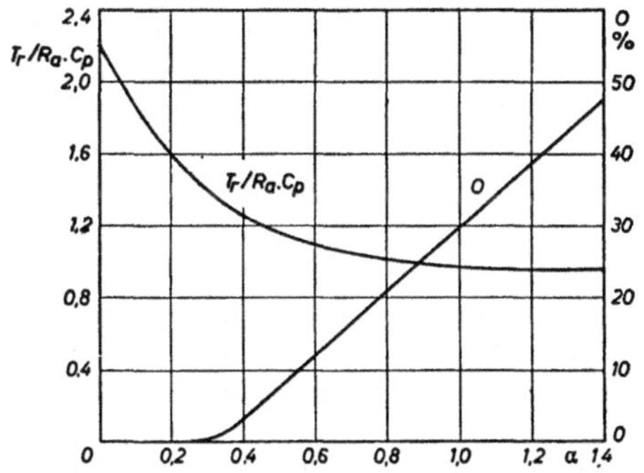

Fig. 5-28 Rise time T_r and overshoot O for a circuit as in Fig. 5-25*a* at various values of α

coupled amplifier — without L-resonance correction — which can be expressed by the equation:

$$B = \frac{0.35}{T_a}. \tag{5.58}$$

If the upper cut-off frequency of the network under consideration is increased by the use of inductance, then, as already shown, overshoot can occur as indicated in Fig. 5-27*c*.

This may be an overshoot of amplitude in the positive direction (A_{+d}), as well as, at higher inductances, an overshoot in the negative direction (A_{-d}). The formula applies only when the overshoot does not exceed 5%. The extent to which the overshoot increases after introduction of the factor α mentioned in the preceding section, is shown in the curves of Fig. 5-28.

The overshoot — according to Kerkhof and Werner — is calculated on the basis of the equation:

$$0 = \sqrt{\alpha} \cdot \exp - \frac{\pi - \arctan \sqrt{4\alpha - 1}}{\sqrt{4\alpha - 1}} \cdot 100\%. \tag{5.59}$$

From the curves it can be seen that in practice by means of L-resonance the rise time cannot be brought below $1.2 \times R \times C$ (see also the oscillograms in Fig. 5-26). The relationship between bandwidth B_1 (equivalent to upper cut-off frequency) and rise time then becomes about $B_1 = 0.2/T_r$.

If several similarly amplifier stages are connected in cascade, the overshoot of the whole arrangement does not increase significantly, if the overshoot per stage is equal to or less than 5% ($\alpha < 0.5$). The total rise time is then simply obtained as the geometrical sum of the individual rise time.

Thus:

$$T_{r\,\mathrm{tot}} = \sqrt{T_{r1}^2 + T_{r2}^2 + \ldots . T_{rn}^2}. \qquad (5.60)$$

If the overshoot per stage is greater than 5%, the overall overshoot increases rapidly when several such stages are connected in cascade. This case is of little interest here, as such amplifiers are unsuitable for use in oscilloscopes. Eq. (5.60) would then no longer be valid, and the measurement of the rise time of signals whose rise time is of similar magnitude as the rise time of the vertical amplifier becomes impossible. If, however, the amplifier corresponds to the conditions described earlier ($\alpha < 0.5$, $O < 5\%$), the total rise time $T_{r\,\mathrm{tot}}$ can be measured and the rise time of the signal can be sufficiently accurately calculated from Eq. (5.60).

The step response curves, as has been indicated in Fig. 5-27*c*, also show the phase delay time τ (Eq. (5.23)). At the lower frequency limit there is, furthermore, a certain fall-off in voltage after a rapid voltage rise, as is shown in Fig. 5-27*d* "The rating of coupling networks for AC voltages with DC component" and "The distortion of a rectangular voltage by phase shifting". According to the demands to be met, the time taken for this drop time is regarded as the decline from the maximum amplitude by 2%, 10% or even 50%. Thus, from the unit step response curve the behaviour of an amplifier can be read, not only as regards frequency response (amplitude in dependence on frequency), but also as regards phase relationships (delay time). This characteristic is of great importance in assessing the performance of wide-band amplifiers, and a study of the comprehensive literature on the subject is strongly recommended [20] [21] [22].

5.18 Feedback

5.18.1 IMPORTANCE OF FEEDBACK IN THE DEFLECTION AMPLIFIER

The voltage at the output of every amplifier is proportional to the signal voltage at the input. If this output voltage is transferred, either wholly or in part, to the input, as, for example, in Fig. 5-29*a*, this amplifier is said to be "feed back". Feedback couplings are employed for the production of undamped oscillations and cyclical pulse sequences. What is of main interest here, however, is what is known as negative feedback, in which the feedback voltage opposes the input voltage. Negative feedback makes possible considerable improvements to the amplifier characteristics. It improves, among other things, the linearity between input and output voltage. In some amplifier circuits the input voltage is not so important as the control magnitude of the input current [43]). Thus, as is shown in the circuit in Fig. 5-29*b*, the feedback can be so arranged that a current proportional to the output voltage is fed back to the input circuit. Negative feedback coupling in this case improves the linearity between output voltage and input current. In any case, for certain purposes, it is not so much the control of the output voltages as the control of the output current which is important. In such cases, feedback as shown in circuits *c* and *d* of Fig. 5-29, can be applied, according whether the input voltage or input current is to be taken as the controlling magnitude. Negative feedback then, can improve either the linearity between output current and input voltage or between output current and input current. Considered from the output side of the circuit, a distinction can be made between feedback of

[43]) This is the case in a grounded grid-input stage or in transistors and in special non-electronic amplifiers (e.g. magnetic amplifiers). The input resistance is then generally not constant.

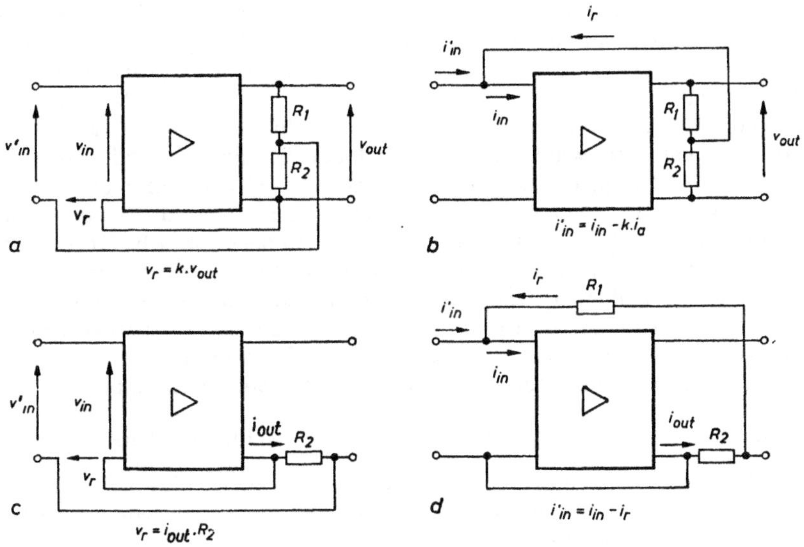

Fig. 5-29 Basic feedback circuits

a) Feedback of output voltage in series with the input voltage

b) Feedback of a current proportional of the output voltage to the input current, shunt feedback

c) Feedback of a voltage proportional to the output current to the input voltage in series with the input voltage

d) Feedback of a current proportional to the output current to the input current, shunt feedback

the output voltage and that of an output current (see circuits *a* and *b* in Fig. 5-29). At the input side, the voltage fed back can be connected in series with the input voltage (circuits *a* and *c* in Fig. 5-29), –series feedback– or a part of the output current is connected in parallel with the input current –shunt feedback– (circuits in Fig. 5-29 *b* and *d*).

In practice two or more of these forms of feedback circuit, and their effects, can be combined. Fig. 5-30 shows a combination the four basic circuits, the individual influences of which may be variously rated.

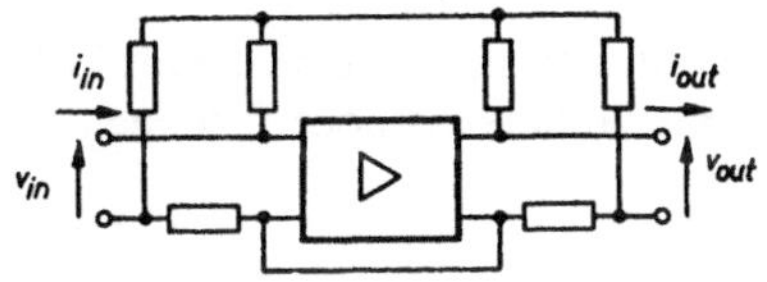

Fig. 5-30 Combination of the four basic types feedback

Among other things, negative feedback can considerably reduce the effects of operating voltage fluctuations. With negative feedback, however, the gain is reduced. If an amplifier is so rated that its gain G' with the required negative feedback is equal to G, then compared with another amplifier with the same gain without negative feedback, it will be less responsive than the latter to supply voltage fluctuations, and in addition there will be less distortion. The importance of negative feedback for deflection amplifiers is obvious from this very important fact. Feedback can also be applied at very low input voltages, as it does not affect the signal to noise ratio.

Feedback can be used to increase greatly the effective amplifier input resistance and thus permit measurements with extremely low input power. It can also be used to produce a circuit component with extremely low output resistance (a cathode follower is a valve to which negative feedback is applied), and can be used as an impedance-matching device. Finally, negative feedback can reduce interference by extraneous voltages (main hum) and make the replacement of valves easier by making unnecessary the circuit re-adjustment which sometimes has to be done.

Of course, the feedback need not necessarily to be the control grid of the input valve. It is also possible to feed back to the auxiliary grid of a multi-grid valve or to the cathode resistor (or to a part of this not bridged by a capacitor). Feedback is possible from "zero" frequency onwards — also in DC amplifiers — up to the highest signal frequencies at which correct relationships can still be maintained.

5.18.2 CHANGING THE PROPERTIES OF A DEFLECTION AMPLIFIER BY MEANS OF FEEDBACK

If series feedback, e.g. according to the circuit shown in Fig. 5-29*a*, is applied, a different value of input voltage V'_{in} as compared with V_{in}, is required for the same output voltage V_a. This voltage is:

$$V'_{\text{in}} = V_{\text{in}} \mp V_r = V_{\text{in}} \mp k \cdot V_a . \tag{5.61}$$

The factor k shows here which proportion of the output voltage is fed back ($k = \dfrac{R_2}{R_1 + R_2}$ in Fig. 5-29*a*). It is therefore called the feedback factor (positive or negative feedback factor). According to whether the voltage fed back is in phase with the input voltage (even number of amplifier stages) or in anti-phase with it (odd number of amplifier stages), the input voltage V'_{in} must be larger or smaller than the original input voltage V_{in}. Thus the gain G' with feedback will be less or more than the gain G without feedback. In the general term "feedback", a distinction must be made between "positive feedback" and "negative feedback". In Eq. (5.61) the "—" sign is to be used when input and output voltages are in phase (positive feedback). The "+" sign is used for anti-phase input and output voltages (negative feedback). As is known, positive feedback leads to self-excitation when $k \cdot G = 1$. This means that the amplifier begins to generate oscillations. Positive feedback (even if $G < 1$) should be avoided in principle in deflection amplifiers; it almost always leads to instability, fluctuating gain coefficients and often to interfering transient phenomena as well. On the contrary, negative feedback offers important advantages.

The output voltage without feedback is

$$V_a = V_{\text{in}} \cdot G . \tag{5.62}$$

While the gain without feedback is

$$G = \frac{V_a}{V_{\text{in}}} , \tag{5.63}$$

with feedback it is

$$G' = \frac{V_a}{V'_{\text{in}}} . \tag{5.64}$$

If the right-hand part of Eq. (5.61) is substituted for V'_{in}, then

$$G' = \frac{V_a}{V_{\text{in}} \mp k \cdot V_a}. \tag{5.65}$$

$G \cdot V_{\text{in}}$ can be substituted for V_a according to Eq. (5.62), so that the gain G' with feedback is

$$G' = \frac{G}{1 \mp k \cdot G}. \tag{5.66}$$

Corresponding equations can be derived for the other circuits of Fig. 5-29 in a similar way. Thus, with negative feedback:

$$G' = \frac{G}{1 + k \cdot G}. \tag{5.67}$$

The greater k is made, the smaller G' becomes as compared with G, and the greater becomes the stabilizing and anti-distortion effect of the negative feedback. In the border case of very high values of $k \cdot G$ with considerable negative feedback, the term 1 can be neglected as compared with $k \cdot G$, and thus we obtain:

$$G' = \frac{1}{k}. \tag{5.68}$$

This relationship means that the gain factor of an amplifier with such strong negative feedback no longer depends on the valve characteristics ($G = g_m \cdot R_a$). The resulting gain G' is determined only by the feedback circuit, i.e. by the resistances (and their constancy); therefore it is almost entirely insensitive to voltage fluctuations and changes in valve data. Very great stability thus results [23] [24] [25].

The curves in Fig. 5-31 give as an example the gain of a two-stage amplifier without feedback (curve G_m), with a positive feedback factor 5 (curve G'_{m1}) and with a

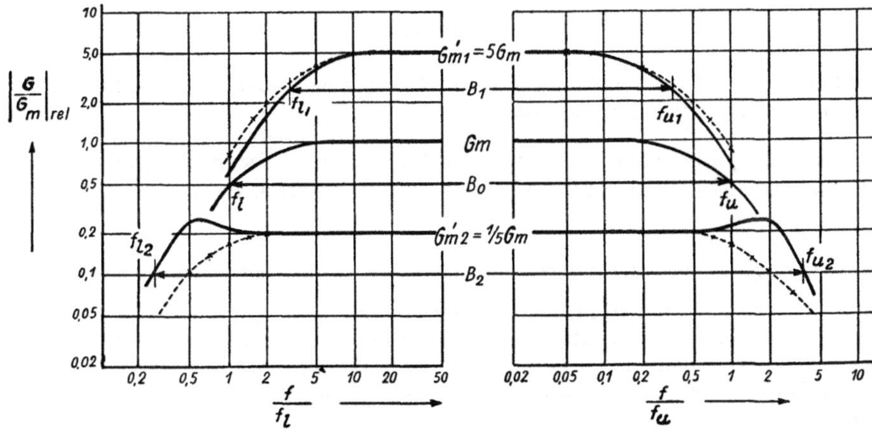

Fig. 5-31 Relative gain of a two-stage amplifier without feedback, with fivefold positive feedback positive feedback and fivefold negative feedback as a function of the frequency

negative feedback factor of 1/5 (curve G'_{m2}). (For the present only the dotted line curves will be considered. [44])

When positive feedback is used, the gain is certainly several times greater, corresponding to the feedback factor, but the spacing between the frequency limits f_{l1} and f_{u1} — the frequency response characteristic — becomes smaller. With negative feedback, on the other hand, although the gain is less, the frequency response characteristic is correspondingly wider [45]).

For this reason wide-band amplifiers make great use of negative feedback. The middle curve G_m refers to the amplifier without feedback coupling. As two coupling elements have been included, the drop in gain at the cut-off frequency is $0.707 \times 0.707 = 0.5$. The dotted curves show the gain response of a feedback-coupled amplifier in the conventional way. To calculate this, only the amount of the relative gain $\left|\frac{G}{G_m}\right|$ as in Figs. 5-9 and 5-22 has been inserted in the expression $\frac{1}{1 \mp k \cdot G}$. This does not always take into account, however, that an increasing phase shift of the output voltage takes place at the ends of the frequency response characteristic (Figs. 5-10 and 5-23). This means that in that region of the response curve the vector of the feedback voltage not only becomes smaller, but also suffers displacement. As a result, the effect of the feedback is changed.

The most usual arrangement of feedback is so dimensioned that a certain, arbitrarily chosen phase displacement $\Delta\varphi$ occurs between input and output voltages. If the arrangement indicated in Fig. 5-29a is adopted, however, the simple relationship in Eq. (5.67) becomes a complex quantity:

$$G' = \frac{G_c}{1 + k \cdot G_c}, \tag{5.69}$$

in which $G_c = G_m \cdot e^{j\varphi}$, $k_c = \frac{R_{c2}}{R_{c1} + R_{c2}}$, and G_c, G', $\Delta\varphi$, R_1 and R_2 are functions of $\omega = 2 \cdot \pi \cdot f$. The self-excitation is obtained from the Barkhaus condition, i.e.

$$k_c \cdot G_c = 1.$$

5.18.3 INFLUENCE OF PHASE SHIFT CAUSED BY COUPLING NETWORKS ON THE FREQUENCY RESPONSE OF AMPLIFIERS WITH FEEDBACK

These effects are particularly pronounced in the case of a feedback amplifier with two coupling networks [46]). In a region beyond the linear part of the gain characteristic

---

[44]) In order to represent the great difference in gain ($\frac{G_{m1}}{G_{m2}} = 25$) clearly on the same graph, a logarithmic scale has been used for the relative gain also.

[45]) Similar results could also be obtained by increasing or reducing the anode resistance. The feedback method is chosen mainly if a particular frequency range is required. Moreover, negative feedback offers certain additional advantages (reduction of distortion and greater stability).

[46]) The number of circuit elements which cause this phase shift at both ends of the frequency response characteristic must not be mutually equal or equal to the number of amplifier stages.

the phase shift approaches the value $2 \times 90° = 180°$ at the cut-off frequencies, as can be seen in Figs. 5-10 and 5-23. Within this region, if the corresponding gain and a sufficiently large feedback factor produce a feedback voltage worth mentioning, this voltage will be fed back shifted 180° in phase with respect to the mid frequency range. In the circumstances the feedback will have exactly the opposite effect to the one intended; positive feedback becomes negative feedback and vice versa. In the transition area between these extreme frequency ranges and the mid range, there is, with positive feedback, a corresponding reduction in relative gain in the vicinity of the cut-off frequency, and, with negative feedback, an increase in gain.

The actual curves resulting from this are shown in full line in Fig. 5-31. With positive feedback the amplification range B_1 becomes even narrower than it was found to be without taking the phase shift into consideration. With negative feedback B_2 there is not only a considerable extension of the range, but, at the limits of the linear portion, even a rise in the gain curve. The position of this rise depends on the number of phase-shifting networks, and their peak value depends on the negative feedback factor and overall gain [26] [27].

In rating an amplifier with feedback, particularly one with negative feedback, the total frequency range from 0 to well beyond the upper cut-off frequency must be taken into consideration. To overcome the difficulties mentioned above, opposed frequency-dependent networks can be connected in the amplification or feedback circuits. This results however in new, additional phase shifts which must likewise be taken into account. At least a fundamental knowledge of phase conditions is essential for working with feedback amplifiers, if their operation is to be clearly understood and controlled [28] [29].

5.18.4 THE STABILIZING EFFECT OF NEGATIVE FEEDBACK

The gain of an amplifier is not necessarily constant; in fact it depends on a whole series of working conditions in the amplifier. If, for instance, the heater voltage of the amplifier valve (or the anode or auxiliary grid voltage) is altered, then the gain also changes. The difference between the gain before and after such a voltage change (e.g., casual fluctuation in the heater supply of the valves) is designated ΔG, and the following equation applies:

$$\Delta G' = \frac{\Delta G}{(1 \mp k \cdot G)^2}. \tag{5.70}$$

When the feedback is positive the denominator is $(1 - k \cdot G)^2$ and when negative, $(1 + k \cdot G)^2$. Assuming that the feedback to be positive but not sufficient to produce self-oscillation (say $k \cdot G = 0.9$), then G' is indeed ten times greater than G but, on the other hand, the change in gain of the fed back amplifier $\Delta G'$, is a hundred times greater than that without feedback. If the gain G without feedback is altered by only $1\% \left(\frac{\Delta G}{G} = 0.01\right)$, then, according to the equation (relative change in gain):

$$\frac{\Delta G'}{G'} = \frac{1}{(1 \mp k \cdot G)} \cdot \frac{\Delta G}{G}, \tag{5.71}$$

so that the gain with positive feedback changes by 10% under these conditions. This shows that positive feedback, in spite of an increase in gain, is in general unsuitable for deflection amplifiers.

If, on the other hand, negative feedback is provided (say with perhaps $k \cdot G = 9$), then, although the gain is only one tenth of the original value, yet the change in gain is only a hundredth part of that without feedback. According to Eq. (5-71), this means that a gain change ΔG of 10% $\left(\frac{\Delta G}{G} = 0.1\right)$ can be reduced only to 1% in this example.

Negative feedback therefore leads to increased stability in an amplifier; stability increases at a greater rate than the gain drops, namely quadratically in the ratio $1/(1 + k \cdot G)$. If an extra stage is added to the negative feedback amplifier to compensate for the loss of gain, then this whole arrangement is much more stable than an amplifier without negative feedback but otherwise equal gain.

If, for example, there are three amplifier stages available in all, each with a gain of 10, then (with a maximum permissible working voltage fluctuation) a variation of gain of $\pm$ 10% per stage must be expected, and, if an overall gain of 100 is desired, three possibilities are available, viz.:

1) to select two such stages, thus obtaining a hundredfold gain $\pm$ 20%,
2) use all three stages and feed back each at $k \cdot G = 1.15$. Each stage then has a gain of 4.65 $\pm$ 4.7% and three stages give a gain of 100 $\pm$ 14%,
3) use all three stages, with overall feedback $k \cdot G = 9$ ($G = 1000$, $k = 0.009$).

According to the above equation, a gain of 100 $\pm$ 3% is obtained.

The least favourable case is naturally that in which there is no feedback. If the feedback is distributed over three stages (case 2), then a considerable improvement is immediately obtained. The most favourable of all is case 3, in which feedback is applied to the complete amplifier.

Feedback over a high-gain electronic layout cannot always be carried out so as to allow for sufficient ease of inspection; such amplifiers are best divided up into several sections (single stages) which are fed back separately. Since phase-shifting elements are present in every electronic layout, a positive rather than a negative feedback might result in certain frequency ranges, so that in the case of heavy feedback ($k \cdot G \geqq 1$) self-oscillation could easily set in. This danger is overcome if each separate stage or even large amplifier sections are fed back relatively weakly ($k \cdot G < 1$).

In the following sections these conditions will be investigated in more detail.

5.18.5 REDUCTION OF DISTORTION BY NEGATIVE FEEDBACK

Both linear and non-linear distortion are reduced in the same ratio as an amplifier with feedback becomes more stable. Linear distortion arises when the frequency curve (Figs. 5-7 and 5-31) is non-linear. For frequencies close to one another in the working range f_l to f_u there is a difference in gain $G_2 - G_1 = \Delta G$, which is reduced by negative feedback according to

$$\Delta G' = \frac{\Delta G}{(1 + k \cdot G)^2} \tag{5.72}$$

$$\frac{\Delta G'}{G'} = \frac{1}{1 + k \cdot G} \cdot \frac{\Delta G}{G} \,. \tag{5.73}$$

In positive feedback there is a minus sign between the two terms "1" and "$k \cdot G$".

This difference is reduced by negative feedback, and because $\frac{1}{(1 + k \cdot G)} < 1$, the linearity of the frequency characteristic improves.

Negative feedback also reduces phase distortion. Bearing in mind that, according to Eq. (5.72), the input and output voltages of an amplifier are not precisely in phase opposition but are shifted by a required phase φ, then the complex notation is used, in which the meanings of the various terms are as follows: $G_c = G\,(\cos\varphi + \mathrm{j}\sin\varphi)$, G = gain factor (as before), φ = phase angle of the amplifier without feedback, φ' = phase angle with feedback. Eq. (5.73) then becomes:

$$\Delta\, G'_c = \frac{\Delta\, G_c}{(1 + k \cdot G_c)^2}\,, \tag{5.74}$$

in which, for the sake of simplicity, the voltage divider is taken as purely ohmic. An equation for the phase angles φ' and φ is derived therefrom as follows:

$$\tan\varphi' = \frac{\sin\varphi}{\cos\varphi + k \cdot G}\,. \tag{5.75}$$

This formula is valid only for negative feedback. Input and output voltages of the amplifier are approximately in anti-phase and φ only indicates the departure from true phase opposition. If φ is less than 45° $\left(\frac{\pi}{4}\right)$, then by approximation

$$\tan\varphi' = \frac{\tan\varphi}{1 + k \cdot G}\,, \tag{5.76}$$

and for $\varphi < 6°$ (or $\frac{\pi}{30}$) the equation can even be simplified to:

$$\varphi' = \frac{\varphi}{1 + k \cdot G}\,.$$

It can be seen from this that in negative feedback the phase angle of the amplifier is reduced by the same factor as the gain.

It is therefore easy to understand that phase distortion in an amplifier is also reduced, for this occurs because neighbouring frequencies have differing transit times through the amplifier. The Eq. (5.75) can easily be rewritten for the transit time τ' since $\varphi/2 \cdot \pi \cdot f = \tau$. Thus:

$$\tau' = \frac{\tau}{1 + k \cdot G}\,. \tag{5.77}$$

The transit time difference $\Delta\,\tau$ of neighbouring frequencies f_1, f_2 is thus reduced by the factor $(1 + k \cdot G)$ [47]) (in example 3 on p. 210 by the factor 10).

[47]) The gain G is the same for both frequencies f_1 and f_2 with pure phase distortion; k is therefore assumed to be independent of phase. Any arbitrarily chosen phase angle ψ in which $k_c = k \cdot e^{\mathrm{j}\psi}$ might equally well be considered.

Non-linear distortions are likewise reduced by negative feedback. They arise at the bends of the valve characteristic which result in input voltages of different amplitudes being amplified to varying extents. For two different input amplitudes V_1, V_2, for which the respective gains are G_1 and G_2, the following formula applies:

$$G'_2 - G'_1 = \Delta G' = \frac{1}{(1 + k + G)^2} \cdot \Delta G . \tag{5.78}$$

The way in which the distortion-reducing influence of negative feedback works in practice can be seen from the valve characteristics recorded with oscilloscopes and shown in Figs. 5-33 and 5-35 *a* and *b*. [48]) The circuit used when obtaining the oscillogram of Fig. 5-33 is shown in Fig. 5-32. The negative feedback was obtained by disconnecting the cathode resistor by-pass capacitor. The alternating anode current then also flows through the cathode resistor and produces an alternating voltage which reaches the control grid via the grid resistor R_g at the same time as the input voltage V_{in}. The feedback voltage is in this way proportional to the output current. In such cases one usually simply speaks of "current" feedback. This arrangement is equivalent to that in Fig. 5-29*c*, i.e. feedback of a voltage proportional to the output current, a series feedback, in fact.

The characteristic I of Fig. 5-33 was obtained with the cathode capacitor switched in, i.e. without negative feedback. In this characteristic, only a limited region on either side of the working point of 23mA could be regarded as sufficiently straight to be suitable for driving the anode current. The sinusoidal cycle I′ at the right is the oscillogram of the output current corresponding to this variation. With negative

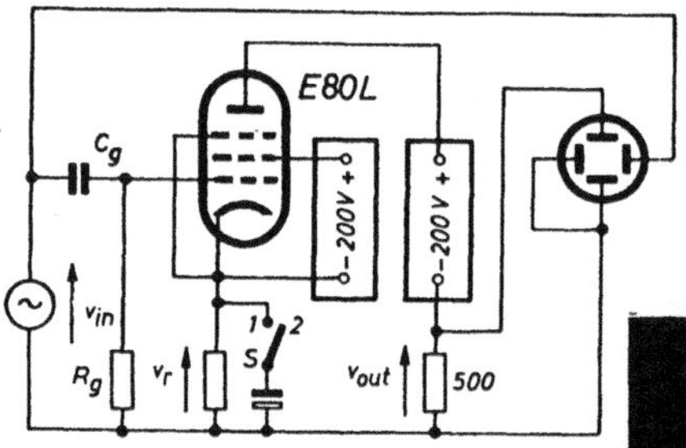

Fig. 5-32 Circuit for recording the characteristics in Fig. 5-33 showing the influence of a "current" negative feedback

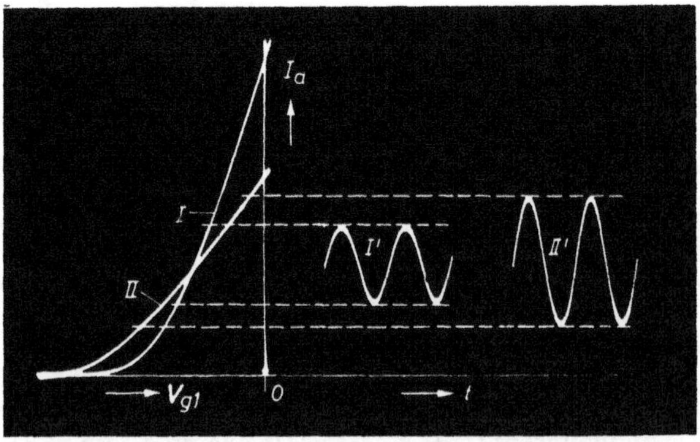

Fig. 5-33 Influence of a current-proportional negative "voltage" feedback on the operating range of a valve characteristic

[48]) It should also be pointed out that hum which does not make its entry into the amplifier at the first stage, but only at the heavy current-consuming output stages, can be reduced by the factor $\frac{1}{1 + G \cdot k}$ [30] [31] [32].

feedback — cathode capacitor switched off — the flatter characteristic II was obtained. Although it has a lower slope, a considerably greater part of the characteristic is sufficiently linear. Oscillogram II shows the output current. Both oscillograms I′ and II′ shows that the waveform is obviously sufficiently free from distortion.

These pictures also confirm, however, that the negative feedback alters the valve data.

With negative current feedback a lower mutual conductance is obtained and the internal resistance is correspondingly increased, but the amplification factor remains unaltered. (The change in the internal resistance will be discussed in the following sections).

The circuit in Fig. 5-34, used when making the oscillograms in Fig. 5-35, correspond to case *b* in Fig. 5-29, that is to say a voltage proportional to the output current is fed

Fig. 5-34 Circuit for recording characteristics in Fig. 5-35 showing the influence of negative "voltage" feedback

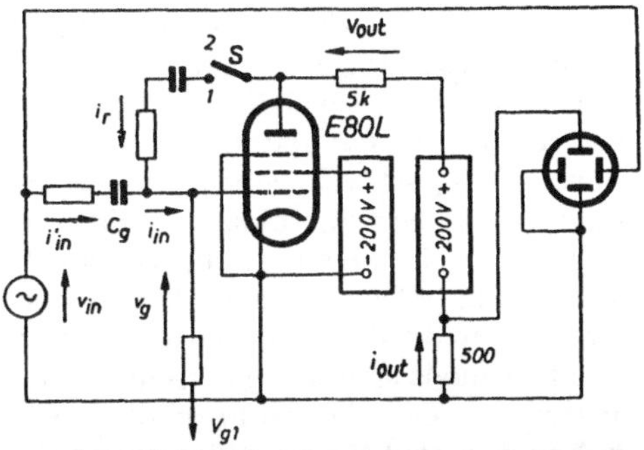

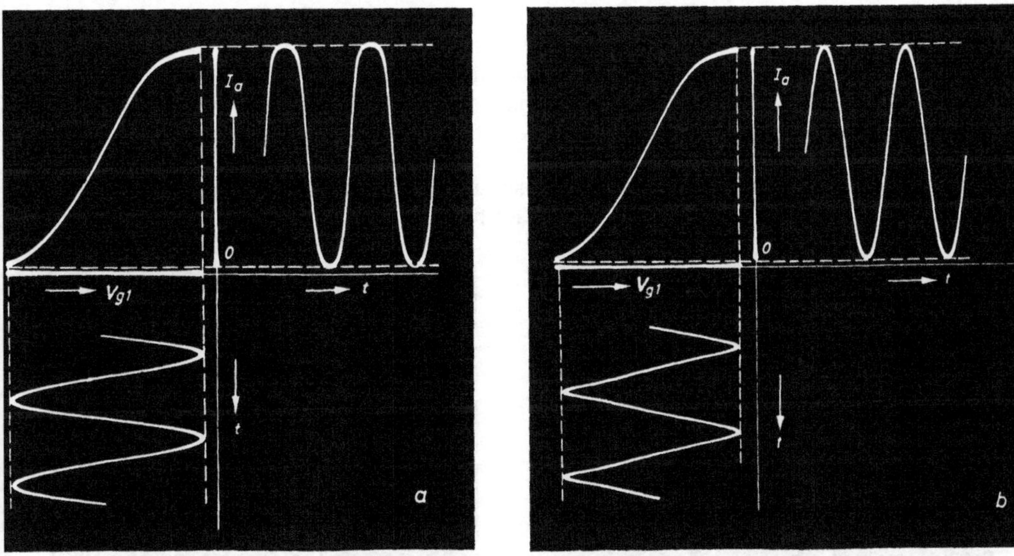

Fig. 5-35 Representation of the influence of a voltage-proportional negative current feedback on the operating range of the dynamic valve characteristic
a) Without feedback, output current distorted
b) With feedback same current amplitude as in *a*), now undistored

back to the input circuit (shunt feedback). This arrangement is usually referred to as voltage feedback.

The oscillograms in Fig. 5-35*a* show a part of the dynamic characteristic without feedback, together with the corresponding sections of the co-ordinates, the modulating voltage at the grid and the waveform of the output current which, due to overdriving, is somewhat distorted.

If switch *S* is placed at position 1, then an alternating current i_r proportional to the anode alternating voltage flows in the grid circuit in anti-phase with the input current i'_{in}, so that the effective current i_{in} which produces the control voltage across the grid resistor is reduced. The values of the circuit elements were such that the feedback factor was about 12. To obtain the same control voltage as without feedback, the input alternating voltage had therefore to be increased by a factor of 12. The results are shown in Fig. 5-35*b*. It will be observed that the valve characteristic has the same slope and form as when operating without negative feedback; yet the output waveform is undistorted. The reason that the output is not distorted is not readily apparent. It can however, be explained by taking into account the individual waveforms of both the control voltage and the output current which supplies the feedback. With a sinusoidal signal voltage and no negative feedback the anode alternating current waveform is non-sinusoidal, i.e. distorted, as shown in Fig. 5-35*a*. But a feedback voltage of this waveform superimposed upon the original sinusoidal input voltage results in an effective input which is non-sinusoidal. This non-sinusoidal input waveform is now so distorted that owing to the curvature of the valve characteristic the resulting anode alternating current is in fact sinusoidal. In other words, the reduction of distortion is entirely due to pre-distortion of the grid input voltage. The final effect is, however, that the amplifier operates as it would if the valve characteristic had in fact been linear. Even so, absolute freedom from distortion cannot be achieved in this way.

These oscillograms again confirm, among other things, *that a valve with negative voltage feedback has the same mutual conductance as a valve without negative feedback, although its amplification factor becomes greater and its internal impedance is reduced.*

In this connection an important fact should be pointed out here. Even though negative feedback can have the above described linearizing effect over the working

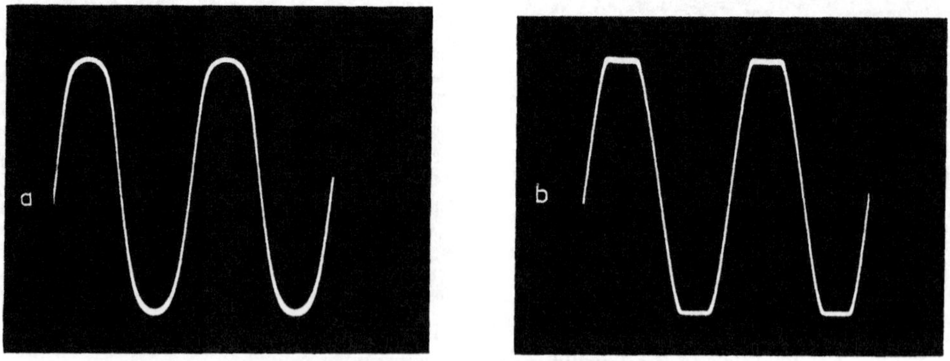

Fig. 5-36 Operating mode of the anode current of an amplifier stage according to a circuit as in Fig. 5-33. *a*) Without negative feedback. *b*) With negative feedback

range, it is certainly not possible to increase the control range by employing negative feedback. The oscillograms of Fig. 5-36 are intended to make this clear. The screen image in Fig. 5.36*a* shows the waveform of the output current without negative feedback but with somewhat increased drive as compared with Fig. 5-35. The sine peaks clearly show some rounding and are therefore distorted. If, however, as in Fig. 5-36*b*, the same amount of drive is applied to this stage under conditions of negative feedback, then, although the trend of the voltage curve is better reproduced up to the limits of the original driving, a sudden flattening occurs at that point. The real driving range cannot therefore be much increased by negative feedback.

5.18.6 FREQUENCY-DEPENDENT FEEDBACK

As has been stated, it is possible to use frequency-dependent circuit elements in the feedback path to modify the feedback voltage very considerably. A reduction of negative feedback at a certain point on the frequency response curve produces an increase in the resultant gain. If, for example, the cathode resistor of an amplifier is bypassed by only a small capacitor, then the negative feedback which occurs at the lower frequencies will be less than that at the higher end of the frequency range. This provides a very convenient method of raising the upper cut-off frequency, for which reason it is very frequently used (Fig. 5-53).

It should not be overlooked, however, that in contrast to the method of increasing the gain by inductive resonance, negative feedback produces no increase in the undistorted output voltage over this frequency range. This method is therefore particularly useful in the initial stages of an amplifier, but not in output stages where a specified voltage amplitude is required.

It has to be accepted that in the extended part of the range only a smaller undistorted output voltage can be produced. The linearity of the frequency response can, however, be modified to suit a particular application. [33] [34].

When calculating the negative feedback circuit in a wide-band amplifier, attention must be paid to the limits set by the response of the amplifier to suddenly rising or falling voltages (e.g., rectangular pulses) [35]. Unsuitable circuit values could otherwise, even without inductance in the anode circuits, lead to overshoot.

5.18.7 AMPLIFIER INPUT RESISTANCE IN NEGATIVE FEEDBACK OPERATION

It is always stipulated in the data given by valve manufacturers that, in order to keep the valve in good working condition, the grid leak resistance of a particular electronic valve may not exceed a certain limit. For this reason the value of the input resistance of a normal amplifier cannot be chosen arbitrarily, since the input resistance of an amplifier loads the voltage sources to which it is connected. The higher the input resistance, the smaller the load. By means of negative feedback in "series connection" (Fig. 5-29*a* and *c*) it is possible to increase the input resistance very considerably (see Ch. 5.29.3 "Cathode follower probes").

If R_{in} is taken as the input resistance of the amplifier without feedback, and R'_{in} as the effective input resistance with negative feedback, then in a series circuit according to Fig. 5-29*a* or *c* the following expression will hold:

$$R'_{\mathrm{in}\,1} = R_{\mathrm{in}} \cdot (1 + k \cdot G)\,. \tag{5.79}$$

If the highest possible value $k = 1$ is chosen (the negative feedback will in this case perhaps only have effect on the input stage), then the input resistance can be increased by the factor $(1 + G)$. At the same time — and this is often of special importance — the effective input capacitance C_{in} is reduced by the same amount, i.e.:

$$C'_{in\,1} = \frac{C_{in}}{1 + k \cdot G}. \tag{5.80}$$

An important application of this is discussed in the Section 5.39.3 "Cathode Follower". Negative feedback with the voltage "connected in parallel" (Figs. 5-29*b* and *d*) has the opposite effect: the input resistance is decreased. The decrease is as if a resistance $R_{in} \cdot (1 + G)$ were connected in parallel with R_{in}. In this way, with shunt feedback the following input resistance is obtained:

$$R'_{in\,2} = \frac{R_{in}}{1 + k \cdot G}, \tag{5.81}$$

and the input capacitance is:

$$C'_{in\,2} = C_{in} \cdot (1 + k \cdot G). \tag{5.82}$$

These facts must be borne in mind in applying the type of feedback shown in Fig. 5-29b and d. For example, if capacitive coupling is employed between a signal source and such an amplifier, the reduced input resistance shifts the lower band frequency limit to considerably higher frequencies. Simultaneously, the input capacitance is increased and thus the upper band limit is lowered.

A variant of the "connection in parallel" shown in Fig. 5-29*b* is the Miller circuit for producing the effect of a very large capacitance (Fig. 4-35). In this arrangement the feedback path is only capacitively coupled; in this way the capacitance can be increased by a factor $(1 + G)$. The circuit in Fig. 4-35 thus represents an important application of the feedback principle.

5.18.8 AMPLIFIER OUTPUT RESISTANCE WITH NEGATIVE FEEDBACK

Every amplifier has a certain internal resistance, with the result that its output voltage depends on the load in the output circuit. In the case of many wide-band amplifiers without feedback, this internal resistance is approximately equal to the anode resistance of the last stage.

As can be seen in Fig. 5-29*a*, a change in the value of this load also brings about a change in the feedback voltage v_r. This means that a different voltage reaches the amplifier input via the feedback connection, and the amplifier input in its turn brings about yet another change in the output. Feedback thus modifies the expression for the output resistance of an amplifier. If $R_{out\,1}$ is the output resistance with a feedback proportional to the output voltage, and R_{out} is the output resistance without such feedback, then the following equation is valid:

$$R'_{out\,1} = \frac{R_{out}}{1 \mp k \cdot G}, \tag{5.83}$$

irrespective of whether series or shunt feedback (Fig. 5-29*a* or *b*) is used. (The plus sign is for negative feedback, the minus for positive feedback).

With negative feedback proportional to the output current (Fig. 5-29*c* or *d*) the output resistance is obtained from

$$R'_{out\,2} = R_{out} \cdot (1 \mp k \cdot G). \tag{5.84}$$

Much use is made of these possibilities in practice; the amplifier (with feedback) is used as an "impedance transducer". The single stage cathode follower is one of the most important applications of this principle.

5.18.9 THE CATHODE FOLLOWER

An amplifier with only one stage provides the phase reversal between the input and output voltages required for negative feedback. If the total output voltage is applied as a feedback voltage to the input ($k = 1$) in series connection, as shown in Fig. 5-29a, the so-called cathode follower circuit is obtained.

Fig. 5-37 shows a practical circuit of this sort. This amplifier has a gain which is always somewhat less than 1 in accordance with the following formula:

$$G' = \frac{g_m \cdot R_k}{1 + g_m \cdot R_k} = \frac{G}{1+G} \text{ [49]} \qquad (5.85)$$

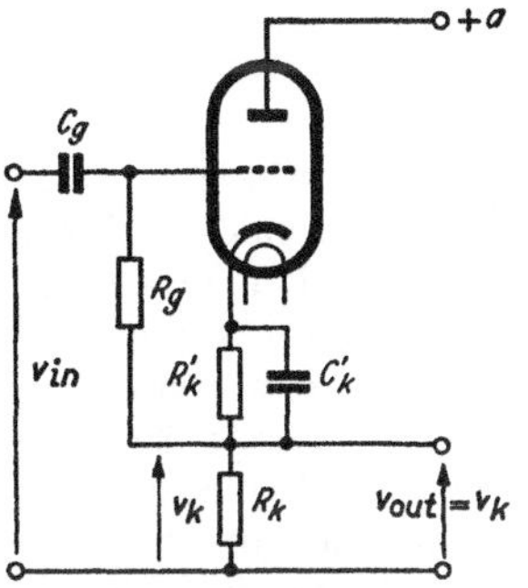

Fig. 5-37 Basic circuit of a cathode follower stage

No voltage gain is produced, and the cathode follower now operates as a 1 : 1 voltage transformer. The effective output resistance of a single stage, non-feedback amplifier is $R_{out} = R_i \cdot R_k / (R_i + R_k)$ (R_i = valve internal resistance, R_k = external resistance in the cathode circuit).

In the feedback described, since $G = g_m \cdot R_k$, a new output resistance, R'_{out}, is obtained from Eq. (5.83):

$$R'_{out} = \frac{R_{out}}{1+G} = \frac{R_{out}}{G} \cdot \frac{1}{1 + 1/G} = \frac{1}{g_m} \cdot \frac{1}{1 + 1/G}. \qquad (5.86)$$

The term $1/G$ is usually so small compared with unity that it can be neglected in the denominator. The equation can then be simplified to:

$$R'_{out} = \frac{1}{g_m} \qquad (5.87)$$

An EF 42 valve connected as a triode has a mutual conductance g_m of 10 mA/V (1/100 A/V). If this valve is connected as a cathode-follower the output resistance $1/g_m = 100\ \Omega$. According to Eq. (5.79) the input resistance is:

$$R'_{in} = R_g \cdot (1 + G) = R_g \cdot (1 + g_m \cdot R_{out}) \qquad (5.88)$$

As the product $g_m \cdot R_{out}$ can easily be made equal to the amplification factor μ (10 or even more) then –even with this single-stage cathode follower– the input

---

[49]) It is important to note that here too G represents the gain without feedback. Attention is drawn to this point, because in some publications the same term is used to denote gain with feedback – actually G' – in equations for the input and output resistors (or capacitors). Thus these equations represent a different picture, although, of course, they lead to the same results.

resistance can be increased more than tenfold. It is understandable that with so high an input resistance even slight currents resulting from stray voltage can produce interfering changes in the input voltage. The screening of the input socket and the grid connection must therefore be carried out with great care in all such cases. It is advisable to connect the screening to the cathode.

According to Eq. (5.80) the input capacitance of this arrangement becomes reduced in proportion in so far as only the grid-cathode capacitance and not including the grid-earth capacitances are concerned. The effective input capacitance of a cathode follower is thus:

$$C'_{\text{in}} = C_{\text{in}} \cdot \frac{1}{1 + G} + C_{ga} \,. \tag{5.89}$$

The influence of the grid-anode capacitance C_{ga} cannot be reduced by negative feedback since, although it effectively appears between grid and anode, the anode in an amplifier must be regarded as earthed. The grid-anode capacitance must therefore be added to the input capacitance. If a pentode is used in a cathode follower stage, the suppressor grid should be connected to the cathode and the second grid decoupled to the cathode. The capacitances of these electrodes to the anode are then in parallel with the low output resistance of the cathode follower.

As the cathode voltage "follows" the input voltage except for a slight difference of amplitude, the influence of these capacitances is reduced considerably. A more detailed account of these conditions is given in the description of a probe with a "tandem" cathode follower Fig. 5-66.

For the EF 80 valve a grid resistance of 2 MΩ is permissible. If $R_k = 5$ kΩ and $R_i = 10$ kΩ, then $R_{\text{out}} = \frac{5.10}{5+10} = 3.3$ kΩ. Without negative feedback the gain, when $g_m = 5$ mA/V, would be $\frac{5}{1000} \cdot 3300$, approximately 16.7. The effective input resistance R'_{in} with negative feedback, therefore, according to Eq. (5.88) is $2 \cdot (1 + 16.7) = 33.3$ MΩ. Because of the strong negative feedback, practically no distortion of any kind occurs, nor any instability due to fluctuations in working voltages. Moreover, with this simple arrangement, the negative feedback is free from phase shift from the lowest to very high frequencies, so that there is no difficulty in attaining bandwidths of from 0 tot 30 Mc/s [36] [37] [38] [39].

The great importance of these qualities when using the cathode follower in probes, as impedance transformers or as a "power stage" in a voltage amplifier and elsewhere, has been pointed out several times in this book.

It should be made clear that in principle each of these properties can be augmented by applying complete negative feedback ($k = 1$) to a three or five stage amplifier. This, however, involves a considerably increased cost; furthermore a reduction in bandwidth is usually unavoidable, so that such multi-stage cathode followers are generally used only for special applications. Pentodes can also be used in cathode follower stages. They are then either connected as triodes, or the screen grid is decoupled to the cathode by a capacitor. The low output impedance of the cathode follower makes it possible to use a variable potentiometer as a cathode impedance. In this way, the signal voltage can be made continually variable over a wide frequency range up to very high frequencies (< 30 Mc/s).

For application in oscilloscope amplifiers it is also important to know the maximum permissible input voltage $V_{\text{in max}}$. The maximum permissible alternating control grid voltage $V_{g\text{ max}}$ can be obtained from the family of characteristics of the valve. The highest permissible input voltage is then calculated from the following equation:

$$V_{in\text{ max}} = V_{g\text{ max}} \cdot (1 + G)\,. \tag{5.90}$$

If, therefore, the valve in the above example is adjusted so that $V_{g\text{ max}} = 0.6\ V_{\text{rms}}$ and, as in the example quoted above, $(1 + G) = 17.7$, then $V_{\text{in max}} = 10.6\ V_{\text{rms}}$. In this calculation the distortion-reducing effect of the negative feedback has not yet been taken into account. This means that a maximum input signal of about 14.0 to 15.0 V_{rms} can be considered acceptable.

In another form of the cathode follower circuit the grid resistor is not connected to cathode, but, as can be seen from Fig. 5-38, to chassis "0". As in this circuit the voltage drop across the cathode resistor supplies too high a negative voltage for the grid working point, it has to be reduced by simultaneously applying a suitable positive voltage (similarly to the cathode-coupled balancing stage (Fig. 5-41)).

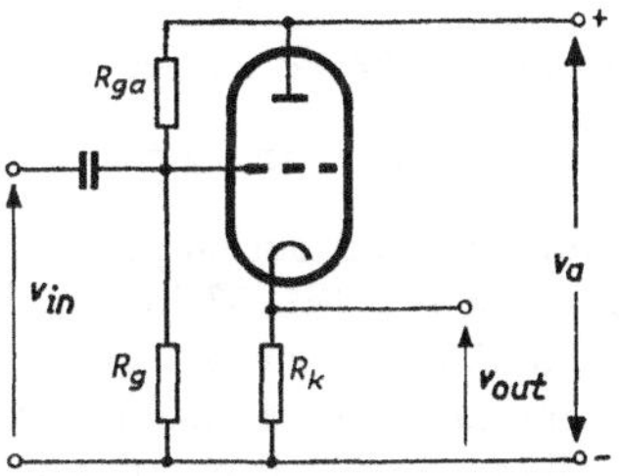

Fig. 5-38 Cathode follower circuit with operating point adjusted by a positive counter-voltage

Fig. 5-38 indicates how this can be done by connecting the grid to the anode voltage source via a resistor. The ratio of the resistors $(R_g + R_{ga})/R_g$ then determines the portion of the anode voltage which works as the positive voltage in opposition to the negative bias. Since this voltage is determined only by the ratio of resistances and can thus be regarded as "fixed", but the negative voltage from the cathode resistor amounts to several times the value of the normal grid voltage, the operating point of this valve always adjusts itself "automatically" to the set value of the cathode current. The time constant of the grid circuit is determined in this case by the coupling capacitor and the parallel circuit of R_g and R_{ga}.

Whereas in a cathode follower stage according to Fig. 5-37 the output impedance is dependent on the resistance of the input voltage source and can increase considerably with it [in Eqs. (5.84) and (5.86) this influence is not taken into account and the source impedance is assumed to be zero], it is possible to design circuits similar to that shown in Fig. 5-38, in which the output impedance remains constantly low [40] [41] [42].

5. 19 Output voltage requirements

The output stage of the amplifier must be able to supply the voltage necessary for the required deflection without introducing amplitude distortion. In this it differs fundamentally from the pre-amplifying stages which are required to deal with only small voltage amplitudes. As, according to Eq. (5.50), the value of the anode resistance must not exceed the calculated amount for a specified upper cut-off frequency, the required value of the anode alternating voltage can be obtained only as the result of a sufficiently large anode current swing. But it should be remembered that the anode current cannot be driven by control grid voltage from zero to twice the value of the quiescent current. In general, the maximum anode current change ($\Delta I_{a\,max}$) will be only from 1.5 to 1.8 times the quiescent anode current. Thus:

$$\Delta I_{a\,max} = (1.5 \ldots 1.8) \cdot I_{a\,0}\,. \tag{5.91}$$

With heavy negative feedback, however, it is possible to approach the factor 2. To achieve a satisfactorily large change in current, high-power output valves must be used in the output stage of a deflection amplifier, and their less favourable characteristics as compared with the pre-amplifying pentodes (larger electrode capacitances) will have to be tolerated.

With a deflection sensitivity of $DS_{=}$, a deflection voltage V_Y is necessary for the voltage applied between the Y-plates to produce a given deflection Y_M. Thus:

$$V_Y = \frac{Y_M}{DS_{=}}\,. \tag{5.92}$$

If, for instance, the deflection sensitivity is $DS_{=} = 0.5$ mm/V, then for a beam deflection of 60 mm, the peak-to-peak voltage required will be: $\frac{60}{0.5} = 120\ V_{pp}$. In a symmetrical output stage, each valve will have to supply 60 V_{pp}. In the family of curves shown for PL 83 in Fig. 5-39, the load line has been drawn in for an anode resistance of 2 kΩ, where the screen grid voltage is 200 V, the mains voltage is 280 V and the quiescent anode current 30 mA.

For a change in anode current of $\Delta I_a = I_{a\,pp} \cdot R_a$ the output voltage is:

$$V_{a\,pp} = I_{a\,pp} \cdot R_a\,. \tag{5.93}$$

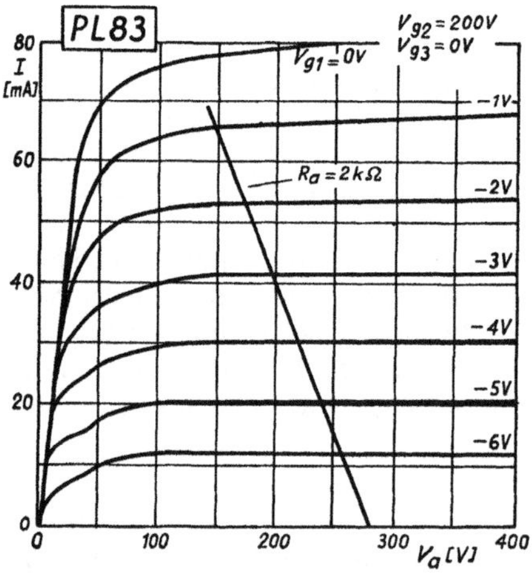

Fig. 5-39 Anode current versus anode voltage characteristics of the PL 83 valve with load line corresponding to $R_a = 2$ kΩ

An output voltage of 60 V_{pp} is obtained by driving the anode current from about 15.0 mA to about 45 mA. Any slight asymmetry is compensated by connecting the output valves in push-pull. In this case a grid alternating voltage of about 3.3 V_{pp}, i.e. approximately 1.2 V_{rms} is sufficient.

In oscilloscope amplifiers with a high upper cut-off frequency the capacitance of the deflection plates of the input leads presents a relatively very low impedance. This necessitates a correspondingly low output resistance for the output stage. For such amplifiers, valves connected as cathode followers are therefore often used in the last stage because they make it relatively easy to obtain the required low output resistance. The high input impedance of a cathode follower stage (usually symmetrical) is particuarly favourable because of its low capacitive loading of the pre-amplifying stage. It is possible to use higher anode resistance with these valves than if the deflection plates were directly connected to their anodes. Although the effective gain of the added cathode follower output stage is less than unity, it is possible in this way to obtain a greater overall amplification or a greater bandwidth (or a somewhat greater bandwidth and a some what greater gain), and also, if the usable grid base of the valves is sufficient, a greater output voltage, than without a cathode follower.

5. 20 Circuits for balancing the output voltage

One pole of the signal voltage is usually earthed, so that for symmetrical operation of the deflection plates (Ch. 2.6 "Connecting the deflection plates") balancing is required in the amplifier circuit. One way of achieving this is by using a voltage divider to obtain from the output circuit of one of the output valves a suitable voltage to drive the other output valve. In the example in Fig. 4-29 of the sweep unit triple pentode circuit a capacitance-compensated voltage divider is used for this purpose. (This arrangement is dealt with in more detail in the next Section "Setting the deflection amplitude"). Fig. 5-40 shows another way of doing this. Here the voltage division takes place across the anode resistor of one valve. The resistor R'_{a1} must be so set that the ratio of the total alternating anode voltage V_a to that portion $p \cdot V_a$ applied to the grid of the other valve is equal to the gain G of the first valve. Thus:

$$G = \frac{V_a}{p \cdot V_a} = \frac{1}{p} . \tag{5.94}$$

From this the resistance ratio is found to be

$$p = \frac{R'_{a1}}{R_{a1} + R'_{a1}} = \frac{1}{G} \tag{5.95}$$

and

$$R_{a1} = \frac{R_{a1} + R'_{a1}}{G} . \tag{5.96}$$

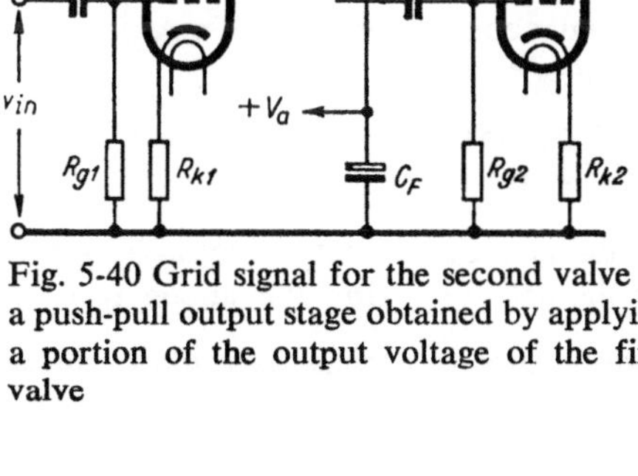

Fig. 5-40 Grid signal for the second valve of a push-pull output stage obtained by applying a portion of the output voltage of the first valve

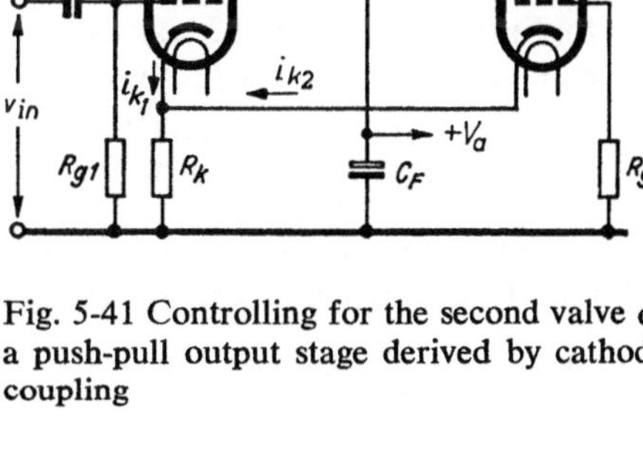

Fig. 5-41 Controlling for the second valve of a push-pull output stage derived by cathode coupling

If, as has already been calculated for the PL 83, $G = 18$ when $R_a = R_{a1} + R'_{a1} = 2\ \text{k}\Omega$, then R'_{a1} must $= \frac{2000}{18} = 111\ \Omega$. If an inductance is used to bring about an extension of the upper frequency range, then of course, a corresponding portion of the inductance will also have to be connected in series with R'_{a1}.

The disadvantage of the two circuits just discussed is that an additional *RC*-network must be used. Naturally, this renders the lower cut-off frequency of the second output valve higher than that of the first, so that there is lack of balance in the load at this end of the frequency range. A phase inversion stage, the basic circuit of which is shown in Fig. 5-41, avoids this disadvantage, as the second valve is driven by the alternating voltage arising at the common cathode resistor. This circuit is particularly useful for amplifiers for operation at extremely low frequencies and for DC-amplifiers.

The operating principle of the circuit shown in Fig. 5-41 is as follows: the input voltage V_{in} results in an alternating anode current i_{a1} which also flows through the cathode resistor R_k. The alternating voltage drop thus occurring across the cathode resistor is in anti-phase with the input voltage at the control grid of valve 1 (negative feedback). At the same time, however, this voltage is also connected between grid and cathode of valve 2 (the grid of which is in this case connected to earth); in this way the valve is driven in anti-phase as desired. An alternating current arises in the anode circuit of valve 2 and this also flows through the common cathode resistor. The alternating voltage drop across this resistor due to this current is, however, in anti-phase to the alternating cathode voltage due to the current of valve 1, thus reducing the control voltage for valve 2 and which in turn produces a decrease in the voltage drop caused by the anode current of this valve. This tends, however, to bring about an increase in the voltage drop across the cathode resistor, so that valve 2 is again more strongly driven. Ultimately a state of equilibrium is attained and the equilibrium condition can be set for a particular valve type by suitable choice of R_k. If, in certain circumstances, this makes the direct negative grid voltage (grid bias) too great, then the working point can be simply re-established by connecting the leak resistor R_{g1} and R_{g2} to a suitable positive voltage. Valve 2 is thus driven by the difference in alternating voltage caused by the cathode currents i_{k1} and i_{k2}. It is also clear that complete symmetry is unattainable in this way. A simple calculation shows that the out-of-balance remains sufficiently small if the product $G_m \cdot R_k$ is large compared with unity.

The ratio of the anode alternating currents is given by the equation

$$\frac{i_{a1}}{i_{a2}} = \frac{1 + \beta \cdot g_m \cdot R_k}{\beta \cdot g_m \cdot R_k} \text{ . [50]} \tag{5.97}$$

If, for instance, two Pl 83 valves with a mutual conductance of 10 mA/V and a cathode resistance of 500 Ω are used in this stage, then

$$\frac{i_{a1}}{i_{a2}} = \frac{1 + 1.14 \times 10 \times 10^{-3} \times 500}{1.14 \times 10 \times 10^{-3} \times 500} = 1.175 \text{ .}$$

The difference of the anode currents and thus also of the output voltages is then 17.5%. If necessary, it can be completely balanced by giving the anode resistors

[50] The factor β expresses the difference between cathode current and anode current (roughly 1.1 to 1.3).

unequal values. This circuit can be used in pre-amplifying stages. In this case the gain is generally approximately only half of that obtained without negative feedback via the cathode resistor (e.g., with fixed grid voltage and earthed cathode).

This circuit is often used in DC amplifiers. With a particularly high-rated cathode resistor it becomes a "difference amplifier", the characteristics and applications of which are discussed in more detail in conjunction with Fig. 5-57.

If, as shown in Fig. 5-42, coupling resistors of the same value are connected in the anode and cathode leads of a valve, the voltages between the grid and the cathode and anode respectively can also be used to drive a push-pull pair. These voltages will be balanced and in antiphase. The gain between the input and each output is, of course, always less than 1, as the entire voltage from the cathode resistor is fed back in antiphase. The disadvantage of this circuit is therefore, that the maximum amplitude of output voltage is somewhat less than twice the input voltage, so that to drive large output stages it is necessary either to use a valve with sufficiently large anode current swing or to interpose a further symmetrical stage consisting of two valves. This offers the advantage of a high input impedance and a low output impedance (heavy negative "current" feedback). In this way both the output valves may be driven entirely symmetrically.

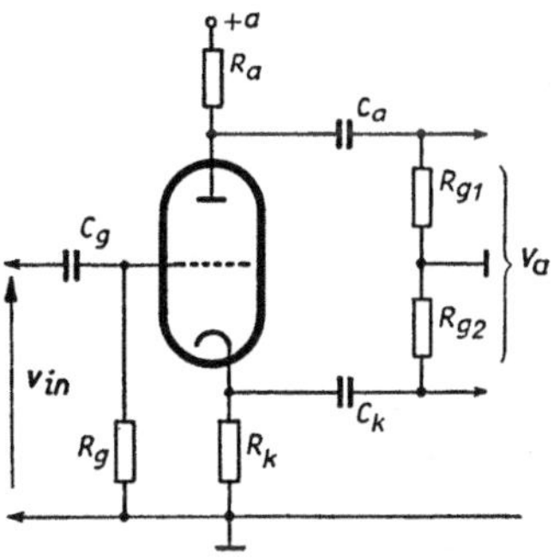

Fig. 5-42 "Catodyne" circuit for balancing the output voltage from the amplifier

Other phase-inverter circuits for driving push-pull stages are in use [43] [44]. Thus, for instance, the so-called anode follower is used in oscilloscopes to balance the time base voltage [45]. This is a grounded-cathode stage in which the negative feedback is made so strong that the gain is unity.

5. 21 Setting the deflection amplitude

Fundamentally there are two possible ways of setting the amount of vertical deflection of the beam. Either the gain of the valves can be varied or a voltage divider may be connected in the input or output circuit of the amplifier. The latter method, which permits the amplifier to operate continuously at maximum gain, is the system most widely used at present time. The voltage division can be effected at the input, between the individual amplifier stages, or at the output. An input potentiometer, as adopted for use with AF amplifiers, would seem to be particularly advisable. Since, however, the upper cut-off frequency is usually much higher in deflection amplifiers for oscilloscopes than in electro-acoustic amplifiers, attention must be paid to the influence of unwanted capacitances on the voltage division. Fig. 5-43a shows a circuit of this sort and Fig. 5-43b shows the equivalent circuit of an input potentiometer. It can be seen that the influence of capacitance C, which is composed of the self-capacitance of the potentiometer, stray wiring capacitance and the input capacitance of the valve, is greatest at the central setting of the potentiometer. The voltage division ratio $\frac{V_2}{p' \cdot V_1}$ is then:

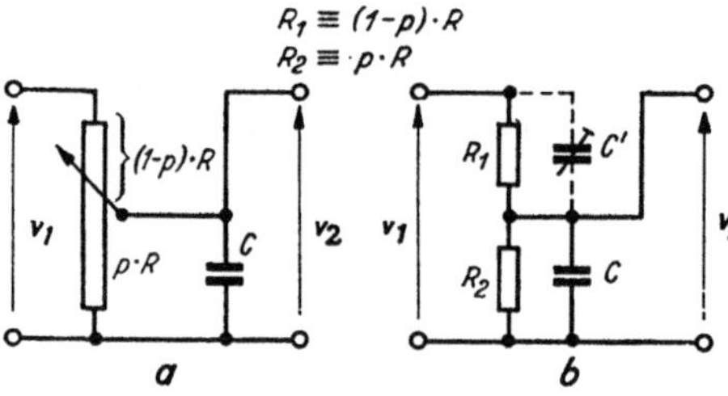

Fig. 5-43 Circuit diagram illustrating the influence of the circuit capacitance at a voltage divider

$$\frac{v_2}{p' \cdot v_1} = \frac{1}{\sqrt{1 + \frac{\omega^2 \cdot C^2 \cdot R^2}{16}}} \ [46]\,. \tag{5.98}$$

If, for example, $R = 10{,}000\ \Omega$, $C = 20$ pF and $\omega = 2 \cdot \pi\ 10^5$ ($f = 10^6$ c/s), then:

$$\frac{v_2}{p' \cdot v_1} = \frac{1}{\sqrt{1 + \frac{4 \cdot 9.85 \cdot 10^{12} \cdot 400 \cdot 10^{-12} \cdot 10^8}{16}}} = \frac{1}{\sqrt{1 + 0.0985}} = 0.953.$$

In other words, as a result of the shunt capacitance C, the output voltage v_2 at a frequency of 1 Mc/s is smaller by about 4.7% than it should be. If $R = 100$ kΩ the error would be 69%, which, of course, could not be tolerated. For this reason such types of controls must have a relatively low ohmic value, but if they are used in the input they impose an excessive load on the voltage source. The difficulty can be overcome by shunting a capacitor — C' in Fig. 5-43b — across the "upper" part of the potentiometer. As the frequency increases, a larger current flows through C; thus compensating the influence of capacitance C, the current through which also increases with the frequency. In this way it is possible to build high-impedance voltage dividers which are to a great extent independent of frequency. In this case the following relationship must be ensured:

$$R_1 \cdot C' = R_2 \cdot C\,. \tag{5.99}$$

Thus at the low frequencies, voltage division is effected by the resistances, at the high frequencies by the capacitances. In the example quoted, compensation could only apply to that particular setting of the potentiometer. It has not so far been used for continuous voltage adjustment since the capacitor C' would likewise have to be variable and thus necessitate complicated control components. Much use of this system is made in step voltage dividers, however (see also Part III, Ch. 25) [46].

In general, the control of gain by varying the direct voltages on the first or second grid cannot be recommended. The voltage jumps which are unavoidable with this procedure are transferred in amplified form to the deflection plates, especially where amplifiers with a very low frequency limit are concerned. It is, however, very convenient to control amplification by means of variable negative feedback. Every other form of adjustment is equivalent to a deliberate reduction of the maximum amplification obtainable, while the frequency response of the amplifier remains unaltered. If, on the other hand, the gain is reduced by increasing the negative feedback, then, as pointed out earlier in Ch. 5.18 "Feedback", the frequency range is extended at the same time. When using variable negative feedback with frequency-dependent circuit components in the feedback circuit, it must be ensured that no disturbing rise in gain or tendency to oscillation occurs near the frequency limits.

5. 22 "Corrected" cathode follower

The time constant of the output circuit of a cathode follower results from the fact that the output resistance (internal resistance) of the valve is shunted by the load capacitance. Because the output resistance is low, a small time constant results, so that the cathode follower stage seems particularly suitable for the transmission of steep-flanked pulses. The data given for input and output impedance are, however, only valid so long as the control voltage excursion takes place in the linear part of the dynamic characteristic. This condition is not always fulfilled. Particularly when cathode followers in output stages of oscilloscope amplifiers have to supply the output voltage for the deflection plates, such large input voltages have to be handled that the instantaneous values of the anode current traverse the whole range of the dynamic characteristic between the start of grid current to the cut-off point. So long as the valve is operating in the straight portion of the characteristic, the average mutual conductance g_m can be used for calculations and the approximate output resistance is obtained from Eq. (5.87), i.e. $R'_{out} = 1/g_m$, and thus the time constant of the output is

$$T'_{out\ 1} = \frac{C_{out}}{g_m} \,. \qquad (5.100)$$

Taking into consideration the shunting effect of the cathode resistance on the output resistance of the valve, the following equation is obtained:

$$T'_{out} = \frac{C_{out}}{g_m + \frac{1}{R_k}} \,. \qquad (5.101)$$

Usually, however, calculations can simply be made according to Eq. (5.100) [51]). If the grid voltage is positive-going, then the rising current charges the output capacitance rapidly, and the leading edge is well transmitted. Large negative-going control voltages, however, encroach on regions in which the characteristic is curved, and the mutual conductance thus becomes progressively less. At the cut-off voltage the anode current is completely suppressed and the mutual conductance is zero. In this case the time constant of the output is determined only by the parallel circuit of the output capacitance and the cathode resistor; thus:

$$T'_{out\ 2} = R_k \cdot C_{out} \,. \qquad (5.102)$$

As the cathode resistance is always many times greater than the output resistance –at low control voltage– the time constant fluctuates at higher negative-going voltages in accordance with the shape of the valve characteristic, between values according to Eq. (5.100) and according to Eq. (5.102).

With a mutual conductance of 10 mA/V and an output capacitance of 100 pF the time constant can be calculated according to Eq. (5.100) as, for example:

$$T_{out\ 1} = \frac{100 \times 10^{-12}}{10 \times 10^{-3}} = 10^{-8}\,\mathrm{s} = 10\,\mathrm{ns}\,.$$

---

[51]) C_{out} is now the output capacitance of the cathode follower, comprising the cathode – filament capacitance and connected load capacitance (circuit capacitance and inter-electrode capacitance of the following valve).

With a cathode resistance of 2 kΩ, however, in the cut-off state of the valve, according to Eq. (5.102) we obtain

$$T_{\text{out }2} = 2 \times 10^3 \times 100 \times 10^{-12} = 2 \times 10^{-7} = 200 \text{ ns}.$$

This time constant is therefore twenty times greater than for low control voltages. This means that with pulses having a steeply rising and decaying waveform of output voltage, the leading edge is reproduced with a low time constant but the trailing edge decays more slowly. The sketches in Fig. 5-44 indicate this more clearly. This disadvantage can be overcome if a valve, preferably a pentode, is used instead of a simple cathode resistor. The use of valves as coupling elements offers many advantages. They make possible DC coupling elements having only a low impedance for DC but a high AC impedance. Although the use of thermionic valves or transistors as coupling elements was suggested some considerable time ago [47], it is only recently that they have come into increasing use for deflection amplifiers [48] [49].

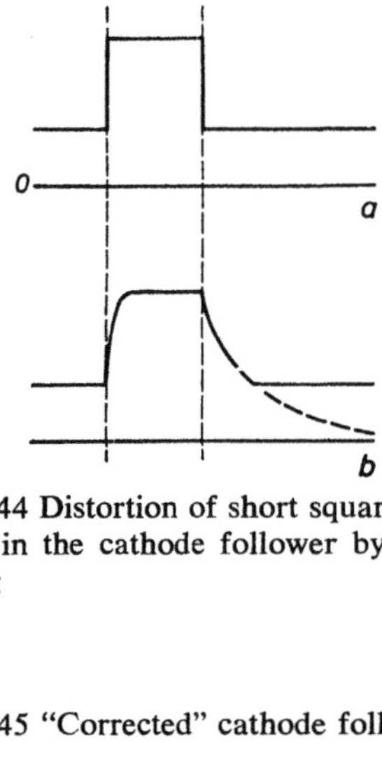

Fig. 5-44 Distortion of short square wave pulses in the cathode follower by heavy driving

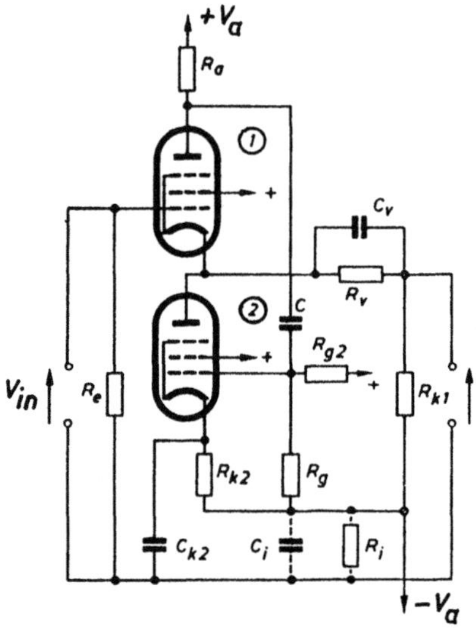

Fig. 5-45 "Corrected" cathode follower

The circuit of such a "corrected" cathode follower ("White cathode follower") [50], as described by M. J. Goddard [51], is shown in Fig. 5-45. It works with DC coupling and is intended as an input circuit (probe). Valve 1 is the actual cathode follower stage. The "corrector" valve 2 is connected in its cathode circuit with a voltage divider consisting of R_v and R_{k1} in parallel with it. This voltage divider is so dimensioned that the voltage at the output terminals is 0 volts when the input voltage is zero. The cathode follower stage in this circuit has, in addition to the cathode impedance, also an impedance (R_a) in the anode lead. Its value is relatively low, so that the dynamic mutual conductance is only slightly lower than the static value. The alternating voltage at the anode of valve 1 is coupled via capacitor C to the grid of valve 2, since the effect of the corrector valve is of importance only for the alternating voltages. If the cathode follower is only slightly driven, then the cathode "follows" the grid, so that the anode current variations of valve 1 are only small. Thus the alternating voltage at the anode of this valve is correspondingly low and the control grid of valve 2 receives only a small alternating voltage signal. If, however, the control grid of the cathode follower is controlled by a negative-going voltage so far that the cathode current becomes almost zero and in consequence the time constant of the output circuit now increases, then the voltage at the anode increases and the grid of the corrector valve (valve 2) receives a positive voltage pulse. This, however, causes the current in the corrector valve to rise, the voltage at its anode drops and thus the cathode of the cathode follower stage quickly "follows" the voltage pulse applied to its grid, which is, of course, exactly what is required. In this way satisfactory transmission of pulse voltages with steep edges and large amplitudes is possible. As a result of using a corrector valve, the output impedance over the linear region of the characteristic becomes considerably lower than that of a normal cathode follower. To a first approximation:

$$R'_{\text{out }k} = \frac{1}{g_{m1}} \cdot \frac{1}{1 + g_{m2} \cdot R_a} \,. \tag{5.103}$$

Here, the mutual conductance of the cathode follower valve is g_{m1}, that of the corrector valve is g_{m2} and the anode load of the cathode follower valve is R_a. For an anode load $R_a = 500\Omega$ and a mutual conductance of the corrector valve 10 mA/V, the output resistance is one sixth of that in a simple cathode follower circuit.
By using electronic valves or transistors as coupling elements in the anode circuit of a cathode follower to provide the anode voltage to control the "corrector" value, circuits with particularly high input impedance can be designed [48] [49].

5. 23 Delay lines; signal delay networks

Every time base unit requires a certain time to trigger the deflection sawtooth even though that time be very short. If, however, rapid pulses are to be studied, then, if the time base unit is directly triggered with this voltage (internal triggering), that period at the start of the voltage rise which corresponds to the triggering time of the deflection unit will be invisible. With signal pulses which are triggered by means of a separate source it is possible to trigger the time base unit of the oscilloscope by means of this source directly and the signal pulse with phase delay. In this way –as will be described in Part II, Ch. 13– any required portion of the oscillogram before the voltage rise under observation can be displayed. For the purpose of observing random phenomena, and especially for obtaining a complete picture of non-recurrent phenomena, it is necessary to delay the signal voltage for the required time on its way from the amplifier input to the deflection plates. Delay lines, they are sometimes termed "signal-delay networks", are used for this purpose. The fundamental properties of these circuit elements will be dealt with in due course, in so far as this seems necessary for understanding the working of modern oscilloscopes. For theoretical and detailed study the reader is referred to the bibliography under [3] [50] [51] [52] [53] [54] [55] [56] [57]. Any pair of lines serves to delay each pulse transmitted because of the line inductance and the capacitance between the conductors. The line can be regarded as a large number of inductances connected in series and the same number of capacitances connected in parallel, as shown in Fig. 5-46*a*. One of the most important properties of such a cable is its characteristic impedance *Z*; it is given approximately by

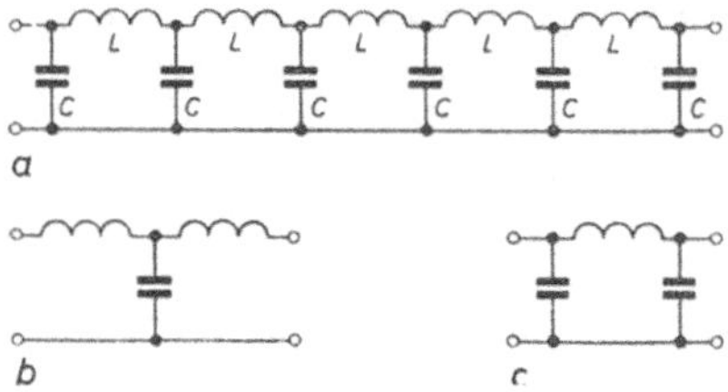

Fig. 5-46 Delay lines
a) Delay lines consisting of *Π*-links
b) *T*-link
c) *Π*-link

$$Z = \sqrt{\frac{L}{C}}. \tag{5.104}$$

In order to prevent the occurrence of undesirable reflection, such cables must be terminated with an ohmic resistance equal to *Z*. Delay cables consist essentially of a flexible magnetic core on which the conductor is spirally wound. In this way considerable delay times can be attained even with short leads (0.13 . . . 1.9 μs/m).

For technological reasons, ordinary delay cables having a sufficiently high cut-off frequency also have a relatively low characteristic impedance. Moreover, it is difficult to make delay cables the delay of which will be sufficiently constant over the frequency range required in oscilloscopy. Since, therefore, no entirely suitable cables are available, it is preferable to obtain the desired delay by using "signal-delay networks". Fig. 5-47 shows the type and arrangement of the symmetrical delay line in the Philips "GM 5602" oscilloscope. For satisfactory transmission of the signal the same demands must be made on the upper cut-off frequency as on the rating of wide-band amplifiers. The upper cut-off frequency of such signal-delay networks (to some extent also in delay cables) is given for the individual network by the expression:

$$f = \frac{1}{\pi \sqrt{L \cdot C}}. \tag{5.105}$$

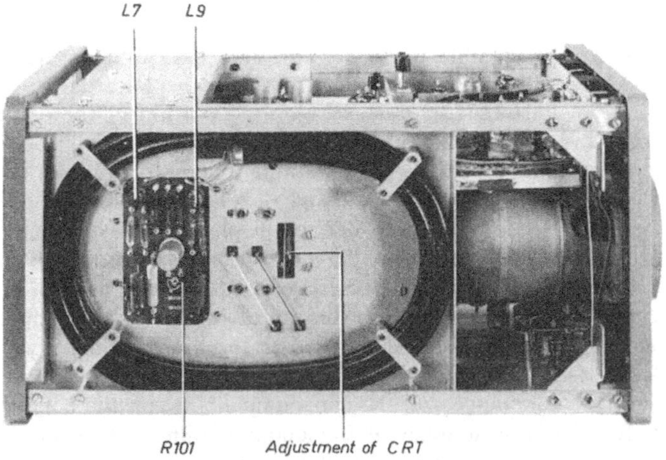

Fig. 5-47 Delay cable in the Philips "GM 5602" oscilloscope L_7 and L_9 compensating inductances at the termination of the cable

From this it can be seen that C and L must only be small for a given high upper cut-off frequency. Thus, with such a network the attainable time delay can also only be small. It can be obtained (per link) from the equation

$$T_v = \sqrt{L \cdot C}. \tag{5.106}$$

For a required delay time, therefore, a corresponding number of such networks –T or Π networks as required, de-coupled or coupled to one another– are connected in series. The firm of Tektronix makes use of a delay network of 48 links in the vertical amplifier of the "540" and "545" oscilloscopes giving an upper cut-off frequency of 30 Mc/s at a delay of 0.2 μs. The delay networks are generally connected either between the pre-amplifier and the main amplifier or in the output circuit — that is to say in the leads to the deflection plates. When connected in the output circuit it is advisable to have symmetrical networks. Fig. 5-48 shows some of the last links of such a delay chain as used in the Tektronix oscilloscope "545". It consists of delay networks with air trimmers and the capacitances of the deflection plates as shunt capacitances. All components are mounted on a transparent plastic panel. One chain of coils is placed below the panel and the other above it.

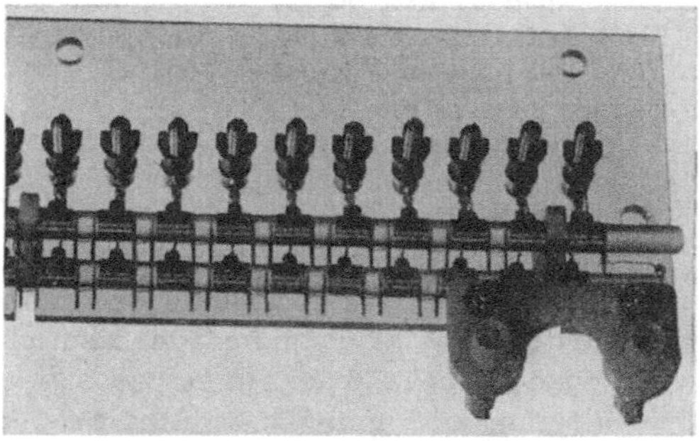

Fig. 5-48 Part of the end links of the delay line of the Tektronix "545" oscilloscope

5. 24 Measuring the response time of the time base unit and the delay time in the Y-channel

The oscillograms in Fig. 5-49 and 5-50 show how the rise response of the time base unit and the delay time of the circuit elements in the Y-channel can be easily determined. A series of voltage pulses with a width of about 0.9 μs and a repetition frequency of about 238 kc/s (4.2 μs periodic time) is fed to the Y-input of the oscilloscope under study. A voltage of the same waveform is used for unblanking. (Owing to the way in which the unblanking circuit of this particular oscilloscope operated, this voltage had to be negative-going.). The Philips "GM 2314" pulse generator is particularly suitable as it supplies symmetrical output voltages. As the time required for unblanking is negligibly short compared with the times to be measured, comparison of the brightened portions of the oscillograms with the displayed waveforms will indicate the desired times with sufficient accuracy. Fig. 5-49 shows two successive pulses with a time coefficient of 1 μs/cm. It can at once be seen that the brigthened portions of the oscillogram occur in time before the image of the corresponding voltages pulse. This time difference obviously corresponds to the signal voltage delay in the Y-channel brought about mainly by the delay cable.

It can also be seen that a small part of the time base portion is missing before the first

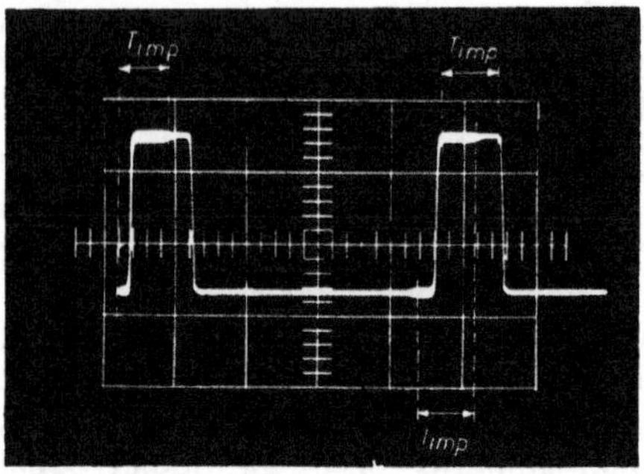

Fig. 5-49 Oscillogram for determining the response time of the time base and delay in the *Y*-amplifier

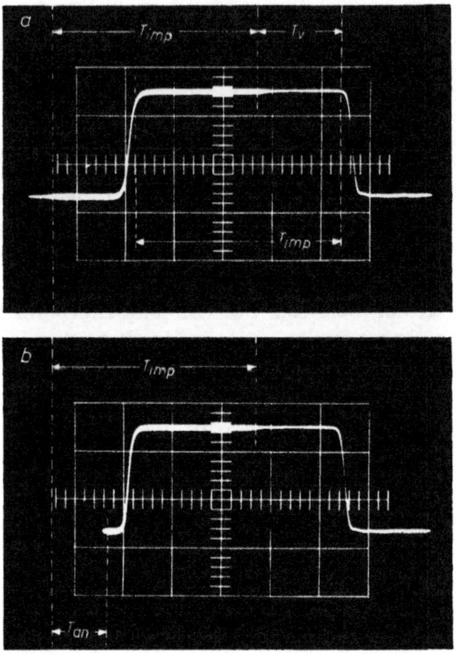

Fig. 5-50 Pulse oscillogram of Fig. 5-49 each expanded fivefold
Time scale: 0.2 μs/cm
a) Second pulse of Fig. 5-49
b) First pulse of Fig. 5-49
$T_p = 0.9\ \mu s$, $T_{an} = 0.24\ \mu s$, $T_V = 0.36\ \mu s$

pulse (the leading edge of which triggers the time base), but not before the second pulse. The length of the missing time base portion corresponds to the response time of the time base. To ensure better interpretation, the oscillogram of Fig. 5-49 was magnified fivefold for Fig. 5-50*a* and *b* (time coefficient 0.2 μs/cm). From this the delay time T_v was found to be 0.36 μs. The response time of the time base is found by comparing the oscillograms of Fig. 5-50*a* with those 5-50*b*, and is 0.24 μs (at the trigger level adjustment selected in this case).

5. 25 Transmission line amplifier (distributed amplifier)

As has been shown in detail in the Chapters 5.16 and 5.17 "Improving gain frequency curve at the upper frequency limit by L-resonance" and "Unit step response of an amplifier", the upper frequency limit in a conventinal amplifier circuit using a particular amplifier valve is given theoretically by the figure of merit $g_m/2 \cdot \pi \cdot C_R$. At this limit only a gain of 1 is possible [see also Eq. (2.11)]. For greater gains a correspondingly lower value of the upper cut-off limit is obtained. Fundamental improvement of such an amplifier is only obtainable by means of circuits in which the effect of valve and circuit capacitances can be considerably reduced or eliminated altogether. One might now be tempted to obtain more favourable conditions by connecting a number of valves in parallel. The mutual conductance is certainly increased by this means, but so are the capacitances, with the result that the g_m/C ratio remains the same. A device which goes very far towards solving the problem is a distributed amplifier.

The basic principle is to use the input capacitances C_g and the output capacitances C_a of several valves as shunt capacitors in a delay-line network. In the example of such a circuit shown in Fig. 5-51, the links of this network are of the *T*-type.

Fig. 5-51 Basic circuit diagram of a delay line (distributed) amplifier

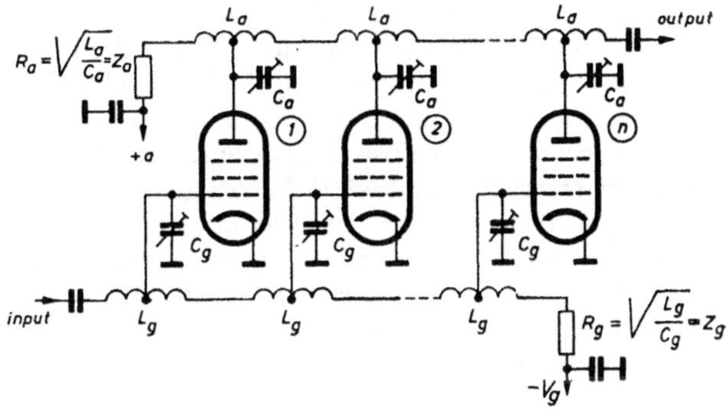

The input signal passes along the delay network at the grid side (L_g) and controls the individual valve grids in turn. The anode currents undergo a corresponding change in the same sequence. In doing so they split up into two components, one flowing to the output and the other to the terminating resistor R_a. Since the delay times are the same on the grid side and on the anode side, the changes in the currents flowing to the output as a result of the alterations to the grid voltages are additive. The reflected current surges are dissipated in the terminal resistance. If Z_a is the terminal resistance of the anode network, then the effective lòad impedance across the anode of the individual valves is $Z_a/2$ and the total gain G_k of the distributed amplifier with n valves of mutual conductance g_m is equal to

$$G_K = n \cdot g_m \cdot \frac{Z_a}{2} \,. \tag{5.107}$$

In contrast to conventional amplifiers connected in cascade, in which the total gain is the *product* of the gains of the individual stages, here the total gain is the *sum* of the gains of the individual stages. On the other hand, however, bandwidths of from 0 to about 500 Mc/s (rise times down to 1 ns) can be obtained [50] [53] [58] [59] [60] [61]. KALLMANN, in his "Components Handbook" has given very useful directions for carrying out calculations and has included very useful examples of practical time delay networks [62]. Distributed amplifiers were first used for output stages, thus making it possible to connect a number of valves in parallel without incurring direct interference from the valve capacitances. Of course, this also permits of the construction of push-pull stages for symmetrical control of the deflection plates. Thus a transit time delay of the signal occurs in the amplifier corresponding to the delay of the delay networks used. The desired total delay can be obtained by extending the networks by the required number of links. This delay system cannot in general just be simply switched off. The output of each stage is connected to the input of the following one. The number of valves and other components used is, of course, correspondingly high. Such amplifiers are therefore only found in oscilloscopes of the highest price range. For example, the "High-Speed" oscilloscope, type "517" of the firm of Tektronix has the following valve complement:

1st and 2nd pre-amplifiers, six 6AK5 valves each;
3rd pre-amplifier, seven 6CB6 valves;
phase invester stage, three 6CB6 valves;
control amplifier, twelve 6CB6 valves and
output stage, twenty-four 6CB6 valves; 54 valves in all.

The rise time of these amplifiers is given as 7ns, which corresponds roughly to an upper cut-off frequency of 50 Mc/s. This extremely high cost is only necessary when the utmost demands are made on the bandwidth. For upper cut-off frequencies up to 50 Mc/s (rise time about 7 ns) completely satisfactory solutions have been found even without recourse to the use of distributed amplifiers. Thus the driving of a wide-band amplifier stage by a cathode follower gives greater band width and gain than direct coupling of the individual amplifier stages. The load impedance of a cathode follower stage can, as has been shown, be much higher than that of a normal amplifier stage, so that the preceding valve is loaded with a many times smaller capacitance. "Corrected" cathode followers have been used for the output stage of such amplifiers with a considerable succes (Solartron "CD 518" and "CD 568").

5. 26 Some examples of AC amplifiers

An example of a simple, two-stage amplifier with a push-pull output stage is that used in the Philips "GM 5655/03" small oscilloscope. In this apparatus two identical amplifiers are provided, one for each deflection direction. The sawtooth voltage generator is connected to the input of its amplifier. Both amplifiers are thus constantly in operation.

As both amplifiers are identical, the gain characteristic and phase of their output voltages are also practically the same, so that, for example, phase measurements over a large frequency range can be carried out. The lower cut-off frequency of this amplifier is given as 3 c/s, so that a phase shift of $+45°$ occurs at this frequency with each amplifier. Nevertheless, the phase difference of two sinusoidal voltages can be measured as a Lissajous figure down to 3 c/s without any difficulty, as the phase difference between the two output voltages at this frequency is considerably less than 1° and therefore not noticeable.

The circuit of this amplifier is shown in Fig. 5-52 — actually the vertical deflection amplifier. The signal voltage is connected between the earth socket (3) and the grid of the input valve either directly via socket (4) and the input potentiometer (R_5) or via socket (5) which inserts a voltage divider — (R_7 and R_8). The input capacitances are then either 35pF or 6pF. The voltage division is 10 : 1. Sockets (3) and (4) may be used for signals up to 100 V_{rms} (283 V_{pp}) and sockets (3) and and (5) for signal voltages of up to 400 V_{rms} (1113 V_{pp}). For "wide-band" transmission of all frequencies within the range of amplifiers, the voltage divider is compensated by a capacitor (C_5). This compensation is, however, only available when the slider of potentiometer (R_5) is in the "upper" position. However, in order that the quality of the transmission of the divider should not noticeably deteriorate at other settings of the potentiometer, a relatively low value was selected (100 kΩ). This, of course, imposes a certain load on higher impedance voltage sources; this had to be accepted, however, because an oscilloscope of such small dimensions ($11.6 \times 24.5 \times 30$ cm; weight approx. 6.5 kg) can only be built economically by making some concessions to less important qualities. The EF 80 HF pentode is the pre-amplifier valve. Its cathode resistor is not bypassed by a capacitor, so that current feedback occurs. The amplified voltage appearing at the anode of this valve directly drives the grid of one system of valve (2) of the output stage which consists of both the systems of a double triode (ECC 81). Both systems are coupled by the common cathode resistor (R_{26}) (cathode coupling, see Fig. 5-41 and the accompanying text). The second system (2) is thus driven in anti-phase — the grid is AC earthed via a capacitor (C_{14}) — by the cathode alternating voltage. The

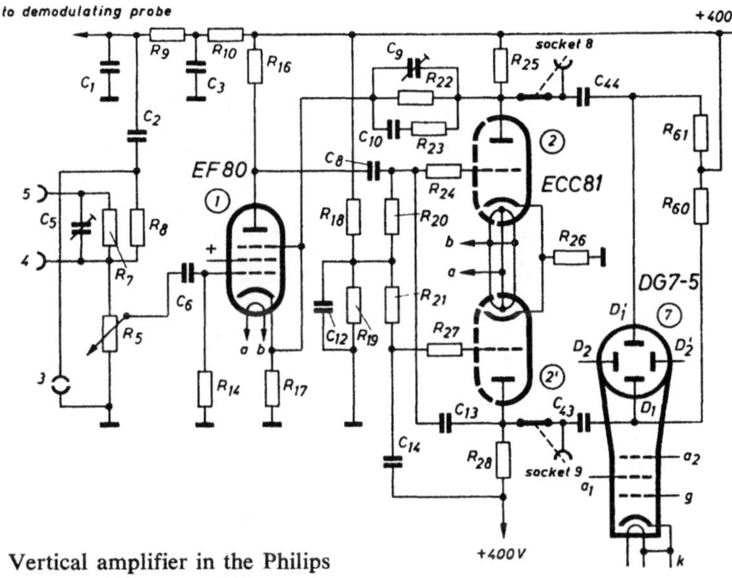

Fig. 5-52 Vertical amplifier in the Philips small oscilloscope "GM 5655/03"

Components of the amplifier	R_{16} = 56 kΩ	R_{26} = 8.2 kΩ	C_5 = max. 6 pF
	R_{17} = 390 kΩ	R_{27} = 1 kΩ	C_6 = 0.27 μF
Resistors	R_{18} = 4.7 MΩ	R_{28} = 62 kΩ	C_8 = 0.1 μF
	R_{19} = 0.47 MΩ	R_{61} = 4.7 MΩ	C_9 = 12.5 pF
R_5 = 0.1 MΩ/lin.	R_{20} = 1 MΩ	R_{62} = 4.7 MΩ	C_{10} = 0.15 μF
R_7 = 0.75 MΩ	R_{21} = 1 MΩ	Capacitors	C_{12} = 10 nF
R_8 = 0.56 MΩ	R_{22} = 0.82 MΩ		C_{13} = 1.8 . . 5.6 pF
R_9 = 0.22 MΩ	R_{23} = 0.33 MΩ	C_1 = 5.6 nF	C_{14} = 0.1 μF
R_{10} = 22 kΩ	R_{24} = 1 kΩ	C_2 = 3.3 nF	C_{43} = 0.1 μF
R_{14} = 1 MΩ	R_{25} = 62 kΩ	C_3 = 0.1 μF	C_{44} = 0.1 μF

relatively high value of the cathode resistance required for this coupling would result in too high a negative bias for the control grid. The correct working point is therefore obtained by feeding a positive counter-voltage to the control grids via the leak resistors (R_{20} and R_{21}). This voltage can be adjusted by the voltage divider (R_{18} and R_{19}). From the anodes of both valve systems in the output stage the output voltage reaches the deflection plates of the cathode ray oscilloscope via the CR couplings (C_{43}/R_{60} and C_{44}/R_{61}). The deflection plates are also accessible via the switch sockets (8 and 9) from outside, so that higher voltages can be observed with only the small load impedance of the leak deflection plates and the leak resistors (30 pF/4 MΩ). In this way, for example, modulation measurements on HF transmitters whose frequency can be higher than the upper frequency limit of the amplifier ($<$ 100 Mc/s) can be carried out.

As the cost of this set had to be kept at a minimum, linearizing the frequency response by means of inductances was dispensed with, but on the other hand, frequency-dependent negative feedback by an *RC*-circuit (C_9, C_{10}, R_{22} and R_{23}) from the anode of the upper system of the output stage (2) to the cathode of the preamplifier stage has been provided. The amplifier circuit in the Philips "GM 5662" HF oscil-

Components of the amplifier

Resistors

$R_9 =$ 1 kΩ
$R_{101} =$ 120 Ω
$R_{102} =$ 4.7 MΩ
$R_{104} =$ 560 Ω
$R_{105} =$ 5.6 kΩ
$R_{106} =$ 1.2 kΩ
$R_{107} =$ 120 Ω
$R_{108} =$ 12 Ω
$R_{109} =$ 270 kΩ
$R_{110} =$ 10 kΩ
$R_{111} =$ 10 kΩ
$R_{112} =$ 68 Ω
$R_{113} =$ 470 kΩ
$R_{114} =$ 750 Ω
$R_{115} =$ 4.7 kΩ
$R_{117} =$ 68 Ω
$R_{118} =$ 100 kΩ
$R_{119} =$ 470 kΩ
$R_{120} =$ 68 Ω
$R_{121} =$ 750 Ω
$R_{123} =$ 1 MΩ
$R_{124} =$ 560 kΩ
$R_{125} =$ 180 Ω
$R_{126} =$ 1.2 kΩ
$R_{127} =$ 1.8 kΩ
$R_{128} =$ 22 kΩ
$R_{130} =$ 470 Ω
$R_{131} =$ 270 kΩ
$R_{132} =$ 150 Ω
$R_{133} =$ 2.7 kΩ
$R_{135} =$ 1 MΩ
$R_{137} =$ 680 Ω
$R_{138} =$ 1.5 MΩ
$R_{139} =$ 680 Ω
$R_{140} =$ 22 kΩ
$R_{147} =$ 1.2 kΩ
$R_{148} =$ 6.8 kΩ
$R_{150} =$ 6.8 kΩ
$R_{151} =$ 50 Ω
$R_{152} =$ 100 Ω
$R_{153} =$ 100 Ω
$R_{154} =$ 180 Ω
$R_{155} =$ 3.3 MΩ
$R_{156} =$ 1.2 MΩ
$R_{157} =$ 180 kΩ
$R_{448} =$ 0.82 MΩ
$R_{449} =$ 1 MΩ

Capacitors

$C_{101} =$ 0.5 μF
$C_{102} =$ 50 + 50 μF
$C_{104} =$ 50 μF
$C_{105} =$ 12 nF
$C_{106} =$ 0.25 μF
$C_{107} =$ 25 + 100 μF
$C_{108} =$ 25 μF
$C_{109} =$ 0.25 μF
$C_{110} =$ 0.5 μF
$C_{111} =$ 175 pF
$C_{112} =$ 22 pF
$C_{113} =$ 0.5 μF
$C_{114} =$ 0.25 μF
$C_{115} =$ 25 + 100 μF
$C_{117} =$ 0.5 μF
$C_{118} =$ 33 pF
$C_{119} =$ 0.5 μF
$C_{125} =$ 25 pF
$C_{131} =$ 68 pF
$C_{132} =$ 33 pF
$C_{419} =$ 0.1 μF
$C_{420} =$ 0.5 μF

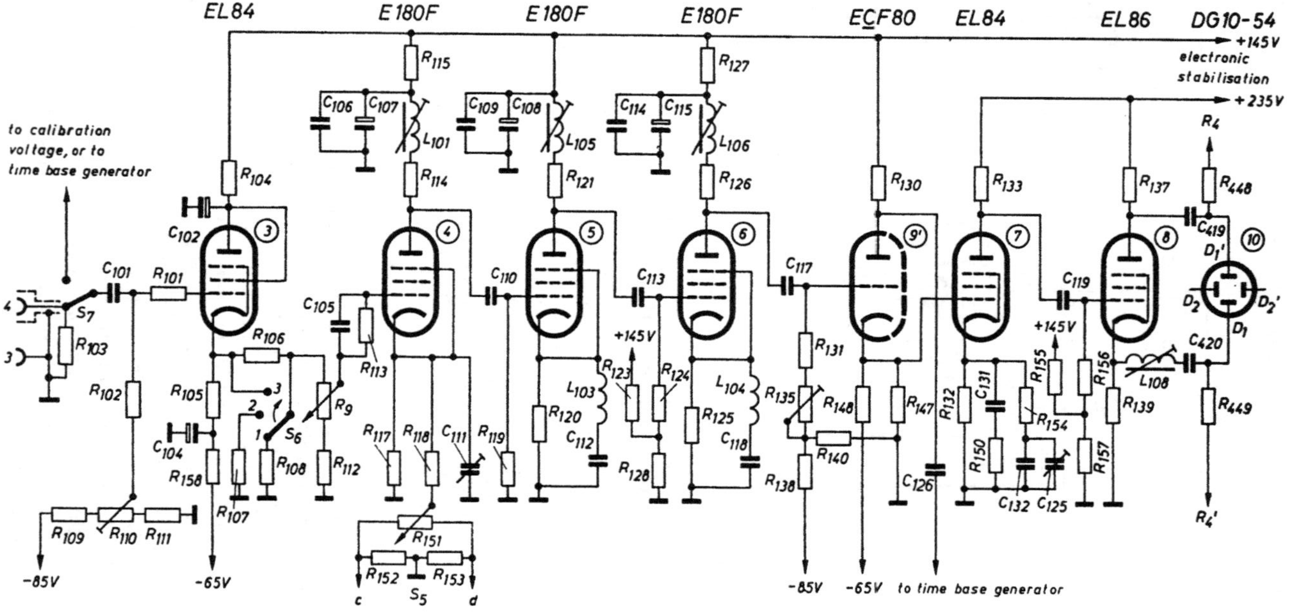

Fig. 5-53 Vertical *AC*-amplifier of the Philips "GM 5662" HF wide-band oscilloscope

loscope serves as an example of a particularly efficient alternating voltage amplifier of moderate cost (Fig. 5-53). The lower cut-off frequency of this amplifier is about 3 c/s, the upper is greater than 14 Mc/s (rise time 20 . . . 25 ns), and the deflection coefficient is 40 mV_{pp}/cm, so that this oscilloscope is particularly suitable for use in the fields of television and pulse technique. In all, seven valve systems are used. Before the output valve which supplies a balanced output voltage, there are a driver stage, synchronization separator stage, three pre-amplifier stages and an output pentode connected as a cathode follower triode functioning as the input valve. The input capacitance is low, being only 15 pF.

By means of a key near the signal voltage socket (Fig. 1-2) the input circuit can be connected by switch (S_7) of small capacitance, either to the signal voltage or to both voltages for the calibration of the vertical deflection and of the time coefficients, or also to the output voltage of the sawtooth generator for the time base (this is done vertically, while the horizontal DC amplifier — upper cut-off frequency 800 kc/s — can be used for deflection by the signal voltage).

The EL 83 "output type" valve (3), connected as a cathode follower triode, is used as the input valve of the amplifier. Such a valve is necessary, because only resistors of low value can be connected in the cathode lead, in order to ensure that the upper frequency limit remains sufficiently high and that an undistorted voltage of a given minimum amplitude shall be available at the output. The cathode current of this valve does not flow to the voltage divider in the cathode lead, but to a point in the power supply stabilized at —65 V, so that the actual cathode resistors draw no current when correctly adjusted. This ensures that the screen image does not jump when the range selector (S_6) is switched to another position, nor when the potentiometer for continuous control (R_9) is operated and gives rise to voltage surges. With the range selector in the cathode lead, the switch (S_6) can be used to provide division ratios 100 : 1, 10 : 1 and 1: 1. Between this stage voltage divider and the input grid of the first pre-amplifier stage still another low-ohmic potentiometer (R_9) makes possible a continuous adjustment in the ratio 12 : 1, so that the input voltage, which must not exceed 35 V_{pp}, can be adjusted over a complete range of about 1200 : 1. For higher voltages a voltage divider probe should be used (Ch. 5.29 "Probes", Fig. 5-64).

The coupling between cathode follower and first pre-amplifier stage is direct (DC coupling — the next Section 5.36 "DC voltage amplifiers" contains further details of this), so that the working point of the first pre-amplifier valve (4) must be adjusted according to the value of the cathode resistance by suitable setting of its cathode voltage. To compensate for hum interference, an alternating current of desired phase and adjustable in amplitude (R_{151}), taken from a transformer winding (S_5), is applied across the cathode resistor of this valve (R_{117}) via a series resistor (R_{118}). Next follow the two other pre-amplifier stages (5, 6) whose circuit elements are so rated in the way already described, i.e. by inductance correction in the anode current circuit and by frequency-dependent negative feedback in the cathode, that the required wide band transmission is assured. The coupling capacitors between these stages must be as large as possible, in order that low frequency signals can also be successfully transmitted. On the other hand, to ensure that there is no restriction of the upper frequency limit, their capacitances to earth must be small, and, for the same reason, their inductances must also be low. Capacitors with a synthetic dielectric are particularly suitable. To improve transmission of low frequencies (at 50 c/s the roof slope of square pulses with a duration T_d = 10 ms should be less than 2%), an RC-network (R_{115} and the capacitors C_{106} and C_{107}) is connected in the lead to the anode resistor of the first pre-amplifier valve (4), as shown in the circuits of Fig. 5-19*a* and *b* and described in detail in the Section 5.13 "Improvement of amplifier characteristics at the lower end of the frequency range".

The grid leak resistor (R_{135}) of the separator valve (9′) following the pre-amplifier stages can be adjusted for precise balance of the time constant of the amplifier for good reproduction of low frequency pulse voltages. The triode portion of an ECF 80 valve (9′) is used in the separator stage. Coupling resistors are provided in the cathode and anode leads of this system. The voltage taken from the cathode is transferred to the grid of the next valve, so that the separator stage functions here as a cathode follower. The amplified voltage arising at the anode is fed to the time base unit, where it can serve for "internal" triggering or synchronization. In this way, good separation between signal amplifier and time base unit is obtained, and feedback from the latter to the amplifier via the sync terminal is practically impossible. In the same way as in the coupling between cathode follower and the first pre-amplifier valve, the connection between this separator stage and the driving valve (7) for the output stage (8) which follows it is a DC coupling. In this way the lower cut-off frequency is kept low. Moreover, the build-up time of the amplifier remains acceptable. If *CR* networks were used in these two couplings, it would be impossible to obtain the desired good reproduction of a square pulse of $T_d = 10$ ms (symmetrical : 50 c/s) with acceptable coupling capacitances and a permissible build-up time. It would, moreover, be impossible with larger coupling capacitances to attain an upper cut-off frequency of $>$ 14 Mc/s because of the higher capacitance to earth.

An output pentode EL 86 is used as the output stage. By means of coupling resistors in both the anode and cathode leads it supplies earth-symmetrical output voltages. As such a stage has only a gain of less than 2 (Fig. 5-42 and accompanying text), a voltage equal to half the required amplitude of the output voltage is required to drive this valve. Thus, this driving voltage can only be supplied by a "power" valve (7), in this case type EL 84. In this stage no *L*-correction is provided in the anode current circuit, but the frequency dependence of the gain is decreased in the cathode lead by frequency-dependent circuit elements. The output stage is so rated, that (with an upper cut-off frequency of 14 Mc/s for the whole amplifier) without special post-acceleration voltage at the cathode ray tube (10) driving up to 50 mm image height is linear.

The DG 10-54 cathode ray tube of high deflection sensitivity (Ch. 2.13 "Cathode ray tubes with particularly high deflection sensitivity of the *Y*-plates for wide-band oscilloscopes") making possible a deflection of 55 mm in any case, is used, so that the ability to drive to 50 mm is quite sufficient. Only the unstabilized voltage of +250 V in the H.T. part of the power pack is used for the post-acceleration voltage. If desired, the post-acceleration electrode can be switched to a point in the E.H.T. chain giving + 1300 V, so that then the total acceleration voltage for the cathode ray tube is about 2500 V. The height of the image which can be obtained free from distortion is then limited to 40 mm.

5. 27 DC voltage amplifiers

5.27.1 REQUIREMENTS

Oscilloscope amplifiers giving linear response down to very low frequencies are required for the examination of changes of phenomena which take place slowly, e.g. biological, thermal, chemical or mechanical processes. For such tasks, AC amplifiers with a suitable response at very low frequencies have been developed [14], but a build-up time corresponding to the time constants of the necessary coupling elements cannot, be avoided, and this causes a certain amount of distortion in the case of rapid voltage changes. Moreover, the static components of the change of phenomena under observation — or changes taking place extremely slowly — are not transmitted by such AC amplifiers, so that a true picture of the variable of state cannot be obtained. Apart from carrier frequency amplifiers specially developed for certain tasks, only the DC voltage-coupled amplifier is available for general application in this frequency range.

In the very early days of radio engineering, when efforts were being made to improve the reproduction in radio receivers and amplifiers, LOFTIN and WHITE [63] gave particular attention to this problem, and direct-coupled amplifiers, which were already familiar, became known as Loftin-White amplifiers. It became clear, however, that radio amplifiers of the AC voltage amplifier type could be quite satisfactory, since their transmission range was far superior to the reproduction range of other elements in the transmission chain, for instance, the loudspeaker. For this reason the use of directly coupled amplifiers was restricted to radio receivers for auxiliary purposes, such as automatic volume control, automatic tuning and similar controls.

In contrast to the AC voltage amplifier, the DC voltage amplifier suffers from the disadvantage that it is not possible to isolate the signal voltages from the "rest" potentials, so that every change of the direct input potential of the amplifier (change of direct grid voltage) is amplified at the output as a corresponding change in voltage. To the factors limiting the gain in the AC amplifier already discussed, the D.C. amplifier presents the additional problem of ensuring constancy of the operating point, particularly in the input stage. Depending upon the purpose for which the amplifier will be used, therefore, more or less elaborate and therefore costly means must be provided for keeping the operating point — and the emission of the amplifier valves — constant.

This is the main reason why DC voltage amplifiers, which have long been in use, principally in measuring equipment such as valve voltmeter circuits, have only recently been applied to oscilloscopes to any great extent. However, means have now been found which make it possible to obtain gains up to 30,000 under stable conditions and at acceptable cost.

5.27.2 COUPLING THE INDIVIDUAL STAGES OF A DC VOLTAGE AMPLIFIER

When directly connecting the anode of the input valve to the grid of the following amplifier valve, care must be taken that the grid of this valve is at the required negative potential with respect to its cathode. If the H.T. supply is derived from a single source, a circuit, shown in simplified form in Fig. 5-54, would result. The anode current source has been indicated by two parallel lines as a voltage divider, in the same way as in G. Kessler's comprehensive and detailed work on the problems of the DC amplifier [64]. This circuit, which corresponds to the one used initially by Loftin-White, has several disadvantages. The total operating voltage is high and the load on the power supply is equal to the sum of the anode currents of the individual valve stages. The heater current for each valve has to be provided by a separate winding, otherwise a high voltage, corresponding roughly to the sum of the anode voltages of the preceding valves, would occur between the heater and cathode of the last stage. A re-

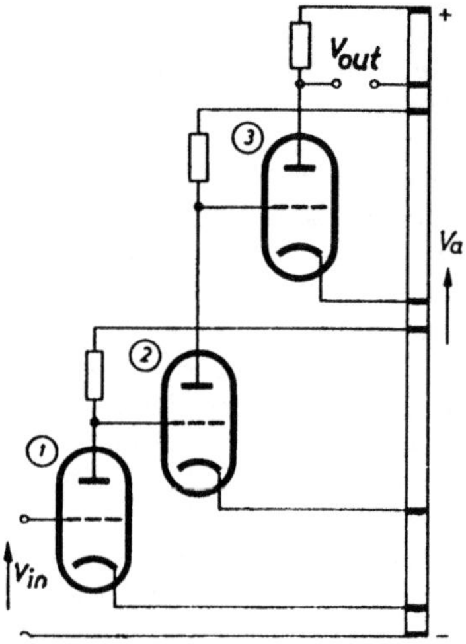

Fig. 5-54 Amplifier stages directly coupled in cascade

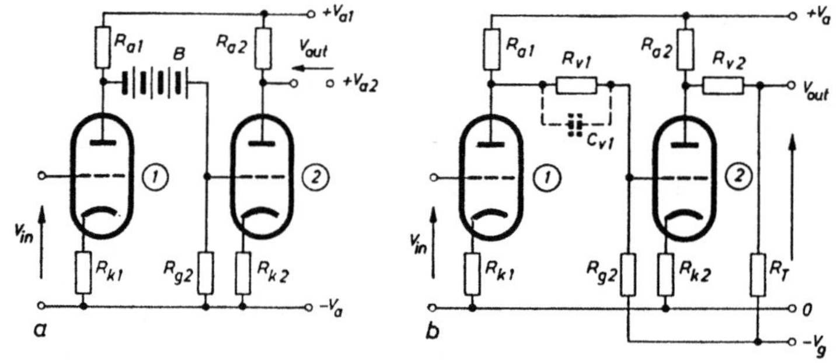

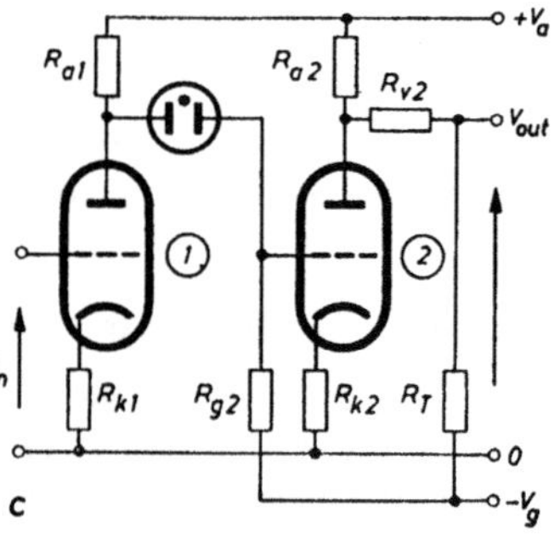

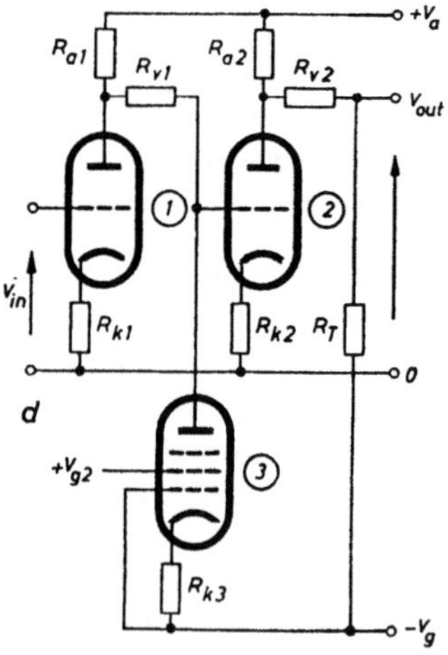

Fig. 5-55 Possible methods for direct coupling of amplifiers with compensation of quiescent DC voltage

a) Compensation with insulated battery

b) Compensation voltage by means of voltage divider with resistors connected to negative voltage source

c) Compensation voltage from a discharge-tube voltage divider

d) Compensation voltage from a voltage divider having a pentode instead of a divider resistor duction of the total supply voltage is possible by compensating the bias of each valve against that of the following valve in some way. The obvious solution would be to connect between the anode of the one valve and the grid of the following valve a battery, well insulated from earth and supplying a voltage equal but of opposite polarity to the direct potential at the anode of the first valve. Fig. 5-55 *a* shows an example of this circuit.

There is no need to advance reasons for the advisability of replacing this battery by a mains-fed voltage divider circuit. (Apart from the usual disadvantages of a battery, its capacitance to chassis would greatly lower the attainable upper cut-off frequency in the same way as in the AC voltage amplifier. If it is only a matter of amplifying a low frequency range, then this circuit can be used, e.g. for a laboratory experiment.) A voltage divider circuit can be seen in Fig. 5-55*b*. An additional current flows through the anode resistor (R_{a1}) of the first valve, through a series resistor (R_{v1}) and through the grid resistor of the following valve (R_{g2}), to a point of negative potential in the power pack. If this negative voltage is of a suitable value, the grid voltage of the second valve can be set at the correct value by correct choice of the divider resistors (R_{v1} and R_{g2}). The voltage change brought about by the input signal voltage V_{in} at the anode of the first valve also reaches the grid of the following valve (2) correspondingly reduced. The loss of gain occasioned by the divider is slight if R_{v1} is very much less than R_{g2}. (This division must not limit the band width, and it is therefore necessary to apply compensation by shunting the series resistor R_{v1} by a capacitor). The greater the negative voltage ($-V_g$), the larger can R_{g2} be as compared to R_{v1}. It has also been suggested, as shown in Fig. 5-55*c*, that a discharge tube be used instead of a series resistor [65] [66]. Such a tube has, as is well known, a low internal resistance, so that the "series resistor" (previously R_{v1}) is low and the voltage division for voltage variations above and below the rest value is more favourable than the direct voltage division. Discharge tube coupling is generally used only for output stages, however, as in the first stages random variation of operating voltage can cause undesirable changes of the working point. A pentode can also be used in the voltage divider chain as shown in Fig. 5-55*d*. A pentode has the advantage of a low DC resistance, so that the negative voltage need not be particularly high, while the AC resistance (the "differential resistance") is high. As shown in the circuit of the vertical amplifier of the Philips "GM 5666" NF oscilloscope in Fig. 5-62, it is also possible to avoid the voltage divider altogether, by suitable gradation of the cathode voltages, particularly in the case of amplifiers having few stages.

In a circuit according to Fig. 5-55*a* the output voltage is superimposed on a direct voltage equal to that at a point in the H.T. supply source and corresponding to the "no-signal" value of the anode voltage. This, however, has the disadvantage that the output has a considerable potential with respect to chassis. This disadvantage is not very serious in so far as use in oscilloscopes is concerned. It is only necessary to take steps to ensure that the last acceleration electrode before the pair of deflection plates — a_2, for instance — is at the same potential (i.e. not zero). Otherwise considerable astigmatism would occur. Alternatively, this difficulty can be overcome by coupling the output valve to the deflection plates in a similar way to that in which the consecutive valve stages are coupled, i.e. by connecting a voltage divider from the anode to a negative voltage point as indicated in the circuits 5-55*b*, *c* and *d* (R_{v2}, R_T). In this way a similar voltage division can be made for the output junction, so that under no-signal condition there is no potential difference between the output and chassis. Naturally the same

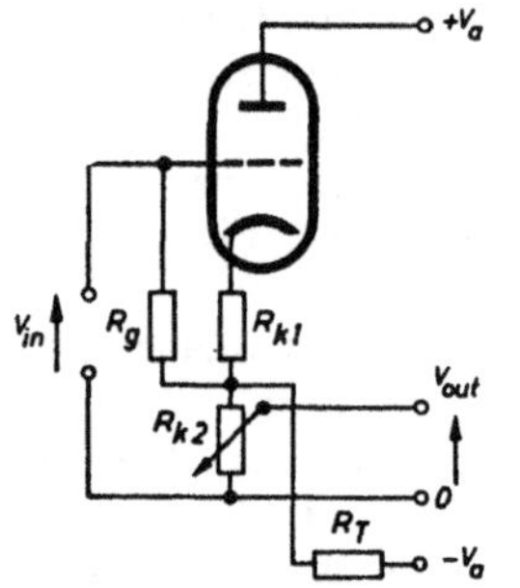

Fig. 5-56 Cathode follower with DC coupling

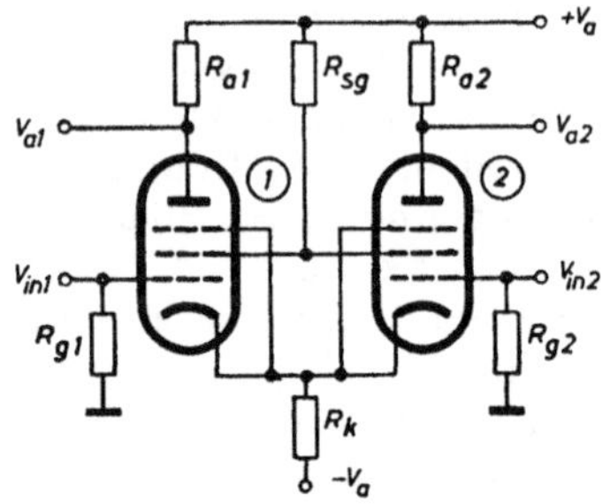

Fig. 5-57 Difference amplifier with DC coupling

considerations as previously set out regarding the coupling of the valves to each other apply to this voltage divider. If a valve is used instead of R_T in Fig. 5-55*d*, it is best connected as a cathode follower, which, by the way, is also possible when coupling the individual valve stages to each other. Thus loading of the anode resistor by the divider current is reduced, and, in addition, a low output resistance is obtained. This also makes it possible, as shown in the circuit in Fig. 5-56, to adjust the amplitude of the output voltage continuously. In such a cathode follower stage in direct voltage coupling the compensation of the DC rest potential is best carried out by allowing the cathode current to flow not to chassis, but via a divider resistor (R_T), to the negative supply voltage source. If the resistances are suitably chosen, it is possible to ensure once more, that the output rest potential to chassis is zero. The input cathode follower is connected in this way in the AC amplifier of the Philips "GM 5662" H.F. oscilloscope — Fig. 5-53 —. By combination of the fundamental circuits shown in Fig. 5-55*a* to *d* for coupling the individual stages of a DC amplifier, many variants of the circuit are possible, thus permitting practically all normal requirements to be met. This is dealt with in detail in the previously mentioned work of Kessler [64]. Cathode coupling is also suitable for coupling, for example, a cathode follower stage to an amplifier stage. Double system valves are best used for this [67] [68].

5.27.3 BALANCED DC AMPLIFIERS - DIFFERENCE AMPLIFIER

Fundamentally the same circuits as have already been discussed for AC amplification (Figs. 5-40 and 5-41) can be used for obtaining symmetrical output voltages in the DC amplifier. Only those points remain to be considered that have already been brought up in connection with Figs. 5-55 *a*, *b* and *c* in the Section 5.27.2 on "Coupling the individual stages of a DC voltage amplifier". Balanced circuits in which each pair of valves is connected by cathode coupling are now very frequently employed (Fig. 5-41). Such circuits have the advantage, among other things, that they are to a large extent self-balancing, which greatly facilitates the stabilization of amplifier characteristics. The larger the common cathode resistance the better the balance and the more stable the ciruit. With this arrangement it is also possible to amplify earth-symmetrical voltages so that only the potential difference between the two input terminals is amplified. As any stray in phase voltages (e.g. hum interferences from the mains) are amplified a few thousand times less [53]), it is possible to construct DC amplifiers with particularly high gain and good stability, as required for instance for physiological investigations [69] [70] [71]. The mode of operation of the difference amplifier will therefore be examined somewhat more closely with reference to Fig. 5-57.

Here again the cathode resistor (R_k) is not connected to the chassis, but to a high negative voltage in the power pack (e.g. —300 V). The adjustment is, like that described for the DC cathode follower circuit (Fig. 5-56) so selected that under no-signal conditions the potential difference between grid and cathode is practically zero. The individual valves can thus be considered as cathode followers.

The current through the cathode resistor is approximately:

$$I_k = \frac{V_a}{R_k}. \tag{5.109}$$

When there are changes in the input voltages — $V_{\text{in}1}$ and $V_{\text{in}2}$ — at the control grids of the valves, there is little change of cathode current because of the cathode follower effect. It is therefore essential that the cathode current (I_k) remains absolutely constant even when the input voltages change (see also the oscillogram of Fig. 4-70e). This condition can be approximated to in practice, if, at a sufficiently high negative voltage, the cathode resistance is so selected as to enable the correct rest current of the valves to be obtained. If both valves are identical, then in the rest state, half of this current flows through each. If a voltage of $+v$ [V] is applied to both grids simultaneously, then the cathode voltage does indeed increase by the same value, but the cathode current scarcely changes at all, so that no voltage change appears at the anodes. In other words, in-phase signals at both grids are not amplified at all.

If, however, for example, a voltage $+v$ is applied to the grid of the left-hand valve (1) and a voltage $—v$ to the grid of the right-hand valve (2), then the current through the one valve (1) rises and the current through the other valve (2) drops to the same extent. The current through the cathode resistor does not change and neither does the cathode voltage. Between the anodes a voltage difference appears which, if R_a is very much less than R_i, is

$$V_{a1} - V_{a2} = V_{\text{in}} \cdot g_m \cdot R_a. \tag{5.110}$$

The gain of the difference amplifier stage with two valve systems corresponds to the gain with one valve in the usual type of circuit. It is, moreover, interesting to discover what output voltage and gain are obtained if the voltage V_{in} is applied only to the grid of one valve (1), the other grid having no applied signal voltage. Assuming first that $+ {}^1/_2\, V_{\text{in}}$ is applied to each grid. The cathode voltage would also rise by the same amount; no voltage difference, however, would occur between the anodes. Now, if an additional voltage of $+{}^1/_2\, V$ is applied to the first grid (1) $+{}^1/_2\, V_{\text{in}}$ and an additional voltage of $—{}^1/_2 V$ is applied to the other grid, the same conditions are obtained as if the voltage v_{in} alone had been applied to the one grid and no signal to the other. At one anode (1) a voltage change of $—{}^1/_2\, V_{\text{in}} \cdot g_m \cdot R_a$ will appear and at the other (2) a change of $+{}^1/_2\, V_{\text{in}} \cdot g_m \cdot R_a$. The total effect will be that the cathode

[53]) The ratio of the desired gain of the difference voltage to the gain of undesired interference voltage (rejection ratio) can be from 100 : 1 up to a maximum of 30,000 : 1, according to the quality of the amplifier. For the high ratios, of course, a high degree of circuit symmetry is essential as regards both component values and valve characteristics.

voltage rises by $^1/_2\ V_{in}$ and a balanced signal of $V_{ni} \cdot g_m \cdot R_a$ occurs between the anodes. This confirms that this circuit only amplifies the voltage difference between the two grids and furthermore that the –symmetrical– output voltage across the anodes resulting from this — asymmetrical — input voltage is equal to a first approximation to

$$V_{a1} - V_{a2} = V_{out} = (V_{in1} - V_{in2}) \cdot g_m \cdot R_a \,. \tag{5.111}$$

The gain at each anode in this circuit is therefore half the gain which such a valve would have in the conventional circuit. The output voltage, i.e. the voltage difference between the two anodes, is equal to the voltage which would be obtained in a normal circuit with one valve and the same anode resistance. It is immaterial here whether the input voltage is connected symmetrically between the two grids or between one grid and chassis. Only the cathode voltage remains constant in the case of a symmetrical input voltage, whereas with unbalanced voltage it shifts by $\frac{V_{in}}{2}$ Of course, two different voltages can be connected to the two inputs; in either case, the difference between the instantaneous values of these two voltages appears in amplified form between the two anodes.

Hitherto, the mode of operation of the difference amplifier has only been considered in those cases in which the input voltages are so small that they only drive over a linear portion of the valve characteristic. Assuming, however, that the grid voltage of the left-hand valve (1) starting from rest goes constantly more positive, but the voltage at the other grid is unchanged, then the anode current of the left valve (1) keeps rising, while the current in the right-hand valve drops. The voltages across the anodes have the opposite trend, as can be seen from the curves in Fig. 5-58. Ultimately, the left-hand valve (1) takes over the whole current and the right-hand valve (2) is cut off. However, the cathode voltage still rises so long as the grid voltage rises. In other words, the cathode voltage always follows that of the more positive grid. It must also be pointed that, if the mutual conductances of the two valves are different, the gain of the individual valve becomes equal [50]

$$V_{(gm1 \neq gm2)} = \frac{1}{2} \cdot \frac{g_{m1} \cdot g_{m2} \cdot R_a}{g_{m1} + g_{m2}} \,. \tag{5.112}$$

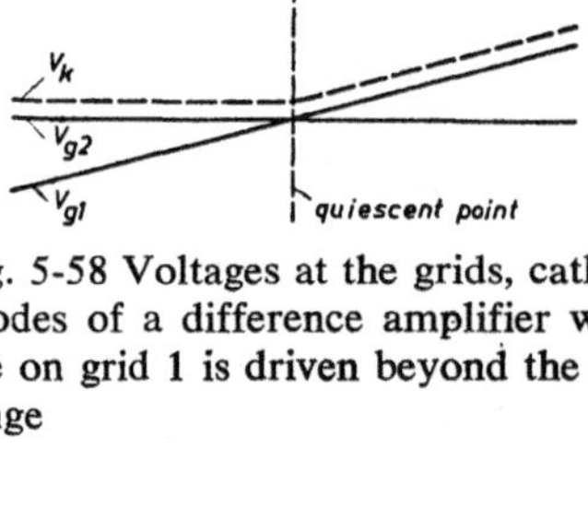

Fig. 5-58 Voltages at the grids, cathodes and anodes of a difference amplifier when voltage on grid 1 is driven beyond the operating range

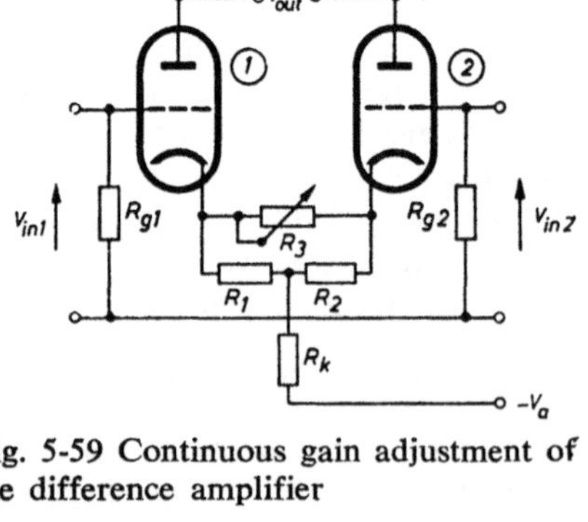

Fig. 5-59 Continuous gain adjustment of the difference amplifier

This circuit is known as the long-tailed pair circuit because the cathode circuit is extended by connecting the cathode resistor to a point of high negative voltage. It is not always necessary, however, for the cathode resistance to be connected to a very high negative voltage. In general it is sufficient if $g_m \cdot R_k \gg 1$. If it is essential that the cathode voltages remain very constant, then it is advisable to use a "constant current valve" in the cathode lead instead of a resistor (Fig. 3-7). By so doing DC amplifier circuits can be built which will meet particularly high requirements as regards constancy. [72] [73] [74].

To go further into this interesting circuit and its possible applications — it can also be used profitably for other purposes such as valve voltmeters, high sensitivity wide band zero indicators, etc. — would be to exceed the scope of this book. The interested reader should refer to the following works given in the bibliography [75] [76] [77] [78] [79].

5.27.4 ADJUSTING THE GAIN

In DC amplifiers all means for adjusting the gain, either in steps or continuously, must be so designed that when the adjustment is altered there is no shift of the working points. This means that the voltage divider or the variable resistor for this purpose must be connected between two points in the circuit, which at rest, have no voltage difference. This is the case, for instance, in the circuit shown in Fig. 5-56 for the cathode follower stage for the potentiometer R_{k2}. (This circuit is used in the practical examples shown in Figs. 5-53 and 4-69.)

In the case of the asymmetrical stages of Fig. 5-55, the gain-control between the anode of the output valve (2) and the voltage point V_{a2} — which must carry the same voltage as the rest potential across the anode — should be connected as in 5-55a. In the circuits *b, c* and *d* of this figure, it would have had to be connected between the output socket and chassis. In push-pull amplifiers, particularly in difference amplifiers, gain adjustment is usually carried out by altering the feedback. In Fig. 5-59 this type of gain adjustment is shown in a difference amplifier stage. Between the common cathode resistor (R_k) and the cathode terminal of each valve a resistor is connected (R_1 and R_2) by means of which current-proportional negative feedback voltage is obtained. The amount of negative feedback, and hence the gain, can be continuously adjusted by means of the variable resistor (R_3). If this resistor is short-circuited, the negative feedback disappears, while at the highest setting of R_3 feedback is at its maximum and thus the gain at its minimum. As has been explained in detail in Ch. 5.18 "Feedback", gain adjustment by means of negative feedback has the great advantage that when the gain is decreased a greater band width and better linearity are obtained than with full gain. It is, of course, possible to apply this change of negative feedback for gain adjustment over several stages — at any rate in amplifiers in which the upper frequency limit is not too high. A wider range of adjustment is thus obtained. Usually coarse adjustment is performed with compensated voltage dividers at the amplifier input, and in the amplifier itself the continuous gain control over a range of 10 or 15 : 1 is provided in the way described above.

5.27.5 PROCEDURES FOR ENSURING THE REQUIRED STABILITY AND FOR REDUCING INTERFERENCE VOLTAGES

As every voltage change between the electrodes of a valve, whether spontaneous and of short duration or permanent, results in a variation of the output voltage level depending upon the gain of the amplifier, it goes without saying that the voltages must be stabilized without inertia. Electronic stabilization must certainly be used for the anode current source. For maintaining the constancy of the heater currents it is sufficient if the regulating time constant of the stabilizer is shorter than the heating time constant of the cathodes. For this purpose magnetic stabilizers with a time constant equivalent to the duration of several mains cycles can also be used with indirectly heated valves. Occasionally in the initial stages, directly heated valves are used, preferably valves taking only a small filament current which is kept constant by electronic stabilization. Sub-miniature valves have proved to be very suitable for this [80] (see also the description of the amplifier of the Philips "GM 5666" NF Oscilloscope, Fig. 5-62). With gains of 10,000 or over it is no longer sufficient merely to keep the heating current of the valves constant in order to maintain the constancy of the valve currents. Because of random processes in the emission layer of the valve cathodes, slow, and also sudden anode current changes (flicker effect) also occur, causing undesirable changes of the zero level at the output of the amplifier. Various circuits have been suggested to reduce this. One example is the circuit put forward by BOUSQUET and KESSLER [64] (Fig. 5-60), in which a "virtual" constant emission cathode is created in a heptode by current distribution control. At the first grid there is only the bias potential due to the voltage drop across the cathode resistor R_{k1} by the current I_k. If, for example, the cathode current rises above the adjusted value, then the first grid receives a higher negative voltage, so that the current passing this grid actually shows only a very small increase. This "compensating" effect of the first grid augmented by the effect of the series resistance in the second grid lead, since an increase in current results in a voltage drop at the second grid (sliding screen-grid voltage).

The third grid serves as the control grid by controlling the current division between screen grid and anode; the average value of the cathode current remains approximately constant. All fluctuations of the cathode current are balanced, if

$$\frac{g_{m\,1\text{-}a}}{g_{m\,1\text{-}k}} \approx g_{m\,3\text{-}a} \cdot (R_k + D_{2\text{-}1} \cdot R_{g2}) \,, \tag{5.113}$$

in which: $g_{m1\text{-}a}$ = mutual conductance of the first grid to the anode, $g_{m1\text{-}k}$ = mutual conductance of the first grid with reference to the cathode current, $g_{m3\text{-}a}$ = mutual conductance of the third grid, $R_k = R_{k1} + R_{k2}$, $D_{g2\text{-}g1}$ = penetration factor of grid 2 to grid 1 and R_{g2} = the series resistance in the second grid lead.

The effectiveness of this compensation is limited by non-linearity of the valve characteristic and by noise introduced by variations in the current distribution between

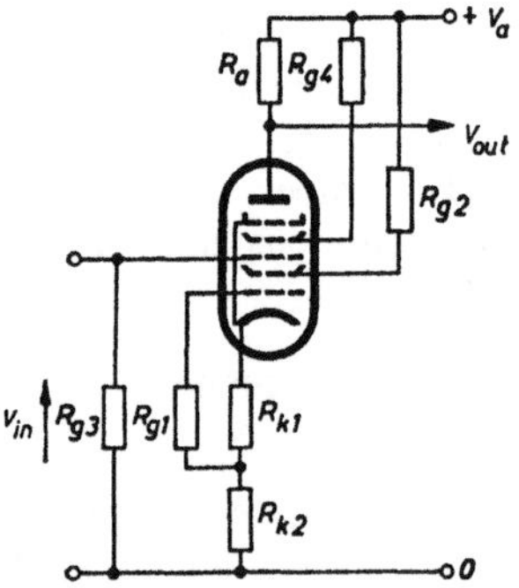

Fig. 5-60 Stabilizing the anode current of the input stage of a DC amplifier by current distribution control and formation of a "virtual" cathode

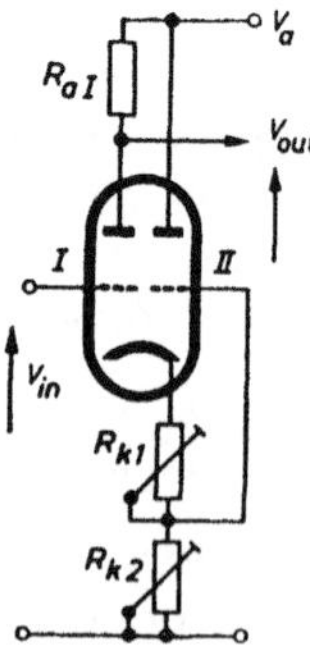

Fig. 5-61 Miller circuit for stabilizing the average anode current of a valve by means of a compensating stage

anode and screen grids. When operating without screen grid resistance ($Rg_2 = 0$) and with $G = g_{m3\text{-}a} \cdot R_a$, a reduction in anode current fluctuation from I_a to approximately 1/40 of the original value is quoted in the above-mentioned work of Kessler. A circuit of this kind is used in the large Philips "GM 4530/01" DC amplifier, which has a gain of 3000 at high constancy, or 10,000 at slightly lower stability.

An interesting compensation circuit which is often used for maintaining the cathode current of a valve constant was put forward by ST. E. MILLER in the description of a stable DC amplifier of high sensitivity (minimum 0.35 mV for full driving — about 100 dB) [74] [80] [81]. Fig. 5-61 shows the basic principle of this circuit. A double triode with common cathode is used in each amplifying stage. Amplification is by means of system I, while system II is used for compensation. Assuming the anode resistance to be large as compared with the internal resistance of this system ($R_{a\mathrm{I}} > R_{i\mathrm{I}}$), the undesirable current fluctuation is least when $R_{k2} = 1/g_{m2}$. The gain of this circuit is given by the following equation:

$$G = \frac{\mu_1 \cdot R_{a\mathrm{I}}}{R_{a\mathrm{I}} + R_{k1} + R_{k2}} \,. \tag{5.114}$$

With a voltage gain of from 30 to 45, it is stated that current fluctuations in the amplifier system are reduced to about 1/50 by the compensator. Such compensated amplifier stages can, of course, also be incorporated in difference amplifiers.

It should also be pointed out that the change of cathode current ΔI_k resulting from a change of heater current ΔI_h, i.e. the ratio $\Delta I_k/\Delta I_h$, can differ for individual valves. It is therefore an advantage to use double valves in a symmetrical amplifier and to connect them so that the two systems work symmetrically in one stage. In these circumstances the emission fluctuations, particularly in difference amplifier circuits, are to a large extent compensated. The use of self-compensating stages of the Miller circuit type (Fig. 5-61) is then only necessary when the requirements as to stability are exceptionally stringent.

5. 28 Some DC voltage amplifiers

Signals with a DC component can be studied with AC amplifier oscilloscopes, incorporating an AC amplifier, supplemented by a DC voltage amplifier.

Fig. 5-62 gives an example of an oscilloscope amplifier with high DC voltage amplification (approximately 8000). It shows the Y-amplifier circuit of a Philips "GM 5666" oscilloscope. This oscilloscope is intended for tasks requiring, above all, high amplification up to approximately voice frequency range. The upper frequency limit is $>$ 40 kc/s and the deflection coefficient 3 mV_{pp}/cm. The input voltage applied between the connecting sockets (3) and (4) is applied to the grid of one input valve (1) either directly or via a compensated voltage divider which can be set to ratios of 10:1, 100:1, 1000:1 of 10.000:1 by a switch ($S_{6.1 \text{ and } 2}$). By operating a press button the input can, if desired, be connceted, by means of a change-over switch (S_7), to a clipped 50 c/s voltage for calibrating the vertical deflection.

Both input valves are directly heated valves of the subminiature type DL 67; the heater current is obtained, via series resistors (R_{131} and R_{132}), from the +170 H.T. line, which is maintained constant electronically [80]. To protect the input valve from accidental excessive input voltages, a resistor (R_{119}; 560 kΩ) is included between the grid and the signal source. To compensate the input capacitance of this valve, this

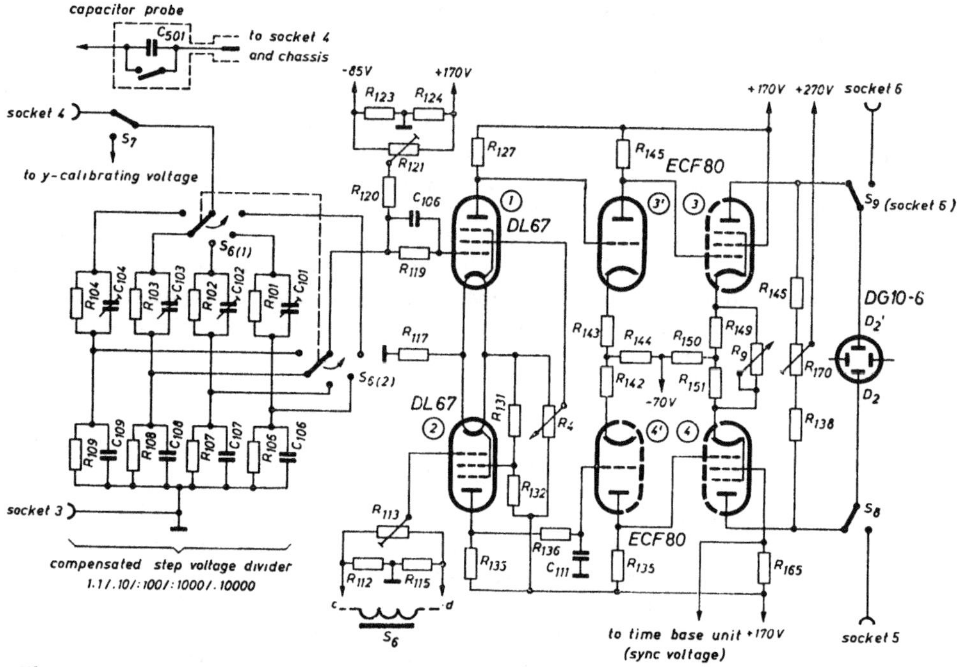

Fig. 5-62 DC-amplifier of Philips "GM 5666" LF oscilloscope ($G \approx 8000$)

resistor is bypassed by a capacitor (C_{106}), so that wide-band transmission is ensured. Amplification in this stage is provided by the upper valve (1). The lower valve (2) only serves as a balancing valve to stabilize the working point and to compensate interference voltages. A 50 c/s voltage, adjustable in amplitude and variable as to phase for compensating residual hum interference is fed to the control grid of this valve from a transformer winding via R_{113} (S_6). The amplified voltage at the anode of the first stage valve (1) is applied to the grid of the upper valve (3′) of the second stage. By cathode coupling (R_{144}) the valve system (4′) shown below it in the circuit diagram is controlled in anti-phase. The cathode potentials and anode resistors are so rated that the correct direct negative grid bias voltages are obtained without a voltage divider. The anodes of the valve systems in the 2nd stage can therefore also be directly connected to the grids of the output stage systems (3 and 4). The cathode resistance, common to both systems of this stage, is relatively high, so that this stage can be regarded as a difference amplifier. A variable negative feedback (R_9) serves for continuous adjustment of gain. The method of operation of this portion of the circuit has already been explained in detail in Fig. 5-59. With the maximum negative feedback the upper frequency limit rises to about 80 kc/s. The deflection plates of the oscilloscope tube are connected directly to the anodes of the output stage. The pair of plates (D_2, D'_2) nearer to the screen, and usually used for the time base, is used here for the vertical deflection. This arrangement was adopted for the following reasons: the potential of the last acceleration electrode and the average potential of the adjacent deflection plates must be equal. This means, however, that this electrode (here a_1) must not be at chassis potential, but must be given a direct voltage equal to this average plate voltage. With this amplifier there was a suitable potential division and a simple circuit, since the anode current for the output stage was taken from the +270 V voltage (which need not be stabilized). But this would have introduced certain difficulties in connection with the adjustment of the anode voltage of the oscilloscope tube, so that it was found preferable to connect the output stages of the horizontal amplifier to the deflection plates (D_2, D'_2) nearer the anode. They can be fed from the +170 V voltage terminal. The anode of the oscilloscope tube is then adjusted to the same voltage as the quiescent potential at the anodes of the horizontal amplifier valves by means of a control (R_{403}; not shown in this circuit).

For internal triggering or synchronization, the second grid of the lower valve (4), which has a coupling resistor (R_{165}) in its lead, is connected to the grid of the pre-amplifier valve in the time base unit. The signal amplifier can also be switched off by means of switch-sockets (5 and 6) and an external voltage directly connected to these deflection plates. Like the horizontal amplifier, the Y-amplifier is very conservatively rated. It is thus possible to "expand" the vertical deflection up to ten times. By adjusting the screen grid voltage (R_4) of the input valve (1) it is possible in addition to shift the image vertically and thus move partial images of the amplified voltage into the centre of the screen, so that the oscillogram can be studied with its details magnified tenfold in the vertical direction as well. If at the same time the time coefficient can be reduced by magnification by a factor up to 10, screen images are obtained which have a surface area 100 times as great as the normal image (Ch. 4 and Part II. Ch. 6). Similarly, relatively small changes of state superimposed on a direct voltage level can be studied in a magnified image. If, however, it is only the alternating quantity which is of interest, or if the ratio of DC to AC is extremely high, the DC components can be removed from the input of the oscilloscope by switching on the capacitor (C_{501}) provided in the probe, for which purpose the contact point of the probe is unscrewed. The oscilloscope then works as with an AC voltage amplifier with a lower cut-off frequency of about 0.3 c/s.

The circuit diagram in Fig. 5-63 shows a simple two-stage DC amplifier which can be switched so that its upper cut-off frequency is 400 kc/s at high gain or 4 Mc/s at reduced gain. This circuit was developed for the small Philips "GM 5650" oscilloscope, intended primarily for use in television servicing. Gain adjustment is carried out in this unit by a switch, in stages of 1:1, 1:3, 1:10 (0 . . . 400 kc/s) and/or 1 : 10, 1 : 30, 1 : 100 (0 . . . 4 Mc/s). In addition, three input sockets are provided by means of which the input voltage can either be fed directly to the switch for the stage voltage divider, or can be first attenuated in the ratio of 30:1 or 100:1 [54]). Moreover, a voltage divider probe is also available, which makes it possible to attenuate the signal in the ratio of 10:1. The input impedance is then 10 MΩ/7pF.

The amplifier consists of an input stage and a symmetrical output stage. The anode load of the input valve can be varied, thus making it possible to choose between high gain and smaller bandwidth or reduced gain and high upper frequency limit. In the narrow-band position the upper cut-off frequency is 400 kc/s and the deflection coefficient $< 42\text{mV}_{pp}$ (15mV_{rms}/cm). In the wide-band position the deflection coefficient is $< 280\text{mV}_{pp}$ (100mV_{rms}/cm) at an upper cut-off frequency of 4Mc/s.

The signal voltage is applied via the input dividers to the control grid of the pre-

[54]) Only one divider is shown in the circuit diagram for the sake of clarity.

Components of the horizontal amplifier

Resistors

R_1 = 100 kΩ/lin.
R_2 = 50 kΩ/lin
R_5 = 50 kΩ/lin
R_{14} = 1 MΩ
R_{15} = 100 kΩ
R_{20} = 200 Ω
R_{21} = 1 MΩ
R_{22} = 100 Ω
R_{24} = 2.2 kΩ
R_{25} = 100 Ω
R_{26} = 220 Ω
R_{27} = 150 kΩ
R_{28} = 100 Ω
R_{29} = 150 kΩ
R_{30} = 270 Ω
R_{31} = 220 kΩ
R_{32} = 120 kΩ
R_{33} = 100 Ω
R_{34} = 220 kΩ
R_{35} = 6.8 kΩ
R_{36} = 10 kΩ
R_{37} = 100 Ω
R_{38} = 5.6 kΩ
R_{40} = 10 kΩ
R_{41} = 18 kΩ
R_{42} = 18 kΩ
R_{44} = 68 kΩ
R_{45} = 82 kΩ
R_{46} = 100 Ω
R_{47} = 1 MΩ
R_{83} = 390 kΩ
R_{84} = 82 Ω
R_{85} = 82 Ω
R_{86} = 47 kΩ

Capacitors

C_5 = max. 6 pF
C_6 = 15 pF
C_{12} = max. 400 pF
C_{13} = 0.1 μF
C_{14} = 0.1 μF
C_{15} = 0.1 μF
C_{16} = 3.9 nF
C_{17} = 0.47 μF
C_{20} = max. 20 pF
C_{30} = 68 nF
C_{53} = 100 pF

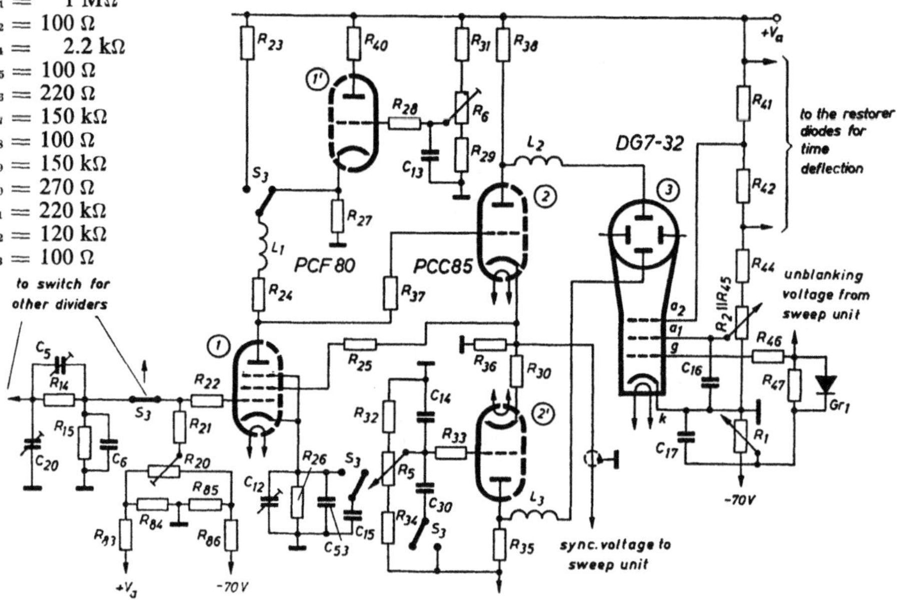

Fig. 5-63 DC-amplifier of Philips "GM 5650" HF small oscilloscope. Upper cut-off frequency 0.4 Mc/s or 4 Mc/s

amplifier valve (1). The anode of this valve is directly connected to the grid of the upper valve (2) of the symmetrical output stage. The second valve system (2′) of this stage is again controlled by cathode coupling (R_{36}). The direct control grid voltage of this valve is adjusted, by means of a voltage divider between the H.T. positive supply terminal and chassis, to about the same value as that at the grid of the other valve system (2) from the anode of the pre-amplifier valve. By adjusting the grid bias of (2′) by means of the potentiometer (R_5), the working point of the amplifier can be altered and the screen image thus shifted vertically to enable asymmetrical voltages to be studied. It should be noted, however, that the control of this amplifier is barely sufficient for providing an image occupying the whole of the C.R.T. screen, so that the picture should not be extended. In order to increase stability and improve the frequency response curve of the amplifier, a (DC) negative feedback is applied from the cathodes of the output stage to the screen grid of the input stage. An improvement in gain in the upper frequency range is also obtained by means of inductance in the anode lead of the pre-amplifying valve as well as in the connections from the output stage to the deflection plates. Moreover, the negative current feedback effect produced by the cathode resistor of the input valve is reduced for the higher frequency ranges by connecting small capacitors in parallel. In order that optimum correction shall be applied for both frequency ranges of the amplifier, the capacitances in the cathode lead are switched over when the anode load is changed (switch S_3). The voltage for internally triggering the time base unit by the amplified signal voltage is taken from the cathode resistor.

The deflection plates of the oscilloscope tube are connected — via inductances (L_2 and L_3) to compensate the plate capacitances — to the anodes of the two valve systems in the output stage. Thus the deflection plates are at the quiescent positive potential of the anodes of (2) and (2′). To ensure that there is no distortion of the picture (astigmatism) due to this, the anode of the oscilloscope tube (a_2) is held — via a resistive voltage divider (R_{41}, R_{42}, R_{44} and R_{45}) — at the same potential as that of the anode of the output stage, and the cathode of the oscilloscope tube is earthed. Direct connection from a_2 to the deflection plates is impossible, as the potential at a_2 is usually below earth potential. It should have the same average potential (about +350 V) with respect to chassis if astigmatism is to be avoided. When changing the anode load of the pre-amplifier stage for wide-band operation ($f_u > 4$ Mc/s), the anode resistance is relatively small. It is therefore important that the inernal resistance of the power supply for this valve should be particularly low; otherwise an undesirable frequency dependence of the gain at low frequencies would arise due to the internal resistance of the power pack. Moreover, the effective voltage at the anode of this valve for both frequency ranges must be the same. When switching to the wide-band position, the current is therefore supplied by a triode (1′) in "constant current circuit" (Fig. 3-7). Such a valve offers the required low source resistance and gives a constant voltage. A circuit of this sort, as is well known, tends to keep the current and hence also the voltage across this resistor constant due to the effect of its cathode resistor. The value of the cathode current can be varied by adjusting the grid voltage (R_6).

5. 29 Probes

It is not always possible to connect the signal voltage source directly to the input of the signal amplifier. If, for instance, the signal is greater than the permissible input voltage, then attenuation by a compensated divider is necessary. This divider is best mounted at the outer extremety of the lead connecting the oscilloscope of the signal source.

In case the upper frequency limit of the oscilloscope amplifier is not high enough to connect the signal voltage directly, satisfactory observation can be obtained by rectifying the high frequency voltage and feeding only the rectified voltage (or the modulation portion) to the oscilloscope amplifier.

Often a screened cable is required between signal voltage source and oscilloscope input. This, however, would usually impose too great a capacitive load on the voltage source. If the signal voltage is sufficiently high, a cable with a divider probe, the input impedance of which can be correspondingly higher, can be used. For low voltages, however, for which the oscilloscope deflection coefficient barely suffices in any case, division is impossible. In such cases a cathode follower stage must be placed at the outer end of the cable. Because of the low output impedance of a cathode follower, a connecting cable having a significant amount of capacitance may then be used

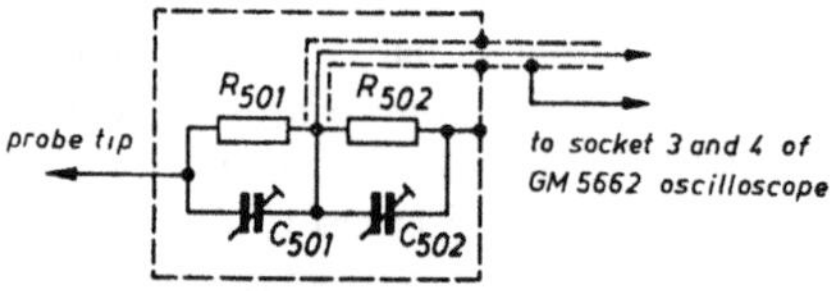

Fig. 5-64 Voltage divider probe for "GM 5662" HF wide-band oscilloscope

between probe and input without any disadvantage. If the load on the signal voltage source must be very small, it is possible to use an electrometer valve at the input of this cathode follower. In all these cases, therefore, the oscilloscope input is connected by a connecting cable carrying at its outer end a probe containing the adjustment device, usually in a cylindrical casing. Some examples of such probes are described in detail in the following pages.

5.29.1 VOLTAGE DIVIDER PROBE

Signal voltages greater than those permissible for the input of the amplifier can be attenuated by a divider probe. The attenuation makes it possible, moreover, to increase the input resistance to a corresponding extent, thus reducing the load on the signal source at the same time. For this reason too, probes are supplied with all high grade HF oscilloscopes. Of course, the attenuation must be uniform over a wide frequency band, corresponding to the frequency transmission range of the oscilloscope. The divider must therefore be well compensated. In general, the circuit is simple. Fig. 5-64 shows the divider probe for the Philips "GM 5662" HF wide-band oscilloscope. The attenuation ratio obtained is equal to the resistance ratio of the parallel connection of the "lower" resistance (R_{502}) and the oscilloscope input resistance, to the upper resistance (R_{501}), which is 10 MΩ. The input resistance of the oscilloscope amplifier is 1.5 MΩ (4.7 MΩ and 2.7 MΩ ||), the "lower" resistance (R_{502}) is 820 kΩ and the parallel circuit is therefore about 0.5 MΩ. In this way an attenuation ratio of 20:1 is obtained. The effective input resistance is thus about 10 MΩ and the input capacitance < 8 pF. Because of the stray capacitances it is difficult to reduce the input capacitance much further and still retain clear-cut definition. The front view of the oscilloscope in Fig. 1-2 shows an example of this divider probe. It should also be pointed out that the connecting cable from the probe to the oscilloscope input has a characteristic frequency determined by its capacitance and inductance and can therefore be excited to oscillation. Such oscillations are even possible at signal frequencies of $^1/_4$ to $^1/_3$ of the resonant frequency. Usually they cause no interference, as their natural frequency is generally considerably higher than the upper frequency limit of the amplifier. For particularly wide-band oscilloscope amplifiers a special cable must be used to connect the probe, the core of which is a high resistance wire. The natural frequencies are sufficiently damped by this means (see Part III, Ch. 25).

5.29.2 DEMODULATOR PROBE

A simple demodulator probe is supplied for the "GM 5655" small oscilloscope, in

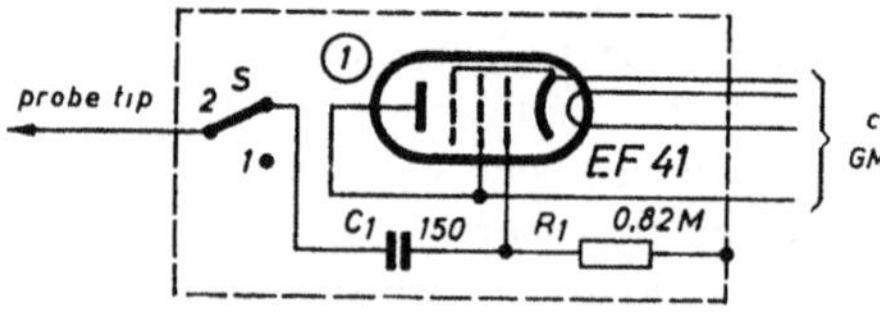

Fig. 5-65 Demodulator probe "GM 4575" for "GM 5655/02" small oscilloscope

which the signal amplifiers have an upper cut-off frequency of 150 kc/s, for studying HF voltages of up to 30 Mc/s. As can be seen from the circuit in Fig. 5-65, the probe contains an EF 41 valve connected as a grid-rectifier triode. The anode resistor (R_9) across which the alternating anode voltage is developed, is shown at the input of the Y-amplifier (circuit in Fig. 5-52). The voltage reaches the input potentiometer from the demodulator via the coupling capacitor (C_2) and the resistor (R_8). The supply voltages (heating and direct anode voltage) are taken from the oscilloscope via the connecting cable. By means of switch (S) in the probe, connection to the measuring point can be made via a 150 pF capacitor (position 2) or, if only a very small load is permissible, via the switch capacitance (position 1). In position 1 the capacitance is about 5 pF, and a deflection coefficient of about 3 V_{pp}/cm is obtained (for 30% modulation depth of the input HF voltage). In position 2 the input capacitance is about 12 pF and the deflection coefficient 300 mV_{pp}/cm. In order to keep the load on the signal source as low as possible, a demodulation valve was chosen. A crystal diode would indeed have made the circuit much simpler, but would have increased the load on the signal source, so that only signals from a low resistance source could be studied.

5.29.3 CATHODE FOLLOWER PROBES

In order that the high input impedance offered by a cathode follower stage may be retained even when a screened lead-in cable is used, it is necessary to place a cathode follower stage at the input of this cable. Taking the grid/anode capacitance C_{ga} into account, the input capacitance of a cathode follower is:

$$C'_{\text{in}} = C_{\text{in}} \cdot (1 - G') + C_{ga} \text{ [55]} . \qquad (5.115)$$

By using a pentode in which the third grid is connected to cathode and the second grid is also decoupled to cathode (not to chassis), the grid-anode capacitance, which is not reduced by the cathode follower effect, can be reduced to a practically negligible value, the effective capacitance being only that between the first and second grids ($C_{g1\text{-}g2}$). This, however, is also reduced by negative feedback. The effective input capacitance is thus:

$$C'_{\text{in}} = (C_{\text{in}} + C_{g1\text{-}g2}) \cdot (1 - G') . \qquad (5.116)$$

The effective input capacitance depends very much on the gain G'. As will be shown, there is a considerable difference in the qualities of a cathode follower if its gain $G' = 0.90$ or 0.95. The gain of a cathode follower [Eq. (5.85)] can also be obtained from equation:

$$G' = \frac{g_m \cdot R_k}{1 + g_m \cdot R_k} . \qquad (5.117)$$

From this it is clear that for a given mutual conductance the gain G' increases with R_k, and that the higher the cathode resistance can be made, the more closely does the gain approach unity. If, however, a screened cable is to be connected to the output of a cathode follower, it must be terminated with a resistance equal to the characteristic impedance of the cable. However, all useful cables have characteristic impedances in the order of magnitude of only about 100 Ω, so that the effect of the cathode follower would be very imperfect. A solution can be found by connecting

[55]) In this equation, in contrast to Eq. (5.89), (1-G') has been substituted for $\frac{1}{1+G}$.

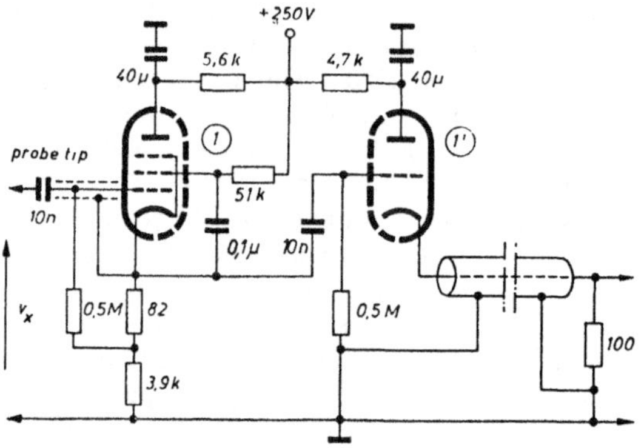

Fig. 5-66 Tandem cathode follower for matching a high input impedance to a low resistance cable

two cathode followers in cascade to the probe. The first cathode follower has a high cathode resistance and thus a high gain (approximately unity), and the cable is then connected to the second cathode follower. The cathode resistance of the first valve is chosen as high as possible in view of the output circuit time constant permissible for a good frequency transmission range. In a circuit described by Roberts [83] and shown in Fig. 5-66, with a cathode resistance of about 4 kΩ, the gain $G' = 0.95$ when $g_m = 5$ mA/V. Thus the dynamic input capacitance of the valve ($C_{in} + C_{g1-g2}$) according to Eq. (5.116) is reduced to 5% of the static value. If, as in this circuit, the grid resistance is connected to the cathode, the input resistance of the circuit is correspondingly raised (5.79) to about 10 MΩ, although the leak resistance is only 0.5 MΩ [56]).

In the publication quoted above, a cable having a characteristic impedance Z = 100 Ω was used. The maximum output voltage is then 2 V_{pp}. When a 200 Ω cable is used it is 8 V_{pp}. The probe cable screening is likewise connected to the cathode, so that it is also affected by the voltage changes at the cathode (it must therefore be insulated from the casing). Its capacitance is thus also reduced by the effect of the cathode follower stage. In the circuit described, the 6 U 8 double valve was used, corresponding approximately to the European valve type PCF 82. It is claimed in the report that the frequency range up to 10 Mc/s has a flat characteristic curve. In a comparison of television test pictures which were observed, first directly and then via this probe, no noticeable difference could be discerned. In practice, the input capacitance is about 3 pF, and the total gain of the probe is 0.5.

For the study of voltages from particularly high impedance sources, probes have been described in which the so-called cascade circuit with an electrometer valve connected as a cathode follower has enabled extremely high input resistance to be obtained. One such probe used in the investigation of cold cathode tubes and in which the load on the signal source had to be limited to a current of only 10^{-10}A or $2 \cdot 10^{-13}$A, has been described by Crowther in a publication [84]. On this and other points concerning the cathode follower (on the behaviour of pulse and sawtooth voltages [85]) further material will be found in the bibliography [86] [87] [88]. With such circuits the use of the oscilloscope is possible for tasks in which the conditions are extremely onerous.

[56]) It should be borne in mind that the input resistance of cathode followers in general, and of such circuits in particular, can become negative under certain circumstances, due to the capacitive voltage divider formed by the grid - cathode capacitance and the cathode filament (chassis) capacitance. With the inductances of the lead at the input an oscillating circuit of the Colpitts type is obtained, which can oscillate at very high frequencies. In such cases the tendency to oscillate must be suppressed by inserting a series damping resistance of about 100 Ω as close as possible to the first grid.

Part II

GENERAL MEASURING TECHNIQUE

CHAPTER 6

OPERATING THE OSCILLOSCOPE; SETTING UP AND PRELIMINARY ADJUSTMENTS

6. 1 Setting up the oscilloscope

Although displays of quite considerable brilliance can be obtained on the screens of modern cathode ray tubes, it is nevertheless desirable to set up the oscilloscope where no direct light, either natural or artificial [57]), will fall on the screen. To improve the contrast, a filter having a transmission characteristic corresponding to the spectral distribution of the light from the screen (yellow-green, orange or even neutral gray), is usually placed in front of the fluorescent screen.

It is inadvisable to set up the oscilloscope near a window, which might make it necessary to adjust picture brilliance to an undesirable high value. To prevent stray light from falling on the screen, the use of a visor is recommended, as shown in Fig. 6-1 (see also Fig. 1-2). By so doing, the operator will not be distracted by extraneous light, and will automatically keep the spot and hence the width of the electron beam small, thus achieving the best pattern definition on the screen. These measures are of particular importance when studying greatly expanded oscillograms of short pulses with low repetition frequency, as then the picture brilliance is very much less than normal.

Fig. 6-1 Front panel of oscilloscope with visor

6. 2 Switching on, brilliance and focus adjustment

In general, no precautions are needed when switching on. The brilliance control should, however, be turned right down beforehand, to avoid overloading the cathode

[57]) The light from the gas discharge of fluorescent lamps, with its strong ultra-violet component, can be particularly unfavourable as it excites the whole screen into fluorescence, so that it appears brighter than accords with its natural whiteness.

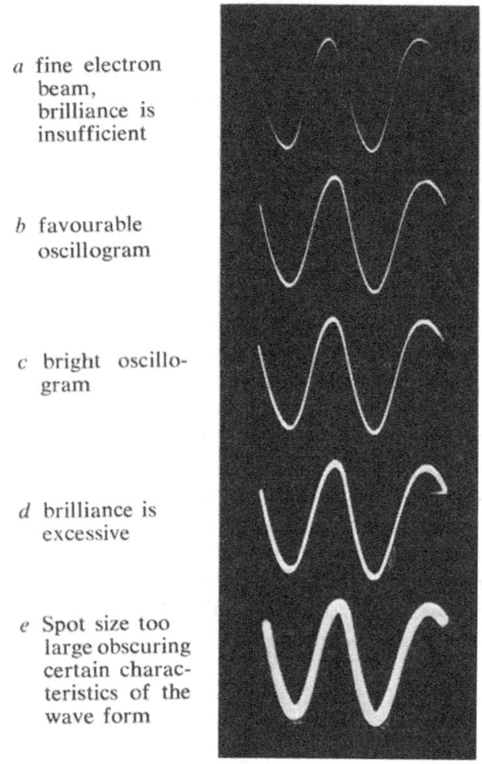

Fig. 6-2 Oscillograms showing different degrees of brilliance

while the filament is warming up. The filament voltage should have been on for at least one minute before the brilliance control is turned up. If there is no deflection voltage on either of the pairs of deflection plates, what appears on the screen is the reduced image of the cathode, its dimensions depending on the brilliance adjustment. Since the smaller the spot, the better the picture definition, the voltage on anode a_1 (focus) must be adjusted by the focus control to produce the smallest possible spot on the screen. During this process the spot changes in size, as has already been shown in Fig. 2-7. When the input voltage is applied to the *Y*-plates simultaneously with the time base voltage on the *X*-plates, the resultant oscillogram should be adjusted for optimum definition with the brilliance and focus controls, as shown in Fig. 6-2. These oscillograms also show the influence of the brilliance adjustment on the accuracy of picture definition. In Fig. 6-2*a* the trace of the spot, although fine, is too faint to allow the whole picture to be photographed satisfactorily. The most favourable brilliance is seen in Fig. 6-2*b*, where the whole oscillogram can be observed distinctly without loss of detail. If a very bright presentation of the image is required, *c*) is the right oscillogram. In *d*) and *e*) the brilliance is too great and the spot size too large, so that certain characteristics of the waveform may well be obscured. Excessive brilliance involves the risk of the trace "burning in" on the screen, and the danger is greater the smaller the surface covered by the spot, that is, the greater the specific load on the screen. This fact should be borne in mind particularly when there is possibility of the deflection voltage cutting out for any reason. When large-area waveforms, e.g. of a modulated RF voltage, are under investigation, the brilliance must be turned up relatively high. If the signal now cuts out, the high power of the beam will be concentrated on the horizontal trace. This is one of the most frequent causes of burnt-in zero-lines on cathode ray tubes. In triggered operation of the time base generator this problem does not arise, as in that case the spot is only unblanked when both vertical and horizontal deflection take place simultaneously. It should not be forgotten that, when assessing the performance of the oscilloscope, a triggering voltage of some sort must be applied; otherwise no image will appear. If no other voltage is available, it is advisable to switch to mains triggering.

6. 3 Astigmatism

If an alternating voltage is applied to the Y-plates only, a vertical trace will appear. When adjusting the sharpness of this trace, it will be found that the setting of the focus control differs from that required for the horizontal trace, the amount of difference depending upon the tube used. This is due to an imperfection in the electron lens of the cathode ray tube and is known as "astigmatism" by analogy with the corresponding error in light optics.

In Fig. 6-3, three sections of oscillograms, enlarged about twice, show the results of astigmatism at the junction of the vertical and horizontal traces. In *a*), the horizontal trace is sharply focused and in b) the vertical trace. (A tube with marked astigmatism was chosen for the purpose of illustration.) With such tubes a compromise usually has to be made between the two settings (Fig. 6-3*c*). The smaller the difference between them, the better will be the overall focus of the pattern.

a *b* *c*

Fig. 6-3 Co-ordinate traces of an astigmatic cathode ray tube

In certain oscilloscopes, particularly those with DC amplifiers, a control is provided by means of which the mid-potential of the deflection plates can be made equal to that of the final accelerating anode. This avoids the additional astigmatism which might otherwise result from the potential difference between these electrodes.

6. 4 Picture width; picture height

The width of the screen image can be adjusted by varying the voltage on the X-plates. It is clear that the "time resolution" of the signal being observed will be greater the wider the oscillogram can be made (the time base frequency remaining unchanged, of course). Nevertheless, there are certain limitations in the case of some simple oscilloscopes. Fig. 6-4 shows the influence of the picture width on the pattern on the screen.

In Fig. 6-4*a* the time base voltage applied was so high that the waveform passed beyond the edge of the screen. Curvature of the edges of the screen therefore distorts these parts of the picture. Such great width is limited to oscilloscopes in which provision is made for reducing the time scale. For general purposes the most suitable adjustment is that shown in Fig. 6-4*b*, since there the whole screen width is used for the time display of the phenomenon, but the flat part of the screen is not exceeded. For some purposes the setting in *c*) may be preferable, although some of

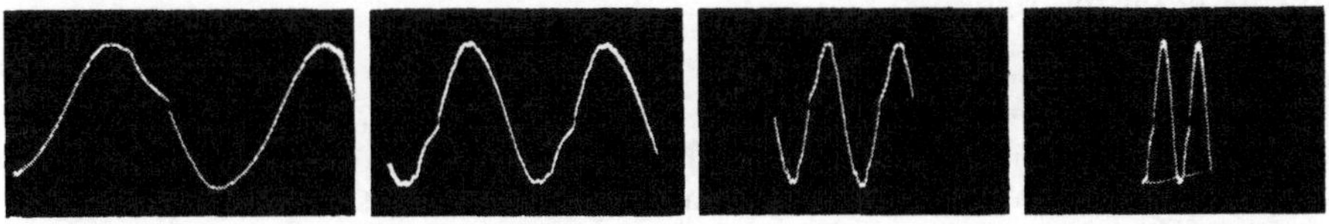

a *b* *c* *d*

Fig. 6-4 Choice of best picture width

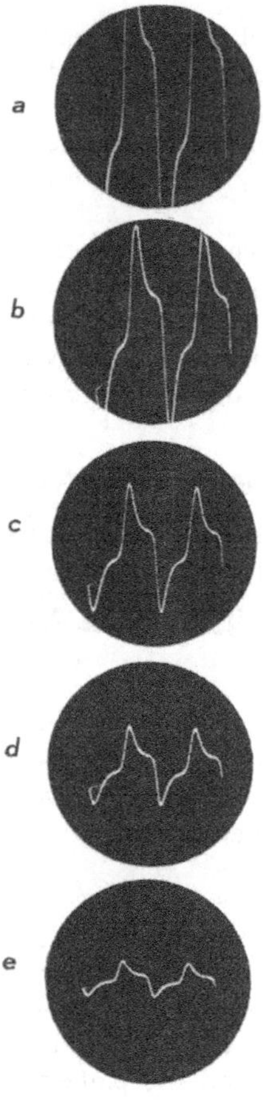

Fig. 6-5 Choice of best picture height. The oscillograms are shown in the same positions on the screen to emphasize their actual relationships

the finer points of the picture are lost through compression along the time axis. Adjustment according to Fig. 6-4*d* is, of course, the most unsuitable.

Similar considerations apply to the choice of suitable picture height. This too can be adjusted by varying the signal amplitude applied to the *Y*-plates.

Fig. 6-5 shows a number of oscillograms from which the most suitable setting can readily be seen. In *a*) and *b*) the useful surface of the screen is once more exceeded. These adjustments will only be useful for special investigations [58]). Waveforms *c*) and *d*) are clearly the most advantageous; the best setting will depend upon requirements. In *e*) insufficient details are visible in the *Y*-direction.

6. 5 Synchronization (self-oscillating time sweep)

Numerous applications of the oscilloscope are concerned with the observation of cyclically recurring phenomena. If the time base frequency is adjusted to be an integral fraction of the frequency of the signal, in each case the patterns obtained on the screen show, in inverse proportion, one or more whole cycles of the phenomenon under investigation. If the frequencies are identical, one cycle appears on the screen; if the time base is half the signal frequency, two cycles appear; if one third, three cycles are shown, and so on.

Only when the time base frequency is a mathematically exact integral fraction of the frequency of the input signal will the oscillogram be perfectly stationary, neither drifting to the left or the right, for only then does the spot describe the trace of successive cycles of the signal in exactly the same path on the screen. If there is no synchronization, the traces described by the individual cycles of the base no longer appear exactly superimposed but by side, resulting in blurred oscillograms. In Fig. 6-6*a* the time base frequency is much too high; however synchronization was nevertheless attempted, but the sloping edge typical of over-synchronization can be seen, particularly on the right.

In Fig. 6-6*b* the base time frequency approaches that of the signal, however, but

[58]) In this way, it is, of course, possible to observe details in the *Y*-direction which would correspond to observation on a larger type of tube. In Fig. 6-5a, for instance, the centre of the oscillogram appears enlarged. In so far as a greater vertical shift of the oscillogram on the screen is possible, the other parts (top and bottom) can also be observed by shifting them to the centre of the screen. This is particularly suitable for DC amplifiers which can be overdriven (Part II, Ch. 6.2 "Electric magnification of the pattern on the screen").

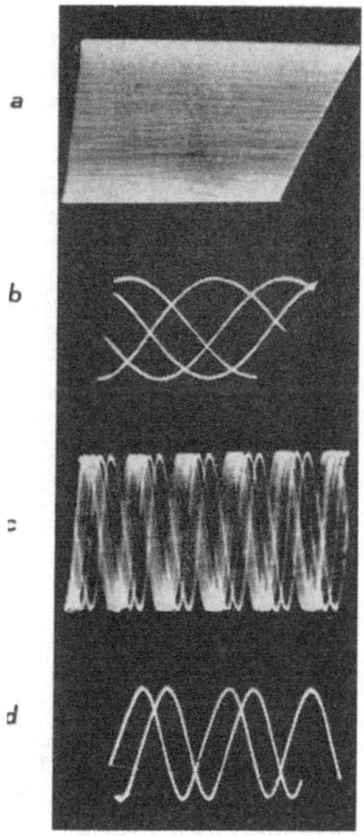

Fig. 6-6 Oscillograms showing the effects of poor synchronisation

the sync pulses were too early and too strong, giving rise to the sloping edge on the right-hand side.

Fig. 6-6*c* is likewise the result of premature and excessive synchronizing pulses. If the time base frequency had been reduced slightly a steady oscillogram of six cycles would have been obtained.

In Fig. 6-6*d*, correct synchronization with 2 or 3 cycles of the input signal is almost attained.

With every good oscilloscope it is possible to synchronize the time base generator in three ways: with the voltage on the *Y*-plates (internal synchronization), with a voltage from an external source (external synchronization) and with the mains frequency (mains synchronization). Proper use of these methods of synchronizing the time base opens up many valuable fields of application which should not be overlooked. Experience shows, however, that in many cases only internal synchronization is used. If use is also made of external synchronization, it is possible with a given voltage to synchronize in a given phase relationship. If this synchronization is maintained and, in place of the first voltage, a second voltage is applied with the same frequency but of different phase, it will appear on the screen displaced along the time axis in a way corresponding exactly to its phase relationship with the first voltage [59]).

If these two voltages are then applied alternately to the *Y*-plates by means of an electronic switch, pictures of both voltages in their correct phase relationships will appear one after the other. More than two voltages can of course be observed or recorded in this way. By photographing each voltage in turn on the same piece of film, the phase relationships and behaviour of any number of voltages can be accurately determined [60]). It is important that the camera should be capable of taking multiple pictures without feeding the film forward.

6. 7 Choice of the most suitable relationship between input frequency and time base frequency

The four oscillograms shown in Fig. 6-7 give an idea of these relationships.

---

[59]) The oscillograms for Figs. 5-12, 5-13 and 5-15 were obtained in this way.
[60]) See also Figs. 14-6, 14-8, 14-9 and 14-11.

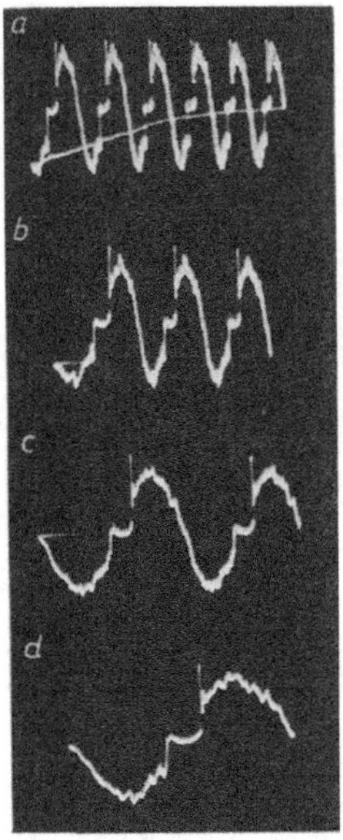

Fig. 6-7 Various possibilities of time expansion by a suitable choice of time base frequency

The oscillograms are of 50 cycles mains voltage, which was not, however, directly applied. Waveforms of this sort are obtained when a short length of wire is inserted into the input terminal of the vertical amplifier, the other end of the wire being left free. Due to the capacitive coupling of this wire with the mains, a current flows in the input impedance of the amplifier, setting up a corresponding alternating voltage on the grid of the first valve. Everyone who has worked with oscilloscopes is familiar with this phenomenon, often only too well, as the stray coupling can often lead to difficulties when small voltages from high impedance voltage sources are to be studied. Since the coupling capacitance for this current is small, the fundamental frequency (50 c/s) is weakly coupled, while the harmonics which are always present in the mains voltage appear exaggerated. This is therefore a simple method of bringing out harmonics and distortions in the display.

Very few details can be perceived in Fig. 6-7*a*, as the six cycles are bunched too closely together. Even here, however, short peaks are visible at the crests, which become more distinct when the number of cycles displayed is reduced (lower time coefficient). But even with only one cycle it is still not possible to ascertain the trend of the voltage at this point. [61])

The observation of details within the voltage waveform of a cycle of the input signal is on the other hand easily possible with oscilloscopes having a triggerable time base and where the pattern on the screen can be electronically expanded.

Oscilloscopes in which it is possible to "trigger" the sweep unit (triggering the time base by means of a positive or negative voltage pulse) are very suitable for such investigations. If the time base is triggered only by the output of the vertical amplifier, the picture on the screen begins mostly on the zero-line with the positive or negative half-cycle, according to choice (Fig. 4-38). If, on the other hand, the part of the input voltage for triggering the time base is taken from an external source via a phase shifter, a greatly expanded image of any desired part of one cycle of the phenomenon under investigation can be obtained. (Part I, Ch. 4.28 "Phase-delayed triggering of the time base" (Fig. 4-73)).

6. 8 Triggering

In "Triggered operation" (4.18.3), the method of operation and the various possible

[61]) Synchonization is achieved by arranging for the peak value of every second, third etc. cycle of the input signal to trigger the flyback of the time base voltage. If the frequency is lower than the time base frequency, several cycles of the time base frequency will coincide with one cycle of the input frequency, but only the peak value of that cycle of the time base corresponding to the peak value of the input voltage can be triggered by the latter. The cycles in between will not be influenced (applies only to relaxed oscillation time base generators).

applications of this type of time base circuit were described in detail.

The essential feature is that the time deflection is initiated at a certain threshold voltage of the triggering voltage, which, in addition, can also be adjustable (see Part I, Ch. 4.28 "Phase-delayed triggering of the time base", and Figs. 4-53 and 56 together with Ch. 4.18.9.2 "Method of operation and possible applications of the trigger level adjustment"). In contrast to synchronization in self-oscillating working, there is complete independence of the input voltage frequency as regards the time scale. By changing the deflection speed in steps or continuously, the time coefficient can be varied in the same way. Sometimes the time coefficients are switched only in fixed steps; they can then be more accurately adjusted than if an additional variable adjustment is demanded. In triggering, switching the time coefficient in steps is sufficient as it is not a condition that the adjusted screen image must be an integral multiple of the duration of cycle of the input. Thus, for instance, 2/3 or 1 3/4 cycles of the signal could be displayed just as easily as integral fractions of a cycle.

The time base can be triggered, just as the synchronization, either by the amplified input, by a voltage from an external source or by means of the alternating mains voltage. Sometimes a device is provided for shifting the phase of the triggering voltage. It is thus possible to display any desired portion of one cycle of the signal waveform greatly expanded. In larger apparatus a special circuit is used for adjusting what is in practice an arbitrarily selected time delay between the occurrence of a given threshold voltage in the input and the commencement of the time deflection. In this way the advantages offered in time expansion by triggering can be fully exploited. Where such a device is not provided in the oscilloscope itself, additional units (trigger-delay units) or, for certain applications, even the ordinary pulse generator may be used.

For instance, a rectangular voltage generator can be incorporated in a circuit by means of which the commencement of the rectangular pulse is triggered by an external voltage and the time deflection commences with the (differentiated) trailing edge of the rectangular pulse. By varying the pulse width the time delay can then be adjusted. For the accurate investigation of mains-synchronous processes the time delay can be obtained by means of a simple *RC* phase shifter which is sometimes built into simple and medium-sized oscilloscopes (Fig. 4-69).

Triggering is usually adjusted by setting the time base circuit to just below the response threshold. The time deflection then always sets in as soon as the input exceeds this voltage threshold. There are various ways of carrying out the adjustment, depending on the way the oscilloscope is equipped. The manufacturer's operating instructions should be consulted for the correct use of these units.

In many oscilloscopes a switching position "automatic triggering" is provided. In this case the time base unit is switched as for triggered operation, but matters are so arranged that it is loosely coupled self-exciting in a low frequency (usually 25 or 50 c/s). However, as soon as control pulses occur from the trigger voltage, the time base is triggered at this frequency only. This adjustment has the advantage that the moving spot is visible even when triggered without a signal voltage, and that the triggering sensitivity (usually known as "stability") need not be specially set.

6.9 Electric magnification of the pattern on the screen

It was shown in Part I, Ch. 4.18.8 "Adjusting the time coefficients", how details of

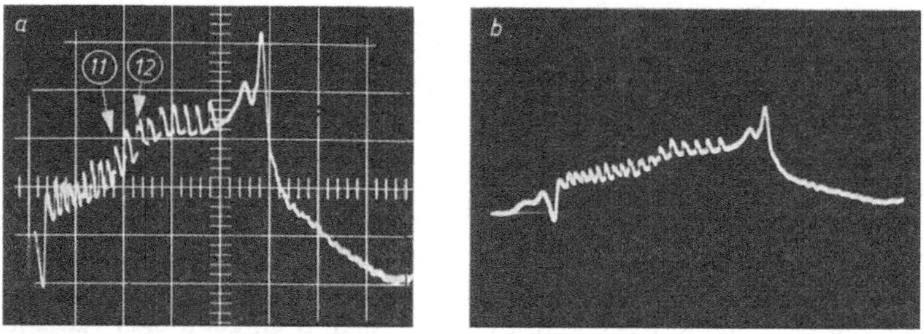

Fig. 6-8 Waveform of an AC cycle to indicate electrical enlargement
a) oscillogram on the observation oscilloscope
b) oscillogram on the auxiliary oscilloscope

the input signal could be clearly displayed along the time axis by amplifying the time base voltage by means of a DC amp!ifier.

It is possible similarly to expand the screen image "vertically" by using DC vertical amplifiers which can be sufficiently driven, and, by shifting the middle working level of these amplifiers, observe an arbitrarily chosen section of the waveform in magnified form. If this process is used simultaniously for both horizontal and vertical deflection, the screen image may be observed, for instance, at a magnification 100 times greater than the normal adjustment. The possibilities thus offered for accurate observation of details in the signal waveform will now be explained in more detail with the aid of a number of oscillograms.

The highly harmonic voltage curve of mains frequency previously in Figs. 4-37 and 4-38 to explain triggering has been taken once more as the input signal. Fig. 6-8*a* first shows approximately one cycle of this voltage triggered by the negative-going part of this voltage. For the better understanding of the processses taking place during the picture enlargement, the *whole* waveform on the *Y*-plates of the Philips "GM 5666" oscilloscope, including the portion going beyond the screen, was also recorded by a

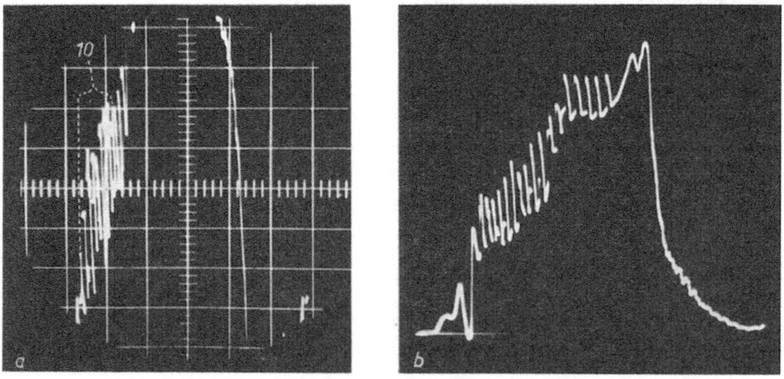

Fig. 6-9 Enlargement of the oscillogram by vertical gain
a) oscillogram on the observation oscilloscope
b) total voltage waveform at the output of the vertical amplifier

Fig. 6-10 Enlarged section of signal amplitude fluctuations of oscillogram *a* in Fig. 6-8 (centre-left)

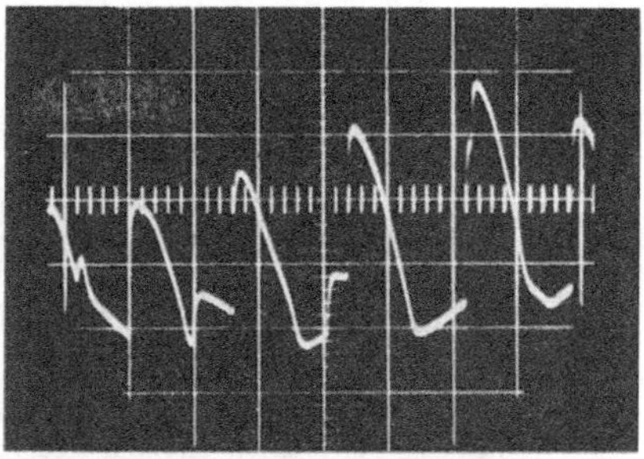

second oscilloscope (Philips "GM 5654", oscillogram *b*). The output voltage of the vertical amplifier of the first oscilloscope was then attenuated by means of a symmetrical (compensated) divider, so that, at maximum amplifier gain, the image on the second oscilloscope still remained well within the screen area. The second oscilloscope had a repeating time base unit, synchronized with the mains frequency. As the mains frequency synchronism voltage differed from the triggering voltage of the first oscilloscope ("GM 5666") the phase relations are different. This is not important here, however, as the phase of the oscilloscope under observation was, in any case, constantly adjusted by shifting the voltage level in the horizontal amplifier, in order to ensure that the interesting proportion of the waveform was visible approximately in the centre of the screen. In Fig. 6-8, oscillogram *b*) is the picture corresponding to oscillogram *a*). If the gain is increased, then the middle portion of the oscillogram is enlarged, though the peaks are compressed due to the limited driving out range of the amplifier. This waveform recorded on the auxiliary oscilloscope can be seen in Fig. 6-9*b*. On the screen of the observation oscilloscope the lower peak voltage is no longer visible, but on the left-hand part of this picture all the small voltage rises visible in the centre of Fig. 6-9*b* appear in enlarged form. (In Fig. 6-9*a* the phase of the picture was shifted somewhat to the right by varying the adjustment of the working point in the horizontal amplifier). If, for instance, the smaller voltage rises of the oscillograms are to be observed in enlarged form, this portion of the waveform is moved to the centre of the screen and the time expansion is increased by increasing the horizontal gain. The result is shown in oscillogram 6-10. Further reduction of the time coefficient by increasing the deflection speed (or even by further increasing the horizontal gain) allows details of the oscillogram that are obscure in the "normal" picture to be displayed very much

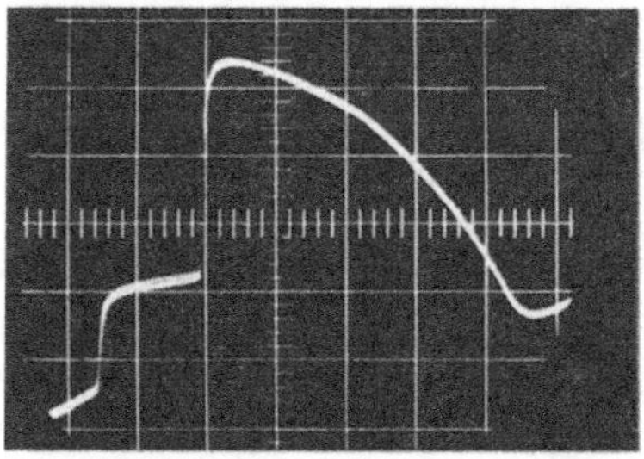

Fig. 6-11 Individual voltage cusp (11) in Fig. 6-8*a* greatly enlarged

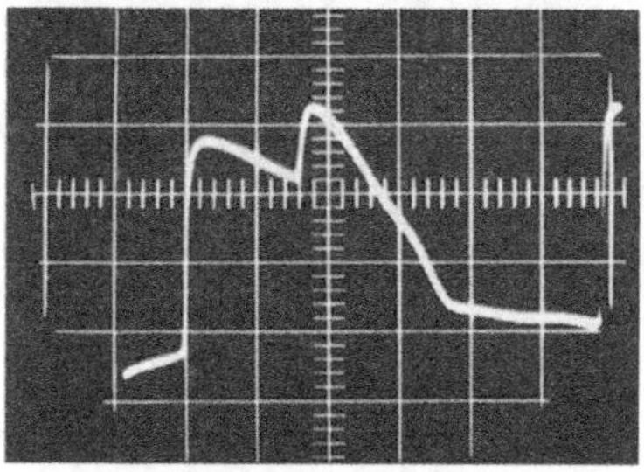

Fig. 6-12 Other cusp (12) from fig. 6-8*a*

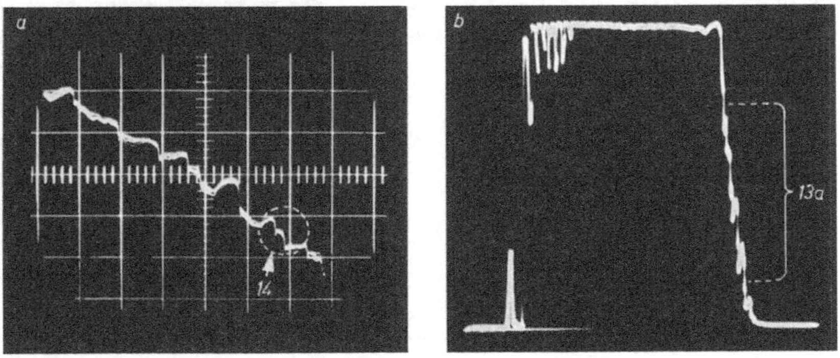

Fig. 6-13 Enlargement of descending portion of curve in Fig. 6-8*a* (bottom right)
a) time expanded oscillogram on observation oscilloscope;
b) total amplitude waveform at the output of vertical amplifier

enlarged. In Figs. 6-11 and 6-12, for instance, two of the voltage peaks indicated at 1 and 2 in the oscillogram of Fig. 6-8*a* were enlarged still more than is shown in Fig. 6-10. That this is the case of purely "electrical" enlargement can be gathered from the recordings with the graticule photographed in. A comparison with the oscillogram in Fig. 6-8*a* clearly shows the facilities offered in this way. The oscillograms of Fig. 6-13*a* and *b* and of Fig. 6-14 show how it is possible to observe other portions of the waveform of a cycle by shifting the trigger phase and the working points in the amplifiers. In this case the vertical gain was made so great that even the small fluctuations in the portions of the descending curve to the right in Fig. 6-8a are considerably expanded. Fig. 6-13*a* shows this waveform with the corresponding increase in time expansion. How far this enlargement can be taken with the "GM 5666" oscilloscope is demonstrated finally by the oscillogram in Fig. 6-14. It shows the section indicated in Fig. 6-13*a*, still further enlarged. In the same way the waveform at the peaks of the curve should be displayed in enlarged form.

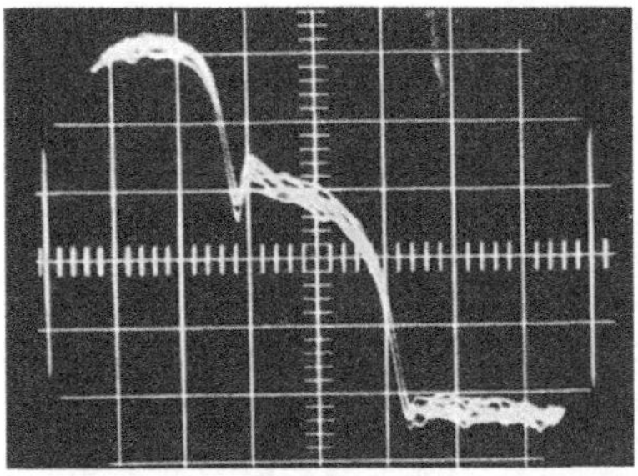

Fig. 6-14 Enlarged section of oscillogram in Fig. 6-13*a*

As has been demonstrated in the above examples, the electrical enlargement of sections of oscillograms makes possible a very detailed analysis of all sorts of changing phenomena with the aid of the cathode ray oscilloscope.

CHAPTER 7

AMPLITUDE MEASUREMENTS

7. 1 Nature of display

A given deflection of the spot on the screen presupposes a commensurate amplitude of the signal voltage applied to the deflection plates. Since the leak resistors for the deflection plates (and also the input resistors of the amplifiers) are in the order of 1 to 10 megohms, the current flowing, and thus the load on the circuit under test, are extremely small [1].

In contrast to this, the familiar indicating instruments, with a few exceptions (e.g. static voltmeters) are actually *current* meters, as an appropriate value of current is a prerequisite of indication. Users of oscilloscopes should always bear these facts in mind, for they open up wide fields of application and lead logically to the correct use of the instruments. With an increasing frequency of the voltage under measurement, the capacitance of the deflection plates represents a rising capacitive conductance. If the capacitance of the deflection electrodes (including connection) is about 10 pF, the impedance will be only about 160 Ω at 100 Mc/s. However, in many cases it will be possible to lump this capacitance with the circuit capacitance, so that it will not appear as a load in this sense.

7. 2 Accuracy of the display and limits of measurement

The extraordinary advantages offered by the oscilloscope even in the determination of the absolute value of the input signal are being exploitated to an ever-increasing extent in practice.

It is necessary, of course, to know the extent to which the spot deflection is proportional to the voltage applied, what the smallest ascertainable change of spot deflection is (accuracy of definition) and what influence fluctuations of the mains voltage have on the results of the measurement, etc.

7. 3 Linearity of the display

The deflection plates of modern cathode ray tubes are constructed and arranged in such a way as to ensure a very good linear relationship between the voltage applied and the deflection of the spot over the whole area of the screen. If, for instance, for 10 volts DC between the two plates a deflection of 4 mm is obtained (deflection sensitivity 0.4 mm/V), then for 25V the deflection will be 10 mm, for 50 V it will be 20 mm, and so on. This applies to both vertical and horizontal deflection directions, of course, as can be seen in Fig. 7-1. In this case a voltage linear with time was applied to one pair of plates each time, and a direct voltage was applied to the other pair and raised in steps of 10 V. A photograph was then taken of each such position with regard to the others on the same negative. (Fig. 7-1*a* thus consists of 37 and Fig. 7-1*b* of 41 separate photographs.) In Fig. 7-1*a* these deflections were recorded in such a

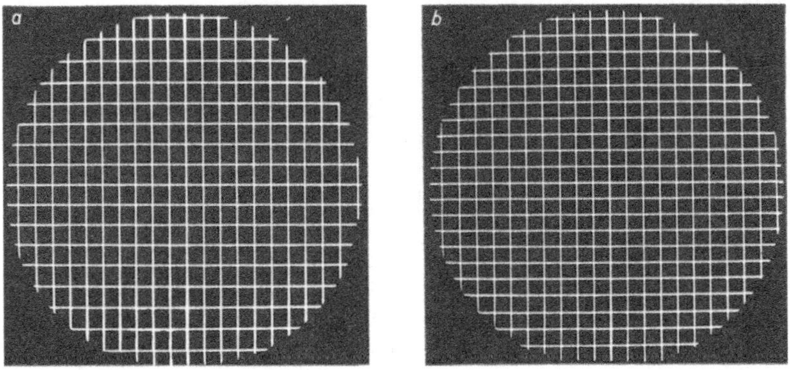

Fig. 7-1 Co-ordinate formed by deflecting a linear trace with direct voltages in steps of 10V
a) without sensitivity correction of the *Y*-plates
b) deflection sensitivity of the *Y*-plates matched to the *X*-plates by voltage division; DG 10-74 oscilloscope tube

way that they corresponded to the different deflection sensitivities of both pairs of plates. (The vertical pair has a higher sensitivity.) As the deflection voltage for the production of the trace was chosen high enough to cover the whole screen, it is immediately clear in which area a linear display can be expected and where distortions will appear. This impression is somewhat clearer in Fig. 7-1*b*. In this case the deflection sensitivity of the *Y*-plates was reduced to the same value as that of the *X*-plates, resulting in a quadratic graticule. The limit of error in linearity of deflection in modern tubes is $< 2\%$ and sometimes even $< 1\%$.

7.4 Reading off the display

The simplest way of reading off the display is to use a strip of transparent millimetre squared paper. A transparent grid with lines 5 mm or even 1 mm apart is often used for this purpose (Fig. 7-2). In high-grade oscilloscopes illuminated graticules are used. The scale is marked on a strip of thick transparent material illuminated from the sides by electric lamps of variable intensity. As an example, Fig. 7-3 shows the scale of As an example, Fig. 7-3 shows the scale of the Philips "GM 5660" oscilloscope with the oscillogram of a pulse 1 μs wide.

A graticule of this sort has the great advantage that it does not obscure parts of the screen like a dark grid in the photograph, but is perfectly bright and visible together

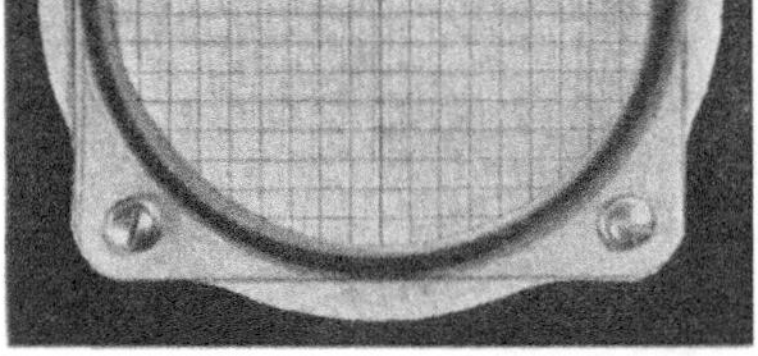

Fig. 7-2 Calibration lattice on the screen of an oscilloscope tube for amplitude measurements

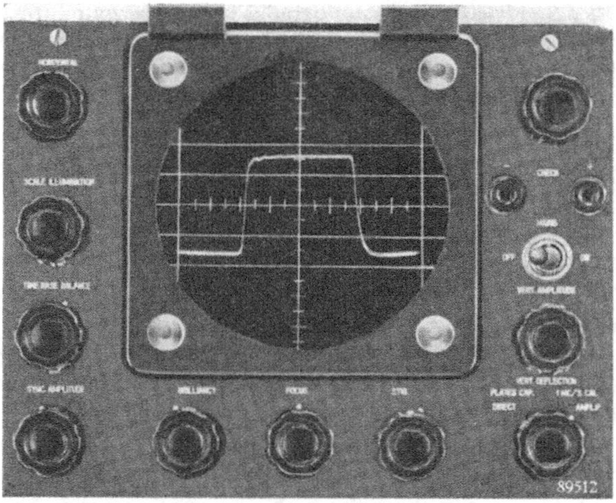

Fig. 7-3 "GM 5660" oscilloscope with floodlit graticule

Fig. 7-4 Observing small changes in amplitude

with the image on the screen. When the illumination is reduced, however, it becomes almost imperceptible and is not troublesome during tests requiring no amplitude measurement. With the illumination full on, such a scale indicates a most satisfactory way whether or not the oscilloscope is switched on.

If it is desired to observe small changes in amplitude, a good method is to cover the rest of the display with a sheet of stout paper (thin pressboard). The peaks which exceed the initial amplitude can then be clearly seen (Fig. 7-4).

It is advisable to create an electrical graticule as shown in Fig. 7-1*b*. By means of such a grid all non-linearities (including those of the oscilloscope tube) are taken into account. Reading off is completely free from parallax errors. A process has been described by ROBINSON and VAN ALLEN, by means of which simple circuits can be used to make such graticules [2].

7. 5 Accuracy of reading

The oscillograms in Fig. 7-5 are intended to illustrate the maximum accuracy of reading obtainable on the screen. They reproduce sections from the horizontal trace enlarged about $1^1/_2$ times. The top trace is a portion of the trace in the zero position. The next picture actually represents two traces, namely the trace in the zero position and a trace vertically deflected by a direct voltage of 0.5 V. In the remaining of oscillograms the zero trace is represented together with its position when vertically deflected by direct voltages increased successively by 0.25 V. It is thus easy to ascertain the voltage at which the two traces are only just distinguishable from each other. This is

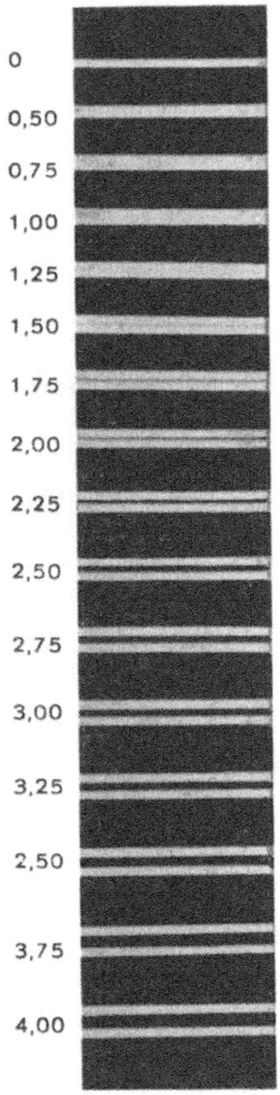

Fig. 7-5 Enlarged sections of photographs of the horizontal trace, first without vertical deflection and then with vertical deflection increased in each case by ΔV, to determine the accuracy of reading from the screen (Deflection sensitivity $DS_{=} = 0.4$mm/V)

clearly the case in the seventh picture from the top, i.e. at 1.75 V [62]).

The spacing between the spot centres is therefore somewhat greater than the diameter of the spot (in this case, 0.7 mm). This is confirmed in the third picture from the bottom. This recording was taken at a voltage difference of 3.5 V (twice the amount previously mentioned). This dark intermediate space is now roughly equal to the thickness of the trace.

The recordings in Fig. 7-5 were made with a DG 9-3 tube. Recently tubes with much superior spot definition have been developed. In order to show the high degree of accuracy here obtaineable, Fig. 7-6 shows recordings similar to those in Fig. 7-5 but made with the DG 10-74 tube. In these oscillograms in Fig. 7-6, for a deflection of 40 mm the zero lines are also recorded on the screen. This was done in order to give an impression of the ratio of the accuracy of spot deflection at the normal picture height obtained with 10 cm tubes. Oscillogram 7-6*a* shows only the zero lines and a line deflected by 40 mm. Fig. 7-6*b* includes a third recording of this line now shifted by $^1/_2$% (0.2 mm). Both these lines, however, coalesce to form a still thicker line. At a distance of 1% (0.4 mm) as in Fig. 7-6*c* the two lines are barely distinguishable. As these details are not clearly recognizable on a scale of 1:1, a portion of the double line shown in Fig. 7-6*c* has been enlarged (see Fig. 7-7). It confirms that for this tube operated at an acceleration voltage of about 2000 V and a post-acceleration voltage of 1000 V accuracy of definition to 1% can certainly be claimed. Finally, the oscillogram in Fig. 7-6*d* shows the lines with a deflection interval of 2% (0.8 mm); here the separation is very clearly defined. This degree of accuracy in definition can be generally expected.

In this connection it must be pointed out that magnetically focused cathode ray tubes which have a spot diameter of only a few μm have been described and are used for special tasks [3] [4]. The accuracy of reading and wealth of detail to be found in the oscillograms made with these tubes are correspondingly great. In the second publication here quoted, the tube works as a travelling wave tube; it is used for the oscilloscopy of extremely evanescent phenomena. Because of their cost and scope such tubes are, however, only of importance to a limited circle of users.

[62]) The practical procedure is adopted here. To obtain a better physical definition of the smallest just readable change of voltage, that voltage value might be determined at which the blackening between the two bright traces of the spot is at most $1/e = 0.37$ times the blackening of the remaining background of the picture.

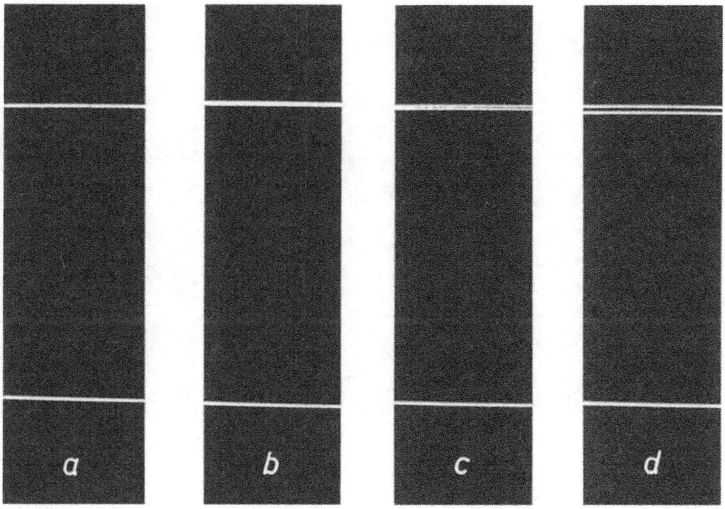

Fig. 7-6 Deflected zero lines with varying deflection difference. *a*) Zero line and deflected line. *b*) two deflected lines with $^1/_2\%$ deflection difference. *c*) deflected lines with − 1% deflection difference. *d*) deflection difference — 2%.

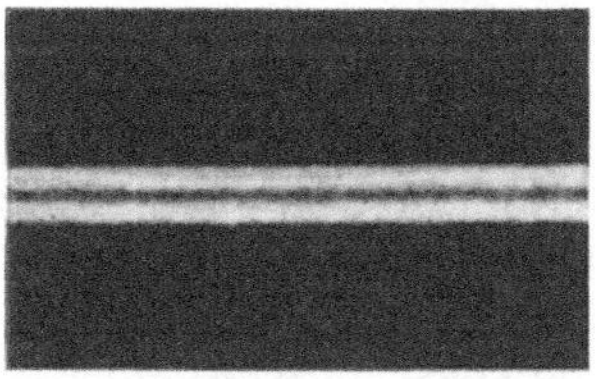

Fig. 7-7 Enlarged section of oscillogram in Fig. 7-6*c*; deflection difference 1%

7. 6 Influence of amplifier on linearity of display

It is understandable that there is a practical limit to measures taken to achieve good linearity combined with adequate output voltage. Thus, for example, the lowest possible values of anode resistors must be used if high frequencies are to be uniformly amplified, but this inevitably limits the undistorted voltage drive to the output stages of the amplifier. Particularly in the case of amplifiers with a high upper cut-off frequency one is obliged to accept this limitation of the undistorted control voltage amplitude (Part I, Ch. 5.19 "Output voltage requirements").

As an example, Fig. 7-8 shows the dynamic characteristics of the amplifier in the "FTO 2" oscilloscope together with the lattice of the floodlight scale [5]. Characteristics of this sort are easily obtained by feeding a suitably chosen portion of the time base voltage of the oscilloscope to the vertical amplifier. For various values of the time base voltage in the horizontal direction the spot simultaneously traces out the corresponding deflection in the vertical direction due to the amplified voltage from the amplifier under investigation. If the gain has been adjusted so that the X- and Y-deflections are equal, then with an ideal amplifier, the spot must trace out a straight line at an angle of 45°. The drive limitation of such an amplifier, due mainly to the low anode resistances, is manifested by a more or less pronounced S-band at the ends of the characteristic.

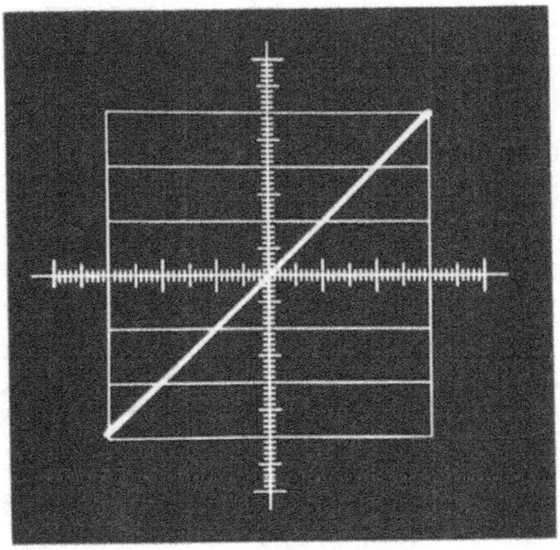

a

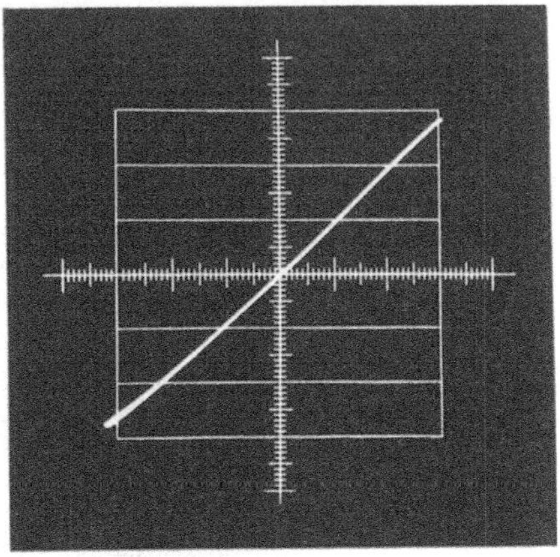

b

Fig. 7-8 Dynamic characteristics of the vertical amplifier of the "FTO 2" oscilloscope. Total acceleration voltage: for *a*) 1.3 kV, for *b*) 2.2 kV

In Fig. 7-8*a* the spot was deflected vertically to the maximum permissible "useful drive" for 60 mm picture height. It can readily be seen that up to 40 mm picture height this characteristic may be accepted as sufficiently linear to meet even high demands. Up to 60 mm the deviation still represents less than 3% distortion (acceleration voltage 1.3 kV). Such slight distortion in the oscillogram of a sine wave could be detected only by very experienced observers. If a total acceleration voltage of 2.2 kV and 1.2 kV post-acceleration voltage are applied in this oscilloscope, an approximately 30% higher deflection voltage is required, which introduces a greater amplitude distortion for the same picture height. As can be seen in Fig. 7-8*b*, however, the characteristic up to a Y-deflection of 40 mm may be regarded as sufficiently linear even under such conditions.

7. 7 Dependence of display sensitivity on mains voltage

For practical purposes it is important to know the influence of mains voltage fluctuations on the display when using a non-stabilized supply.

Fig. 7-9 shows the change in sensitivity of the cathode ray tube alone — curve α — and the sensitivity curve (β) when the vertical amplifier of the "FTO 2" oscilloscope is used. As expected, the deflection sensitivity of the cathode ray tube is inversely proportional only to the mains voltage. If, therefore, the mains voltage falls by 5%, the deflection sensitivity will rise by the same amount and vice versa [Part I, Ch. 2.3 "Deflecting the beam", (Eq. 2.3)]. When the deflection amplifier is used, however, it naturally exerts an influence on the deflection sensitivity. If the mains voltage rises, the mutual conductances of the amplifier valves become greater and vice versa, so that, according to the number of amplifying stages (or the gain for which the amplifier is adjusted), there is a measure of compensation, within certain limits, for changes in the sensitivity of the cathode ray tube. Curve β shows that, for slight fluctuations of the mains voltage, smaller changes in sensitivity occur. But even the influence of larger fluctuations of mains voltage is smaller than for the cathode ray tube alone.

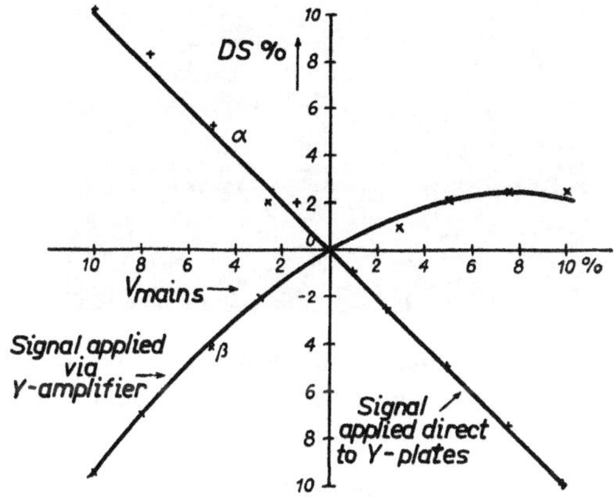

Fig. 7-9 Dependence of sensitivity of "FTO 2" oscilloscope on changes in mains voltage

Most of the cathode ray tube oscilloscopes put on the market in recent years are provided with power packs in which not only the H.T., but also the E.H.T. is electronically stabilized (Fig. 3-9). Normal mains fluctuations affect such units but little. Only in the case of small sets, where light weight is essential and the price must be kept low, are elaborate stabilization measures omitted.

7. 8 Relation between deflections due to direct voltages and those due to alternating voltages

It has been shown that the deflections over the whole screen are linearly proportional to the voltage applied. This is true for direct as well as for alternating voltages. With alternating voltages, the maximum deflection due to a voltage whose rms value has the same amplitude as a given direct voltage is considerably greater than the deflection due to the direct voltage. As the spot always follows the instantaneous value of the frequencies under consideration, it will always be deflected to the *maximum* positive and negative value of an alternating voltage.

The rms value I_{rms} of a sinusoidal alternating current with a maximum value I_{max} is that current which generates the same amount of heat in a resistance as is generated by the corresponding value of direct current measured with a direct current instrument. This relation is expressed by the following familiar equations:

$$I_{max} = I_{rms} \cdot \sqrt{2} = I_{rms} \cdot 1.414 \tag{7.1}$$

and

$$I_{rms} = \frac{I_{max}}{1.414} = I_{max} \cdot 0.707\,. \tag{7.2}$$

The peak swing of the deflection with alternating voltage will therefore be $2\sqrt{2}=2.828$ times the deflection due to a direct voltage of the same value. This means that the deflection sensitivity for alternating voltage is correspondingly greater. The following equation thus applies:

$$DS_{\sim} = DS_{=} \cdot 2.828\,. \tag{7.3}$$

These conditions are illustrated in the oscillogram in Fig. 7-10. With a 50 V direct voltage the spot was deflected vertically only once (*a*) and in (*b*) along the time axis. Above this the spot was traced out with an alternating voltage of 50 V_{rms}.

For measuring alternating voltage the time base deflection can also be dispensed with, but then a correspondingly long vertical trace is obtained. If the time base voltage is synchronized to the signal frequency, the distance between the crests of the curve ("peak-to-peak" V_{pp}) can be measured. If it is only a matter of determining the

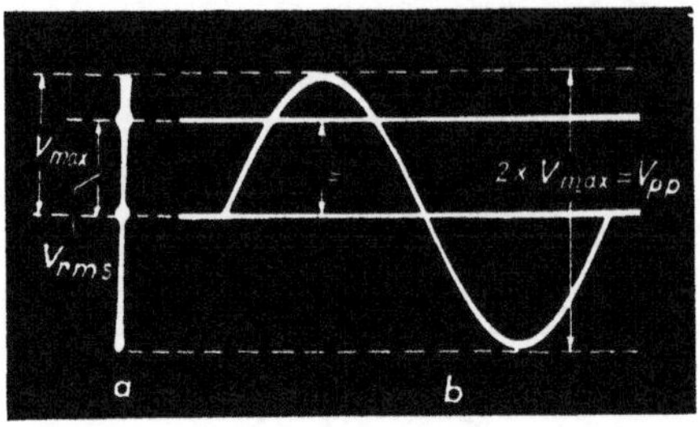

Fig. 7-10 Oscillograms illustrating beam deflection with direct and alternating voltages

voltage, it may be advisable to select the time base frequency considerably higher or lower than the input signal frequency. This produces a luminous area, the height of which corresponds to the voltage to be measured (similar Fig. 11-22*c* and 11-29*c*).

7. 9 Direct voltage measurements

When the voltage amplitude to be measured is sufficiently great, it can be applied directly, that is, without amplification, to the *Y*-plates. As has been shown in Fig. 7-6, with a deflection sensitivity of $DS_= = 0.45$ mm/V and a trace breadth of approx. 0.3 mm, a voltage difference of 1.0 $V_=$ can be read off quite easily. If a maximum deflection of 40 mm from the zero position is taken (corresponding to 90 $V_=$), this value can be read off with an accuracy of about 1%. For smaller deflections the degree of uncertainty will, of course, be greater, but with 10 cm tubes one should be able to read with an accuracy of from 1 to 3%. For such measurements it should be possible to dispense with deflection in the *X*-direction; in general, it is an advantage to have a time base voltage of about 100 . . . 1000 c/s and giving a short horizontal trace. The direct voltage will then move this trace just like the pointer of an indicating instrument. A scale can be calibrated based on the deflection sensitivity. Fig. 7-11 shows a scale of this sort and the picture on the screen with a deflection voltage of 43 V. In this way it is possible to measure not only all anode voltages and screen grid voltages in radio receivers and amplifiers, but also A.V.C. voltages, grid bias etc., so long as they are not too small. The polarity of the *Y*-plates is such that a positive voltage will produce an upward deflection and a negative voltage a downward deflection. Fig. 7-12 shows the circuit of a device by means of which, using a C.R.T. with $DS_= =$ 0.4 mm/V, DC voltages can be read off from the minimum perceptible value to 1300 V. Up to 100 V, intermediate values must be read from a scale on the screen. The accuracy of the reading decreases with diminishing voltage. If a high input impedance is desired, the switch *S* is set in position 2 and resistor R_1 adjusted to its highest value. For voltages above 100 V the position of switch *S* should be chosen as indicated in the diagram. Within the given voltage range, the setting of potentiometer R_1 should now be such that the spot deflection will be just

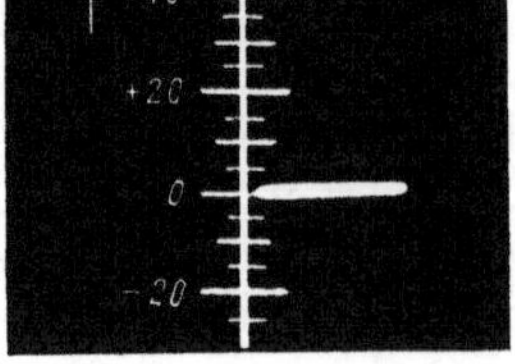

Fig. 7-11 Scale for DC measurements; positions of the trace with 0 V and 43 V on the *Y*-plates

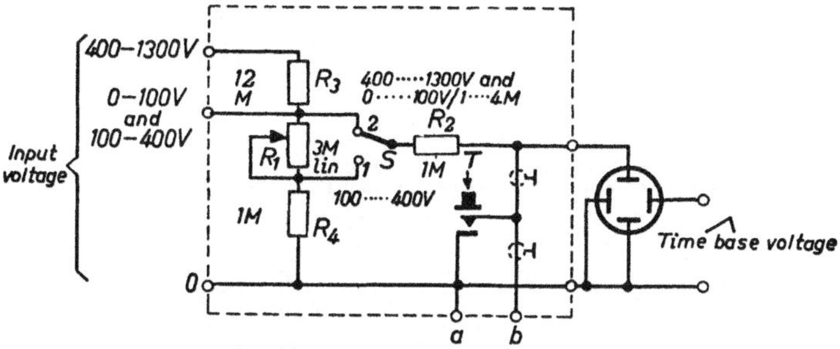

Fig. 7-12 Circuit of auxiliary device for measuring direct voltages in the ranges of 0-100 V, 100-400 and 400-1300 V.

as large as with direct connection of 100 V. From the position of the pointer on potentiometer R_1 the value of the voltage under test can be read off directly if the angle of rotation has been previously calibrated (the resistance value and the voltage scale run counter to one another). The great advantage of this is that it is possible to make measurements with the maximum accuracy, namely $\approx 1\%$.

By depressing the short-circuiting button T, the zero starting position can be restored at will, so that the deflection difference can be observed accurately. (The resistor R_2 prevents short-circuiting of the input voltage source.) It may be useful to remake this contact at rapid intervals by means of a relay in conjunction with an electronic switch, so that both positions of the spot can be observed simultaneously. A device of this sort can be connected to terminals *a* and *b* [63]). In this way very small deflections of the spot can be read distinctly.

The cathode ray tube is *actually a voltage indicator.* Measurements of current are therefore only possible by measuring the voltage drop across a resistance of known value included in the circuit. As this drop in voltage must be kept small, such DC measurements are generally only possible with the aid of a DC voltage amplifier.

7. 10 Alternating voltage measurements

The applications of the oscilloscope for alternating voltage measurements are very numerous. The vertical amplifier enables an extremely wide range of voltage to be investigated, and, with a suitable amplifier, measurements of voltages down to a few millivolts can be made over a wide range of frequencies. With the majority of oscilloscopes the maximum sensitivity of indication is known when the fine adjustment is fully turned up; intermediate values can be estimated according to the position of the control. For more exact quantitative analysis, comparison with a known voltage (e.g. 50 c/s), possibly calibrated against a measuring instrument is very advisable.

A facility of this kind is provided in oscilloscopes specially intended for the quantitative evaluation of oscillograms. Should this not be the case, it is advisable to have

[63]) The input terminals and the corresponding series resistors must be properly insulated and made safe. When measuring DC voltages, auxiliary voltages for correcting the zero position should be switched off. As these are fed in via the high-ohmic leak resistors, the effective voltage on the deflection plates would depend on the internal resistance of the input voltage source, and the beam deflection would be less clear in consequence.

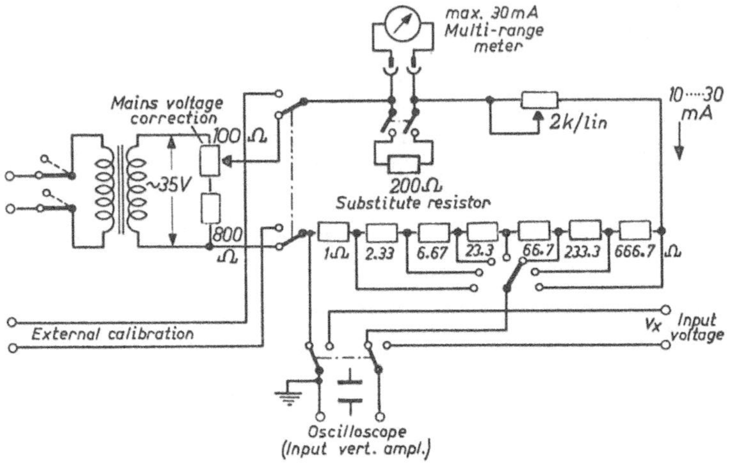

Fig. 7-13 Circuit of a calibration unit for alternating voltages

a calibration instrument made. Fig. 7-13 shows the circuit of such an instrument and Fig. 7-14 a photograph of an actual apparatus. Measurement is carried out by alternately applying the unknown voltage and the calibration voltage to the input terminals of the oscilloscope.

By means of the 2 kΩ intermediate control, the amplitude of the calibration voltage is adjusted to produce the same peak value as the unknown voltage. Thus it is possible with the calibration unit described to adjust the reference voltage to any value between 10 mV_{rms} and 30 V_{rms}. The instrument shown can be replaced by a resistor having the same value as the instrument resistance, but in this arrangement variations in the mains voltage will also be included in the measurement. A calibration voltage with a square waveform is particularly advantageous. If synchronization with the time base voltage is then deliberately avoided, two horizontal traces will be obtained on the screen and these greatly facilitate amplitude calibration. A square wave voltage of

Fig. 7-14 Practical model of a calibration unit for alternating voltages based on the circuit in Fig. 7-13

this kind can be generated quite simply by means of a gas discharge tube, connected via a resistor across a 50 c/s mains frequency alternating voltage source. The peaks of the sine wave will then be clipped at the level of the operating voltage (see oscillograms of Figs 7-15, 7-16 and 11-18).

If a stabilizing tube with a very constant operating voltage is used, as for example, the 85 A 1 or 85 A 2, the calibration voltage will be largely independent of the mains voltage and extremely constant over the life of the tube.

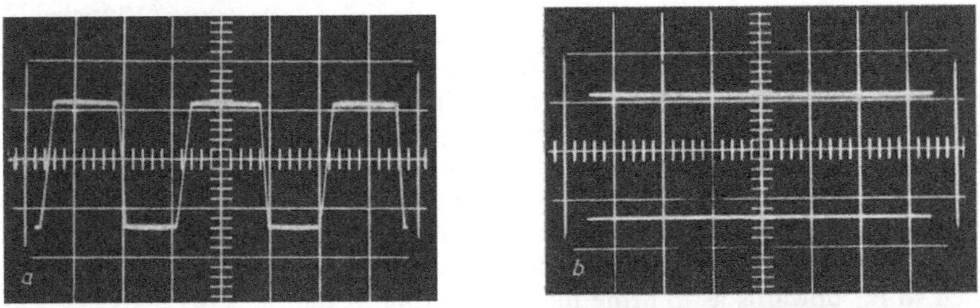

Fig. 7-15 Square calibrating voltage
a) triggered to represent waveform of the curve
b) time base unit relaxed oscillation, not synchronized

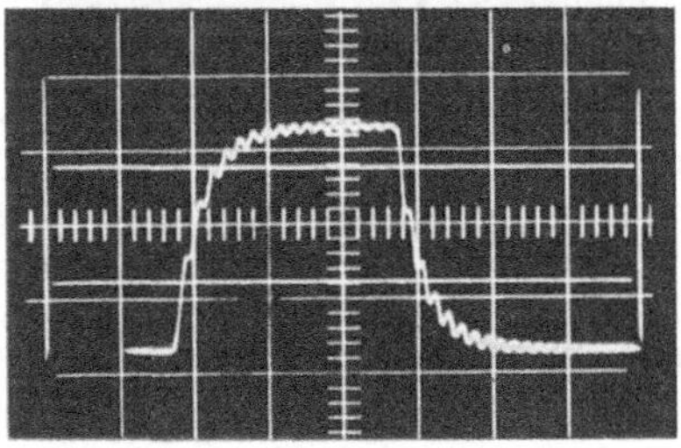

Fig. 7-16

Another circuit for clipping a 50 c/s voltage in which the voltage amplitude is determined and likewise stabilized by vacum diodes, is used in the Philips "GM 5662" and "GM 5666" oscilloscopes. This portion of the circuit is included in the wiring diagram of Fig. 4-100. A 50 c/s alternating voltage occurs here and its waveform is shown in the oscillogram in Fig. 7-15*a*. In this case the mains frequency was so triggered that a stationary picture was obtained. If the time base unit is working with a frequency other than 50 c/s, two horizontal traces are obtained for amplitude calibration as in Fig. 7-15*b*. These calibration traces can be recorded simultaneously with the actual oscillogram by pressing a button, thus obtaining a simple scale of amplitude as can be seen, for instance, in Fig. 7-16.

For measurements of alternating currents, it is necessary, as with direct currents, to connect in circuit a low value resistor of known resistance. The voltage drop can then be displayed on the screen. The greater the gain of the vertical amplifier, the smaller the resistor or the current can be.

7. 11 Determining the voltage amplitudes of an oscillogram with any waveform by displacement of the pattern and measurement of the direct shift voltage

In the preceding section it has been shown how a voltage corresponding to the amplitude of an oscillogram can be measured by comparing it with the peak-to-peak value of a known alternating voltage (sinusoidal or square). In many applications of the oscilloscope it is of particular importance to determine the voltage at particular points of the waveform (in the Y-direction) simply and clearly. The procedure about to be described fulfils this purpose very conveniently; moreover, given a sufficiently high voltage, the accuracy of definition can even be increased.

7.11.1 THE MEASURING PROCESS

In this method, amplitude is measured by determining the value of the direct shift voltage required to shift the oscillogram vertically through the difference in amplitude to be measured. The starting point can be the time base ar any other reference point. Fig. 7-17*a* shows first of all a normally adjusted oscillogram of a sinusoidal voltage with very pronounced peaks at the crests. The time base is also depicted. In Fig. 7-17*b* the oscillogram has been shifted so far upwards as to bring the lowest point of the sine wave to rest on the time base. For this a direct shift voltage of +15.5 V was required. A shift voltage of 65 V was needed to adjust the lower point of the peaks to the time base (Fig. 7-17*c*). In Figs 7-17*d* and 7-17*e* the appropriate shift voltage settings for the positive rest of the sine wave and the positive voltage peak were —15.5 V and —65 V respectively. It is thus seen that the amplitude of the sinusoidal voltage is $V_{\sim} = 31\ V_{pp}$ and that of the peaks is $V_p = 130\ V_{pp}$. This procedure has proved itself particularly useful in intepreting pulse-shaped oscillograms such as frequently occur, for example, in the investigation of blocking oscillators and multi-vibrator circuits and the like. The Philips "GM 5660" pulse oscilloscope is particularly suitable for this type of interpretation; the part of the circuit essential for this work is to be seen in Fig. 7-18. It is seen that the direct voltage for the vertical shift is led from the potentiometers R_{55} and R_{56} to sockets 11 and 12. A high resistance voltmeter (Philips "P 817") may be connected to these sockets for measuring the shift voltage. Very often the voltage to be investigated is of sufficient amplitude to produce satis-

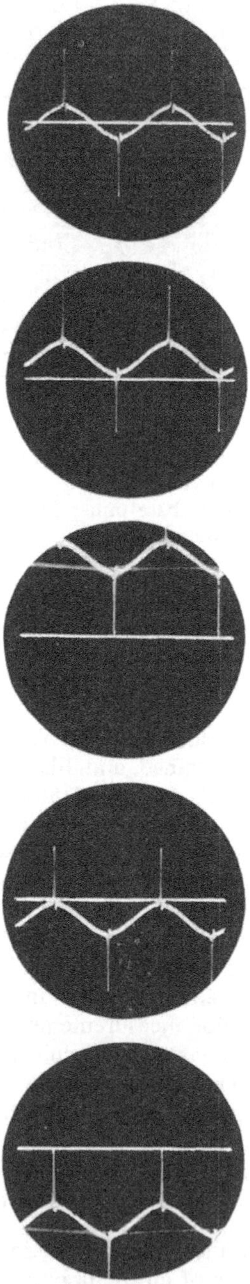

a

b

c

d

e

Fig. 7-17 Oscillograms for evaluating various amplitude sections on patterns on the screen by means of vertical shift
a) normal setting
b) determining the negative crest of the sine wave
c) determining the negative peak
d) determining the positive crest of the sine wave
e) determining the positive voltage peak

factory deflection when directly connected to the plates. A vertical amplifier, whose frequency limits always restrict the measuring range, can then be dispensed with. This is of great advantage, of course, with pulsed voltages with short rise times, as then the faithful reproduction of the waveforms is limited only by the capacitances of the deflection plates and circuit. If the voltage source has a relatively low output resistance, the picture obtained on the screen will be very close to reality. Moreover, if the signal under investigation is applied directly (without coupling capacitance), any DC component present will be visible in the oscillogram, and this component can be evaluated in the manner described.

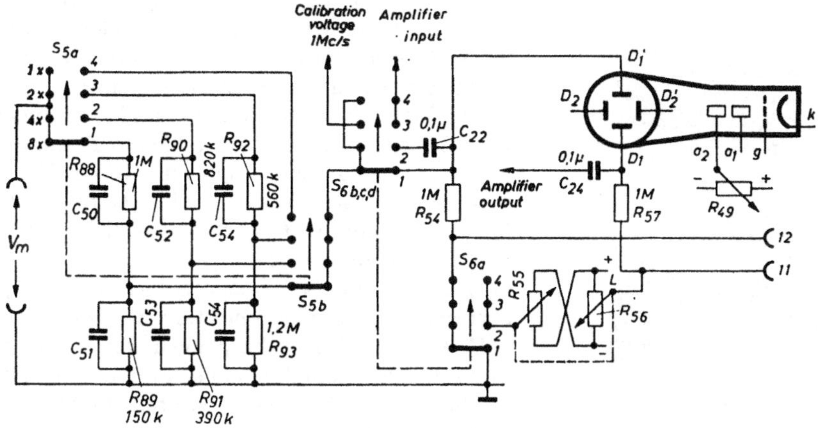

Fig. 7-18 Part of the circuit of the Philips "GM 5660" pulse-oscilloscope

On the "GM 5660" pulse oscilloscope, the first position of the selector switch for the vertical deflection is the one intended for such measurements (Fig. 7-18). The voltage under examination reaches the "upper" Y-deflection plate D'_1 via the compensated voltage divider ($R_{88} \ldots R_{93}$ and $C_{50} \ldots C_{54}$). The direct shift voltage from potentiometer R_{56} reaches the other plate D_1 via the leak resistor R_{57}. The leak resistor of plate D_1 is therefore tied to chassis in this position of switch S_6. Regarded individually, each of these voltages is applied asymmetrically to the Y-plates. (In the other positions of switch S_6, the vertical shift is effected by symmetrical direct voltages. A corresponding opposing direct voltage then reaches plate D'_1 from R_{55} via R_{54} in switch positions 2, 3 and 4.) Since, however, the oscillogram is approximately in the centre of the screen, the voltage on both plates (with opposite polarity) will be roughly equal and thus the mid-potential of the plates approximately zero. There will therefore be no astigmatism.

But even with the greater shifts required for evaluating the waveform in the way described, the potential on anode a_2 can be adjusted by means of potentiometer R_{49} to equal the mid-potential of the deflection plates and thus maintain a well focused image. This is of particular advantage when, as will be described in the following paragraph, the signal amplitudes are increased to several times the diameter of the screen in order to improve the accuracy of reading.

7.11.2 IMPROVING THE ACCURACY OF READING BY INCREASING THE SIGNAL AMPLITUDE AND SUPPRESSING THE ZERO POINT

Considerably greater accuracy can be attained in the measurement of voltage amplitudes if the vertical deflections can be made to exceed the useful area of the screen.

The pattern shift needed for determining a certain potential difference in the voltage curve is correspondingly greater, which leads to a corresponding increase in the accuracy of reading. The accuracy of reading is thus determined in practice only by the accuracy of the DC instrument used to measure the shift voltages.

Fig. 7-19 shows some oscillograms taken with the "GM 5660" pulse oscilloscope, with the switch in position 1 as shown in Fig. 7-18. Under observation is the voltage at the anode of an amplifier valve when negative pulses are applied to the control grid.

Fig. 7-19*a* shows the oscillogram in the conventional way; the zero level has been added to all these pictures to render the direct component visible. For this recording the voltage was divided in the ratio 2:1 (switch $S_{5a,b}$ in position 3). For the rest of the recordings the voltage was applied to the Y-plates undivided (switch $S_{5a,b}$ in position 4). The oscillogram in Fig. 7-19*b* corresponds to the zero position; the shift voltage was 0 V. For Fig. 7-19*c* the pattern was shifted so as to bring the horizontal voltage level to the time base. In this case 19 V_{pp} was measured. In Fig. 7-19*d* the upper edge of the oscillogram just touches the zero line (to some extent with the zero point suppressed). A shift voltage of 150 V was necessary for this. The total amplitude of the measured voltage is thus 150 V_{pp} with a level portion at 19 V_{pp}. Using this process, which gives an accuracy of reading equivalent to that on a screen several times larger, it is also possible to evaluate accurately the individual modulation levels of a modulation levels of a modulated HF voltage. This method can also be used with simple oscilloscopes; the only condition is that the Y-plates must be readily accessible. The schematic circuit of such an arrangement is given in Fig. 7-20. The shift voltage, V_{comp},

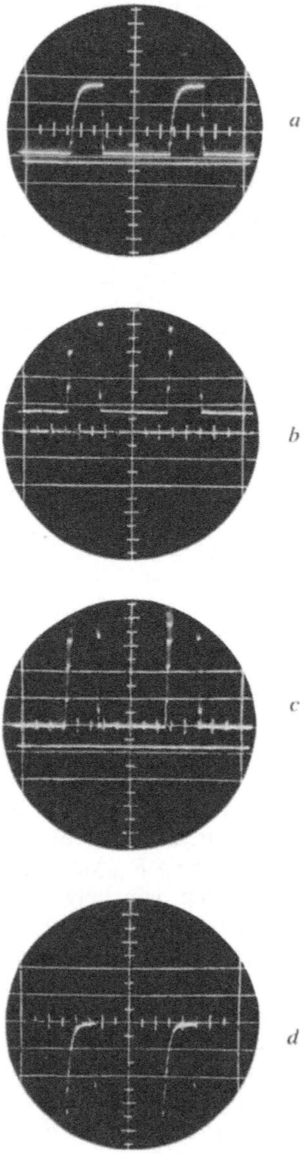

Fig. 7-19 Oscillograms for accurately reading the voltage by enlarging and vertical shift
a) normal oscillogram
b) zero setting at twice picture height
c) determining the direct voltage level
d) determining the voltage crest

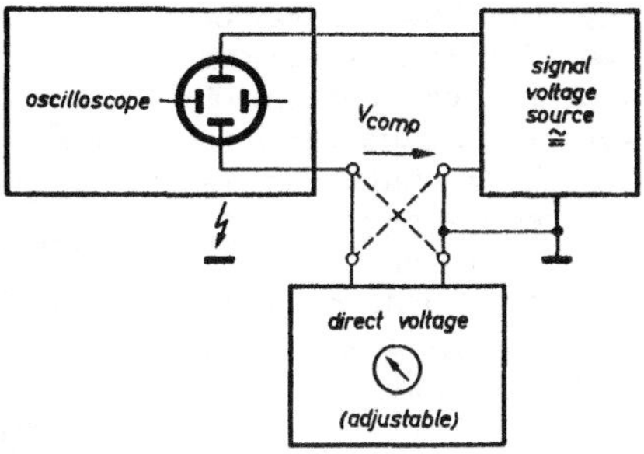

Fig. 7-20 Circuit for evaluating oscillograms obtained with a normal oscilloscope and an external voltage source for the shift ("electric lift")

must be supplied from an external voltage source. One deflection plate is connected directly to the shift voltage, while the output voltage of the phenomenon under test is applied between the other deflection plate and the direct volage source. By adjusting the shift voltage, various amplitude values of the oscillogram can also be determined in the way already described.

In this circuit, the greatest care must be taken that the oscilloscope chassis is at the same direct potential as the compensation voltage V_{comp} with respect to earth. Appropriate *precautionary* measures must therefore be taken. As the pattern on the screen moves up and down while the compensation voltage is being adjusted, the circuit is also known as the "electronic lift".

Optimum use of the "electronic lift" is made in the difference amplifier comparator plug-in, type Z made by Tektronix. This pre-amplifier is particularly proof against overloading, and the voltage amplitude can be increased to an apparent picture height of 20 m. The vertical shift of the picture can be accurately adjusted within a limit of error of 0.2% by means of a helical potentiometer. In this way reading the voltage amplitudes in a correspondingly enlarged oscillogram is possible with the same degree of accuracy within the frequency limits of the amplifier (5 . . . 13 Mc/s, depending upon the oscilloscope used).

7. 12 Determining the voltage amplitude and the time scale by shifting the image

The interpretation of the image on the screen by calibrated shifting with a direct voltage can be used not only to determine the voltage amplitude but also to measure the time scale.

This is done in the TF 1330 Marconi oscilloscope, the upper part of which is to be seen in Fig. 7-21. In this unit the starting position of the picture can be chosen in the usual way by means of adjusting knobs for the vertical and horizontal directions. The oscillogram can also be shifted by two adjusting knobs with large scales (*A* and *T* in Fig. 7-21).

The oscillogram to be measured is first set at a suitable starting position by the uncalibrated adjusting knobs. It is then shifted by one of the calibrated, large-scaled adjusters until the point at which the amplitude is to be measured is brought to rest at the same position on the graticule (in a similar way to the measurement of band width in Fig. 18-7). Knob *A* (amplitude) is used for the vertical shift and knob *T* (time measurement) for the horizontal shift.

The multiplication of the scale reading by the adjusted range of the deflection coefficient or of the adjusted time scale gives the required measurement.

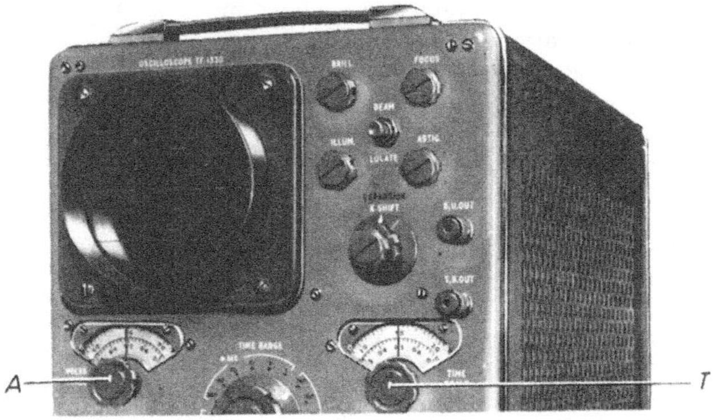

Fig. 7-21 Marconi oscilloscope TF 1330. Oscillogram interpretation by means of calibrated picture shift
A Shift of pattern in *Y*-direction, amplitude measurement
T shift of pattern in *X*-direction, measurement of time scale

7. 13 Plotting the amplitude with electronic switch

If an input voltage is applied to only one input circuit of a dual trace switch (see Ch. 9), a record image in the form of a horizontal trace can be caused to appear on the screen.

By suitable adjustment of the vertical positions of the two patterns, this horizontal trace can be shifted vertically over the image. If the circuit of the dual trace switch permits this shift to be calibrated in volts, it becomes possible, by shifting the trace, for the amplitudes of the pattern on the screen to be easily plotted (without parallax error). Fig. 7-22 shows as an example four oscillograms of pulse sequences in which this "measuring line" was moved to four different positions, thus permitting the amplitude at four interesting levels to be determined. These photographs were made with the Philips "GM 5666" oscilloscope and the "GM 4580" dual trace switch, both channels working in DC coupling.

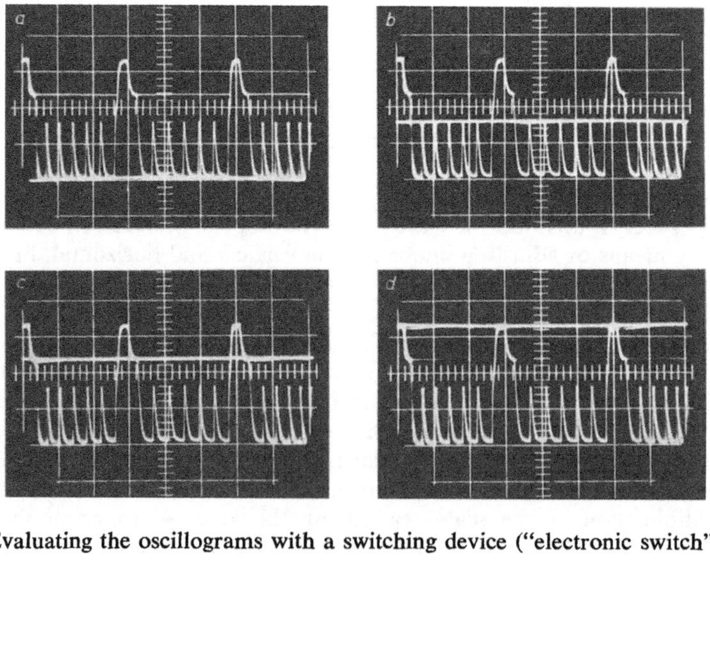

Fig. 7-22 Evaluating the oscillograms with a switching device ("electronic switch")

Schroeder [6] has described a similar process in an auxiliary apparatus for measuring pulse peaks in television investigation, in which the switching is done by a relay switch. In the device there described the relay is driven by rectangular voltages from a valve oscillator. The switch frequency is adjustable between 40 c/s and 250 c/s.

7. 14 Digital oscillogram interpretation

The firm of Du Mont has found a very interesting way of interpreting the time and amplitude measurements of the oscillograms made with its type 425 oscilloscope and of recording the results in digital form. Fig. 7-23 shows part of the front plate of this oscilloscope and the adjusting knobs essential for this form of interpretation. This device, known as "Display Logic" alternating switching, makes use of two auxiliary luminous spots which are added into the oscillogram. The Lissajous-image of a square voltage is locked in phase with both the *X*- and *Y*-deflection voltage in a way similar to the electronic switch. One of these auxiliary spots is represented by the index dot, *I*, and the other serves as the scaling dot, *M*. The scaling dot can be moved vertically and horizontally to any desired point of the screen by means of two sets of multi contact switches operated by knurled knobs. When all these knurled knobs are set at 0, scaling dot *M* and index dot *I* coincide. They can be moved simultaneously to any desired point on the screen by the "index" adjuster (*K1*) which can be shifted both vertically and horizontally.

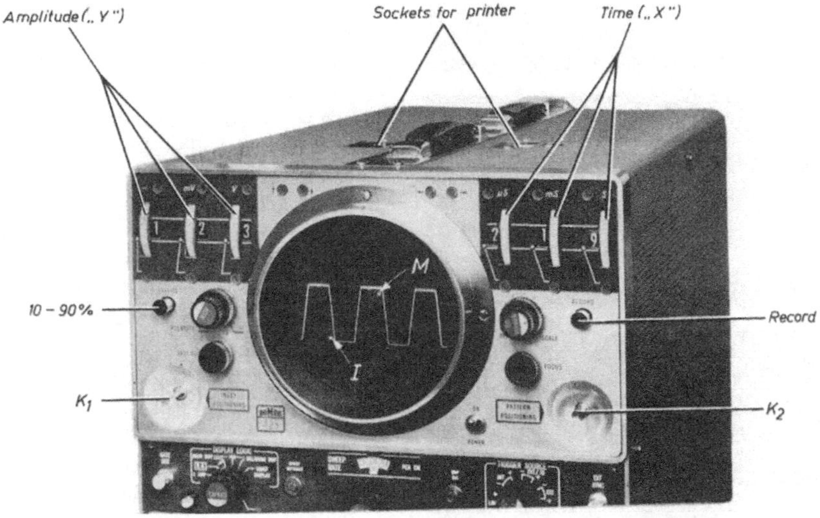

Fig. 7-23 Du Mont oscilloscope 425. Digital interpretation of the screen image in amplitude and time direction

The oscillogram which it is desired to interpret can also be adjusted in the required position on the screen with a second adjuster (*K*2). To measure the oscillogram the index and scaling dots must first be made to coincide by turning the amplitude and time step switches to 0. These dots must now be moved, by means of the index adjuster *K*1, to the point on the oscillogram at which the measurement is to commence. Next, the scaling dot *M* should be moved by means of the knurled knobs to that point of the oscillogram at which the measurement is to cease, the index dot *I* remaining stationary at its original position. The measurement result in time and amplitude values can then be read off directly at these adjusters. One of the pilot lamps lights up under these adjusters according to the ranges selected, indicating even the decimal point with unmistakable clarity. No calculation is therefore required for interpretation.

As the voltages for generating the auxiliary luminous spots are taken from a point in the circuit prior to the amplifier output stages, the non-linearities liable to occur in these stages, particularly when overdriving is considerable, have no influence on the result of the measurements. As the luminous spots also occur in the same plane as the oscillogram, there are no errors in measurement due to parallax. The limit of error in evaluation is stated to be $< 2\%$.

These measurement results can be manipulated still further. Two forty-pole plug connections are provided in the upper part of the oscilloscope to which a number printer — a perforated tape — or punched card printers can be connected. These printers are supplied with the measuring results in decimal-coded form. When evaluation on the fluorescent screen is finished, printing can be started by means of the knob marked "Record". At the other output connections there are also voltages available which are in a corresponding ratio to the positions of the luminous spots. If a *X* - *Y* recorder with a maximum rating of 100 mV (e.g. Philips PR 2210A) is plugged into these connections, it is possible, by following the image on the screen with the scaling point, to record the corresponding enlarged oscillogram as well.

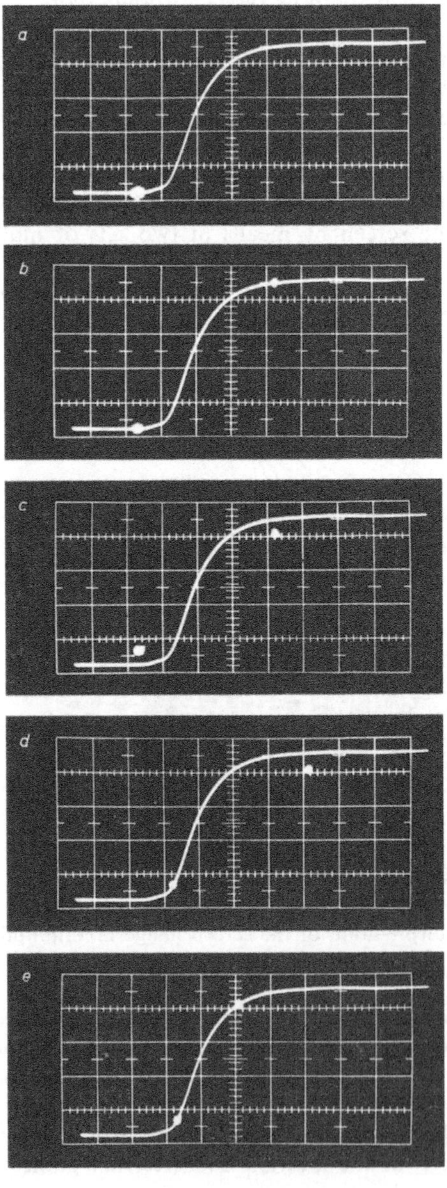

Fig. 7-24 Measurement of the rise time of a pulse with the oscilloscope of Fig. 7-23

The 10—90% knob serves to facilitate the measurement of pulse rise times. As can be seen in oscillograms *a—e* in Fig. 7-24, the evaluation is carried out as follows: first of all index dot and scaling dot are made to coincide at the start of the pulse on the pulse datum line (Fig. 7-24*a*). All knurled-disc adjusters are now at 0. At this point the scaling dot is set to the pulse peak as desired (*b*). The pulse height can be read off at the adjusters for amplitude as an interim result. If button "10—90%" is now pressed, the index dot now rises by 10% of the total amplitude, and the scaling dot drops 10% (*c*).

The vertical distance between the dots now corresponds to the amplitude range as defined for determining the rise time. Now both dots must be shifted with adjusting stick *K*1 in a precisely horizontal direction until the index dot coincides with the rising flank of the pulse image at this height (*d*). Now the scaling dot need only be shifted with the knurled knob switch to the left until it too coincides with the ascending flank of the pulse (*e*). Now the rise time can be directly read off at the adjusters for the time direction. If a printing device is connected for this purpose, the value can immediately be recorded in digital form by means of the knob marked "Record". It is, of course, also possible to record additional data by means of suitable extra signs (e.g. date, number of measurement, etc).

It is evident that such processes, especially for long series of measurements, offer considerable advantages as compared with conventional screen image interpretation.

Oscilloscope 567 of Tektronix, the "Readout Oscilloscope", enables the image of the voltage waveform displayed in an analogous way with double image and intensity marking, to be interpreted automatically in the horizontal direction in digital form. As it is possible to select automatically determined types of measurement for this (rise time, decay time, delay time, etc.), this oscilloscope with cabinet racks 3 S76 and 3 T77 for sampling oscilloscope technique (rise time 0.45 ns), is particularly suitable for short time series measurements of transistor switching times, for measuring the dynamic behaviour of magnetic cores, the properties of delay lines, and for similar tasks.

7. 15 Resistance measurements

As a typical voltage measuring instrument, the oscilloscope is also very suitable for measuring high resistance with a direct voltage. Fig. 7-25 shows the basic circuit.

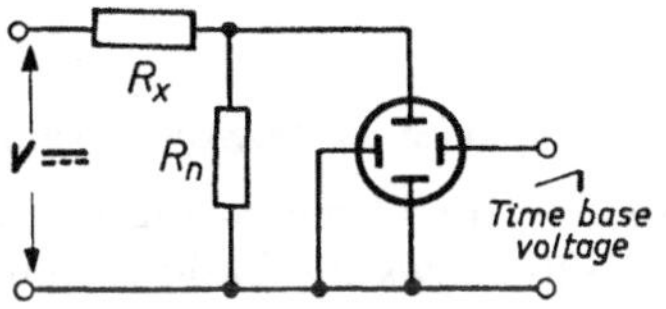

Fig. 7-25 Circuit for measuring high resistances

A small amplitude of time base voltage is advisable on the *X*-plates as for direct voltage measurements, in order to produce a short horizontal trace. The value of the unknown resistance R_x is found from the leak resistance — symbolized here as R_n — and from the input voltage $V_{=}$, according to the equation:

$$R_x = R_n \cdot \left(\frac{V_{=} \cdot DS_{=}}{Y} - 1 \right). \tag{7.4}$$

If the first term in the bracket is considerably greater than 1, it is permissible to simplify the equation to:

$$R_x = R_n \cdot \frac{V_= \cdot DS_=}{Y}, \tag{7.5}$$

in which $DS_=$ is the deflection sensitivity of the cathode ray tube for direct voltages and Y is the deflection of the spot read on the screen. If, for instance, the 300 V supply voltage of the vertical amplifier is used for this purpose, then, when R_n = 3 MΩ, $DS_=$ = 0.4 mm/V and Y = 0.5 mm, a value for R_x = 700 MΩ only just distinguishable can be read off the screen.

This method is particularly suitable for testing the insulation resistance of block capacitors. For measuring even higher resistances the anode voltage of the cathode ray tube of the oscilloscope (or a DC amplifier for amplifying the input voltage) could be used. In view of the high voltages the necessary safety precautions would have to be taken to avoid accidents, for example by inserting a series resistor of at least 1 MΩ, which, if necessary, must be deducted from the result, In this way, resistances up to ≈ 2400 MΩ can be measured [64]).

As an amplifier is usually available, smaller resistance values can also be measured. If a sufficiently small resistance is included in the circuit and, as described, the voltage drop across it, and hence the current, are measured at a known voltage, then impedance measurements by comparison with a correspondingly adjusted ohmic resistance can be carried out.

7. 16 Power measurements

The following is a very simple method of power measurement: the voltage of the circuit under test is applied via the vertical amplifier to the Y-plates. A voltage proportional to the current, taken across a low value resistor in the circuit to the unknown impedance, is used, via the horizontal amplifier, to deflect the spot in horizontal direction. As current and voltage are generally not in phase, an ellipse appears on the screen. By calibrating the amplitude along both co-ordinates, current and voltage can now be determined as described. The distance between the points where the co-ordinates intersect is, moreover, a measure of the phase angle φ or cos φ (see Ch. 11 "Phase measurements", particularly Fig. 11-7b and Fig. 11-10, with descriptions). The power dissipated in the impedance can therefore be found from the equation:

$$W_x = V_x \cdot I_x \cdot \cos\varphi. \tag{7.6}$$

7. 17 Capacitance measurements

A special case of measuring AC impedances is the measurement of small capacitances. Fig. 7-26 shows the basic circuit. The value of the reference capacitance and of the input frequency must be so high that $1/\omega C_n \ll R_p$.

The unknown capacitance C_x is then found from the equation:

$$C_x = C_n \cdot \frac{V_n}{V_m - V_n}. \tag{7.7}$$

---

64) Just in the same way as small DC voltages can be measured with an oscilloscope with the aid of a DC amplifier, it is, of course, also possible in the way described to measure correspondingly small resistances too.

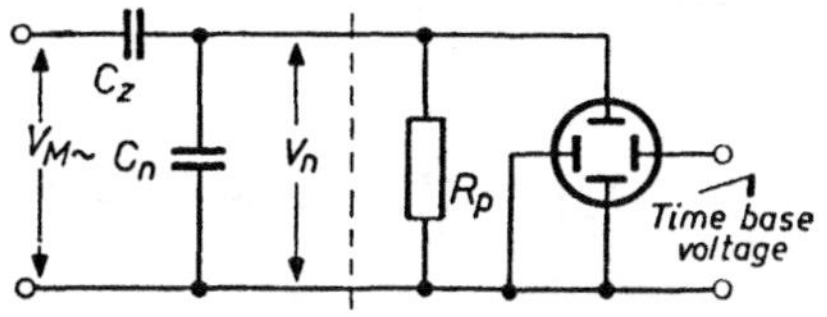

Fig. 7-26 Circuit for measuring small capacitance values

If $V_n \ll V_m$, the equation can be written:

$$C_x = C_n \cdot \frac{V_n}{V_m} \,. \tag{7.8}$$

For $R_p = 3\ \text{M}\Omega$ and $f = 5000$ c/s, C_n can be 1 nF. If $V_m = 30$ V and $V_n = 10$ mV are still readily measurable, this corresponds to a minimum value for $C_x = 1/3$ pF [65]).

This method is therefore particularly suitable for measuring small capacitances (valve capacitances, etc.). Since the voltage on V_n must always be amplified for the display, a vertical amplifier should be assumed to be inserted at the dotted line position in Fig. 7-26. The exact value of this voltage is determined as described in Ch. 7.10 "Alternating voltage measurements".

Summarizing, it can be said that the foregoing examples show that the cathode ray oscilloscope can be used for amplitude measurements just as an AC or DC valve voltmeter (high input resistance).

In the case of input voltages which require to be amplified, the pass band of the vertical amplifier determines the frequency application of the oscilloscope.

---

[65]) It may be necessary to take into consideration the input capacitance of the oscilloscope with respect to capacitance C_n, so that it must accordingly be rated smaller.

CHAPTER 8

NULL-INDICATION IN AC BRIDGE CIRCUITS

The cathode ray oscilloscope, a voltage indicator of extreme sensitivity, can be used with great advantage as a null-indicator for AC measuring bridges. Its high input impedance is of particular value in this connection.

8.1 Simple null-indicator

Fig. 8-1 shows in its simplest form the application of the C.R.O. as a null-indicator in a "sliding-wire" bridge. The horizontal plates are not employed, so that the out-of-balance voltage, amplified by the vertical amplifier, produces linear vertical deflections on the screen, as can be seen in the oscillogram of Figs. 8-2*a* and *b*. If the bridge is balanced, the out-of-balance voltage and therefore the deflection disappear, and all that is left on the screen is the spot *O*. The value of the unknown impedance Z_x is given by the familiar expression:

$$Z_x = \alpha \cdot Z_n, \tag{8.1}$$

where the factor α is the ratio of the potentiometer arms a:b, as shown in Fig. 8-1.

Traces I and II appear as the result of arbitrarily varying the setting of potentiometer *P* to the right and left of the null position. In Fig. 8-2*b* these deflections have been recorded superimposed on one another, as they appear to the observer.

This procedure, however, has the disadvantage common to all other conventional indicating methods in AC bridges (with the exception of such instruments as are provided with phase-dependent rectifiers), be they headphones, magic eyes, vibration galvanometers or such like, that if the bridge is not in balance, there is no way of telling whether the correct position of the potentiometer *P* should be sought further to the right or to the left. At high indicating sensitivities it becomes very difficult to find the bridge minimum.

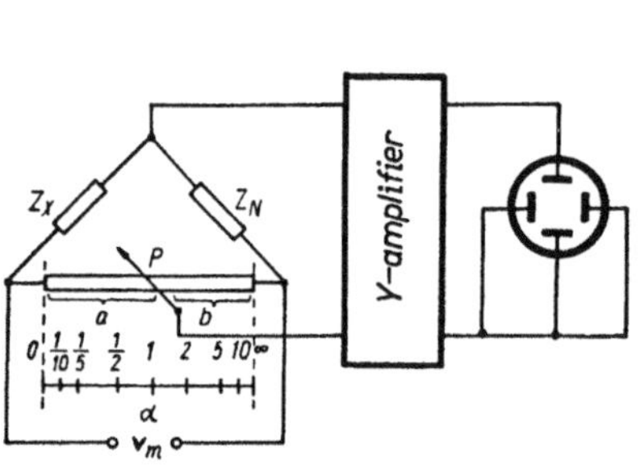

Fig. 8-1 AC bridge circuit with oscilloscope as null-indicator

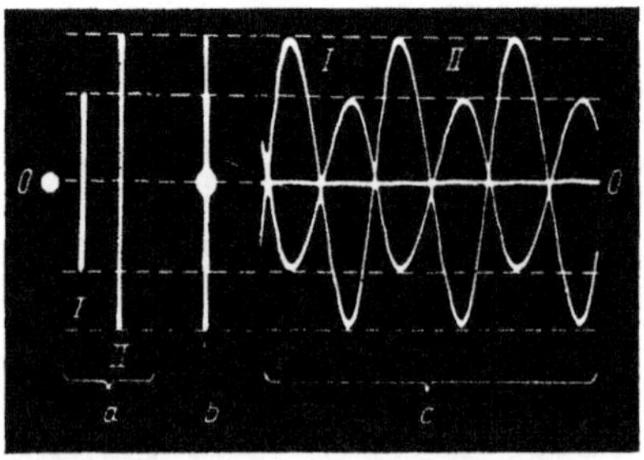

Fig. 8-2 Oscillograms obtained with the oscilloscope as null-indicator. *a*) and *b*) without horizontal deflection. *e*) horizontal deflection voltage synchronized at $1/3\ f_m$

8.2 Phase-dependent indication by synchronizing the time base with the bridge voltage

The above disadvantage can be avoided by applying the linear time voltage to the *X*-plates and synchronizing it accurately with the bridge supply voltage as shown in the circuit in Fig. 8-3. The picture on the screen is now the trend of the null voltage with respect to time. In the oscillogram shown in Fig. 8-2*c*, the frequency of the voltage applied to the *Y*-plates was 50 c/s and the time base frequency was 16 2/3 c/s, which resulted in an image of three cycles.

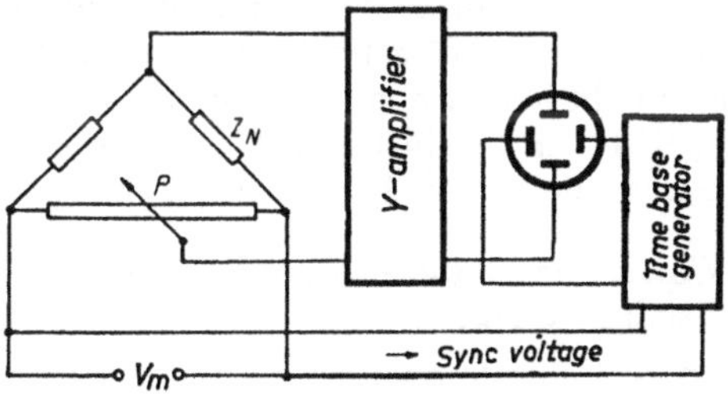

Fig. 8-3 Bridge circuit with oscilloscope; time base synchronized with the frequency of bridge supply voltage

As the time base voltage was firmly synchronized with the frequency, the oscillogram not only shows the amplitude of the out-of-balance voltage but also, from the phase relations, indicates the direction in which the unbalance lies. When the bridge potentiometer is operated, the phase position of the out-of-balance voltage reverses when passing through the point of balance, as can be seen from the oscillograms I and II of Fig. 8-2*c*.

8.3 Null indication by means of a rotating trace produced by horizontal deflection with the bridge voltage

A still more accurate null indication is obtained by applying the bridge supply voltage directly as shown in Fig. 8-4*a*.

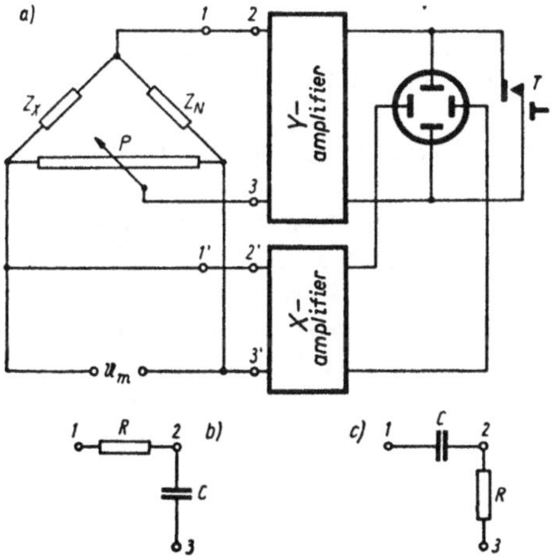

Fig. 8-4 Bridge circuit with oscilloscope for null-indication with rotating trace. Horizontal deflection by bridge voltage

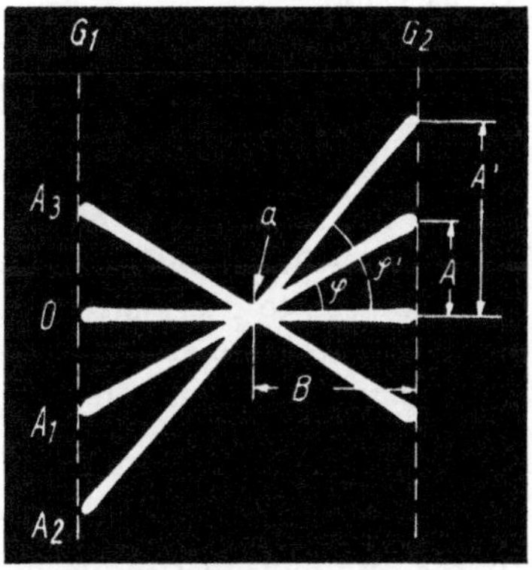

Fig. 8-5 Oscillograms of null indication with rotating trace

Since the voltage V_m applied to the bridge itself is in general hardly sufficient to bring about a horizontal deflection of 30 ... 60 mm, it must be brought to the required amplitude by a suitable amplifier. If the impedances Z_x and Z_n are of the same type, without phase differences and the amplifiers cause no phase shift, the spot deflection in both directions will be in phase and will produce straight-line images.

When the bridge is balanced the spot will be deflected horizontally only, giving the horizontal trace O in Fig. 8-5. With an unbalanced bridge the spot is also influenced in the Y-direction, so that, when the potentiometer is operated, the trace turns about the point a. The extremities of the traces which result will lie on straight lines in the vertical direction, provided the flat portion of the screen is not exceeded. These straight lines are indicated by G_1 and G_2 in Fig. 8-5.

Traces A_1 and A_2 in Fig. 8-5 correspond to different degrees of bridge unbalance on the one side, while A_3 indicates a change of balance in the other direction.

8.4 Correction of the phase relationship between bridge voltage and horizontal deflection voltage

It is also possible to obtain a sufficiently large deflection voltage for the X-plates by stepping up the bridge voltage with a transformer. Phase shifts occur during this process, so that the pattern on the screen is no longer a simple trace but a more or less open ellipse. If the mains voltage is used for measurement, the 50 c/s X-deflection position of the selector switch provided on some oscilloscopes for deflecting along the horizontal axis can be used. In this case too, a phase difference with respect to the bridge supply voltage will occur. When the vertical amplifier also causes a phase shift or when the shift differs as between X- and Y-amplifiers, its effect must be eliminated by turning back the phase to a corresponding extent. This can be done by incorporating a phase-shifting network in the out-of-balance voltage lead to the vertical amplifier (points 1, 2, 3) or in the deflection voltage lead to the X-plates (points 1′, 2′, 3′) as shown in the layout in Figs. 8-4*b* and *c*. This network must be so rated that, when comparing impedances without phase difference, a simple trace will again appear on the screen. (For preliminary adjustment it is better to compare ohmic resistances.) According to the ratio $R : 1/\omega C$, the phase shift, when $R = 1/\omega C$, will be 45°. At the same time the voltage at the output of the network will fall to 0.707 of its value.

The output voltage of the network in Fig. 8-4*b* lags behind the input voltage, and that of Fig. 8-4*c* leads the input voltage.

If considerable phase correction is necssary, anti-phaseshifting networks can be incorporated simultaneously between points 1, 2 and 3 on the one side and 1′, 2′ and 3′ on the other. When $R = 1/\omega C$, the phase difference obtainable with both networks will be 90°. The essential point here is that the horizontal deflection voltage need to necessarily originate direct from the one bridge diagonal, but can be obtained from the voltage source in other ways. Once the correction has been made, the measurements of complex resistances can be carried out in a manner which will be described later.

8.5 Bridge sensitivity

If the minimum deflection coefficient of the available oscilloscope is, for example, 3.0 mV_{pp}/cm and the mean diameter of the spot is 0.3 mm, a change in the position of the extremety of the indication due to an out-of-balance voltage of 0.1 mV_{pp} will still be readable on the screen. By using a specially tuned amplifier, considerably higher null sensitivities can be achieved.

The observation of particularly small deflection amplitudes is much simplified by short-circuiting the voltage being measured, either by a key T (Fig. 8-4*a*) or periodically by mechanical means or by an electronic switch. This results in the appearance on the screen of both the null position 0 and the pattern deflected by the out-of-balance voltage.

This idea can be taken further; one can, for example, switch over to two pairs of resistances corresponding to certain plus and minus tolerances. Four pictures then appear on the screen:

1. null trace,
2. trace with minus tolerance,
3. trace with plus tolerance,
4. moving trace corresponding to the amplitude of the voltage being measured.

Since the indication of the bridge ratios is displayed by a rotating trace on the screen, it is of interest to consider the dependence of the angle of rotation φ upon the out-of-balance voltage. The horizontal deflection of the spot by the bridge voltage is constant. (In Fig. 8-5 it equals $2 \times B$.) The deflection of the extremity of the trace is determined by the (amplified) out-of-balance voltage. Here the ratio A/B is the tangent of the angle φ. The angle of rotation is therefore found from the equation

$$\varphi = \text{arc tan} \frac{A}{B} . \tag{8.2}$$

By plotting the dependence of the angle of rotation upon the relation $\frac{A}{B}$, a curve is obtained of the form shown in Fig. 8-6*a*. It corresponds to the arc tan curve for the range $\pm\ 90°$. As larger angles than $\pm\ 90°$ are impossible in this arrangement, the other branches of the curve of more than $\pm\ \pi/2$ of the arc tan function need not be considered.

From the curve in Fig. 8-6*a* emerges the important and valuable fact that the

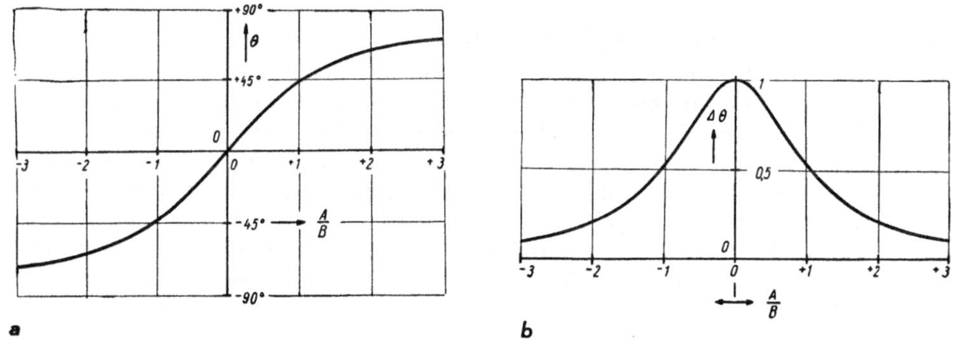

Fig. 8-6 Curves showing indication sensitivity of oscilloscope with rotating trace
a) Angle of rotation in dependence on A/B (see oscillograms in Fig. 8-5)
b) Change of angle in dependence on A/B (= tan of angle of rotation φ).

change of angle is greatest for a certain small change of the out-of-balance voltage near the point of bridge balance since in this region the curve is at its steepest part. (It is determined by the deflection sensitivity of the tube, the amplification and the value of the bridge supply voltage.)

This picture is still clearer if the change of angle itself is plotted against the ratio A/B, as shown in Fig. 8-6*b* [66]).

One of the essential advantages of this method of measurement is that the acute sensitivity of the out-of-balance indication declines rapidly with bridge unbalance. This, however, represents a insensitivity to overload scarcely equalled by any other null-indicating device.

8. 6 Direct measurement of bridge unbalance

An additional advantage of this procedure is that the rotating indicator trace can also be used for direct measurement of bridge unbalance. The straight lines joining the extremities of the traces G_1 and G_2 in Figs. 8-5 and 8-7, for instance, can be fitted with a scale which may be calculated from the mathematical conditions already considered, but in practice it is simpler to produce certain known values of unbalance by means of test resistors and then calibrate the scale accordingly, as has been done in Fig. 8-7 in which are shown eight different positions of the rotating trace. It is now no longer necessary to balance the bridge; the unbalance can be read direct and thus the impedance Z_x determined according to Eq. (8.1). This procedure is particularly useful when, for instance, it is required to sort a number of components within certain manufacturing tolerances. As the out-of-balance voltage on the straight part of the curve in Fig. 8-6*a* is lenearily proportional to the unbalance, the divisions of such a percentage scale are linear, as can be seen in Fig. 8-7.

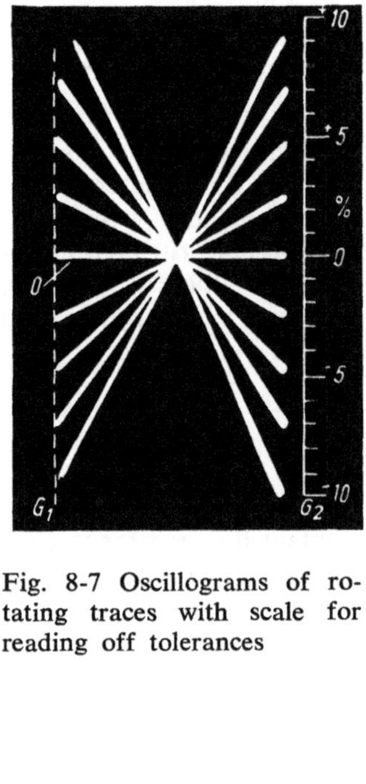

Fig. 8-7 Oscillograms of rotating traces with scale for reading off tolerances

8. 7 The measurement of complex impedances

A further advantage of the rotating trace method as compared with other methods of null-indication in AC bridges appears in the measurement of complex impedances. In such applications the out-of-balance voltage is also complex, that is to say, it consists of two voltage vectors corresponding to the different types of impedance.

If, as in Fig. 8-8, for example, an electrolytic capacitor is to be tested, the pattern on the screen when the bridge is off balance will not be a simple trace but an ellipse, as is seen in Figs. 8-9 and 8-10. If bridge potentiometer P is operated the bridge balance can be restored. The ellipse then rotates to the position marked O_{Ph}, thus balancing the two voltage vectors corresponding to the capacitance ratio $C_x : C_n$, but

---

66) Mathematically this curve represents the differential quotient of the arc tan curve. The equation $\dfrac{d\,(\text{arc tan}\, A/B)}{d\, A/B} = \dfrac{1}{1 + (A/B)^2}$

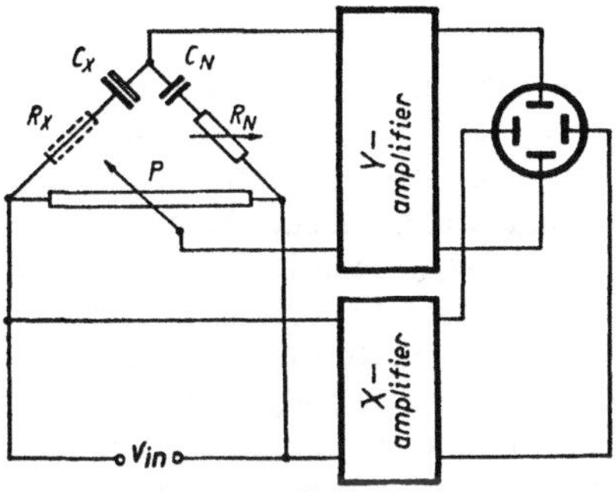

Fig. 8-8 Circuit for measuring complex impedances

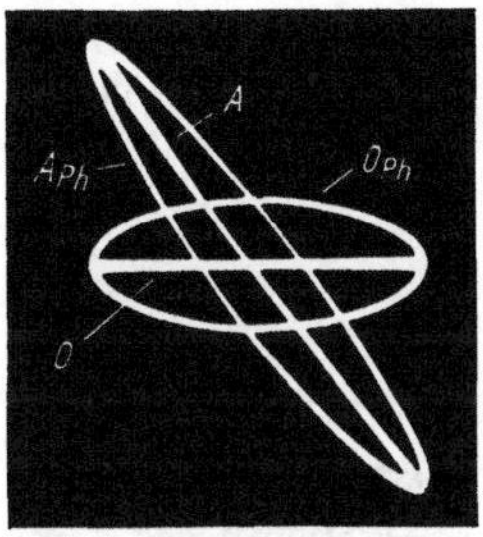

Fig. 8-9 Oscillograms obtained in measuring complex impedances

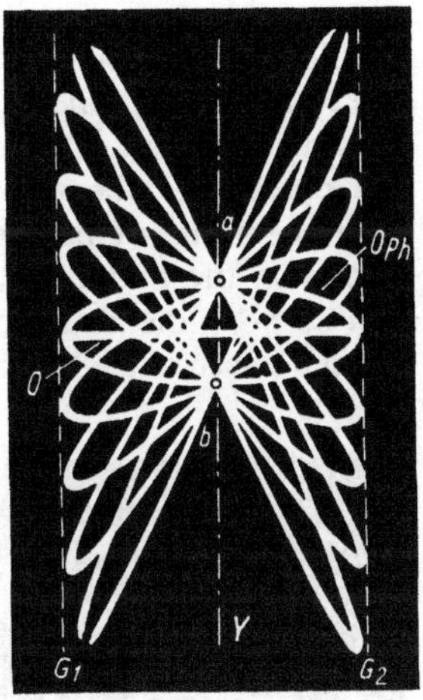

Fig. 8-10 Family of oscillograms based on Figs. 8-8 and 8-9, with points of intersection *a* and *b*

because of the ohmic "series resistance" there will still be a residual voltage to deal with. As the vector of this voltage is at right angles to the capacitive voltages of C_n and C_x, the ellipse will remain open even when the bridge is balanced. Only when a variable resistor R_n in series with C_n is adjusted to the corresponding value will the ellipse collapse to the null line *O*.

The value of the "leak resistance" R_x can be found from the following equation:

$$R_x = \frac{1}{\alpha} \cdot R_n \quad , \tag{8.3}$$

in which α is again the bridge ratio of the potentiometer *P*.

It is interesting to note that "bridge balance" and "phase balance" are now completely independent of each other, as in this procedure the influence of each balance can be separately seen. Thus, in Fig. 8-9, when the bridge is unbalanced — A_{Ph} — the phase balance can first be adjusted to obtain the indicator *A*, and then the bridge can be balanced to obtain the indicator *O*, and vice versa.

This is a unique advantage compared with all other methods used for measurements of this kind, which always require alternate adjustment of "bridge balance" and "phase balance" until the minimum is reached.

With the method described here, however, both adjustments can be rapidly and accurately made in complete independence of one another.

Fig. 8-10 shows null trace *O*, the null ellipse O_{Ph} and eight different oscillograms of ellipses with an unbalanced bridge to illustrate the practical working of this procedure.

8. 8 Direct reading of the loss angle without balance

In Fig. 8-10 one is struck by the fact that all figures without phase balance have the intersection points a and b in common with respect to the Y-axis. At a certain bridge sensitivity, the distance between these points is a measure of the ohmic component R_x of C_x. The larger this component, the wider will be the ellipses and the greater the distance between a and b. In Fig. 8-11 an attempt has been made to illustrate these relationships by means of recordings of two measurements with a different ratio of the resistive component. It is evident that phase balance is unnecessary for measuring the loss angle of a capacitance. It is sufficient to balance the bridge and read off the opening of the out-of-balance ellipses at a_1 and b_1 and a_2 and b_2 respectively.

If the value of the unknown quantity is not important and it is just a matter of obtaining the loss angle, the bridge need only be roughly balanced. As can be seen in Fig. 8-11, the points of intersection on the ordinate always make it possible to read off the phase angle or one of its functions, even for the deflections A_1 and A_2 [1] [2].

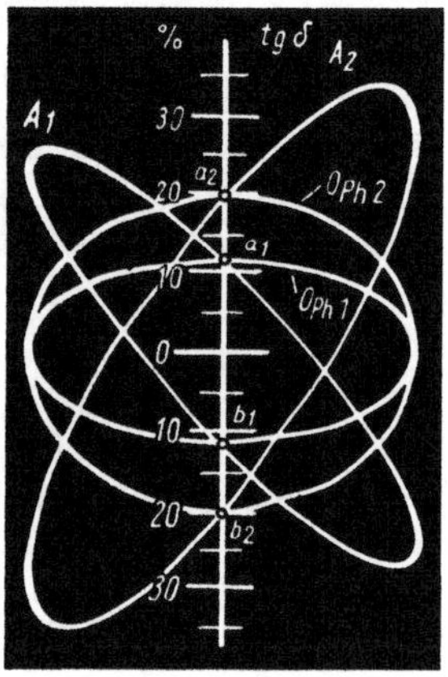

Fig. 8-11 Oscillograms for reading the loss angle of a capacitor

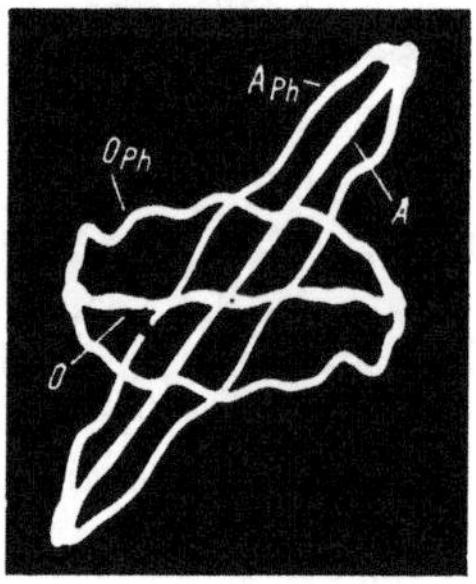

Fig. 8-12 Oscillograms obtained during measurement of complex impedances with voltage containing harmonics

So far, only sinusoidal voltage measurements have been considered, those being the most suitable for this type of measurement. But it occasionally happens that measurements have to be made with voltages containing harmonics. Fig. 8-12 therefore shows a number of screen patterns resembling those in Fig. 8-9 but using a test voltage taken direct from the mains without filtering. Measurements are still clearly possible here, whereas with other methods of null indication no satisfactory results could be produced. Moreover, in distortion measurements using bridge circuits the harmonics can thus be separated from the fundamental.

8. 9 Impedance measurements by voltage comparison

The cathode ray oscilloscope can often be used to advantage in other methods of measurement based on voltage comparison, apart from actual bridge circuits. Fig. 8-13 shows a circuit with which R, C and L impedances can readily be compared. The unknown Z_x and the known Z_n impedance are connected in series across the voltage source v_{in}. Each of the alternating voltages across these components is fed to one or other pair of deflection plates. As the current through both impedances is equal, the voltage drops are proportional to the impedances. To obtain satisfactory deflections V_m should be approximately 100 V. This means that unless an amplifier is used this procedure is only suitable for comparing high impedances. If suitable amplifiers are available for both directions of deflection, then small signal voltages can also be used and correspondingly small impedances compared. The internal resistance of the measuring voltage source has been assumed to be zero. If impedances whose series connections are not very large in comparison to those of the internal resistance are to be compared in this way, then the input voltage of the circuit must be kept constant during this measurement.

If the higher deflection sensitivity of the Y-plates is compensated by a voltage divider P, the pattern on the screen when both impedances are equal will be a trace at an angle of 45°. If $Z_x = \infty$, the full voltage will appear on the X-plates and the trace will be horizontal. When $Z_x = 0$, the full voltage will appear across Z_n and thus on the Y-plates, giving a vertical trace. The trace on the screen will thus turn according to the ratio $Z_n : Z_x$ and, when $Z_n = Z_x$, will be exactly 45°. Patterns like this are shown in the oscillogram in Fig. 8-14. The extremities of these pointers lie on a straight line joining the end points of the X- and Y-traces.

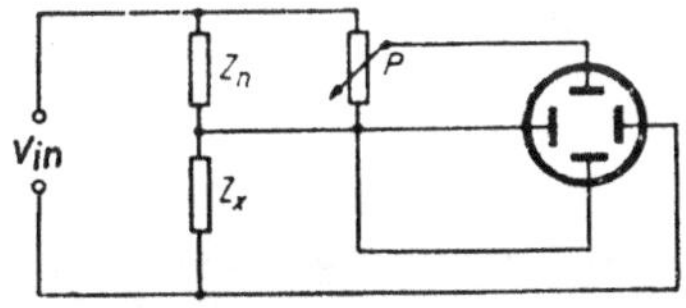

Fig. 8-13 Circuit for comparing impedances

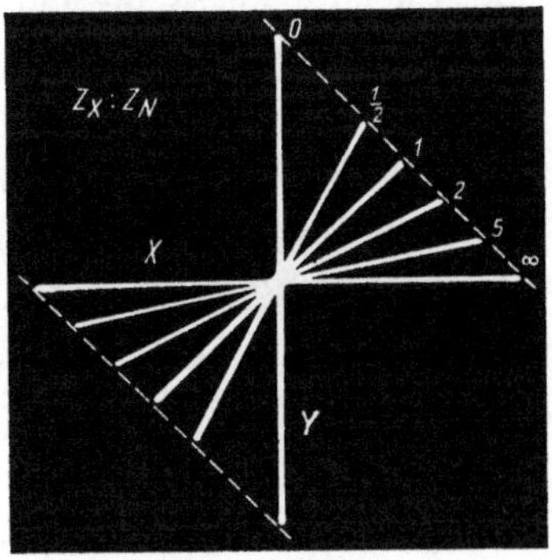

Fig. 8-14 Oscillograms based on circuit in Fig. 8-13

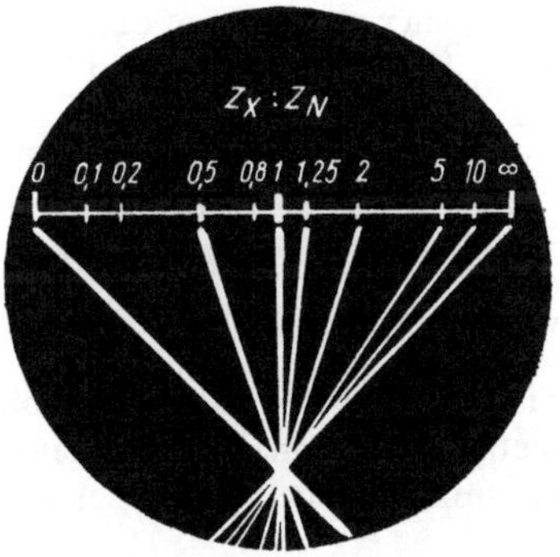

Fig. 8-15 Family of oscillograms for impedance measurement using circuit of Fig. 8-13; pivotal point shifted to edge of screen and tube turned 45°

In contrast with the bridge method previously described, it is not possible with this circuit to increase the accuracy of indication by overloading the amplifiers. The available screen surface can, however, be better exploited and maximum reading accuracy achieved if the pivot of these traces is displaced to the edge of the screen (or even beyond) by appropriate direct bias voltages, as is always possible in good oscilloscopes. This process can be taken a step further by turning the cathode ray tube round in such a way that the pivot lies at the bottom centre of the screen as in Fig. 8-15. For a 1 : 1 ratio the indicator will be the central and vertical, and the extremities will move over a straight horizontal line which can be fitted with a scale.

This arrangement, too, has the advantage that sensitivity is greatest in the centre and decreases with increasing values of $Z_x : Z_n$ or of $Z_n : Z_x$. This arrangement is very suitable for rough sorting and, as already mentioned, for comparing high resistances and inductances as well as small capacitances where $\frac{1}{\omega \cdot C_x}$ is high at a given measuring frequency.

Where there is a phase angle between the two impedances, an ellipse appears on the screen in place of a linear trace. Thus, the power factor of a capacitor can be accurately determined by comparison with a loss-free capacitor as previously described for the bridge circuit. This will be considered in greater detail when dealing with phase measurements (Ch. 11).

It should be pointed out here that this procedure can also be used to determine the resonance resistance of an LC-circuit. For this measurement the frequency of the measuring voltage source is adjusted to the natural frequency of the circuit. Resonance can then be recognized not only by the amplitude maximum, but much more clearly by making the phase angle zero [3].

The uses here described of the oscilloscope in bridge circuits have only been able to touch upon the basic possibilities of application. It is quite clear, however, that there are many other special applications in which it can be similarly employed with advantage [4] [5] [6].

8. 10 Bridge circuit for sorting core plates

An interesting application is a bridge circuit for sorting core plates. The appropriate circuit is shown in Fig. 8-16, but for the sake of simplicity the amplifiers have been omitted.

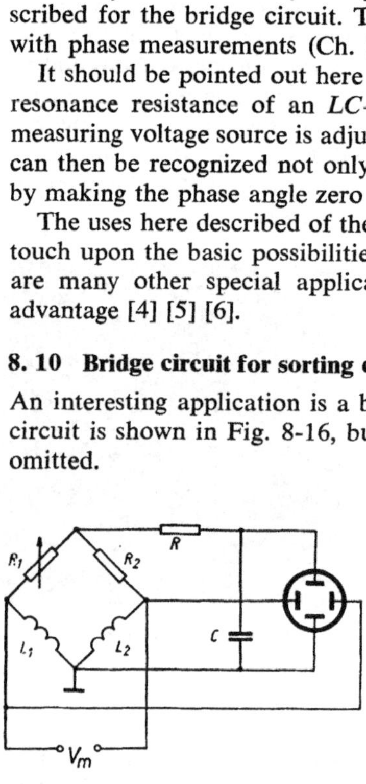

Fig. 8-16 Bridge circuit for magnetic sorting

Coils L_1 and L_2 are included in two arms of the bridge and specimen core plates of equal size are placed in them. The coils consist of 15,000 turns of 0.1 mm copper wire. The other two arms of the bridge consist of resistors R_1 and R_2, R_1 being used to balance the bridge. Due to resistor R and capacitor C, integration of the out-of-balance voltage takes place ($\frac{1}{\omega \cdot C} \ll R$), so that the spot is deflected vertically by a voltage which is directly proportional to the integral of the magnetic fluxes (not the flux change $\frac{d\Phi}{dt}$) [67])

In order to obtain a balance, plates of equal dimensions, made of the same material with known characteristics, are placed in both coils and the bridge balanced by adjusting R_1. If one of the plates is then replaced by an unknown core plate of equal dimensions, the pattern on the screen will be seen to change. With some experience this simple method can be used to obtain rapid information on the properties of the unknown material [7] [8].

---

[67]) For further details see Part III, Ch. 16 (also footnote 78 on page 330).

CHAPTER 9

THE ELECTRONIC SWITCH

9. 1 Method of operation

In some investigations it is required to indicate not merely the dependence of one magnitude on time (or some other magnitude) but also to observe the trend of two or even more processes simultaneously. If these are cyclic processes, a multi-ray oscilloscope is not essential; a single-ray oscilloscope can be employed. If, for example, photographic records are desired, it is merely necessary to switch in the various processes in succession and to take a photograph of each. It is then an essential condition that, when the dependence on time of the processes is being shown, the time deflection in all photographs is always rigidly locked in phase with some reference value. Fig. 9-1, which will be discussed in more detail later, shows this circuit. To facilitate observation it is essential to switch from one signal to the other so rapidly that discontinuities of the patterns are not perceptible or, at least, not obtrusive. The impression of continuity can be much improved by using oscilloscope tubes of very long persistence. Tubes with *N* (*P*2) and *P* (*P*7) screens such as DN 10-78, DP 10-78 or DN 13-79 are particularly suitable. As the stimulation light of the *P*-tubes is blue-white, but the afterglow is yellow, the light fluctuations at low switching frequencies can be reduced by means of an orange-yellow filter. Thus it is possible to display two processes simultaneously by means of a simple switching relay. Rotary switches have also been used, particularly if the waveforms of several magnitudes are to be displayed at the same time. An important example is the display of several characteristic curves intended to show the dependence of one magnitude on another and also the influence of a parameter [1].

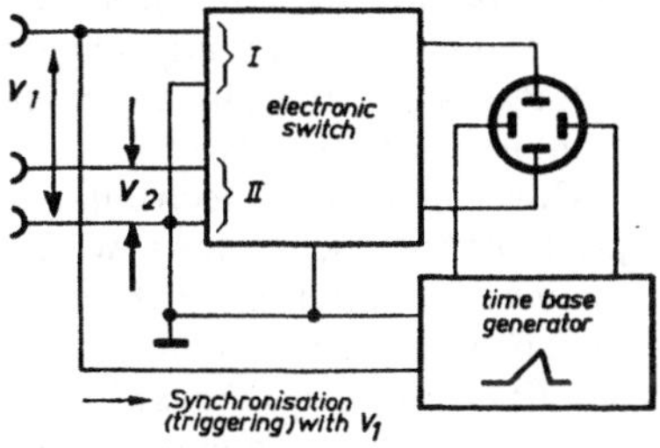

Fig. 9-1 Circuit of an electronic switch and the time base generator to obtain correct synchronization

A particularly elegant solution is electronic switching by means of diodes or amplifier tubes, when it is a simple matter to build up devices with which a considerable number of voltages can be observed simultaneously. Generally, however, this device is limited to the switching of two processes. It is thus possible to switch alternately signals of frequencies from zero to about 100 Mc/s using comparatively inexpensive apparatus. If amplifiers are used in such switches, the devices are known as "electronic switches". [68])

[68]) The author suggests the use of the term "amplifier switch" in place of electronic switch, as the former expresses the fact that the stream of electrons not only changes path but is also amplified. The later term, he says, should be used only when the simple switching of a stream of electrons is referred to [2].

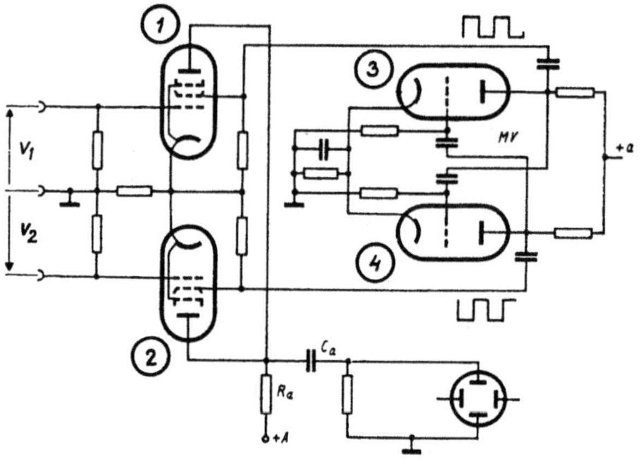

Fig. 9-2 Fundamental circuit diagram of an electronic switch

The principles of an electronic switch can be followed by reference to Fig. 9-2; the two valves 1 and 2 represent two normal (e.g. wide-band) amplifier stages with a common anode load R_a. The two input voltages V_1 and V_2 are applied to the control grids of these valves. The screen grids are not supplied with a constant DC voltage but with an alternating square voltage generated by multivibrator circuit *MV*. The screen grids of the amplifier valves therefore go positive alternately so that one or the other valve is conducting in turn. The amplified voltages therefore appear across the anode resistor in turn. These voltages are applied to the deflection plate of the oscilloscope tube via coupling capacitor C_a. It is, of course, possible to produce these voltages symmetrically as explained in Part I, Ch. 5, and DC coupling could be employed if necessary.

One such type of electronic switch is available for direct connection to the deflection plates of the oscilloscope, while another type is intended to be connected in front of the vertical amplifier, it thus operating as a "two-channel pre-amplifier". In the units intended for direct connection to the deflection plates the characteristics of the amplifier tubes in the switching device alone determine the frequency range and gain. In the other type identical gains are obtained in both channels provided that the subsequent oscilloscope amplifier will transmit the two signals satisfactorily. In some models with direct connection to the deflection plates it is sometimes possible to connect the oscilloscope amplifier to one channel only, so that it serves as a pre-amplifier. In this arrangement, therefore, one channel has a high gain and in the other the gain is limited by the characteristics of the amplifier stage in the switch.

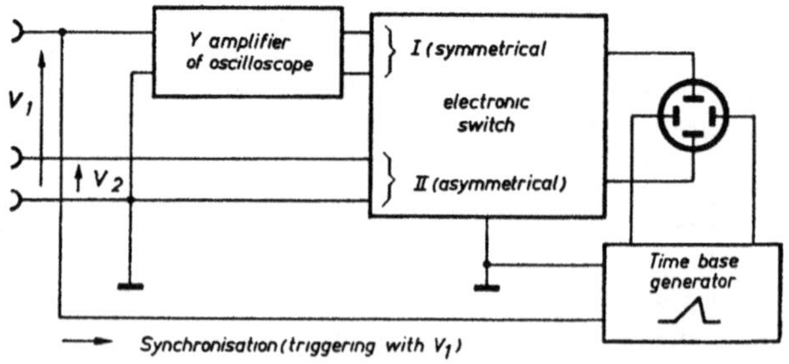

Fig. 9-3 Series connection of oscilloscope amplifier to one input of electronic switch

In this way, for instance, the DC amplifier of the Philips "GM 5666" DC oscilloscope can be connected in series to the "GM 4580" electronic switch and can, in addition, still employ DC coupling. A channel with a gain of about 8000 is thus obtained with a frequency range from 0 . . . 250 kc/s. Fig. 9-3 shows this type of circuit. The input of the electronic switch to which the output of the oscilloscope amplifier is connected must then be switched to symmetrical input. The gain of this channel in the actual switch is unity. If the switching times of the electronic switch valves is variable, the brilliance of each screen pattern can be adjusted independently [3] [4].

The oscillograms in Fig. 9-4*a* and *b*, recorded with the "GM 5666" oscilloscope together with the electronic switch "GM 4580" will serve to illustrate the working principle of the electronic switch. In oscillogram *a*) the synchronization was "internal" and, in addition, the switching frequency was chosen purposely to be nine times that of the input frequency. It can clearly be seen that the spot described portions of the waveform curve of one voltage and changes over abruptly to show portions of the other waveform in the intervals. Whenever one curve is described there is automatically a gap of equal size in the pattern of the other. These are the patterns that are always obtained when incorrectly synchronized (or triggered) with the *Y*-voltages on the plates and the switching frequency is an exact multiple of the input frequency. In principle, therefore, when the electronic switch is used, synchronizing and triggering should always be externally applied, as shown in the circuit in Fig. 9-1. By this means, oscillograms can obtained like that in Fig. 9-4*b*, in which both curves are traced without discontinuities. It is, however, essential that the ratio of the input frequency to switch frequency should differ at least by 20 c/s. Under these conditions the switching remains invisible.

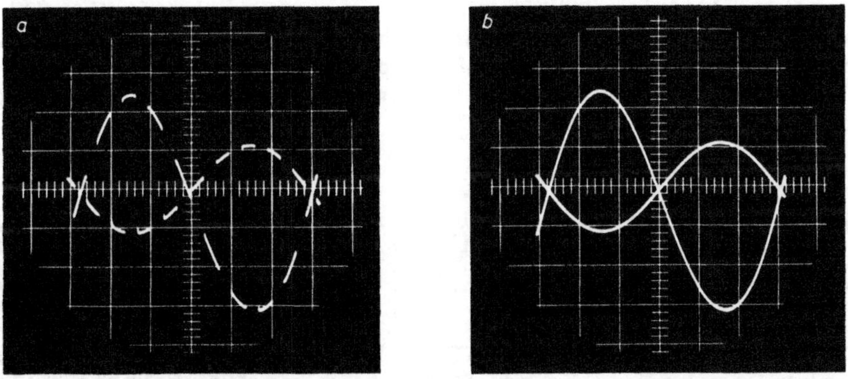

Fig. 9-4 Oscillograms showing method of working and correct synchronization
a) Switching frequency in integral ratio to signal frequency and internally synchronized
b) externally synchronized and switching frequency sufficiently differentiated from signal frequency

Electronic switches have been described in which the switching can take place as required, and as described above, either by means of self oscillation or triggered by the time deflection voltage. In that case the change from the pattern one signal to that of the other always takes place at the end of a time deflection, that is to say, during the flyback [5] [6]. Thus the switching remains invisible, as the flyback is always blanked in the oscilloscope. In such switches, self oscillation operation — with high switch frequencies up to 500 kc/s — is used for the lowest and the highest input frequencies. For signals of medium frequencies the switching is triggered. This is of advantage in view of the fact that the switching requires a certain time. At low switching frequencies this time is short compared with the duration of the cycle, so that the switching process is invisible since its deflection speed is much greater than the writing speed of the oscillogram. But when smaller time scales are used, the switching duration approaches the duration of the signal cycle it is desired to study. In such cases a troublesome hiatus would be interposed between the two patterns. By blanking the flyback in triggered working, this hiatus is avoided.

9.2 Special applications

Should it be necessary to observe three magnitudes simultaneously, a second electronic switch may be connected in front of one channel of the first electronic switch. This second unit connects the second channel of the first switch alternately to two voltages. Further details may be gathered from the sketch in Fig. 9-5. In this case too, the time base united can be triggered by one of the three voltages as required. An oscillogram obtained in this way can be seen in Fig. 9-6. The top trace is a sawtooth curve from a time base unit, the middle one shows the first derivative (differentiated voltage) of the upper curve. The lowest curve is a sinusoidal voltage of known frequency serving as a time calibration. When using the two electronic switches connected one behind the other like this, it should be ensured not only that the switching frequencies stand in no rational ratio to the input frequency, but that they should differ from one another so as to obtain clear pictures without surges or other irregularities. If an electronic switch receives an input voltage at only one input, then a horizontal line is obtained via the other channel and this may be used as a reference line ("null" line). It has already been said in Chapter 7 "Amplitude measurements" that if the shift voltages are calibrated, this reference line can be used to measure the other oscillogram in the amplitude direction (Fig. 7-21).

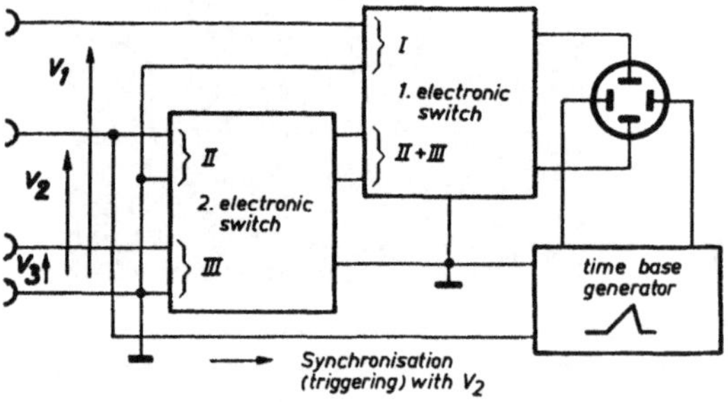

Fig. 9-5 Connecting two electronic switches to an oscilloscope to display three voltages

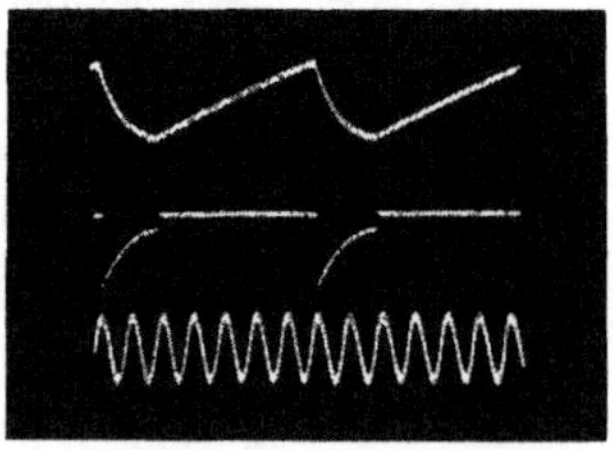

Fig. 9-6 Triple oscillograms obtained with the circuit of Fig. 9-5

If an electronic switch is applied to each pair of deflection plates, a number of different combinations can be obtained thus showing their mutual dependence (characteristic) [7], the curve co-ordinates being obtained at one and the same time (Part III, Ch. 16).

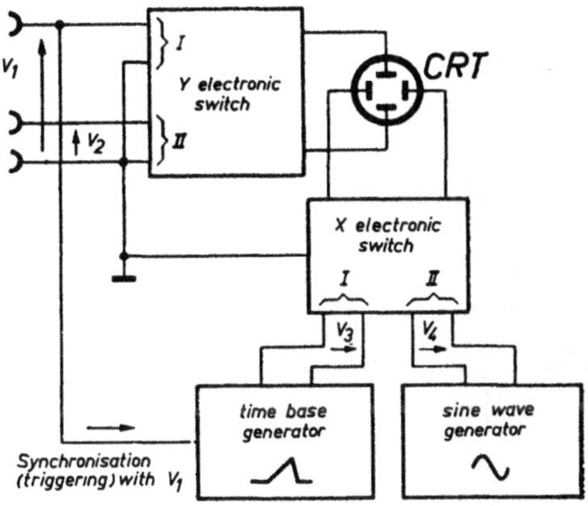

Fig. 9-7 One electronic switch per direction of deflection to display four oscillograms

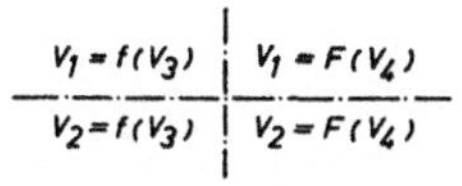

Fig. 9-8 Representation of the dependencies in a screen pattern obtained with the circuit in Fig. 9-7

This circuit makes it possible, for example, to display four different oscillograms by means of a single-beam oscilloscope. Fig. 9-7 shows the principles underlying this circuit. An electronic switch is used for each direction of deflection i.e. vertical and horizontal. While two input voltages are applied to the switch for vertical deflection, both inputs for horizontal deflection are each fed with two voltages with respect to which the dependence of the input voltages is to be demonstrated. As an example, a sawtooth voltage was applied to the left-hand horizontal input to show the dependence of both input voltages on time; a beat frequency oscillator with a sinusoidal alternating voltage was connected to the right-hand horizontal input. In this way four different pictures are obtained on the screen, which can be conveniently shifted into four quadrants by the positioning control. As shown in Fig. 9-8, the following traces are obtained. In the top-left corner the dependence on time of the voltage V_1; bottom-left the dependence on time of the input voltage V_2 (with the same time scale as above), top-right the Lissajous figure of the trend of the input voltage V_1 as a function of the sine voltage V_4 (at frequency ω_4) and bottom-right the dependence of the waveform of the input voltage V_2 on the sine voltage V_4. An oscillogram obtained in this way is shown in Fig. 9-9. The output voltage of the Philips "PR 9260" pick up for mechanical vibration, mounted on top of a running motor, was used for the input voltages V_1 and V_2. The voltage V_1 was taken directly from the pick up. It is obtained by moving a coil in a magnetic field. The output voltage is therefore proportional to the speed of movement ($\frac{\mathrm{d}x}{\mathrm{d}t}$). The voltage V_2 is obtained from V_1 by integration in an *RC*-network in the so-called "PR 9250" calibrating unit; it corresponds to the vibration magnitude (x), that is, to the oscillation amplitude. The left-hand pictures show the time dependence of the two voltages. The right-hand oscillograms can be used for the determination of frequency, it being known that the frequency of V_4 was $7^1/_2$ c/s. This shows that the number of revolutions giving rise to the oscillations was 450 rpm. This also determines the duration of the cycle of the bottom-left curve as being 0.13 s. From this, for instance, the frequency of the oscillation speed (top-left) is found to be 60 c/s. In place of the sinusoidal voltage V_4, the output voltage of an additional vibration pick up might be connected here, and from the right-hand pictures conclusions drawn as to the phase relationship of the oscillating parts. This example is

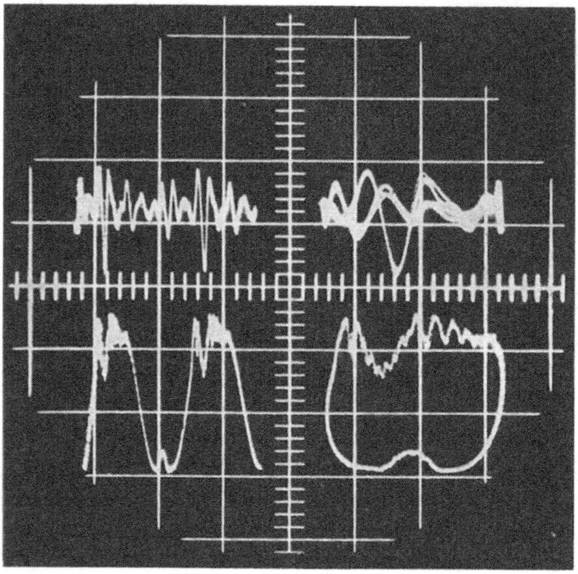

Fig. 9-9 Fourfold oscillograms as in Fig. 9-9 but right-hand oscillograms in different phase position

intended to show that by the appropriate use of potentialities to the single-beam oscilloscope, a great variety of results can be obtained. The principle of the electronic switch, apart from this application, is also similarly used in general measuring technology for many special tasks [8] [9] [10]. Special electronic switches have been described for a multiplicity of measuring points (up to 12) in seismic investigations [11] or in the study of servo-mechanisms [12]. This circuit principle is also used for measuring processes other than oscilloscopes [13]. A publication by DORSMANS and DE BRUIN can be consulted for studying the history of the development of this device [14].

It is also possible to show the dependence on time of several variables of state by means of the intensity modulation of a grid traced on oscilloscope tubes (similar to a television picture) [15] [16].

9. 3 Practical form of electronic switch

Electronic switches are produced both as independent units and as multi-channel pre-amplifiers (usually two-channel plug-ins, Fig. 1-5). Switching to the two channels generally takes place in one of the input stages. It is now only seldom that the electronic switch itself supplies the large deflection voltages required for the direct connection to th edeflection plates of the oscilloscope tube.

Fig. 9-10 shows the Philips PP1071 electronic switch as an example of an independent unit. Two wide-band DC amplifiers with a frequency range of 0—15 Mc/s are switched in this unit. A choice is offered between self-oscillating or triggered operation. This unit is intended for use with oscilloscopes with deflection coefficients > 50 mV/cm. The total gain from input to output is about 2. There are potentiometers to adjust the gain so that, in conjunction with the oscilloscope amplifier, the same deflection coefficients are obtained as in single-channel operation without an electronic switch.

Fig. 9-10 Philips electronic switch PP 1071
Two-channel display 0....... 15 Mc/s

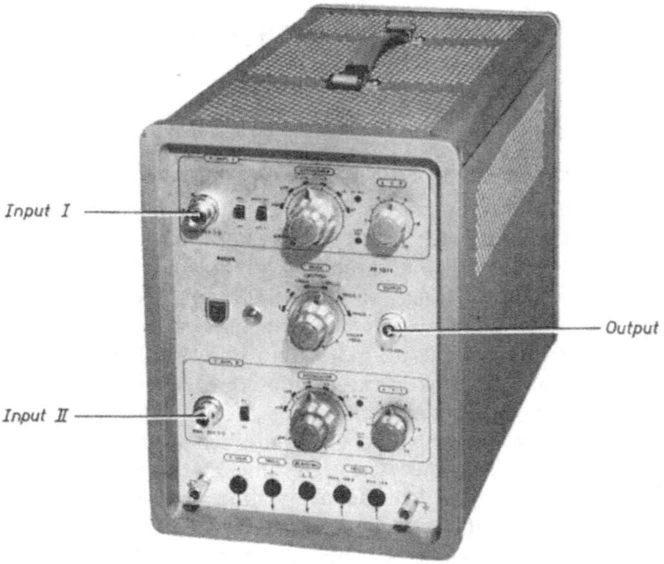

The firms of DuMont, EMI, Ribet-Desjardins, Hewlett-Packard and Tektronix supply electronic switches as plug-in units. Hewlett-Packard supplies not only oscilloscopes with built-in switch for displaying double images (type 122A/AR) but also units with two-channel plug-in units (types 150/AR, 160 A and 187A/187B, see Fig. 4-97). Tektronix makes a number of different two-channel plug-in units and has recently brought out the *M*-rack unit for 4-channel display in the frequency range of 0—15 Mc/s at deflection coefficients of 20 mV/cm upwards.

The provision of an electronic switch in the form of an independent unit has the advantage that it can be used in conjunction with any of the various oscilloscopes to be found in a laboratory. It is, of course, more expensive than multi-channel pre-amplifier plug-in units, as the cost entailed in the supply portion, which must be built up separately, is considerable. The plug-in units can be supplied from the power pack. Their disadvantage is that their use is limited to large oscilloscopes built for the plug-in unit system.

CHAPTER 10

USES OF INTENSITY MODULATION

10. 1 Rating the circuit components; time marking

In most measurements with the oscilloscope, the electron beam, and thus the spot, is deflected by both pairs of deflection plates in co-ordinates at right angles to each other -the Y- and the X-axes. It is, however, sometimes necessary to make a third quantity, namely time, visible on the fluorescent screen.

As the result of experience with moving coil oscillographs, where there is usually no difficulty in making more than one simultaneous measurements, it is often considered that in cathode ray oscilloscopes a second trace should also be provided e.g. by using a dual-beam oscilloscope or electronic switch. So long as it is not a matter of displaying the course of a second phenomenon, but only of time marking, this extra trace is usually unnecessary. A time marking voltage can be used to modulate the intensity of the electron beam by superimposing this voltage on the negative bias of grid g (Wehnelt electrode) of the C.R.T. This voltage (usually alternating) can be applied, via capacitor C_g, between the grid and chassis as shown in Fig. 10-1. The current from this voltage source reaches at the cathode via capacitor C_2 and the leak resistor R_g via C_1. The alternating voltage across R_g then controls the grid voltage. It should be noted that the grid g is connected via resistors R_g, R_1 and potentiometer P to the side of the filter capacitor C_2 which is at a high potential with respect to the chassis. Coupling capacitor C_g must therefore be suitably insulated.

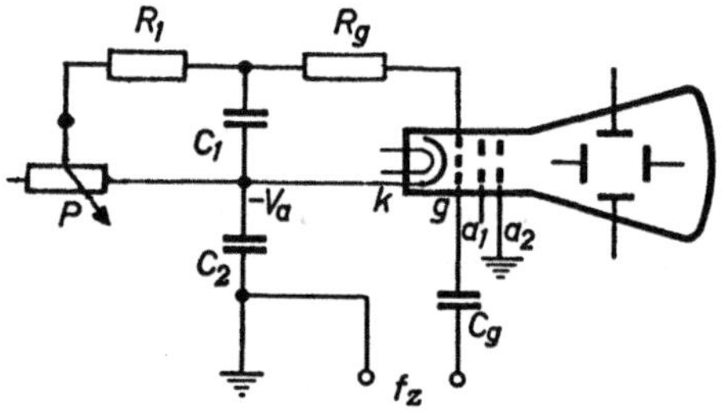

Fig. 10-1 Circuit for modulating the intensity of the beam

In this arrangement, the spot on the screen will be made brigther or fainter during each half cycle of the control voltage, i.e. brightness modulation occurs. This effect can be seen in the oscillogram of Fig. 10-2. In parts *b*), *c*), *d*) and *e*) the alternating voltage *a* was raised in steps and the grid bias increased in such a way that the brilliance peak remained roughly the same.

Initially the variations in brilliance are slight but become increasingly pronounced later. This third method of influencing the spot is often called the Z- axis, although other measures are required for the actual representation of three-dimensional oscillograms, and intensity modulation here serves only to produce an impression of space.

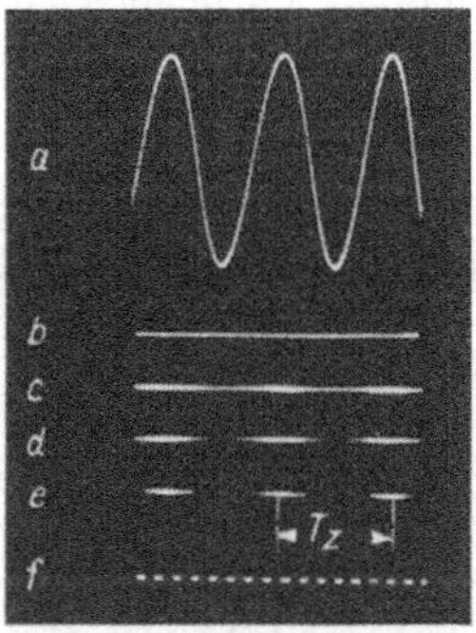

Fig. 10-2 Sinusoidal brightness modulation with traces of different intensity

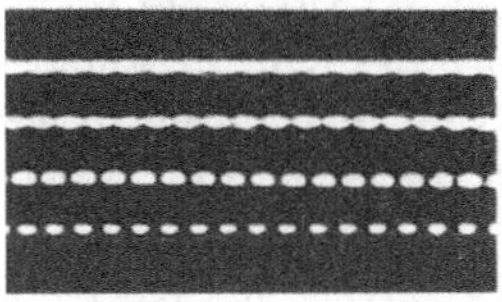

Fig. 10-3 Brightness modulation with the markings mere close together as in Fig. 10-2*b* to *e*

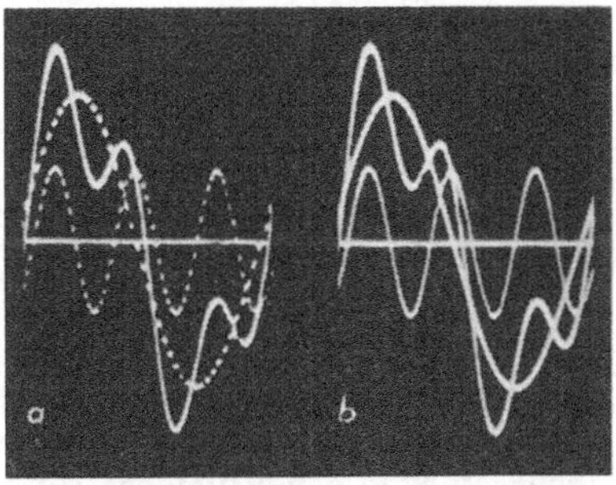

Fig. 10-4 Brightness modulation to clarify the presentation of multiple oscillograms

Although there may be no objection to the relatively large intervals between the brilliance markings in these oscillograms, since it is still possible to determine the centres of the markings or of the darkened trace, such interruptions are often undesirable, as they could mean the loss of details in the oscillogram. The markings are therefore kept as close together as possible (Fig. 10-2 f). Fig. 10-3 shows four such traces, in which, as a result of varying modulation voltage and correspondingly higher grid bias, only slight variations in brilliance and good intensity modulation were obtained. The spacing of the bright points (or of the centres of the darkened trace) corresponds to a time difference T_Z which is the reciprocal of frequency f_Z, thus:

$T_Z = \frac{1}{f_Z}$, so that a frequency of 1000 c/s the internal between points $\frac{1}{1000}$ s = 1 ms.

For the sake of clarity, the traces shown in Fig. 10-3 have been somewhat enlarged as compared with those in Fig. 10-2.

The time marking thus achieved is satisfactory for many purposes.

Oscillograms of several different quantities, which can easily be recorded one after another on one picture, can also be clearly distinguished by varying the degrees of intensity modulation. The time base generator must then be synchronized in fixed phase with the reference voltage. An example is given in Fig. 10-4*a* showing the fundamental frequency of an alternating voltage strongly punctuated and the third harmonic (shifted in phase) less strongly punctuated, together with the unmodulated curve of the sum of both voltages. For comparison, the same curves are shown without intensity modulation in Fig. 10-4*b*. An intensity-modulation frequency of 2200 c/s was used to modulate the fundamental (50 c/s); for the smaller harmonic (150 c/s) the modulation frequency was 1250 c/s.

When intensity modulation is used for time marking, it is almost invariable bright-dark modulation (see lower trace in Fig. 10-3). Such a time mark will only be adequate if the deflecting speed of the beam is reasonably constant. Sudden fluctuations could

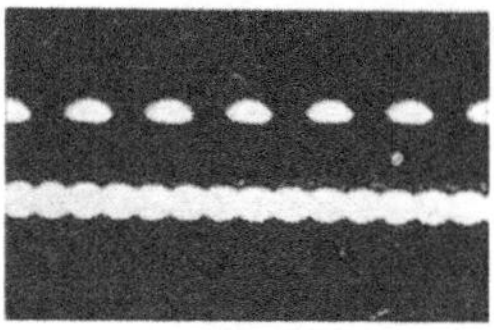

Fig. 10-5 Magnified traces with bright-dark modulation and with closely bunched modulation marks

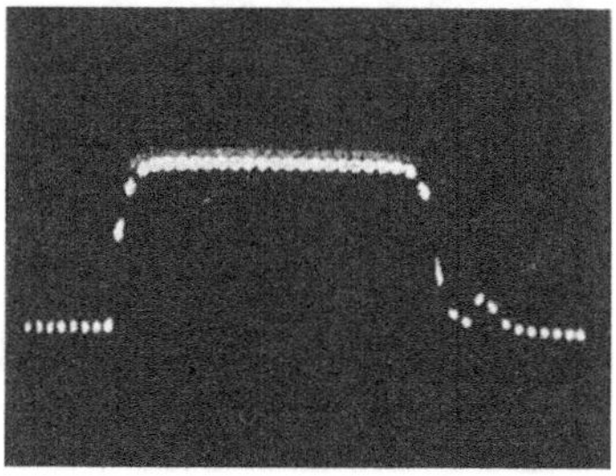

Fig. 10-6 Opening behaviour of a between-lens shutter with time marking by brightness modulation $f_z = 1000$ c/s

cause gaps. It is clear, however, from the enlarged traces in Fig. 10-5 (as seen under a magnifying glass) that it is still possible to count the time markings with reasonable accuracy even when they are closely crowded together. The section reproduced shows that the bright-dark trace contains six time-marking spaces, whereas the closely modulated trace contains fourteen.

An example of the practical application of this procedure is given in Fig. 10-6. Even in the steep portions of the oscillogram the time markings can still be counted. This example represents the opening behaviour of a between-lens camera shutter. The opening and shutting time amounted to 3 ms (time marking 1000 c/s). The shutter was open for 20 ms ($\frac{1}{50}$ s; 20 time markings). It is of interest to note that, perhaps due to rebound, the shutter opened again for about 3 ms roughly 1 ms after closing [69]). [1]

10. 2 Synchronous intensity modulation

As long as differences, however small, exist between the frequency of the intensity modulating voltage and corresponding multiples of the frequency under observation, the modulation markings will drift along the pattern on the screen. The interpreting and photographic recording of such images can be very difficult. If such tasks are frequent, it is advisable not to take the modulating voltage from a self running source but to trigger it by coupling to the time deflection (Figs. 4-100 and 4-104).

This is achieved as follows: a tuned, lightly damped oscillator circuit is released by the flyback of the time deflection voltage. During the forward trace of the spot this circuit decays at its characteristic frequency. This voltage can therefore be used (if required, after rectification) as a time marking for bright (or blank) modulation of the pattern on the screen. As the moment of commencement of these oscillations is determined by the flyback of the time deflection, these time markings are also stationary in the oscillogram irrespective of whether there is a whole number frequency ratio between time base frequency and time marking frequency or not. If the frequency ratio varies, then the distribution of the time markings along the oscillogram merely varies (see Part I, Ch. 4.35 "Calibrating the time scale").

---

[69]) We are dealing in this case with the shutter of an older type camera. Other similar oscillograms made with more modern shutters can be found in Part III, Ch. 28.

10. 3 Short brilliance markings without gaps or short blank-markings

As can be seen in the oscillograms of Fig. 10-2*c* to *f*, in sinusoidal intensity modulation voltage the ratio between the brilliance markings and their intervals is always constant. It may, however, be necessary to have brilliance markings at greater intervals perhaps, but sharply defined, without blanking pauses. Short, sharp voltage pulses are required for this.

There are a great number of circuits for generating voltages of this kind. They are used, for example, in radar, electronic computers and television engineering. These voltages are usually generated by the differentiation of a square wave. For the following oscillograms a type of electronic switch was used as the voltage source, which not only supplies rectangular voltages over a wide range of repetition frequencies but also permits them to be adjusted symmetrically or asymmetrically, as desired.

Differentiation can be effected by a pulse transformer or simply by a *CR*-network, as shown in Fig. 10-7. Whereas $\frac{1}{\omega \cdot C} \leqq \frac{1}{100} \cdot R$ must apply for passing a square wave properly, good differentiation can be achieved when $\frac{1}{\omega \cdot C} \gg R$. Good practical results were obtained when $\frac{1}{\omega \cdot C} = (3 \ldots 10) \cdot R$. The cut-off frequency is that at which $\frac{1}{\omega C} = R$, [2] [3] [4].

This process is illustrated in the oscillograms shown in Fig. 10-8. Fig. 10-8*a* shows three cycles of a rectangular voltage and *b*) the waveform after application to a *CR*-network of the type shown in Fig. 10-7, in which $\frac{1}{\omega \cdot C} = 3 \cdot R$. During the rise a positive peak occurs; as the wave decays a negative peak is formed. If the screen trace is modulated by this voltage, the resulting pattern is as shown in Fig. 10-8*c*, brightness modulation alternating with blanking. In order to obtain brightness modulation only, full-wave rectification of the modulating voltage is necessary. Two time markings then occur in each cycle of the rectangular voltage. Fig 10-9 shows a circuit of this type for operation at a frequency

Fig. 10-7 *CR*-network for differentiating alternating voltages

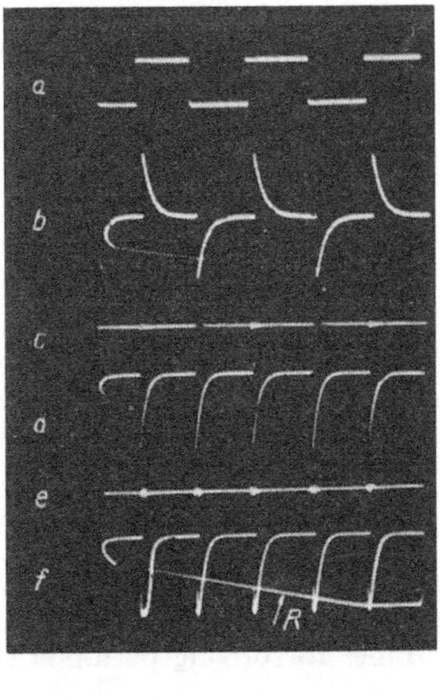

Fig. 10-8 Waveform of voltages in a circuit according to Fig. 10-9 with corresponding traces

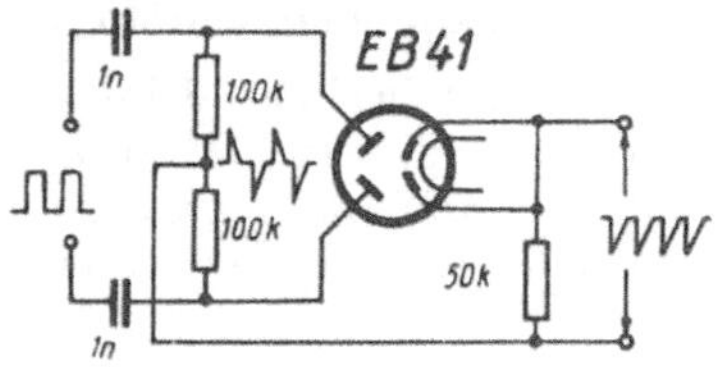

Fig. 10-9 Circuit for full-wave rectification of differentiated voltages ($f = 1000$ c/s)

of 1000 c/s. Oscillogram 10-8*d* shows the curve of the output voltage. The brigtness-modulated trace of such a voltage is to be seen in 10-8*e*. One time mark per cycle can be obtained by connecting a single diode (crystal diode) with series resistors, between the two poles of the modulation voltage, instead of the full-wave rectifier.

With suitable modulating voltage, semi-punctuated and other types of oscillograms can be obtained. Fig. 10-10*a* shows an example of a damped oscillation intensity-modulated by the time-marking trace in 10-10*b*. The differences in spacing give a clear impression of the decrease in speed of the spot. By reversing the polarity of the modulating voltage short blank markings can be obtained, as shown in Figs. 10-11 *a* and *b*.

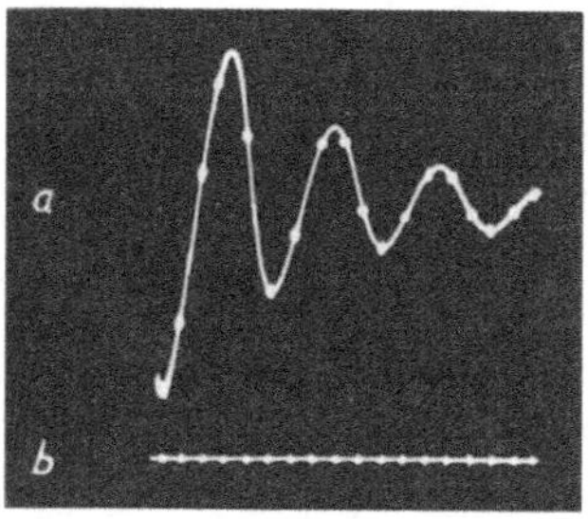

Fig. 10-10 Time-marking by trace brightening

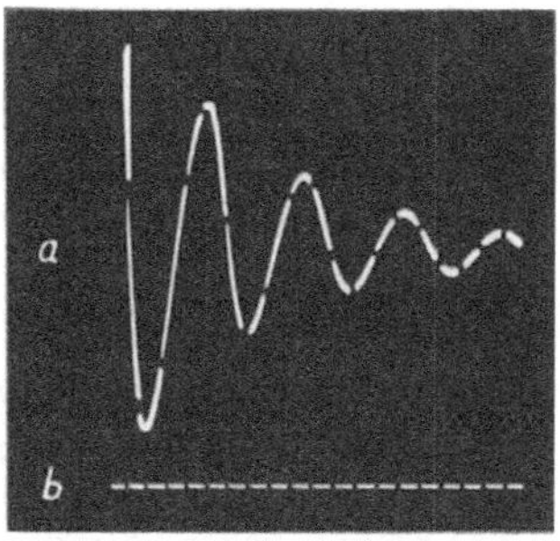

Fig. 10-11 Time-marking by trace blanking

10. 4 Intensity modulation proportional to deflection speed (Automatic brilliancy control)

Oscillograms in which the deflection speed is constant are seldom met with in practice and a trace of varying intensity has to be accepted. But, even when displaying a simple sine wave, it may be undesirable in the vicinity of the zero axis, where the speed of the spot is greatest, for the brilliance to be considerably less than at the crests of the curves. In this example, the speed trend is displayed by 90° according to the cosine function.

The aim must therefore be to modulate the intensity so as to obtain a uniform trace. A circuit designed for the purpose can be seen in Fig. 10-12. The signal voltage V_Y is fed via an *RC*-network to the vertical amplifier, and the output applied to the *Y*-plates. At the same time V_Y is fed via a *CR*-network to a normal AF transformer for driving push-pull stages, the secondary of which is centre-tapped. After

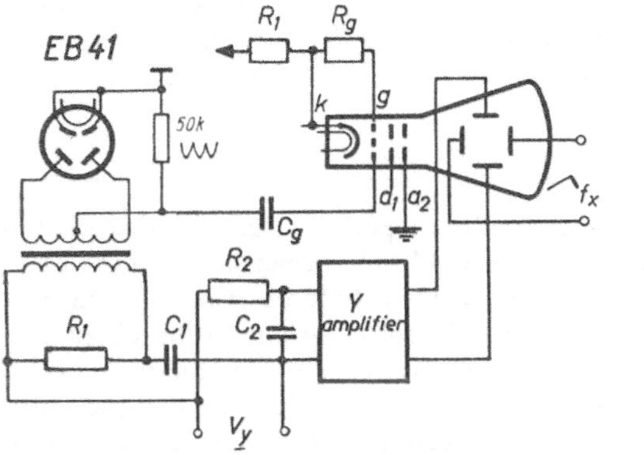

Fig. 10-12 Circuit for speed-proportional brightening of sinusoidal voltages

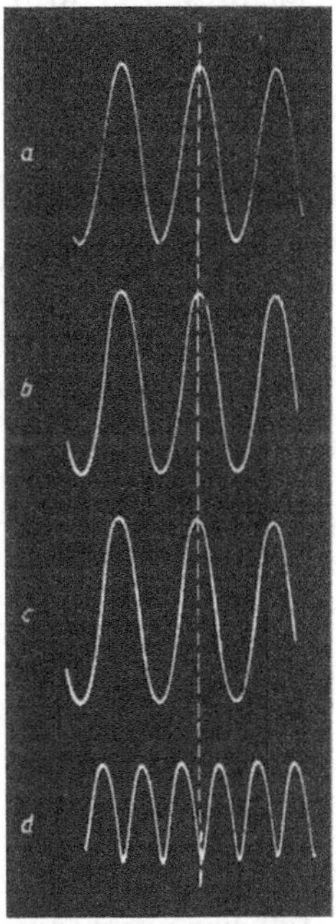

Fig. 10-13 Patterns showing the effect of speed-proportional brightening with circuit of Fig. 10-12

full-wave rectification the output obtained is as shown in the oscillogram in Fig. 10-13*d*. The required phase difference can be obtained by C_1, R_1 (voltage leading). To avoid too great a voltage drop during this process, the voltage on the Y-plates can at the same time be counter-displaced, so that the required phase difference of 90° is achieved in each case with a voltage loss of only 30%.

Figs. 10-13*a*, *b* and *c* show the effect of this circuitry. Fig. 10-13*a* shows the normal sine wave without corrective intensity modulation [70]). The differences in intensity between the peaks and the zero transition points can be recognized clearly. Intensity modulation was introduced in Fig. 10-13*b*, so that the zero transition approximately equals the peaks in brilliance. Modulation was somewhat exaggerated in Fig. 10-13*c* to emphasize this effect. All waveforms show the same brillance at the peaks, but different degrees of brilliance on the other parts of the curve. (The reference line allows of easier comparison of the time spacings.) Other waveforms can also be considerably improved by intensity modulation dependent on speed. Fig. 10-8*f* shows an example of modulation voltage as in Fig. 10-8*d*. Used at the same time for intensity with Fig. 10-8*d* it resulted in a considerable improvement in the display of the peaks. The flyback, indicated by ***R*** is now emphasized as compared with Fig. 10-8*d*. The simple circuit shown in Fig. 10-12 is only suitable for sinusoidal voltages. For more complex tasks a more elaboratic arrangement would be required [5] [6] [7].

It is, for example, desirable in radiometric techniques to be able to determine the pulse amplitude with the utmost accuracy on moving film. As, however, the pulse rise is the same at the greatest recording speed, it is difficult to record it satis-

[70]) See also Fig. 6-2*a* p. 256.

factorily. For such tasks brightening the ascending flank by the use of valve circuits is possible [8]. As the voltage rises or falls a Schmitt-trigger is activated, releasing an intensity modulation to the oscilloscope tube.

10. 5 "Switching" the brilliance

When observing and recording single transients, it is especially desirable that the time during which the spot is visible on the screen should not exceed the duration of the transient itself. If the stationary spot, of sufficient intensity to describe a bright trace, were to appear beforehand on the screen, it would certainly dazzle the observer or produce blackening of the recording material by its strong halo. It might even produce burns on the screen. These disadvantages can be avoided by blocking the grid with a suitably high negative voltage until before the recording is to be made. This can be done by including a change-over switch *S* in the lead to the arm of potentiometer *P*, as indicated in Fig. 10-14. If the switch is set in position 2, the grid receives from point *a* a biasing voltage sufficient to cut off the tube. In position 1 the actual voltage on the potentiometer arm appears on the grid. Some Philips oscilloscopes provide another means of switching. The voltage from a 45 V dry battery is applied to terminals *A* and *B* via a relay contact. (Switch *S* can, of course, be a relay change-over contact.) Since the internal resistance of the battery is always smaller than the steadying resistor of 0.5 MΩ, the battery will determine the operating point [71]). In this way, by means of further relays or valve circuits, single transients can be recorded in the following sequence:

1) the spot appears,
2) the time base deflection begins,
3) the transient is released,
4) the transient ends,
5) the time base reaches the end of the screen,
6) the spot disappears.

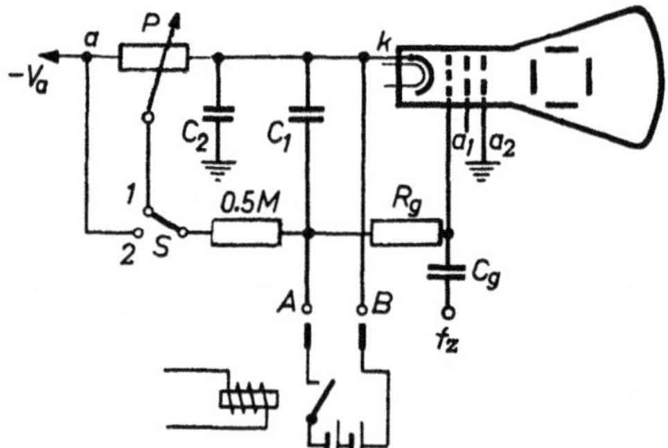

Fig. 10-14 Circuit for electrical switching of the brightness

Fig. 10-6 was recorded in this way. The spot brightness was modulated in addition by an AF voltage as time marker (see also Figs. 28-1, p. 495 and 28-7, p. 501 and Part I, Ch. 4.21 "Unblanking circuits in oscilloscopes with triggered time base units".

[71]) It should always be taken into consideration that all these circuit components are to chassis under E.H.T.

10. 6 Further applications of intensity modulation

In this chapter it has not been possible to do more than deal with the rudiments of intensity modulation. Its importance in television engineering needs no emphasizing since it comprises the picture signal itself. Blanking the flyback has already been dealt with in the section on time base generators. Intensity modulation is also used for displaying vector locus diagrams and three-dimensional oscillograms [9] [10] [11] [12] [13] [14] [15].

The "stroboscopic" oscilloscope (sampling oscilloscope) for frequencies up to 3000 Mc/s is partly based on intensity modulation as well ([16] and Chapter 4, page 154). It can, moreover, be applied for recording valve characteristics [17]. It is possible in this way to display several variables of state simultaneously [18]. It is clear from the foregoing that the appropriate use of intensity modulation — influencing the spot in the third dimension — extends the range of application and greatly widens the scope of the cathode ray oscilloscope.

CHAPTER 11

PHASE MEASUREMENTS

Fundamentally, there are two distinct types of investigation involving phase measurement:

1) the determination of the phase difference between two or more voltages or currents, and
2) investigation of the characteristics of a circuit unit comprising a fourpole network (circuit components, amplifiers, etc.) by measuring the phase shift of the output with respect to the input by means of a signal voltage of given frequency.

It is important to bear in mind that *every phase measurement amounts, in fact, to a measurement of time difference.*

Thus every measurement of phase difference permits at the same time the determination of time differences, and vice versa. The term "phase" means nothing more than the relative time difference between the phenomena under observation.

The most important of the great number of procedures for phase measurement developed up to the present time will be dealt with in this chapter.

11. 1 Phase measurement by multiple oscillograms

When the waveforms of two voltages are shown simultaneously on the screen of the oscilloscope, the mutual phase difference, or time difference between corresponding points of the voltage curves, can be read off directly. To read off the phase relations it is best to observe the points at which the voltages pass through zero; the maxima do not give so clear an indication. The spacing of the zero points of one cycle corresponds to 360°, i.e. to the duration (T) of one alternation (the time taken to complete one cycle) or 2π.

A multibeam oscilloscope is not essential for investigation of this sort. The vertical deflection plates of a single beam oscilloscope can be switched to two or more voltages

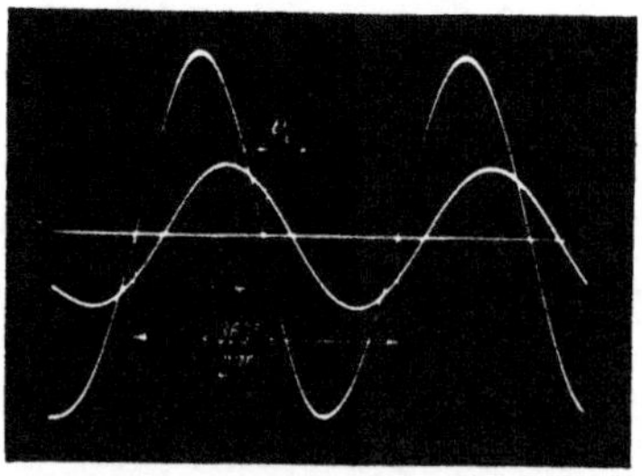

Fig. 11-1 Determining the phase difference between two voltages by a double oscillogram

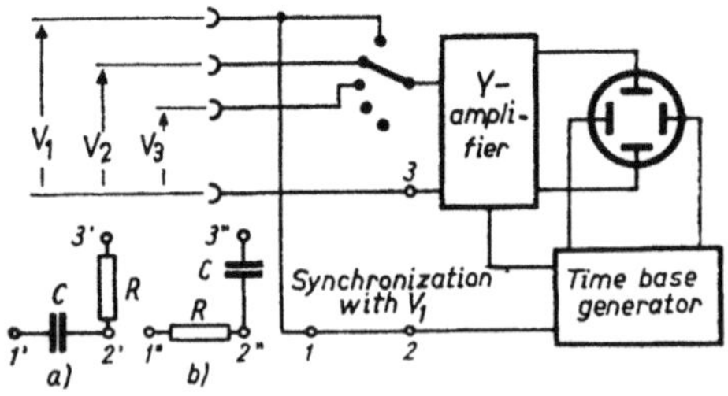

Fig. 11-2 Circuit diagram for measuring the phase difference between a number of voltages. Time base firmly synchronized with reference voltage V_1

in succession. What is essential is that during this process the time base generator is always locked in phase synchronism with the reference voltage, as shown in Fig. 11-2 (external synchronization with voltage V_1). This ensures that the waveform traces of the other voltages will faithfully represent their true phase relationship [72]).

A special switch device is not absolutely necessary for the photographic recording of multiple oscillograms. As described at the outset, the time base unit is locked in synchronism with one of the voltages to be observed and this synchronism is retained for the subsequent recordings. The individual phenomena are now adjusted on the screen at the appropriate amplitude and recorded one after the other on the same photographic material.

This was the method followed for recording the oscillograms of Figs. 5-12, 5-13, 5-15 and 14-6, 14-8, 14-9 and 14-11.

11.2 Measurement by means of a phase mark

With the method just described the actual waveform of the phenomena can be observed at the same time on the screen. There are some tasks, however, in which this is not essential and all that is required is information on the phase difference (time difference).

It is advisable for this purpose to take off phase-dependent pulses from the voltages in question. This can be done magnetically by means of "pulse transformers" or by means of valve circuits (Fig. 11-3). The input voltage to be measured is amplified by valve 1, and in valve 2 the peaks are clippel by anode and grid rectification in such a way that a rectangular voltage appears in the anode circuit. By correctly rating the following CR coupling network ($\frac{1}{\omega \cdot C} > 3R$), differentiation is obtained, resulting in the indicated voltage peaks.

As the grid of valve 3 is maintained at a high negative potential, only the positive pulses are able to cause a change in anode current (anode rectification). Thus a voltage

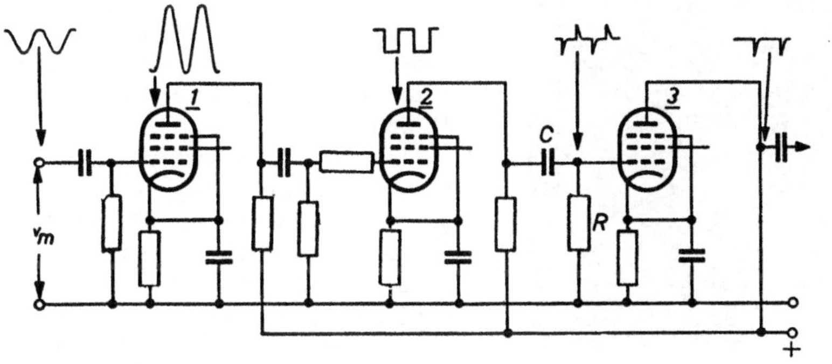

Fig. 11-3 Circuit diagram for generating pulses from a sine wave for phase measurement

[72]) To displace the whole picture sideways and adjust a certain overall phase position, a corresponding correction is possible by means of phase-shifting networks in the sync lead as shown in Fig. 11-2*a* and *b*.

pulse appears at the output during each cycle and is used for indicating the phase [73]). For phase measurement the time base generator is now adjusted to the frequency of the input signal and synchronized with the reference voltage (Fig. 11-4). The voltage pulses from V_1 and V_2 are applied simultaneously to the input of the vertical amplifier. Mutual reaction can be avoided in this type of circuit very simply by resistors R_1 and R_2. If necessary, a mixing valve can be incorporated (see Fig. 12-15).

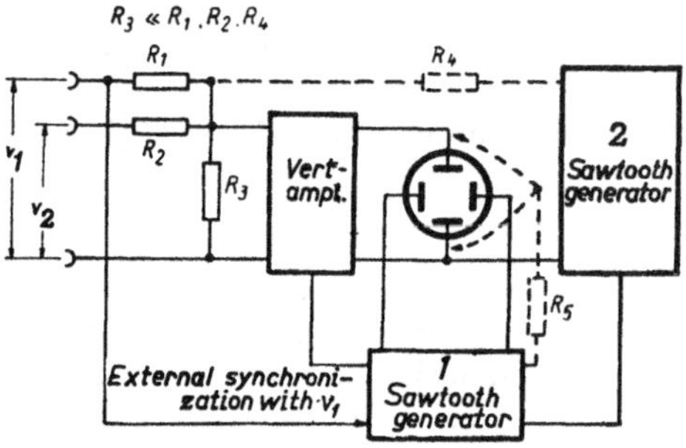

Fig. 11-4 Circuit diagram for phase comparison with pulses according to Fig. 11-3

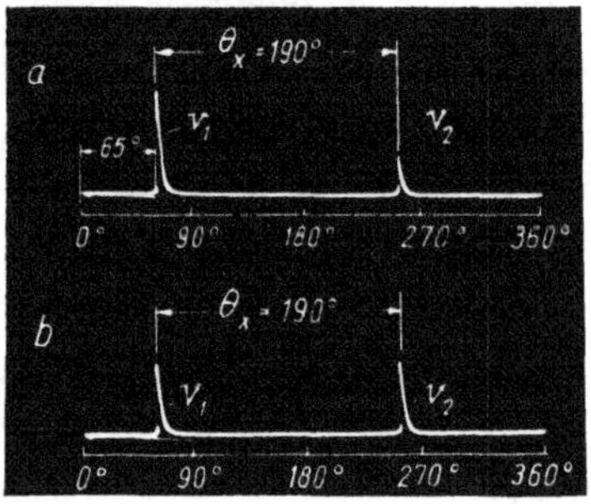

Fig. 11-5 Oscillograms of phase comparison with pulses
a) pulses of different amplitude,
b) pulses of equal amplitude

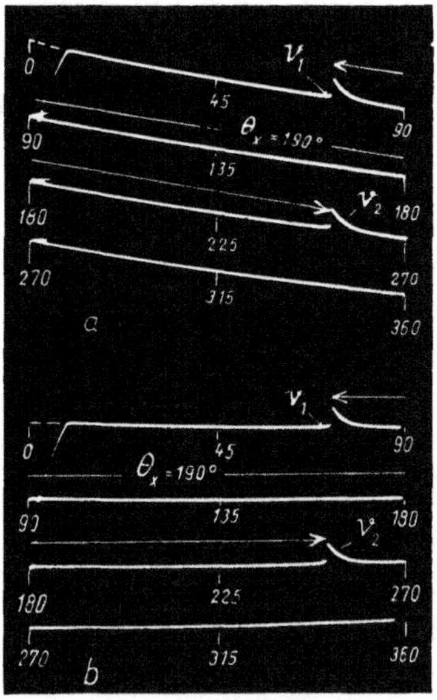

Fig. 11-6 Phase comparison with pulses extended to 4 lines
a) tilted traces without correction,
b) horizontal lines after correction

The pulse from V_1 synchronizes the time base. Since the sweep frequency is equal to the signal frequency, the total picture width, as shown in Fig. 11-5, corresponds in phase to one cycle, that is to say, 360°. The time base voltage rises linearly with time, so that the phase scale over the whole picture width is also linear and the phase difference can easily be determined by measuring the distance between the markings V_2 and V_1.

However, the accuracy with this procedure, as with that previously described, is not very great, being only between 5 and 10%. It can be improved by displaying several traces. The linear output from a second sawtooth generator, the input pulse frequency of which is synchronized, is applied to the *Y*-plates via resistor R_4 (Fig. 11-4).

If the time base frequency of the oscilloscope is now adjusted, let us say, to four times the second sawtooth frequency, four lines will

[73]) Phase correction is also possible here by connecting phase-shifting networks (*RC* or *CR*) to the input of this circuit, thus retaining a required output position of the pulse phase.

appear on the screen (Fig. 11-6). Each line corresponds to a phase range of 90°, so that the reading accuracy is now within 2°; flyback times must, however, be taken into consideration.

From Fig. 11-6*a* it can be seen that the traces are tilted. This blemish can be remedied by applying to the appropriate *Y*-plate a part of the time base voltage which must be found by trial and error. As is shown in Fig. 11-4, this is effected via resistor R_5. A display corrected in this way is to be shown in Fig. 11-6*b*.

11. 3 Phase measurement by Lissajous figures (ellipses)

Measurement of the phase difference between two sinusoidal voltages by the representation of an ellipse is very widely adopted. For such measurements each of the two voltages is simply applied to a different pair of deflection plates (Fig. 11-7*a*).

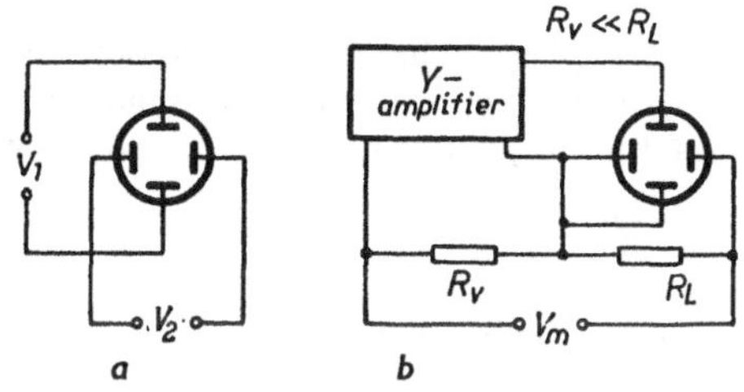

Fig. 11-7 Phase comparison by deflecting the spot in both co-ordinates to produce an ellipse (Lissajous figure);
a) basic circuit of the C.R.T.
b) layout for measuring the phase difference between current and voltage on a reactive load R_L

In order to indicate the operation of this method, a number of oscillograms are given in Fig. 11-8 to show how the spot behaves under the influence of in-phase voltages on the deflection plates. In both this figure and also in 11-9, in which there was a phase difference of 30° between the two voltages, 13 points were selected at equal distances on the time axis, and the formation of the reference image, which was

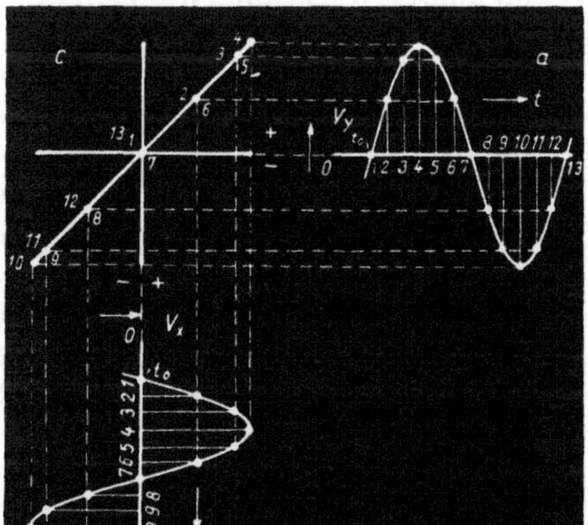

Fig. 11-8 Trace of the spot (*c*) under the influence of two sinusoidal voltages in phase (*a*) and (*b*)

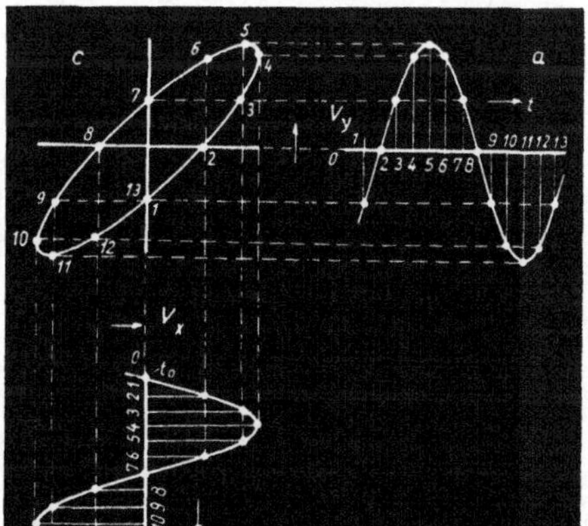

Fig. 11-9 Trace of the spot (*c*) under the influence of two sinusoidal voltages (*a*) and (*b*), with voltage (*a*) lagging

also recorded photographically, is shown by projecting these points from the corresponding points of the voltage curve.

Whereas in Fig. 11-8, the spot travels up and down along a straight diagonal in the course of one cycle, in Fig. 11-9 it describes an ellipse, the minor axis of which increases with increasing phase difference.

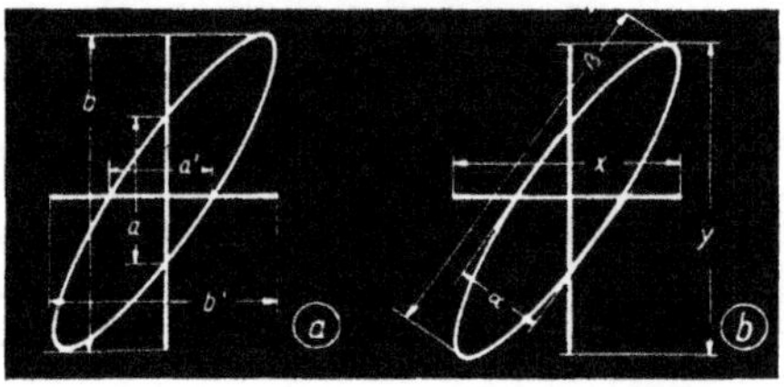

Fig. 11-10 Determining the phase angle from the ellipse

The phase angle can now be determined as shown in Fig. 11-10*a* or *b*. The ratio of the part of the axis (either ordinate or abscissa) intersected by the ellipse to the length of the projection to the corresponding axis is equal to the sine of the phase angle.

$$\sin \theta = \frac{a}{b}, \tag{11.1}$$

and the angle θ is thus:

$$\theta = \arc\sin \frac{a}{b}. \tag{11.2}$$

The lengths a and b must therefore be measured and the angles corresponding to the quotient $\frac{a}{b}$ then obtained from the trigonometrical tables.

Even this slight trouble can be avoided if a suitably calibrated scale on transparent material is fixed over the screen (Fig. 11-11). For measurement, the deflections must be large enough for the edge of the grid to be touched by the ellipse in both directions. The phase angle can now be read off directly [74]). The phase angle can also be obtained by measuring the axes a and β of the ellipse and their

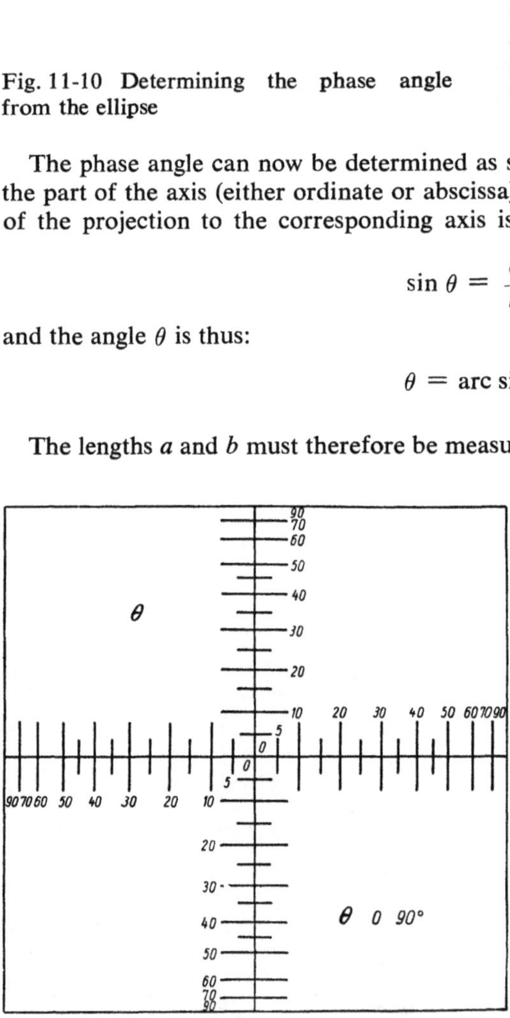

Fig. 11-11 Scale for phase measurement for $\theta < 90°$

[74]) This figure has been purposely reproduced in its present size to enable a copy of it to be made on a similar scale, if required (photostat on transparent material). If the oscilloscope is fitted with a removable floodlit scale, it is advisable to have this grid made for phase measurements.

projections X and Y, and applying the equation:

$$\theta = \arcsin \frac{a \cdot \beta}{X \cdot Y} \tag{11.3}$$

(see Fig. 11-10b).

Examples of the patterns resulting from the different phase angles is given by the oscillograms in Fig. 11-12. The opening of the ellipse can already be recognized at 1°.

In these and in the oscillograms to follow, it has been assumed that the sinusoidal voltages compared have either been applied directly to the deflection plates or via amplifiers having the same number of stages and the same phase characteristics in each case. If one of the amplifiers has a stage more than the other, however, an extra phase difference of 180° is obtained. If the phase trend of the amplifiers is not the same in each case, it can be adjusted before the entry to the "better" amplifier by the choice of an appropriate *RC*-network (Fig. 11-2*b*). In the Philips "GM 5666" oscilloscope this is achieved, for instance (before the X-input), with $R = 11\ \text{k}\,\Omega$ and $C = 100\ \text{pF}$.

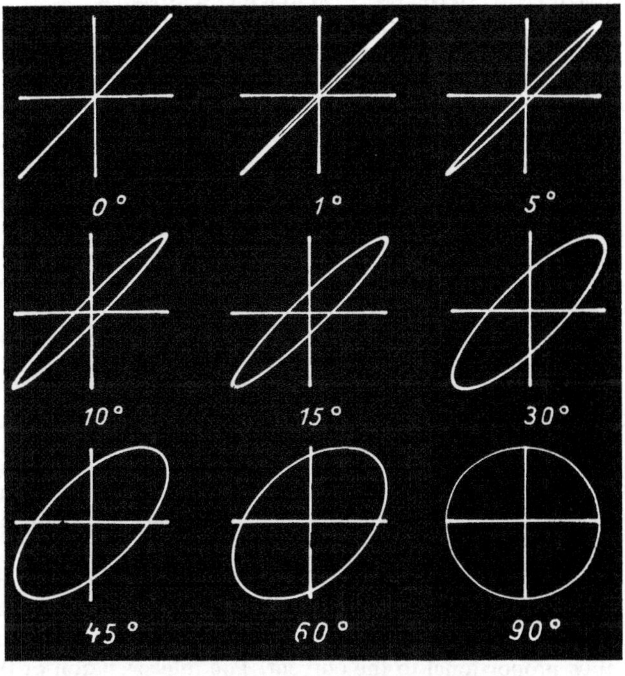

Fig. 11-12 Nine characteristic oscillograms showing various phase differences

In the oscillograms shown so far the deflections were equal in both directions so that the major axis of the ellipse always showed a tilt of 45°. This need not be the case, however. In Figs. 11-13 *a*, *b* and *c* are three oscillograms with a phase angle of 30° in which the amount of deflection in the two directions differed. It is the ratio between the lengths *a* and *b* that indicates the phase relationship.

The phase angle may not always be of direct interest, however. Very often a function of the phase angle — $\tan\theta$ or $\cos\theta$ — is required. This can also be read off

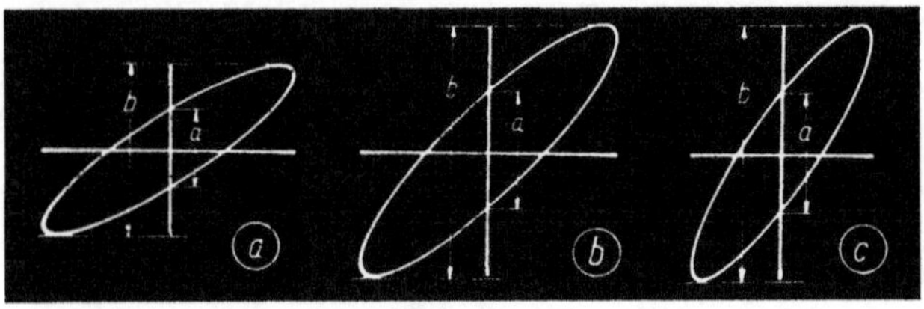

Fig. 11-13 Examples of the independence of phase measurement on deflection amplitude
a) *Y*-amplitude smaller than *X*-amplitude.
b) *Y*-amplitude equal to *X*-amplitude.
c) *Y*-amplitude greater than *X*-amplitude.

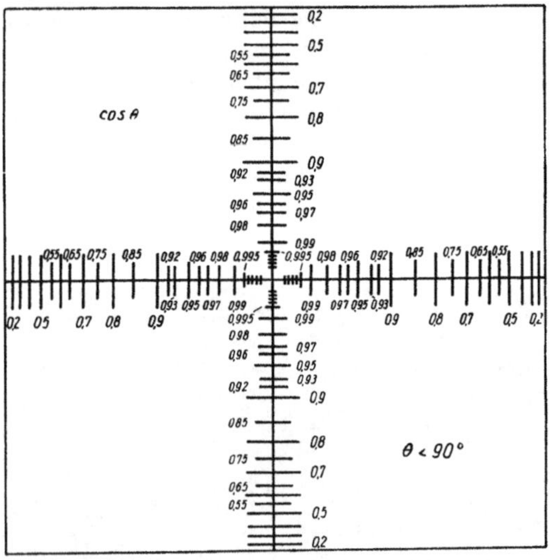

Fig. 11-14 Scale for direct reading of cos θ

immediately. A scale calibrated in terms of $\cos\theta = 0 \ldots 90°$ is given in Fig. 11-14, similar to that in Fig. 11-11, but $\cos\varphi$ can be calculated from Fig. 11-10*a* according to equation

$$\cos\theta = \sqrt{1 - \left(\frac{a}{b}\right)^2}. \tag{11.4}$$

It is worth noting in this connection that $\cos\theta$ values above 0.5, which are of main interest in practice, become increasingly easy to read as they approach unity.

The phase difference between current and voltage in a reactive load R_L can be measured as shown in Fig. 11-7*b*. The voltage across R_L is used for horizontal deflection. The total current flows through a resistor R_v which is small compared with R_L, and the resultant voltage drop is used, after amplification, for the vertical deflection, which is thus proportional to the current. The interpretation of the ellipse on the screen makes it possible, in the way described, to determine current, voltage, $\cos\theta$ and thus the power on the circuit at the same time.

11. 4 Determining the sign of the phase angle

The phase difference of two voltages can be ascertained with considerable accuracy from the ellipse formation. These oscillograms, however, do not immediately show whether the voltage under observation leads or lags behind the reference voltage.

One way of determining the sign of the phase angle is to include in the connection of the voltage of unknown phase value a phase shifting network of one of the types shown in Figs. 11-2*a* and *b*, which will then introduce a phase change of known direction. In Fig. 11-2*a* the output voltage leads, and in 11-2*b* it lags.

If, as a result of this network, the ellipse is widened, then the phase shift of the unknown voltage will have the same direction, but if it is narrowed, it will have the reverse direction. It is best to use a *CR*-network for this purpose (Fig. 11-2*a* - output voltage leading). The resistance *R* can be formed, in the AF band at least, by the input impedance of the oscilloscope, so that in practice it is only necessary to connect a suitably rated capacitor in series with the voltage and test by short-circuiting it.

If the phase difference is greater than 90°, the ellipse inclines towards the left of the screen, as in the oscillograms in Fig. 11-15. At 180°, a straight diagonal forms as at 0°, but inclined towards the left i.c. is a mirror reflection of the trace for 0°. If the phase difference is greater than 180° the ellipse opens still more. At 270° the patterns are the same as between 90° and 180° unless special measures are taken; similarly, the patterns between 270° and 360° would be the same as those between 0 and 90°.

Various measures have been suggested to avoid this ambiguity. That put forward in the circuit in Fig. 11-16 is particularly simple. A portion of the reference voltage (not of the voltage whose phase changes) is fed to the grid of the cathode ray tube, via shifting networks, in such a way that with a phase difference between 0 and 180° the right side of the pattern on the screen is brightened and the

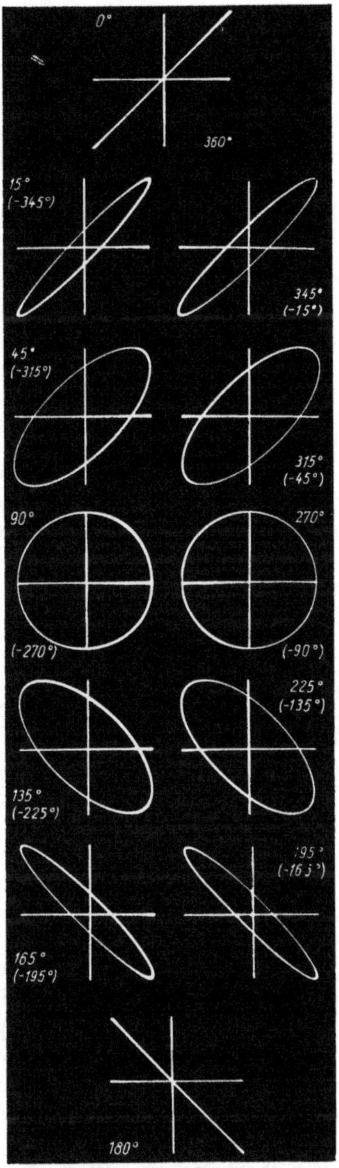

Fig. 11-15 Oscillograms for phase measurement with brilliancy modulation for interpreting clearly all patterns between 0° and 360°

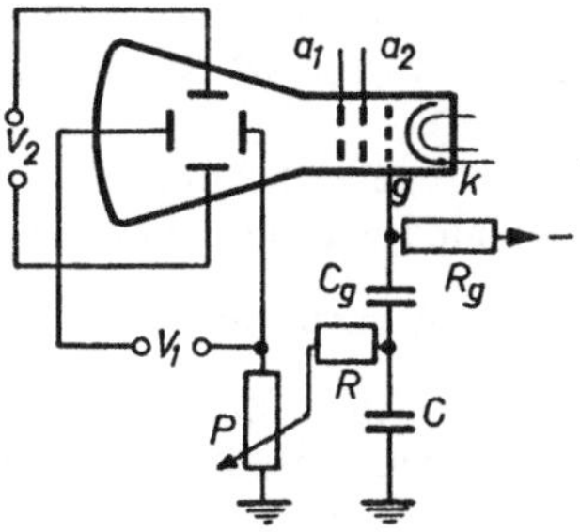

Fig. 11-16 Circuit for unequivocal interpreting ellipses in phase measurement from 0° to 360° by brilliancy modulation with the reference voltage

left is dimmed. This was the method used for the oscillograms in Fig. 11-15. If the phase exceeds 180° the situation is reversed and the left side is brightened. Thus, in the example shown in Fig. 11-15 it is easy to distinguish between the patterns representing phase differences between 15° and 345°, 45° and 315°, 90° and 270°, 135° and 225° and 165° and 195°.

The most favourable value of the phase shifting components R and C depends on the values of the grid network R_g, C_g, and must be found by experiment for a given signal frequency. The intensity modulation can be varied by potentiometer P if required. In this way it is possible to see from the oscillogram whether the phase relationship is leading or lagging, i.e. whether θ is less than or greater than 0° or 360°. The reading accuracy for phase measurements depends, of course, also on the sharpness of the spot. A phase angle of 1° is still clearly readable. If voltage curves are distorted, measurement will become less accurate [1].
Another interesting way of showing without ambiguity the phase angles obtained with sinusoidal voltages is shown in the circuit in Fig. 11-17.

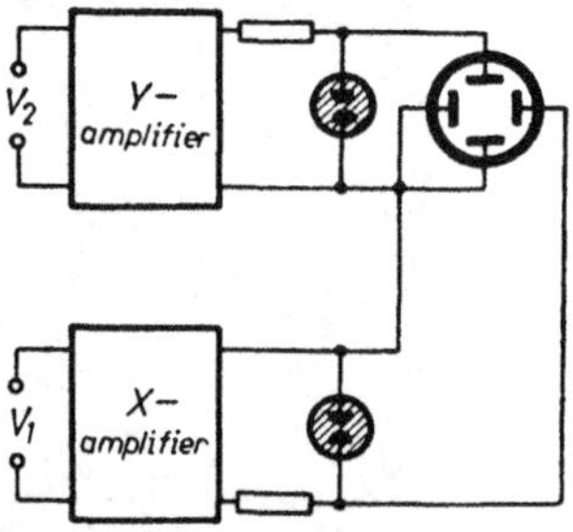

Fig. 11-17 Phase measurement with voltage curves flattened by gas discharge tubes

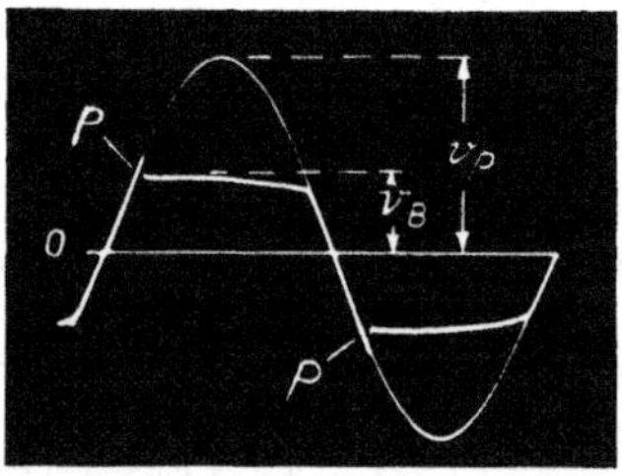

Fig. 11-18 Operating voltage waveform of a gas discharge tube circuit as in Fig. 11-17

Both voltages, the reference voltage and the input voltage, are raised to an amplitude higher than the ignition potential of a gas discharge tube by amplification. The voltage on both glow-tubes is switched alternately to both deflection plates. When the voltage on the glow-tube circuit during one half cycle reaches ignition potential, the voltage, after ignition, drops to the "burning" value (v_B) as shown in the oscillogram of the time trend of this voltage in Fig. 11-18, so that the peak P is produced each time. If the value of the peak voltage V_p is at least $2 \times v_B$, the remaining portion of the sine wave will be almost level. As a result of such voltages on the deflection plates, angular patterns with peaks appear on the screen. The length of their sides is a measure of their phase difference. At the same time the positions of the peaks at the end of the figure indicate the direction of the phase difference. Fig. 11-19 shows the two characteristic patterns for $\theta = 90°$ and $\theta = 270°$ (or—90°).

Brightness modulation is another way of obtaining an accurate indication of the phase angle. If, as in Fig. 11-20, the ellipse is intensity-modulated with an alternating voltage whose frequency is $\frac{360}{5} \cdot f_x$, a brightness marking will occur at every 5°, making 72 points in all. To determine the phase angle if it is less than 90°, it is only necessary to count the points occurring in the second or fourth quadrants. The number of points × 5° gives the phase angle. In this case it is $4^1/_3 \cdot 5 = 21^2/_3°$.

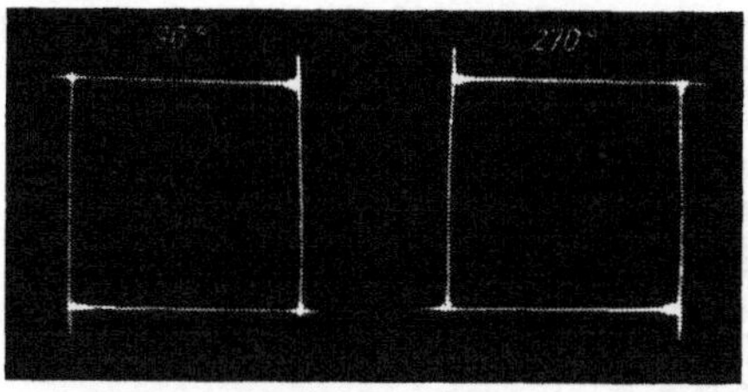

Fig. 11-19 Patterns on the screen for $\theta = 90°$ and $\theta = 270°$ in measurements with a circuit as in Fig. 11-18

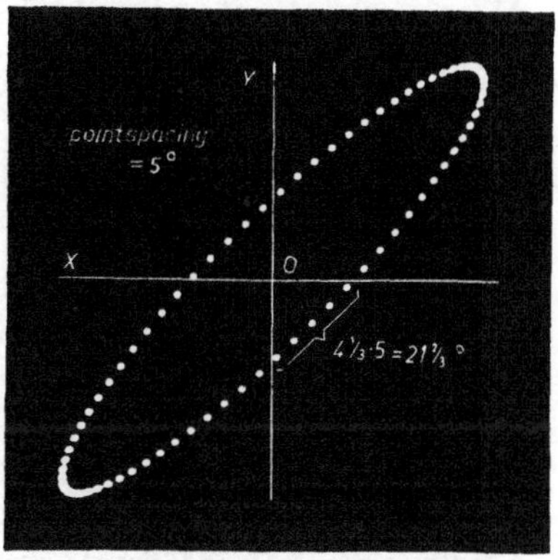

Fig. 11-20 Phase measurement with brilliance marking

These modulation markings are best produced by multiples of the input frequency. In this way the markings are stationary and the pattern can be easily interpreted.

11. 5 Measuring the phase with a bent sine wave

This interesting method, particularly suitable for measuring small phase angles, has been described several times [2]. The input signal is applied across two cross-connected diodes (valves or crystal diodes) which receive a small bias from battery *B* (Fig. 11-21). Without bias, both half-cycles of the sine wave would be passed through unattenuated. The bias prevents current from flowing for a short period in the vicinity of the zero transition points. The waveform of the voltage across the output resistor for $\theta = 0°$ will be as shown at the the top-left corner of Fig. 11-22.

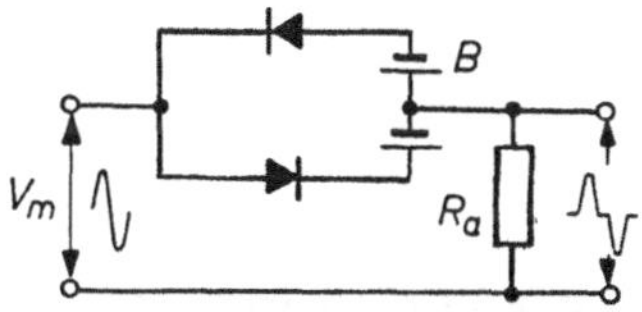

Fig. 11-21 Circuit for generating a "bent" sine wave

If this voltage is now applied to a circuit element (*CR*-network, amplifier etc.) in which the phase change under observation occurs, then these "steps" become displaced, the displacement *A* being a measure of the phase change. In Fig. 11-22 the oscillograms *a* show corresponding patterns for phase angles of 0°, 1°, 5°, 15° and 30°. No comparison voltage with a reference phase is required here. Other possibilities present themselves when this idea is extended. For instance, without horizontal deflection (Fig. 11-22*b*) the bends in the curve cause two bright spots to appear in the vertical trace, the distance between them being likewise a measure of the phase difference.

In Fig. 11-22*a* the horizontal deflection frequency was half the input frequency to produce the waveform observed. But even when the time base frequency is deliberately not synchronized with the vertical deflection frequency, but is considerably higher or lower, interpretable patterns can still be obtained, as can be seen from the

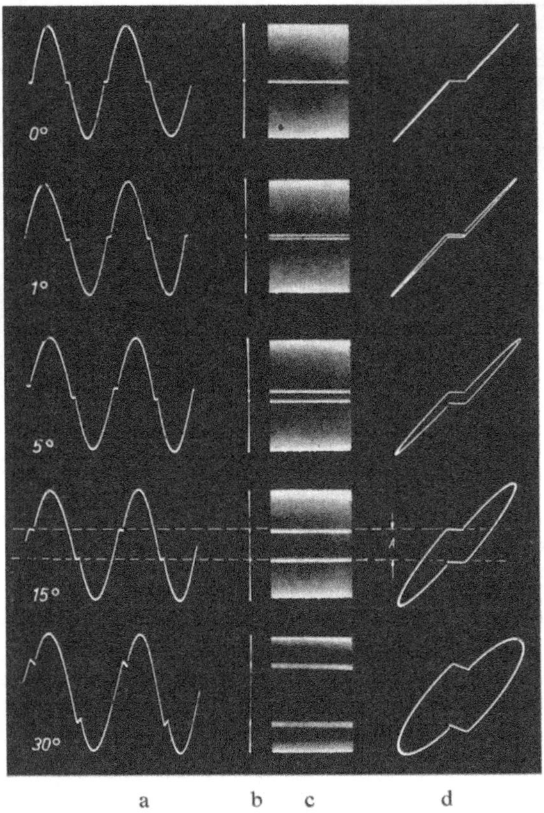

Fig. 11-22 Oscillograms of phase measurements with a "bent" sine wave

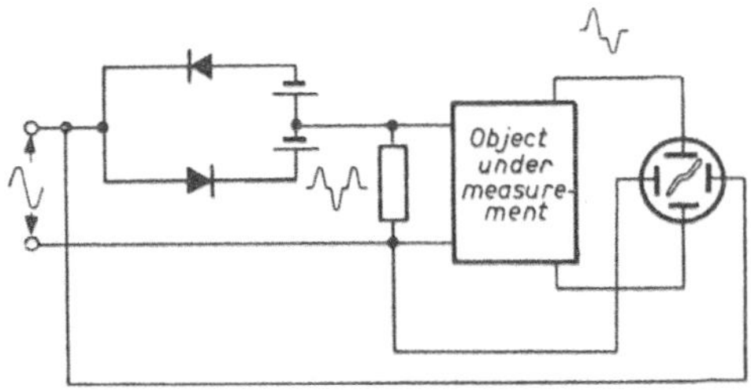

Fig. 11-23 Circuit for phase measurement with bent sine waves for patterns as in Fig. 11-22d

series of oscillograms in Fig. 11-22*c*. On a luminous area, which becomes darker towards the middle, two bright horizontal traces appear, the spacing of which can also serve to indicate phase difference. It is, however, also possible to dispense with the time base generator altogether and carry out the horizontal deflection by means of the "unbent" input signal (amplified if necessary), as shown in the circuit in Fig. 11-23. The "bent" ellipses shown in Fig. 11-22*d* with their short, horizontal portions in the middle are then obtained. The distance between these portions of the curve is once more a measure of the phase.

11. 6 Phase measurement on a circular scale

Indication of the phase angle between two alternating voltages on a circular scale has the valuable advantage that the readings can be made directly in degrees of arc. The following is one way of achieving this. The reference voltage V_1 is applied in such a way to both pairs of deflection plates that the component voltages have a phase difference of 90°. The generally recommended circuit is that shown in Fig. 11-24, in which the comparison voltage V_1 is applied across the series arrangement of a resistor R with a capacitor C. If $1/\omega C = R$, then the phase difference between the voltages V_C and V_R is almost exactly 90°. If the deflection amplitudes in the X- and Y-directions are equal, the pattern appearing on the screen at exactly 90° phase shift is a circle.

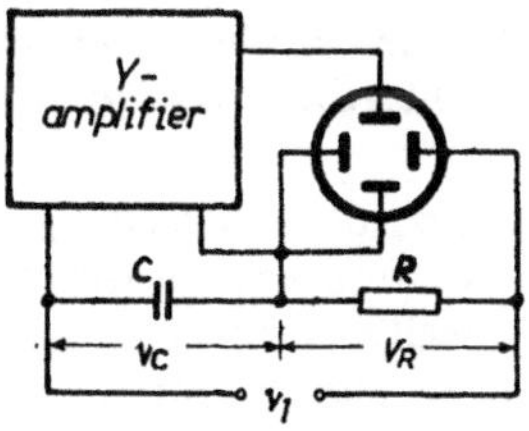

Fig. 11-24 Circuit for generating a circular trace

The pre-requisite in this circuit is that V_1 is symmetrical with respect to earth. As this is seldom the case, the circuit usually employed is that shown in Fig. 11-25. Here the deflection voltages for the pairs of plates are phase-shifted by an *RC*- or a *CR*-network, one by minus 45° and the other by plus 45°. In the process the voltages drop to 0.707 of the amplitude of the input voltage. For symmetrical working of the deflection plates identical, symmetrical *RC*- and *CR*-networks can be connected in the lead into each of the plates of the deflection system. In this example the mains voltage is employed for V_1. As a variable auto-transformer with one winding is used at the input, attention must be paid to correct polarity; alternatively an automatic make-and-break transformer — Philips type "RTT 54" — sould be used.

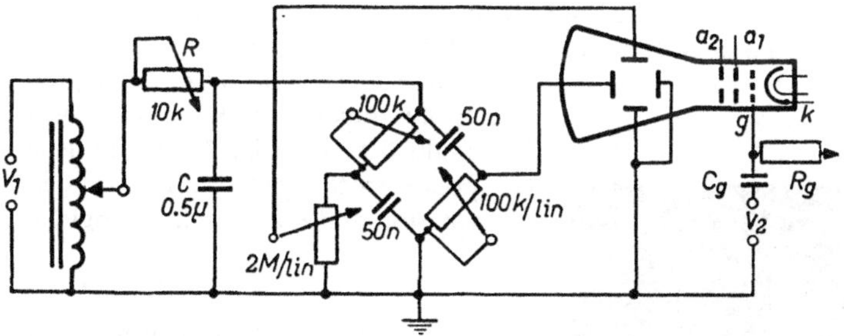

Fig. 11-25 Circuit for a circular trace. V_1 earthed on one side

The harmonics always present in mains voltages must be filtered out by an *RC*-network to prevent severe distortion of the circle. (The circle trace is an ideal criterion for harmonics.) The diameter of the circle can be adjusted by the input transformer. The phase and thus the circle itself can be adjusted by the two 100 kΩ variable resistors, and the pattern symmetry can be adjusted by the 2 MΩ potentiometer.

A pulse is now generated as described in Fig. 11-3, from voltage V_2, whose phase difference with respect to V_1 is to be measured, and this pulse is led to the grid of the

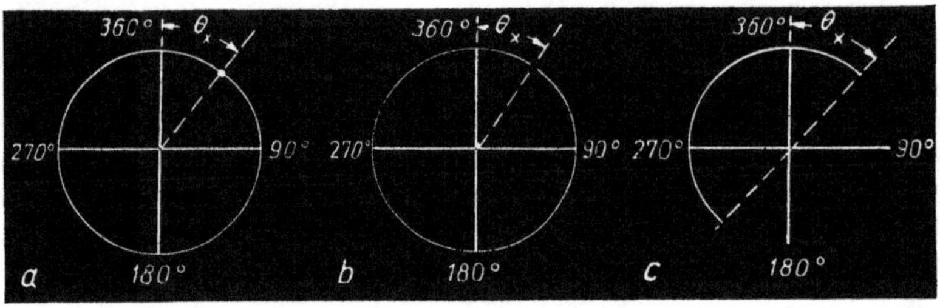

Fig. 11-26 Oscillograms of phase measurements with a circular trace; *a*) brilliance marking, *b*) blank marking, *c*) brightening of semi-circle

cathode ray tube, so that, according to polarity, a bright marking or an interruption appears on the circumference of the circle, as shown in Fig. 11-26*a* and *b*. The marking for $\theta = 0$ can easily be ascertained by connecting V_1 itself to the pulse-generating source (Fig. 11-3). The output position on the circumference of the screen can be corrected as described by means of phase-shifting networks at the input of the pulse-former. The distance of the measuring mark from the zero mark on the circumference gives the phase angle directly up to 360°. The oscillogram in Fig. 11-26*c* indicates a similar method. Here a rectangular voltage is taken from the second valve in the circuit of Fig. 11-3 and applied to the grid of the cathode ray tube. This brightens exactly one half of the circle. The chord connecting the end points of the semi-circle also gives a phase scale in degrees of arc. The degree of uncertainty of interpretation with this method is about 2° in standard cathode ray tubes [8] [9].

A number of circuits, known as "phase bridges" have been developed, which produce a trace which rotates in dependence on the phase difference between the two voltages [5] [6] [7] [8]. (An identical circuit is described in the next chapter on the representation of cycloids.) The amplitudes of these voltages must, however, be equal. The phase angle can then be read off directly in degrees on a circular scale: Fig. 11-27 shows this in a number of oscillograms for various phase angles.

It can be seen that phase angles are indicated up to maxima +360° or —360°. Twice the angle described by the rotating pointer corresponds to the phase angle. If α is the angle of rotation and θ is the phase angle, then:

$$\theta = 2 \cdot \alpha \,. \qquad (11.5)$$

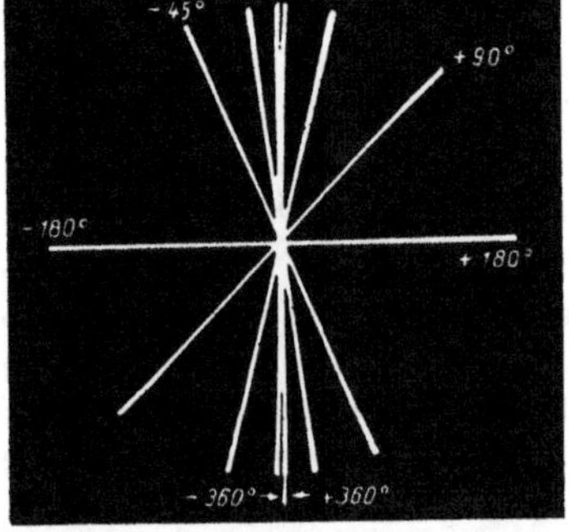

Fig. 11-27 Phase measurement with rotating pointer

The relevant circuit is shown in Fig. 11-28. In this example the phase difference between voltages V_1 and V_2 at mains frequency was observed. The voltage is taken from the mains after adjusting by Tr_1; isolation from the mains is provided by Tr_2 and smoothing by R and C. For the elements of the bridge circuit, $\frac{1}{\omega \cdot C}$ must be equal to R. The values given in the circuit provide satisfactory results.

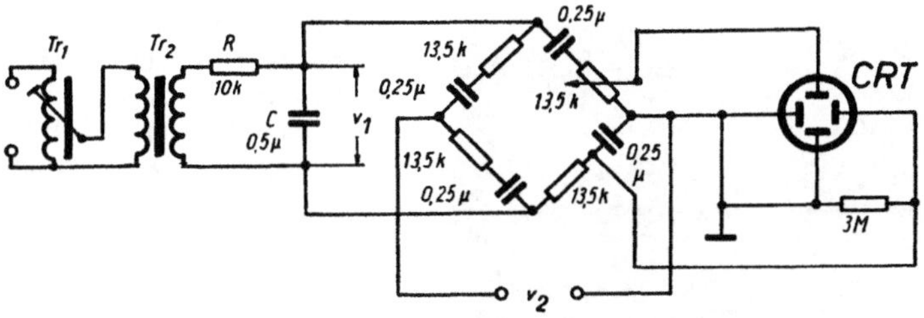

Fig. 11-28 Bridge circuit for producing a rotating pointer

This indication is ambiguous, however, for the trace for 0° could also be 360°, that for 90° could be —270° and that for 180° could be —180°. This method is most suitable for measurement of large phase angles. The reading uncertainty over the whole range (< 360°) is constant at about 5°.

If the two voltages V_1 and V_2 are taken from different AC generators, then, unless the generators are in perfect synchronism, a rotating trace is obtained. Observation of this trace greatly simplifies the operation of synchronizing the generators (see Bibliography [11] on Ch. 12). It should be noted that where sufficient voltage is available, only a few inexpensive circuit elements (no vertical amplifier and no time base generator) are needed for the purpose, apart from the cathode ray tube and its mains supply. It should further be noted, that it is possible in this way to carry out phase measurements on voltages having frequencies up to 1 Mc/s [9].

The circuit of the phase bridge is, as in most measuring arrangements given here, rated for a given frequency. If, however, phase measurements are to be carried out over a certain frequency range, it is necessary to include a mixer valve in each voltage lead for frequency transformation. The frequency-determining elements of this stage should be so rated that the output frequency is always equal to the bridge frequency.

11.7 Phase measurement with rectangular voltages

The phase shifting effect of circuit elements or of a whole amplifier, particularly when the phase differences are small, can be satisfactorily investigated with the aid of a symmetrical rectangular voltage. Alternating voltages of this sort can be generated in multivibrator circuits, in a square-wave voltage generator (e.g., Philips "GM 2314" or "PP 1122") or by clipping a sinusoidal voltage.

This voltage of suitable amplitude is applied to the element or circuit under observation and the waveform examined on the oscilloscope. The amplifier must, of course, be able to pass this waveform faithfully at the required frequency, and this should be checked in advance by connecting the input signal directly to the oscilloscope, since square waves make very great demands on an amplifier, as was described in Part I, Ch. 5 (Fig. 5-1). The behaviour of the voltage in relation to time is then observed (preferably over two cycles) with a voltage linear with time on the *X*-plates. The resultant patterns of the most important gradations of phase difference at the lower frequency limit of a four-pole are shown in the series of oscillograms of Fig. 11-29*a*. Viewing these pictures from the side, it is seen that even at $^1/_2$° the straight portion of the waveform is slightly tilted. (The vertical quick transients cannot be

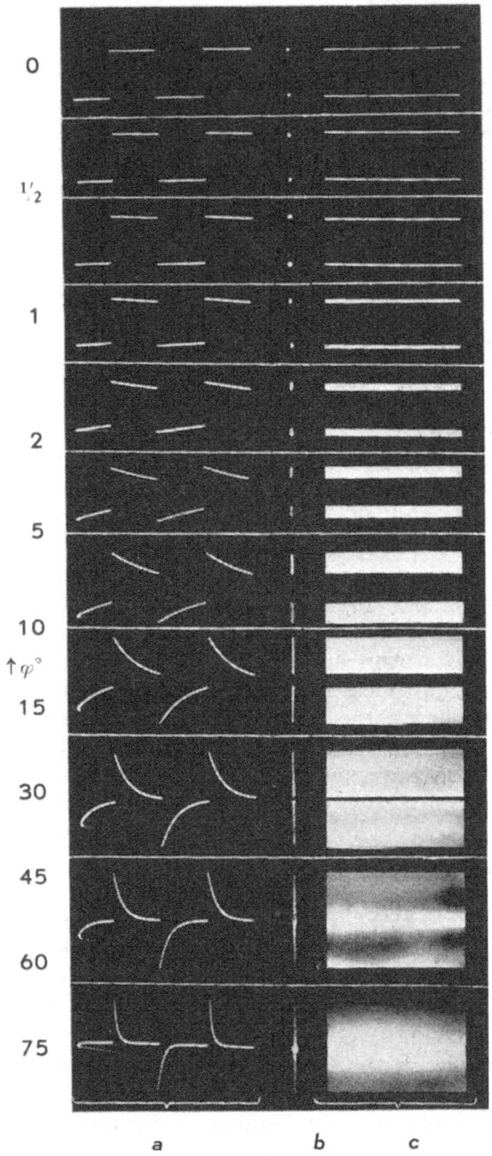

Fig. 11-29 Oscillograms of phase measurements with square waves; *a*) Patterns of two cycles, *b*) without time base, *c*) time base frequency higher than signal frequency and not synchronized

recorded owing to the extremely high deflection speed.) In addition to this wellknown method, two other possibilities are indicated, in oscillogram series *b* and *c*. In series *b* the voltage on the *X*-plates was switched off; the time base generator can be dispensed with entirely for this purpose. Line traces are obtained, whose height in relation to the total picture height is a measure of the phase difference. In series *c* the time base generator was in use, but its frequency was much higher than the input frequency and not synchronized. There now appear traces or strips of different thickness, whose height in relation to the whole picture height can once again be used as a measure of the phase difference.

11. 8 The distortion of a square wave by phase shift

To be able to judge the oscillogram thus obtained it is necessary first to consider the causes of the distortion. According to Fourier, a square wave can be built up from a fundamental sine wave and an appropriate number of sinusoidal harmonics in certain phase positions.

For a square voltage beginning with a positive half-cycle the following equation applies:

$$y = \frac{4 \cdot A}{\pi} \cdot \sum_{n=1}^{n=\infty} \frac{1}{2n-1} \cdot \sin(2n-1)x, \qquad (11.6)$$

or

$$y = \frac{4 \cdot A}{\pi} \cdot [\sin x + \frac{1}{3} \sin 3x + \frac{1}{5} \sin 5x + \ldots \frac{1}{n} \sin nx] \, . \qquad (11.7)$$

The amplitude relationships are as follows:

Fundamental:	amplitude	1
3rd harmonic:	,,	1/3
5th ,, :	,,	1/5
7th ,, :	,,	1/7
9th ,, :	,,	1/9
11th ,, :	,,	1/11, and so on.

This means that a voltage of this shape contains the 101st harmonic with an amplitude equal to 1% of the fundamental amplitude.

In a circuit element which determines the low frequency end, the output voltage leads the fundamental frequency of the input voltage. In a circuit element which determines the upper frequency end, the output voltage lags behind the input voltage. Since at the low-frequency end the harmonics, depending on their frequency, are appreciably less shifted in phase, and since, moreover, their "wave-length" is only a fraction of that of the fundamental, the instantaneous values change in such a way that the horizontal portions of the voltage curve become tilted in the time direction. In a circuit element determining the upper frequency limit the harmonics are influenced more strongly than the fundamental (attenuation occurs as a rule too) and the horizontal portions of the curve are rounded counter to the direction of the time axis (Fig. 11-34).

The change in waveform at a given phase shift could be investigated by drawing the fundamental together with its harmonics and constructing the sum curve for the characteristic cases. But this would require a great deal of time. The connection between phase difference and change of the square wave can, however, be obtained from the following consideration.

At the commencement of each half cycle of the signal, capacitor C (Fig. 11-30*a*) is charged up [75]).

Until the arrival of the following half cycle, capacitor C can discharge through resistor R. The voltage decay is given by the time constant $\tau = R \cdot C$, to which the familiar equation for the discharge curve of a CR-circuit applies:

$$v_E = v_0 \cdot e^{-t/RC} \, . \qquad (11.8)$$

As indicated in Fig. 11-30*a*, v_0 is the voltage at the beginning and v_E the voltage at the end of the half cycle.

In this case (symmetrical signal) the time t is equal to half the reciprocal of the frequency of the rectangular voltage, that is to say:

$$t = \frac{1}{2 \cdot f} \, . \qquad (11.9)$$

[75]) For the lower frequency limit, the corresponding circuit elements could be combined in such an equivalent circuit of a CR-network. It is usually possible to describe the upper frequency limit in the same way by connecting a corresponding resistor and an "equivalent" capacitor in parallel.

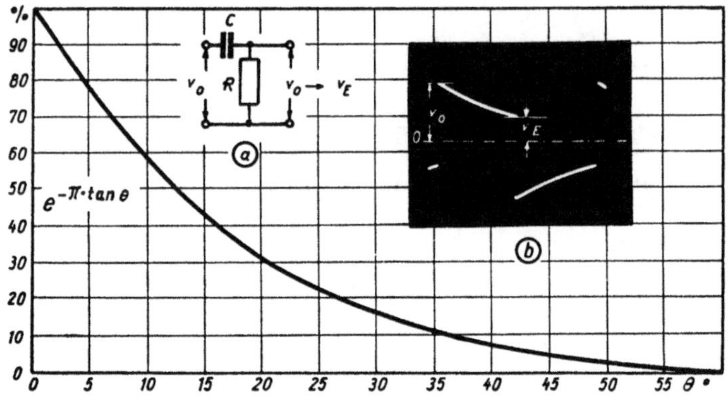

Fig. 11-30 Dependence of the attenuation factor $e^{-\pi \cdot \tan\theta}$ on phase angle θ. *a*) *CR*-network, *b*) oscillogram distorted by phase shift

It is now necessary to show the voltage v_0 and v_E on the one hand and the phase angle on the other in their mutual dependency. Here we should recall the equation $\cot\theta = \omega \cdot R \cdot C$, (Eq. (5.24) in Ch. 5 of Part I derived for this type of coupling elements.

If Eqs. (11.9) and (5.24) are newly inserted in Eq. (11.8) and $\frac{1}{\tan\theta}$ is substituted for $\cot\theta$ we obtain:

$$v_E = v_0 \cdot e^{-\pi \cdot \tan\theta}. \tag{11.10}$$

Fig. 11-30 shows the curve for the dependence of the attenuation factor $e^{-\pi \cdot \tan\theta}$ on the phase angle θ, calculated according to Eq. (11.10). From this, for a given phase angle, the voltage decay at the end of the half cycle, or, conversely the phase angle corresponding to a certain fall in amplitude, can be ascertained. The series of oscillograms in Figs. 11-29*b* and *c* are particularly suitable for this measurement, as they are easily interpreted [76]).

These oscillograms likewise satisfy previously mentioned conditions and are therefore suitable for use as a kind of "scale" for practical comparison. For this reason a comprehensive series of oscillograms was selected showing the essential stages in the change of pattern. The reader is thus provided with a basis of comparison in any given case, and it is recommended that these characteristic patterns be memorized.

In Fig. 11-29, particularly in the lower pictures of series *a*, it can be seen that the total picture height increases with increasing phase-shifting influence of the circuit elements present.

The process can be explained as follows:

If the *CR*-constant is so great that no noticeable phase change appears, then the voltage amplitude is such that, as a result of a "re-charging", the spot jumps from the given negative deflection to the positive deflection and vice versa (Fig. 11-31*a*).

---

[76]) Of course, this curve can be used to construct a suitable scale, so that the phase angle can also be read off directly on the screen. The picture height should be adjusted so that the end markings are reached both above and below.

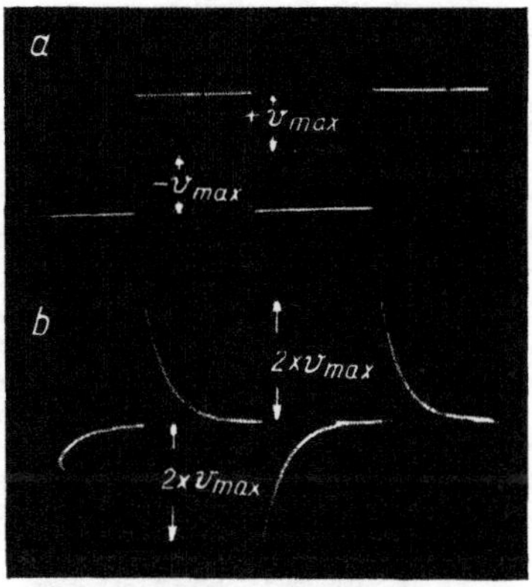

Fig. 11-31 Oscillograms to explain the increase of amplitude at large phase angles. *a*) Phase angle $\theta = 0°$ *b*) $\theta = 60°$

The voltage source thus overcomes the voltage difference $2v_{max}$. If, however, the time constant is so small that the zero line is reached *before* the end of the half cycle, as shown in Fig. 11-31*b* (the capacitor can be almost completely discharged within this time), then the voltage source again allows the output voltage to rise to the value $2v_{max}$, but this time from the zero line. At the next half cycle it falls again by the same amount after having first of all become zero once more, so that the amplitude of the voltage peaks is approximately twice what it was before. This applies for $\theta < 60°$. Where $\theta > 60°$ (75° and so on), the decay of the amplitude of the fundamental is noticeable and the total pitcure height becomes somewhat smaller. If the value of the *RC*-network is such that the capacitor can partly, but not wholly discharge, then, of course, corresponding intermediate values of the amplitudes are found, as can be seen in the oscillograms in Fig. 11-29 for $\theta \geqq 10°$ onwards.

11.9 Electrical differentiation

With increasing phase shift the oscillogram corresponds less and less to the wave-form of the input voltage itself, and more and more to its change or to the slope of the curve at each instant. As a curve of this kind represents the differential quotients of the original curve, the possibility of electrical differentiation arises. For the picture in which $\theta = 60°$, $\frac{1}{\omega \cdot C}$ was equal to $1.8R$ for $\theta = 75°$, $\frac{1}{\omega \cdot C}$ was equal to $4R$. It can be seen that from $\frac{1}{\omega \cdot C} = 3R$ onwards, this effect becomes progressively more pronounced. In this way, from all voltages corresponding to any change of state, an oscillogram can be obtained representing mathematically the first derivative of the process [77]).

[77]) If, in a circuit according to Fig. 11-30a, $\frac{1}{\omega \cdot C} \gg R$, the current i flowing through these circuit elements is determined mainly by C. The voltage v_c across this capacitor is practically equal to the input voltage v_o; thus $v_o \approx v_c$. The current through a capacitor is known, however, to be the voltage change $\frac{dv}{dt}$ and hence the current i here is proportional to the voltage change $\frac{dv_o}{dt}$. The voltage v_E is thus obtained from:

$$i \cdot R = v_E \approx R \cdot C \cdot \frac{dv_o}{dt}$$

The output voltage v_E is obtained in a circuit so rated from the product of a constant ($k = R \cdot C$) and the differential quotient of the input voltage. It is thus proportional to the first derivative of the input voltage [10].

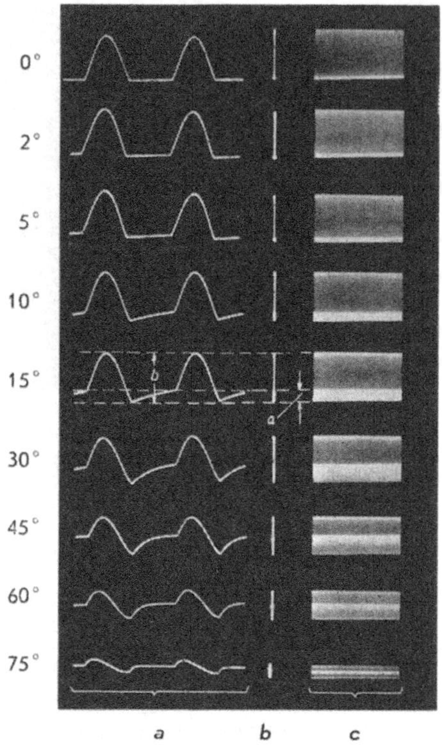

Fig. 11-32 Oscillograms of phase measurements with half-wave rectified sine waves. *a*) With time base, *b*) without time base, *c*) with high, non-synchronized time base frequency

11. 10 Phase measurement with half-wave rectified voltages

Since there may not always be a voltage source available for producing square waves of variable frequency, it should be pointed out that similar measurements are possible with a half-wave rectified alternating voltage (during alternate half cycles the voltage waveform is horizontal). In Fig. 11-32*a* a series of such oscillograms are set out similar to those for square waves in Fig. 11-29. Here too, horizontal deflection can be dispensed with patterns as obtained in Fig. 11-32*b*. With a high, non-synchronized time base frequency, patterns like those in Fig. 11-32*c* are obtained. With a voltage curve of this sort, however, phase changes to be readable must be somewhat greater than in the case with square voltages, that is, they must be at least 2°.

11. 11 Investigation of circuits with lagging phase

The phase change produced in a sinusoidal voltage by an *RC*-network (Fig. 11-33*a*) can also be investigated by means of a square wave. Alternatively the permissible distortion of rectangular pulses is also a measure for rating such circuit elements, the phase change being used as a scale. Fig. 11-34 shows a series of oscillograms for such a case. The voltage across the capacitor now reaches its maximum value after some delay. The greater the time constant *RC*, the slower the voltage rise. As the influence of the circuit in Fig. 11-33*a* is equivalent to the one in Fig. 11-33*b* under certain circumstances which apply in amplifying technique [11] [12], these results can also be applied to investigations into the upper frequency limits of circuit elements or amplifier stages. As shown Fig. 11-34, no rise occurs, but a decrease always takes place in the deflection amplitude — in contradistinction to a *CR*-network.

Fig. 11-33 *RC*-circuit (*a*) and equivalent circuit with resistor and capacitor in parallel (*b*)

Fig. 11-34 Oscillograms of phase measurements at negative phase angle

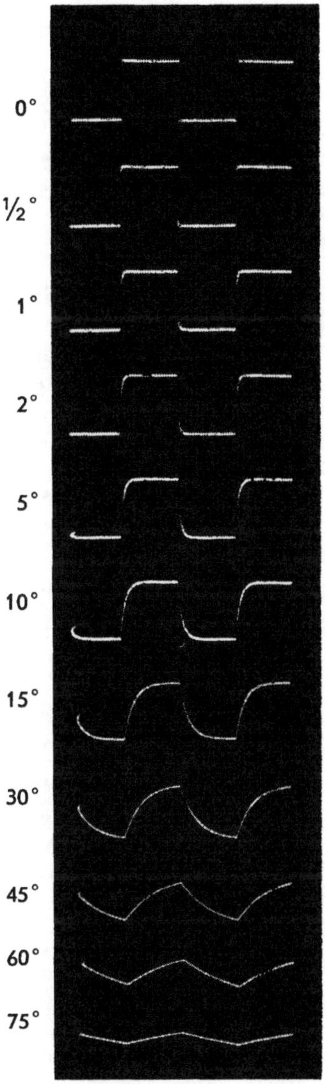

The relationship between the ratio of the permissible rise time T_r, pulse duration T_d and phase angle θ (corresponding to the amount of distortion of the oscillograms in Fig. 11-34), is given in the following equation:

$$\frac{T_a}{T_d} = \frac{2.2}{\pi} \cdot \tan\theta = 0.70 \cdot \tan\theta. \qquad (11.11)$$

With a given pulse duration T_d, for a certain permissible distortion (expressed by the phase angle θ) the required rise time is:

$$T_r = 0.70 \cdot T_d \cdot \tan\theta. \qquad (11.12)$$

For a symmetrical square voltage, f is equal to $\frac{1}{2\,T_d}$, so that

$$T_r = \frac{0.35}{f} \cdot \tan\theta. \qquad (11.13)$$

For the cut-off frequency f_{cu}, $\tan\theta = 1$, the rise time is

$$T_r = \frac{0.35}{f_{cu}}. \qquad (11.14)$$

This equation is identical with Eq. (5.58) in Part I, Ch. 5. If the quotients $0.35/f_{cu}$ from Eq. (11.14) are inserted for T_a in Eq. (11.11), the following new equation is obtained:

$$f_{cu} = \frac{1}{2 \cdot t_d \cdot \tan\theta}. \qquad (11.15)$$

From this the required upper cut-off frequency of the fourpole can be calculated for a given permissible pulse distortion. For a phase error of 2° (Fig. 11-34) f_{cu} must be equal to $14.3 \cdot \frac{1}{T_a}$, for a pulse duration of, for instance, 5 μs, the cut-off frequency is found to be 2.86 Mc/s. As a pulse duration of 5 μs corresponds to a symmetrical rectangular voltage of 100 kc/s, this means in other words, that the upper cut-off frequency of the four-terminal network must be at least 2.86 Mc/s, if the distortion of this voltage may not be greater than in the oscillogram for 2° in Fig. 11-34. If, however, the distortion is to remain barely perceptible, then the upper cut-off frequency is obtained from f_{cu} ($^1/_2{}^\circ$) $= 57.5 \frac{1}{T_d}$. For a rectangular pulse of 5 μs duration (or a symmetrical rectangular voltage of 100 kc/s) this means an upper cut-off frequency of about 11.5 Mc/s.

Eq. (5.2) was based on these considerations. If symmetrical rectangular voltages are to be reckoned with, then the required cut-off frequency for a given permissible distortion is obtained from:

$$f_{cu} = f \cdot \cot \theta \,. \tag{11.16}$$

For very stringent requirements, e.g. $\theta = {}^1/_2$; $\cot \theta = 115$. In most cases, however, the rating for $\theta = 2^\circ$ is sufficient, i.e. $\cot \theta = 28.6$.

It should be noted particularly, however, that with pulses of higher frequency, the rise time is already greater at the voltage source than would correspond, for instance, to a phase angle of $^1/_2{}^\circ$. But since the rise times of several networks only add quadratically [Eq. 5.60], there is generally little point in making such stringent demands on the oscilloscope amplifier in the highest pulse frequency ranges, as would be required according to Eqs. (11.15) and (11.16). For economic reasons, therefore, it is advisable to be content with an upper cut-off frequency of the oscilloscope which stands in a reasonable ratio to the shortest rise time of the pulses to be measured. However, a prerequisite is that the overload does not exceed 5%. Otherwise Eq. (5.60) no longer applies.

11. 12 Electrical intregation

As the picture for $\theta = 75^\circ$ in Fig. 11-34 shows, the square wave has become an almost perfect triangular wave. During every half cycle of the square voltage the output voltage rises at a uniform rate and falls likewise during the ensuing half cycle. Since this corresponds to the integrated curve of the input voltage, a circuit like that in Fig. 11-33*a* makes possible electrical integration. For this purpose R must be $> 3 \cdot \frac{1}{\omega \cdot C}$ [78]).

---

[78]) If, in this circuit according to Fig. 11-33a, R is selected as $\gg \frac{1}{\omega C}$ then the current i flowing through these circuit elements is practically determined by the value of R. In a first approximation $i = \frac{v_0}{R}$. The voltage v_E arising from the alternating current i across the capacitor C is, however, now given by the equation $v_E = \frac{1}{R \cdot C} \int i \cdot dt$. (The capacitance integrated the current i over the time t.) Assuming $i = \frac{v_0}{R}$, the voltage across the capacitor is obtained from the equation $v_E = \frac{1}{R \cdot C} \int i_0 \cdot dt$. It corresponds to the integral of the voltage v_0 over the time t [13] [14].

Fig. 11-35 Phase measurement with half-wave rectified sine waves with negative phase angle for the fundamental

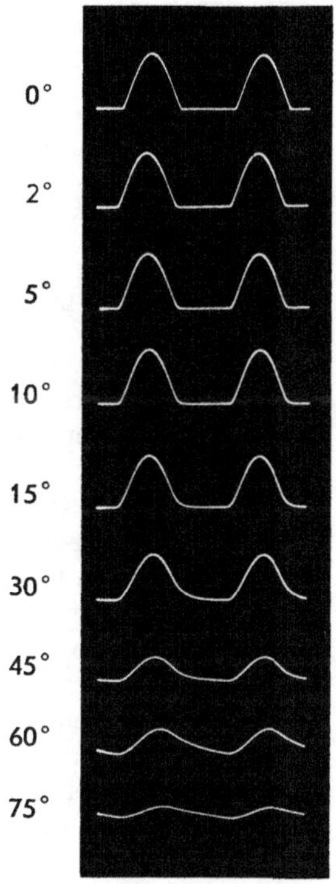

In this way the voltage corresponding to any other change of state (for example, the velocity of a physical movement) can be converted into a voltage proportional to the time integral of this voltage (path amplitude). (See also the bibliography to Ch. 10, [2] [3] [4].)

The oscillograms in Fig. 11-35 show that half-wave sinusoidal voltages are less suited to the investigation of circuit elements with a phase lag.

A rounding of the trailing edge of the voltage peak and a displacement of the position of the peak due to increasing phase lag of the fundamental are particularly noticeable. These changes are not so marked, however, as with a square wave.

11. 13 The use of square waves for assessing the properties of electrical transmission systems

In Chapter 5.17, Part I "Unit step response of an amplifier", it was stated that a very c'ear general idea of the essential characteristics of a deflection amplifier can be obtained by observing its behaviour during sudden periodical increases or decreases of input voltage.

It has been shown in the present chapter how phase changes can be studied with the aid of square waves. The combination of the above investigations shows that square waves, which can be regarded as periodic sequences of switching on and off, can provide information on all the essential characteristics of the circuit under measurement, i.e., frequency response, phase distortion, overshoot, etc.

Asymmetrical rectangular voltages are sometimes used in order to allow the system under study a longer decay time after the shorter part in accordance with its time constants.

Contrary to frequency response measurement with sine waves, which requires a whole series of derivations, a single derivation is often sufficient to discover whether a given square wave is undistorted, or distorted to a certain known extent, thus making it possible in practice to form a sufficiently accurate judgment of the transmission system under observation. A special oscillographic technique has been developed from this which finds particular application in the field of pulse technique and television. It is also used for locating faults in cables and overhead lines [15] and for ultrasonic testing for defects in manufactured parts [16].

The reader is referred to the extensive literature on the subject [17] [18] [19] [20] [21] [22] [23] [24] [25] [26] [27]. A work by J. Müller [22] deserves special attention in view of the excellent survey it provides (see also Ch. 25. Adjustment of high impedance wideband voltage dividers by square pulses or symmetrical square voltages).

CHAPTER 12

FREQUENCY MEASUREMENTS

12. 1 Frequency measurements; frequency comparison

The cathode ray oscilloscope is not itself a device for measuring frequency, but it makes possible frequency comparison to a degree of accuracy which can hardly be achieved by other means. (Except "Absolute frequency measurement with rotating pointer", p. 360). The accuracy of reading is limited only by the "absolute" limit at which changes in the pattern on the screen can be discerned with sufficient precision. The higher the frequency, the higher the accuracy of reading.

In view of the basic character of the oscilloscope, i.e. that it is voltage-operated, the following pages will deal mainly with the investigation of electrical voltages. It should be emphasized at this point that, in general, every change of state, be it mechanical, optical, thermal, etc., can be converted by an appropriate transducer into a proportional voltage and thus observed. For example, very accurate measurements of rotation speeds can be obtained in this way. The rotating parts can be picked-up by a light beam and a photocell, so that no mechanical connection or loading is necessary. The time-keeping of watches can be checked and regulated in this way in a surprisingly short space of time [1] (see also Part III, Ch. 30 and 31).

12. 2 Frequency measurement by comparison with time base frequency

As the time base frequency is continuously adjustable over the individual ranges, its use in determining the unknown frequency is obvious. Generally, however, the scale adjustment only approximates the actual time base frequency (it can be dependent on mains voltage, amplitude and so on), so that further measures must be taken if really reliable results are to be obtained [79]).

However, accurate measurements are possible immediately if the time base frequency is locked in synchronism with a suitable standard frequency. The standard frequency should be equal to, or a small integral multiple of the time base frequency required for the measurement.

The first step, therefore, is to adjust the frequency of the time base generator so that it is equal to, or a fraction of the standard frequency. When synchronization has been firmly established, the accuracy of the time base frequency is equal to that of the standard frequency.

The voltage of unknown frequency, amplified if necessary, is then applied to the Y-plates. Fig. 12-1 shows the circuit for frequency comparison of this kind.

---

79) The ideas expressed in these chapters dealing with time base generators only refer directly to the relaxed-oscillation synchronized working. The triggered circuit is unsuitable, as then the flyback time would have to be taken into consideration. It is, however, not exactly known and, moreover, varies with the different adjustments.

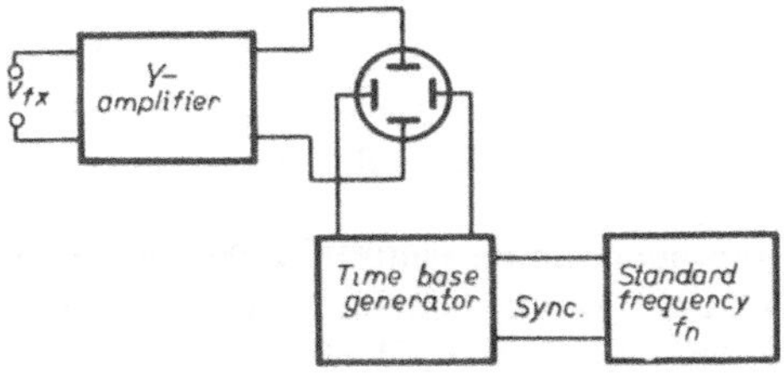

Fig. 12-1 Frequency measurement by comparison with the synchronized time base

$f_x : f_n =$	a	b	c	d	e	f	g	h	i
	$^1/_{10}$	$^1/_5$	$^1/_3$	$^1/_2$	1	2	3	5	10

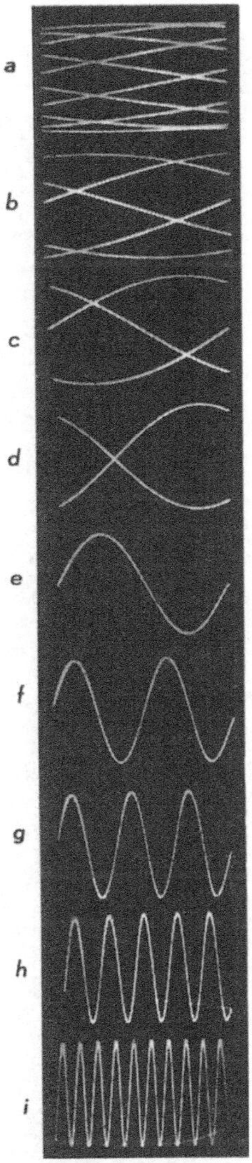

Fig. 12-2 Oscillograms obtained with a circuit as in Fig. 12-1

The voltage of unknown frequency, amplified if necessary, is then applied to the Y-plates. Fig. 12-1 shows the circuit for frequency comparison of this kind.

The waveform of the voltage of known frequency now appears on the screen. Sinusoidal voltages will be used generally in the following examples. Zig-zag waveforms and the like may also be used, however, and can be interpreted in the same way.

If the number of cycles appearing on the screen is N_{fx}, and if f_n is the known frequency, then, if the frequency ratio is an integral number, the unknown frequency will be:

$$f_x = N_{fx} \cdot f_n \cdot p . \qquad (12.1)$$

The factor p represents the ratio of the time base frequency to the reference frequency. If they are equal, the ratio is unity; otherwise it will always be a fraction or an integral multiple of 1.

In the series of oscillograms in Fig. 12-2, oscillograms e to i represent cases in which the unknown frequency is exactly equal to, or an integral multiple (1, 2, 3, 5 and 10) of the time base frequency.

If, on the other hand, the unknown frequency is only an integral fraction of the time base frequency, then during each time base cycle only a portion of the waveform of the unknown frequency will appear. A complete image of one cycle of this voltage now appears during a certain number of time base cycles in sections one above the other, as seen in Figs. 12-2a to 12-2d for integral frequency ratios of $^1/_{10}$ to $^1/_2$.

The unknown frequency is found from frequencies of this kind as the quotient of the frequency f_n and the number of curve portions N_c of the unknown frequency, i.e.:

$$f_x = \frac{f_n}{N_c} \cdot p \,. \tag{12.2}$$

It is still possible to interpret the oscillograms in cases where the frequency ratios are not whole numbers or integral fractions.

Fig. 12-3 reproduces nine arbitrarily chosen figures of the innumerable ones obtainable with ratios intermediate between 2 : 1 and 3 : 1 (Fig. 12-2 between *f* and *g*). If the comparison frequency f_n = 50 c/s (mains) and the time base frequency is synchronized in the ratio of 1 : 1, then these patterns correspond to the values of f_x set out in the last column of table 12-1. The frequency ratios for these patterns are given in fractions and decimals to show how, with a single comparison frequency, any required number of intermediate values in the frequency adjustments can be found, which is extremely useful for continuous calibration.

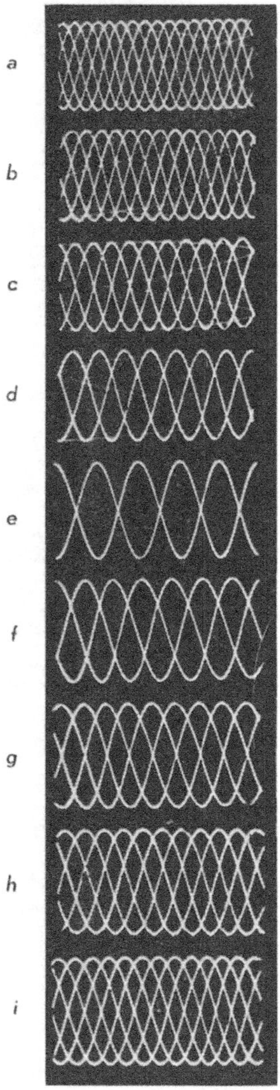

Fig. 12-3 Oscillograms with non-integral frequency ratios between 3 and 2, Table 12-1

TABLE 12-1

FREQUENCY RATIOS FOR THE OSCILLOGRAMS OF FIG. 12-3

Symbol	$f_x : f_n$	f_x [c/s] (for f_n = 50 c/s)
a	$17/6 = 2\,\frac{5}{6} = 2.833$	141 2/3
b	$14/5 = 2\,\frac{4}{5} = 2.800$	140
c	$11/4 = 2\,\frac{3}{4} = 2.750$	137 1/2
d	$8/3 = 2\,\frac{2}{3} = 2.667$	133 1/3
e	$5/2 = 2\,\frac{1}{2} = 2.500$	125
f	$7/3 = 2\,\frac{1}{3} = 2.333$	116 2/3
g	$9/4 = 2\,\frac{1}{4} = 2.250$	112 1/2
h	$11/5 = 2\,\frac{1}{5} = 2.200$	110
i	$13/6 = 2\,\frac{1}{6} = 2.167$	108 1/3

To interpret such patterns it is best to consider the figure as enclosed in a rectangle and then to count the number of points of contact of the loops and the horizontal tangent (above or below) and of the loops and the vertical tangent (left or right) as shown in Fig. 12-4. The unknown frequency is now obtained from:

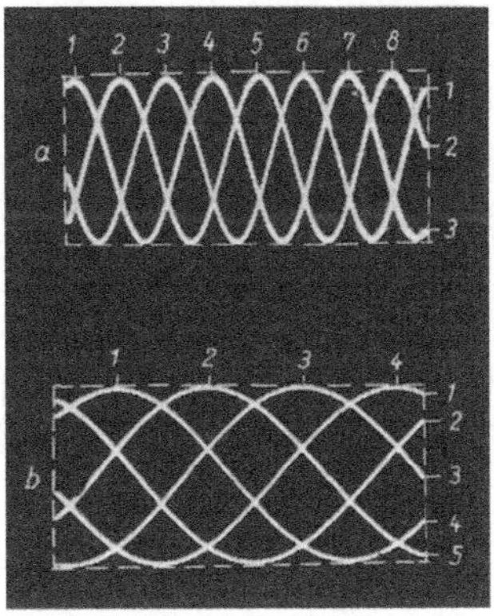

Fig. 12-4 Counting the patterns at frequency ratios $f_x : f_n$. *a*) $8/3 = 2^2/_3$; *b*) $4/5$

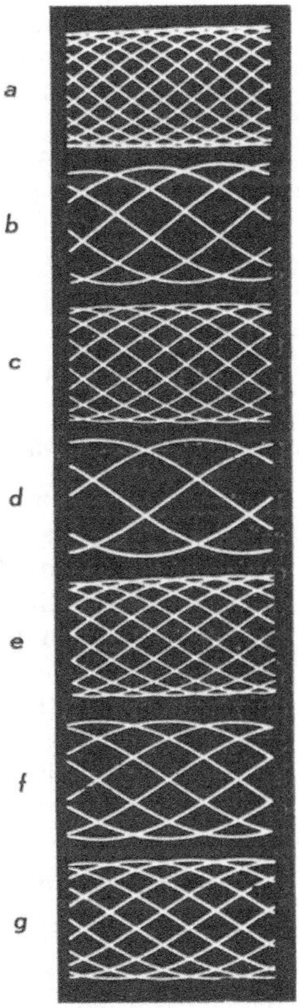

Fig. 12-5 Oscillograms with non-integral frequency ratios between 1/2 and 1/3, Table 12-2

TABLE 12-2 FREQUENCY RATIOS FOR THE OSCILLOGRAMS OF FIG. 12-5

Symbol	$f_x : f_n$	f_n [c/s] (for $f_x = 1000$ c/s)
a	7/16 = 0.438	437 1/2
b	3/7 = 0.428	428 4/7
c	5/12 = 0.417	416 2/3
d	2/5 = 0.400	400
e	5/13 = 0.385	384 8/13
f	3/8 = 0.375	375
g	4/11 = 0.364	363 7/11

$$f_x = f_n \cdot \frac{N_l}{N_c} \cdot p, \tag{12.3}$$

in which N_c is again the number of curved portions and N_l the number of loops touching the tangents.
In Fig. 12-4*a*, $f_n = 50$ c/s, $f_x = 50 \cdot {}^8/_3 = 133\,{}^1/_3$ c/s.

Eq. (12.3) also applies to frequency ratios at which $\frac{N_l}{N_c} < 1$.

In the example given in Fig. 12-4*b*, $\frac{N_l}{N_c} = 4/5$ and $f_n = 50$ c/s; $f_x = 50 \cdot 4/5 = 40$ c/s.

The arbitrarily selected oscillograms in Fig. 12-5 show that many intermediate values can also be found for frequency ratios less than 1 in this instance between the frequency ratios 1 : 2 and 1 : 3. Assuming once more that the time base is synchronized in the ratio 1 : 1, with $f_n = 1000$ c/s, frequency ratios and intermediate values are obtained as set out in Table 12-2.

The ability to count the number of loops N_l touching the tangents sets the limit to the interpretation of these patterns. It should also be noted that if the number of curve sections is too large, some of them may be lost during the flyback time. The patterns in Figs. 12-3 and 12-5 may at first appear somewhat confusing and difficult to interpret. With some practice, however, it will soon be found that patterns such as those in Figs. 12-3*e* and 12-5*d* stand out very clearly. It is usually not even necessary to count them in this way. When calibrating a scale, it is found by varying the unknown frequency that such characteristic patterns appear quite distinctly, so that such points are readily picked out as calibration marks between the integral frequency ratios. These oscillograms have been reproduced and discussed at such length mainly to draw attention to these possibilities of the oscilloscope which are not always exploited to the full.

12. 3 Frequency division

It should be mentioned here that a time base unit locked in synchronism with a reference frequency can be readily used as a frequency divider. If, for example, 10 cycles of the voltage are made to occur during one cycle of the reference frequency, it is assured that the time base voltage is working with a frequency of $f_n/10$.

If desired, a number of time base units may be connected in series and a large ratio frequency division obtained. This hint may be useful for laboratory work entailing only temporary frequency division and where it is highly desirable to avoid the acquisition of special frequency dividers for which there would be no further need. The sawtooth voltage with divided frequency can then be used as described either directly or differentiated.

12. 4 Frequency measurement with line traces

The oscillograms discussed above and shown in Figs. 12-2, 12-3, and 12-5 will have made it amply clear that for large frequency ratios the screen should be able to accommodate as widely extended a picture as possible. For this purpose, as already described in the section on phase measurement, it is possible with two linear time base voltages to trace horizontal lines in the screen and superimpose on them the voltage of unknown frequency. The relevant circuit is shown in Fig. 12-6 and greatly resembles the circuit in Fig. 11-5. The oscillator for the "picture" frequency is synchronized with

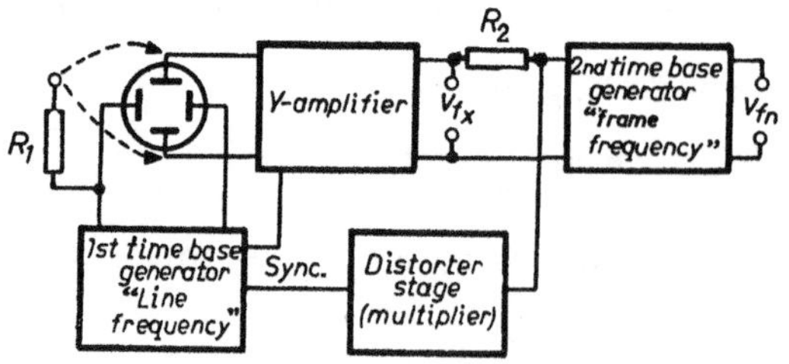

Fig. 12-6 Block diagram for frequency comparison with several frame lines

the standard frequency. A distorting stage (shaper) is used to generate and filter multiples of the "picture" frequency and to synchronize the line frequency oscillator. If the standard frequency is 50 c/s, for instance, it is possible to observe in five lines one hundred times the standard frequency, i.e., 5000 c/s. By suitable choice of lines it is possible, with one "picture" frequency and various line frequencies, to cover a wide band of frequencies. Conversely, if a standard frequency equal to or a multiple of the line frequency is available, it can be used to synchronize the first sawtooth oscillator, and from this the second can be synchronized. The shaper with its filter elements can then be dispensed with. Figs. 12-7*a* and *b* show typical patterns. In order to obtain horizontal lines, a portion of the line frequency voltage was here again fed to the Y-plates via resistor R_1.

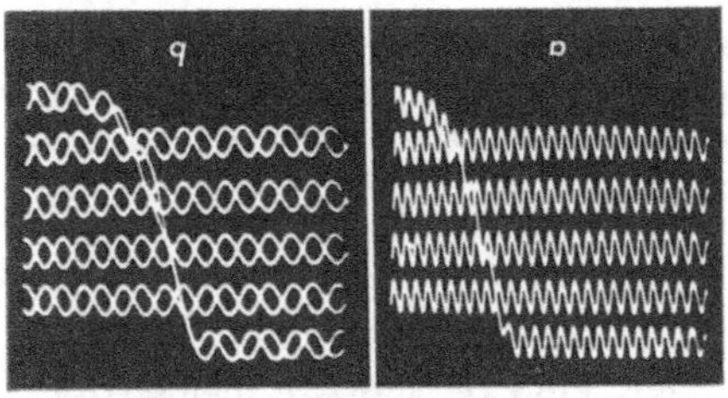

Fig. 12-7 Oscillograms with several frame lines for frequency comparison.
a) 5 traces; $f_x : f_B = 127$; *b*) 5 traces; $f_x : f_B = 83/2 : 1 = 41^1/_2$, f_B = picture frequency

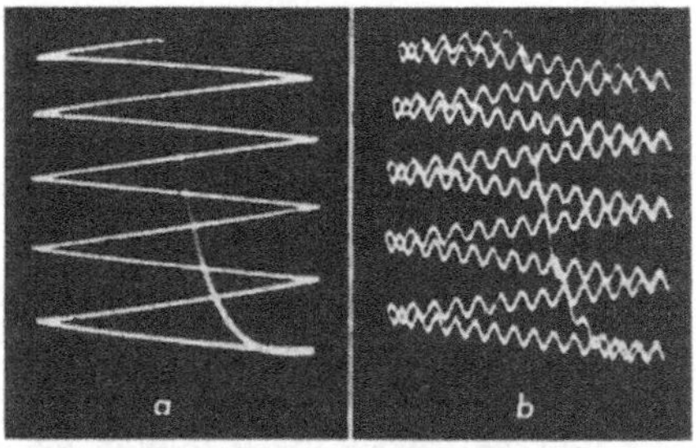

Fig. 12-8 Spot trace with triangular voltage for frequency measurement.
a) Pattern of the base line.
b) Base line with signal frequency $f_x : f_B = 141$

It should be noted here that, when there are a large number of lines, a considerable part of the unknown frequency pattern is lost in the flyback. It is best in such cases to use a triangular and not a normal sawtooth line voltage. Such voltages can be obtained by integration of a rectangular voltage [2] (see also Part III, Ch. 28 and Figs. 28-2*c* and 28-3). Instead of the first sawtooth voltage oscillator, an arrangement comprising a square wave generator, an integration network and, may be, an amplifier should be employed in the circuit of Fig. 12-6. Examples of the results then obtained are shown in Fig. 12-8. Trace and flyback are now equalized and the patterns can be evaluated in both directions without any portions remaining unseen.

12. 5 Frequency comparison by double oscillogram

If a twin-beam oscilloscope or a single-beam oscilloscope with an electronic switch is available, frequency comparisons can be made by direct comparison of the patterns of the two voltages. The time base must again be synchronized with the standard frequency. With a twin-beam oscilloscope the same time base voltage can be used for both systems, as it must be in any case with the single-beam oscilloscope. The voltage of the comparison frequency is applied to one input of the electronic switch, and the voltage of the unknown frequency to the other. The waveform of the known frequency then remains stationary and shows directly whether the time base frequency

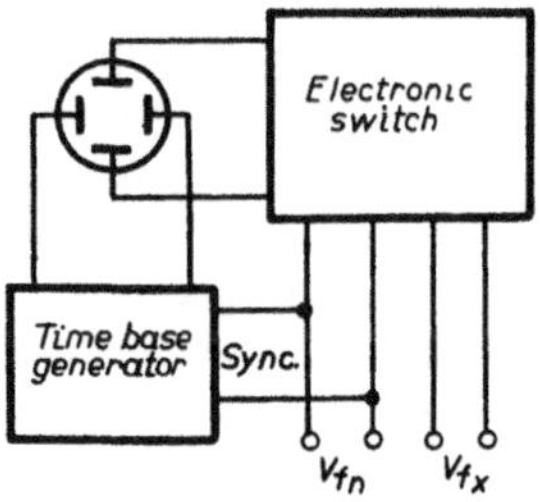

Fig. 12-9 Block diagram for frequency comparison with electronic switch

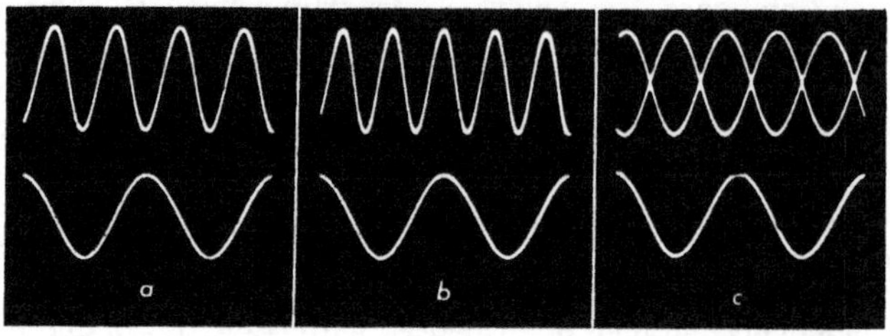

Fig. 12-10 Double oscillograms with the electronic switch; *a)* $f_x : f_n = 2$; *b)* $f_x : f_n = 5/2 = 2^1/_2$; *c)* $f_x : f_n = \frac{5/2}{2} = 1^1/_4$

is equal to or a fraction of the comparison frequency. Fig. 12-9 shows the basic layout for this method of frequency comparison. By comparing the number of cycles of the unknown frequency N_{fx} with the number of cycles of the known frequency N_{fn}, the unknown frequency is:

$$f_x = f_n \cdot \frac{N_{fx}}{N_{fn}} . \tag{12.4}$$

Fig. 12-10*a* shows a double oscillogram of this kind (f_x above f_n below) for an integral frequency ratio of 4 : 2. In Fig. 12-10*b* the frequency ratio was not integral but 5 : 2.

If the waveform of the reference frequency is not a simple figure but appears like the oscillograms in Fig. 12-3, then according to Eq. (12.3) the factor p must be introduced for N_{fn} (Fig. 12-10*c* top). The same applies when the waveforms of the voltage of unknown frequency are not complete cycles (Fig. 12-5).

12. 6 Frequency comparison by anode voltage modulation of a circular trace

For this method a circular trace is obtained, as has already been described, by applying two voltages having a phase difference of 90° to the two pairs of deflection plates. The basic circuit is again shown in Fig. 12-11.

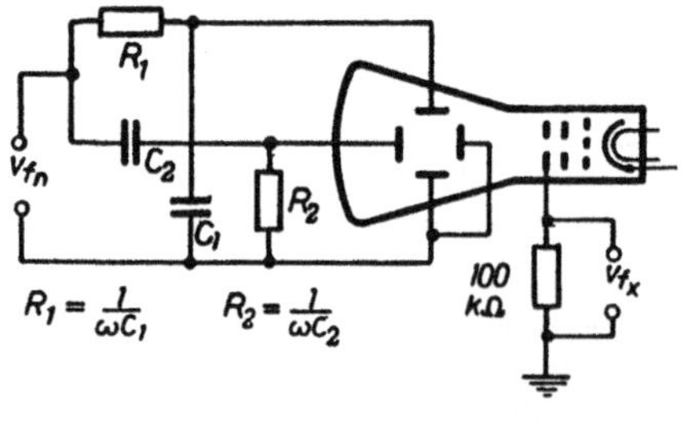

Fig. 12-11 Circuit for frequency measurement by deflecting a circular trace with anode modulation

Fig. 12-12 Radial deflected circular trace by a circuit as in Fig. 12-11

If the anode voltage of the cathode ray tube is now modulated sufficiently by the voltage of unknown frequency, the sensitivity of the deflection plates will vary in such a way that the diameter of the circle will rhythmically increase and decrease. For this purpose the anode of the cathode ray tube is not directly earthed but is connected to chassis via a resistor. The voltage of unknown frequency is applied to this resistor resulting in waveforms as shown in Fig. 12-12. It should be added that modulation of the anode vo!tage also affects the sharpness of the picture. (The sharpness adjustment corresponds only to those parts of the trace where the circle is of mean diameter.) The applied voltage should therefore not be too large. Good spot focus and the possibility of better modulation can be obtained by simultaneously modulating the other electrodes (a_1, g_2) by an appropriate part of the voltage of unknown frequency in proportion to their share of DC voltage. The unknown frequency is then found from the number of displacements N_{fx} at the circumference of the circle, i.e.

$$f_x = f_n \cdot N_{fx}. \tag{12.5}$$

(The "circle" frequency is then equal to f_n.)

12. 7 Frequency comparison by mixing the voltage of unknown frequency with the comparison frequency voltage

Here both voltages are applied together to the input of the oscilloscope amplifier as shown in Fig. 12-13. In order to avoid mutual reaction on the voltage sources, at least two series resistors, R_{v1} and R_{v2}, should be provided as indicated, and the common input resistor R should be small with respect to them. For the oscillograms of Fig. 12-14 $R_{v1} = R_{v2}$ was 800 kΩ and R was 170 kΩ. Mixing the

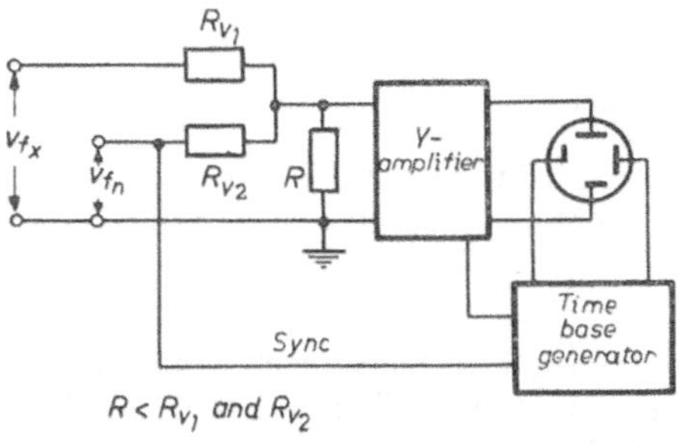

Fig. 12-13 Circuit for frequency comparison by mixing V_{fx} and V_{fn}

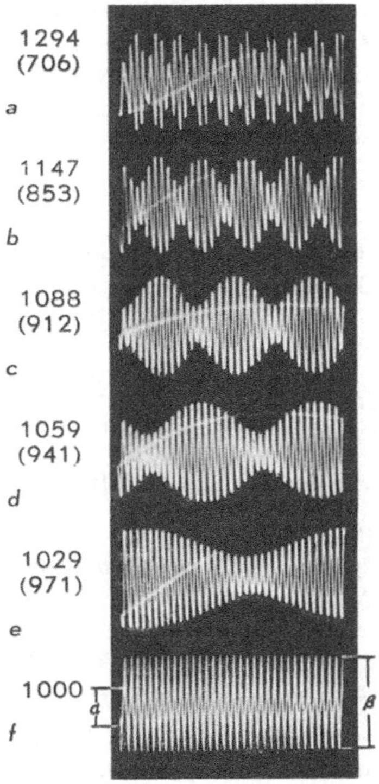

Fig. 12-14 Oscillograms for frequency comparison by mixing. The numbers beside the patterns apply to f_n = 1000 c/s and $f_n : fx = 34$

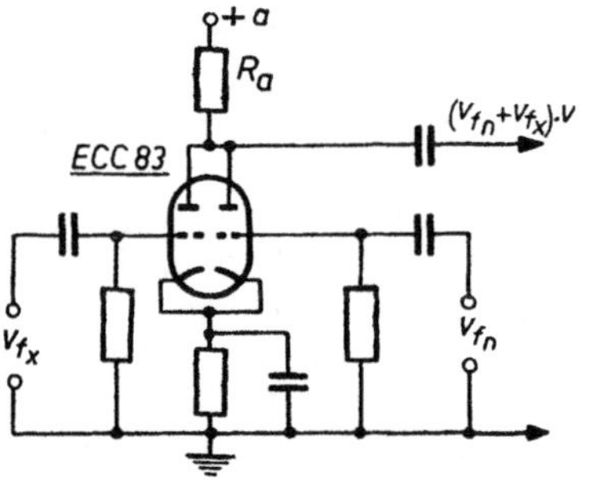

Fig. 12-15 Circuit for mixing by means of a double valve

voltages by means of double valves is a still better method, however. This is shown in Fig. 12-15 using a double triode (e.g., ECC 40 or ECC 83). As the two valve elements have a common load resistor R_a, the voltages V_{fn} and V_{fx}, amplified in both systems are added in the anode circuit. The risk of reaction is thus extremely slight. If necessary, it could be reduced still further by use of pentodes. The input voltages for comparison should preferably have the same amplitude. The time base generator should be locked in synchronism with the reference voltage V_{fn} so that its frequency will not change if, during measurement, amplitude fluctuations appear on the deflection plates (external synchronization).

As long as the two frequencies do not stand in any integral relation to one another the patterns on the screen will drift one through the other. When frequency equality is attained, however, the pattern of the reference voltage appears again, but the amplitudes fluctuate in the rhythm of the frequency difference, in other words interference occurs. The oscillogram in Fig. 12-14*f* illustrates this; the amplitude fluctuations from α to β were photographed by means of a long exposure with a small aperture.

If on the other hand the unknown frequency differs from the comparison frequency, symmetrical amplitude variations appear superimposed on the waveform of the standard frequency, and at certain frequency differences they appear as stationary beating patterns (amplitude modulation). Oscillograms of this nature are shown in Fig. 12-14*a-e*. If N_{fn} is the number of time base cycles corresponding to the quotient of the comparison and time base frequencies, thus:

$$N_{fn} = \frac{f_n}{f_X}, \tag{12.6}$$

then the unknown frequency f_x, which, together with the reference frequency, causes the beating, is found from

$$f_x = f_n \pm f_n \cdot \frac{N_{\text{beat}}}{N_{fn}}, \tag{12.7}$$

in which N_{beat} is the number of beating patterns in the picture. The interesting fact emerges from this that in this way it is easy to obtain fixed frequency points in the vicinity of a standard frequency at a spacing of $\pm f_n \cdot \frac{N_{\text{beat}}}{N_{fn}}$. By the choice of N_{fn} (the number of cycles of reference frequency adjusted on the screen), the frequency spacing can be chosen within very wide limits. In the oscillograms in Fig. 12-14, N_{fn} was equal to 34.

At larger frequency spacings the patterns are not always so easy to interpret, but after a little practice, accurate results can still be obtained from them. For instance, in Fig. 12-16*a*, $f_x = {}^1/_2\, f_n$ [80]); *b*, $f_x = 1/3\, f_n$; *c*, $f_x = 1\,{}^1/_2\, f_n$ and *d*, $f_x = 2\, f_n$.

[80]) Otherwise expressed, the oscillogram represents the picture of a sine wave with the first harmonic in a certain phase position.

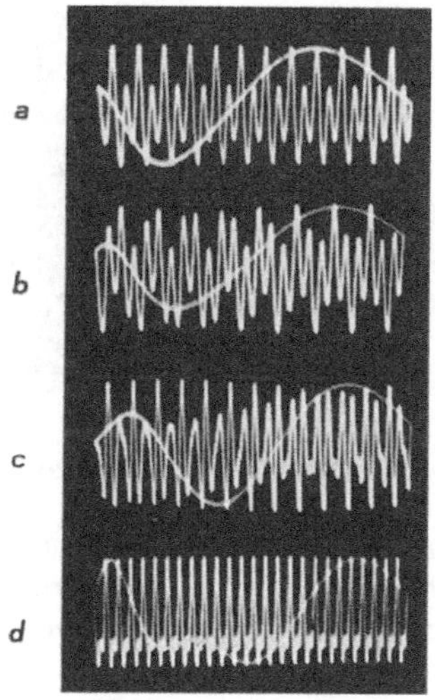

Fig. 12-16 Oscillograms with large differences between the known and unknown frequencies

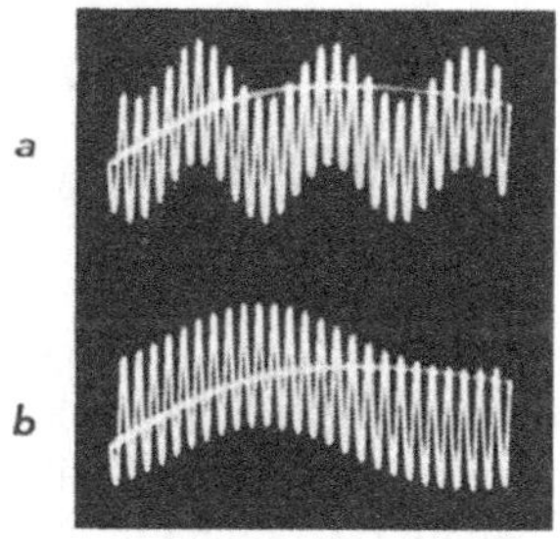

Fig. 12-17 Oscillograms when f_x is a small integral multiple of f_X

Useful results can also be obtained when the unknown frequency is so low that it is equal to or a multiple of the time base frequency. The amplitude now fluctuates in the rhythm of $f_x : f_X$. The frequency can be determined in the same way as has been described in Ch. 12.2 "Frequency measurement by comparison with time base frequency".

Here, however, the ratio $f_x : f_X$ can be read straight off the screen from the number of cycles N_{fx}. The unknown frequency is found from the equation:

$$f_x = \frac{N_{fx}}{N_{fn}} \cdot f_n \,. \tag{12.8}$$

The factor N_{fx} thus represents the number of fluctuations due to f_x. In Fig. 12-17*a* the unknown frequency f_x was equal to $3f_X$ and in b, f_x equalled f_X. If $N_{fn} = 26$ and $f_n = 1000$ c/s, the unknown frequency is 115 5/13 c/s or 38 6/13 c/s.

These examples show how simple it is to make accurate frequency measurements over wide ranges with one standard frequency.

12. 8 Frequency comparison with Lissajous figures

When one of the two voltages whose frequency ratio is to be measured is applied to the Y-plates and the other to the X-plates of the cathode ray tube, they will cause the spot to describe patterns on the screen, called Lissajous figures after their discoverer.

Fig. 12-18 shows the fundamental circuit for this. If the two frequencies are equal, a simple figure without cross-over appears, as already described in the section on phase measurement and illustrated in the elliptical oscillograms of Figs. 11-10 and 11-13. The resultant patterns are ellipses, whose inclination and width depend on the phase relationship and phase difference of the two voltages. The analytical evaluation of such figures has been dealt with elsewhere, and the interested reader is referred to these publications mentioned in the bibliography [3] [4].

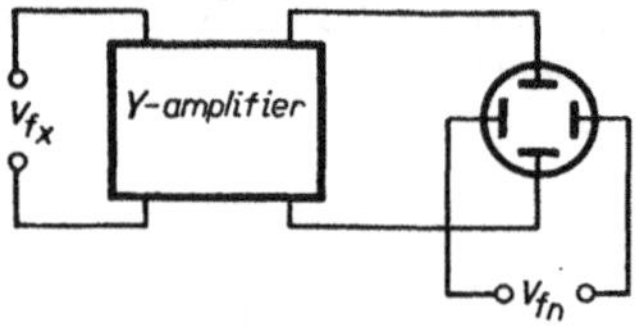

Fig. 12-18 Circuit for frequency comparison by Lissajous figures

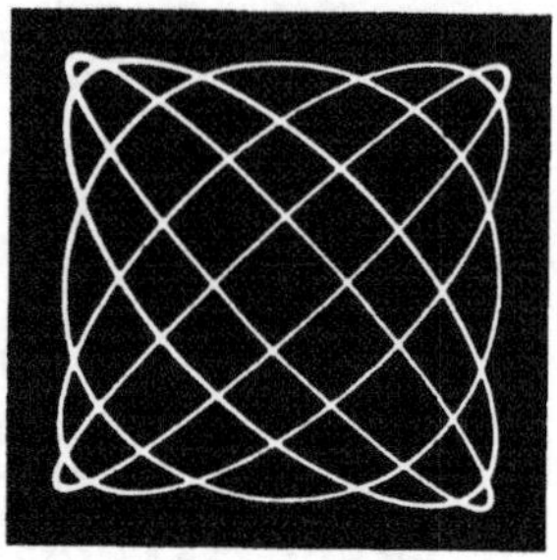

Fig. 12-19 Composite oscillogram for voltages of equal frequency but having some phase differences;
Phase difference of individual figures, 15°, 45°, 90°, 105° and 135°

When there are small differences in frequency the oscillogram passes through all the patterns for phase relationship from 0° to 360° in a certain space of time. Fig. 12-19 shows the composite oscillograms for phase differences of 15°, 45°, 90°, 105° and 135°, where the deflection was equal in both directions. The frequency difference Δf, if small, can be determined by using a stop watch to check the time taken for the phase to pass through 360° (one cycle). The reciprocal of this time T (in seconds) is then exactly equal to this frequency difference:

$$\Delta f\ [\mathrm{c/s}] = \frac{1}{T\ [\mathrm{s}]}\ . \tag{12.9}$$

In this way, very slight frequency differences, amounting to small fractions of one cycle per second, can be accurately determined.

If the frequency ratio is an integral multiple, the familiar Lissajous figures with crossed loops appear, which, depending on the initial phase relationship of the two voltages, can also pass through a great variety of forms. A set of original oscillograms are arranged in Fig. 12-20 to illustrate how different patterns are produced for different phase positions [81]) of the voltage on the Y-plates (higher frequency), when the frequency ratio is 2 : 1.

Certain patterns can be seen to repeat themselves and can thus be interpreted in more than one way. The figure for a phase difference $\varphi = 0°$ is also that for $\varphi = 180°$, that for $\varphi = 45°$ is also that for $\varphi = 135°$, etc.

The oscillogram for 90°, which repeats itself inversely at 270° and also appears in a correspondingly transformed shape at higher even-numbered frequency ratios, when the curves coincide, is identical with what is known as the "Tschebyscheff function" [5]. Perfectly clear patterns can also be obtained in this case by using intensity modulation with a reference voltage, as shown in Fig. 11-15. An asymmetrical voltage waveform of a higher frequency is the most suitable for this purpose. A sawtooth voltage waveform can be used.

Figs. 12-21*a* and *b* show the effect of intensity modulation where the frequency ratio is 2 : 1 at phase difference of 0° and 180°. A wedge-shaped marking appears, the direction of which, when the patterns for 0° phase difference are known, allows the other phase positions to be unambiguously distinguished.

---

[81]) The term "phase position" refers here to the initial phase of the voltage with the higher frequency. A phase difference of 360° is, therefore, a cycle of the voltage with the higher frequency.

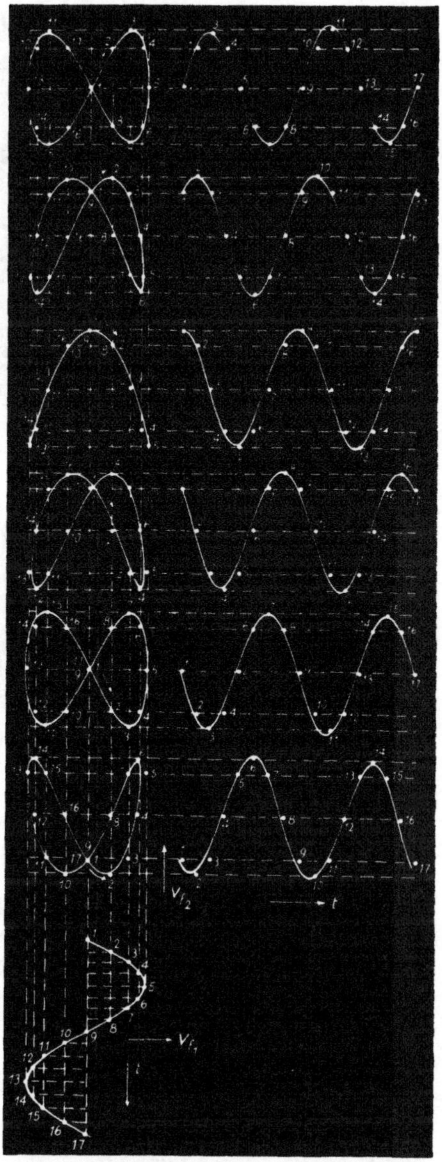

Fig. 12-20 Lissajous figures at a frequency ratio of 2:1 for varying initial phase of the higher frequency. The oscillograms are set out in steps of 45° between 0° and 225°. The direction of movement of the spot from the starting point is indicated by the arrow

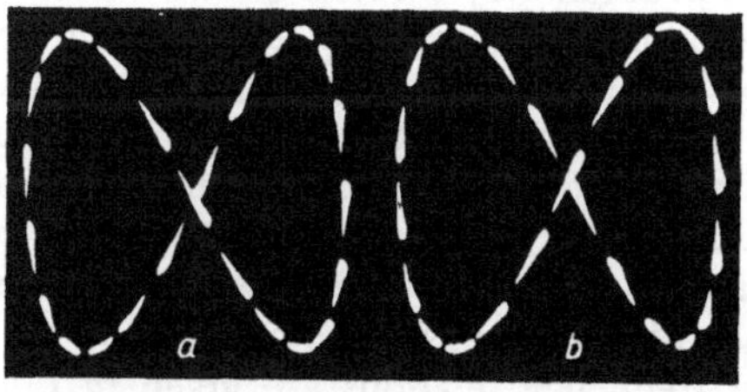

Fig. 12-21 Non-linear brilliance modulation for clarifying ambiguous patterns. *a)* 0°; *b)* 180°

Of course, at higher frequency ratios too, the oscillograms pass through the patterns corresponding to the various phase differences. An example of this is the series of oscillograms in Fig. 12-22, in which the frequency ratio is 3 : 1 and the phase differences are given in steps from 45° to 360°.

The reader is referred for an analysis of such figures to the Bibliography [6]. Typical Lissajous figures for the ratios 1, 2, 3, 5 and 10 are shown in Fig. 12-23.

A satisfactory method of interpreting these patterns is, as has already been de-

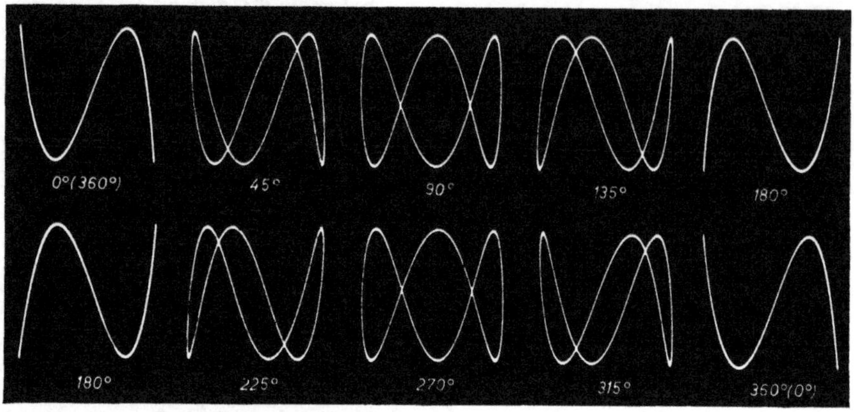

Fig. 12-22 Lissajous figures for 3:1 frequency ratio at different phase positions

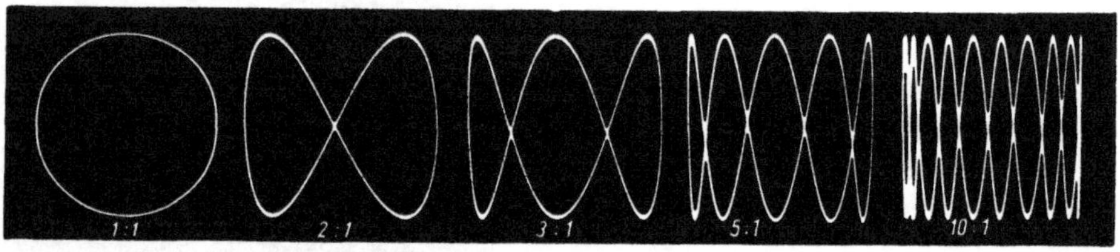

Fig. 12-23 Patterns for integral frequency ratios 1:1, 2:1, 3:1, 5:1 and 10:1

scribed, to think of the oscillogram as being surrounded by a rectangle whose sides are tangents to the pattern on the screen. The number of points of contact with the adjacent tangents gives the frequency ratio of the deflection voltage in a particular direction. For instance, in Fig. 12-24*a*, the ratio of the vertically deflected to the horizontally deflected voltage frequency is 3 : 1. But frequency ratios that are not even numbers can also be determined by Lissajous figures, provided these ratios correspond to rational fractions. The oscillograms in Figs. 12-24*b* and *c* show patterns of this sort for the frequency ratio 3 : 2 for two different phase relationships.

A number of oscillograms for non-integral frequency relations are given in Fig. 12-25. Oscillograms *a*), *b*), *c*) and *d*) show patterns with open loops representing frequency ratios of 5 : 2, 8 : 3, 10 : 3 and 7 : 2. Oscillograms *e*) and *f*) represent Lissajous figures for the same frequency ratios as in *a*) and *b*), but at instants when the loops were coinciding. In such cases the number of loops (*A*) on two adjacent tangents must be counted twice and the points of contact at the beginning and end of these figures (*B*) counted once.

The patterns in Figs. 12-19, 12-20, 12-21 and 12-22 are shown as "within" a square. The other ones, however, are shown as within a rectangle, corresponding to the frequency ratio. These patterns are easier to interpret, as may be seen by comparing the two oscillograms in Fig. 12-26. In all the oscillograms so far discussed the frequency of the vertical deflection was higher than that of the horizontal deflection. If the reverse

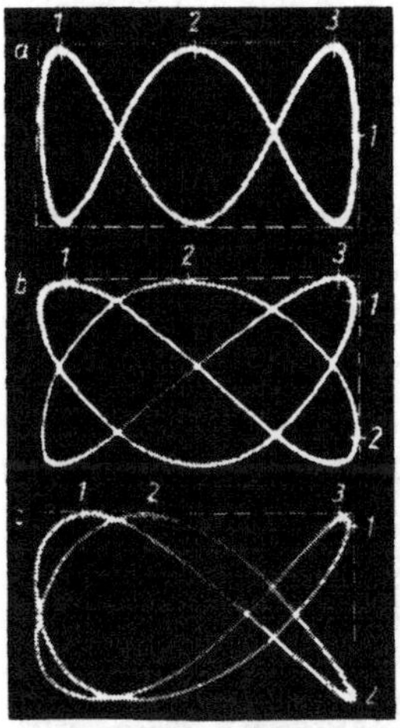

Fig. 12-24 Counting on Lissajous figures. *a*) 3:1, *b*) and *c*) 3:2

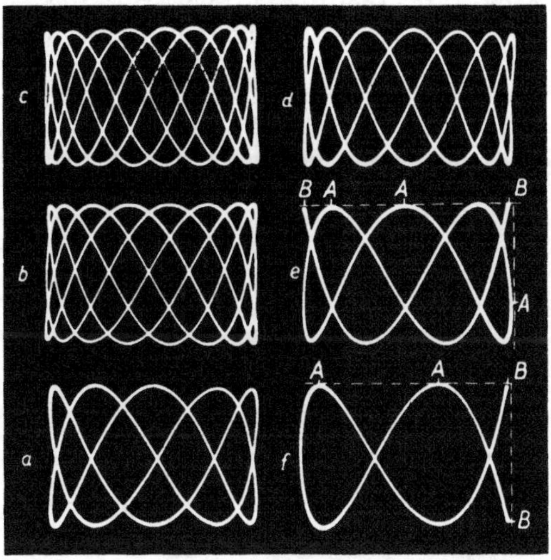

Fig. 12-25 Patterns for non-integral frequency ratios ($f_n = 50$ c/s) *a*) $5/2 = 2^1/_2$, $f_x = 125$ c/s; *b*) $8/3 = 2.2/3$, $f_x = 133.1/3$ c/s; *c*) $10/3 = 3.1/3$, $f_x = 166.2/3$ c/s; *d*) $7/2 = 3^1/_2$, $f_x = 175$ c/s; *e*) as *b*) 8/3 but different phase position; *f*) as *a*) 5/2 but different phase position

Fig. 12-26 Choice of the most suitable presentation of the oscillogram corresponding to the frequency ratio. *a*) unfavourable; *b*) correct (frequency ratio 5:2)

is the case, patterns will be obtained which, although they conform to the shape of the others, are turned by 90°. Fig. 12-27 shows a series of oscillograms for frequency ratios of 1 : 2, 1 : 3, 1 : 5 and 1 : 10.

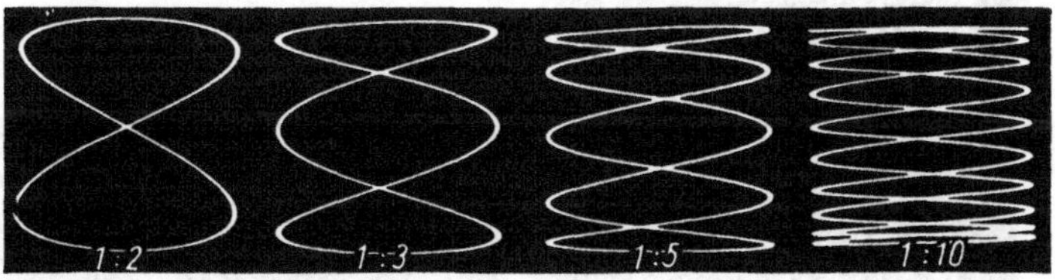

Fig. 12-27 Lissajous figures for frequency ratios < 1

12. 9 Lissajous figures with elliptical base line

With increasing differences in frequency, particularly when the frequency ratios are not whole numbers, it becomes more and more difficult to interpret the patterns obtained. The range of measurement can be considerably increased if the horizontal paths of the spot traced from left to right and back from right to left are separated by applying a part of the Y-deflection voltage to the X-plates but shifted in phase by 90°. The forward and return traces will not then move in a straight horizontal line but will describe an ellipse, so that the paths of the spot which would otherwise intersect each other will appear one above the other.

A suitable circuit for this type of measurement is given in Fig. 12-28. The low frequency comparison voltage, taken in this case as the reference frequency f_n, is used for the horizontal deflection. From this voltage, two components, each phase-shifted by 45°, are produced by means of the circuit components R_1, C_1 and R_2, C_2.

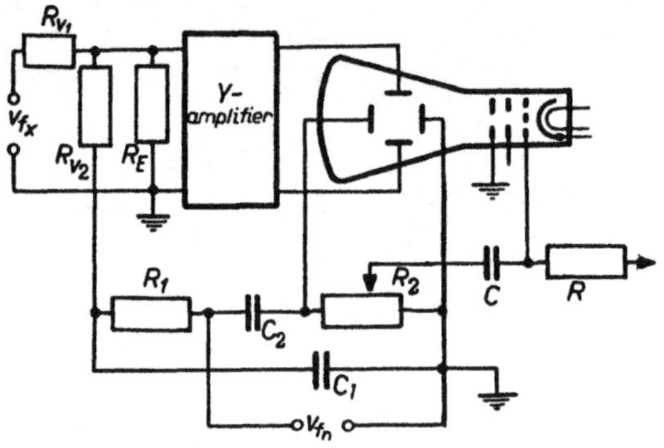

Fig. 12-28 Circuit for frequency comparison by Lissajous figures with elliptical basic trace

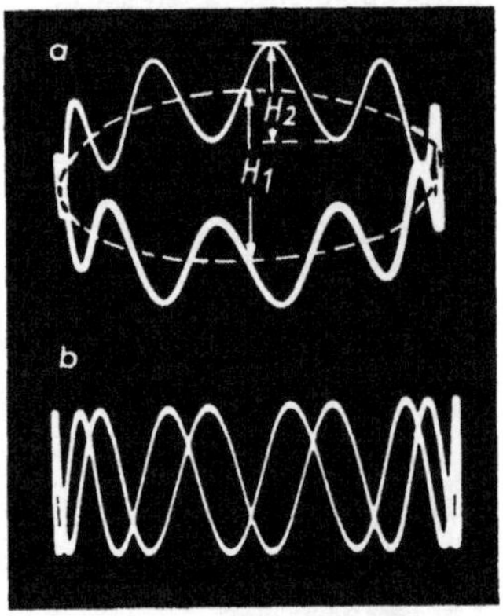

Fig. 12-29 Improvement of readability by using elliptical trace. *a*) Deflection on elliptical (or circular) trace at same frequency ratio as *b*), in this case only horizontal deflection for basic line has taken place

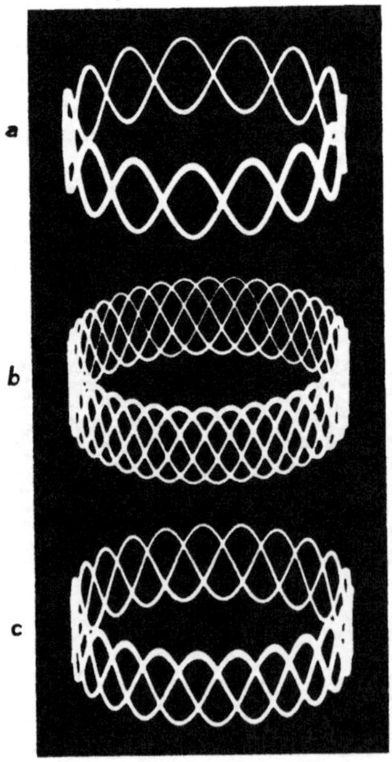

Fig. 12-30 Determination of non-integral frequency ratios with Lissajous figures on an elliptical trace (f_n = 50 c/s). *a*) 15/2 = 7 1/2, f_x = 375 c/s; *b*) 37/5 = 7 2/5, f_x = 370 c/s; *c*) 23/3 = 7.2/3, f_x = 383 1/3 c/s

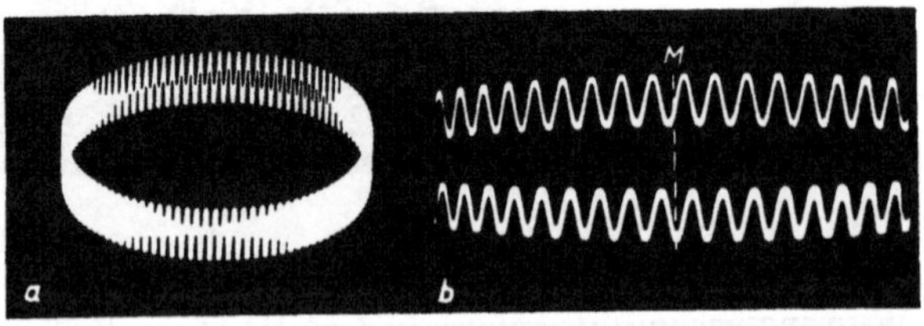

Fig. 12-31 Determining the frequency difference by increasing the horizontal deflection voltage

The voltage from R_2 is then applied to the X-plates, while the 90° phase-shifted component from C_1 is fed, together with the frequency to be measured, via resistors R_{v1} and R_{v2}, and the vertical amplifier to the Y-plates. By controlling the gain of the amplifier the minor axis of the reference trace (H_1 in Fig. 12-29), which now forms a horizontal ellipse, can be adjusted as required. The height of the curve traced by the voltage with the unknown frequency (H_2) is best adjusted by this voltage itself. The result can be seen in Fig. 12-29. With a frequency ratio of 10 : 1 the figure is far easier to interpret on an elliptical base line than in the conventional form.

In addition, as shown in the circuit in Fig. 12-28, a portion of the X-deflection voltage can be fed to the intensity modulation electrode of the cathode ray tube in such a phase that one half of the pattern will be brightened, as shown in Fig. 12-29*a*. This gives the oscillogram a certain "relief", so that the observer has the impression that he can distinguish between foreground and background, which makes for greater ease of interpretation.

The three oscillograms in Fig. 12-30 show how non-integral frequency ratios can be determined by this method.

If only the frequency difference is to be determined, and if the frequency ratio is large, good interpretation is possible by simply increasing the deflection voltage on the X-plates. The horizontal movement of the spot may then increase to several times the diameter of the screen.

This effect is clearly shown in Figs. 12-31*a* and *b*. At small frequency fluctuations the pattern shows some horizontal drift, the top and bottom parts moving in opposite directions. The duration of a cycle may be checked by a time mark, indicated at M in Fig. 12-31*b*, and a stopwatch, and the frequency difference determined.

At faster frequency changes, the number cycles passing within a given time can be counted in order to arrive at frequency difference. This method can be used for checking the frequency of broadcasting transmitters against a standard frequency, and is used for time measurements in general.

Particularly large frequency ratios can be bridged by means of an auxiliary frequency f_a. A beat pattern is then formed between f_n and f_a, in one of the ways previously described, and a second beat pattern between f_n and f_x. For this type of measurement two oscilloscopes can be used, or a twin-beam oscilloscope or an oscilloscope with an electronic switch. In this way it is possible to compare frequencies in a ratio of $F^2_{\max}$ ($F_{\max}$ being the maximum frequency ratio measurable by this method [7].)

Methods of frequency measurement employing radial deflections are ideal for circular screen surfaces. Anode voltage modulation cannot be used, however, because of the periodically recurring lack of definition, as can be seen from Fig. 12-12.

Owing to the bunching together at the ends of the Lissajous figure, however, there is an upper limit beyond which interpretation becomes uncertain. For instance, the inexperienced would find it rather difficult to count the voltage peaks at these positions in Figs. 12-30*b* and *c*.

Contrary to the often-expressed opinion that only polar co-ordinate cathode ray tubes are suitable for radial deflection (and only these can be used for this purpose), it should be pointed out that, using relatively simple circuits, it is possible to obtain "polar" oscillograms with standard cathode ray tubes having two pairs of plates for deflection in rectangular co-ordinates. Very good results are obtained from a simple circuit when representing cycloids.

12. 10 Frequency measurement with cycloids on a circular trace

The "function" of a quantity to be measured can also be represented in such a way that the length of a radial vector is plotted against its angle of rotation, so that with such "polar co-ordinates",

$$r = f(\varphi)\,. \tag{12.10}$$

These curves generally originate at a centre point with zero amplitude. There are many known circuits [8] with which oscillograms of this kind may be obtained using standard cathode ray tubes with two pairs of deflection plates. However, they generally require the use of special deflection amplifiers. On the other hand, very excellent results can be obtained by frequency measurements using cycloid traces, which can also be described as circular oscillograms. They require very simple circuitry, consisting only of resistors and capacitors, but nevertheless allow a wide range of measurements of whole-number frequency ratios as well as an exceptionally precise determination of non-integral frequency ratios.

The following paragraphs are based on a very comprehensive work by W. Bader [9], in which the geometric analysis of these patterns is also dealt with in detail.

12. 11 Circuits for frequency comparison with cycloids

The basic circuit for this type of measurement is shown in Fig. 12-32. Each of the two voltages, V_{f1} and V_{f2}, to be compared is applied to a resistor and a capacitor in series. The capacitor is so rated that for the frequency concerned the capacitive reactance $\frac{1}{\omega \cdot C}$ equal to the value of the resistance in series with it. Hence the voltages applied to the pairs of deflection plates are 90° out of phase with each other in every case. This means that each of these voltages alone will describe a circular trace on the screen with a circumferential velocity determined by its frequency and amplitude. If both voltages are applied simultaneously, the spot will describe a path representing the sum of two circles. Fig. 12-33 shows the vectors of the two voltages V_{f1} and V_{f2}. As the vector with the lower frequency, in this case V_{f1}, rotates, the vector of the voltage with the higher frequency is rotating around the peak of the first vector. The path of the spot corresponds to that of a point on the radius of a circle described by the vector of the voltage with the higher frequency, which, in turn, revolves around the circumference of the circle described by the lower frequency vector. The pattern

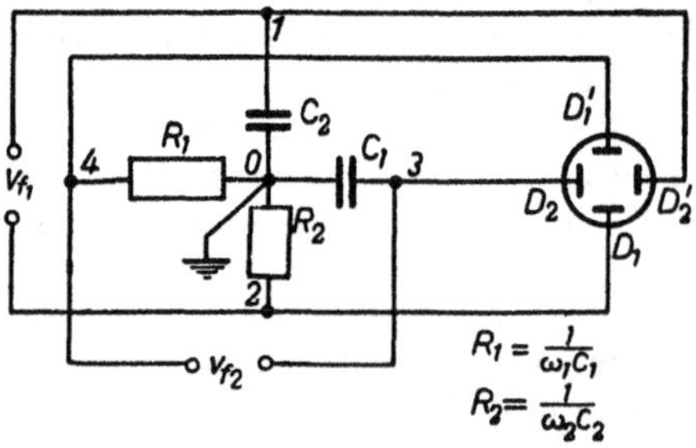

Fig. 12-32 Basic circuit for producing cycloid patterns

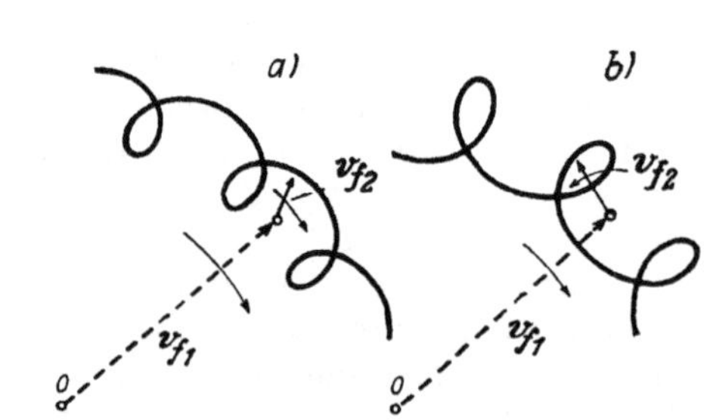

Fig. 12-33 Formation of a cycloid pattern by the addition of two rotating vectors. *a*) Epicycloid; *b*) Hypocycloid

on the screen is thus a basic circular path with loops or peaks. If the faster vector rotates in the same direction as the slower, then, in single line figures, one loop less will be described than corresponds to the frequency ratio (the point on the small circle runs counter to the large circle), and the peaks or loops point inward. Such curves are known as epicycloids (Fig. 12-33*a*). If the faster vector rotates counter to the slower one (the point on the smaller circle leads the larger circle) then, in single line figures, one additional loop is described during one rotation of the slower vector than corresponds to the frequency ratio. The loops or peaks point outwards. This type of curve is known as hypocycloid [82]).

In the circuit of Fig. 12-32 the *Y*-plates are connected to points 2 and 4 and thus receive the "ohmic" component of both voltages from the resistors R_1 and R_2. At the same time the *X*-plates receive from points 1 and 2 the 90° phase-shifted component of these voltages from the capacitors C_1 and C_2. If the *Y*-plates are connected across 1 and 4, they will then receive the "ohmic" component (R_1) of voltage V_{f2} and the capacitive component (C_2) from V_{f1}. The *X*-plates then, however, receive from 2 and 3 the capacitive component from voltage V_{f2} from C_1 and the non-phase-shifted component of V_{f1} at R_2. Provision for reversing polarity is made in the circuit used in carrying out these tests, as shown in Fig. 12-46. Here the voltages of the "ohmic" components can be switched over, thus allowing a choice of either hypercycloids or epicycloids.

12. 12 Interpreting cycloid patterns

The oscillograms in Fig. 12-34 illustrate the process described. The two initial circles, *A* and *B*, as well as the oscillographic sum *C* of these two patterns, are shown superimposed. (The ratios of the circles to each other in these two recordings were not equal.) The frequency ratio concerned is $f_x : f_n = 4$ [83]). Fig. 12-35 gives a series of cycloid patterns for whole-number frequency ratios of $f_x : f_n = 2, 3, 5, 10, 21, 30$ and 51.

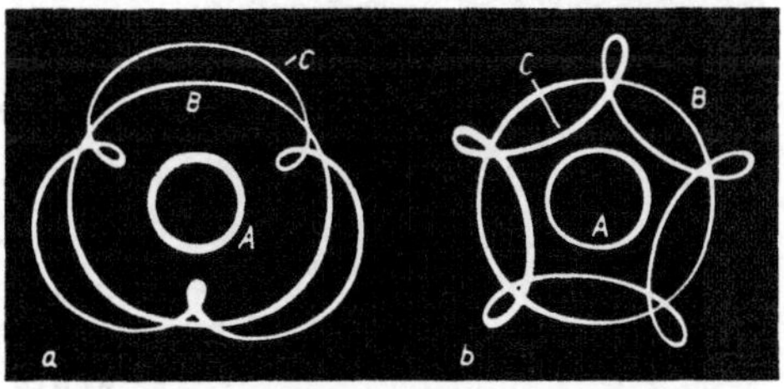

Fig. 12-34 Formation of the circular traces to cycloid patterns *a*) Epicycloid; *b*) Hypocycloid

If the known frequency is equal to the reference frequency or differs only slightly, then, in the epicycloid, a circle appears which "beats" with the frequency difference.

In Fig. 12-36*a* the two initial circles for this are shown — the signals to be measured were not equal — together with the background pattern.

The limits of this area are given by the sum and difference of the peak values of both

[82]) In American literature on the subject, such screen patterns are known as "roulette patterns".

[83]) In the following, the lower frequency will always be taken as f_n. For all oscillograms recorded it was 50 c/s. Assuming this, f_x will also be given from time to time to emphasize the possibilities offered for counting by this method.

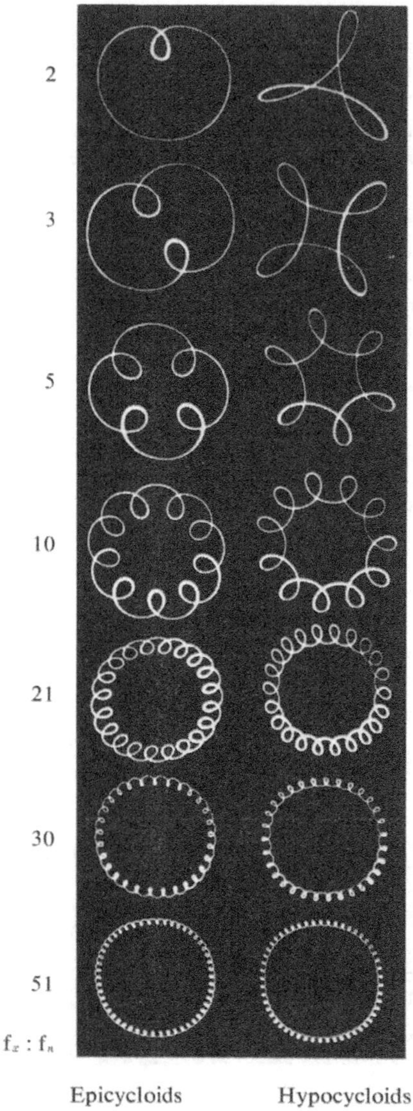

Fig. 12-35 Frequency comparison by cycloids; "single" figures

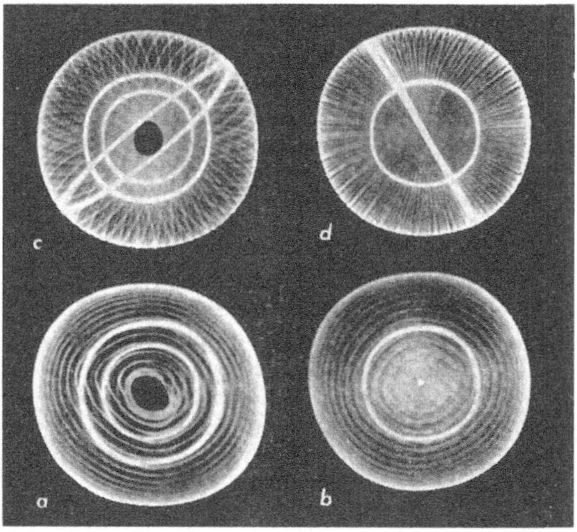

Fig. 12-36 Cycloid patterns at approximately equal frequencies.
a) Epicycloid, comparison voltages of different amplitudes;
b) epicycloid, comparison voltages equal in amplitude;
c) hypocycloid, comparison voltages different in amplitude;
d) hypocycloid, comparison voltages equal in amplitude

voltages. If the voltages are equal in amplitude, there will, of course, be only one initial circle, and the limits will be found at the centre or at twice the diameter, as shown in Fig. 12-36*b*.

In the case of hypocycloids for unequal amplitudes an ellipse will generally appear, which will rotate according to the difference frequency. If both voltages are equal, there will be a straight-line trace rotating in the same way. This trace can also be used as an indicator of the phase difference. The phase angle is equal to twice the rotation angle of the trace (Fig. 11-27).

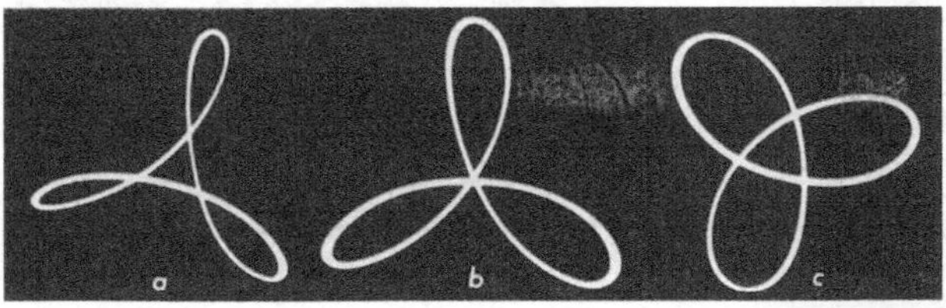

Fig. 12-37 Influence of amplitude of comparison voltage on hypocycloid trace

Fig. 12-38 Cycloid patterns showing possibilities of reading in the vicinity of frequency ratio 3

The patterns in Figs. 12-36*c* and *d* are composite photographs showing the initial circles, the rotating traces and the area covered by the traces in a hypocycloid circuit [10]. Such an arrangement can serve for checking phase in machines connected in parallel [11].

The advisability of having a smaller signal amplitude for the higher frequency than for the lower one can be seen from the oscillograms of Fig. 12-37. In oscillogram *a*), the higher frequency is of smaller amplitude than the lower frequency; in Fig. 12-37*b*, however, the signal amplitude are equal, while in *c*) the amplitudes of the signal with the higher frequency is greater than that of the signal with the lower frequency. Figures such as *b*) or *c*) could under certain conditions lead to confusion with the patterns of non-integral frequency ratios. The choice of voltage amplitude gives the pattern the form most favourable for correct interpretation. In general, loops appear on the screen, but if the amplitude of the voltage with the higher frequency is reduced, peaks are produced (Fig. 12-38), which are obviously easier to read. In some cases even figures with straight edges are pro-

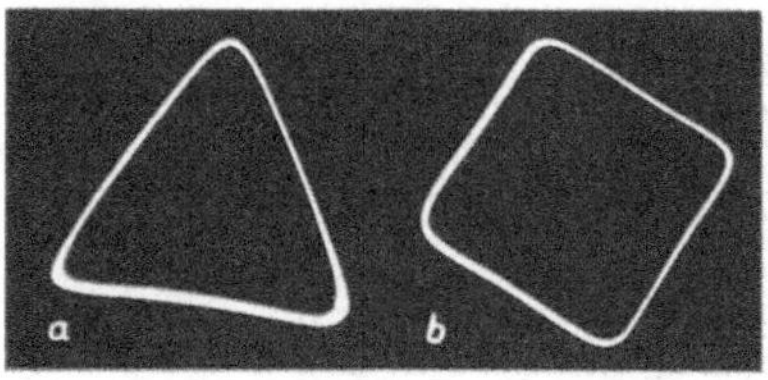

Fig. 12-39 Hypocycloid figures with straight sides obtained at certain voltage and frequency ratios

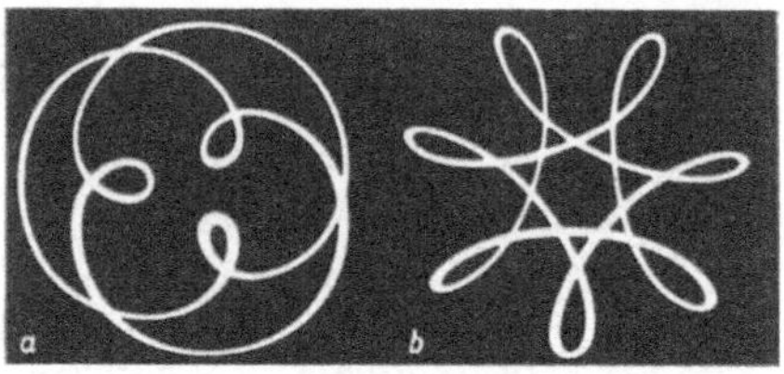

Fig. 12-40 Patterns with the frequency ratio of 5/2. *a*) epicycloid, *b*) hypocycloid

duced, as in the oscillograms of Figs. 12-39*a* and *b*. (These are hypocycloids with frequency ratios of 2 and 3.)

If the figures are not stationary, the frequency difference from that of the ratio which would correspond to the stationary pattern, can be determined by counting the number of peaks (backwards or forwards) passing a given point in unit time, as in Lissajous figures with an elliptical trace [9].

This process is particularly suitable for the determination of non-integral frequency ratios. Fig. 12-40 shows the oscillograms of the epicycloids and hypercycloids for the frequency ratio $m = 5 : 2 = 2^1/_2$. In general terms, the frequency ratio m_H or the frequency f_{xH}, when s is the number of loops or peaks and p indicates whether the figure is traced in a single or multiple form, for hypocycloids is determined from the equations:

$$m_H = \frac{s - p}{p} \tag{12.11}$$

and

$$f_{xH} = f_n \cdot \frac{s - p}{p} \,. \tag{12.12}$$

For epicycloids the equations are:

$$m_E = \frac{s + p}{p} \tag{12.13}$$

and

$$f_{xE} = f_n \cdot \frac{s + p}{p} \,. \tag{12.14}$$

	a	b	c	d	e
$f_x/f_n =$	$^8/_7 = 1^1/_7$	$^6/_5 = 1^1/_5$	$^4/_3 = 1^1/_3$	$^3/_2 = 1^1/_2$	$^5/_3 = 1^2/_3$
f_x[c/s] $=$	$57^1/_7$	60	$66^2/_3$	75	$83^1/_3$

Fig. 12-41 Straight-sided hypocycloids for non-integral frequency ratios

In this way it is possible, with a fixed standard frequency, to achieve exceedingly delicate gradations in the determination of unknown frequencies. Fig. 12-41 shows a number of oscillograms for frequency ratios between 1, 2 and 5 which were obtained in this way. It can be seen from patterns *a* and *b*, as well as from oscillograms of hypercycloids in Figs. 12-38*h* and *m*, that it may be difficult in the case of multipleline hypocycloids to determine the factor *p* (the number of figures outlined). Epicycloids are more suitable in this respect as can be seen from the series of oscillograms in Fig. 12-38. This applies particularly when the peaks are drawn towards the centre, as in Figs. 12-38*a* and *e*. For this reason epicycloids will generally be considered the more suitable [84]).

For frequency ratios in which $f_x : f_n < 1$, epicycloids are less suitable, as is illustrated in the two oscillograms in Fig. 12-42, whereas hypocycloids can be interpreted just as well as before. This is illustrated by seven oscillograms in Fig. 12-43. The asymmetry that can be seen at extremely low frequencies at the lower limit of the working range of the voltage source is due to relatively slight distortions of the waveform of the comparison voltage.

Some examples of unfavourable settings are shown in Fig. 12-44. The hypocycloid in oscillogram *a* had a non-integral frequency ratio, and both signals were of equal amplitude; since the inner loops rotate in the centre, interpretation is impossible.

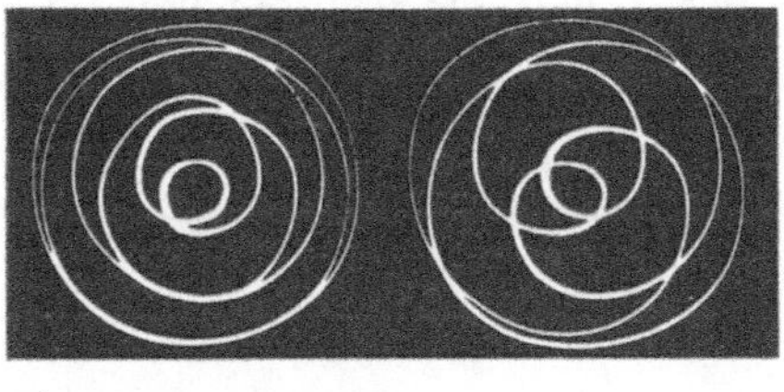

Fig. 12-42 Epicycloids at $f_x : f_n < 1$ (unfavourable)

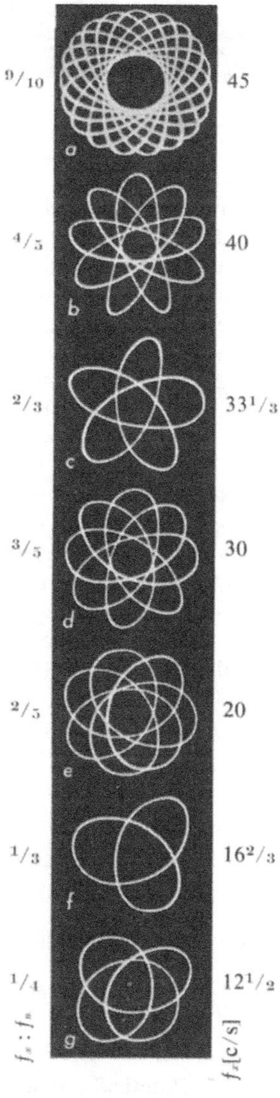

Fig. 12-43 Hypocycloids at $f_x : f_n < 1$ (easily interpreted)

[84]) To determine the factor *p*, it is best to start counting from the centre of the figure and move out along a radius cutting the points of intersection. (At these points the number to be counted is, of course, 2.)

Fig. 12-44 Unfavourable adjustment of voltage ratios

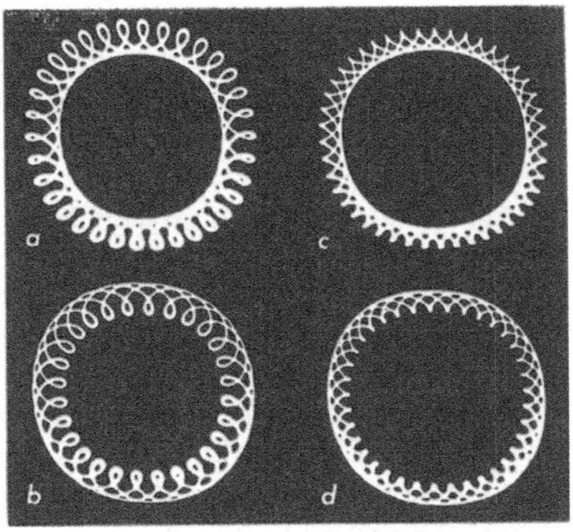

$m = 10^1/_3$ $f_x = 516^2/_3$ c/s $m = 10^1/_4$ $f_x = 512^1/_2$ c/s

Fig. 12-45 Oscillograms at a frequency ratio of 10 : 1 indicating finely graded readings

For the epicycloid pattern in Fig. 12-44*b* the amplitudes of the signals were also equal, so that the loops needed for counting are seen to overlap in the centre. The epicycloid in Fig. 12-44*c* can, in fact, be interpreted as $m = 25$; $p = 1$, $f_x = 1250$ c/s for $f_n = 50$ c/s, but could very easily be confused with a double pattern ($p = 2$). A smaller amplitude of the signal with the higher frequency is always preferable.

The oscillograms in Fig. 12-45 illustrate how fine steps of reading can be obtained at higher frequency ratios. The three-fold hypocycloid pattern in Fig. 12-45*a* contains 34 loops and the corresponding epicycloid (*b*)28. According to Eqs. (12.11), (12.12), (12.13) and (12.14) the frequency ratios are:

$$m_H = \frac{34 - 3}{3} = 10\ 1/3 \text{ or } m_E = \frac{28 + 3}{3} = 10\ 1/3,$$

and the unknown frequency is 516 2/3 c/s ($f_n = 50$ c/s).

In the fourfold hypocycloid oscillogram in Fig. 12-45*c* there are 45 peaks; the epicycloid in oscillogram *d* has 37. The frequency ratio is thus $m_H = \frac{45 - 4}{4} = 10^1/_4$ or $m_E = \frac{37 + 4}{4} = 10^1/_4$; the unknown frequency is $512^1/_2$ c/s.

12. 13 Practical circuit arrangement for cycloids

It has already been said that the reference frequency used in these examples was 50 c/s. Fig. 12-46 shows the circuit with the ratings of the components used, based on the circuit in Fig. 12-32. To filter out the mains harmonics, an *RC*-network of 10 kΩ and 1 μF is incorporated. The input voltage is applied via a transformer to a 50 kΩ potentiometer. This allows the signal amplitude to be adjusted. Two linear potentio-

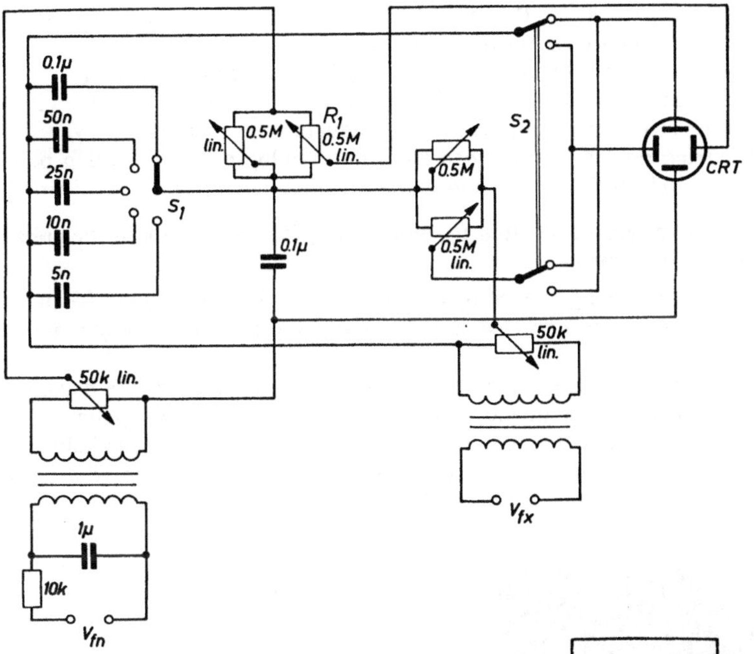

Fig. 12-46 Practical circuit for producing cycloid patterns

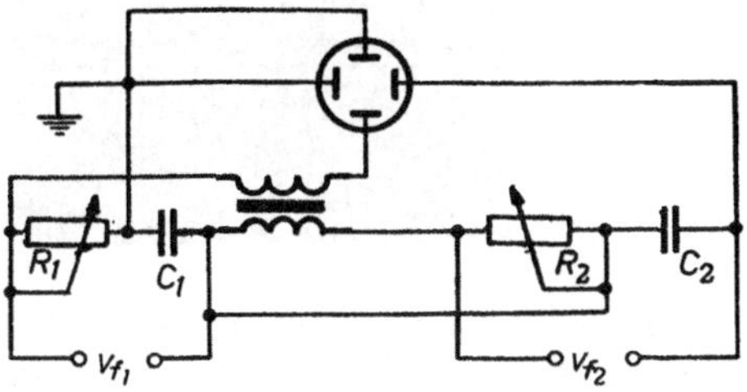

Fig. 12-47 Circuit for representing cycloid patterns with deflection plates earthed at one pole

meters, each of 0.5 M Ω, are connected in parallel as resistor networks in each branch of the comparison voltages. One potentiometer in each pair is connected as a variable resistor $R = \frac{1}{\omega C}$ to adjust the circular pattern. Since a reference frequency of only 50 c/s is used, one capacitor of 0.1 μF is used for C_1. If the whole audio frequency range is to be covered, it must be possible to select various values for C_1 in the same way as C_2. Two switches, not shown on the circuit, can be used to switch off the signal alternately as well as to connect directly the appropriate pair of deflection plates. In three switching operations it is thus possible to adjust the two circle individually one after the other and then switch over both signals to produce the sum pattern. The switchable capacitors in the branch for V_{fx} make it possible to fulfil the condition $R_2 = \frac{1}{\omega_2 \cdot C_2}$ over the whole audio frequency range. Change-over switch S_2 offers a choice between epicycloids and hypocycloids. Fig. 12-47 illustrates another circuit in which asymmetrically operated deflection plates (or amplifiers) are used. It is identical with the circuit described by Reich in 1937 [12] [13]. Rangachari had already drawn attention to this type of circuit in 1928 [14] [15].

Frequency comparison with cycloid patterns has been repeatedly shown to offer a number of advantages over other methods. The construction of the simple circuit required is therefore strongly recommended if frequency measurements over wide ranges are often carried out. As can be seen from the oscillograms, calibration points up to 50 times the standard frequency and fractions of it can be read off distinctly.

12. 14 Frequency measurement by intensity-modulating the oscillogram of the voltage with the unknown frequency

Brightening the trace, by using the comparison voltage with the higher frequency to intensity-modulate the beam, can also serve in many ways for frequency measurement.

The most obvious method is to use intensity modulation to brighten the waveform of the voltage of the unknown frequency. To achieve this the signal is applied in the usual way to the Y-plates and the pattern adjusted on a time base of suitable frequency. The comparison frequency voltage is then applied between the grid of the cathode ray tube and the oscilloscope chassis (Fig. 10-1).

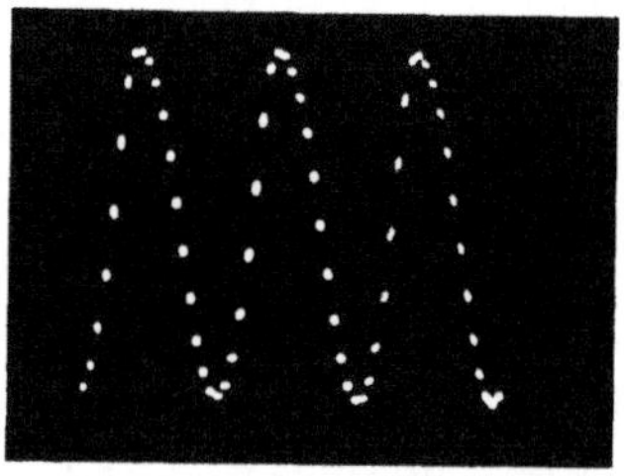

Fig. 12-48 Brilliancy-modulated oscillogram of the voltage of unknown frequency

If the comparison frequency is adjustable (e.g., by an AF generator), almost stationary points can be obtained, which can be counted, as illustrated in the oscillogram in Fig. 12-48. For exact measurements, particularly at large frequency differences, photographic recording is required. From the number of points in the oscillogram of the voltage with the unknown frequency, this frequency is calculated:

$$f_x = \frac{N_{fx}}{N_p} \cdot f_n, \qquad (12.15)$$

in which N_{fx} is the number of cycles of the unknown frequency displayed and N_p the number of brightness markings. In Fig. 12-48, $f_n = 100$ c/s, $N_{fx} = 3$ and $N_p = 60$, which, according to Eq. (12.15), gives $f_x = 5$ c/s. From the example it is clear that this method is very suitable for the measurement of fairly low frequencies. In the case under consideration, an ordinary AF generator, whose frequency range usually begins at 20—30 c/s, was used to measure a frequency of only 5 c/s.

12. 15 Intensity modulation of a circular trace with the second frequency

For this measurement the circuit illustrated in Fig. 12-49 is used. In the way previously described, a circular trace is produced by applying the voltage of unknown frequency, v_{fx}, to one pair of deflection plates, and the same voltage, displaced in phase by 90° to the other pair.

The voltage with the comparison frequency, v_{fn}, is applied between the grid of the cathode ray tube and chassis. Care should be taken that the time constant of the *CR*-coupling network for this latter voltage is sufficiently large to deal with low frequencies of V_{fn}.

If the waveform of V_{fx} is not sufficiently sinusoidal, the harmonics should first be filtered out. This need not be taken too far, however, as slight irregularity of the circular trace has no effect on the measurement. The circular trace now shows regularly spaced bright markings, remaining stationary if the frequency ratio is an exact integral. The number of bright markings is a direct indication of the frequency ratio itself. If the frequency ratio is not quite integral, the bright marks will move around the circle. Figs. 12-50*a* — *f* show typical sinusoidally intensity-modulated oscillograms for $f_n : f_x$ ratios of 1, 2, 3, 5, 10 and 103. If f_x is greater than f_n, the inputs V_{fx} and Vf_n must, of course, be changed over. The ratio $f_x : f_n$ is then found directly from the number of bright markings. Thus one standard frequency is sufficient to cover wide frequency ratio ranges (at least 1 : 10 and 10 : 1, i.e. a total of 1 : 100). In addition to integral frequency ratios, intermediate values can also be seen from the oscillograms of Fig. 12-51 obtained with a ratio in the region 4 : 1.

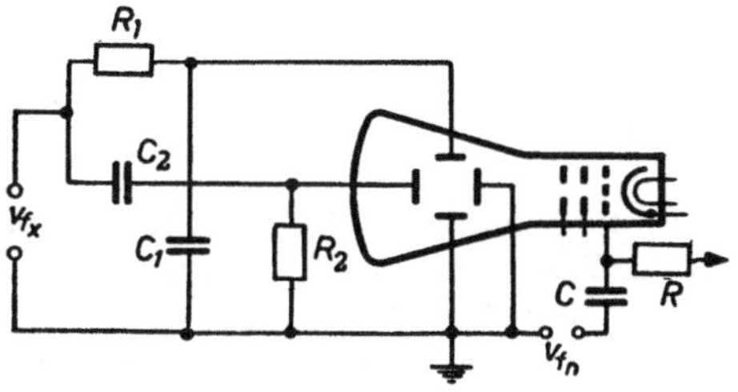

Fig. 12-49 Circuit for brilliancy modulating a circular trace by a voltage with known comparison frequency

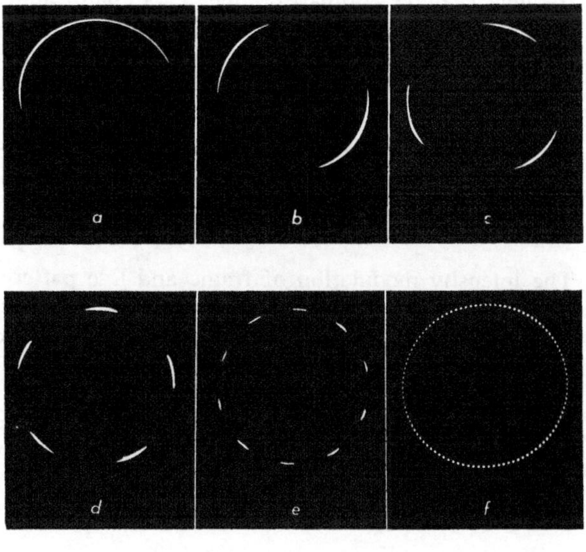

Fig. 12-50 Oscillograms of frequency measurement by brilliancy-modulation of a circular trace

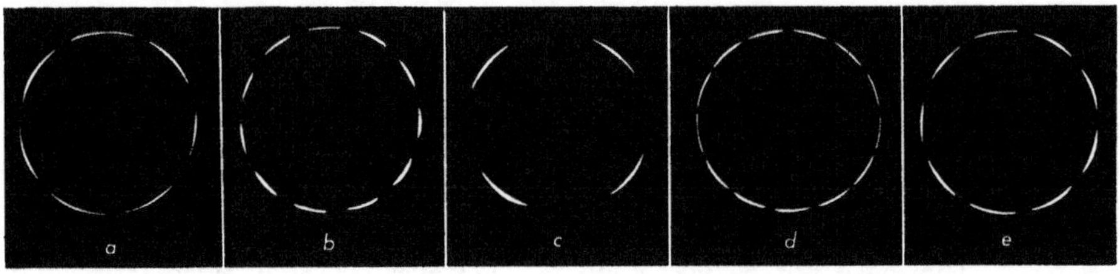

Fig. 12-51 Intensity-modulated circular traces for non-integral frequency ratios (except *c* which corresponds to $f_x : f_n = 4$)

TABLE 12-3 FREQUENCY RATIOS FOR THE OSCILLOGRAMS OF FIG. 12.51

Pattern	Frequency ratios $f_x : f_n$	f_x at $f_n = 50$ c/s
a	7/2 = 3.500	175 c/s
b	11/3 = 3.667	183 1/3 c/s
c	4 = 4.000	200 c/s
d	13/3 = 4.333	216 2/3 c/s
e	9/2 = 4.500	225 c/s

The table shows that the number of bright markings on the circle is the numerator of the fraction which indicates the frequency ratio. The denominator, however, cannot be found from the number of loops or suchlike, as in frequency measurement by comparison with the time base frequency or by means of Lissajous figures. However, if, as usually done, calibration is carried out by progressively varying one of the frequencies under comparison, the denominator number can be found exactly by means of a simple experiment. There will always be only one whole number which will produce a possible frequency ratio. These patterns differ from those for integral frequency ratios by the small spacing between the bright parts of the picture. The spacing diminishes the more the frequency ratio deviates from a whole-numbered ratio, so that patterns are ultimately obtained in which the circle is completely filled in. Such patterns, cannot, of course, be interpreted.

12. 16 Intensity modulation of a line pattern

The intensity modulation of frame and line patterns (similar to television scanning) with the voltage of unknown frequency makes possible frequency comparison over a very wide range. Two sawtooth generators are required for this. The appropriate circuit is shown in Fig. 12-52. The sawtooth oscillator for the vertical (frame frequency) deflection frequency (low) is synchronized with the horizontal (line) deflection frequency. The oscillator for this (which generates a frequency which is a multiple of the vertical frequency) is synchronized with a standard frequency which can generally be up to a ten times the horizontal frequency.

If the unknown frequency is equal to or an integral multiple of the vertical fre-

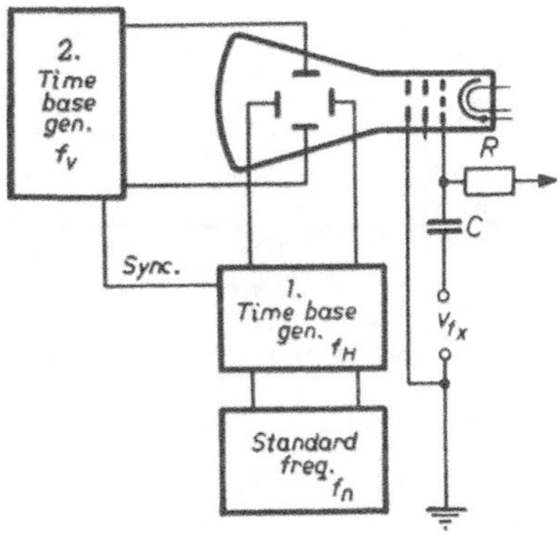

Fig. 12-52 Circuit for frequency measurement by brightness modulation of line patterns

quency, then one or correspondingly more bright patches will appear during one pattern in a horizontal direction, as shown in the series of oscillograms of Fig. 12-53. If the unknown frequency is, however, equal to or an integral multiple of the horizontal frequency, then one or a number of patches of brightness will appear during the trace of one line — vertical — as can be seen in Fig. 12.54.

If, in the oscillogram illustrated, the vertical frequency f_V is equal to 100 c/s and 60 lines are traced, that is, $f_H = 6000$ c/s, the patterns shown in Figs. 12-53*a*—*e* correspond to 100, 200, 300, 500 and 1000 c/s and those in Figs. 12-54*a*—*e* to 6, 12, 18, 30 and 60 kc/s respectively.

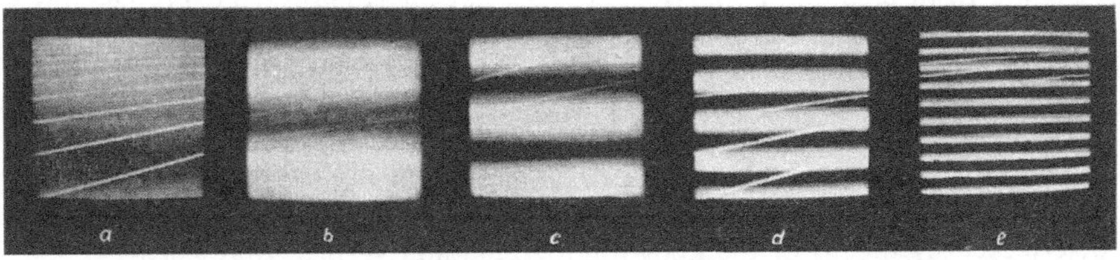

Fig. 12-53 Oscillograms at multiples of vertical frequency

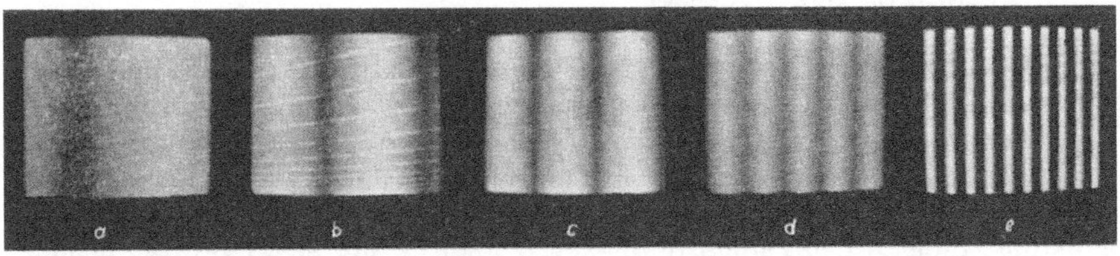

Fig. 12-54 Oscillograms at multiples of horizontal frequency

It is interesting to note that in this way two separate frequency ranges can be determinated with one standard frequency. There is a gap between the two, which can be selected within wide limits by changing the ratio $f_H : f_V$. This method should thus be eminently suitable for adjustment at the frequency limits of a system under examination (amplifier of suchlike). The frequencies under investigation can thus be read off quickly and clearly in fixed, pre-determined steps.

If the unknown frequency is not in an exact integral relationship with the horizontal or vertical frequency, the bright patches will no longer appear exactly above or beside

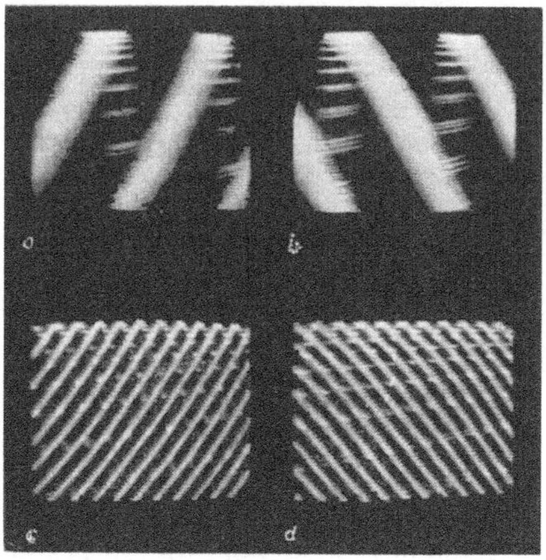

Fig. 12-55 Oscillograms of intensity-modulated line patterns when the unknown frequency is somewhat less or more than a multiple of the horizontal frequency
a) and *b*) unknown frequency lies in the vicinity of twice the horizontal frequency; *c*) and *d*) unknown frequency eleven times the horizontal frequency

each other, but will drift or, at slight deviations, will appear at a slant. This can be seen in the oscillograms of Fig. 12-55, where *a* and *b* represent ratios of $f_H : f_V$ deviating slightly from 2, and in *c* and *d* where they deviate slightly from 11. The unknown frequency in each case was somewhat lower or higher than the indicated frequency ratio. So long as the frequency ratio differs from the comparison frequency by less than one complete cycle per second, there is an additional phase difference. The slant of the pattern markings can then serve as a direct measure of the phase.

This section, in addition to giving a description of a method of frequency measurement, also shows how an interfering frequency in a television picture and its cause can be determined by means of the known horizontal and vertical frequencies.

12. 17 Absolute frequency measurement with rotating pointer

12.17.1 METHOD OF MEASUREMENT

All the frequency measuring methods described so far are based on the comparison of the unknown frequency of an alternating voltage with the known frequency of an independent voltage source. Other methods have been described, however, by which direct indication of the unknown frequency is obtained.

According to a circuit patented by E.T. Jaynes, described by H. Dorf [16], the alternating voltage of unknown frequency is applied, via phase shifting elements, to both pairs of deflection plates of the cathode ray tube, in such a way that the phase difference of these voltages is 180°. This can be done with *RC*- and *CR*-networks in a way similar to that already described in these pages for a circular trace of the spot with two alternating voltages of identical frequency and displaced with respect to one another by 90° (Figs. 11-25, 12-11 and 12-28).

12.17.2 CIRCUIT

Fig. 12-56 shows the circuit and examples of component ratings. The components are so rated that each phase-shift element produces a phase shift of 45° for 200 c/s (cut-off frequency). As two such elements are provided in each branch, the voltage arriving

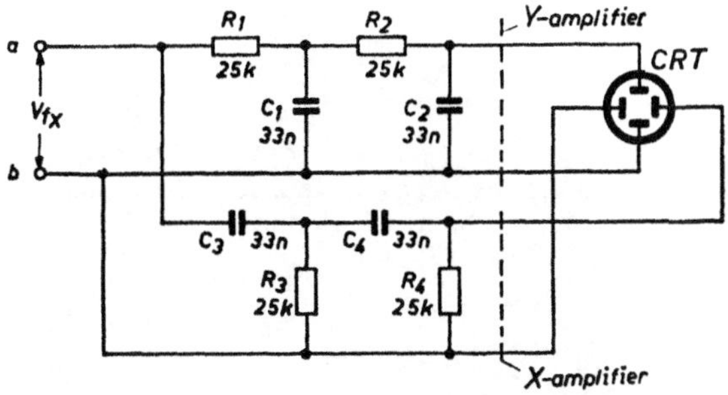

Fig. 12-56 Circuit for absolute frequency measurement with rotating pointer

at the imput of the Y-amplifier is displaced by —90° and has rather less than half the amplitude of the input voltage. The voltage at the input of the X-amplifier is attenuated by the same amount but has a phase shift of + 90°. Since the phase shift elements are directly connected, some slight corrections due to reaction are required to obtain the desired phase difference of 180°. The only correction needed in the circuit shown in Fig. 12-56 was to increase the capacitor C_2 to 36 nF.

In general, the voltage with unknown frequency must be amplified both vertically and horizontally to obtain a sufficiently large image on the screen. The "GM 5654" oscilloscope was therefore used for the measurements about to be described, as this permits of amplification in both directions. As the input impedance for the X-deflection in this oscilloscope is 50 kΩ, R_4 was likewise made 50 kΩ, so that the two in parallel gave the specified value of 25 kΩ at the end of this filter.

12.17.3 THE OSCILLOGRAMS AND HOW THEY ARE EVALUATED

By means of this circuit a diagonal trace is obtained on the screen of the cathode ray tube from top left to bottom right (see Fig. 11-15, bottom oscillogram, for 180°). At the cut off frequency of the circuit components and with equal deflection amplitude in both directions, this line will slope at an angle of 45°. Input frequency changes give rise to changes in phase and amplitude of the output voltage of the filters. Circuit elements R_1, C_1 and R_2, C_2 form a low-pass filter; its output voltage decreases with increasing frequency (Fig. 5-22*b*). Capacitors C_3, C_4 and resistors R_3, R_4, on the other hand, form a high-pass filter whose output voltage increases with increasing frequency (Fig. 5-9). The phase difference remains constant at 180° (Figs. 5-10 and 5-23). The result is always a line which rotates in accordance with the amplitude changes as described in Fig. 11-27 for phase measurement with a rotating trace. The end points of the trace do not describe a circle but a hyperbolic curve, as can be seen in Fig. 12-57. To illustrate this process as clearly as possible, the oscillograms were photographed with a small aperture (f = 11) and with the frequency varied as steadily as possible from 25 c/s to 2500 c/s during an exposure time of 4 secs. The deflection amplifiers were so adjusted that the almost vertical 25 c/s trace was of the same length as the almost horizontal 2500 c/s trace. The trace inclined at 45° corresponds to 250 c/s at the gains selected.

It should be emphasized that an absolute frequency indication is obtained in this way, its accuracy depending only on the accuracy and constancy of the circuit elements — apart from the constancy of amplification. Thus, after the frequency range has once been calibrated, no voltage source with a reference frequency is required. The reading accuracy can be improved if the pivotal point of the trace is

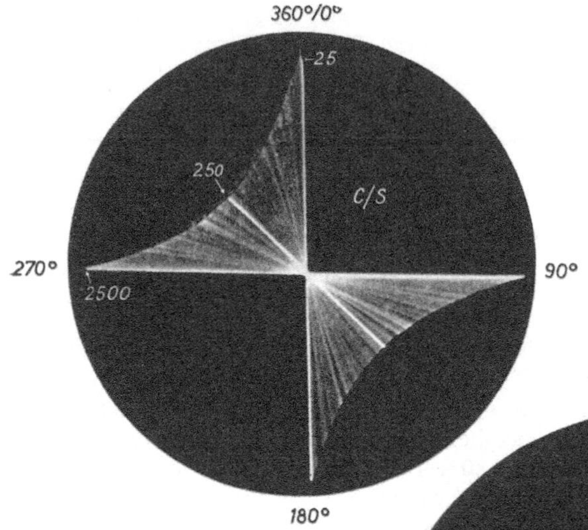

Fig. 12-57 Oscillogram for absolute frequency measurement using the circuit in Fig. 12-56 without special measures

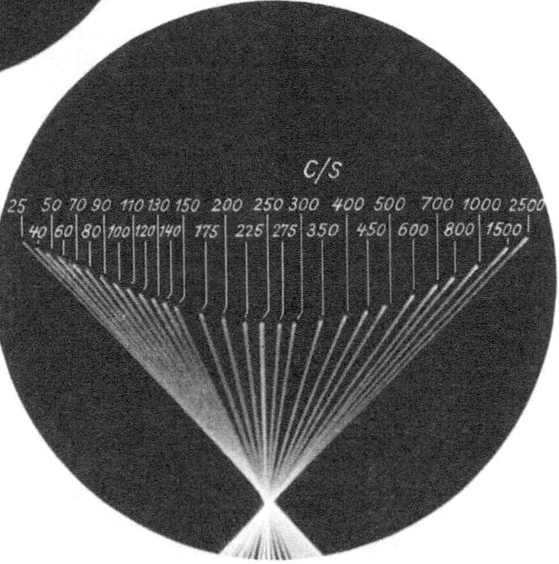

Fig. 12-58 Pivotal point of rotating trace displaced to edge of CRT; tube turned 45°

moved to the edge of the screen. The reading is further made easier if the cathode ray tube is rotated by 45° to the right.

Such a pattern is shown in Fig. 12-58. The trace positions of 29 different frequency settings are shown.

Using the circuit as in Fig. 12-56, Fig. 12-57 shows a frequency scale moving from top centre to left bottom. By reversing the connections of the X-plates a frequency scale which increases in the usual way from left to right can, of course, also be obtained.

Fig. 12-58 and all subsequent oscillograms in this section are simply enlargements of the negatives reversed.

12.17.4 CHOICE OF MEASURING RANGES

With a relatively large frequency range as in Fig. 12-58, the scale obtained leads to crowding of the range at the ends. The accuracy of reading is greater in the middle

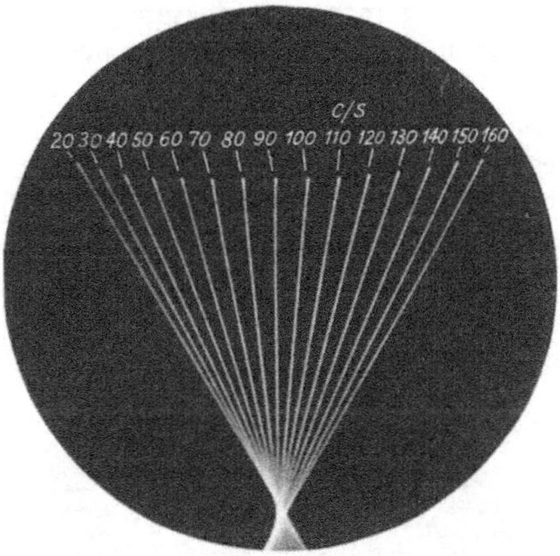

Fig. 12-59 Almost linear frequency reading

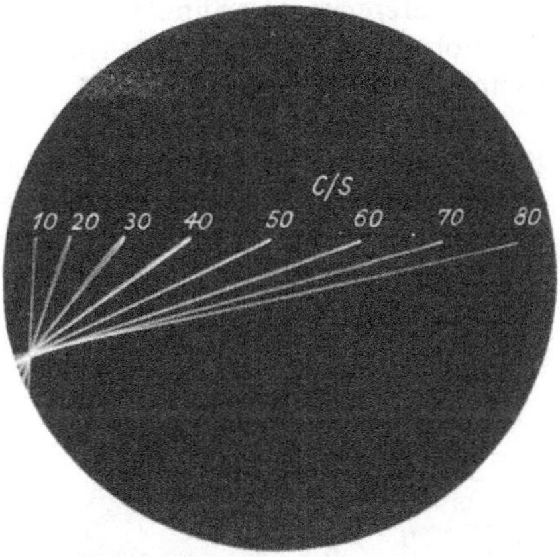

Fig. 12-60 Accuracy of reading increases with frequency

(about 2 %). By adjusting the gain, the range can be made more open, and the accuracy of the reading correspondingly improved. Fig. 12-59, which was also taken with the circuit shown in Fig. 12-56, illustrates an almost linear scale in the range from 20 to 160 c/s with an accuracy of reading of about 1 %. Fig. 12-60 is another variant, giving a frequency range from 10 to 80 c/s, in which the accuracy of reading (about 1 %) increases with frequency.

12.17.5 SIMULTANEOUS MEASUREMENT OF SEVERAL FREQUENCIES

It is also possible in the same way to determine simultaneously the frequencies of two or more alternating voltages. If two alternating voltages with different frequencies and equal amplitudes are applied to the input sockets *a, b* for V_{fx}, the image obtained on the screen is a slanting Lissajous figure as shown in Fig. 12-61 [85]). (The voltage with the higher frequency serves for the horizontal deflection in this example.) The inclinations of the sides of this pattern are measures of the two frequencies, just as was the case with the rotating trace having a single frequency. The amplitudes of these frequency components must, of course, be taken into account. It is advisable to adjust them to equal size.

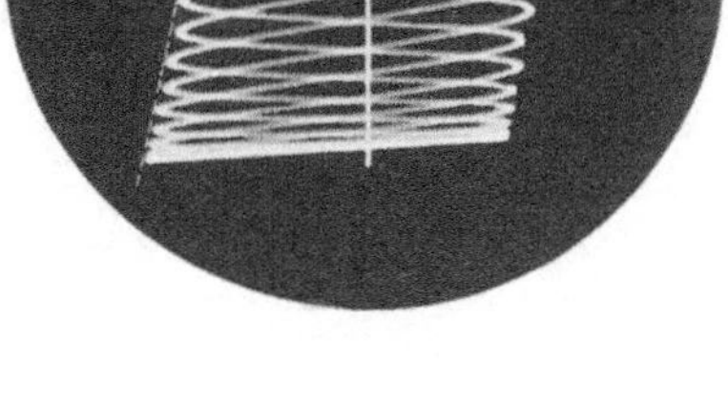

Fig. 12-61 Simultaneous determination of two frequencies: $f_1 = 60$ c/s, $f_2 = 650$ c/s

---

[85]) Two or more voltages can be added very simply by means of a resistance *T* network, as described in Fig. 4-87 on p. 149 and in Fig. 12-13 on p. 339

The reference co-ordinates in these oscillograms were recorded separately to facilitate observation. For this purpose the voltages on the pair of X- and Y-plates were switched off alternately.

As a rule the frequencies do not stand in a whole-numbered or rational relationship to one another. In that case Lissajous figures with recognizable contours are not obtained on the screen but instead a straight-sided luminous area, as illustrated in the oscillogram in Fig. 12-62. Three alternating voltages of different frequencies produce a hexagon as in Fig. 12-63. Here too, the inclinations of the sides are measures of the corresponding frequencies.

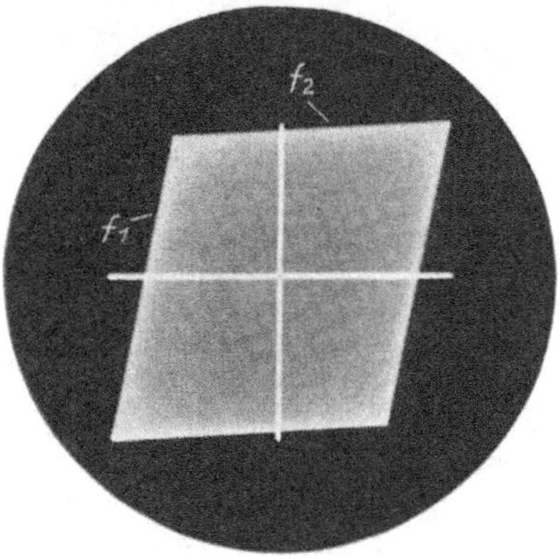

Fig. 12-62 As in Fig. 12-61 but frequencies not in an exactly rational relation to one another

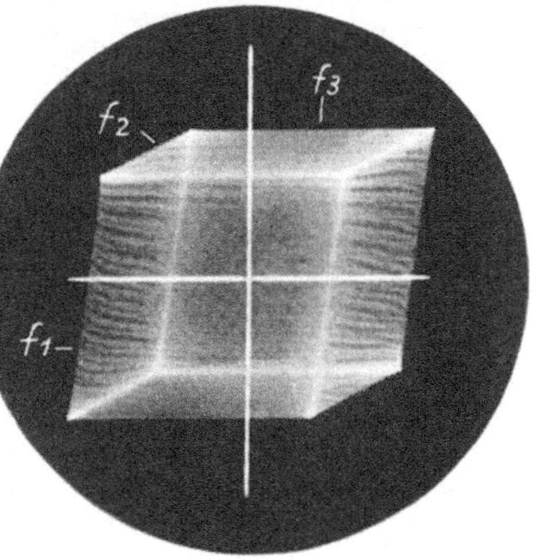

Fig. 12-63 Simultaneous determination of three frequencies: $f_1 = 50$ c/s, $f_2 = 150$ c/s, $f_3 = 1000$ c/s

12.17.6 SPECIAL ADVANTAGES AND APPLICATIONS OF THIS METHOD

Since both frequency and amplitude may be read off from the oscillogram, this process, which has been dealt with only very briefly here, will no doubt be of great advantage in frequency analysis.

As has been pointed out at the beginning of this chapter, all the methods of frequency measurement described can, with the aid of suitable transducers, be used for measuring mechanical vibrations, speeds of revolution and other non-electrical variables.

The absolute indication it affords makes the method described above especially suitable for observing mechanical vibrations or the speed of a rotary engine.

CHAPTER 13

REPRESENTATION OF THE RISING FLANK OF PULSE-SHAPED SIGNAL VOLTAGES WITH OSCILLOSCOPES WITHOUT DELAYING ELEMENTS IN THE VERTICAL AMPLIFIER

Elsewhere in this book attention has been drawn to the significance of the measuring process whereby the distortion of rectangular pulses in a four-terminal network makes it possible to evaluate the characteristic properties of the latter (Ch. 11.13 "The use of square waves for assessing the properties of electrical transmission systems"). In all such investigations it is desirable to be able to observe the behaviour of the quadripole after a sudden voltage rise and its building-up behaviour with the greatest possible accuracy.

In the simplest type of such investigations symmetrical square voltages are used. However, short input pulses of low repetition frequency can also be used. In all such cases it is important to observe the rising flank of the voltage surge over as extended a period of time as possible. Every time base generator requires a certain repeating value of the triggering voltage to release its sawtooth voltage. As this value only reaches its circuit elements after an interval determined by the circuit time constants, this circuit must have a certain response time. Before the time deflection begins, the input voltage has already attained a certain value. The oscillogram in Fig. 13-1 is of a somewhat distorted pulse 10 μs in breadth in "self-triggering" (with the voltage applied to the *Y*-plates), and it is seen that because of this response time, the rising flank, which is of prime interest in this case, remains invisible. A complete picture can be displayed only when the time deflection can be initiated before the vertical deflection of the spot is started by the input signal. This, however, means that the input signal must be delayed by some means [86]).

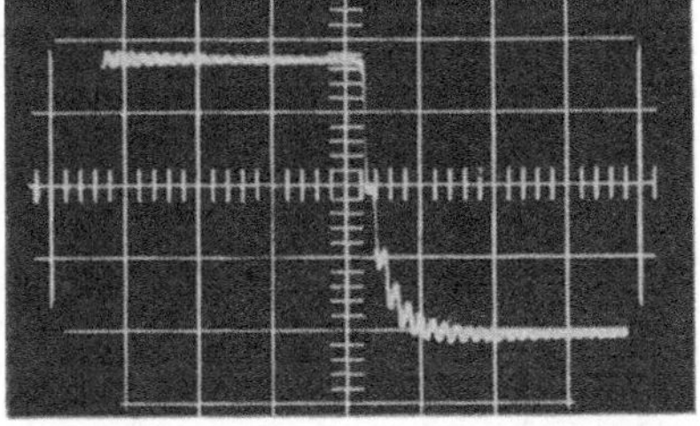

Fig. 13-1 Distorted rectangular pulse of 10 μs duration, internally triggered

[86]) With signal pulses which occur cyclically it is also possible, by means of suitable circuit arrangements, to use the previous signal pulse to advance the setting in of the time deflection by a previously adjusted period of time, so as to be able to study the subsequent pulse. By this means it is possible to observe not only the initial flank of the triggering pulse, but also any desired portion of the subsequent pulses greatly expanded in time. A simple example of this method is the Philips time base expansion unit "GM 4584" (Part I, Ch. 4 p. 140 and 149).

13. 1 Delaying elements in the amplification system

The required time delay can be achieved in such a way that the time deflection is directly released by the input voltage, but delaying circuit elements must then be inserted in the path of the input voltage to the Y-plates. Only then, after being suitably delayed, can the input voltage proceed to carry out the vertical deflection. This can be achieved by means of so-called delaying cables or a delay network of π links consisting of inductances and capacitances (see "Delay lines" Part I, Ch. 5; p. 227).

At the same time, these delaying networks must not recognizably change the waveform of the input voltage. Satisfactory solutions involve considerable cost for multi-link delay lines or expensive delay cable, and even then a drop in the upper cut-off frequency can never be completely avoided. If, moreover, delay cable of average quality is used, a noticeable debasement of the quality of the amplifier characteristics will have to be accepted. This method of delaying is always needed for observing non-recurrent or irregular phenomena.

For most types of measurement, in which each has its own input voltage source, which can be controlled as required, a different process, described below, is more suitable in many respects. An auxiliary alternating voltage controls the time deflection directly while the input voltage pulses are released after having been delayed by suitable means.

13. 2 Delayed release of the input voltage pulses

Any signal voltage generator which can operate with external control can be used for this. Such generators as for instance the Philips "GM 2314" pulse and square voltage generator, usually comprise a sinusoidal voltage generator which directly controls the other circuit of the generator in the case of the "GM 2314", a Schmitt-trigger, a multivibrator and an output stage.

As, however, the rectangular pulse voltage can be released by means of an external generator, it is possible to trigger the time base unit of the oscilloscope directly by means of this generator, while the release of the input pulses can occur later via phase-delaying circuit elements. (Filters or delay line cables can also be used for time delaying [1].) This is explained more fully by reference to the block diagram of Fig. 13-2 in which a 10 μs pulse is released by including an RC-network with a variable capacitor. With a 20 kc/s controlling voltage from a sine-voltage generator (Philips

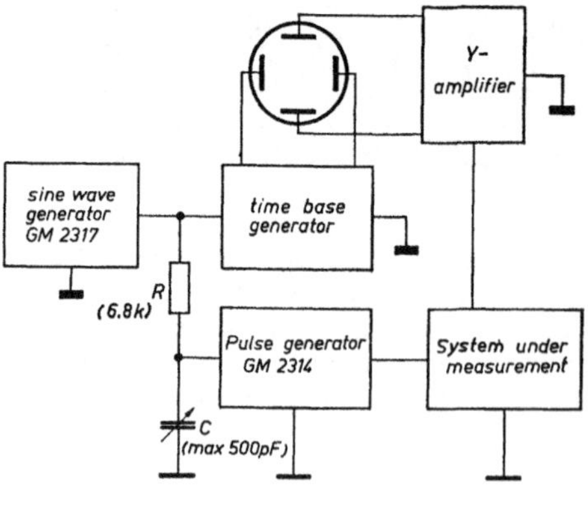

Fig. 13-2 Delayed triggering of signal pulses

Fig. 13-3 Adjusting picture position by varying delayed triggering of signal pulse; capacitor C in Fig. 13-2 was varyingly adjusted

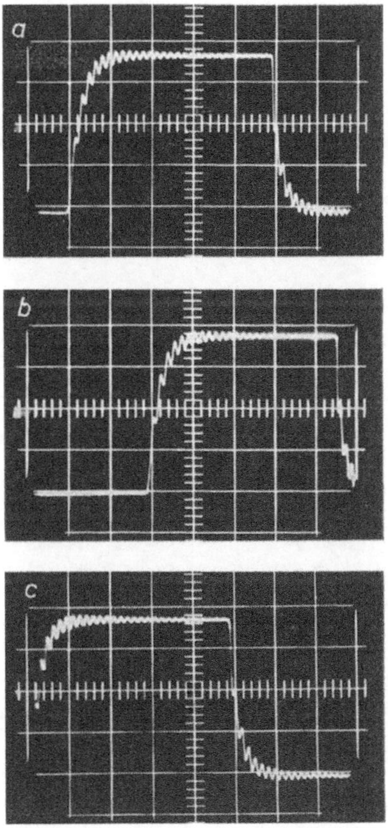

"GM 2317") a phase delay of 37° (45° corresponding to the cut-off frequency) corresponds to a time delay of about 5 μs. As is confirmed in the pulse oscillograms in Fig. 13-3, this time suffices to permit the pattern to be adjusted horizontally to any required position on the screen. For greater time shifts a lower frequency of the auxiliary alternating voltage and correspondingly higher rated circuit elements for the phase shifts would be required.

This process makes possible time-expanded observation not only of the leading flank of pulses with a rapid voltage rise, but also of the leading flank of low frequency pulses using medium priced oscilloscopes having no delay elements in the amplifier, so that the required delays cannot be obtained in the amplifier circuit. As has already been emphasized, this method can be used only for the observation of cyclic phenomena. The leading flank of irregular or non-recurrent phenomena can be shown only by oscilloscopes having circuit elements in the vertical amplifier for delaying the transit of the input voltage.

Part III

PRACTICAL EXAMPLES

CHAPTER 14

RECORDING THE WAVEFORMS OF LUMINOUS FLUX, CURRENT AND VOLTAGE OF FLUORESCENT LAMPS

14. 1 General

It is particularly important to know the nature and extent of the variations of the luminous intensity of fluorescent lamps during one cycle of the mains voltage (stroboscopic effect). As the oscilloscope is actually a voltage indicator, a photocell for supplying a voltage proportional to the luminous intensity or to the variation of luminous intensity has to be used to observe the behaviour of light. The basic circuit for this type of investigation is shown in Fig. 14-1.

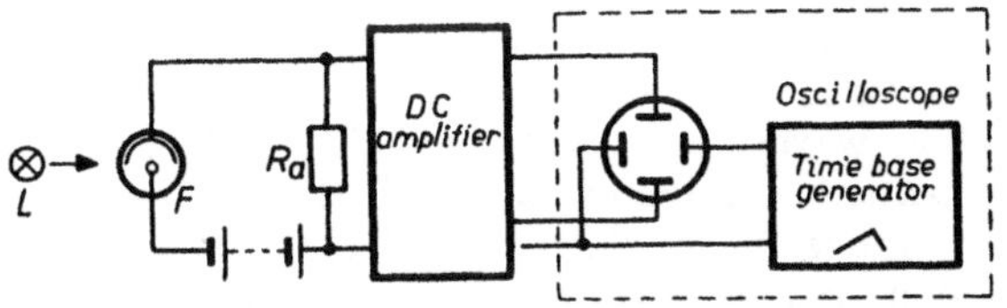

Fig. 14-1 Arrangement for displaying the luminous flux of sources of illumination on the oscilloscope

The distance between the light source L and the photocell F must be so adjusted that the maximum permissible intensity of radiation for the photocell is not exceeded. This only applies in practice to incandescent lamps, however, as the radiation intensity of fluorescent lamps is relatively low.

For the purposes of the investigation, it is essential to know the relationship between "direct light" and "alternating light". Care should therefore be taken that the direct voltage components from the photocell are also displayed. As there was no DC oscilloscope available for the recordings dealt with here, the built-in amplifier was connected to an external DC amplifier. A deflection coefficient of 10 mV/cm is quite sufficient for this purpose. Such a coefficient is now available in the Philips "PM 3206" oscilloscope.

14. 2 Incandescent and fluorescent lamps

To provide a practical scale of comparison for investigations on fluorescent lamps examples are first shown of the patterns thus obtained of luminous flux variations of 40 W and 200 W incandescent lamps during one cycle of the 50 c/s mains (Fig. 14-2). In these, as in the subsequent patterns, the controls were adjusted to give approximately the same amplitude for the average brightness in all the oscillograms. In this way a clear picture is obtained of the relative proportion of flux

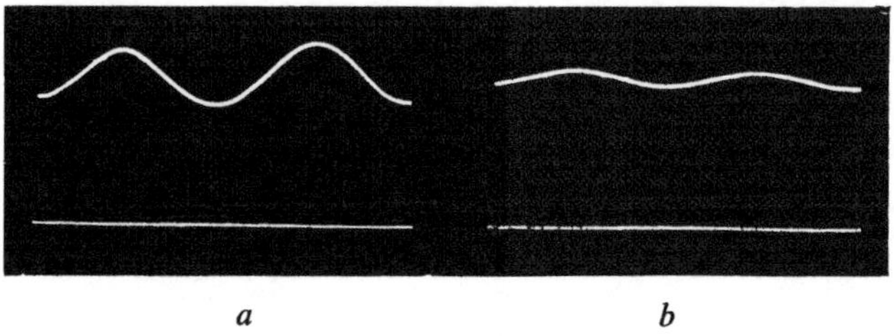

Fig. 14-2 Luminous flux of incandescent lamps during one cycle of the mains voltage
a) 40 W/220 V lamp;
b) 200 W/220 V lamp

variation in the light from various sources. These oscillograms cannot, however, be a measure of the absolute values of the luminous flux.

In all cases two cycles of the mains voltage are shown, extended over the whole width of the screen. For simple and clear comparison, those parts of the oscillogram extending beyond one cycle have been blanked out. The luminous intensity of the 40 W lamp fluctuates around the mean value by about ± 19% and that of the 200 W lamp by about ± $5^1/_2$%. Fig. 14-3 first shows the waveform of the luminous flux of the mercury vapour light of a lamp without fluorescent material. The luminous flux fluctuates by about ± 90% around the average value. The light thus follows the course of the current almost perfectly but, of course, with twice the frequency. Fig. 14-3*b* shows the luminous flux waveform of a "TL" "Daylight" lamp. The fluctuation of the luminous flux is about ± 52%.

The luminous flux waveform of the "white" "TL" lamp can be seen in Fig. 14-3*c*. Here the fluctuation amounts to ± 33%. Fig. 14-3*d* shows the waveform of the so-called "warmtone" lamp which has a ripple of ± 27%.

It is immediately clear, among other things, how the fluctuation of the luminous flux can be influenced by the fluorescent material. Development is in constant

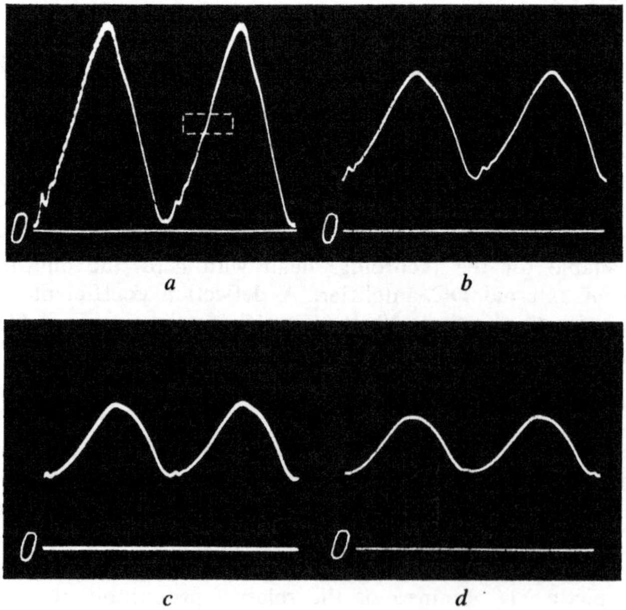

Fig. 14-3 Luminous flux of fluorescent lamps
a) Lamp without fluorescent material
b) "Daylight" lamp
c) "White" lamp
d) "Warmtone" lamp

progress in this field and considerable advances may be expected [87]. A further interesting detail emerges from the oscillogram of the fluctuation of the luminous flux of the mercury vapour light in Fig. 14-3*a*. Using time base expansion and photographic enlargement of a section of this waveform, it can be seen from Fig. 14-4 that the whole oscillogram contains a small, superimposed low-frequency flux variation of about 22,000 c/s. This is due to certain processes taking place during gas discharge [1] [2].

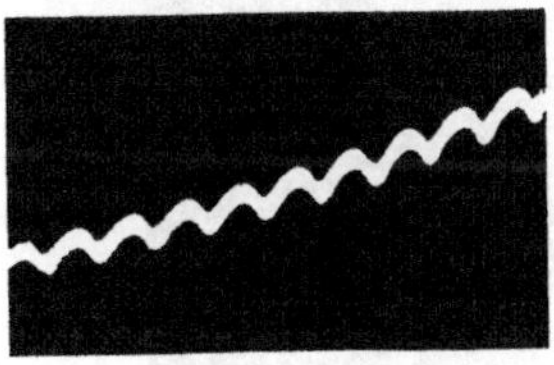

Fig. 14-4 Time-expanded and photographically enlarged section of waveform from Fig. 14-3*a*

14. 3 Current and voltage waveforms of fluorescent lamps

As is known, fluorescent lamps are usually employed with a ballast choke. This gives rise to an undesirable phase difference between current and voltage, so that the power factor cos φ in this arrangement is only from 0.5 to 0.6.

Fig. 14-5 shows a circuit, with which, according to the position of switch *S*, waveform patterns of the current, the voltage accross the lamp and the mains voltage can be produced on the oscilloscope one after another. To obtain the current waveform, a resistor R_v of about 10Ω is connected in the main circuit. The voltage drop across this resistor is applied to the vertical amplifier and amplified sufficiently to produce the required deflection.

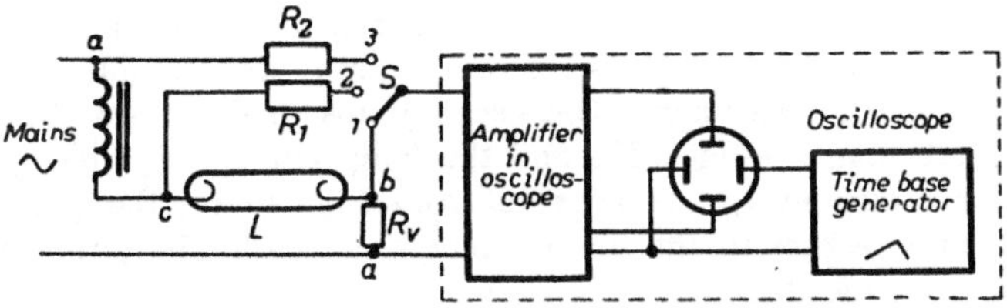

Fig. 14-5 Arrangement for displaying the current and voltage curves of fluorescent lamps

In the second position of the switch the lamp voltage appears across resistor R_1 at the input of the amplifier, and in position 3 the mains voltage appears across the input of the amplifier via R_2. The resistors should be chosen such that the desired deflection is obtained on the screen without having to adjust the gain control further. A resistor of 2 MΩ is appropriate if, as here, the input impedance of the amplifier is 150 kΩ. If R_1 and R_2 are equal, the oscilloscope gives a direct impression of the relationship between mains voltage and lamp voltage. If the time

---

[87]) These oscillograms were recorded some considerable time ago.

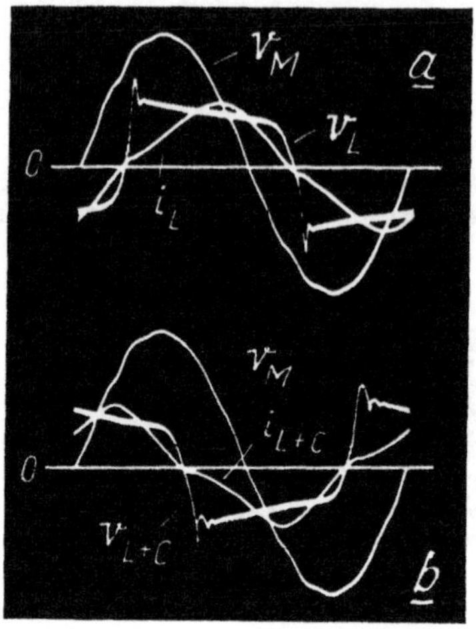

Fig. 14-6 Current and voltage waveforms
a) Lamp with inductive ballast
b) lamp with capacitive ballast

base frequency is locked in synchronism with the mains voltage, the waveforms of lamp current, lamp voltage and mains voltage appear on the screen in such a way that their displacement with respect to each other corresponds to their actual phase relationships. (In most oscilloscopes a special switching position is provided for investigations of this kind.) The result of this measurement is shown in Fig. 14-6*a*.

14. 4 Fluorescent lamps connected in duo

Attempts were soon made to improve the poor power factor of fluorescent lamps and to reduce the variations of luminous flux and thus the stroboscopic effect. The "duo" arrangement of fluorescent lamps brought about a considerable improvement [3]. One lamp functions in the normal circuit with a series-connected choke. The current and voltage relationships for this arrangement have already been discussed with reference to Fig. 14-6*a*. Lamp current and voltage lag behind the mains voltage.

A combination of capacitor and choke precedes the second lamp. This combination is so rated that the current is determined by the capacitance and thus leads the mains voltage. These relationships can be seen in the oscillogram in Fig. 14-6*b*. (The mains voltage waveform shows considerable harmonics.) Fig. 14-7 shows the due arrangement. The oscillograms in Fig. 14-8 show the relationship between the mains voltage V_M and the lamp currents i_L and i_{L+C} individually and of the total current i_T. As appears from Fig. 14-8*c*, the phase shift of the total current with respect to the mains voltage is small. The power factor for this arrangement is given as 0.95.

This is not the only advantage offered by the duo arrangement, however. The

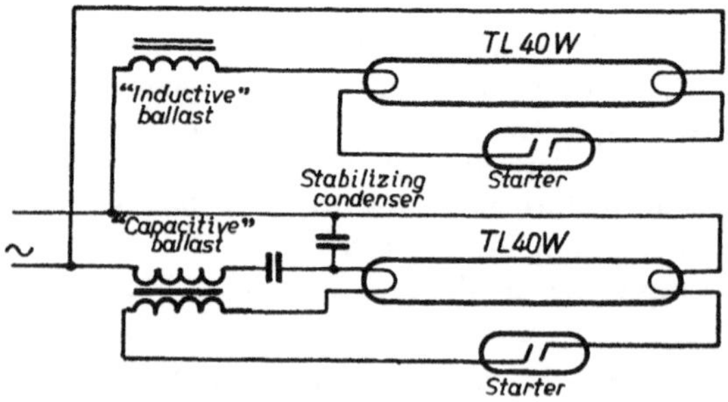

Fig. 14-7 "Duo" arrangement of fluorescent lamps

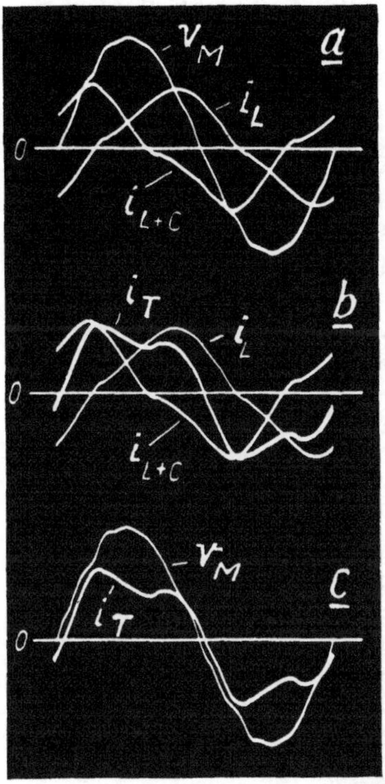

Fig. 14-8 Current and voltage waveforms of fluorescent lamps in "duo"
a) Mains voltage V_M and currents i_L and i_{L+C} of lamps individually
b) currents of individual lamps and sum current i_T
c) waveforms of mains voltage and total current

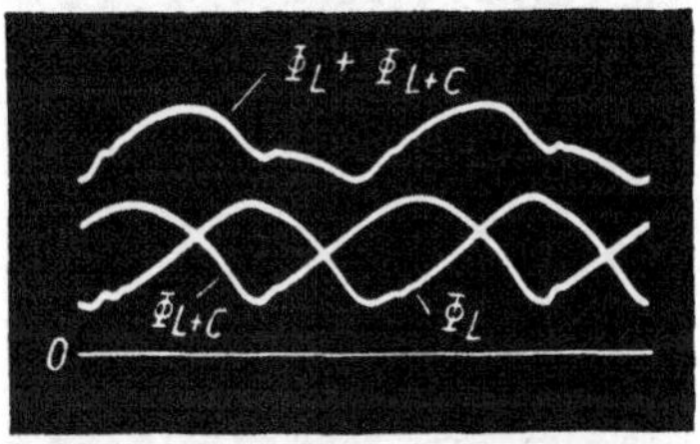

Fig. 14-9 Luminous flux of two "daylight" lamps, individually Φ_L and Φ_{L+C}, and their sum $\Phi_L + \Phi_{L+C}$ (duo arrangement)

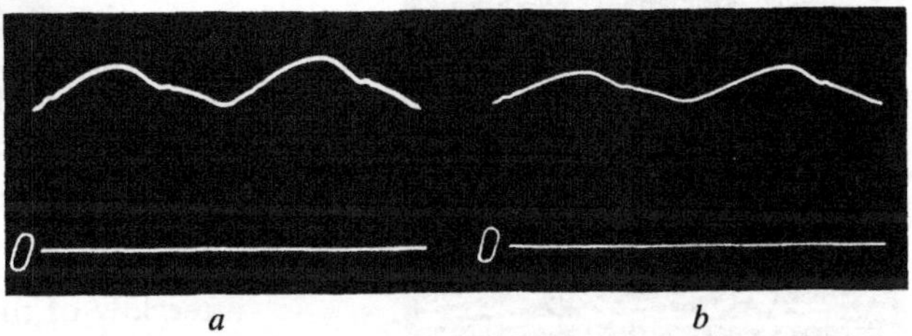

Fig. 14-10 Luminous flux of fluorescent lamps in duo; *a*) "white"; *b*) "warmtone"

mutually phase-shifted currents of both lamps also result in a corresponding phase shift of the variations of luminous flux. Thus, when the flux of one lamp declines, the flux of the other correspondingly increases, and vice versa. These results are shown in the oscillograms of Fig. 14-9 for "daylight" lamps. The luminous flux of each lamp is shown individually — Φ_L and Φ_{L+C} — and also the sum of the two luminous fluxes $\Phi_L + \Phi_{L+C}$. The brightness variation of the combined light of the two lamps now only amounts to $\pm$ 18%; it is thus no more than that for a 40 W incandescent lamp, as opposed to $\pm$ 52% in a single fluorescent lamp. Fig. 14-10 shows the oscillograms representing the waveform of the luminous flux in the duo arrangement for two "white" lamps (*a*) and two "warmtone" lamps (*b*). The flux variations were found to be $\pm$ 16% and $\pm$ 11% respectively. These variations are even less than those of a 40 W incandescent lamp. The oscillograms also show that a power factor of 0.95 is not difficult of achievement with a duo arrangement [4].

14. 5 Lamp current and luminous flux waveforms of electronically controlled lamps

For many purposes it is often necessary to control the brightness even of fluorescent lamps. As these lamps have a relatively high ignition potential and a practically constant working voltage, it is impossible to exert control by voltage variation as with incandescent lamps. The problem can be solved by connecting an antiparallel arrangement of two thyratrons in series with the fluorescent lamp and its ballast. By changing the phase of the grid voltage responsible for ignition, the current is made to flow only during an adjustable portion of each half cycle. In this way the average current of the lamp can be reduced considerably without extinguishing the lamp [5] [6].

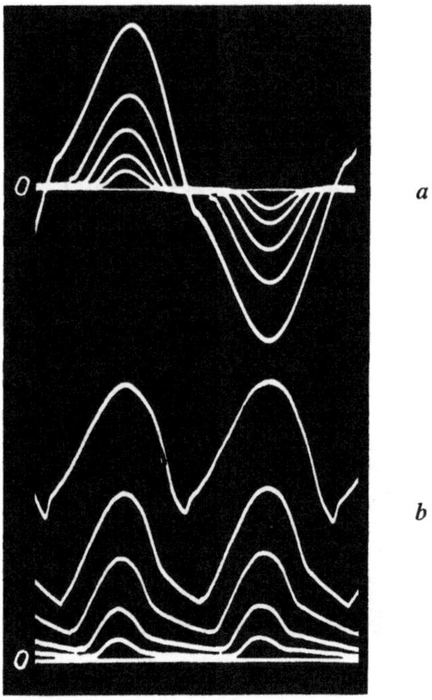

Fig. 14-11 Current and luminous flux waveforms for fluorescent lamps, with electronic control in five different stages of adjustment

The current waveforms of a lamp so controlled can be seen in the oscillograms of Fig. 14-11*a*. The settings were 1200, 600, 300, 150 and 75 mA (measured with a standard multi-range meter calibrated in rms values). They were recorded in the way already described, with the time base synchronized in fixed phase with the mains frequency. In this way the same reference phase position was maintained for all composite photographs.

The corresponding oscillograms of the luminous flux are shown in Fig. 14-11*b*. Since the reduction of the current means a shorter time of current flow, it might be expected that the stroboscopic effect would be increased. Experience shows, however, that the light fluctuates just as little at reduced brightness as at normal working current. The explanation for this strange behaviour can be found in the flux waveforms in Fig. 14-11*b*. It can be clearly seen that the afterglow of the fluorescent material, which at normal current (top curve) is hardly noticeable, represents a progressively increasing part of the total light at decreasing current.

The foregoing serves to show that the oscilloscope can offer a thorough insight not only into the behaviour of the lamps themselves, but also into the entire field of thyratron control.

CHAPTER 15

SWITCHING PHENOMENA WITH ELECTRIC LIGHT BULBS

What happens in electronic appliances, radio receivers, amplifiers and the like after they have been switched on is fairly well known. It is becoming more generally recognized that a cathode ray oscilloscope may be used to obtain all the required information. Less interest is shown, however, in what takes place when actually switching on and off, and yet one could cite many examples of occasions when exact knowledge of switching phenomena could lead either to greater operating safety or to greater economy. The next few oscillograms of the switching phenomena observed on electric light bulbs furnish convenient examples [88]).

Recordings were made of the course of the current through the lamp as well as of the luminous flux at the moment of switching on [89]). The basic circuit used is shown in Fig. 15-1. The lamp current is supplied, either from an accumulator or from the secondary of a mains transformer, via contact k_3 and resistor R_v. The value of this resistor must be small in relation to the cold resistance of the lamp in order not to mask the switching effects.

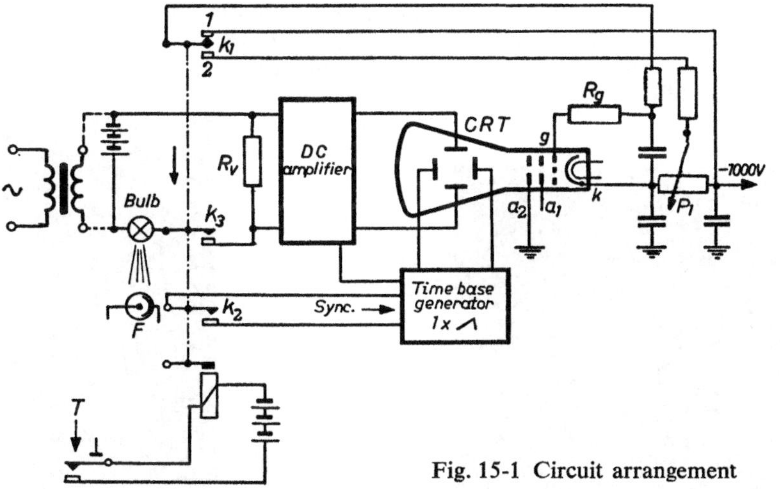

Fig. 15-1 Circuit arrangement

---

[88]) The results apply of course by analogy to incandescent lamps generally.

[89]) Naturally, both processes could be recorded simultaneously with a twin beam oscilloscope e.i. the Philips oscilloscope PM 3236. But as they are repeated in the same way more than once, it was possible to put the recordings together one by one later as the same time scale and equipment were used in each case.

The voltage drop across this resistor is then proportional to the current through the bulb. After amplification in the DC vertical amplifier, it provides the beam deflection voltage on the Y-plates of the cathode ray tube. For this measurement the time base generator must be adjusted for a single sweep.

To avoid premature exposure of the film and to ensure the correct sequence of the recording, the required switching processes are carried out by means of a relay with several contacts. The relay is switched on for recording by button T. The armature of the relay is then actuated and closes the contacts k_1, k_2 and k_3 in succession. This sequence is arranged by suitably adjusting the contacts. In the quiescent state no spot is seen on the screen as the control grid of the cathode ray tube is at full cut-off potential. When k_1 closes, the grid receives only a partial voltage from potentiometer P_1. The closure of k_2 triggers a single time base and finally k_3 switches on the current to the bulb. This contact should be a double one for preference, so as to avoid "chattering" and other disturbance when the relatively high peak current is switched on (see also Fig. 10-14).

In oscilloscopes with a "real", triggered single time sweep (for instance the Philips "PM 3236") these measures are much simplified, as then time base and brightening are firmly synchronized. The instructions for use issued with the apparatus give the necessary information.

For recording the behaviour of the luminous flux the light of the bulb is directed during the switching process on to a high-vacuum photocell. The voltage across its load resistance, which is proportional to the luminous flux, is fed to the Y-plates via a DC amplifier not shown in the circuit diagram. An alternating voltage of known frequency was recorded below the waveforms under investigation and provided a time scale. To provide a measure of the current also, a further recording was made, the bulb being replaced by a rheostat in series with an ammeter, and the current to be recorded was adjusted to 500 mA. Fig. 15-2 shows the oscillograms thus obtained in the first ten milliseconds with a bulb of 6.3 V/0.15 A. The time-marking frequency was 2500 c/s.

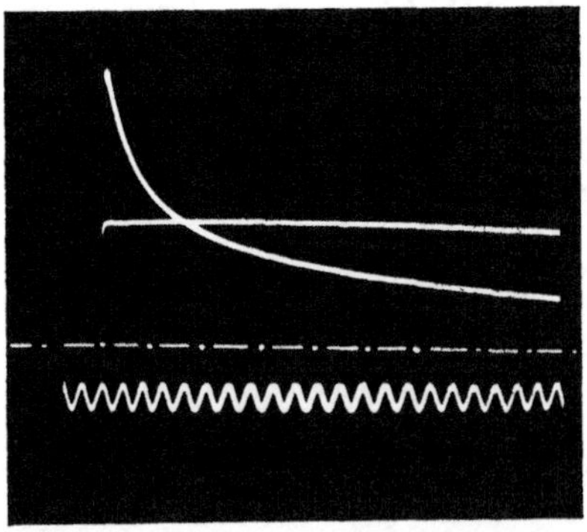

Fig. 15-2 Switching-on current of a 6.3 V/0.15 A bulb in DC operation. The broken line represents zero, the horizontal full line a constant current of 0.5 A. The sinusoidal frequency for the time scale is 2500 c/s

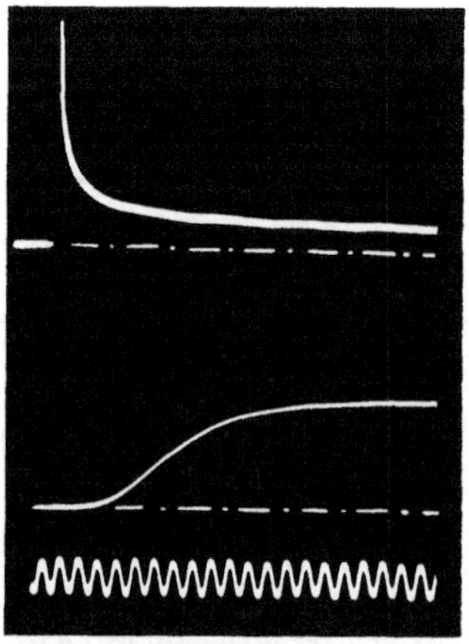

Fig. 15-3 Above: Switching-on current under same conditions as in Fig. 15-2. Centre: Trend of luminous flux. Below: Time scale 200 c/s

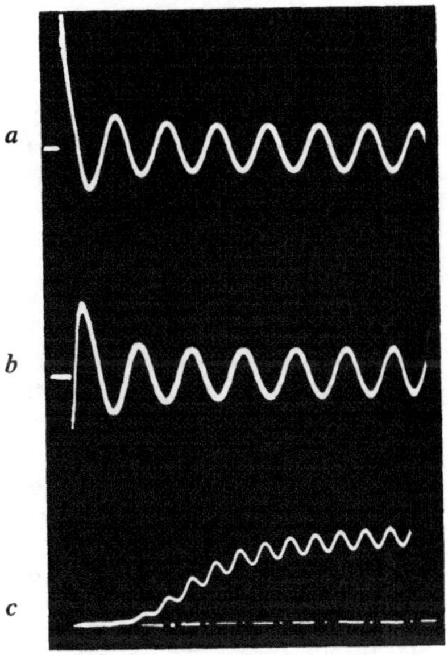

Fig. 15-4 In AC operation the height of the current surge depends on the voltage phase at the moment of switching on.
a) Switching on in the positive peak
b) Switching on at a smaller amplitude
c) Rise of the luminous flux

In Fig. 15-3 the duration of the recording was long enough to allow the lamp to reach a stable condition. The frequency of the scale was 200 c/s. The interesting fact appears that, at the moment of switching on, the current rises to about nine times the normal value. It is further interesting to note that light radiation does not begin until the current has reached its nominal value. The time in the case under consideration is about 15 ms. After a further 50 ms the luminous flux has attained its normal value. Observing the switching-on current with alternating voltage in the oscillograms in Fig. 15-4 it can be seen that the height of the initial peak depends on the phase of the voltage at the moment of switching on.

In Fig. 15-4a it is clear that the moment of switching was in the vicinity of the positive peak; a maximum value was therefore reached similar to that with direct voltage. In oscillogram b), on the other hand, only a small current peak is observed. The luminous flux oscillograms shown in Fig. 15-14c show a ripple of the light of about 15% of the basic light with twice the frequency of the supply current. No time marking was required for these recordings, the time scale being given by the known frequency of the supply current (50 c/s) or of the luminous flux (100 c/s).

CHAPTER 16

THE DISPLAY OF HYSTERESIS LOOPS

In the display of the hysteresis loop of magnetic material the magnetic induction B is represented as a function of the field force H.

If the iron core is sufficiently well closed (no air gap), the field force can be assumed to be proportional to the number of ampere-turns. This means that the primary current I_p can be used to represent the field force. The test core is provided with two windings; a resistor R_v is connected in series with the primary winding and the resultant voltage drop used as a deflection voltage for the X-plates, as shown in Fig. 16-1.

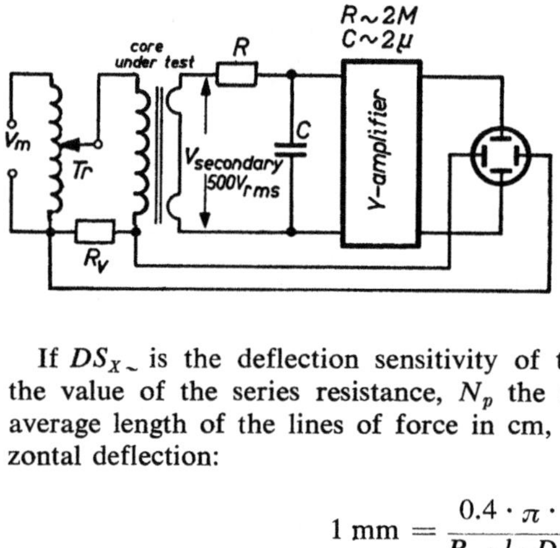

Fig. 16-1 Arrangement for recording hysteresis loops

If $DS_{X\sim}$ is the deflection sensitivity of the X-plates for alternating current, R_v the value of the series resistance, N_p the number of primary windings and l the average length of the lines of force in cm, the following scale applies for the horizontal deflection:

$$1\ \text{mm} = \frac{0.4 \cdot \pi \cdot N_p}{R_v \cdot l \cdot DS_{X\sim}}\ [\text{Oersted}]\,. \tag{16.1}$$

The vertical beam deflection, proportional to the induction B, is obtained from a secondary winding on the test core. The voltage induced in this winding is

$$V_{\text{sec}} = c \cdot \frac{\mathrm{d}B}{\mathrm{d}t}, \tag{16.2}$$

in which c is a constant. If this voltage is fed via a large resistor R to a capacitance, such that $R \gg \frac{1}{\omega \cdot C}$, this voltage is electrically integrated (see Part II, Ch. 11.12 "Electrical integration"). Hence the voltage across capacitor C is

$$V_c = c' \cdot \frac{B}{R \cdot C}\,. \tag{16.3}$$

Integration results in only a small voltage, which must be amplified for display. It is preferable that the secondary winding consists of two parts arranged so that each begins and ends at the outside and each part has roughly the same capacitance to the primary winding and to the core. Screening should be provided between the primary and secondary winding so that the secondary voltage is produced only by magnetic induction. If $DS_{Y\sim}$ is the deflection sensitivity of the Y-plates for alternating voltage, C the value of the capacitance in μF, R the resistance in MΩ, N_s the number of windings of the secondary armature, Q the cross-section of the core and G the gain, then the following scale for vertical deflection (corresponding to B) applies:

$$1\ \text{mm} = \frac{R \cdot C \cdot 10^8}{N_s \cdot Q \cdot G \cdot DS_{Y\sim}}\ [\text{Gauss}]\ . \tag{16.4}$$

The pattern of the hysteresis loop then appears on the screen (Fig. 16-2). By altering the input voltage by means of transformer Tr, various degrees of magnetization and thus correspondingly hysteresis loops can be recorded one above the other. The pictures of the axes are obtained by cutting off the voltage on one pair of plates during maximum magnetization and recording the resultant line in addition.

The vertical amplifier must not show any phase change for this test, or patterns as in Fig. 16-3 will result. It is possible, however, to compensate for small phase errors by means of a network as in Fig. 8-4*b* or *c* at the input of the vertical amplifier.

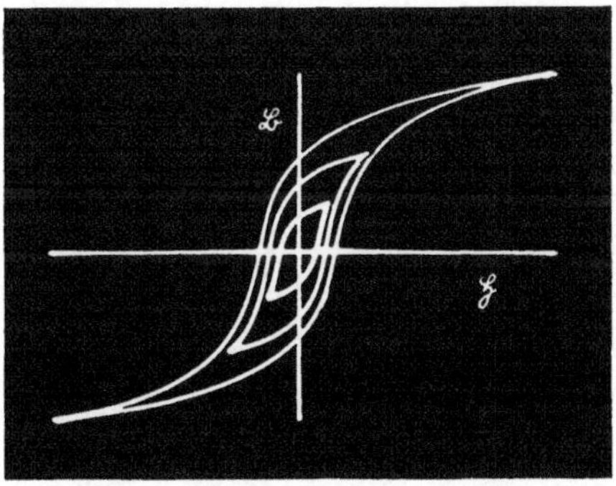

Fig. 16-2 Hysteresis loops recorded according to the arrangement shown in Fig. 16-1

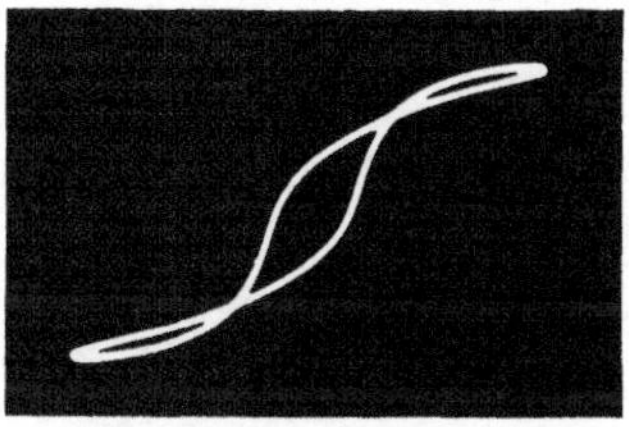

Fig. 16-3 Hysteresis loop, distorted by phase error of vertical amplifier

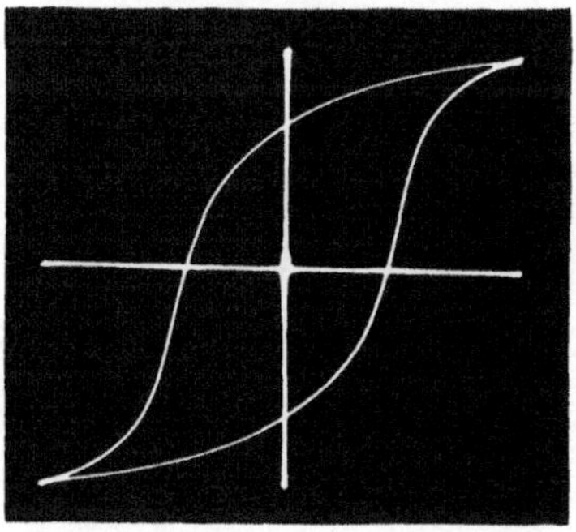

Fig. 16-4 Hysteresis loop with simultaneous display of the coordinates by means of two electronic switches

Using two electronic switches it is also possible to display the hysteresis loop and the co-ordinates simultaneously. The oscillogram in Fig. 16-4 was recorded in this way. A separate electronic switch must obviously be used for each direction of deflection. It switches the deflection plates alternately to the null voltage (line of co-ordinates) and to deflection by means of the input voltage. One such arrangement described by Griessen and van der Zwaag shows how the B/H curve of the first magnetization cycle can be displayed [1]. The actual hysteresis loop remains invisible; only its upper end can be brought out by brightness modulation. If the field (magnetizing) force is then progressively increased from zero to maximum value then this point traces the magnetization curve.

Many other hysteresis phenomena are known. In particular, the manufacture of dielectric with a high ε value has necessitated close study of electrical hysteresis. [2] [3].

The recording of hysteresis loops is not always necessary for judging various types of iron and in general practice the type of sorting bridge described in Part II, Ch. 8 (Fig. 8-16) should suffice, as it also offers the advantage of allowing the object being measured to be changed rapidly.

CHAPTER 17

RECORDING THE CHARACTERISTICS OF CRYSTAL DIODES, TRANSISTORS AND ELECTRONIC VALVES

17.1 Demands on the oscilloscope

The characteristic of a rectifier or amplifying element indicates the dependence of one characteristic magnitude on another, and such characteristics can be displayed by means of the oscilloscope. As in the case of displaying the hysteresis loop, it is not a matter of showing the dependence of amplitude on time, but the mutual dependence of two magnitudes on one another. As such characteristics generally have some irregular parts and, in addition, the range of the measuring circuit must not be overdriven to any considerable extent, this arrangement leads to distorted voltage curves. Since the position of the waveform must be maintained when the parameter changes — a fixed null level is necessary — only oscilloscopes having DC voltage amplifiers for both directions of deflection are suitable for this task.

17.2 Measuring technique

In the usual diode characteristics the current passing as a function of the applied voltage is always measured. This means, therefore, that to display this curve on the screen of the oscillogram the horizontal deflection must be produced by the applied anode voltage and the vertical deflection by a voltage proportional to the current.

This measurement could be made in a similar way to static measurements by means of indicating instruments and varying the direct anode voltage. In this way, however, only a spot moving across the screen according to the voltage variation would be obtained.

The spot trace could, of course, be observed on a screen with a long afterglow or could be photographed, but it is more preferable to superimpose on the characteristic an alternating voltage with an amplitude corresponding to the adjusted picture dimensions, so that the observer has the impression of a static picture. The shape of the curve of this alternating voltage is immaterial; if its speed is not constant the various parts of the curve are just different in brightness. However, the frequency of this alternating voltage must be chosen such that no phase shifts occur either in the circuit or in the amplifier, otherwise double lines (distorted ellipses) are obtained and not a simple waveform. A sinusoidal voltage with a frequency of 50 . . . 500 c/s was used in all the following oscillograms, as this permits of the simplest working conditions. With sinusoidal voltages the ends of the curve are brighter than the middle, according to the speed of the spot. Uniformly bright traces can be obtained with voltages linear with time, such as sawtooth or triangular voltages. Thus, for instance, the horizontal deflection voltages can be amplified via an output stage to drive such a measuring arrangement. (The internal resistance of the signal source must be small so as not to cause distortion of the curve.) Pulse-shaped voltages can also be used. To obtain uniformly bright traces, care must be taken that the spot remains blanked during the pauses. (The brightness is, of course, reduced in comparison with a continuously traced curve, according to the scanning ratio.) Measurement with pulse-shaped voltages makes it possible to display characteristics in ranges unattainable in static measurement, since in static measurement the object under investigation would be overloaded. In pulsed operation the average load can be kept within reasonable limits, as for a given pulse amplitude it is considerably less than with a continuous voltage. Such processes have been described by HEINS VAN DER VEEN, FOSTER, KAMMERLOHER and KREBS [1] [2] [3].

The oscillogram in Fig. 17-1 illustrates how a characteristic of the object under investigation is produced when driven with a sinusoidal voltage. In three photographs it shows the trace in dependence on time (horizontal deflection) of the alternating anode voltage of a diode and, deflected vertically, a voltage curve corresponding to the current then flowing. The waveform of the anode current is flattened at the bottom, as during the period of time in which the anode voltage is negative there is no flow of current. (There was an additional positive direct bias on this anode, so that the curve distortion extends to the centre of the sine trace.) The time dependence of these voltages was displayed by applying sawtooth voltages in the appropriate directions.

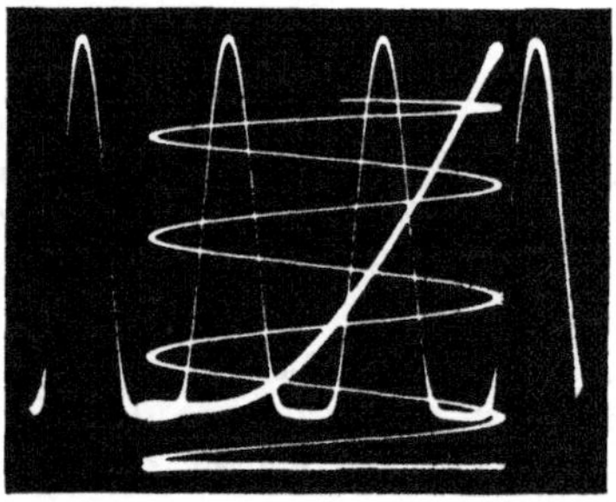

Fig. 17-1 Diode characteristic brought about by the alternating deflection voltages

If the input voltages are applied to both pairs of plates, the spot describes a path in which the vertical deflection corresponds at every moment to the anode voltage at the particular moment. As the horizontal deflection, on the other hand, is proportional to the instantaneous values of the anode voltage, the "characteristic curve" is obtained (the more strongly displayed curve in Fig. 17-1).

17. 3 Diode characteristics

Fig. 17-2 shows the circuit for recording diode characteristics. The input voltage is applied between anode and cathode. A resistance R_a is included in the circuit. It must be small in value compared with the internal resistance of the diode to avoid introducing abnormal conditions. A voltage directly proportional to the anode current appears across this resistor and is fed to the vertical amplifier. The alternating anode voltage is also applied to the input of the X-amplifier for the horizontal deflection. In this way a curve can be obtained as described, which shows the dependence of the anode current on the anode voltage.

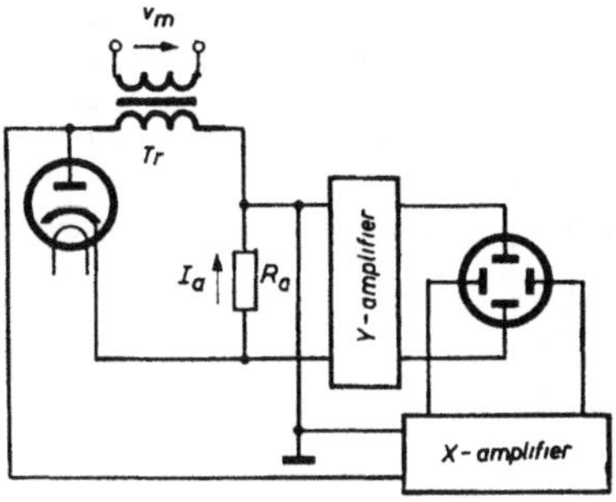

Fig. 17-2 Arrangement for recording diode characteristics

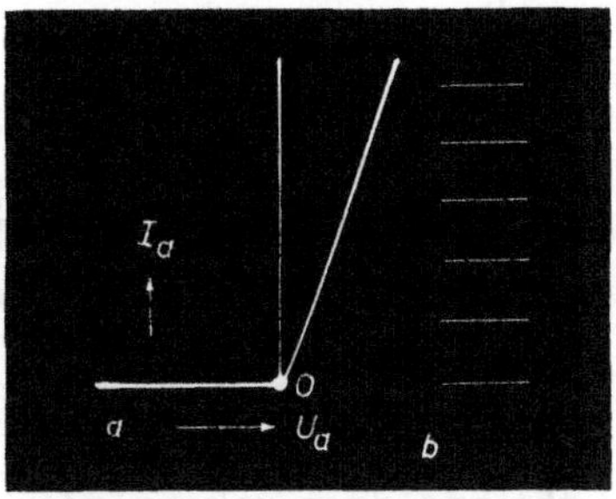

Fig. 17-3 Characteristic of the OA 85 germanium diode; calibrated in steps of 5mA

Fig. 17-3*a* shows the characteristic of the germanium diode OA 85 recorded with the Philips "GM 5656" oscilloscope. On the right near the characteristic curve *b*) a previously recorded current scale can be seen. For this scale the zero-line of the gain required for recording the characteristic curve was deflected by direct voltage steps of 0.5 V; with a resistance of 100 Ω this corresponds to a current scale of 5 mA. The alternating anode voltage was about 3 V_{rms}, the peak value, and hence the highest deflection for the characteristic curve, was about 4.2 V_p. The characteristic of one diode of the valve EAA 91 recorded with the Philips "GM 5666" oscilloscope is to be seen in Fig. 17-4.

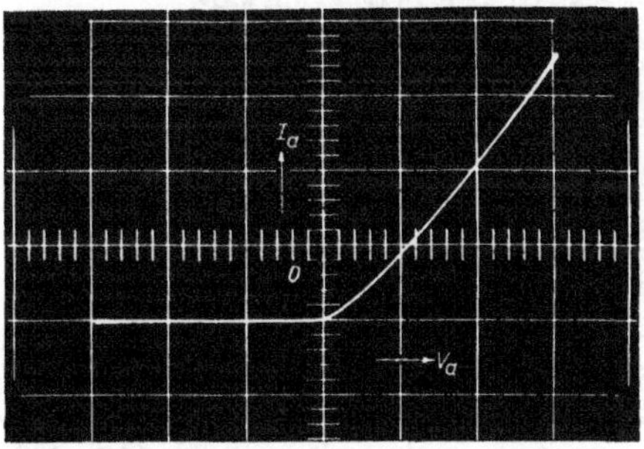

Fig. 17-4 Characteristic of a diode of a valve EAA 91. Scale (on the oscilloscope screen): vertical 6mA/cm, horizontal: 1.5 V/cm

17. 4 Characteristics of amplifier valves

The layout used for the recording of the anode current/grid current characteristic can be seen in Fig. 17-5. Direct voltage sources, preferably mains rectifier units such as the Philips "GM 4561" DC supply unit, supply the direct currents and voltages used to adjust the working conditions of the valve. The internal resistance of these direct voltage sources must be sufficiently low to prevent undesired couplings from introducing errors. A pentode is shown as the valve under investigation in the circuit diagram. For triodes, of course, there is no need for a current source for the screen grid. The horizontal deflection is obtained by means

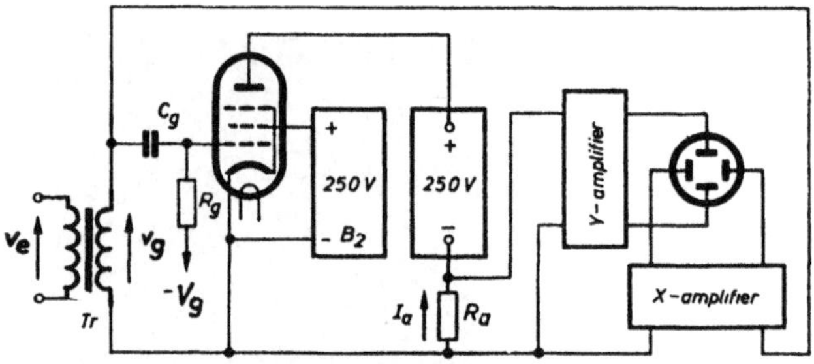

Fig. 17-5 Arrangement for recording the anode current/grid voltage characteristic of amplifier valves

of the grid drive voltage, while the vertical deflection is obtained from a voltage taken from a resistor connected in the anode circuit.

If the direct grid voltage or the anode voltage as parameter is varied in stages a family of characteristic curves is obtained. For Fig. 17-6 a number of curves of one of the triode systems of valve E 80 CC were recorded, the anode voltage being adjusted successively to 100, 150, 200 and 250 V.

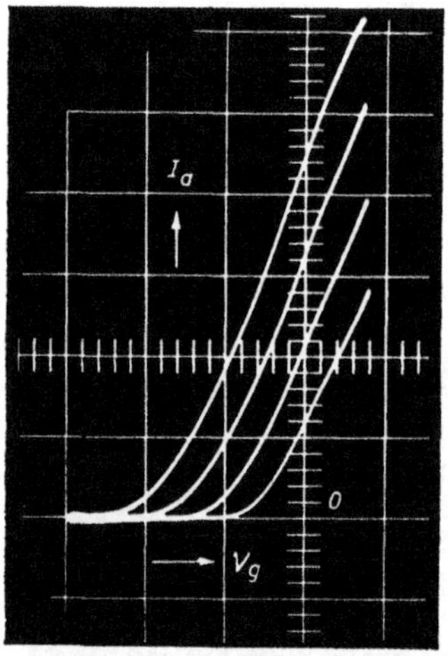

Fig. 17-6 Anode current/grid voltage characteristic of a system of the Valvo E 80 CC valve. Scale: vertical 6 mA/cm, horizontal 4 V/cm; parameter: anode voltage 100, 150, 200 and 250 V

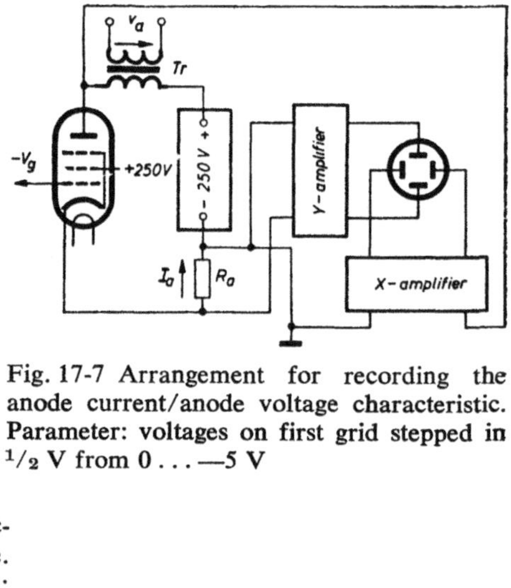

Fig. 17-7 Arrangement for recording the anode current/anode voltage characteristic. Parameter: voltages on first grid stepped in $^1/_2$ V from 0 . . . —5 V

In a similar way other dependences of amplifier valves can be displayed. A circuit for recording the anode current/anode voltage characteristic curve is shown in Fig. 17-7. It corresponds in essentials to that in Fig. 17-2. Now, however, corresponding DC sources are needed for the supply of the valve and for adjusting the working point. The family of characteristic curves of one system of the E 80 CC valve recorded in this way is seen in Fig. 17-8. The topmost curve corresponds to zero grid voltage, for the other curves the direct grid voltage was increased in stages of —0.5 V to —5 V. The dependence of the anode current on the anode voltage in the case of the E 80 L pentode is shown in the oscillogram in Fig. 17-9.

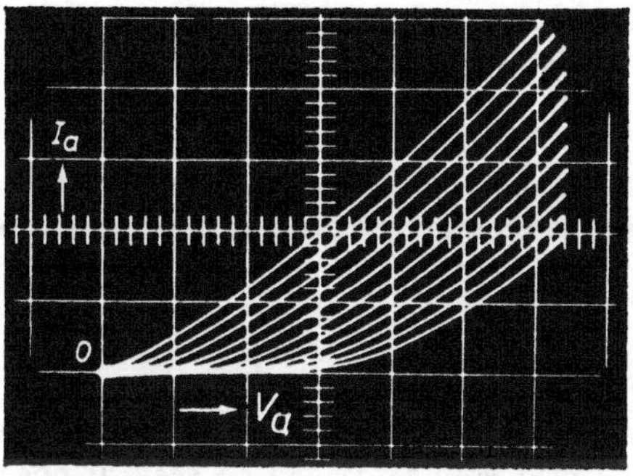

Fig. 17-8 Anode current/anode voltage characteristics of a system of E 80 CC valves.
Scale: vertical 6 mA/cm, horizontal 50 V/cm

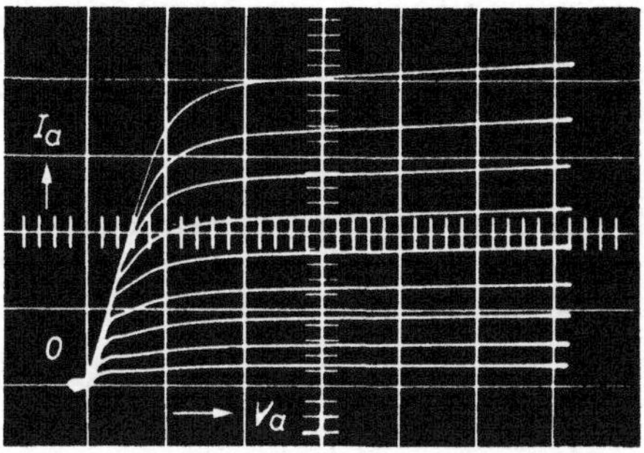

Fig. 17-9 Anode current/anode voltage characteristic of the E 80 L pentode; $V_a = V_{g2} = 200$ V; scale: vertical 20 mA/cm, horizontal 31 $^1/_3$ V/cm; parameter: voltage on 1st grid in steps of 1 V from 0. . –8V

17. 5 Characteristic curves of transistors

The characteristic curves of transistors can be recorded by the same process. An example of this can be seen in oscillograms a) and b) of Fig. 17-10. The dependence of the collector current on the collector-emitter voltage is shown in the two families of curves of the switching transistor OC 76. In normal working the base current was adjusted in stages of 0.3 mA between 0.3 mA and 2.7 mA as parameter [90]).

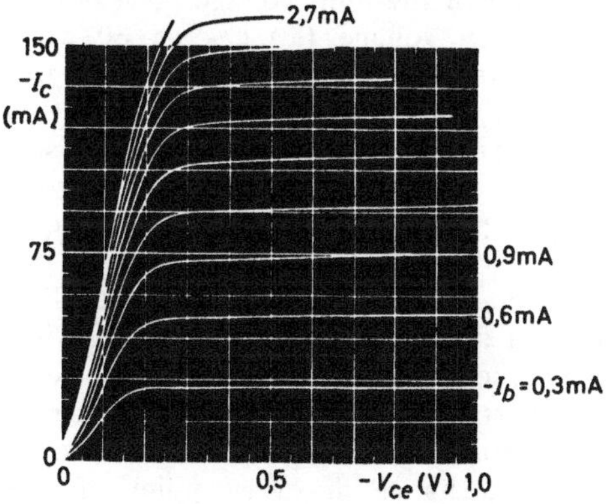

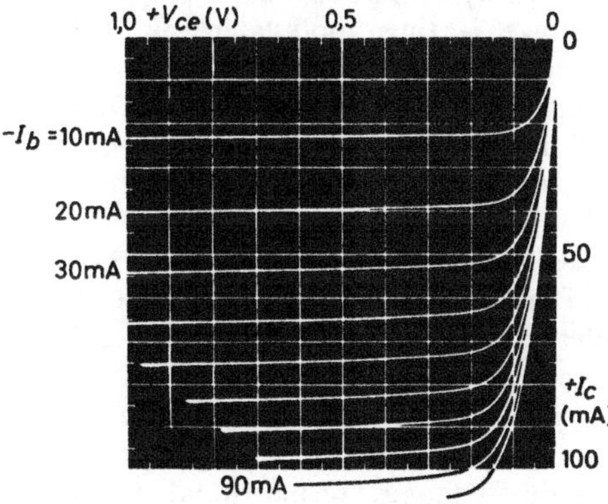

Fig. 17-10 Characteristics of the Philips OC 76 transistor.
a) Normal operation, dependence of the collector current on the voltage between collector-emitter for various basic currents;
b) as *a)* but inverted operation

[90]) These recordings were made in the applications laboratory of Valvo Ltd., Hamburg-Stellingen, with a special recording apparatus. By means of a switch relay, switching is evenly and automatically carried out to the various values of the basic current -as parameter- throughout the open time of the camera shutter.

Fig. 17-10*a* shows the "forward" characteristic curves, while Fig. 17-10*b* represents curves of the "reverse" current.

In contrast to amplifier valves, relatively large currents are needed to measure crystal diodes and transistors at low voltages. This must be taken into account when measuring apparatus of this sort are produced. The internal resistance of these voltage sources must be particularly low. The bibliography contains references to works dealing in particular with the recording of the characteristic curves of transistors [4] [5] [6] [7] [8].

17. 6 Displaying the characteristic curve indicating variation g_m

In judging an amplifier valve it is not always enough to know, for instance, the direct dependence of the anode current on the grid voltage. It can be seen at once from a characteristic curve that the relationship is not linear. It is therefore often an advantage to be able to observe the change in this relationship. The ratio of anode current change to grid voltage change — $\frac{\Delta I_a}{\Delta V_g}$ — is an important value, commonly known as the "slope" or mutual conductance. As has been dealt with in detail in Part I, Ch. 5, the attainable gain in wide band amplifiers is directly proportional to the mutual conductance. For instance, a characteristic curve showing the mutual conductance as a function of the grid voltage is of interest in finding the most favourable working point. If it is necessary to control the anode current over a region of varying mutual conductance, a varying degree of gain results. This means, among other things, that a small HF alternating voltage applied to the grid circuit of a valve in a region where a relatively large hum voltage is also present is not uniformly amplified but is amplified in the rhythm of the hum voltage. The output voltage is thus modulated by the mains frequency voltage (50 c/s or 60 c/s). After demodulation, a hum results, and this is termed *modulation hum.* In the same way any other interfering voltage, for instance, the audio frequency component of a powerful local transmitter demodulated by the curvature of the characteristic curve of the input valve of a radio receiver, can modulate the input signal from the wanted weak transmitter. These important aspects of amplification technique have been exhaustively studied [9]. Mathematically the mutual conductance is the differential quotient or the first derivative of the I_a/V_g valve characteristic. The derivatives of higher degree are also important in distortion effects. The measuring process for recording the mutual conductance characteristic curve is described below.

The starting point here is the fact that an alternating voltage at high frequency but small amplitude over the range of the characteristic, supplies, at the anode, output voltages proportional to the mutual conductance at every point of the characteristic curve. The pre-requisite is, however, that the amplitude of the voltage of higher frequency is chosen so small that its excursion can be regarded as a straight line compared with the relevant part of the characteristic curve. Otherwise only the waveform of the average mutual conductance would be obtained. Fig. 17-11 shows the basic arrangement of a circuit for recording the characteristic curve of mutual conductance. The valve is once more adjusted to a favourable working point by means of direct voltages. Now two alternating voltages — V_{g1} and V_{g2} — are applied to the control grid, connected via transformers for modulation purposes. One of the alternating voltages — V_{g1} — has a low frequency (50 c/s) and such a large amplitude that it more than covers the range of the whole

Fig. 17-11 Arrangement for displaying slope curves of amplifier valves

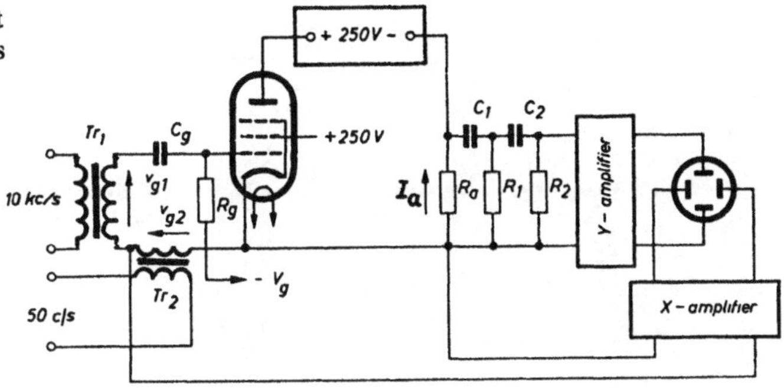

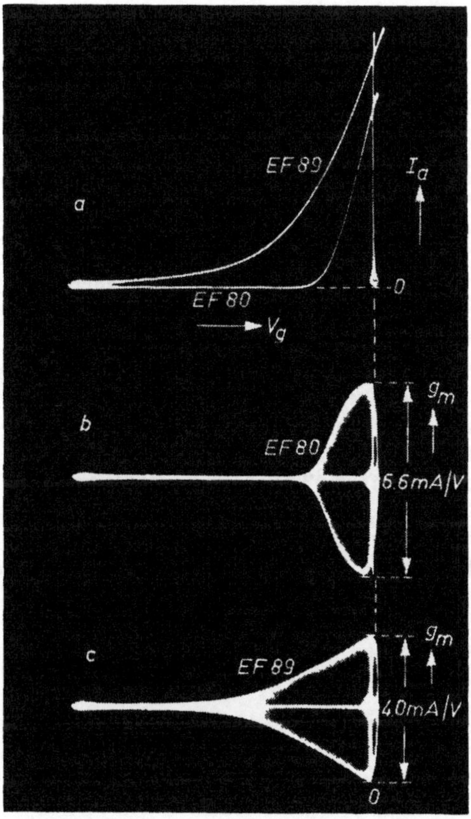

Fig. 17-12 Characteristic and slope curve of a valve with a "straight" characteristic and of one with an "exponential" characteristic. *a*) $I_a = f(V_g)$; *b*) slope curve of EF 80 valve; *c*) slope curve of EF 89 valve

family of curves. The other alternating voltage — V_{g2} — has as small an amplitude as possible and a high frequency (10 kc/s).

The low frequency voltage traversing the family of curves deflects the spot in the horizontal direction via the *X*-amplifier so that its position at any instant corresponds to a particular grid voltage and thus to a corresponding value of anode current. The anode current contains amplified components of both alternating voltages. A high pass filter, admitting only the higher frequency voltage component, is connected between the anode resistor (R_a) and the input of the vertical amplifier. The amplitude of this voltage fluctuates, however, in accordance with the varying slope of the characteristic curve traversed by the low frequency. As the horizontal deflection of the spot also corresponds to the amplitudes of this voltage, a luminous area occurs on the oscilloscope screen, the limits of which constitute a standard of measurement for the mutual conductance.

The oscillograms in Fig. 17-12 show firstly *a*) the curves of two valves with different characteristics, a "straight" curve (valve EF 80) and the curve of an "exponential valve" (EF 89). The curve of the exponential valve moves

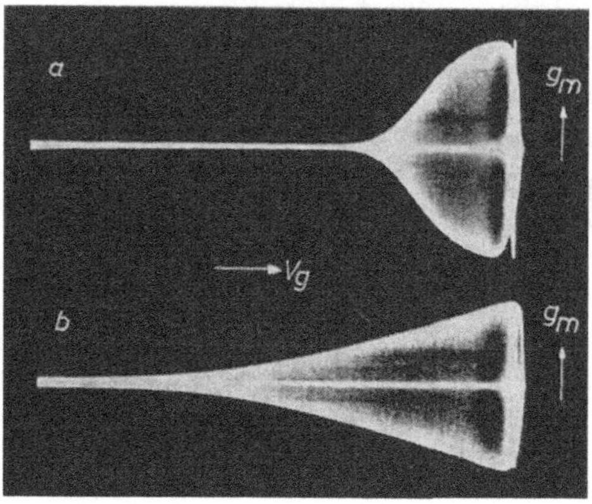

Fig. 17-13 Slope curves in Fig. 17-12*b* and *c* with a more favourable scale

in accordance with such an exponential function that a linear mutual conductance results.

Below are reproduced the slope oscillograms recorded by means of the circuit shown in Fig. 17-11; in the centre is the oscillogram of the valve with the straight characteristic curve *b*) and below it *c*) that of the exponential valve. Vertical lines have also been recorded from the zero point with maximum slope as a scale. In the case of the EF 80 valve this line corresponds to a mutual conductance of 6.6 mA/V, in the EF 89 to about 4.0 mA/V [91]). It can be readily seen that although the mutual conductance of the valve with the "straight" curve attains a higher value, it is only linear over a very small range. The mutual conductance curve of the exponential valve, on the other hand, can be regarded as linear over a wide range.

This becomes still clearer, if, as in the oscillogram in Fig. 17-13, the curve is modulated no more than is just needed for satisfactory observation of the mutual conductance oscillogram. Greater horizontal enlargement expanded this oscillogram to a very clear picture. Oscillograms of the mutual conductance waveform also give an analogous picture of the "intermodulation factor" of the valve [10] [11].

The dynamic behaviour of amplifier circuits can also be studied by means of arrangements for recording valve characteristic curves. Exceedingly clear pictures are obtained if the dynamic curve, i.e. the pattern described by the working point in the modulation range during operation is also photographed in [1].

[91]) It should be pointed out that these recordings have been made with valves which have not been selected. They therefore correspond to the list data only within the usual limits.

CHAPTER 18

RECORDING THE PASSBAND CURVES OF HF CIRCUITS, RADIO AND TELEVISION RECEIVERS

18. 1 Measuring arrangement

In recording the passband curves of HF four-poles, what is required is to show the dependence of the output voltage on the frequency of the input voltage. This means that the horizontal deflection must be proportional to the frequency and the vertical deflection must be proportional to the output voltage. The input voltage frequency must therefore be varied rhythmically to produce the horizontal deflection of the spot. In the process described in Ch. 26, for recording "low frequency characteristic curves", the horizontal deflection proceeded by means of the variable frequency input voltage obtained from a frequency-dependent network. In recording HF characteristic curves it is often the case that the frequency change of the input — the "frequency modulation" — is derived from the X-deflection voltage. It is obvious that this frequency change — "wobbulating" — can be achieved mechanically by periodically detuning an oscillator. In early publications in which the display of resonance curves on an oscilloscope tube was described, the frequency change was carried out by a motor-operated variable capacitor [1]. The slider of a potentiometer was driven by the same spindle, the potentiometer resistance track being connected to a direct voltage source. Between the slider and one end of the resistance track a voltage which increased linearly with the frequency change was obtained. It was used to deflect the spot horizontally, the horizontal position of the spot thus corresponding at each instant to a particular frequency. If the vertical deflection of the spot is carried out simultaneously by the output voltage of the HF four-pole, then the spot path corresponds to its passband curve. As the output voltage is used directly to deflect the spot vertically, the frequency characteristic is seen as the boundary curve of a luminous area, as will be seen later in the oscillograms of Figs. 18-5 and 18-8.

When a motor was used for detuning, some considerable difficulties were encountered. In particular the generation of the voltage for the horizontal deflection by means of a variable potentiometer was found to be far from perfect with the means available at the time. Since the required frequency variation need not exceed 50 to 100 kc/s, it has proved more convenient to modulate the input voltage by means of a so-called "reactance valve".

It is perhaps of interest to point out that motor detuning has recently again been used, particularly in wobbulators for ultrashort waves and VHF.

The circuit for a wobbulator for the frequency range from about 100 kc/s to approx. 30 Mc/s for AM receiver circuits is shown in Fig. 18-1. It is, in essence, a mixing stage (1) with its oscillator (1') which can be detuned by a reactance valve (2) in accordance with a control voltage. (The oscillator is adjusted to about 4 Mc/s.) By using such a reactance valve it is possible to detune an adjusted circuit in accordance with a control voltage. If this is obtained from a heterodyne

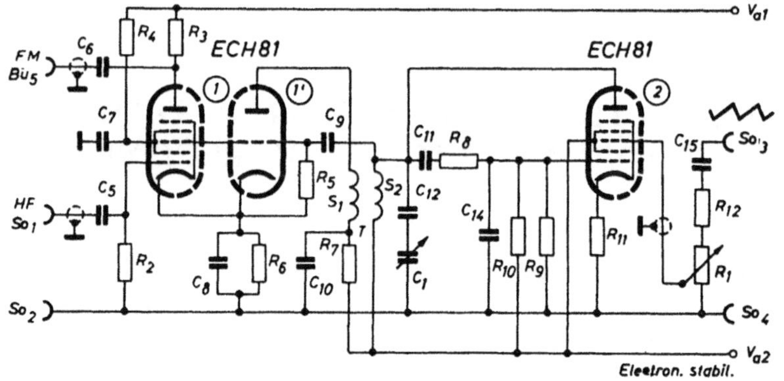

Components:	R_6 = 150 Ω	Capacitors	C_9 = 100 pF
Resistors	R_7 = 12 kΩ		C_{10} = 82 nF
	R_8 = 12 kΩ	C_1 = 500 pF	C_{11} = 10 nF
R_1 = 1 MΩ	R_9 = 12 kΩ	C_5 = 10 nF	C_{12} = 27 pF
R_2 = 1 MΩ	R_{10} = 220. . 330 kΩ	C_6 = 22 nF	C_{14} = 56 pF
R_3 = 270 Ω	(for adjusting	C_7 = 82 nF	C_{15} = 82 nF
R_4 = 34 kΩ	grid voltage)	C_8 = 82 nF	
R_5 = 22 kΩ	R_{11} = 1 kΩ		

Fig. 18-1 Circuit of the Philips "GM 2886" frequency modulator

oscillator stage, the frequency of the oscillator output voltage can be altered electronically. A frequency-modulated voltage is thus obtained.

The circuit can be so chosen that the reactance valve functions as a variable capacitance or as a variable inductance. In the case of the variable capacitor adjustment it is better to use an inductance rather than a capacitance for detuning, as then the frequency change over the frequency range is more uniform. The operating principle of a reactance valve circuit will be explained in somewhat greater detail with reference to Fig. 18-2.

The resonant circuit of the oscillator, as far as AC is concerned, is connected between the anode and cathode of the reactance valve. Parallel to it is connected a voltage divider, consisting of a resistor (*R*) and a capacitor (*C*). The grid of the valve is connected to the junction point of the resistor and the capacitor. In the following explanation only the alternating quantities will be considered and it will be assumed, moreover, that the grid resistance (R_g) is large compared with the impedance of the capacitor (*C*).

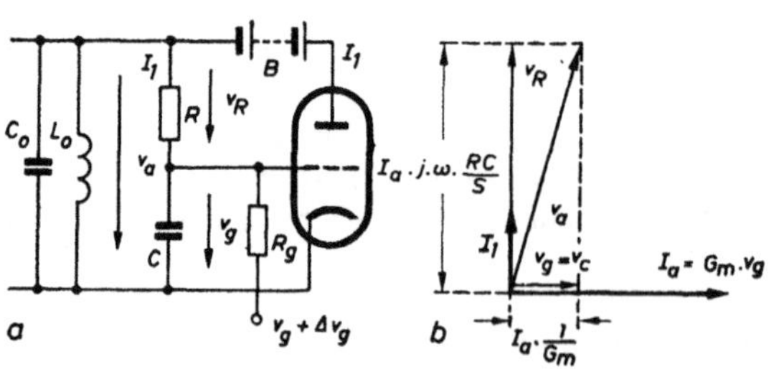

Fig. 18-2 Method of working of a reactance valve circuit

The current I_1, in the series connection of the resistor (R) and the capacitor (C) is:

$$I_1 = \frac{V_a}{R + \dfrac{1}{j \cdot \omega \cdot C}} \,. \tag{18.1}$$

It must be small compared with the alternating current I_a through the valve. As can be seen from the vector diagram (Fig. 18-2*b*), the anode voltage V_a is the resultant of the voltage across the resistor

$$V_R = I_1 \cdot R, \tag{18.2}$$

which is in phase with the current I_1, and the grid voltage (the voltage across the capacitor) which is:

$$V_g = I_1 \cdot \frac{1}{j \cdot \omega \cdot C} = -j \cdot \frac{I_1}{\omega \cdot C} \tag{18.3}$$

and lags by 90°. The anode current I_a of the valve is in phase with the grid voltage and corresponds to the equation

$$I_a = g_m \cdot V_g \,. \tag{18.4}$$

From the vector diagram in Fig. 18-2*b* it can be seen that, when the anode current lags behind the anode voltage by almost 90°, the anode impedance

$$Z_a = \frac{V_a}{I_a} \tag{18.5}$$

is inductive.

If the anode voltage V_a is separated into its two components V_R and V_C ($= V_g$) and each is divided by the anode current, we obtain the impedance represented by this valve at the resonant frequency of the oscillation circuit ($L_0 - C_0$). To make this division possible we express the component V_R in terms of V_g as follows:

$$\frac{V_R}{V_g} = \frac{R}{\dfrac{1}{j \cdot \omega \cdot C}} \,, \tag{18.6}$$

or:

$$V_R = V_g \cdot j \cdot \omega \cdot R \cdot C \,. \tag{18.7}$$

After transformation and substitution of Eq. (18.4) for I_a we obtain

$$Z_a = \frac{V_g + V_R}{g_m \cdot V_g} = \frac{1}{g_m} + j \cdot \omega \frac{R \cdot C}{g_m} = R' + j \cdot \omega \cdot L'. \tag{18.8}$$

The valve thus behaves as if a resistor $R' = \dfrac{1}{g_m}$ were connected in series with an inductance

$$L' = \frac{R \cdot C}{g_m} \,. \tag{18.9}$$

It is, of course, necessary to ensure that the product $R \cdot C$ is large as compared to 1, so that the reactive components of the circuit predominate. As can be seen from Eq. (18.9), the self-inductance L' of the diode corresponding to the grid-anode path, is inversely proportional to the mutual conductance of the valve. It is thus possible to vary the inductance by varying the mutual conductance. This could be achieved by simply varying the grid voltage (in Fig. 18-2*a* $\Delta\ v_g$). Usually, however, multi-grid valves are used. Heptodes are particularly suitable and are used in the circuit of the Philips frequency modulator (see Fig. 18-1). In this circuit the phase-shifting voltage divider consists of resistor R_8 and capacitor C_{14}. Capacitor C_{11} is used to isolate the grid from the direct anode voltage. Its influence on the effect of the voltage divider can be neglected, as the reactance of a capacitance of 10 nF for an oscillator frequency of 4 Mc/s compared with the other resistors is practically zero.

Such circuits are also used for electronic fine tuning of radio receivers [2]. According to the task in hand there is a choice among four basic circuits of a reactance valve [3]. For a relatively narrow frequency band the phase shift of the voltage divider can be maintained constantly at about 90°. If C_0 and L_0 are the elements of the resonant circuit of the oscillator, then with a relatively slight detuning of the circuit with frequency Ω_0 by a reactance valve, the frequency change $\Delta\omega$ is approximately

$$\frac{\Delta\,\omega}{\Omega_0} = \frac{1}{2} \cdot \frac{L_0}{L} \,. \tag{18.10}$$

(In the case of a reactance valve working capacitively, C' should be substituted for L_0 and C_0 for L'.) As the detuning is proportional to the mutual conductance, the slope of the grid voltage/anode current characteristic should be linear for a linear relationship between the detuning $\Delta\omega$ and the modulation voltage Δv_g. This means that the characteristic curve should be a quadratic function [4]. A particularly large frequency variation is possible by using several cathode follower stages connected in cascade [5]. Instructions for building such "wobbulators" have been published on a number of occasions [6] [7].

To ensure that an HF alternating voltage of the desired input frequency occurs at the anode of the converter valve, a signal must be applied to the first grid of this system of valves (1) from one of the usual test oscillators, of such a frequency that the sum — or difference — gives together with the 4 Mc/s frequency of the oscillator a voltage of the required Δ frequency. The heterodyne principle ensures a constant output voltage and goes far to obviate undesirable amplitude modulation. If the X-deflection voltage is fed to the third grid of the reactance valve, then this valve detunes the oscillator in the rhythm of the waveform of the control voltage on the third grid. In general a sawtooth voltage linear with time is used. A frequency change linear with time is then obtained. The oscillogram in Fig. 18-3 shows such a frequency-modulated voltage for two modulation cycles.

The sinusoidal 50 c/s voltage of the supply mains could just as easily be used. The frequency change would then no longer be proportional to time. This is not essential for the measurement, however. The only condition is that the horizontal position of the spot must always correspond to a given frequency. As this frequency is proportional to the voltage amplitude at every moment, it is immaterial whether its waveform is linear with time, or is sinusoidal, or whether it corresponds to some

Fig. 18-3 Voltage frequency modulated linear with time (two modulation cycles)

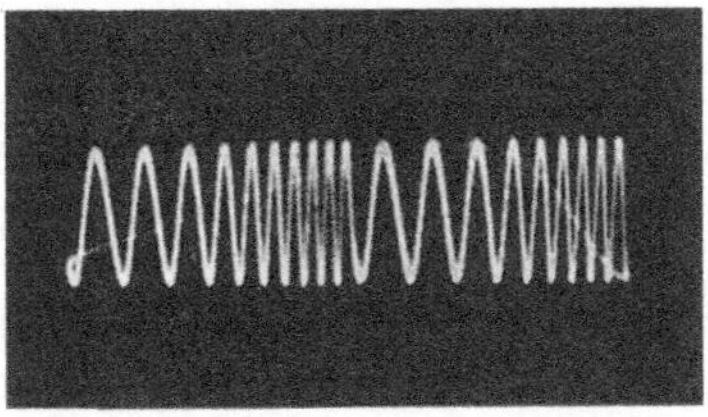

other function. Only the brightness distribution changes when the waveform is not linear with time. With sinusoidal deflection the brightness at the voltage peaks as compared with the centre portions is relatively greater than with deflection linear with time, since in the latter case deflection speed decreases.

The fact that the oscillogram of the tuning curves in Fig. 18-5 is darkened towards the middle even when wobbulation linear with time was used, is due to the speed of the vertical deflection in this part of the pattern being greater than at the sides.

18. 2 Recording the passband curves of HF circuits and AM radio receivers

A measuring circuit similar to that described above, is shown in Fig. 18-4 in the form of a block diagram. The fundamental frequency f_1 is mixed in its mixing valve with the voltage of a HF test oscillator f_2, so that the product of mixing has the desired frequency f_m. For frequencies less than 4 Mc/s the difference has to be taken and for frequencies greater than 4 Mc/s the sum. Frequency modulation control is best obtained via the oscilloscope sawtooth voltage. As it is in most cases too great, it is attenuated at the input of the frequency modulator by a series resistor (R_{12} in Fig. 18-1). This avoids both overloading the sawtooth voltage source and overmodulating the reactance valve. A frequency swing linear with time of about 50 c/s max. can be adjusted in the Philips "GM 2886" frequency modulator, the circuit of which is given in Fig. 18-1, by a potentiometer (R_1), using a sawtooth voltage of about 150 V_{pp}.

The frequency-modulated output voltage of the wobbulator is fed to the component or circuit to be measured. It must be ensured that the output impedance of the wobbulator is low. Coupling to the component or circuit under test must be such that its properties are not altered (loose capacitive or inductive coupling), or it must be ensured that it is working under conditions corresponding to its application in practice. For instance, an artificial antenna should be tapped into receiver input circuits. For circuits used in the amplification path of a receiver (ZF-circuits) it is advisable to apply the wobbulator output voltage to the grid of the

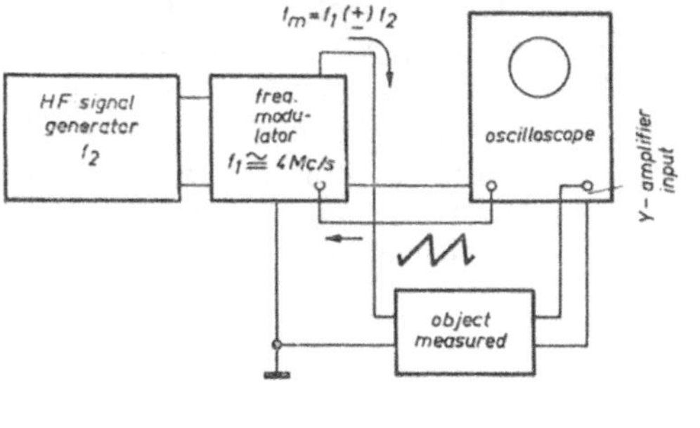

Fig. 18-4 Recording passband curves with the "GM 2886" Frequency Modulator

preceding valve and to take off the output voltage of the filter in the anode of the following valve. As the grid circuit is damped by the low output impedance of the wobbulator, its influence would in general be negligible. If necessary, this circuit can be additionally damped by a parallel resistor or detuned by a parallel capacitor in such a way that its resonant frequency lies well outside the observed frequency range. Similar steps must likewise be taken to ensure that the influence of the means of detuning on the anode circuit of the following valve is sufficiently small. If a probe of sufficiently high input impedance is available, it may also be used to take the voltage directly of the filter unter investigation. Care must always be taken, however, to ensure that the detuning of the filter by the input capacitance of the probe is so small as to be negligible.

If the oscilloscope gain is set at its highest possible value, a very loose capacitive coupling of the probe to the measuring point (between tenths pF and 1 to 3 pF) is generally sufficient to obtain sufficiently large patterns on the screen. The coupling capacitance and the input capacitance of the probe then form a capacitive voltage divider. In any case (be it a compensated divider or a direct connection to the oscilloscope input) a screened cable is required, as otherwise the stray alternating current from the lighting mains would produce intolerable interference with the oscillogram. In difficult cases it is advisable to include a screened high-pass (*CR*-network) directly at the input of the oscilloscope. This has the effect of suppressing the 50 c/s part of the voltage fed in, but permits the unaltered transmission of the HF input signal. As the frequency change affects the movement of the light spot, the passband curve is, of course, obtained as a stationary pattern.

If the bandwidth of the oscilloscope amplifier is sufficient for the purpose, the output voltage of the fourpole network under measurement can be connected in the familiar way to its input. Otherwise rectification is required. The oscillogram in Fig. 18-5*a* shows the HF passband curve of a simple resonant circuit of about 350 kc/s resonant frequency. The output voltage of the wobbulator is coupled inductively in this case.

For perfect display of the passband curve the speed of change of the signal frequency –wobbulating– is not a matter of indifference. The oscillating circuits are excited by the frequency-modulated input voltage and begin to oscillate at their resonant frequency after a building-up time T_b. This oscillation decays slowly or rapidly in inverse proportion to the quality of the circuits. If the woblulation takes place so fast that a noticeable amplitude of the oscillator circuit voltage is present while the input frequency is already at a considerable distance from the resonant frequency, the two HF voltages overlap; interferences and distortions of the curve pattern occur [8] [9]. On the one hand this causes the leading flank of the curve pattern to appear flatter than corresponds to the actual waveform of the passband curve, and on the other hand, from the moment the frequency of the input voltage exceeds the resonance frequency of the circuit, beats occur between the still increasing signal frequency and the decaying fundamental frequency of the circuit.

To determine the most favourable operating conditions, it is best to start from the build-up time T_b. (The decay time can generally be assumed to be equally great.) As has been shown by Küpfmüller, it can be assumed that

$$T_b = \frac{1}{\Delta f_h} \tag{18.11}$$

is a sufficiently acceptable approximation [10].

The denominator Δf_h is the frequency difference at which the passband curve is at half height ("half amplitude width").

Oscillograms *b* to *d* of Fig. 18-5 were taken to display these relationships clearly.

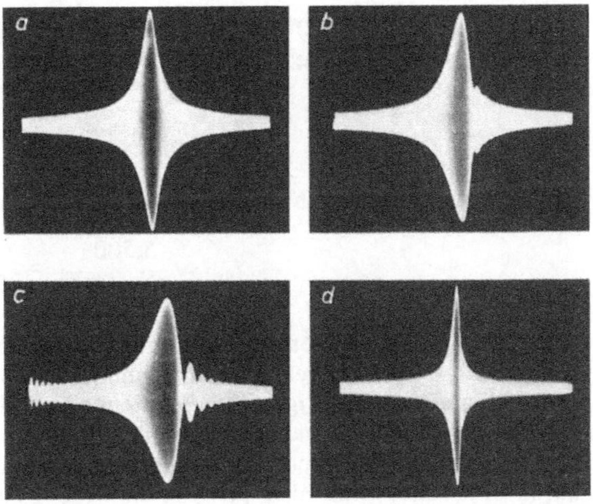

Fig. 18-5 HF passband curves of a simple resonant circuit
a) Wobbulating frequency 50 c/s;
b) wobbulating frequency 175 c/s; wobbulating frequency 460 c/s [frequency swing for *a*) to *c*) about 60 kc/s];
d) wobbulating frequency 50 c/s, but frequency swing about 140 kc/s

While the wobbulation frequency was 50 c/s in *a*) at a bandwidth of 5.5 kc/s (half amplitude width) and a swing of 60 kc/s, the wobbulation frequency in *b*) was increased to 175 c/s and in *c*) to 460 c/s. In both these photographs the decaying oscillation voltage of the circuit is overlapped by the input voltage increasing in frequency with the forward trace of the spot. From this it can be seen that the frequency change must not be more rapid than a maximum value obviously dependent on the quality of the circuit. As is shown in detail in Küpfmüller's book, this duration T_d must at least be such that

$$T_d = 20 \cdot \frac{1}{\Delta f_h} \,. \tag{18.12}$$

In the passband curve of Fig. 18-5*a* the building-up time according to Eq. (18.11) was about 0.18 ms and the duration 1.7 ms. According to Eq. (18.12) it was therefore still too short. On close examination the oscillogram shows that the peaks of the curve are bent a little to the right. In general, such small distortions are permitted to avoid the need for using too low wobbulation frequencies, as in such cases the pattern flickers. There is, however, a general condition that the wobbulation speed (the rate of change of the frequency df/dt) must not exceed a certain limiting value. In the case of a sawtooth waveform of the wobbulation voltage, of frequency f_w and of a frequency swing f_H, the time T_f for every frequency swing is $T_f = \dfrac{1}{f_w}$. The wobbulation speed v_w is thus:

$$v_w = \frac{f_H}{\dfrac{1}{f_w}} = f_H \cdot f_w \,. \tag{18.13}$$

From this the required duration T_v can be determined within the half width of the passband curve Δf_h. It is:

$$T_v = \frac{\Delta f_h}{f_H \cdot f_w} . \qquad (18.14)$$

Introducing Eq. (18.12) we obtain for an assumed frequency swing f_H at a given half width value Δf_h the permissible wobbulation frequency:

$$f_w = \frac{(\Delta f_h)^2}{20 \cdot f_H} . \qquad (18.15)$$

In the filter curve shown in Fig. 18-5*a*, according to Eq. (18.12):

$$f_w = \frac{5{,}500^2}{20 \cdot 60000} = 25 \text{ c/s} .$$

In wobbulation waveforms not linear with time, for instance, frequency modulation with sine voltages, this maximum permissible level must not be exceeded. Oscillogram *d*) in Fig. 18-5 is intended to illustrate from another angle these relationships for wobbulation which is linear with time. It shows the passband curve of the same tuning circuit as in the other oscillograms of this figure. The wobbulation frequency was also 50 c/s, but the frequency swing was now so greatly increased that it corresponds to 140 kc/s for the whole width of the pattern. (At the limits of the modulation range the reactance valve was already overdriven.) Thus the wobbulation speed is about 7 kc/s/ms as against 3 kc/s/ms in Fig. 18-5*a*, and the remaining time in this circuit is only about 0.8 ms (instead of 4 ms as required). According to Eq. (18.15) the modulation frequency for wobbulation should, in this case have been only $f_w = \frac{5{,}500^2}{20 \cdot 140{,}000} = 11$ c/s. Hence the characteristic distortion makes its appearance again at the right-hand side of the curve. The curve peaks also lean slightly to the right. The narrower the passband, the lower the wobbulation speed must be, to keep the measuring errors so small as to be negligible [11]. If particularly low frequencies are examined in this way, it is best to use oscilloscope tubes with long-afterglow screens.

When studying the passband curves of mains-operated apparatus, the output signal voltage is often found to contain slight residues of the 50 c/s mains voltage. If the wobbulation frequency is not an integral multiple of the mains frequency, the pattern of the curve "flickers" in the rhythm of the difference frequency. In such cases it is best to work with a wobbulation frequency (= horizontal deflection in the oscilloscope) which is an integral fraction of 50 c/s (25, $16^2/_3$, $12^1/_2$ . . . . c/s). The oscillogram is then completely stable. Of course, this is only permissible when stray voltages are slight, as otherwise the pattern of the passband curve would be distorted in accordance with the waveform and the amplitude of the stray 50 c/s voltage in addition. When the 50 c/s part is larger ($> 1\%$) a high-pass filter should be connected in the normal way in the oscilloscope input.

The quantitative frequency measurement of the passband curves displayed can be carried out in the following way, among others: the oscillator of the frequency modulator can be detuned by a variable capacitor (C_1 in Fig. 18-1) by $\pm$ 25 kc/s.

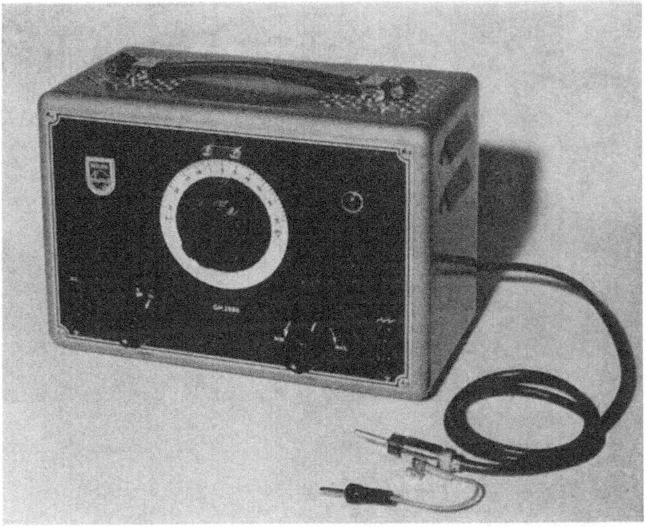

Fig. 18-6 Exterior of "GM 2886" Frequency Modulator; central scale for detuning fundamental frequency of 4 Mc/s by ± 25 kc/s

This detuning can be read off on a calibrated scale on the front of the apparatus (Fig. 18-6). The detuning causes the pattern on the screen to shift horizontally. If it is established what detuning is necessary to shift a passband curve of a given reference height (e.g., 0.71, $^1/_2$ or 1/10 of max. value) in such a way that the other flank appears at the same part of the screen, then the bandwidth can be read off at this spot. The oscillogram in Fig. 18-7 is intended to illustrate this more clearly. It shows the two passband curves which serve to determine the bandwidth at $1/\sqrt{2} = 0.71$ of the max. height.

The photograph of the curve shifted to the right was given only half the exposure time of the left-hand picture for the sake of differentiation. The passband width at this height was found to be 3.0 kc/s.

The photographs of Fig. 18-8 show more pictures of the passband curves of a bandpass filter for various degrees of coupling. The oscillogram in Fig. 18-9 shows the passband curve of such a band filter with coupling as in Fig. 18-8*c*, but with one circuit detuned.

For these photographs the frequency swing was 43 kc/s, and the wobbulation frequency 25 c/s. Since, according to Eq. (18.15), it should not exceed 35 c/s, the working conditions fully comply with this requirement.

As has already been mentioned, the output voltage of the fourpole network or other system under examination can also be rectified. The passband curve is then obtained as an oscillogram which is not in the form of an illuminated area, but a line. Rectification is a necessity if the frequency of the filter under investigation is greater than the upper cut-off limit of the oscilloscope amplifier. As only such curves will be discussed in the following paragraphs dealing with the ultra-short waveband, it is superfluous to introduce more such oscillograms at this point. It should be noted, however, that the time constants of the filter elements in the rectifier circuit must not be too great, as that too can give rise to undesirable distortion.

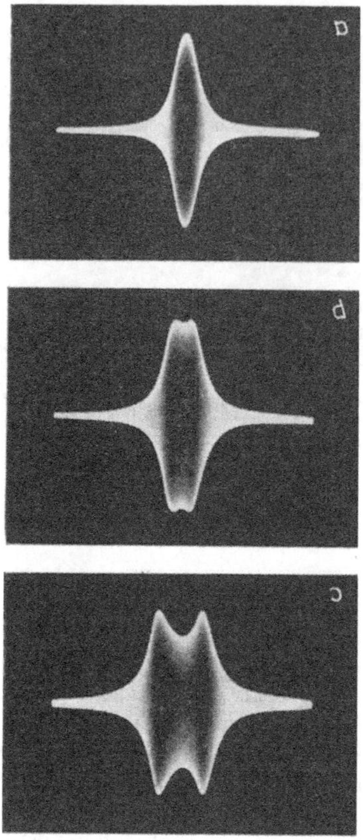

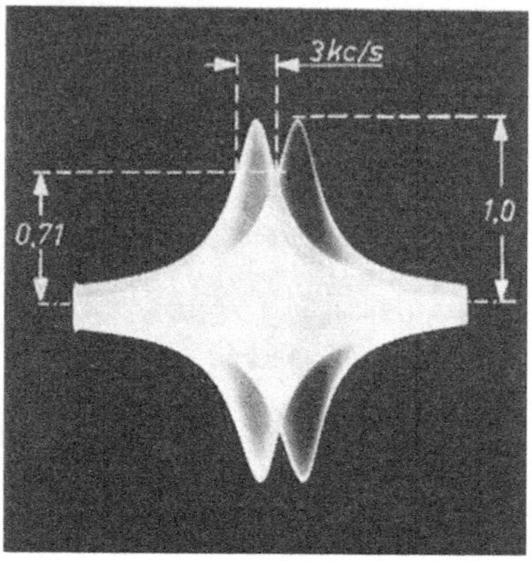

Fig. 18-7 Measuring passband curves by defined shifting of the oscillograms

Fig. 18-8 Passband curves of a ZF bandfilter. *a*) Critical coupling; *b*) over-critical coupling; *c*) too strong coupling (saddle-formation of curve)

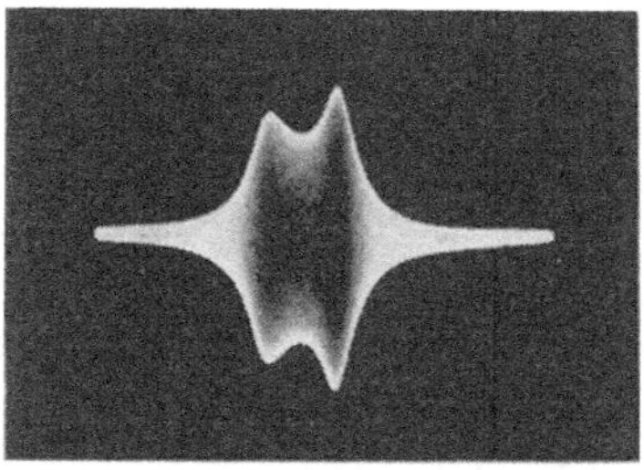

Fig. 18-9 Bandfilter as in Fig. 18-8c, but with one circuit detuned

18. 3 Recording the passband curves for ultra-short wave and television receivers

For such tasks frequency-modulated signals with a considerably greater frequency swing than previously described are required. The frequency modulation is therefore carried out electro-mechanically in most cases. There are, however, ultra-short wave wobbulators, in which the detuning of the oscillator stage is achieved by changing the magnetic bias of an iron core. In the majority of such wobbulator signal generators the frequency modulation occurs because a part of the oscillator circuit capacitance is varied by means of a mechanical oscillatory system. In the Philips "GM 2889" AM-FM signal generator frequency modulation is obtained by an electro-dynamic system which causes the moving element of a plunger-type capacitor to oscillate mechanically. The operating principle of this signal generator will be shortly discussed on the basis of the block diagram 18-10.

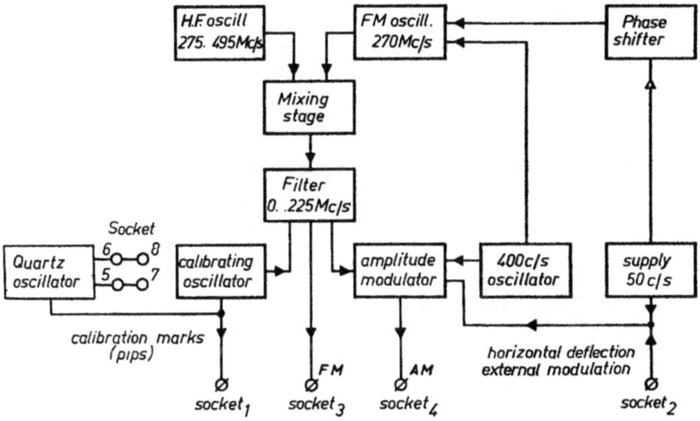

Fig. 18-10 Block diagram of Philips "GM 2889/01" AM/FM generator

The output frequency is obtained, here too, by mixing the voltage of two HF generators. One oscillates at 270 Mc/s; it can be frequency-modulated in the way described. The frequency of the second oscillator can be adjusted by a variable capacitor between 275 and 495 Mc/s. It is thus possible to generate difference frequencies in the mixing stage between 5 and 225 Mc/s. The frequency modulation can be obtained as desired either from the 50 c/s mains frequency or from a built-in 400 c/s oscillator. A phase modifier is connected between the 50 c/s voltage source and the modulation system so that the picture phase can be adjusted. The frequency-modulated voltage can be taken off from socket (S_1) and used for the horizontal deflection of the spot on the oscilloscope screen. A suppression circuit ensures that in the large wobbulation swing of 15 Mc/s maximum the 270 Mc/s oscillator oscillates only during one half cycle of the 50 c/s voltage, in fact, while the rising flank is oscillating. As a result the passband curve is traced only during the forward trace of the fluorescent spot, while during the flyback no HF voltage occurs and hence the zero line appears on the screen.

The output voltage of the mixing stage is led via a low-pass filter which sufficiently suppresses all voltages of the mixed products with frequencies $>$ 225 Mc/s. For certain tasks the means of amplitude modulation is also provided in the form of a germanium diode. An auxiliary oscillator with a fundamental frequency range of 15 - 30 Mc/s is provided for calibrating the passband curve. Strong harmonics are generated by a distortion stage so that precise measurements up to 300 Mc/s can be effected. The voltage of the calibration frequency is combined with the wobbulated signal voltage. In the periods of time in which the frequency of the output voltage is equal to that of the calibration frequency (or one of its harmonics), beating causes characteristic marks ("pips") to occur.

The frequency error of the fundamental frequency is $<$ 50 kc/s, that is to say, rather better than 0.3%. In view of the fact that reading uncertainty in oscillograms is $>$ 1%, this is quite sufficient. Where very great accuracy is requested it is also possible to calibrate the calibration oscillator even more accurately by means of a quartz crystal oscillator which can be inserted in front of the test signal generator (Fig. 18-11). The voltage of the quartz oscillator is combined with that of the calibration oscillator. It is fed to the output voltage of the mixing stage via the output filter. The voltage of the quartz oscillator also reaches the output socket for the calibrating frequencies, so that two calibrating frequencies are available at this point. Two interference marks can be made visible simultaneously on the pattern of the passband curve. It is thus possible, when examining television receivers, to check the picture-sound separation in the passband curve with a 5.5 Mc/s crystal oscillator by a single adjustment of the calibrating oscillator.

Fig. 18-11 Exterior of the "GM 2889/01" generator; right-hand scale: adjustment of the fundamental frequency; left-hand scale: calibration frequency generator; below it, left, both pairs of sockets for the plug quartzes

A measuring system for investigating the discriminator curve of the sound part of a television receiver is shown in Fig. 18-12. It applies equally to the ultra-short wave part of radio receivers, so that a separate example is unnecessary in this instance. The output voltage of the signal generator is applied to the grid of the valve preceding the discriminator filter. The probe (connected 1 : 1) takes the voltage off the discriminator stage. Between the probe lead-in and the oscilloscope input the *RC*-filter indicated has proved suitable. The upper cut-off frequency is reduced by this means so that the overlapping mark due to the marking frequency becomes narrower as the range of the beat frequencies is narrowed by it.

A discriminator curve of the audio channel of the Philips "17 TD 140 A" receiver is shown in Fig. 18-13. The linear middle part of the curve is about $\pm$ 60 kc/s wide.

For certain tasks, e.g. examining the interference suppression of ratio detectors, the input voltage is modulated with sinusoidal voltages or pulse voltages [12]. Fig. 18-14 shows such an oscillogram with additional sinusoidal amplitude modulation where the discriminator was adjusted symmetrically. Passband curves of an USW-MF filter are reproduced in the last two oscillograms of Fig. 18-15. In *a)* the mid-frequency was marked at 10.7 Mc/s. For *b* the marking voltage of 10.7 Mc/s was mixed in addition with the output voltage of a 100 kc/s *RC* generator, so that frequency markings occur at this frequency interval also on the flanks of the passband curve.

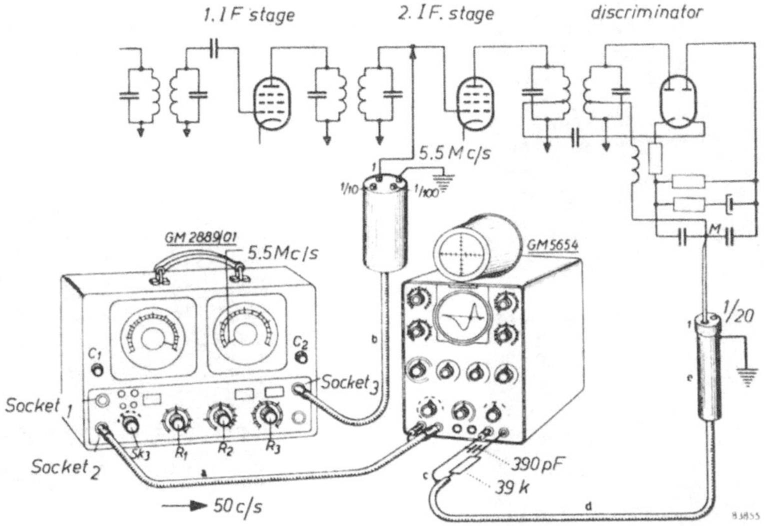

Fig. 18-12 Displaying discriminator curves in the audio part of television receivers

The passband curve of a television receiver can be displayed in a similar way. The frequency swing must be a maximum of 15 to 20 Mc/s. If the measurement is to be over all, the signal generator must be connected to the input of the set and its output frequency adjusted to the required value. In order to examine the MF part the inter-frequency signal voltage is coupled in at a suitable place in the MF portion.

If the whole MF-portion is to be examinated, it is advisable to introduce the signal voltage into the circuit via an auxiliary electrode mounted on the mixing valve with a rubber cap, via the capacitance to the valve anode ("insufflator cap").

The input of the oscilloscope amplifier is either connected to the demodulation valve or to the anode of the video stage i.e. to the cathode of the picture tube, preferably via a voltage divider probe. Fig. 18-16 shows the filter curve of the TV-picture MF amplifier in the Philips "17 TD 140 A" television receiver as an

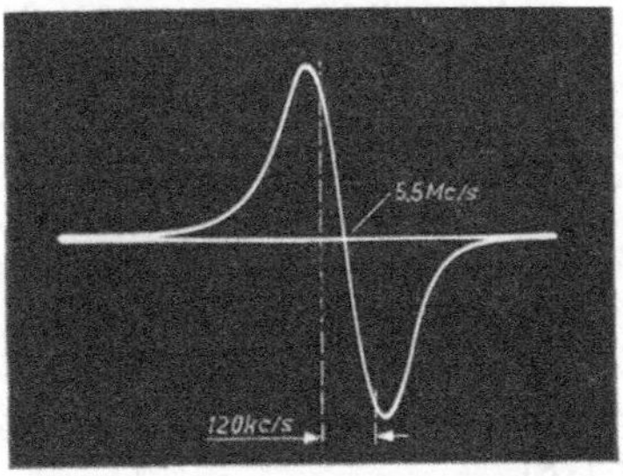

Fig. 18-13 Discriminator curve of Philips "17 TD 140 A" television receiver

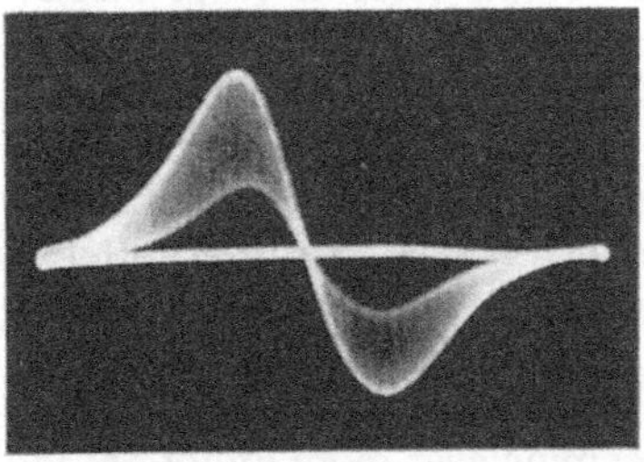

Fig. 18-14 Discriminator curve with AM for adjusting the symmetry of the circuit

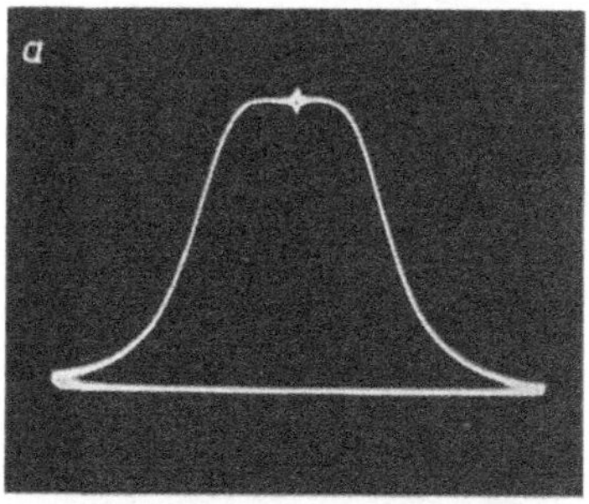

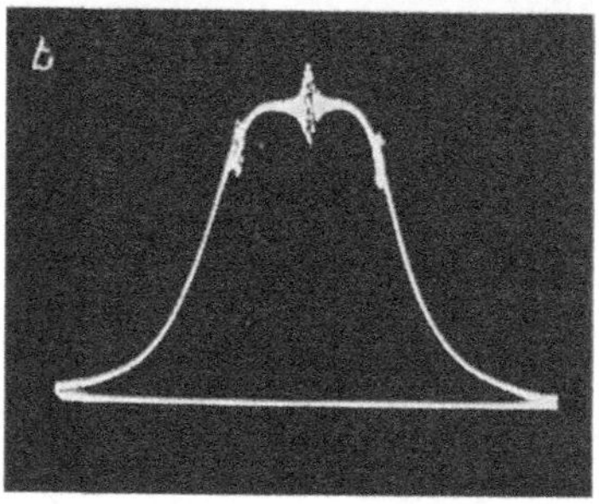

Fig. 18-15 Passband curves of a 10.7 Mc/s filter
a) 10.7 Mc/s mark in the middle of the curve
b) 10.7 Mc/s mark and frequency markings ± 100 kc/s

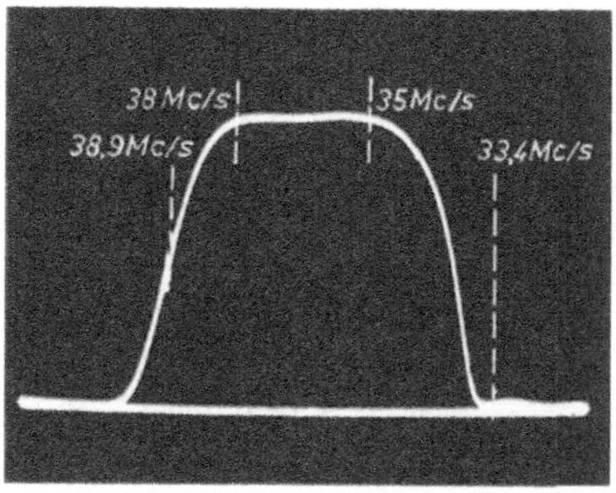

Fig. 18-16 Picture-MF-passband curve of the "17 TD 140 A" Philips TV receiver

example. The middle of the Nyquist flank — the place for the picture carrier — is marked by a marking frequency of 38.9 Mc/s. The passband of the filter for practically constant response from 35...38 Mc/s and the position of the sound carrier in this curve at 33.4 Mc/s were previously measured by the marking generator and then entered on the oscillogram.

The processes described are suitable for all applications and can satisfy stringent demands. For special tasks with special demands as to accuracy and ease in reading, special measuring apparatus have been described, the use of which is restricted to narrow fields of application by reason of its high cost [13].

With the aid of the "GM 2894" phase velocity measuring apparatus as an accessory to the Philips AM-FM "GM 2889" signal generator, it is also possible to display the phase velocity or phase linearity of television receivers as a curve on the oscilloscope screen. As is known, it is not sufficient for the amplitude characteristic of the receiver to be linear within the limits of the transmission range; certain demands must be made on the waveform of the phase characteristic curve. For this reason great importance is attached to observation of the phase characteristic or phase velocity. This is especially the case when the problem arises whether certain "standard" abnormalities in the phase characteristic of the receiver must be corrected at the transmitter, or if this should be taken into consideration in the design of the receiver. The measurement of such characteristics has been the subject of detailed discussion in many publications. Irrespective of which of the two solutions will be favoured in future, measurement of phase velocity is becoming ever more important [14] [15] [16] [17] [18].

CHAPTER 19

INVESTIGATIONS IN TELEVISION ENGINEERING

19. 1 Application of the oscilloscope in TV engineering

In judging the quality and performance of television engineering installations, be it in development, for checking circuitry or in repair work, it is essential to be able to examine the dependence on time of the various voltages and currents much more closely than in other branches of radio engineering. The patterns of these waveforms are usually much more important than the absolute values. For this reason the cathode ray oscilloscope is of particular value in this field, in which it has three fundamental types of task.

In the first of these, the oscilloscope is used for examining the pattern and amplitude of the video signal containing the information appearing as intensity fluctuations of the television picture during the individual picture lines (amplitude modulation), and usually also synchronizing and blanking pulses.

In the second of these applications it is necessary to observe the waveform and currents with which, controlled by the synchronous pulses of the master pulse generator of the TV transmitter, the picture raster is traced in the receiver. A similar task is the observation of the deflection voltages and deflection currents in television cameras and the investigation of similar forms of voltage in the master pulse generator and of the installations in the studio.

In the third type of investigation requiring the use of oscilloscopes, the frequency response or attenuation curve of the transmission channels (in receivers of the high and medium-frequency stages) is displayed directly with a frequency-modulated input signal (by means of a wobbulator). A similar use is the display of standing waves in measurements of matching, as described in detail in Ch. 20. The display of frequency response or passband curves of television receivers has been discussed in section 18, so that the present part can be limited to a treatment of the two first-named applications (observation of video voltage and examination of the deflection voltages).

19. 2 Choice of oscilloscope

The stringency of the demands made on the performance of the oscilloscope depend upon the particular field of application.

The repair or service workshop makes the lowest demands on measuring techniques. Essentially, what is required is the display of the waveforms of deflection voltages, and currents, their frequencies, and the passband curve by means of a frequency-modulated signal derived from a modulator, and the fairly accurate measurement of the other voltages occurring in the set. Observation of the video voltage content individually is very seldom neccessary. Hence only limited demands are made on the frequency range of the Y-amplifier and on the range of the adjustable time scale. This means that the amplifier is called on to amplify square voltages in the frequency range from 50 c/s to about 20 kc/s without appreciable distortion. The picture raster is, of course, traced in accordance with the standards accepted in the various television regions of the world, that is to say frame deflection frequencies of 50 c/s (Europe incl. Great Britain) or 60 c/s (U.S.A.) and horizontal deflection frequencies of 10,125 c/s (Gt. Britain), 15,750 c/s (U.S.A.), 15,625 (continent of Europe, excl. France) or 20,475 c/s (France). In accordance with Eqs. (5.1) and (5.2) the oscilloscope is quite capable of displaying such voltages if the cut-off frequencies of its amplifiers lie in the region of 2 c/s and 2.3 Mc/s. What is most important of all is good pulse reproduction (overshoot $< 5\%$). A range of time scales of about 5 μs/cm to 10 ms/cm is quite sufficient. For horizontal deflection frequencies of 15,625 c/s or 15,725 c/s (Central-European 625-line standard, excluding France and the 525-line standard in the U.S.A.), an upper cut-off limit of only about 1.5 Mc/s would be sufficient. The Philips "GM 5659" oscilloscope, which has cut-off frequencies of 0.8 c/s and 0.8 Mc/s for these purposes, has even been found fully satisfactory as a service oscilloscope.

The extent to which the upper cut-off frequency of the oscilloscope amplifier influences the image of higher modulation frequencies can be seen in Fig. 19-1*a* and *b*. For this oscillogram the output voltage of the Philips television test generator "GM 2887C" was photographed for two line cycles with 11 bar pulses in addition as modulation (11 vertical bars in the picture). In Fig. 19-1*a* the upper cut-off frequency of the oscilloscope was 3 Mc/s, in Fig. 19-1*b* it was 0.8 Mc/s. It is immediately clear that the modulation pulse reproduction is quite good enough, even at a cut-off frequency of 0.8 Mc/s, for most tests on deflection voltages and currents required in repair work. A cut-off frequency of 3 Mc/s would satisfy even more stringent demands.

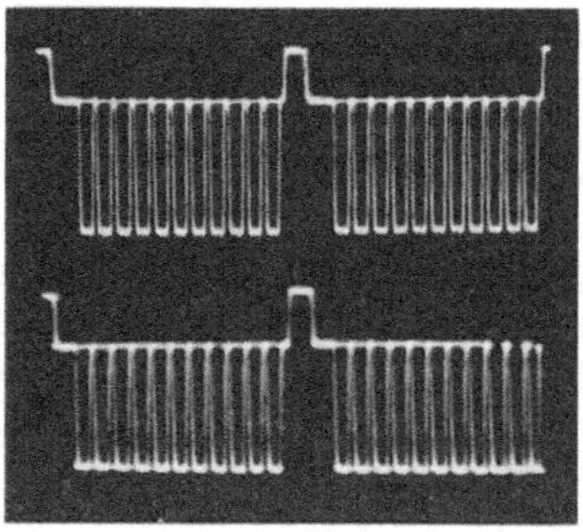

Fig. 19-1 Portrayal of signal from "GM 2887 C" pattern generator for the duration of two horizontal deflections of vertical amplifiers with an upper cut-off frequency of *a*) = 3 Mc/s and *b*) = 0.8 Mc/s

Nowadays the signal amplifiers of such oscilloscopes are usually supplied with a DC amplifier (lower cut-off frequency 0) and an upper cut-off frequency of 5 Mc/s (Philips "GM 5650", Fig. 5-63). The upper cut-off frequencey of 5 Mc/s is, of course, more than twice as high as would seem necessary for the purpose for reasons already given. But it is made very high to make these oscilloscopes as versatile as possible.

Far higher demands are made on oscilloscopes used in TV receiver development. Here the upper cut-off frequency must be at least 5 Mc/s (Philips "PM 3201"), but in general the requirement for such work is that the frequency response should be linear up to 5 Mc/s. This means that the upper cut-off frequency should be

10 Mc/s or, better still, 14-15 Mc/s (Philips "GM 5602" and "GM 5603"). The time scales should thus at least include the range of from 5 μs/cm to 10 ms/cm and image expansion should be possible to at least 0.5 μs/cm. Good, stable triggering with both frame and line frequency pulses is required. For measurements within the circuits of the equipment being investigated, the probes should have as high an impedance as possible. DC-coupled cathode follower probes as supplied, for example, with the Philips "GM 5603" ("GM 4603 D/00") oscilloscopes are especially suitable. Their input impedance is equivalent to a parallel circuit of 0.5 MΩ and 5.5 pF. In selecting such an oscilloscope, even greater store must be set on perfect reproduction of pulses with only slight overshoot ($< 2\%$).

Particularly stringent demands are made on oscilloscopes used for checking transmission systems. Such demands have already been discussed in innumerable publications and have been laid down in the standards of the postal authorities and other competent institutions [1] [2] [3] [4] [5] [6] [7] [8].

It is essential, for instance, that the difference of the deflection coefficients in the Y-direction in the range between 30 c/s and 4.5 Mc/s should be less than 0.5 dB [5]. On the other hand it has been strongly recommended that an oscilloscope for these purposes should have an upper limiting frequency of at least 10 Mc/s and that overshoot should not exceed 2% [2] and [8], Philips "PM 3230".

Similar demands are made on oscilloscopes used in the development of television studio equipment and of equipment for investigating transmission systems. For these, as for the transmission system control equipment, the very greatest demands as to modulation linearity and sharpness of spot focus for the highest possible accuracy in reading must be met. Models with special properties and applications have been brought out as path control equipment and for studio purposes [9] [10]. The oscilloscopes generally produced for work in the field of television engineering often have special switching facilities for providing the equipment with the requisite "standard" properties, if so desired. Thus, for instance, the Tektronix 524 AD oscilloscope, brought out specially for such tasks, has a "FLAT" switching position. In this case the frequency characteristic of the Y-amplifier does not deviate more than 1% from the horizontal in the range between 60 c/s and 5 Mc/s. Apart from other special facilities this equipment also has an arrangement for delaying the time sweep within a range of from 0 to 25 ms. It is thus possible to display details of the voltage pulse sequence or even of the voltage waveform of any desired single line of the television picture. The greatly expanded pulse images shown can be measured with great accuracy in time by brightness modulation, using frequency dots which are even multiples of the line frequency (Fig. 4-103). DILLENBURGER has given a comprehensive survey of the measuring techniques employed in television engineering [11]. MACEK has dealt with the special measurements required for television receivers [12] and television transmitters [13].

19. 3 Measuring processes and voltage sources

When investigating voltage and current sources in the sets, the picture of the process is obtained directly when the probe is applied in the usual way. It is merely necessary to take care that the probe impedance does not impose any impermissible load or detuning.

An additional voltage source must, however, be available for amplifier measurements.

For general studies simple square voltage pulse generators are used. If, at the same time, it is also desired to observe the characteristics of the picture received, then test pattern generators, which generate an electronic test pattern, are required as a voltage source [14], [15]. For observing and adjusting the geometry of the picture itself, an oscilloscope is of course unnecessary. The picture on the television screen is then itself the best criterion. But the moment more thorough knowledge of the transmission behaviour of the system under investigation is required (overshoot and the like), it is only the oscillogram which can give the necessary information. If the performance of a camera is to be investigated then a test pattern of this sort in photographed. Fig. 19-2 shows a test picture displayed in this way by a television receiver. The two oscillograms obtained from this picture, Fig. 19-3, show the voltage waveform across the first video stage grid, that is to say, immediately after demodulation. A second test pattern, often used both for adjusting cameras and receivers, and closely resembling the American RMA test

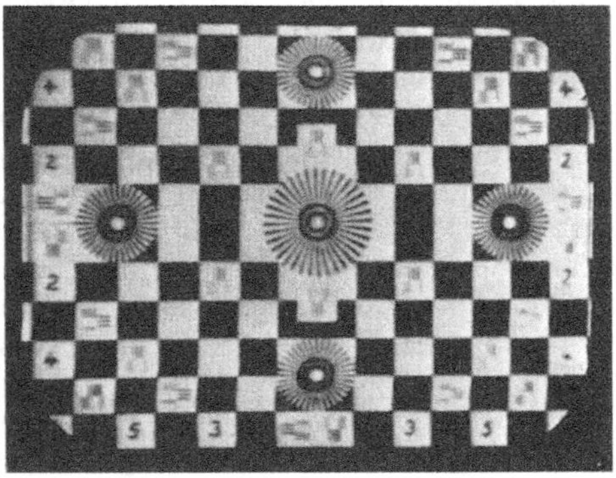

Fig. 19-2 Test-pattern transmitted from West Berlin and used for oscillogram Fig. 19-3

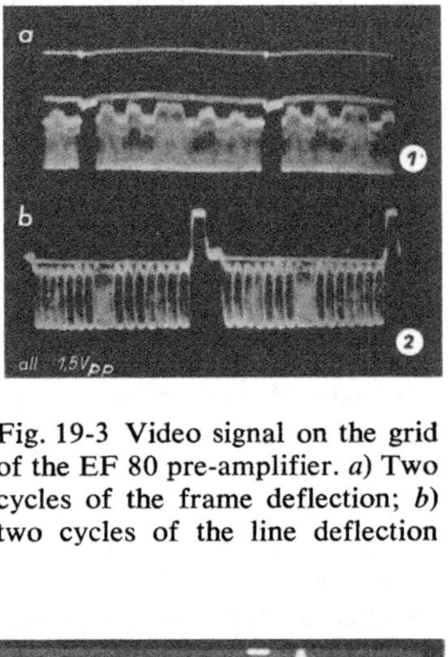

Fig. 19-3 Video signal on the grid of the EF 80 pre-amplifier. *a*) Two cycles of the frame deflection; *b*) two cycles of the line deflection

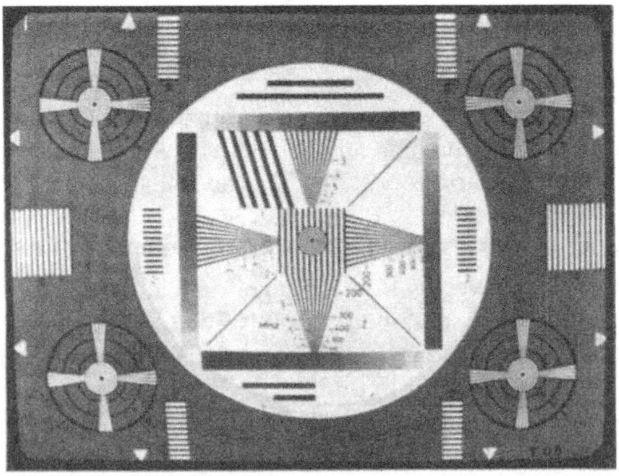

Fig. 19-4 Diapositive test pattern for general checking (studio photograph)

pattern, can be seen in Fig. 19-4. This photograph, like the ones forming part of Fig. 19-5, was taken in the studio before the broadcast [94]).

It is clear that this test pattern is so designed as to provide facilities for examining a large number of different picture details.

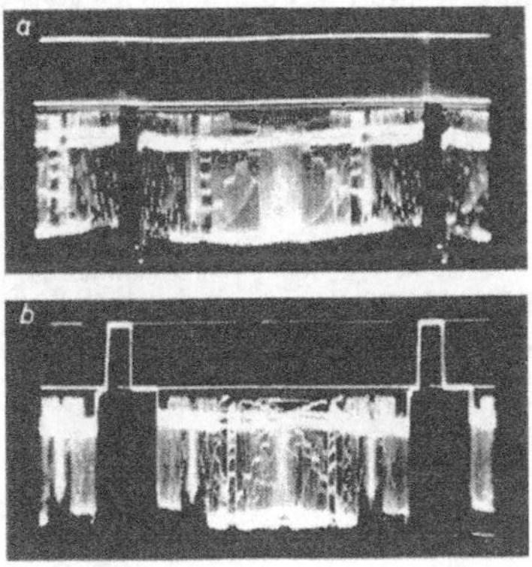

Fig. 19-5 Oscillogram of Fig. 19-4. *a*) one vertical cycle; *b*) one horizontal cycle

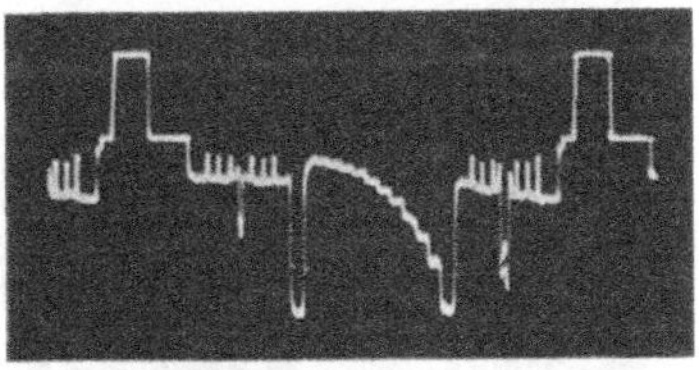

Fig. 19-6 Oscillogram of the lower grey panel of Fig. 19-4

Such pictures (made in this case by a diapositive-scanner) in themselves make possible a large number of investigations such as judging picture geometry and picture definition, and also give a general impression of the behaviour of the transmission system as a whole (flagging, light boundaries due to errors in adjustment and such like). They are not suitable, however, for giving an impression of the linearity of the transmission system, as can be confirmed, for instance, by the oscillogram in Fig. 19-6. This oscillogram of the waveform of a single picture line traversing the horizontal grey scale shows that this step is already transmitted as a curve by the scanner [95]).

Electronically generated test patterns as shown in Fig. 19-7 are better suited for such investigations and are indeed being used to an ever-increasing extent. The voltages for these patterns are so formed and combined that even the simple (unexpanded) oscillograms with the frame or line frequency offer a comprehensive impression of the linearity and general behaviour of the system under study [16].

The rows of vertical lines in Fig. 19-7, going downwards, represent voltage changes with frequencies of 1, 2, 3, 4 and 5 Mc/s. A black panel follows, into which the name of the station can be faded in by a camera or diapositive (see also Fig. 19-15). The white horizontal line in this panel is caused by a needle pulse in the waveform, which can be used to advantage in reflection measurements (aligning television receiver antennas) (Fig. 19-9). The pattern is bounded at the bottom by a grey wedge of graded brightness.

[94]) These photographs and the ones in fig. 19-6, 19-7. 19-8 and 19-9 were made available to the author by the kind permission of the Norddeutsche Rundfunk, Hamburg.

[95]) Adjustement of the contrast control would probably have increased the basic noise intolerably with this type of scanning. The section on "Slow motion oscillograms compared with the relevant lines in the picture on the screen" contains details on oscillographic recording of the waveform of single lines of the picture.

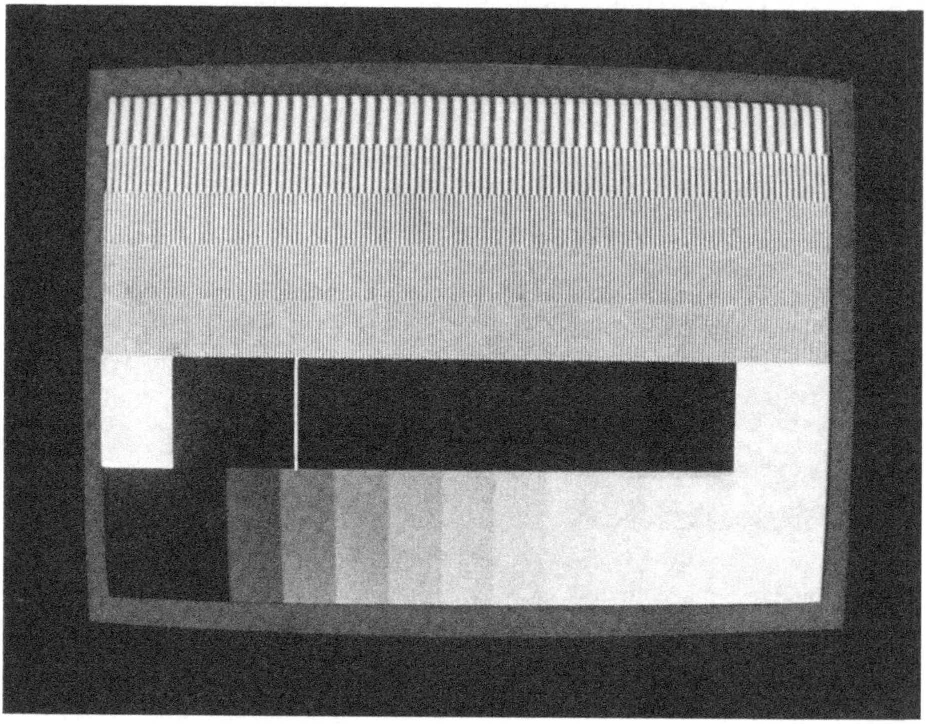

Fig. 19-7 Electronic test pattern. (Studio photograph – Developed by Bavarian Radio)

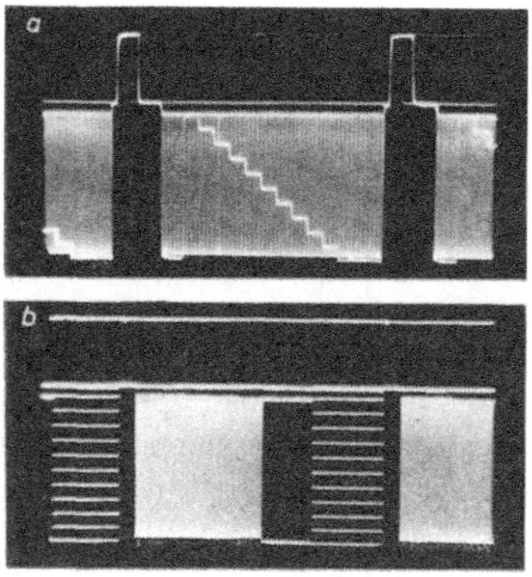

Fig. 19-8 Oscillograms of the electronic test pattern in Fig. 19-7. *a*) one horizontal cycle; *b*) one vertical cycle (studio photographs)

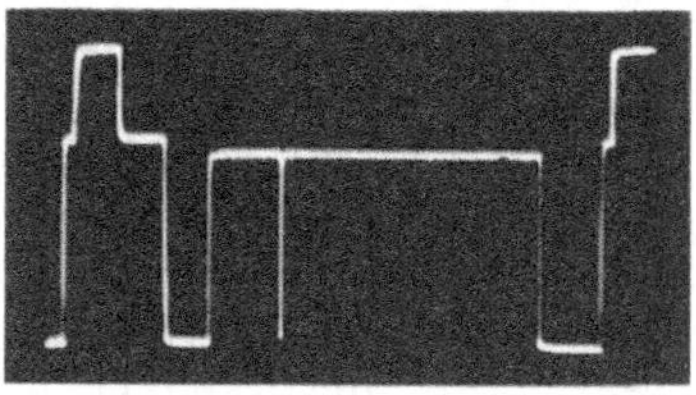

Fig. 19-9 Oscillograms of a line from Fig. 19-7 in the black part of the pattern

In the oscillogram in Fig. 19-8*a* (a line cycle) the rows of frequency verticals, shown one above the other, appear as a uniformly bright surface as all the verticals are equal in height. The multiple line voltages which form the grey scale cause a step voltage to appear above this field of the pattern, giving an indication of the modulation linearity. In the vertical oscillogram in Fig. 19-8*b*, the "staircase" brings about horizontal lines, one above the other, the distances between which are likewise a measure of the linearity. The deflection due to the bar voltages, the frequency increasing from left to right and commencing from the *V*-blanking gap, now appear as rectangular bars, one beside the other. As these are oscillograms of the voltage source (in the studio), these bars are of the same height or the same distance apart. If such an oscillogram had been taken to investigate the behaviour of a frequency-dependent transmission system (Fig. 19-14), then a decrease in height of these bars would be a direct indication of the frequency response.

The oscillogram of Fig. 19-9 shows the voltage waveform of a picture line crossing the black panel, and shows the signal pulse corresponding to the white line (see also the pattern on the receiver screen in Fig. 19-15 and the oscillogram in Fig. 19-16).

Electronic test pattern generators are available in various models both for studio and other use. In conjunction with a standard synchronous pulse generator and an HF generator, these units are often employed as test voltage generators during the manufacture of television receivers. The Philips generators "GM 2657", "GM 2671" and "GM 2681" have been specially introduced for this purpose. (The "GM 2671" supplies the test pattern and synchronous pulses. The "GM 2681" gives 10 crystal-controlled HF signals). For repair work there is little point in making such stringent demands. Test generators which can control the HF signals with simple bar or grid patterns are sufficient. Essentially it is the pattern geometry which has to be examined when the receiver is adjusted. Fig. 19-10 shows the pattern on the receiver screen as obtained with the modulation adjusted to give a fixed block pattern using the Philips television test generator "GM 2892".

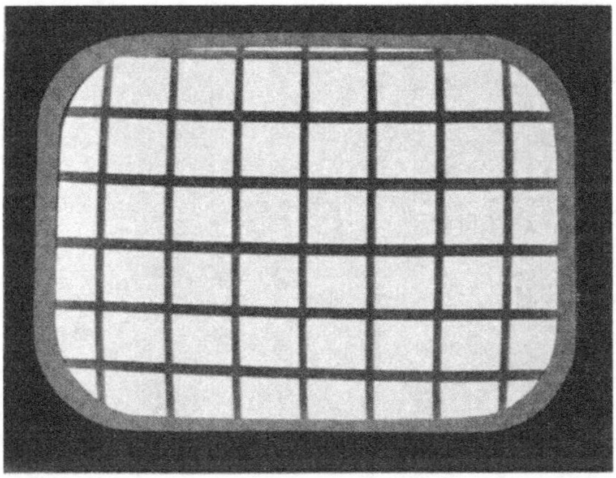

Fig. 19-10 Receiver pattern obtained with the Philips "GM 2892" test generator

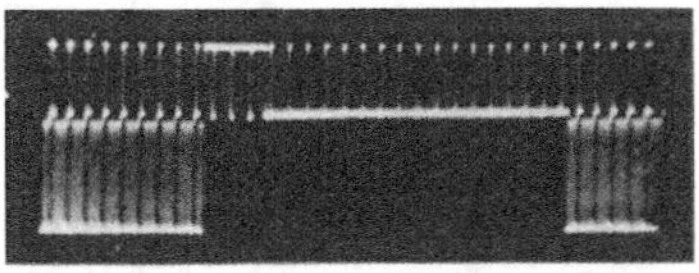

Fig. 19-11 Vertical pulse with blanking gap in Philips "GM 2892" test generator

The synchronous pulses too need only approximate to the standard provided the performance of the receiver corresponds reasonably to standard signals. In the oscillogram in Fig. 19-11 the vertical sync pulse of the "GM 2892" test generator is shown with the blanking gap switched to 50 c/s, mains-synchronized and somewhat expanded. As the horizontal and vertical synchronizing pulses in such units are not synchronized together, the latter do not remain stationary in the vertical frequency oscillogram. A suitable instant had to be seized during which they happened to be stationary long enough to take the photograph.

19. 4 Investigations on television receivers

It is not intended to offer a detailed description of the method of operation of a given set in the following pages. Models of television receivers are too subject to fashion to make this worth while. The reader is referred for this to the relevant servicing manuals. This section merely offers a general survey of the possible applications of the oscilloscope in investigating television receivers. Fig. 19-12 gives as an example the circuit of the video stage of the Philips "TX 1420 U" ("TD 1420 U") television receiver.

A comprehensive body of literature deals in detail with repair techniques [17] [18] [19] [20].

When a test pattern broadcast by a station is received on the screen, as shown in Fig. 19-4, a voltage appears at the first grid of the PL83 valve, the

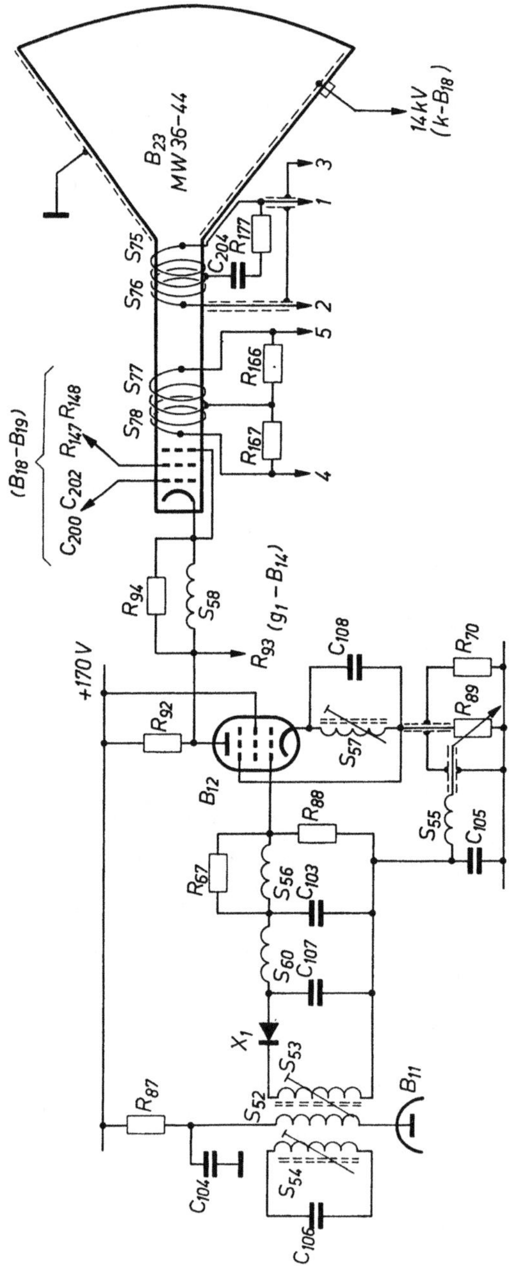

Circuit elements

Resistors	Capacitors
R_{67} = 33 kΩ	C_{103} = 5.6 pF
R_{87} = 1 kΩ	C_{104} = 1.5 nF
R_{88} = 3.3 kΩ	C_{105} = 150 pF
R_{89} = 1 kΩ	C_{107} = 5.6 pF
R_{90} = 330 Ω	C_{108} = 1 nF
R_{92} = 2.2 kΩ	
R_{94} = 6.8 kΩ	

Fig. 19-12 Circuit of video output stage of Philips TX 1420 U (TD 1420 U) receiver

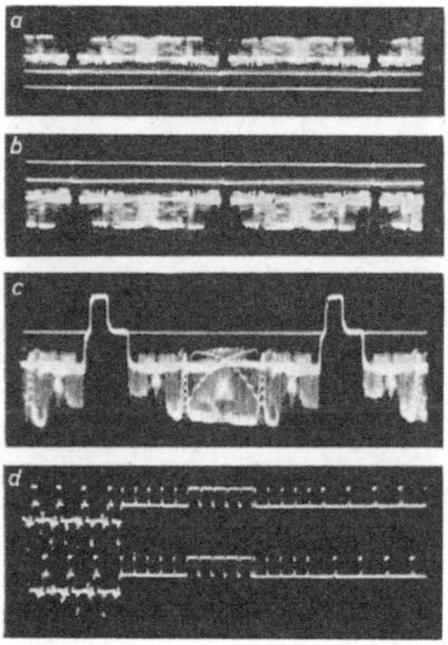

Fig. 19-13 Voltages on video output stage of Fig. 19-12 (Test pattern as in Fig. 19-4).
a) two vertical cycles on 1st grid 7 V_{pp}; *b*) two vertical cycles across anode 90 V_{pp}; *c*) one horizontal cycle across anode; *d*) vertical pulse with blanking gap

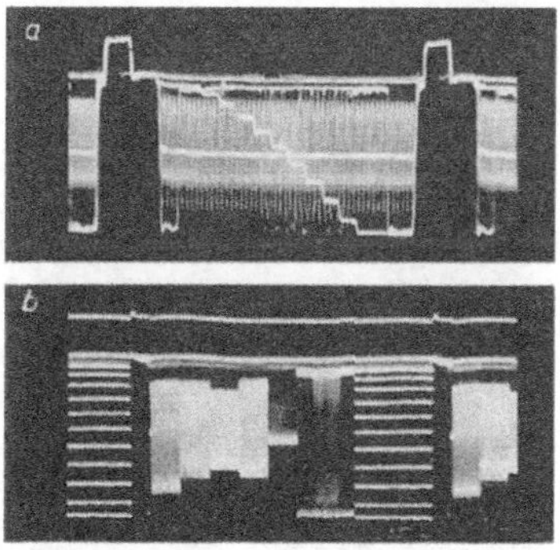

Fig. 19-14 Voltages across anode of PL 83 valve using electronic test pattern as in Fig. 19-7. *a*) one vertical cycle; *b*) one horizontal cycle

waveform of which during two vertical cycles is shown in oscillogram *a*) in Fig. 19-13. The voltage waveform across the anode of this valve during two vertical cycles in shown in Fig. 19-13*b*. Oscillogram 19-13*c* represents a horizontal cycle at the same point as the first two, but somewhat amplified.

The vertical sync pulse is shown as a double oscillogram, for both frame pulse sequences in an expanded section of the start of the vertical blanking gap [96]).

For the purpose of comparison with the corresponding photographs of the voltage pictures of the electronic test pattern in the studio (Fig. 19-8), Fig. 19-14 shows the voltages at the anode of the video output tube in the receiver during both a vertical and a horizontal cycle. The unfavourable influence of the various parts of the transmission chain, especially of the limited frequency range in the receiver, is clearly recognizable.

Fig. 19-15 shows how the electronic test pattern of Fig. 19-7 itself arrives in the receiver. The oscillogram of the brightened line on the black panel showing the name of the station is seen in Fig. 19-9. The amplitude of the needle pulse has been reduced to about 1/3 for reasons already stated.

---

[96]) The section on "Phase-delayed triggering of the time deflection" (figs. 4-76 and 4-77) describes how time-expanded sections of oscillograms can be displayed. The simultaneous display of both series of frame-sync pulses is described in the section on "the time base expansion unit". Some of these photographs were obtained by combining the Philips "PP 1071" electronic switch, the "GM 4585" pulse delay unit and the "GM 5603" oscilloscope. Likewise, the Tektronix type 535 (or type 545) oscilloscope can be used in combination with the "CA" double-channel preamplifier rack unit and the "delayed sweep".

In both cases the input signal voltage should be applied simultaneously to both inputs of the electronic switches and the latter set to "alternate".

Fig. 19-15 Electronic test pattern photographed at the receiver

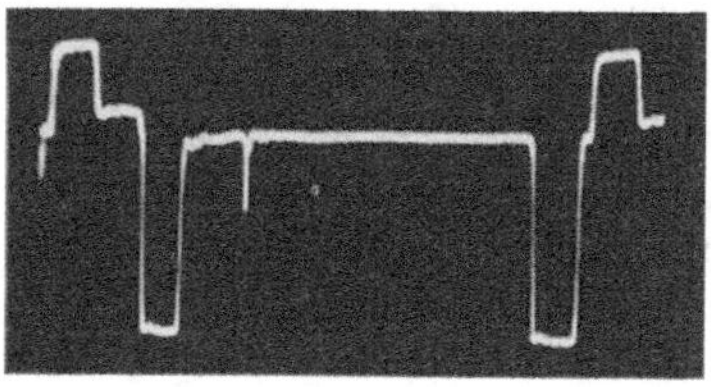

Fig. 19-16 Voltage waveform of the brightened line in the receiver in Fig. 19-15

From the anode of the PL 83 (B_{12}) valve the amplified video voltage controls the cathode of the picture tube (B_{23}). This voltage is also applied to the pulse-separator stage, the so-called amplitude discriminator (B_{14}). In this stage the synchronized pulses for frame and horizontal deflection are separated from the mixed signal consisting of picture signal, blanking signal and synchronization pulses, and further amplified. Fig. 19-17 shows this circuit in the "TX 1420 U" receiver together with the line deflection oscillator.

Commencing with the oscillograms of the voltage at the anode of the video output tube PL 83 (B_{12}), as shown in Fig. 19-13, the following illustrations show oscillograms of the operation of the pulse separator stage and of the amplitude discriminator [21].

The valve system B_{14} serves both as a pulse discriminator and as a pulse amplifier. The oscillogram of the voltage waveform during two frame periods across the grid

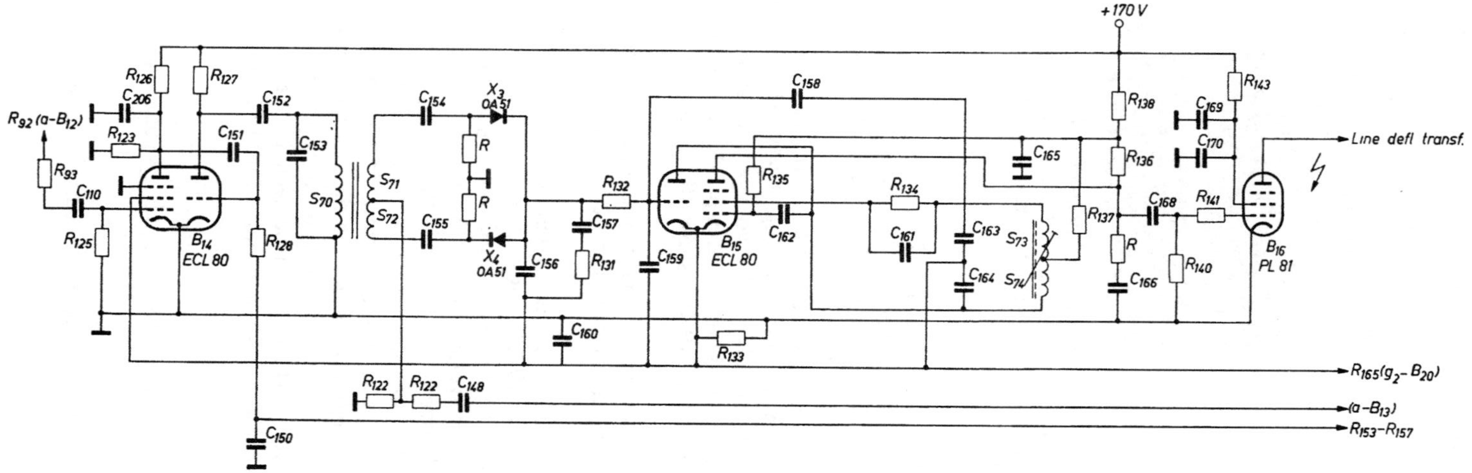

Circuit elements

Resistors

R_{93} =	33 kΩ
R_{121} =	2.2 kΩ
R_{122} =	270 Ω
R_{123} =	39 kΩ
R_{125} =	2.2 MΩ
R_{126} =	220 kΩ
R_{127} =	18 kΩ
R_{128} =	470 kΩ
R_{129} =	120 kΩ
R_{130} =	120 kΩ
R_{131} =	5.6 kΩ
R_{132} =	56 kΩ
R_{133} =	680 Ω
R_{134} =	22 kΩ
R_{135} =	3.3 MΩ
R_{136} =	100 kΩ
R_{137} =	2.7 kΩ
R_{138} =	1 kΩ
R_{139} =	15 kΩ
R_{140} =	560 kΩ
R_{141} =	1 kΩ
R_{143} =	2.2 kΩ

Capacitors

C_{110} =	820 pF
C_{148} =	470 pF
C_{150} =	47 pF
C_{151} =	1.5 nF
C_{152} =	1.5 nF
C_{153} =	180 pF
C_{154} =	1.5 nF
C_{155} =	1.1 nF
C_{156} =	56 nF
C_{157} =	0.47 μF
C_{158} =	68 pF
C_{159} =	82 pF
C_{160} =	0.1 μF
C_{161} =	6.8 nF
C_{162} =	2.7 nF
C_{163} =	10 nF
C_{164} =	3 nF
C_{165} =	50 μF
C_{166} =	2.2 nF
C_{168} =	2.7 nF
C_{170} =	1.5 nF
C_{170a} =	47 nF
C_{206} =	100 pF

Fig. 19-17 Circuit of pulse separating stage and of the horizontal deflection generator

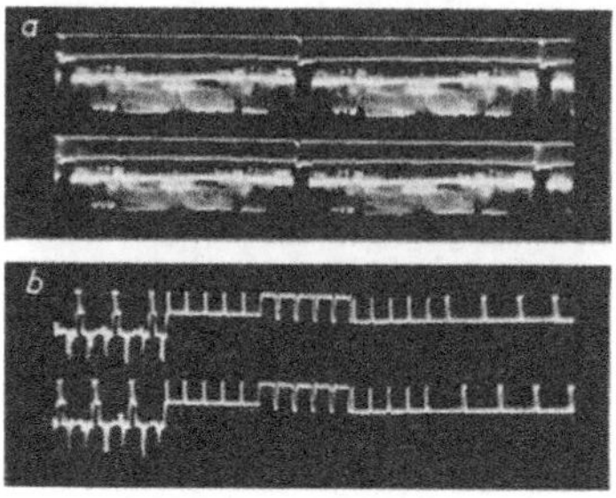

Fig. 19-18 Voltages on 1st grid of pulse separator stage. *a*) two vertical cycles 75 V_{pp} *b*) time expanded oscillogram of the vertical sync pulse

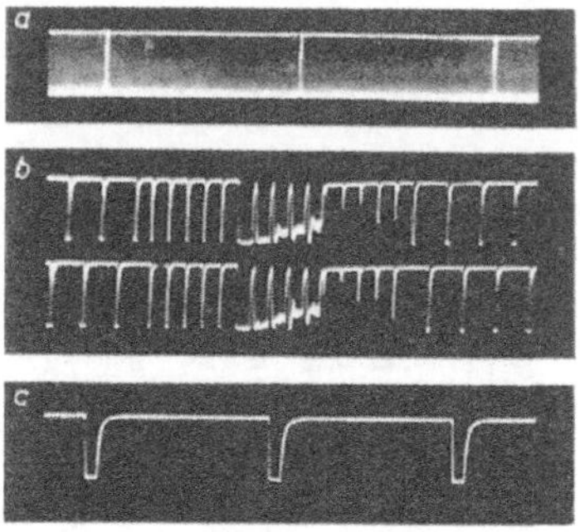

Fig. 19-19 Anode of the pentode of the pulse separator stage. *a*) two vertical cycles; *b*) time expanded picture around vertical pulse; *c*) two horizontal cycles

of the pulse discriminator stage (pentode B_{14}) can be seen in Fig. 19-18*a*. The time-expanded waveform at the commencement of the vertical blanking is shown in Fig. 19-18*b* as a double oscillogram. Only very slight distortion occurs here, which does not significantly affect the working of the circuit.

The voltage waveform across the anode of the pentode of stage B_{14} is shown in the oscillograms of Fig. 19-19. The pattern on the screen in Fig. 19-19*a* shows the voltage waveform during two vertical cycles. Details of the voltage waveform are not recognizable in this oscillogram. It is only by making a time-expanded oscillogram as shown in Fig. 19-19*b* that this is possible. Fig. 19-19*c* shows the voltage waveform during two line cycles.

The voltage at the anode of the pentode system of this valve reaches both the triode grid and the first link of the integrating network (*R*128 - *C*150), accentuating the sync pulse for the vertical synchronization.

The voltage waveform across the anode of the triode system of this stage can be seen in the oscillograms of Fig. 19-20. Fig. 19-20*a* shows the waveform during two frame cycles, and Fig. 19-20*b* shows the time-expanded oscillogram of the voltage waveform with the frame sync pulse. The horizontal cycles with the sync pulses are shown in the oscillogram of Fig. 19-20*c*.

This pulse train is now fed to a phase discriminator stage via the discriminator transformer. The voltage waveform on the primary of this transformer can be seen in the oscillograms in Fig. 19-21*a*, *b* and *c*. Fig. 19-21*a* shows

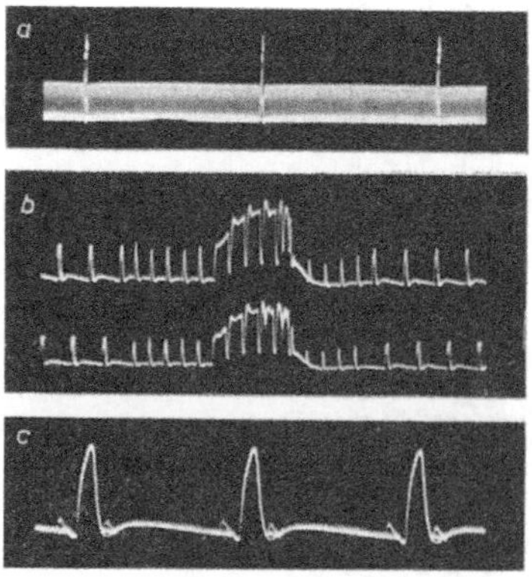

Fig. 19-20 Anode of the triode of pulse separator stage
a) two vertical cycles
b) time expanded picture about the vertical pulse
c) two horizontal cycles

the voltage pattern during two cycles of the frame deflection and *b*) the pattern during two cycles of the line deflection. The time-expanded oscillogram in *c*) shows the pulse train again in the vicinity of the frame sync pulse. In this receiver a control voltage which is dependent on the relative phase position of the sync pulses with respect to a voltage pulse fed back from the line deflection generator is generated by means of two germanium diodes OA 51 (X_3 and X_4) in Fig. 19-17.

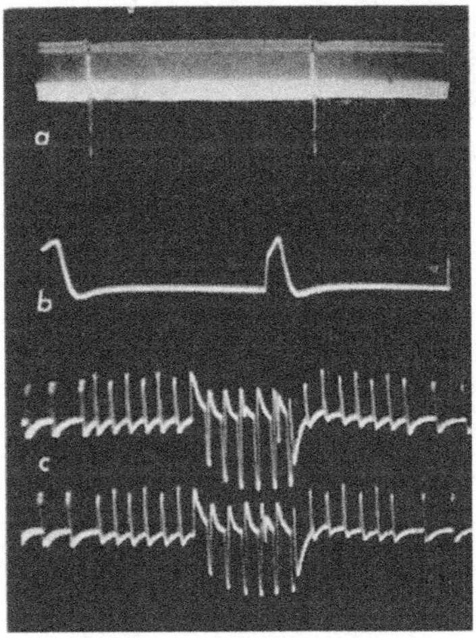

Fig. 19-21 Voltage on primary of discriminator transformer. *a*) Two cycles of frame deflection; *b*) two cycles of line deflection; *c*) time-expanded oscillogram

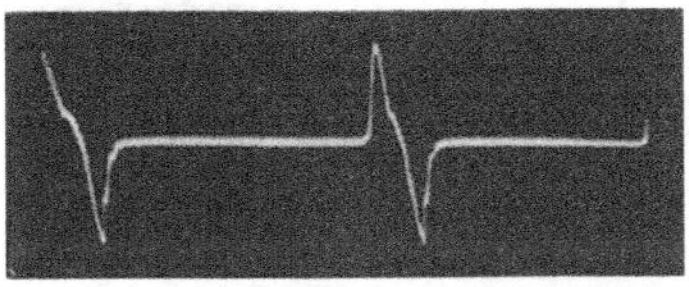

Fig. 19-22 Voltage waveform at centre tap of secondary of discriminator transformer

The germanium diodes are oppositely connected to the symmetrical secondary winding of the discriminator transformer. The voltage waveform at the centre tap of the secondary during two line cycles is shown in the oscillogram of Fig. 19-22. The control voltage, which is the difference voltage of the two diodes, controls the inductive reactance of the triode system of the ECL 80 valve (B_{15}) of the line deflection generator. It thus brings about a frequency change of the sine-wave generator in the pentode system and hence the necessary synchronization between transmitter and receiver. The form of grid voltage required for driving the line output valve PL 81 (B_{16}) is obtained by adding the asymmetrical rectangular voltage generated by the oscillator circuit to a sawtooth produced by integration of a part of the rectangular voltage in an *RC*-network. The waveform of this voltage at the anode of the pentode section of the ECL 80 (B_{15}) can be seen in Fig. 19-23.

The sawtooth for frame deflection is generated in the triode section of an ECL 80, working as a blocking oscillator. The circuit for generating the currents for vertical deflection in the "TX 1420 U" receiver is to be seen in Fig. 19-24.

For synchronization, pulses are again taken from the anode of the synchronization

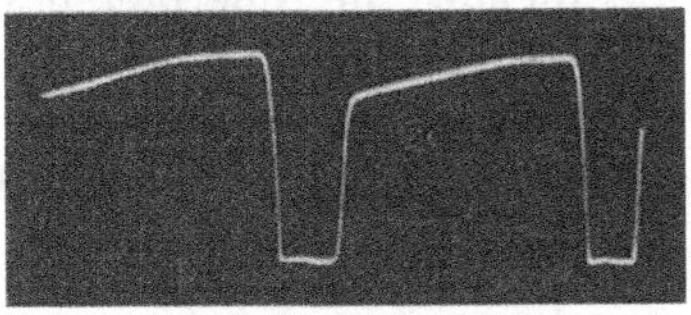

Fig. 19-23 Voltage on anode of the pentode section of line deflection generator

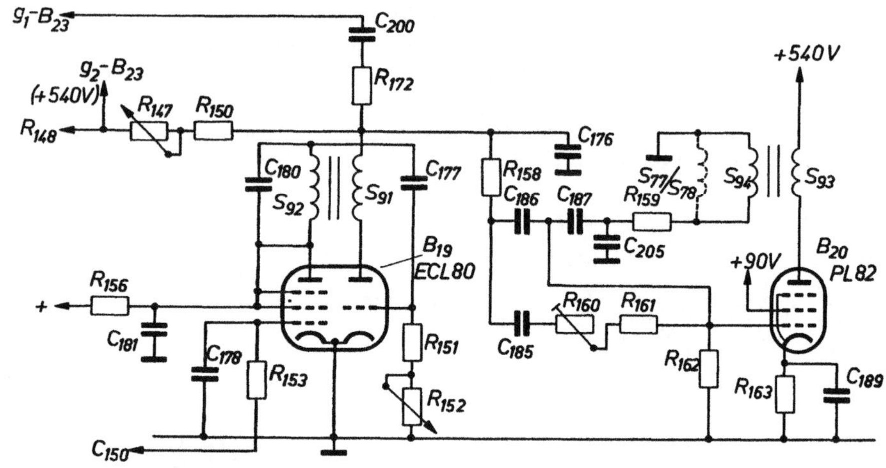

Circuit elements

Resistors

$R_{147} = 1$ MΩ
$R_{150} = 0.68$ MΩ
$R_{151} = 0.33$ MΩ
$R_{152} = 0.5$ MΩ
$R_{153} = 0.47$ MΩ
$R_{156} = 68$ MΩ
$R_{158} = 12$ kΩ
$R_{159} = 3.3$ kΩ
$R_{160} = 1$ MΩ
$R_{161} = 0.39$ MΩ
$R_{162} = 1$ MΩ
$R_{163} = 680$ Ω
$R_{172} = 0.1$ MΩ

Capacitors

$C_{176} = 82$ nF
$C_{177} = 22$ nF
$C_{178} = 220$ pF
$C_{180} = 270$ pF
$C_{181} = 47$ nF
$C_{185} = 56$ nF
$C_{186} = 12$ nF
$C_{187} = 47$ nF
$C_{189} = 100$ μF
$C_{200} = 1.5$ nF
$C_{205} = 0.1$ μF

Fig. 19-24 Oscillator and output stage for vertical deflection

separator triode (ECL 80 -B_{14}) and passed through two *RC*-networks to filter out the line pulses. The voltage waveform on the capacitor of the first network is shown in Fig. 19-25*a* for two frame pulses. The time-expanded oscillogram in Fig. 19-25*b* shows the pulse train of both fields. The corresponding voltage waveform on the capacitor (C_{178}) of the second network can be seen in Fig. 19-26*a* and *b*. This voltage appears at the grid of the pentode section of the ECL 80 valve (B_{19}). Here it is amplified, separated and reversed in phase so that the voltage from the anode of this system appears as positive pulses on a winding of the transformer (S_{92}) in the blocking oscillator and is used for synchronization.

The normal oscillogram of the anode voltage waveform during two cycles of the frame frequency is shown in Fig. 19-27*a*, but the real trend of the voltage pulse only becomes clearly visible after time expansion — shown singly this time — in Fig. 19-27*b*. For the purpose of determining the frequency of the visible ringing, the waveform of a sinusodial voltage with a frequency of 10 kc/s was recorded below the oscillogram. Interpretation of the oscillograms shows the ringing to have a frequency of roughly 175 kc/s. A sawtooth voltage appears once again across a charging capacitor in the blocking oscillator circuit (C_{176}), which, due to the operation of the oscillator, has a negative peak superimposed on it, as shown in Fig. 19-28.

A parabolic component is added to this voltage in the usual way by means of

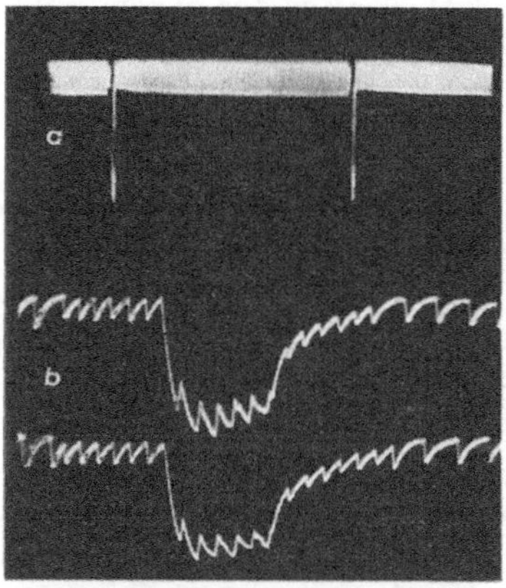

Fig. 19-25 Waveform of vertical sync pulse on first *RC*-network. *a*) Two frame deflection cycles; *b*) strongly expanded oscillogram (25 V_{pp})

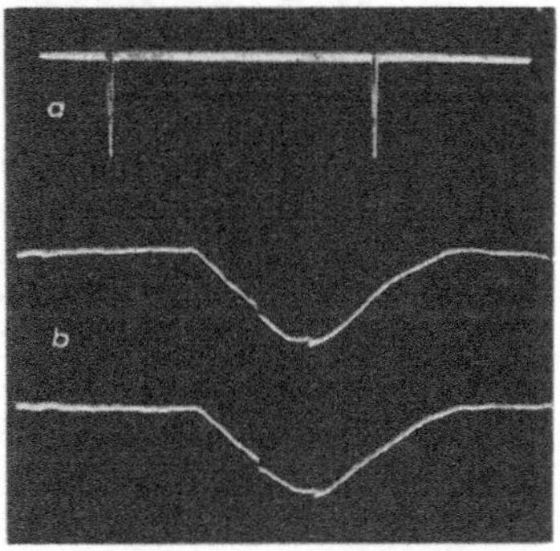

Fig. 19-26 Voltage on grid of vertical sync amplifier. *a*) Two cycles of frame deflection; *b*) greatly time-expanded oscillogram (7.5 V_{pp})

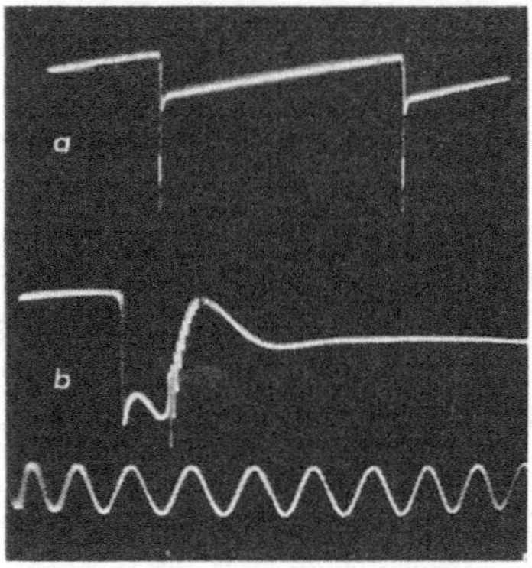

Fig. 19-27 Anode voltage of blocking oscillator for frame deflection (150 V_{pp}). *a*) Two cycles of frame deflection; *b*) greatly time-expanded; time marking 10,000 c/s

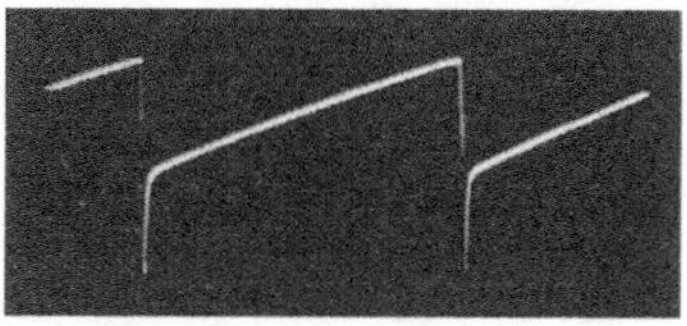

Fig. 19-28 Voltage across the charging-capacitor of the blocking-oscillator-circuit

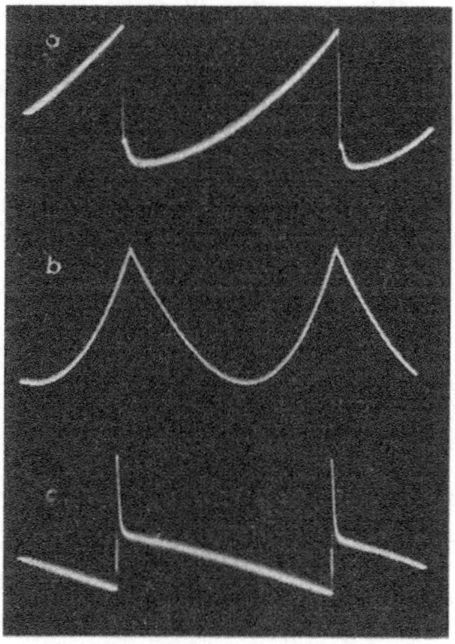

Fig. 19-29 Voltage on frame output stage. *a*) Waveform on the grid (7.5 V_{pp}); *b*) Parabolic component of cathode voltage (0.4 V_{pp}); *c*) Waveform on the anode (1500 V_{pp})

RC-networks in the grid circuit of the PL 82 output stage (B_{20}), so that the actual voltage on the grid is shown by the oscillogram in Fig. 19-29*a*. The approximate waveform of the parabolic voltage can be observed on the cathode resistor (R_{163}) (Fig. 19-29*b*). The waveform of the corresponding anode voltage is seen in Fig. 19-29*c*.

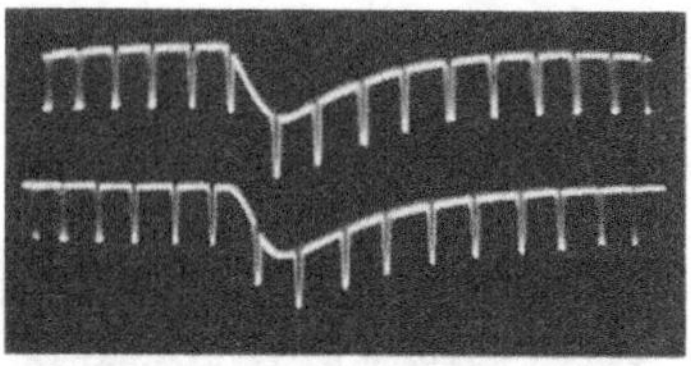

Fig. 19-30 Flyback blanking on grid of picture tube (45 V_{pp}; total pulse height 65 V_{pp})

To blank the spot, flyback traces on the screen of the MW 36-44 picture tube (B_{23} in Fig. 19-12), the grid receives a pulse at line frequency from the line output transformer as well as a pulse at the frame frequency from the anode of the blocking oscillator triode section of the ECL 80 (B_{19}). The time-expanded oscillograms of the pulse trains in Fig. 19-30 represent the waveform of these voltages on g_1 of the picture tube in the vicinity of the vertical blanking pulse. It can clearly be seen that the vertical blanking of both fields is in phase, whereas the line blanking, as desired, is displaced by a distance equal to half the spacing between the pulses.

As has already been said at the outset, there is no doubt that study of the waveforms is of particular importance in television engineering. The oscillograms reproduced here can only be regarded as a random selection from the most important waveforms. For accurate measurement of the currents and voltages employed during the development of deflection circuits, care must be taken that no alteration of the working conditions occurs when switching on the oscilloscope. A suitable procedure for the study of line deflection circuits has been described by Schröder [22].

19. 5 Time-expanded oscillograms when comparing selected lines of the picture

The normal oscillograms which show the composite signal during one or more cycles of line deflection (Figs. 19-5*b*, 19-8*a* and *c*, and 19-31*b*) are usually referred to as "line patterns". Yet it should always be remembered that what is seen is the sum of the voltages of all 625 lines of both fields during the time of perception or of the time required to record the oscillogram. This produces the well-known "veil" between the line pulses, which fluctuates with the instantaneous values of the video signal voltages or moves in a horizontal direction. The similar oscillograms are shown in Fig. 19-31. In this way, among other things, the lateral movement of an object in the television picture can be recognized in an oscillogram of the horizontal deflection cycles. However, by using a time base expansion unit to produce a greatly expanded oscillogram in the rhythm of the frame frequency, a single curve can be observed which represents the actual voltage waveform of the signal-pulse during only one cycle of the line frequency. With a moving scene the voltage curve will, of course, vary in amplitude according to the variations in brightness, and will also move horizontally, corresponding to the lateral movements of the bright or dark patches. None the less, there will be nothing but simple lines all the time. When the pictures are stationary, as, for instance, test patterns, the oscillograms will, of course, also be stationary. Moreover, if at the same time the appropriate lines in the receiver picture are brightened by an unblanking pulse from the time base expansion unit, as described in detail in Part. I, Ch. 4.30 "Time base expansion unit", page 149. (Figs. 4-83, and 4-85 and 4-86), a very clear and useful means of comparing the oscillogram with the corresponding lines of the receiver picture is made available.

The next four illustrations, Figs. 19-32, 19-33, 19-34 and 19-35, will serve to illustrate the possibilities offered by this method. They were recorded during normal reception of the test patterns transmitted by the West Berlin television transmitter.

The voltage fed to the vertical amplifier of the oscilloscope was taken from the anode of the video output stage PL 83. Fig. 19-32*a* shows a double oscillogram, expanded over about twelve cycles of the line frequency, with the frame sync pulse at the beginning of the vertical blanking. Fig. 19-32*b* shows the corresponding picture on the receiver, with the frame frequency adjusted so as to make visible on the screen the gap in the picture during vertical blanking (see also Fig. 4-89*b*). The intensity modulation shows very clearly that the section under observation ends within the vertically blanked gap indicated by arrows in Fig. 19-32*b*.

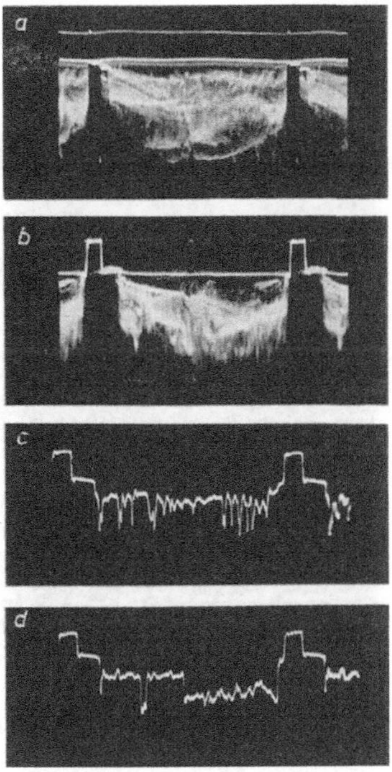

Fig. 19-31 Oscillogram with normal picture modulation. *a*) one vertical pulse; *b*) one horizontal pulse; *c*) and *d*) voltage waveforms of single lines

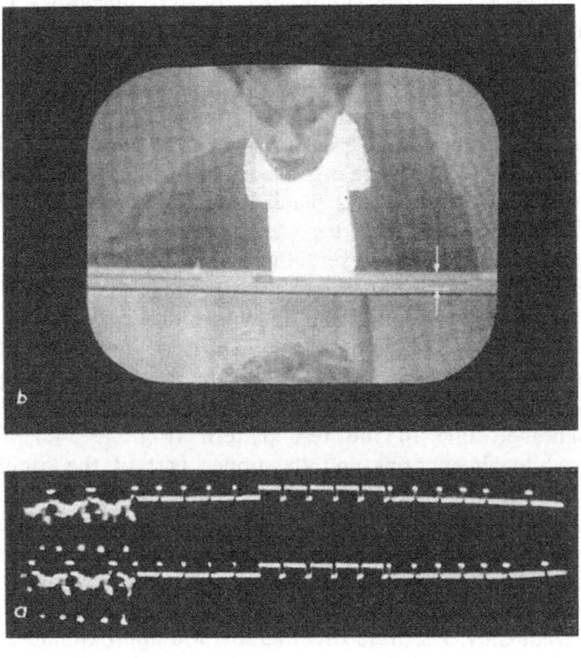

Fig. 19-32 Receiver picture *b*) together with expanded oscillogram *a*) at the beginning of vertical blanking

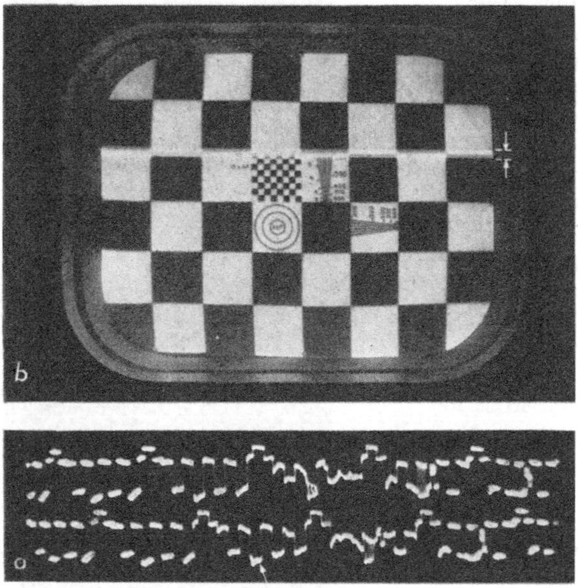

Fig. 19-33 Test pattern *b*) with oscillogram *a*) of five intensity modulated lines of both fields

In Fig. 19-33 a double oscillogram of about five cycles of the line frequency can be observed. Picture *b*) shows how these lines correspond to the lines at the transition point from the simple alternation of black and white squares (second row from the top) to those which are further subdivided for test purposes. The pulse train during the first four line-cycles (counted from the left in the oscillograms of both fields) shows in its essentials only the regular rectangular voltages corresponding to the black-white alternations of the black squares. In the fifth cycle, however, at the point marked with an arrow, the voltage form producing the smaller squares can already be seen (12.5 picture points in width). In the following cycles the amplitude of this voltage increases according to the further lines of the scan, until they lie between the "white" and black level (fourth cycle in the top series of oscillograms). In this case the time base expansion is insufficient to show the details of these voltage forms. However, the "GM 4584" time base expansion unit allows of such extreme expansions along the time axis that it affords clear insight into voltage waveforms of less than half a picture-line in width.

Thus, in Fig. 19-34, with the same test pattern we can now investigate the voltage form of the line which passes over the centre of the small squares, through the 200 lines fan and through the figures 2,0,0. If we compare the oscillogram in Fig. 19-33*b* with the brightened line in the test pattern (Fig. 19-34*c*), the connection between the black-white alternations and the upper part of the oscillogram can easily be distinguished. The voltage fluctuations for scanning the small squares are now quite easy to follow. In these parts of the oscillogram that show the 200 lines fan, or that scan the number "200", it is, of course, clear that they no longer correspond to 100% modulation of the video signal.

In Fig. 19-34, the expanded oscillogram in *b*) is compared with a normal oscillogram in *a*) with the line frequency. As these oscillograms are approximately as

Fig. 19-34 Test pattern *c*) with oscillogram *b*) of one line each of both fields; normal oscillogram during one line deflection

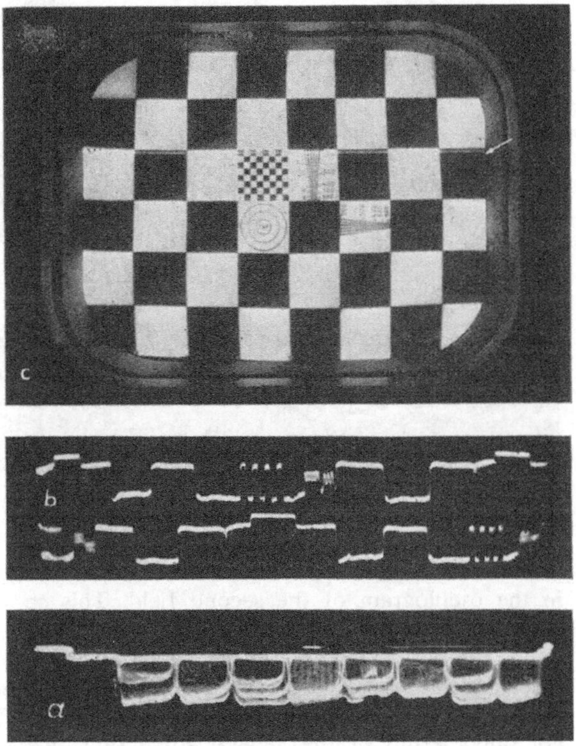

wide as the receiver picture, the corresponding sections along the time axis can be directly compared.

Fig. 19-35 is intended to demonstrate the possibilities as well as the limitations of extreme time base expansion [97]).

The oscillogram in Fig. 19-35*a* shows the voltage waveform during less than half a picture line in both fields, while Fig. 19-35*b* shows the relevant section of the pattern on the screen. It can be clearly seen that the voltage alternations

[97]) Such oscillograms are particularly faint owing to the extreme time base expansion, and must therefore be compensated by the most favourable conditions obtainable during photographic recording. These oscillograms (Figs. 19-37*a* and 19-39*a*) were obtained with the following equipment:

Time base expansion unit:	Philips "GM 4584"
Oscilloscope:	Philips "GM 5654"
Cathode ray tube:	DB 10-6 (blue-fluorescent screen)
Anode voltage of CRT:	$V_k = -1.2$ kV, $V_{a3} = +2$ kV
Camera:	Tenax II, lens/Sonnar 1:2 max. aperture
Scale of picture:	4 : 1
Exposure time:	1/10 s
Film:	21° DIN Pan
Developer:	Rapid (Gevaert 230, Agfa X-ray developer or similar)
Enlargement paper:	Silver bromide, extra hard.

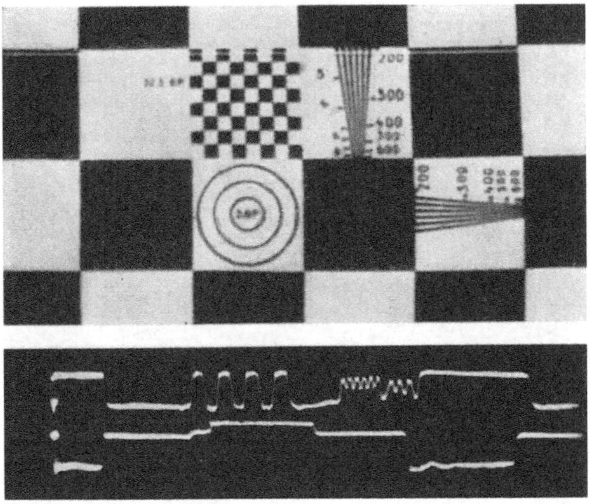

a

b

Fig. 19-35 Oscillogram *a*) with section of test pattern *b*) of Fig. 19-34*c*, showing maximum time base expansion

corresponding to the small squares are still quite rectangular, and the scanning of the 200 lines fan and the numbers 2,0,0 is much clearer than in Fig. 19-33*b*.

The voltage corresponding to the region of the sync pulse is also shown expanded in the oscillogram of the second field. This shows clearly the difference between black and blanking level.

For the observation of the voltage content of single lines of the television picture special oscilloscopes have been developed and are used only as "line selector unit" at the various control points [23] [24] [25] [26] [27]. Line selector unit oscilloscopes not only permit of the general study of single television lines, but have been of particular value in observing the quality of the television transmission from the time input signals were first added to the frame blanking gap for control purposes.

19.6 Checking the transmission characteristics of a television system during the programme by studying test-line oscillograms

For levelling in and for carrying out precise measurements on the transmission line, use is made of clear-cut rectangular pulses [4] or of sawtooth voltages [28] [29] which are modulated in instead of the picture modulation.

For observing the differential modulation linearity the sawtooth is modulated with a HF sinusoidal alternating voltage (4 Mc/s with the 625 line system; 2 Mc/s with the 405 line system). The amplitude of this additional HF voltage is set at 10% of the amplitude of the sawtooth voltage. Behind the transmission link this voltage is then filtered out and studied on test oscilloscopes which are synchronized with the line deflection frequency. If the slope of the modulation characteristic is constant, a uniformly high band is obtained on the fluorescent screen. If nonlinearities occur, the height of the oscillogram varies according to the various slopes. In this way a critical standard for the transmission element is obtained [98] [2].

[98] This process is also eminently suitable for testing the modulation linearity of vertical amplifiers. A clear picture of the change in slope is thus obtained (modulation factor; see also the oscillograms of Fig. 17-12*b* and *c* portraying the valve slope). Deviations in linearity can here be determined much more clearly than by mere observation of the modulation.

The above processes have the disadvantage that they cannot be employed during programme transmission although it is also necessary to be able to check constantly the quality of the various transmission paths during transmission as well.

One method which overcomes this difficulty is the test line process. Signal voltages whose waveforms can be universally observed with a time base expansion unit or a line selector switch are keyed in between two adjacent line sync pulses during vertical blanking. It is also possible, as described by FRÖLING [30], not only to introduce simple signal pulses, but also to indicate the origin of the transmission by its varying composition.

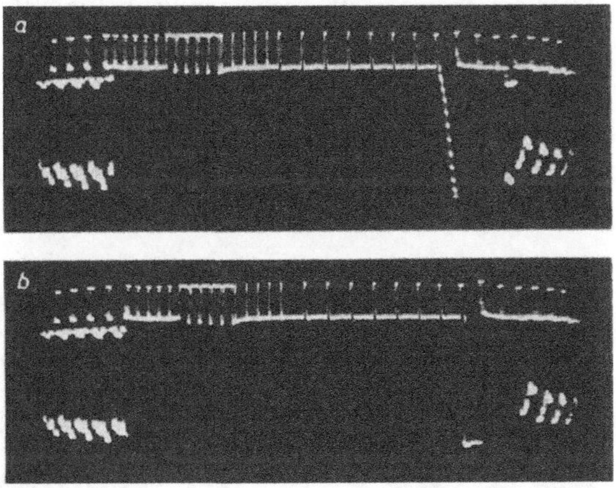

Fig. 19-36 Signal sequences of both fields at the vertical blanking gap with signal voltages in two line gaps. *a*) "staircase"-voltage of the eighth line *b*) the constant white level value of the ninth line

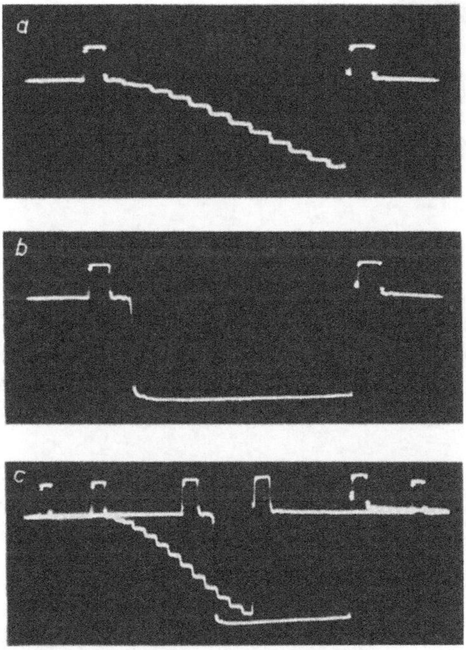

Fig. 19-37 Oscillograms of test lines
a) Stepped test line
b) white level test line
c) oscillograms of *a*) and *b*) together

The two oscillograms in Fig. 19-36 show the voltage waveforms of the signal sequences of both fields at the vertical blanking gap with signal voltages in two line gaps (recorded during test transmissions of the Norddeutscher Rundfunk).

In Fig. 19-36 the eighth line has a "staircase" voltage and in Fig. 19-36*b* the ninth line has a constant white level value. The even more strongly time-expanded oscillograms of Fig. 19-37*a* and *b* were recorded with the Philips "GM 4584" time base expansion unit on a "GM 5654" oscilloscope. The line deflection was triggered at half the frame frequency. If it is intended to study simultaneously the waveform of both test signals in as large a pattern as possible, the frame deflection frequency (50 c/s) can also be used to trigger directly and adjust the oscillogram as is shown in Fig. 19-37*c*.

The corresponding line is, of course, brightened by the signal voltages, particularly by those shown in the oscillogram in Fig. 19-37*b*. As these lines are located at the end of the flyback gap for vertical deflection, this brightening causes no trouble in a receiver which is working properly, as it falls within that part of the picture which is normally invisible. The photograph of a pattern on the receiver screen in Fig. 19-38 shows the position of both test lines quite clearly. This was only possible, however, because the pattern dimensions were purposely so reduced for this recording, that the total area swept by the spot could be studied.

Fig. 19-38 Test pattern of the test signal according to oscillograms in Figs. 19-36 and 19-37

The circuit for generating the currents for vertical deflection in the "TX 1420 U" receiver is seen in Fig. 19-24.

In modern television receivers the deflection circuits are stabilized in such a way that brightness control cannot cause interference by test pulses. It has been the practice for some time in Germany to include the white pulse as well as the "staircase" voltage in the test line, thus making it possible to observe these signal patterns even in the simple oscillogram of the synchronized pulse sequence of a field.

The double oscillogram of Fig. 19-39 shows the vertical blanking gap for both fields when receiving the electronic test pattern as in Fig. 19-7. It can be seen that these voltages are located in the 11th or 12th line behind the equalizing pulses. Exhaustive tests have proved this to be the best arrangement.

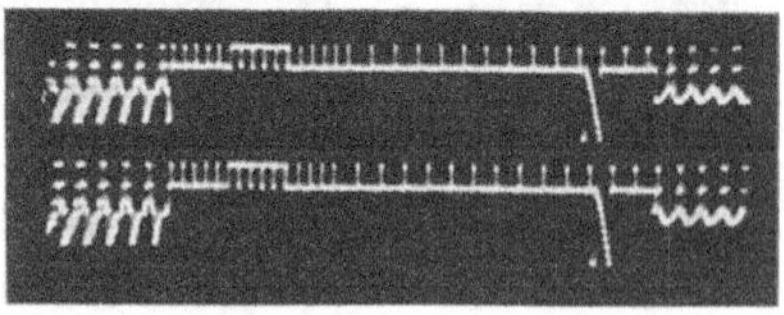

Fig. 19-39 Vertical blanking gap with test signals (sync pulse sequences of both fields)

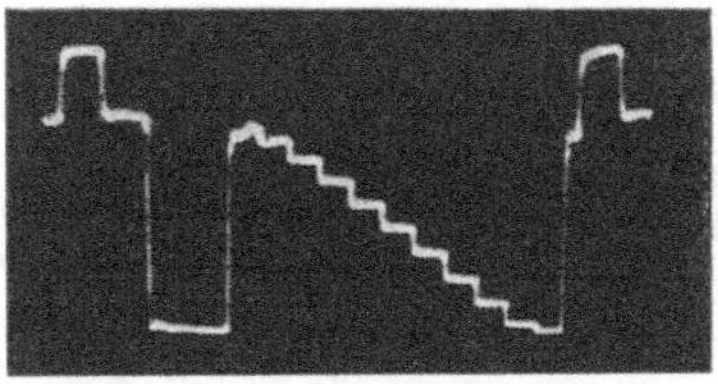

Fig. 19-40 Test signal with white level and staircase, photographed at receiver

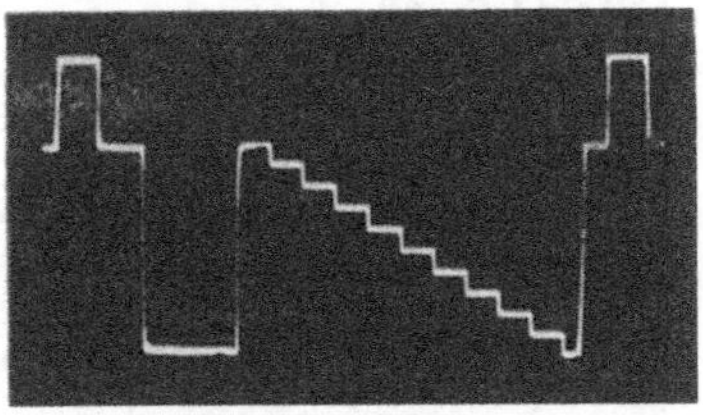

Fig. 19-41 Test signal as in Fig. 19-40, but photographed in the studio

The oscillogram of Fig. 19-40 shows the waveform of a single test-line voltage as obtained by the receiver. Fig. 19-41 shows it as it is when it left the studio sets.

In European broadcasting (Eurovision) the test line is usually transmitted at white level. This makes it possible not only to maintain the modulation at a constant level throughout the transmission path, but also to control the transmitters automatically (at a fixed value). Owing to the great advantages of such a test line, use is made of it everywhere and international standardization has been considered [31].

It is thus seen that the cathode ray tube oscilloscope has a wide field of application in television engineering [32], [33], [34], [35].

CHAPTER 20

INVESTIGATING MATCHING CONDITIONS AND MEASURING IMPEDANCES IN THE USW-BAND BY MEANS OF A LONG TRANSMISSION LINE

20. 1 Investigating matching conditions and measuring impedances in the USW-band in general

The devices used for studying matching conditions and measuring impedances at high frequencies include bridge circuits, transmission lines, and rectifying couplers with their associated accessories. The transmission line system has now come into fairly widespread use. In this method a sliding probe is used to observe the voltage waveform along the line occurring as a result of interference between two alternating voltage waves passing along the line in opposite directions. The one wave originates in the supply generator, while the other occurs by reflection at the end of the line when the terminal impedance differs from the characteristic impedance of the line. Standing waves then occur in the line, which have potential antinodes and potential nodes at an interval of half a wavelength. In order to study the matching conditions and measure unknown impedances, such a transmission line is connected between the voltage source and the system under investigation. The characteristic impedance of the transmission line must obviously be well matched to the generator output impedance in order to avoid the occurrence of new reflections at these points. Examples in which the unknown impedance is connected to the input or output of a transmission line have been described by van Hofwegen among others [1] [2].

The voltage waveform along the line is observed by means of a probe to which a diode is connected. The rectified current through the diode is measured by a sensitive instrument. Position and value of the voltage maxima and minima on the line permit of accurate determination of the extent and type of mismatching at the end of the line. The proportion of the highest potential value to the lowest is termed the standing wave ratio (SWR):

$$s = \frac{V_{\max}}{V_{\min}}; \qquad (20.1)$$

its reciprocal:

$$m = \frac{1}{s} = \frac{V_{\min}}{V_{\max}} \qquad (20.2)$$

is termed the matching factor. (The difference $1-m$ is the mismatching.) From this the reflection factor r can be calculated according to the equation

$$r = \frac{1 - m}{1 + m} \tag{20.3}$$

[3] [4] [5] [6] [7] [8] [9] [10] [11] [12].

Measurements carried out with a transmission line have the advantage of giving very precise results, but the disadvantage is that the transmission lines have to be exceedingly accurately built up and are thus relatively expensive. In addition, the measurement takes some considerable time as the voltage values along the line must be measured point by point. If it is intended to restore given operating conditions by balancing, it may require several series of readings, which can take a great deal of time.

20. 2 Displaying the voltage waves by means of oscilloscopes

It is obviously much more convenient to use the movement of the measuring probe to deflect the fluorescent spot horizontally and synchronously, and the voltage of the probe for vertical deflection. If the detuning and horizontal deflection of the spot can be carried out so rapidly that the impression of a continuous oscillogram is obtained, it is possible to observe the voltage conditions in the transmission line "at a glance". Then, for instance, the desired matching conditions can be quickly found. A solution to this problem has been found in the arc-shaped transmission line, at least for cm waves [13]. Apart from the fact that such ring-shaped transmission line seems suitable only for cm waves, the cost involved by the necessary precision in construction will certainly be no less than for a straight transmission line.

For many practical applications, particularly in television engineering, it is very desirable to be able to carry out such investigations, especially the determination of matching conditions, with measuring instruments already available for normal measuring tasks, not only in laboratories, but also in well-equipped repair workshops. Considerable concessions as to accuracy can be made since the requirements are not usually too stringent.

As BAUER has shown, it is also possible to use a FM generator (ultra-short wave wobbulator) and a transmission line which corresponds to a multiple of the wavelength to display on an oscilloscope the voltage waveform on the transmission line which is dependent on the matching conditions [14]. In a subsequent publication by VAN DEN HOOGENBAND and STOLK it was also shown that even an unknown impedance can be determined as to magnitude and phase [15]. In this work the results of measurements have been subjected to detailed mathematical analysis, so that the work is recommended for study of the exact relationships. The process is shown here in the simplest form by means of original oscillograms and its suitability for matching measurements demonstrated. It is hoped to rouse very wide interest in its use. (In many tasks the absolute value of the impedance per se is of minor interest.)

Measurement with ultra-short wave wobbulator and a long transmission line

The HF voltage source, in this case a frequency-modulated signal generator, is connected to the input of a long transmission line as shown in Fig. 20-1; the detector is connected at the same time. If the frequency of the input HF voltage

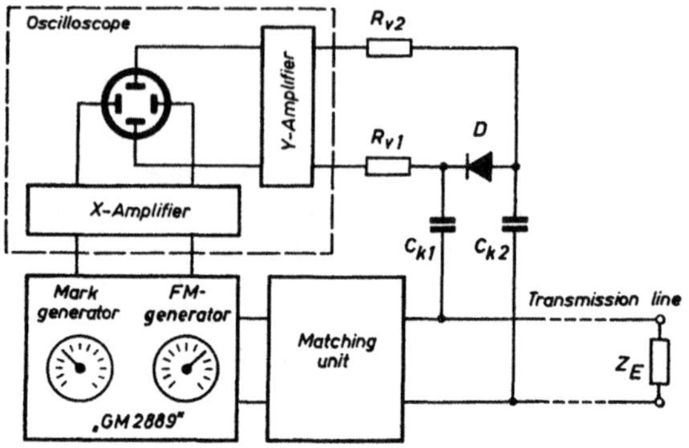

Fig. 20-1 Circuit for determining the reflexion factor of an unknown impedance by portraying the waveform of the voltage at the cable input in dependence on the frequency

is altered by frequency modulation, then the input impedance and hence the input voltage experience the same maxima and minima as those obtained on an adjusted transmission line by shifting the shorting bar.

A cable with as low damping as possible is best used as a transmission line. Depending on the task in hand, it should be either a coaxial cable or a double-cored ribbon cable. In Fig. 20-1, as in all the oscillograms reproduced in this chapter, a symmetrical ribbon cable of some 240 Ω characteristic impedance was used. The length of the cable must be a multiple of the average wavelength, depending on how many standing waves are to be represented on the oscilloscope screen for a given frequency swing. If the number of waves to be displayed is taken to be N, the frequency swing as f_H and the phase velocity of the voltage in the cable as v_{ph}, then the required length l of the cable is obtained from the following equation

$$l = \frac{N \cdot v_{ph}}{2 \cdot f_H}\,. \tag{20.4}$$

The phase velocity v_{ph} (which is less than the speed of light) can be obtained from the equation

$$v_{ph} = \frac{c}{\sqrt{\varepsilon_r \cdot \mu_r}}\,. \tag{20.5}$$

Here c is the propagation velocity of the electromagnetic waves in vacuo ($3 \cdot 10^8$ m/s), ε_r the relative dielectric constant and μ_r the relative permeability. Normally, μ_r can be assumed to be unity. The value of the relative dielectric constant can also be determined with this arrangement, as will be shown later, if it is not stated by the manufacturer. In the case of the ribbon cable it was found that $\varepsilon_r = 1.4$, and the concentration factor

$$\xi = \frac{1}{\sqrt{\varepsilon_r}} \tag{20.6}$$

was 0.85. This shows the extent to which the wavelength in the line is shorter than in vacuo.

According to Eq. (20.4), with a maximum frequency swing of 15 Mc/s and a phase velocity $v_{ph} = 3 \cdot 10^8 \cdot 0.85 = 2.55 \cdot 10^8$ m/s, a cable of $l = 51$ m is required to display 6 voltage waves. Attachment points with insulators give rise to certain non-homogeneities, and it has not been found easy to pack the cable into the smallest possible space and still avoid all electrical contact points and prevent the individual parts of the line from exerting influence on each other. It would be best to slacken the cable. This must not be done in the open, however, as humidity has a strong damping effect. Vertical suspension, for example in a light-shaft, with few attachment points, would be suitable. A solution for the use of such a cable in a laboratory was found by folding it in figure-eight folds, tying the loops tightly with thread and suspending it so that the loops are about 20 cm from one another [100]). Fig. 20-2 shows a portion of the cable "folded" in this fashion. fashion.

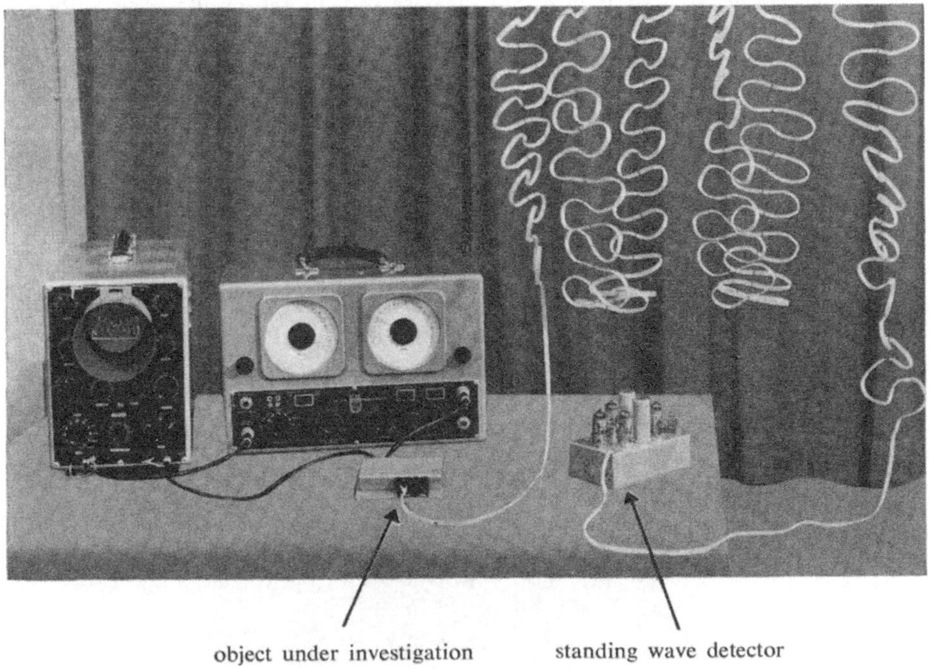

Fig. 20-2 Arrangement according to Fig. 20-1 with USW-wobbulator, oscilloscope, standing wave detector and object under investigation

The operating principle of this arrangement can be explained with reference to Fig. 20-3 which shows the voltage conditions. The output impedance of the HF voltage generator must be so adjusted in relation to the impedance of the measuring cable that the circuit is as free from reflection as possible. The output impedance of the signal generator is usually 70 Ω, though it is very often necessary to carry out measurements with 240 Ω leads also. Thus a corresponding adjustment link for

[100]) See footnote on page 432

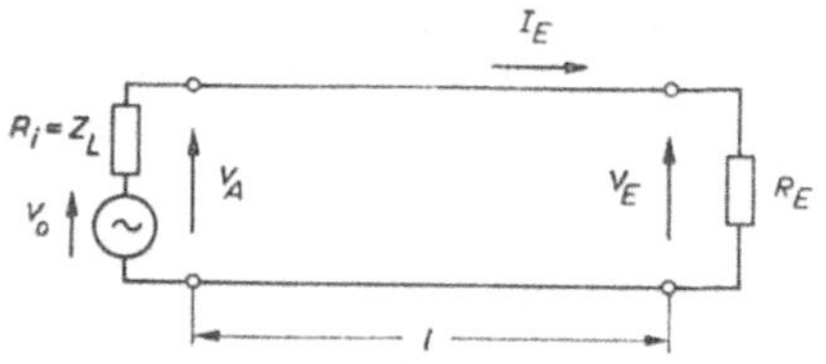

Fig. 20-3 Equivalent circuit diagram to Fig. 20-1

connection to the measuring cable at the generator end would be most advisable. It is necessary for this adjusting and balancing to be done so as to ensure sufficient balanced-to-ground voltages under no-load conditions also. The output voltage of the signal generator which, at a maximum of 100 mV_{rms} still falls within the quadratic portion of the diode characteristic curve, should also be made as high as possible. Then the output impedance of the diodes rises, and, in addition, the direct output voltage is greater, so that it also can be observed on the oscilloscope with a not very small deflection coefficient. In order to meet these requirements, a wideband transformer of simple construction was developed [100]). As can be seen from Fig. 20-4, the primary coil consists of one, and the secondary coil of two hairpin windings, whose "shanks" are enclosed in a ferroxcube tube. This keeps the scatter inductance small. The circuit of this arrangement is shown in Fg. 20-5.

When connecting the terminals marked "300 Ω" to a 240 Ω cable, the reflection factor of the connection between the output of the signal transmitter and the input of the cable, within the frequency range 10 . . . . 225 Mc/s, is less than 0.25. This can be considered satisfactory for this measuring process. In the unit described, three germanium diodes were used in a voltage tripler circuit for rectification, in

Fig. 20-4 View of the standing wave detector

[100]) This suggestion comes from H. RUMÖLLER, who, in the applications laboratory of ELEKTRO-SPEZIAL, Hamburg, also developed the "standing wave detector" which is described in the following pages.

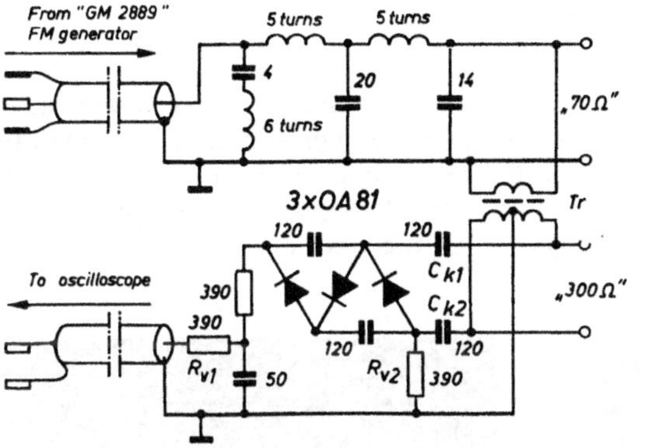

Fig. 20-5 Circuit of the standing wave detector

order to obtain as high an output voltage as possible. This, of course, increases the load capacitance as compared with that of a single diode (it is not tripled, however); obviously the effect of the detector capacitance and the capacitive interference at the connecting terminals are partly compensated by the wiring inductance and the scatter inductance of the matching transformer.

The Philips "GM 2889/01" AM-FM generator employed has a low-pass filter connected to the output, which suppresses interference signals in the range up to 225 Mc/s. For the tasks for which the unit is normally intended, i.e. for selected subjects of study, the effect of this filter is entirely satisfactory. But for the present application, in which the output must be wideband-matched, additional filtering by a further low-pass filter was found to be necessary, as otherwise, *a*) the sensitivity of the demodulator would be reduced by the initial load of the interference signal, and *b*) because the direct voltage occurs at the detector, causing additional shifting of the pattern on the screen. Fig. 20-5 shows this filter connected to the cable from the generator. In the upper portion of Fig. 20-4 the inductances and capacitances of this filter are clearly seen. The whole arrangement, that is to say, the matching unit of the basic diagram in Fig. 20-1 with coupling capacitances (C_{k1} and C_{k2}), demodulator (D) and matching resistors (R_{v1} and R_{v2}), was combined as an auxiliary unit called a standing wave detector [101]). For practical matching measurements it is only necessary to connect the cables of the standing wave detector to the FM generator and the oscilloscope, to connect the measuring cable to the requisite terminals and to connect the system to be measured to the end of the measuring cable. The horizontal deflection of the spot is caused in the usual way by a 50 c/s voltage supplied from the FM generator.

As the frequency modulation of the 50 c/s wobbulator here used is sinusoidal, the frequency scale of the time deflection is almost linear. In order to set the most favourable picture phase it is an advantage if the voltage phase can be varied. The photographs in this chapter were therefore made with the "GM 5666" oscilloscope, which, like the "GM 5662", can be switched to a 50 c/s voltage which can be varied in phase by about 160°. By changing the polarity of the mains plug of the oscilloscope the phase of this deflection voltage can be varied by 180° in addition.

---

[101]) A few number of laboratory models made only.

20. 3 Some practical examples with oscillograms

20.3.1 INFLUENCE OF CABLE DAMPING AND OF DEMODULATOR CHARACTERISTIC

The oscillograms in Figs. 20-6*a* and *b* show first of all the standing wave voltage at the input of the cable for both border cases of total reflection (short-circuit at the end of the cable) and reflection-free termination with an ohmic resistor equal to the characteristic impedance of the cable (in this case 235 Ω) at a mean frequency of 40 Mc/s ($\pm$ 7 Mc/s) [102]. The curve troughs, in the case of short-circuited cable ends, touch the reference zero line.

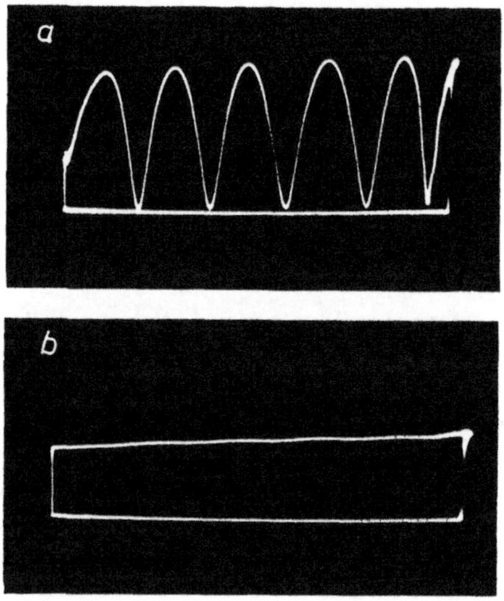

Fig. 20-6 Voltage waveform on cable input at about 40 Mc/s. *a*) Short-circuit at cable end; *b*) termination with a resistor equal to the surge resistance of the cable

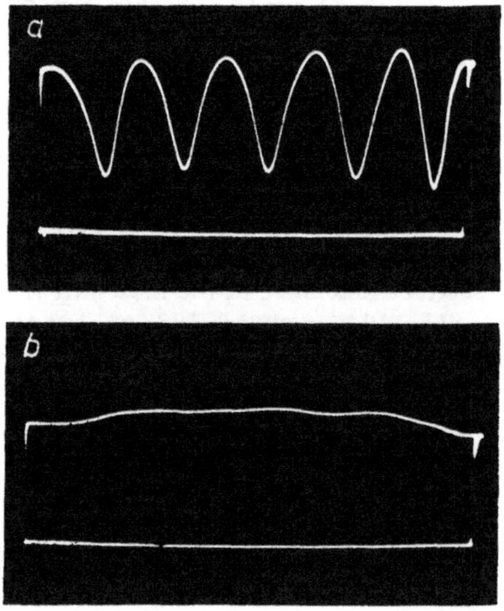

Fig. 20-7 Voltage waveform at cable input at about 210 Mc/s. *a*) Cable end open; *b*) cable end terminated with surge resistor

In accurate observation of the waveform of such voltage curves it should be noted that the demodulator, like precision transmission lines, works at low voltages in the quadratic portion of its characteristic. The minima of the voltage curves are therefore not so peaked in the oscillograms of Figs. 20-6, 20-7, 20-8, 20-9, etc. as could theoretically be expected, but are actually rounded. The possible errors have been calculated in the publication quoted [15] and correction curves supplied.

[102]) That the oscillogram, when matched with the characteristic impedance is not quite smooth and still evinces the slightly irregular ripple, is due to a certain frequency dependence of the surge voltage amplitude and irregularities in the cable. In this connection it must be stressed that the frequency dependence of the voltage amplitude in the "GM 2889" USW-FM generator is relatively slight as a result of the beat principle used in it. Only such voltage sources may be used for the measuring process described, as otherwise the amplitude-frequency characteristic distorts the pattern excessively.

Krausse, too, has pointed out possible corrections of these errors [16]. Using a signal generator with calibrated attenuator and a DC voltmeter (e.g., Philips "GM 6020") or a DC oscilloscope (e.g., "GM 5600" or PM 3206), the published characteristic can also be checked and then taken into account by means of a correspondingly calibrated vertical scale.

For frequency measurement of the curve pattern the unmodulated voltage of the marking signal oscillator is added to the frequency-modulated output voltage of the signal generator. At a suitable setting the well-known interference markings are obtained on the voltage waveform curves at the same frequency each time.

The oscillograms in Figs. 20-7*a* and *b* show similar recordings to those in Fig. 20-6, but with an average frequency of 210 Mc/s. The damping in the cable is now noticeably greater at the higher frequency, so that even with an open cable end the full voltage is no longer reflected, with the result that the troughs no longer reach the reference line, nor are the maxima so high as they would be for 100% reflection. None the less, the oscillogram in Fig. 20-7a means complete reflection; hence, according to Eq. (20.2), a matching factor $m = 0$ or, according to Eq. (20.3), a reflection factor of $r = 1$. It is advisable, in interpreting such oscillograms, not to start by considering the distance of the voltage maxima and minima from the null line, but rather to compare the amplitude difference at nominal total reflection (short-circuit or open cable end) with the amplitude difference when the impedance is unknown. The oscillograms in Figs. 20-8 and 20-9 are intended to illustrate this more clearly. In Fig. 20-8 curve *A* was recorded with an open cable end, curve *B* with unknown impedance and curve *C* with the optimum matching. For better evaluation in the frequency direction (horizontal) the frequency swing for a cable length of about 58 metres was adjusted to only 4 Mc/s.

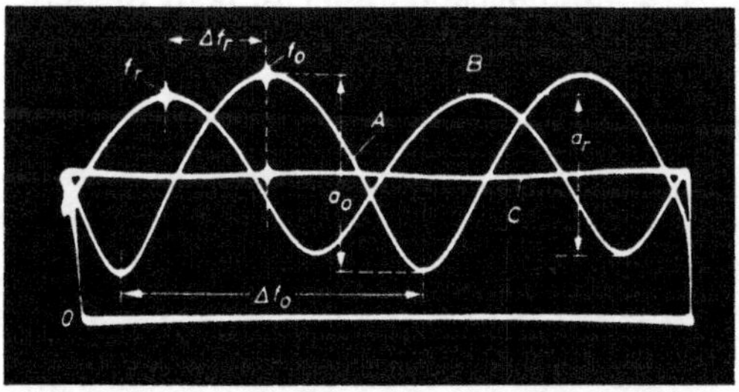

Fig. 20-8 Termination of cable with complex resistor (470Ω; 10 pF). *A* open cable end; *B* termination with unknown (complex) impedance; *C* termination with ohmic resistor = surge resistance of cable

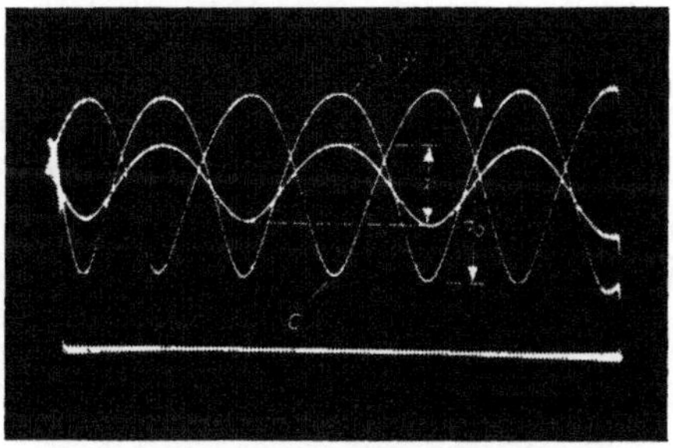

Fig. 20-9 Measuring terminating resistors with ohmic characteristics. *A* short-circuit at cable end; *B* unknown resistance applied; *C* cable end open

In Fig. 20-9 are shown in curve A the voltage waveform for short-circuit at the cable end with a frequency swing of about 7½ Mc/s and an average frequency of about 150 Mc/s, in curve *B* with an unknown impedance, and in *C* with an open cable end. Although the average frequency in Fig. 20-8 was relatively low (about 36 Mc/s), the minima no longer reach the null line. This is still more clearly seen in the oscillograms of Fig. 20-9, where the minima of curves A and C (short-circuit and open cable) occur at some considerable distance from the null line because of cable damping. Only the amplitude differences a_0 are obtained. If the curve with the unknown impedance has an amplitude difference a_r, then the amount of the reflection factor

$$|r| = \frac{a_r}{a_0} . \tag{20.7}$$

From the characteristic impedance Z_L of the line from the terminating impedance R_E to the cable end we obtain the (complex) reflection factor in the general form

$$r = \frac{Z_E - Z_L}{Z_E + Z_L} . \tag{20.8}$$

The unknown impedance is then calculated according to

$$Z_E = Z_L \cdot \frac{1 + r}{1 - r} . \tag{20.9}$$

As has already been mentioned, when Z_E is complex the reflection factor r is also complex. It is then given by

$$r = |r| \cdot e^{j\varphi} . \tag{20.10}$$

This means that Z_E can be indicated in such a way that the terminating impedance is represented as the sum of an ohmic resistance and an impedance, as

$$Z_E = R_E + \mathrm{j} \cdot X_E . \tag{20.11}$$

It must be stressed that Eq. (20.11) is general in application, that is, both for series and parallel connection of resistance and impedances. For practical calculation the corresponding conversions are required with impedances connected in parallel. The unknown impedance is then given, according to Eq. (20.9), by

$$Z_E = Z_L \cdot \frac{1 + r \cdot e^{j\varphi}}{1 - |r| \cdot e^{j\varphi}} . \tag{20.12}$$

Further manipulation gives Eqs.

$$R_E = Z_L \cdot \frac{1 - |r|^2}{1 - 2\,|r| \cos\varphi + |r|^2} , \tag{20.13}$$

and

$$X_E = Z_L \cdot \frac{2\,|r| \sin\varphi}{1 - 2\,|r| \cos\varphi + |r|^2} . \tag{20.14}$$

The interpretation of such measurements is much simplified by the Smith diagram [15] [17] [18].

Eqs. (20.10) to (20.14) are only intended as an indication of the way in which the connected impedance should be calculated. The phase of the reflection factor (not to be confused with the phase angle of the voltages) can be determined in the following way.

The oscillograms are cyclic, as has already been described. The voltage at the start of the cable repeatedly assumes the same value if f_0 increases by such a magnitude Δf_0, that $(4 \cdot \pi \cdot \Delta f_0 \cdot l)/v = 2\pi$. It is thus

$$\Delta f_0 = \frac{v}{2 \cdot l}. \tag{20.15}$$

The minima in the case of unknown impedance are shifted with respect to the short-circuit minima by the distance $\Delta f_r = \varphi \cdot (\Delta f_0)/2\pi$. From this we find the phase to be

$$\varphi = \frac{\Delta f_r}{\Delta f_0} \cdot 2\pi . \tag{20.16}$$

Thus, even for complex resistances, the magnitude $|r| = a_r/a_0$ — and the "modulus" — and the phase φ — the "argument" — of the reflection factor can be obtained and calculations carried out as shown in Eqs. (20.13) and (20.14).

In the example in the oscillograms of Fig. 20-8, at an average frequency of about 110 Mc/s and with the cable terminated with $R_E = 470\ \Omega$ and $C_E = 10$ pF, it was found that $\Delta f_r = -1.0$ Mc/s (as compared with the open cable) and $\Delta f_0 = 2.65$ Mc/s. From this the phase is obtained as $\varphi = -0.38 \cdot 2 \cdot \pi$ or $-136°$. In this example, therefore, the phase of the reflection factor is negative. As the reactance $\left(\frac{1}{\omega C}\right)$ at 110 Mc/s is considerably less than the ohmic terminating resistance (about 145 Ω), the phase is determined mainly by the capacitive load.

As we are dealing here with a parallel circuit, and this should once more be emphasized, Eqs. (20.12) and (20.13) could only be applied after the corresponding conversion.

In the measurement of impedances of an ohmic character much simpler relationships result. (The measuring cable serves as "impedance standard".). Eq. (20.9) then changes in form to

$$(R_E > Z_L) \qquad R_{E1} = Z_L \cdot \frac{1+r}{1-r}. \tag{20.17}$$

As indicated, it only applies for $R_E > Z_L$. If the terminating impedance is less than the characteristic impedance Z_L of the transmission line, then calculation should be according to

$$(R_E < Z_L) \qquad R_{E2} = Z_L \cdot \frac{1-r}{1+r}. \tag{20.18}$$

Which of these conditions applies can be quickly determined by observing the voltage waveform during short-circuit and when there is an open cable end. The quotient $\frac{1-r}{1+r}$ then corresponds to the concept of the matching factor m as previously defined.

In Fig. 20-9, for instance, it can be seen that the value of the unknown terminating impedance lies between short-circuit and the value of the characteristic impedance (235 Ω; see also Figs. 20-6 and 20-7), and therefore $R_E < Z_L$. The reflection factor is found to be $r = \frac{15}{37.5} = 0.40$. In the case of the surge resistance of a transmission line of $Z_L = 235\ \Omega$, the factor m in Eq. (20.18) is found to be equal to $(1 - r)(1 + r) = 0.43$. Thus $R_E = 101\ \Omega$.

The process for determining the surge resistance of a transmission line is similar. The line must in that case be terminated with such an ohmic resistance that the surge disappears completely. For this, normal variable resistors (adjustable potentiometers) are unsuitable, as the influence of the unwanted capacitances and inductances is no longer negligibly small at such high frequencies and prevents satisfactory termination. It is therefore necessary to ascertain the matching by trial and error with small layer resistances. If the precise value of the terminating resistance cannot be found in this way, the characteristic impedance of the cable can also be calculated according to Eqs. (20.17) or (20.18) with a known value of the terminating resistance R_E.

Proper adjustment of the correct terminating impedance can also be obtained by connecting a small layer potentiometer (also called dehummer) in series with a fixed resistor of somewhat less value than seems necessary for the correct termination. The value of this variable resistor can be so small that its reactances are negligible. When the most favourable resistance value has been found it can be precisely determined in the usual way with a low-frequency or DC measuring bridge.

20.3.2 LIMITS OF ERROR OF THIS METHOD

The degree of accuracy to be expected is shown by the three oscillograms in Fig. 20-10. They represent three recordings at mean frequencies of 30 Mc/s, 100 Mc/s and 210 Mc/s respectively, at total reflection and a termination as free as possible from reflection, and with a termination in which the matching impedance was made to deviate by 15% from that of the surge impedance of the cable. The interpretation, in accordance with Eq. (20.7), for the mismatching $(1 - m)$ resulted in values of about 18, 16 and 17%. As the amplitude of the slight ripple (a_r) can be measured only to a similar degree of accuracy, the results are very satisfactory.

20.3.3 MATCHING RECEIVER INPUT CIRCUITS

As a practical example, the matching of the input resonant circuit of a television receiver can be considered. As such circuits are always damped to increase the bandwidth, the range of approximate matching is correspondingly wide. The oscillogram in Fig. 20-11 shows the recording of such a measurement at $\pm$ 7 Mc/s frequency swing. The ripple disappears at perfect match (the parallel resistance was the same as the surge impedance of the cable). The best matching point was marked with the calibrating oscillator.

The oscillogram in Fig. 20-12 shows an interesting discovery. The large curve *A* shows the waveform at total reflection (short-circuit at the end of the wiring), while *B* with the smallest amplitudes shows the waveform on connection to the input (channel tuner) of a television receiver. The receiver was set to channel 6 (182.25 Mc/s). If the input stage (a cascode stage) is "regulated down" on receiving a powerful transmission, the curve of average amplitude *C* is obtained. It shows

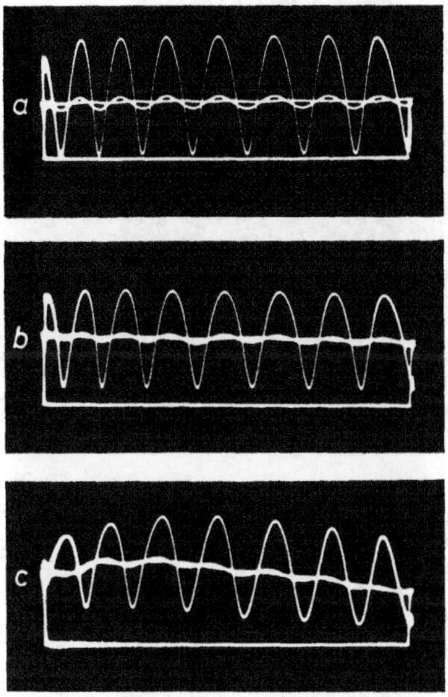

Fig. 20-10 Oscillograms representing the limits of error of this process. *a*) Average frequency about 30 Mc/s; *b*) average frequency about 100 Mc/s; *c*) average frequency about 210 Mc/s

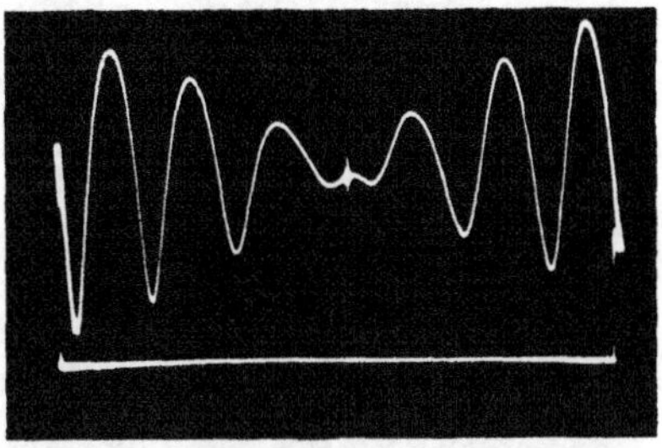

Fig. 20-11 Termination of cable with damped resonant circuit

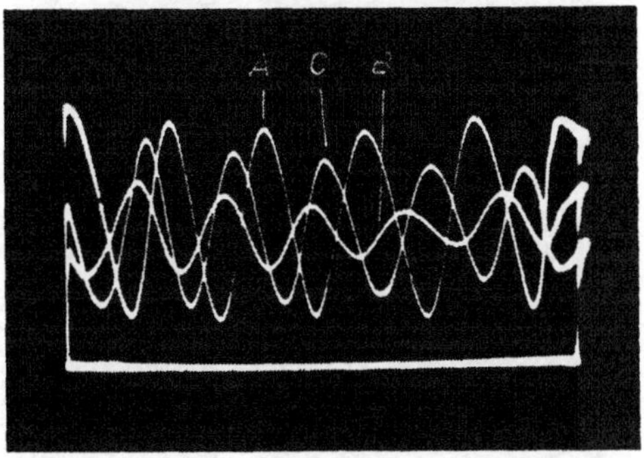

Fig. 20-12 Varying the matching conditions of a television receiver by regulating the input valve. *A* short-circuit at cable end; *B* television receiver input normal; *C* television receiver input valve regulated down

that the matching in the neighbouring channel has been shifted and that thus a considerable increase in reflection must occur in the adjusted channel.

It can be seen that the detuning of the matching in this set (an older type) is alarmingly great. This point has received the necessary attention during further development of television receivers. Förster [19] among others, has described the television picture faults which can be brought about by mismatching and detuning a tuner.

20.3.4 MATCHING SYSTEMS WITH PARTICULARLY WIDE-BANDS (TELEVISION ANTENNAS)

In wide-band elements such as television dipole antennas, the frequency range over which there is satisfactory matching is very considerable, and this is indeed usually the aim. The oscillograms in Fig. 20-13 give an impression of this. These oscillograms show one recording of a total reflection and one on connection with the antenna at 10 Mc/s steps. It can be seen that matching between about 175 and 210 Mc/s is sufficient for the requirements of TV reception. It is most favourable at about 190 Mc/s. This measuring device makes it possible to match TV antennas, particularly complicated arrays, to the characteristic impedance of the cable.

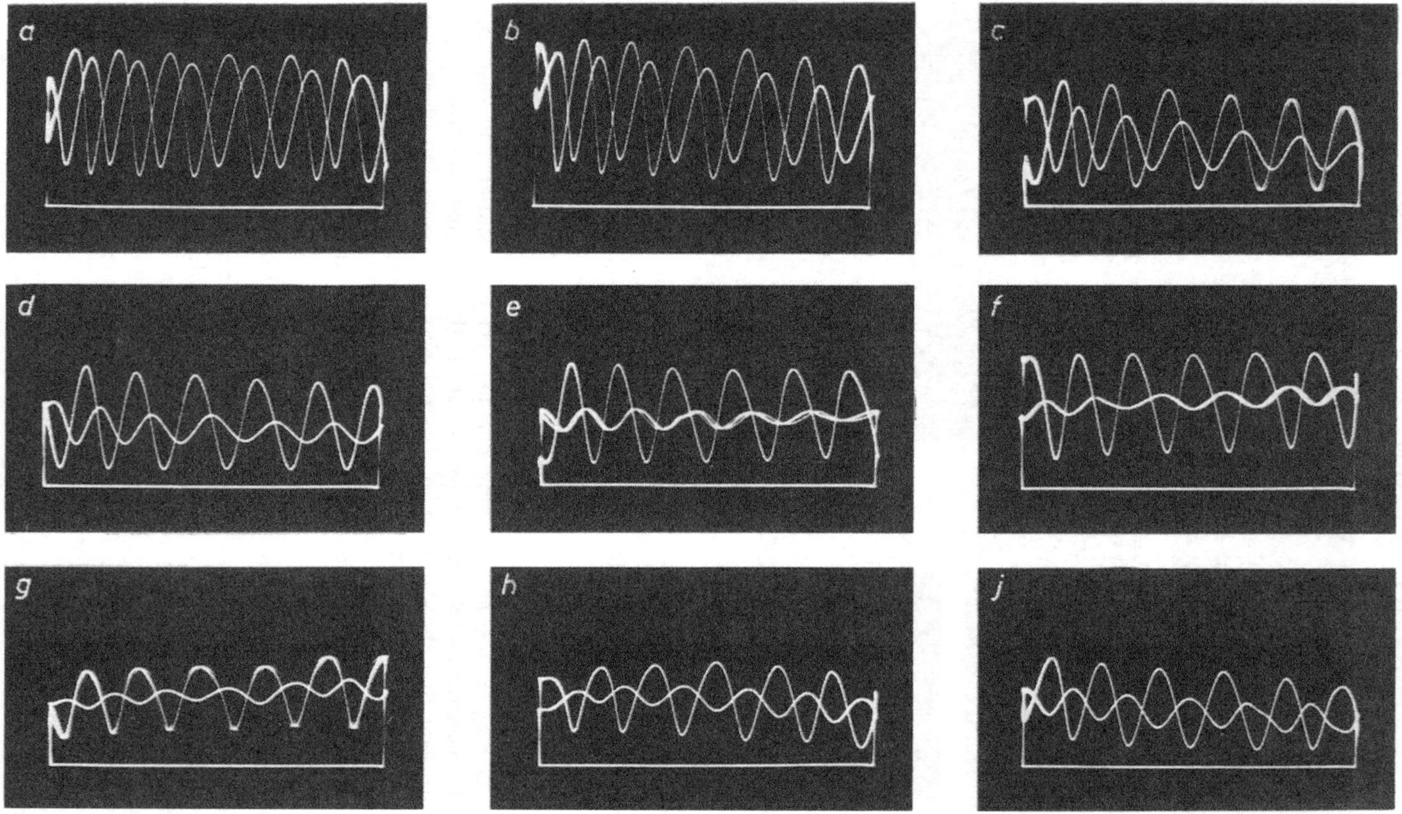

Fig. 20-13 Matching a TV folded-dipole in range 175 Mc/s to about 228 Mc/s to measuring cable with $Z_L = 235\Omega$; average frequency ($\pm\ 7^1/_2$ Mc/s): *a*) 140 Mc/s; *b*) 150 Mc/s; *c*) 160 Mc/s; *d*) 170 Mc/s; *e*) 180 Mc/s; *f*) 190 Mc/s; *g*) 200 Mc/s; *h*) 210 Mc/s; *j*) 220 Mc/s

20.3.5 MATCHING MEASUREMENTS WITH NARROW-BAND NETWORKS

Even the matching of individual narrow-band elements can be clearly observed in an oscillogram. It is essential for the resonant frequency of the element under study to coincide with a frequency at which a voltage maximum or minimum occurs at the start of the cable with an open or a short-circuited cable end. If necessary, the transmission line must be slightly lengthened or shortened, by $\lambda/4$ at most, in order to apply a voltage maximum or minimum to the resonant frequency of the element under study. It is usually both possible and permissible to detune the element under study to a small extent without affecting the measurement to an unacceptable degree. Fig. 20-14 will serve to illustrate this more clearly. Here is shown one waveform at total reflection and one matched. A narrow-band element was then connected, and four other oscillograms were recorded in which the impedance of the narrow-band element (a few hunderd kc/s) was variously adjusted. It can be clearly seen that, in the case of reflection-free termination, the peaked troughs even reach the level of the oscillogram recorded for perfect termination.

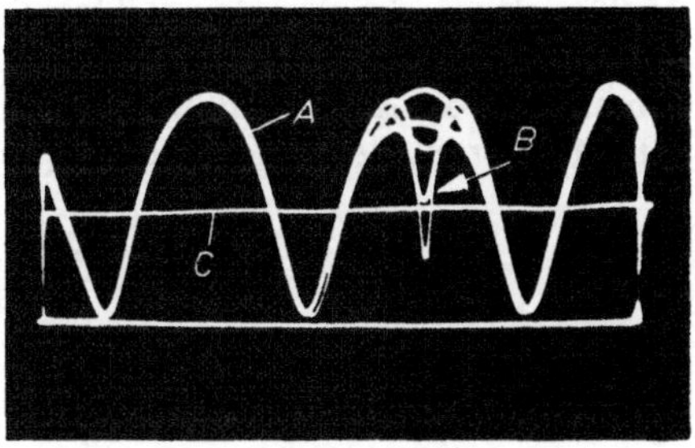

Fig. 20-14 Matching conditions with a narrow-band object. *A* total reflexion; *B* various matchings of object under study; *C* most favourable termination with an ohmic resistor

20.3.6 DETERMINING THE CONTRACTION FACTOR

The contraction factor ξ is determined, according to Eq. (20.6), by the relative dielectric constant ε_r. If the frequency interval of the voltage peaks, or better still, the frequency intervals of the minima, are measured on the curve of the standing wave voltage, and if the geometrical length l of the cable is known, then the contraction factor ξ can be calculated thus

$$\xi = \frac{2 \cdot l \cdot \Delta f_0}{c} \,. \tag{20.19}$$

It is rather awkward to determine the exact length of cable, since, according to Eq. (20.4), the usual maximum possible frequency swing of about 15 Mc/s, which allows for a sufficient number of voltage waves, requires about 50 metres.

If, in addition, the cable is stowed as shown in Fig. 20-2 in order to decrease space requirements, this measurement is impossible. The following procedure was therefore adopted: a three-metre length was cut off the same cable, suspended freely, and connected to the standing wave detector. Owing to the short length simultaneous voltage surges are not obtained; by varying the mean frequency of the signal generator it was possible to adjust a minimum or a maximum, as in Fig. 20-15 (in this example a minimum). The value of Δf_0 can be determined

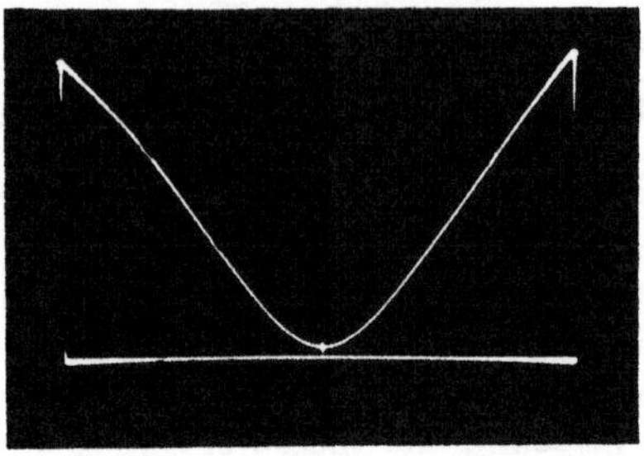

Fig. 20-15 Determining the shortening factor by measuring the frequency spacing of the minima positions

by ascertaining the frequency difference at which a minimum (or a maximum) appears at the same place on the screen each time. In the present case the minima occurred at 63.3, 105.4, 147.5 Mc/s, etc., i.e., $\Delta f_0 \approx 42$ Mc/s. From Eq. (20.19) the value of the contraction factor $\xi = 0.84$ was obtained.

20.3.7 MEASURING THE CABLE DAMPING

If the value m for the mismatching (or the reflection factor r) has been measured in the way described at a given frequency, and if the cable length l is known, the amount of damping β at this frequency can be calculated in nepers per unit of length according to the following expression

$$\beta = \frac{1}{2l} \cdot \ln\left(\frac{1+m}{1-m}\right), \tag{20.20}$$

or

$$\beta = \frac{1}{2l} \cdot \ln\left(\frac{1}{r}\right). \tag{20.21}$$

In this measurement the cable should be short-circuited at the end.

20.3.8 LOCATION OF FAULTS IN CABLES

If the characteristics of a cable are known (in particular the characteristic impedance), its total length or the distance to a contact point (short-circuit or discontinuity) can be measured. It is thus also possible to locate the position of cable faults [20].

It should also be pointed out that in studying the matching of an installed TV antenna the down lead can also be used as a transmission line. The numerous contacts of the attachment points cause additional ripple in the voltage curve, it is true, so that the pattern is not so clear as in the example quoted. None the less, such measurements, particularly when carried out by persons with some experience, can lead to valuable results.

CHAPTER 21

MEASURING TRANSIT TIME AND INVESTIGATING MATCHING CONDITIONS IN CABLES BY MEANS OF PULSE VOLTAGES

21. 1 Methods of measurement

In addition to knowing the voltage waveform at the input of a fourpole network in dependence upon the frequency, it is also often of equal interest to know the transit time of a signal in this network. It is easiest to determine this with a pulse oscilloscope. All other methods of measurement give only indirect results, and hence the reliability of the measurement varies.

In oscilloscopic determination of the transit time, the test pulse is best triggered with a delay in the way described in Part II, Ch. 13, by an auxiliary voltage generator, and the time base generator directly triggered by the auxiliary voltage [1].

If, for instance, the voltage at the cable input to which the test pulses are applied is observed, the trace shows not only the pattern of the test pulse but also a deflection due to the voltage reflected by the cable input or cable end. The value and polarity of the reflected voltage depend on the ratio of the terminating impedance (input R_A, end R_E) to the surge impedance Z of the cable.

The reflected voltage is then determined by the reflection factors r_1 (input) and r_2 (end) [2] [3].

They are given by

$$r_1 = \frac{R_A - Z}{R_E + Z} \tag{21.1}$$

and

$$r_2 = \frac{R_E - Z}{R_E + Z}\,. \tag{21.2}$$

For extreme cases these give the following values: for perfect matching $r = 0$, with no-lead $r = 1$, and for short-circuit, $r = -1$.

21. 2 Circuit for the measuring device

The following oscillograms afford some insight into the conditions which result. They were obtained by means of a device constructed with a piece of delay cable. The signal pulse from the "GM 2314" generator reaches the cable input via an E 83 F buffer valve, as shown in the wiring diagram of this arrangement in Fig. 21-1.

The negative pulses applied to the grid of the valve suppress the anode current temporarily, so that positive pulses occur at the cable input. The buffer valve

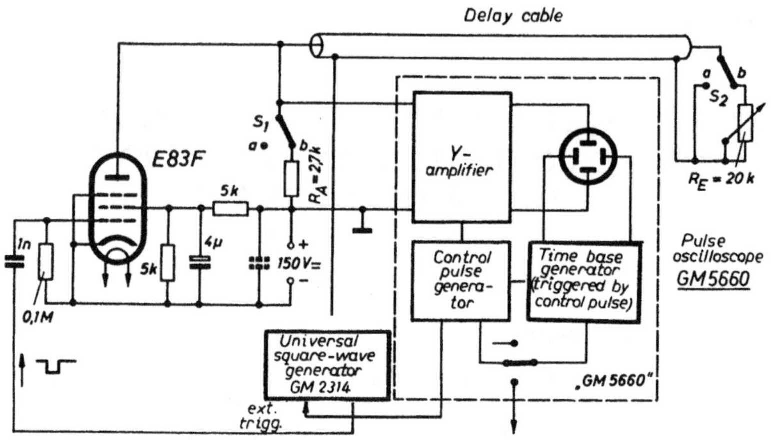

Fig. 21-1 Circuit for measuring transit time and judging matching conditions of a cable

was needed in order to be able to make the cable input highly resistive if desired. The output impedance of the impulse generator is low, so that, if the voltage source was directly connected to the cable, the input impedance would also be low.

The pulse generator was triggered by the control pulse generator of the "GM 5660" oscilloscope at a repetition frequency of about 2000 c/s. As the output pulse of the control pulse generator of the "GM 5660" oscilloscope is delayed with respect to the triggering by 1/3 μs, the time deflection starts about 0.2 μs earlier, so that a complete pattern of the input pulse (starting with a short portion of the null line) is obtained on the oscilloscope screen. The voltage at the input of the cable was to be studied. By means of switch S_1 the input of the cable can be connected to or disconnected from its surge impedance, as desired. Furthermore, by means of switch S_2 the cable output can be either short-circuited (position *a*) or connected to a variable resistor of max. 20 kΩ (position *b*).

If this resistance is set at 2.7 kΩ it is equal to the characteristic impedance of the cable, and matching has been obtained. When set at the maximum value the cable is open as regards matching.

21. 3 Measured results

For the trace shown in Fig. 21-2*a* the switch S_1 was at position *b*), and thus the input of the cable was matched with its characteristic impedance; the switch at the cable output was also at position *b*) and the resistance set at its highest value; hence the cable end was open. After the primary pulse is seen the reflected pulse according to Eq. (21-1) with the same polarity. As a result of damping losses in the cable its amplitude is somewhat less than that of the primary pulse. The evaluation of the base distance τ (at 10% of pulse amplitude) from the time calibration in Fig. 21-2*d* in which 1 Mc/s ($\hat{=}$ 1 μs) shows that the delay is 3.3 μs. The transit time in one direction is accordingly about 1.65 μs. As the transit time is given by the cable manufacturer as 1.9 μs per metre (Hackethal "2500"), it can be calculated that the length of the cable was 0.87 m, which in fact agrees very well with the actual length. If the terminating resistance at the end of the cable is adjusted to the characteristic impedance $Z = 2.7$ kΩ, the reflected pulse disappears, as can be seen in Fig. 21-2*b*. On short-circuiting the cable end the reflected pulse shown in Fig. 21-2*c* is negative, in accordance with Eq. (21.2). The time scale for these recordings was 1.6 μs per cm.

If the cable input is open and the output short-circuited, the pattern shown in Fig. 21-3 is obtained. The pulse energy oscillates between cable input and cable end, until it is dissipated by the cable damping. The time scale, 11 μs/cm, was greater than for the oscillograms of Fig. 21-2, so that the whole of this decay process is recorded.

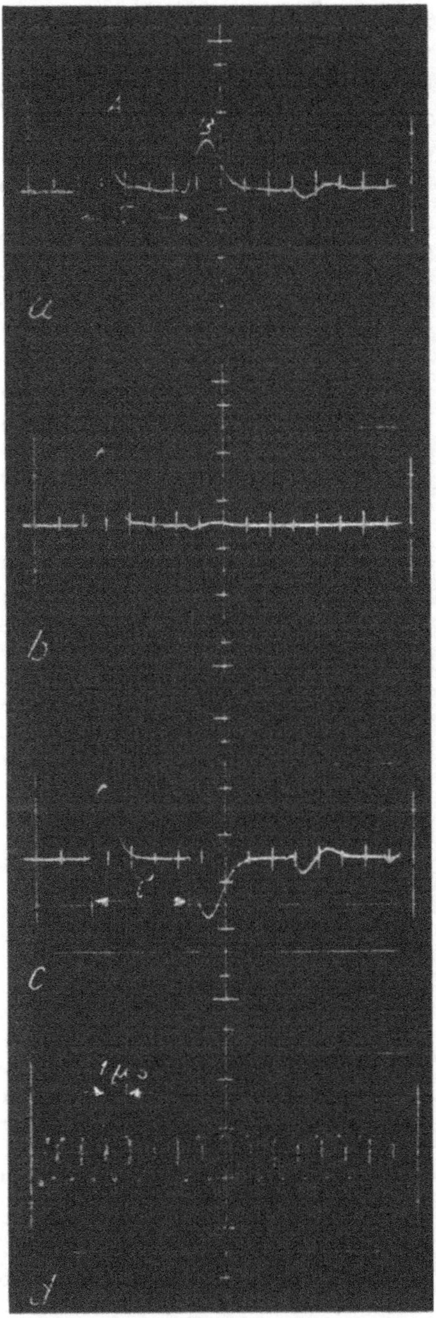

Fig. 21-2 Oscillograms for measuring the cable transit time and judging matching conditions
a) Cable beginning terminated with surge resistor, cable end open
b) beginning and end of cable terminated with a resistor equal to surge resistance of cable
c) end of cable short-circuited
d) time calibration: 1 Mc/s = 1 μs

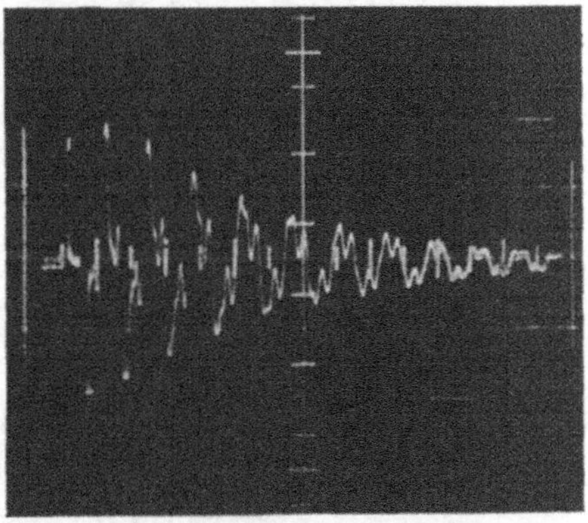

Fig. 21-3 Voltage waveform with open cable input and short-circuited output

21. 4 Oscillograms with a relatively long rectangular pulse

As described in the previous section, it is natural that signal pulses as short as possible are desirable for transit time measurements. Their duration must be short in relation to the transit of the system under observation.

However, if the pulse duration is increased until it is equal to or greater than twice the transit time (double journey of the pulses), overlapping of the input voltage and of the reflected voltage occurs. Form these oscillograms it is possible, just as in the preceding section, to draw conclusions concerning the matching conditions and transit times. In many practical applications, however, such arrangements are used to generate pulses with a desired waveform by means of suitably selected control pulses together with delay cables or networks and certain matching relationships. An oscillogram of one such longish, negative pulse is seen in Fig. 21-4*a*. From the time calibration 1 Mc/s, recorded as a second oscillogram, the pulse duration is found to be 46 μs. This pulse generates in the same circuit

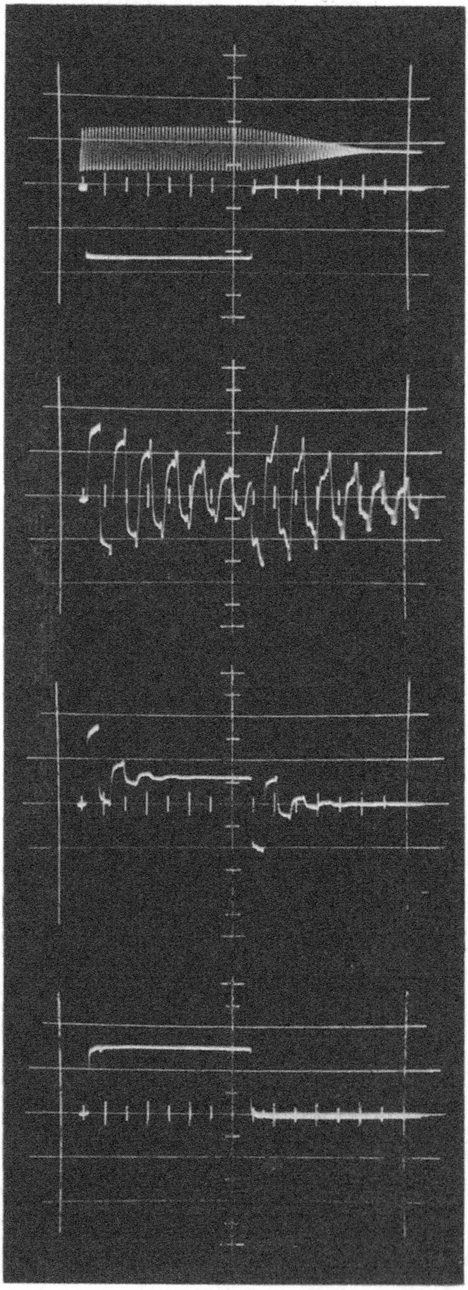

Fig. 21-4 Voltage forms on the cable of circuit in Fig. 21-1 with a square pulse 46 μs wide at input. *a*) Negative drive pulse on grid of valve E 83 F and time calibration of 1 Mc/s = 1 μs; *b*) cable input open, output short-circuited; *c*) cable input open, at output R_E = 0.7 kΩ ($< Z$); *d*) cable input open, output terminated with $R_E = Z$

as illustrated in Fig. 21-1 a correspondingly long positive impulse in the anode circuit of the E 83 F valve, and this pulse is applied to the input of the delay cable. If both switches (S_1 and S_2) are in the positions marked *a*, the cable input is open, and the cable end is short-circuited.

Once again, as shown in Fig. 21-4*b*, a negative pulse occurs at the cable end and appears at the input after the double transit time. As the input end is open, the pulse energy oscillates similarly to and fro during the whole duration of the input pulse, as can be seen in the oscillogram of Fig. 21-3 after a short pulse. At the end of the input pulse a similar line pulse is released and begins with negative polarity but otherwise oscillates in the same way.

If the output impedance has a certain value ($> 0, < Z$), only a part of the incoming voltage surge is reflected. As the oscillogram in Fig. 21-4*c* for an output impedance $R_E = 0.7$ kΩ shows, it becomes overlaid on the pattern of the input pulse which now appears at reduced amplitude. If the output impedance of the cable corresponds exactly to its characteristic impedance, in this case 2.7 kΩ, the reflections disappear, as the input energy is completely dissipated in the terminating resistance. The pulse pattern with the full voltage amplitude [103]) can now be observed at the beginning of the cable, as shown in the oscillogram of Fig. 19-4*d*.

If the output resistance is several times greater than the characteristic impedance of the cable and the input impedance is also high, a voltage surge runs through the cable and its amplitude corresponds to a part of the input voltage determined by the ratio of the characteristic impedance to the internal impedance of the voltage source (in this case the output impedance of the E 83 F valve).

Fig. 21-5*a* shows an example of this with an output resistance of 10 kΩ. The surge coming from the end of the cable has the same polarity. Its voltage builds up again, after the double transit time, to the voltage at the input. Thus the second step in the oscillogram is formed at the beginning of the pattern. So long as the input pulse continues, charge continues to flow into the cable end once again and, after the double transit time causes a voltage rise at the cable input. As the charging current is always determined by the difference between the input voltage and the voltage at the beginning of the cable at any particular instant, steps are obtained, whose mean values rise according to an exponential function.

The discharge of the cable at the end of the input pulse takes place in a similar way. After the double transit time the charge always drops by a certain value. If the cable output is terminated with a multiple of the characteristic impedance ($R_E = 10$ kΩ) and the cable input is terminated with a resistance equal to the characteristic impedance ($R_A = Z$), the waveform is as shown in the oscillogram of Fig. 21-5*b*. At the beginning and the end of the input pulse of the square voltage generator only one step occurs.

If input and output of the cable are terminated with an impedance equal to the characteristic impedance ($R_A = R_E = Z = 2.7$ kΩ), a pattern of the positive pulse similar to that in Fig. 21-4*d* is obtained. However, its amplitude is only half that shown in Fig. 21-4*a*, as the total voltage of that arrangement is divided into

[103]) The vertical deflections in the various oscillograms only correspond approximately to the actual amplitude conditions. For clearly visible pictures gain correction must be carried out to adjust the larger differences in voltage. The main intention here was to show and discuss the voltage waveforms.

a

b

c

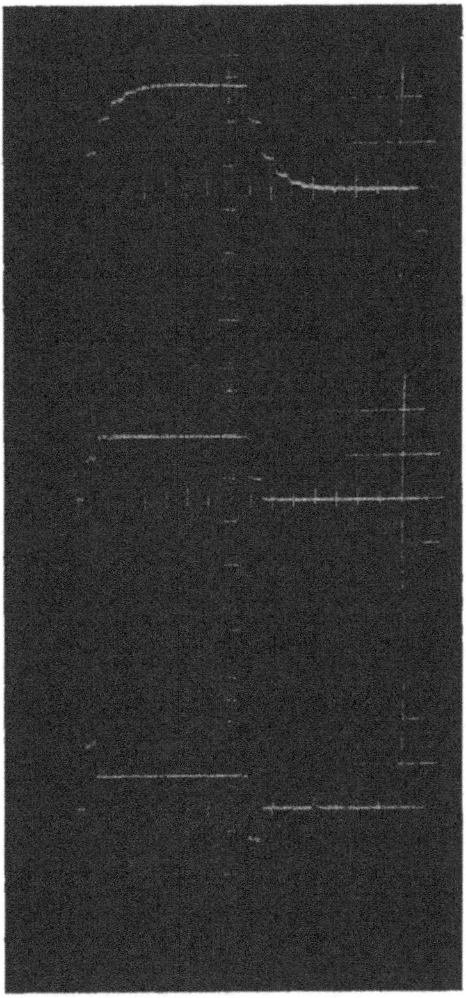

Fig. 21-5 Voltage on cable input with pulse 46 μs wide. *a*) Input open, $R_E = 10$ kΩ; *b*) input resistor $R_A = Z = 2.7$ kΩ, output resistor $R_E = 10$ kΩ ($> Z$); *c*) input resistor $R_A = Z = 2.7$ kΩ, output resistor $R_E = 0.7$ kΩ ($< Z$)

two practically equal parts, one across the input impedance and the other across the output impedance.

If the output resistance is still further reduced to a fraction of the characteristic impedance ($R_E = 0.7$ kΩ), an oscillogram is obtained like that of Fig. 21-5*c*. At the commencement of the pulse pattern there now occurs a positive pulse, and at the end a negative one. Its width corresponds once again to the double transit time. The amplitudes are determined by the ratios of the input impedance to the output impedance, and by the value of the characteristic impedance of the cable.

These examples show how voltage pulses with predetermined and even complex waveform can also be generated by the choice of a suitable generator pulse with a delay cable or delay network and by suitably rating the terminating impedance.

The practical measuring technique only has been described in this section. For the precise quantitative relationships and for further practical applications the reader is referred to the relevant literature [2] [3] [4] [5] [6].

CHAPTER 22

DETERMINING THE CHARACTERISTIC QUALITIES OF RESONANT CIRCUITS AND BANDPASS FILTERS FROM THE PATTERN OF THE DECAYING OSCILLATION AFTER SHOCK EXCITATION

22. 1 The characteristic qualities of resonant circuits and their measurement

The most useful methods of measuring the qualities of resonant circuits are based on recording their selectivity curve or resonance curve. For accurate measurements the voltage waveform appearing across the circuit (discussed here as a parallel resonant circuit) is recorded, when a constant alternating current having a frequency in the vicinity of the resonant frequency is connected to it, via as loose a coupling as possible from the signal generator. For rapid working, the resonance curve can be recorded with an oscilloscope in the way described in Part III, Ch. 18.

The resonant frequency ω_0 is given by the equation

$$\omega_0 = \sqrt{\frac{1}{L \cdot C} - \left(\frac{R_s}{2 \cdot L}\right)^2}. \tag{22.1}$$

Here, L is the circuit inductance, C the circuit capacitance and R_s the loss resistance of the inductance. The damping by the insulation resistance and the dielectric losses in the capacitance which occur in practice and indicate a parallel resistor R_p in Fig. 22-1, are neglected. As the second term under the root is generally so small as to be negligible compared to the first, the resonant frequency f_0, if f_0 is taken equal to $\omega_0/2\pi$ is given by the familiar equation

$$f_0 = \frac{1}{2 \cdot \pi \cdot \sqrt{L \cdot C}} \tag{22.2}$$

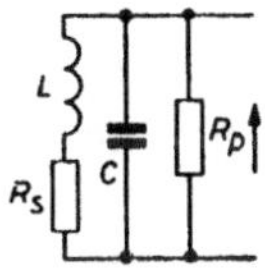

Fig. 22-1 Parallel resonant circuit with loss resistors

(Resonance occurs when the reactances of the inductance and capacitance are equal, thus $1/\omega C = \omega L$). The characteristics of a resonant circuit are variously designated by the damping factor

$$\delta = \frac{R_s}{2 \cdot L}, \tag{22.3}$$

or by the logarithmic damping decrement

$$\vartheta = \pi \cdot \frac{R_s}{\omega_0 \cdot L} = \frac{R_s}{2 \cdot f_0 \cdot L} \tag{22.4}$$

the damping

$$d = \frac{R_s}{\omega_0 \cdot L} \tag{22.5}$$

or the damping reciprocal, (the Q value):

$$Q = \frac{\omega_0 \cdot L}{R_s} \tag{22.6}$$

The relationship between these magnitudes can be seen from Table 22-1.

The resonant resistance of a parallel circuit consisting of inductance and capacitance is

$$R_0 = \frac{1}{R_s} \cdot \frac{L}{C} \tag{22.7}$$

or

$$R_0 = \frac{1}{d} \cdot \sqrt{\frac{L}{C}} = Q \cdot \sqrt{\frac{L}{C}}\,. \tag{22.8}$$

The factor $\sqrt{\frac{L}{C}}$ in this equation is the characteristic impedance Z of the circuit. The circuit quality factor Q can be calculated from equation

$$Q = \frac{1}{R_s}\sqrt{\frac{L}{C}}\,. \tag{22.9}$$

The selectivity or discrimination of a circuit σ can be expressed also by the ratio of the frequency detuning of the circuit which produces a voltage equal to $\frac{1}{\sqrt{2}}$ or 0.71 of the voltage at resonance, to the damping d, thus

$$\sigma = \frac{\Delta f}{d}\,. \tag{22.10}$$

Measurement of circuit characteristics is usually carried out by recording the resonance curve, thus determining at what frequency difference the height of the curve corresponds to 0.71 of the peak value.

It is better, after determining the voltage value at resonance, to raise the output voltage of the signal generator by the $\sqrt{2} = 1.41$ times its value and then find the frequency detuning at which the same voltage as that previously measured at resonance reappears across the circuit.

The quality factor of the circuit is then found to be

$$Q = \frac{f_0}{\Delta f} \tag{22.11}$$

or the damping to be

$$d = \frac{\Delta f}{f_0} . \tag{22.12}$$

In practice it may be preferable not to adjust the frequency, but to observe at what capacitance detuning the voltage obtained is 0.72 of the voltage at resonance. The quality factor is then found to be

$$Q = 2 \cdot \frac{C_0}{\Delta C} . \tag{22.13}$$

Whichever circuit characteristic may be of immediate interest can then be calculated from the equations discussed at the beginning of this section.

22. 2 Generating the pattern of the decaying voltage

In contrast to the measuring process described, the specific characteristics of resonant circuits can also be gathered from the oscillogram obtained when the circuit receives a voltage or current shock ("shock excitation"). There is then obtained a pattern showing the decay of the circuit voltage following this shock, and, if the shock is non-recurrent, the trace can be photographed and then interpreted.

This process is, however, made easier if the shock applied to the test circuit is repeated. If this shock is synchronized with the time deflection a stationary pattern of the voltage waveform is to be seen on the fluorescent screen. The time base voltage itself can be used to excite the circuit.

Its sawtooth waveform makes it particularly suitable for excitation. Synchronism is then automatic and its operation becomes extremely simple. In the oscillograms in Figs. 22-2*a* and *b* are shown, for the sake of clarity, *a*) the sawtooth waveform of the voltage for the time base, and *b*) its voltage curve differentiated by a *CR*-network (see also Fig. 11-30). It can be seen that at each instant of flyback, corresponding to the rapid change of the voltage, a strong negative pulse occurs [104]). If the time base unit operation is non-triggered and at the same time the voltage across the circuit is used for time deflection as described, a stationary

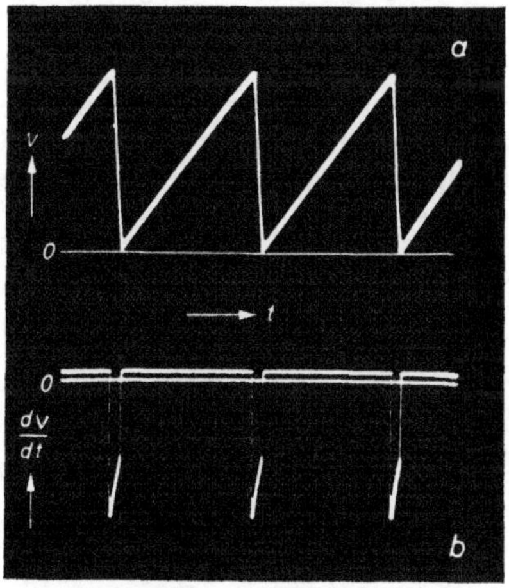

Fig. 22-2 Sawtooth voltage *a*) and its differentiated waveform *b*)

[104]) In this oscillogram the flyback path, contrary to the practice of this book, has been drawn in, as otherwise the pattern of the whole waveform would lack clarity.

picture of the waveform following these shocks will appear. In the practical application of this process the circuit can be quite simple, as can be seen in Fig. 22-3.

The resonant circuit is connected to the terminals of the vertical amplifier. To excite the circuit, the sawtooth voltage is connected to the "hot" pole of the oscillating circuit via a relatively small capacitor (a few pF). On closer examination of this circuit it is easy to recognize that a differentiation circuit is formed by the coupling capacitance (C_k) and the coil inductance L (in series with the loss resistance R_s). If the capacitor (C) is left out of account for the instant of excitation, it can be shown that the voltage across the coil is proportional to a constant ($\frac{L}{R_s}$) and to the differential quotient of the sawtooth voltage $\frac{dv}{dt}$ i.e., and must correspond to the waveform of the oscillogram in Fig. 22-3*b*.

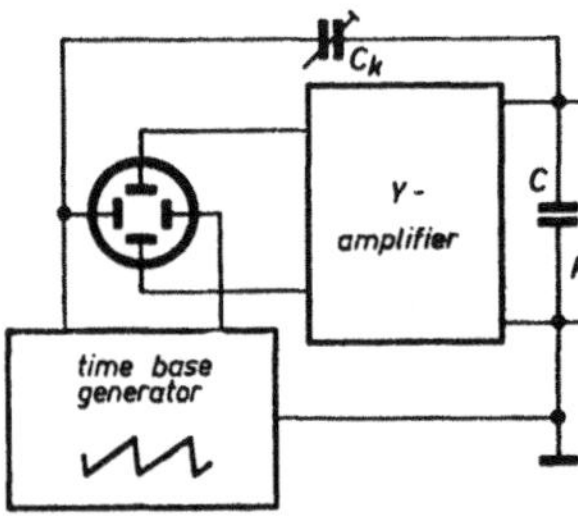

Fig. 22-3 Generating the pattern of the decaying voltage of a resonant circuit by starting with the time base voltage

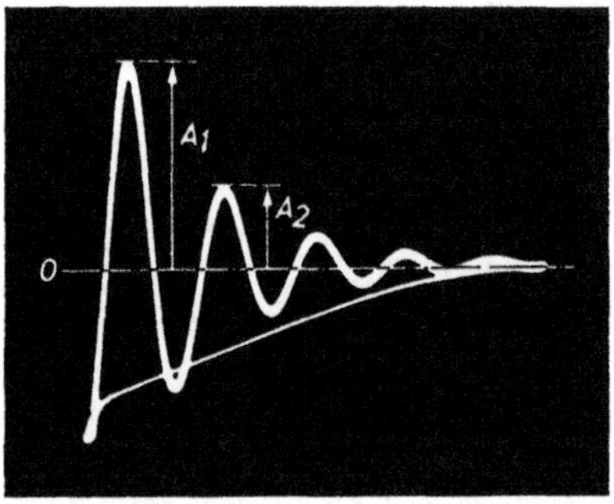

Fig. 22-4 Decaying oscillation of a strongly damped resonant circuit

In a simple resonant circuit the pattern of the decaying oscillation is obtained as is shown in the first example in Fig. 22-4 [105]) [1] [2] [3].

The voltage waveform is then according to the equation

$$v_t = V_0 \cdot e^{-\delta t} \cdot \cos \omega \cdot t \tag{22.14}$$

and in the exponent,

$$\delta = \frac{R_s}{2 \cdot L}.$$

22. 3 Interpreting the oscillograms in simple resonant circuits

The amplitude relationships of successive oscillations determine the circuit characteristics. When examining these circuit characteristics it is customary to term the natural logarithm of the relationship of the successive amplitudes the *logarithmic damping decrement*. It is thus

---

[105]) Such oscillograms were shown by the author in lectures held as far back as 1937/1938.

$$\vartheta = \ln \frac{A_1}{A_2} \tag{22.15}$$

the actual damping *d*, used in most calculations today, is obtained from equation

$$d = \frac{\vartheta}{\pi} = 0.318\,\vartheta\,. \tag{22.16}$$

All other circuit data can be derived from it at a known resonance frequency $\omega_0 = 2 \cdot \pi \cdot f_0$ in accordance with Table 22-1.

TABLE 22-1

	δ	ϑ	d	Q
$\delta =$	—	$\vartheta \cdot f_0$	$\pi \cdot f_0 \cdot d$	$\frac{\pi \cdot f_0}{Q}$
$\vartheta =$	$\frac{\delta}{f_0}$	—	$\pi \cdot d$	$\frac{\pi}{Q}$
$d =$	$\frac{\delta}{\pi \cdot f_0}$	$\frac{\vartheta}{\pi}$	—	$\frac{1}{Q}$
$Q =$	$\frac{\pi \cdot f_0}{\delta}$	$\frac{\pi}{\vartheta}$	$\frac{1}{d}$	—

Table 22-1 Relationship between the individual magnitudes of an oscillating circuit

However, the damping in Fig. 22-4 is abnormally great. In HF circuits with normal characteristics the amplitude difference of the successive deflections is, on the other hand, very small, as can be seen from the oscillograms in Fig. 22-5, so that such interpretation would be far from accurate. A great improvement in accuracy has been described by NENTWIG by determining, not the relationship of the successive oscillation amplitudes, but how many voltage cycles there are for a given amplitude reduction to $^1/_2$ or $1/e$ [4].

The pattern is adjusted by varying the time base frequency, so that a reasonable number of oscillation cycles per deflection cycle can be obtained.

If it is assumed, for instance, that the amplitude after *N* cycles has been reduced to half of the initial value, the following equation applies for the determination of the damping decrement ϑ

$$\ln 2 = \ln \frac{A_1}{A_N} = N \cdot \ln \frac{A_1}{A_2} = N \cdot \vartheta\,. \tag{22.17}$$

From this we obtain

$$\vartheta = \ln 2 \cdot \frac{1}{N} = \frac{0.693}{N} . \tag{22.18}$$

From Eq. (22.16) damping d can be derived from the simple equation

$$d = \frac{0.22}{N} . \tag{22.19}$$

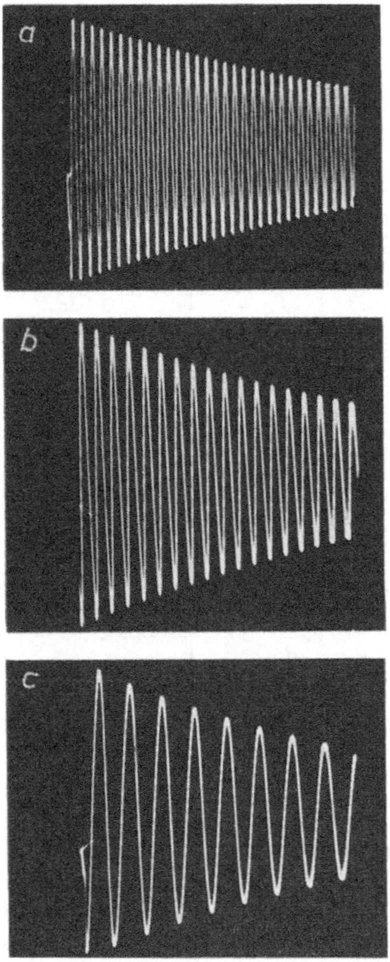

Fig. 22-5 Measuring circuit quality Q during voltage decay to half the initial value. a) $Q = 137$; b) $Q = 77$; c) $Q = 36$

In the oscillograms of Fig. 22-5, for instance, 30,17 and 8 cycles can be counted for a reduction to half in amplitude. According to Eq. (22.19) this corresponds to dampings of 0.73, 1.3 and 2.75%.

The circuit quality factor is the reciprocal of the damping — see Eqs. (22.11) and (22.12) — so that it is

$$Q = \frac{N}{0.22} . \tag{22.20}$$

In the examples in Fig. 22-5 the quality factor is thus 137,77 and 36.

It is clear that resonant circuits with normal quality factors can be satisfactorily measured in this simple way. For instance, 55 voltage cycles correspond to a circuit quality of $Q = 250$, and these can be easily counted on a 10 cm screen.

In the case of particularly high circuit quality (or small damping) a lesser amplitude drop, for example to 0.8, is observed. Then, since $\ln 1/0.8 = \ln 1.25 = 0.223$, the constant 0.071 must replace the constant 0.22. At known resonant frequency f_0, other circuit characteristics can be calculated from the values d or Q. Thus, for instance, the bandwith

$$\Delta_f = \frac{f_0 \cdot 0.22}{N} . \tag{22.21}$$

With a resonant frequency of about 350 kc/s the bandwith for Fig. 22-5*a* is found to be 2.6 kc/s. In the recording of the resonance curve (Fig. 18-7), which could not be very accurately interpreted, a bandwidth of about 3 kc/s was measured, which is in fairly close agreement.

If, in addition, the inductance L is known, the dynamic resistance of the circuit is found from:

$$R_0 = \frac{\omega \cdot L \cdot N}{0.22} \,. \tag{22.22}$$

It should be pointed out that in the simple circuit shown in Fig. 22-3 a certain amount of reaction effect on the time base generator cannot be avoided. (The capacitor C_k is not a reaction-free coupling). It is therefore advisable to connect the circuit under investigation to the anode lead of a pentode, as in the circuit used by BLOK (see also Fig. 22-6) [1].

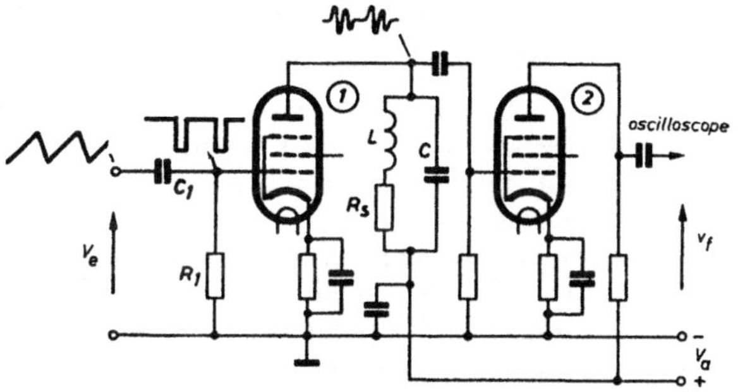

Fig. 22-6 Determining the characteristics of a valve-coupled circuit

The excitation voltage, the waveform of which is that of the sawtooth voltage differentiated by C_1 and R_1 (Fig. 22-2*b*), is applied to the grid of the amplifier valve. The anode oscillating circuit is then excited by means of positive pulses.

If a circuit within another circuit is to be investigated, then the above circuit or a similar one is usually most suitable. The circuit data are then obtained under true working conditions. Care must be taken, however, that the pulse voltage on the grid of the corresponding valve is not distorted in any undesirable way by other circuit elements. Oscillating circuits are best switched off and replaced by ohmic resistors of suitable value. Moreover, care should be taken that the oscilloscope vertical amplifier input is connected to the oscillating circuit in such a way that the characteristics of the former are not noticeably altered. Here too, it can be an advantage not to take the voltage direct from the circuit, but from the anode of the following valve (in a wireless receiver, for instance, from an IF stage). This valve is then connected as a wideband amplifier (with an anode resistance of 1...5 kΩ). It is advisable to switch off any means of tuning which may be present, so that the voltage taken off is not additionally influenced by them. In this way the characteristics of four-pole networks can be generally determined even when a high degree of accuracy is required [5]. As has been shown by VAN SLOOTEN, the excitation of a four-pole network for certain measuring tasks can also be achieved satisfactorily by using a current surge [6] with which not only resonant circuits but also other kinds of four-pole networks can be examined.

It should be pointed out here that not only electrical resonant circuits, but every sort of structure capable of oscillation — even mechanical ones — can be examined as to their characteristics. But for this purpose it is necessary, with the aid of suitable pickups (or transducers), to convert the variable of state into electrical voltage changes. Thus, for example, strain gauges (Ch. 31 "Strain gauges") can be used to measure the material constants of test rods. OLSSON and ORLIK-RÜCKEMANN for example have described a measuring device under the name "Dampometer", which is used to test the mechanical behaviour of aircraft parts. The pattern of the decaying oscillation is in this case not linear with time, but describes a spiral (Ch. 28, Figs. 28-2*f* and 28-5). A photocell placed opposite slits in front of the oscilloscope observes every traverse of the fluorescent spot and transmits the corresponding pulse to an electronic counter equipped with a E1T decade counter tube. The counted pulses are thus a standard of measure for the object under test [7].

22. 4 Measuring the coupling factor in coupled circuits

22.4.1 COUPLED CIRCUITS AND THE COUPLING FACTOR

If two or more structures capable of oscillations (regarded in this case as electrical resonant circuits) are coupled together, certain characteristics can be obtained which are unobtainable with simple circuits alone e.g. their high selectivity at greater bandwidths).

Resonant circuits are generally described as coupled if they are so connected by magnetic or electric fields or even by ohmic resistances, that during oscillation energy is transmitted from one circuit to the other and vice versa. This can be done in a great variety of ways. Four fundamental circuits are shown in circuits *a)* to *d)* in Fig. 22-7. In *a)* is illustrated one type of inductive coupling. It is brought about when two tuning coils L_1 and L_2 are arranged spatially so that the energy transmission takes place through the magnetic field. But it is also possible that part of the inductances L_1 and L_2 is formed by the inductance L_k common to both, as shown in *b)*. The inductances shown in each case consist of the series connection of the large circuit inductance (L_1 and L_2) and a smaller coupling inductance (L_k) common to both.

The coupling factor k showing the amount of reciprocal coupling of the circuits is represented for the circuit diagram *a)* by the equation

$$k = \frac{M}{\sqrt{L_1 \cdot L_2}} . \tag{22.23}$$

The amount M shows the strength of the coupling inductances or, as in *b)*, with a common partial inductance, how big this part of the inductances is in relation

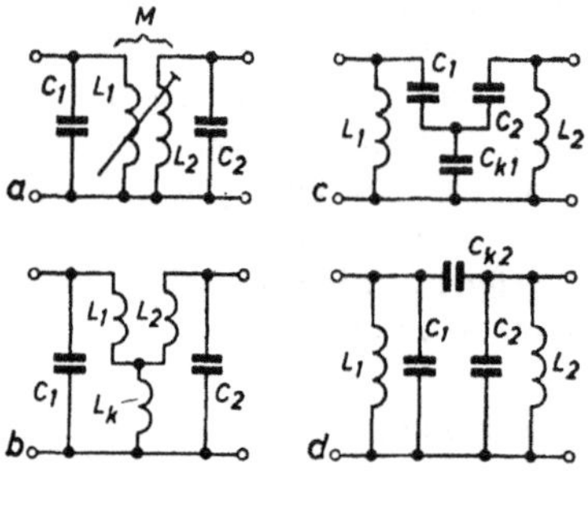

Fig. 22-7 Fundamental coupling possibilities of two circuits. *a)* Coupling by mutual inductance; *b)* inductive "current" coupling; *c)* capacitive "current" coupling; *d)* capacitive "voltage" coupling

to the total circuit inductances. If both inductances are equal, i.e. $L_1 = L_2 = L$, the coupling factor is simply

$$k = \frac{M}{L}. \tag{22.24}$$

In the capacitive coupling shown in *c*), the capacitance of the individual oscillation circuits is formed in each case by the series connection of the actual oscillating circuit capacitors (C_1 and C_2) and a coupling capacitor (C_{k1}), which the coupling capacitor is common to both circuits.

In capacitive coupling according to *c*) the coupling factor is obtained in a similar way to the inductive coupling in *b*) from the proportion of the capacitive reactances of the coupling capacitance to the reactances of the circuit capacitances. It can be calculated from the equation

$$k = \sqrt{\frac{C_1 \cdot C_2}{(C_1 + C_k) \cdot (C_2 + C_{k1})}} \approx \sqrt{\frac{C_1 \cdot C_2}{C_{k1}}}. \tag{22.25}$$

(The approximation applies if $C_{k1} > C_1, C_2$.)

If, in addition, the circuit capacitances are equal, i.e. $C_1 = C_2 = C$, the coupling factor is obtained from the simple equation

$$k = \frac{C}{C_{k1}}. \tag{22.26}$$

In circuits *b*) and *c*), coupling takes place because of a coupling impedance in the part common to both circuits; this is therefore termed *current coupling*. However, similarly to *a*), where the coupling is by the electromagnetic field, it is possible to produce an *electrostatic* coupling by means of a capacitance at the "hot" end of the circuit, as shown in circuit *d*) with the capacitor C_{k2}. Again assuming that the circuit capacitances are equal, i.e. $C_1 = C_2 = C$, and that the coupling capacitor C_{k2} is small as compared with C ($C_{k2} < C$), the coupling for this circuit is:

$$k = \frac{C_{k2}}{C}. \tag{22.27}$$

There are many other ways of coupling circuits. Even those using two or more types of coupling simultaneously are quite common. A selection has been noted in the bibliography [8] [9] [10] [11] [12]. The characteristics of simple oscillatory circuits are dealt with in detail there also.

22.4.2 THE PATTERN OF DECAYING OSCILLATION IN COUPLED CIRCUITS AND THE DETERMINATION OF THE COUPLING FACTOR

If, instead of a simple resonant circuit, the primary of a number of coupled circuits is connected in an arrangement such as in Fig. 22-3 or 22-6, it is possible to observe the pattern of the decaying oscillation in both the primary and the secondary circuit with an oscilloscope. So long as the coupling remains under the critical coupling value, the voltage in the circuits decays in a way similar to that in a simple resonant circuit having the same characteristics. In the secondary circuit the maximum value of the amplitudes is naturally dependent on the coupling factor. When the coupling factor increases, the voltage in the secondary circuit naturally increases also.

If the response curves of two such coupled circuits are studied by the process described in Ch. 18, it can be seen that the peak of the curve first rises with increasing coupling. After critical coupling has been reached it does not rise any further but breadens out. When the coupling is further increased it splits up into two peaks, one on each side of the resonant frequency ω_0 (Fig. 18-8).

The critical coupling factor k_{cr} is equal to the damping d, that is,

$$k_{cr} = d = \frac{R_s}{\omega_0 \cdot L} = R_s \cdot \sqrt{\frac{C}{L}}\,. \tag{22.28}$$

Assuming that the damping and the coupling factor are small ($d < 0.1$ and $k < 0.1$) the voltage v_t occurring across the secondary circuit after a current surge I_{max} and after time t can be calculated according to equation

$$v_t = I_{max} \sqrt{\frac{L}{C}} \cdot e^{-\frac{R_s}{2L}t} \cdot \sin \frac{1}{2} k \cdot (\omega_0 \cdot t - d) \cdot \cos (\omega_0 \cdot t - d). \tag{22.29}$$

The two oscillograms in Fig. 22-8, recorded using the circuit in Fig. 22-3, show the waveform on the primary and secondary circuits of a band filter.

It can be clearly seen how energy is exchanged back and forward between the two circuits periodically according to the coupling factor. In the equation this is indicated by the term $1/2 \cdot k \cdot (\omega_0 \cdot t - d)$. It is also clear that in both circuits the voltage decays in accordance with the term $e^{-\frac{R_s}{2L}t} = e^{-\delta \cdot t}$

The ratio of two successive maxima depends on the ratio d/k. The time T_m between two maxima is given by $1/2 \cdot k \cdot \omega_0 \cdot T_m = \pi$. Thus:

$$T_m = \frac{2 \cdot \pi}{\omega_0 \cdot k} = \frac{1}{f_0 \cdot k}\,. \tag{22.30}$$

If the time T_m is measured in units of duration of oscillation $T_0 = \frac{1}{f_0}$ that is, if we count the number of cycles N_k corresponding to time T_m, we find that

$$k = \frac{1}{N_k}\,. \tag{22.31}$$

This equation shows that from such oscillograms the coupling factor is simply obtained as the reciprocal of the number of oscillation cycles counted between two minima (or two maxima). The tighter the coupling the fewer cycles correspond to this region of the oscillogram.

In Fig. 22-8, for instance, 11 cycles can be counted, so that the coupling factor is 9.1%. It is thus considerably greater than the critical coupling. The oscillograms in Fig. 22-8 are rather difficult to interpret as the individual oscillation cycles lie

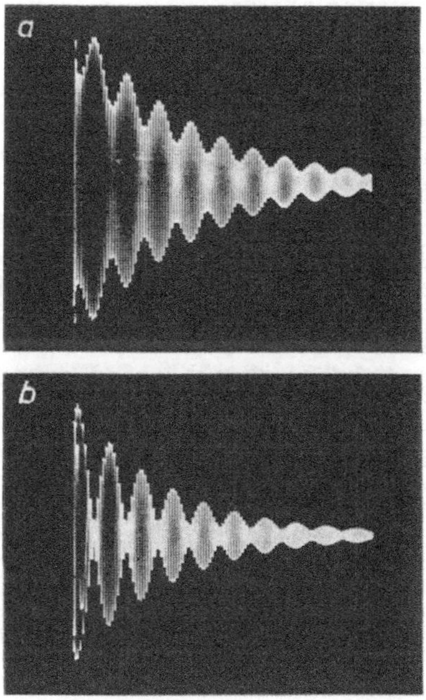

Fig. 22-8 Oscillograms of the decaying voltage of two coupled circuits ($k = 9.1\%$). *a*) Primary; *b*) secondary

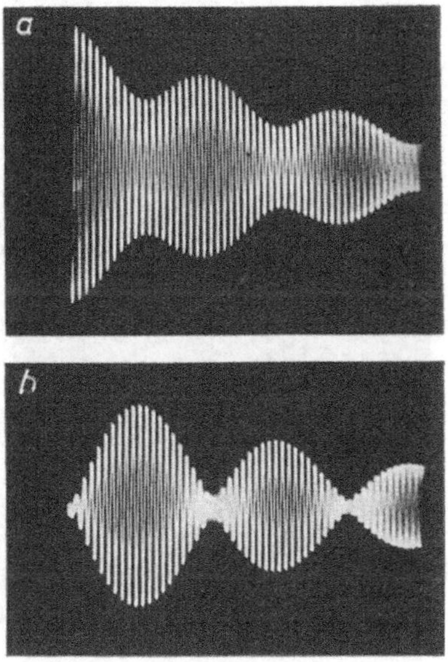

Fig. 22-9 Bandfilter with $k = 5\%$. *a*) Primary; *b*) secondary

close together. This method of representation was only chosen to show the decay process clearly.

For determining the coupling factor it is best to adjust only 3-4 cycles of the amplitude fluctuations, as is the case in the oscillograms of Figs. 22-9, 22-10 and 22-11. Here evaluation results in $N_k = 20$,13 and 7, and the coupling factors were 5, 7.7 and 14.3%. These are relatively tight couplings. The smallest coupling which can be measured in this way depends upon the maximum number of cycles that can be counted over the width of the oscillogram. If the pattern is so adjusted as to contain precisely two maxima (or minima) of the amplitude fluctuations, then a 10 cm oscilloscope tube will certainly enable 50 cycles to be counted. The lowest coupling factor which can be measured by means of this process is thus 2%. But there are patterns which can be evaluated although the number of cycles of f_0 can no longer be counted. According to Eq. (22.29) all that need be done is to read off the time corresponding to the interval between the two amplitude maxima. The ratio a of the successive amplitude maxima is given by the equation

$$a = e^{-\frac{Rs}{2L}\cdot\frac{2\cdot\pi}{\omega_0 k}} = e^{-\pi\cdot\frac{d}{k}}. \qquad (22.32)$$

Thus, at critical coupling, the maximum following decreases with respect to the

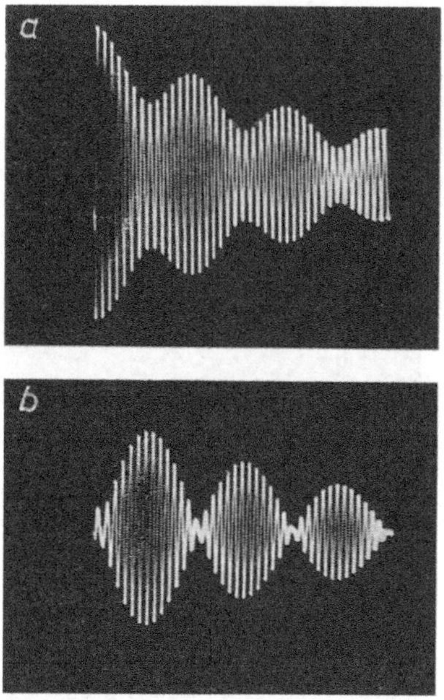

Fig. 22-10 Bandfilter with $k = 7.7\%$. *a*) Primary; *b*) secondary

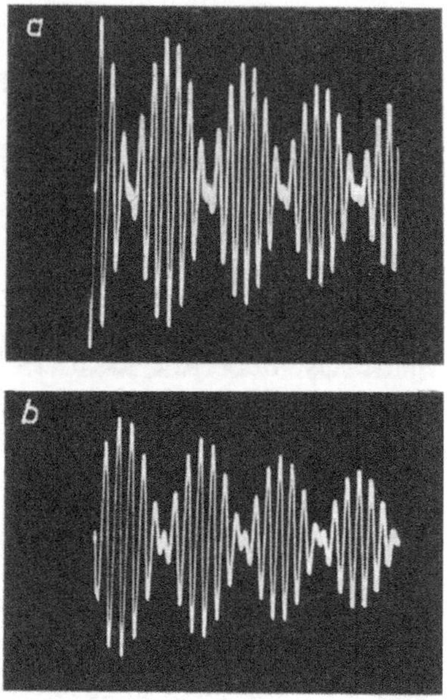

Fig. 22-11 Bandfilter with $k = 14.3\%$. *a*) Primary; *b*) secondary

proceding one only by $e^{-\pi} \approx 1/23$ ($d/k = 1$). The difference is practically impossible to distinguish.

The method of determining the coupling factor by means of the pattern of voltage decay can be used to advantage in many tasks and particularly for instructional purposes because of its outstanding clarity, as changes in the coupling factor can be seen immediately on the pattern.

It should be pointed out in particular that for measurements of circuit characteristics and measurements for determining the coupling factor of coupled circuits no other equipment beyond the oscilloscope and a few circuit elements is needed.

CHAPTER 23

SOME METHODS OF MEASURING THE AMPLITUDE MODULATION OF HF VOLTAGES

23. 1 Modulation in general and measuring the amplitude modulation

For the transmission of all kinds of signals (Morse signals, audio-frequency alternating voltages with speech or music content, television modulation voltages, signal voltages and suchlike) an alternating voltage of higher frequency is modulated by these signals in such a way that the signal content can be separated in the receiver with as little distortion as possible. The HF voltage transmitting the signal is therefore called the "carrier" voltage, and the process of combining this carrier and the signal to be transmitted is known as modulation.

The instantaneous value of the HF oscillation which is to be modulated is given by the equation

$$v_t = V_0 \cdot \sin(\omega_c \cdot t + \varphi) . \tag{23.1}$$

In this equation v_t is the voltage at time t, V_0 is the amplitude of the unmodulated HF carrier, $\omega_c = 2 \cdot \pi \cdot f_c$ is the angular frequency of the carrier voltage and φ its phase angle. According to which of these magnitudes is changed by the modulation, the method of modulation is termed amplitude, frequency or phase modulation. Here only the measurement of amplitude modulation will be considered. If a sinusoidal modulation voltage [106]) is used, the instantaneous value of the modulated carrier voltage is given by the equation

$$v_t = V_0 \cdot \left\{ 1 + m \cdot \sin(\omega_m \cdot t) \right\} \cdot \sin(\omega_c \cdot t) . \tag{23.2}$$

Here, m is the ratio of the amplitude change to the average amplitude, that is to say,

$$m = \frac{\Delta v}{V_0} \tag{23.3}$$

the degree of modulation, and $\omega_m = 2 \cdot \pi \cdot f_m$ is the angular frequency of the modulation voltage [1] [2] [3] [4].

Thus amplitude fluctuations of the HF voltage occur, as can be seen in the oscillogram of Fig. 23-1.

From this it can be gathered that the amplitude modulation of the carrier voltage can, at most, be modulated to the extent that its ampliude fluctuates between zero and twice the value of V_0. To interpret an oscillogram like the one in Fig. 23-1, it is advisable to measure the highest amplitude (a) and the lowest

[106]) Every non-sinusoidal voltage can be split up into a corresponding number of sinusoidal voltages with the fundamental frequency and multiples of it at certain amplitudes and in certain phase relationships.

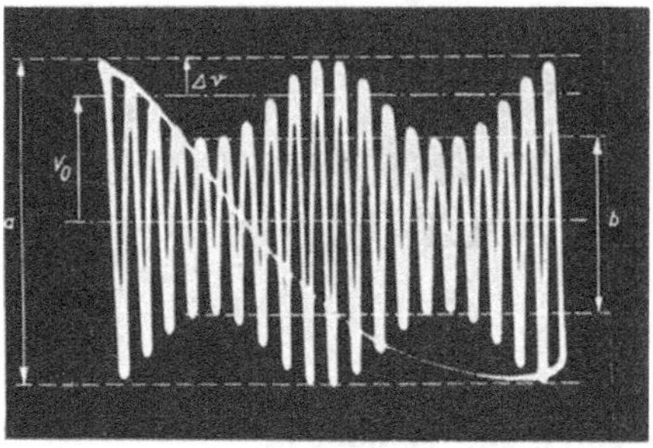

Fig. 23-1 Sinusoidal, amplitude-modulated HF voltage; $m = 30\%$; $f_c : f_m \approx 10 : 1$

(*b*). The degree of modulation is obtained, by analogy with Eq. (23.3), from equation

$$m = \frac{a - b}{a + b} \cdot 100\ [\%]\,. \tag{23.4}$$

If the values from the oscillogram in Fig. 23-1 are substituted, we obtain $m = 0.3$ in this example. As the modulation can only be a fraction of the quiescent amplitude of the carrier voltage, it is customary to express the degree of modulation as a percentage. For this the quotient $(a - b)/(a + b)$ [Eq. (23.4)] need only be multiplied by 100. In the example in Fig. 23-1 we obtain $m = 30\%$.

23. 2 Various circuits for measuring amplitude modulation

To record an oscillogram as in Fig. 23-1, the waveform of the HF amplitude in dependence upon time must be displayed so that one or more cycles of the modulation voltage can be seen. This means that the time base frequency must be equal to or an integral fraction of the modulation frequency. The time base must also be synchronized with the modulation voltage V_m externally, as shown in Fig. 23-2*a* [107]). (By triggering with the modulation voltage, time intervals can be selected and adjusted at will, but at least one modulation cycle must be visible in order to be able to judge the degree of modulation.) In Fig. 23-1 the ratio of carrier frequency to modulation frequency was purposely kept low in order to show clearly the influence of the modulation on the individual cycles of the carrier voltage. In practice, however, the frequency difference is so great that, as in the oscillogram in Fig. 23-2*b*, the carrier voltage cycles appear crowded together as a fluorescent area.

The process as described hitherto is the one usually employed in practice. There are also other alternatives for measuring the degree of modulation, and to these we shall refer in due course as their use, for some tasks, may be more favourable.

The time base unit in Fig. 23-3*a* is adjusted, not as in Fig. 23-2*a*, to a fraction of the modulation frequency, but to a similar fraction of the frequency of the HF carrier. The synchronization can then be "internal", that it to say, it can be obtained from the vertical deflection voltage. As shown in Fig. 23-3*b*, a picture

[107]) Internal synchronization is impossible for such a pattern, as the high frequency voltage lies on the vertical amplifier and thus also on the vertical deflection plates. Should it be possible, however, perhaps with older-type oscilloscopes, to make the pattern stationary, it can only be explained in that non-linearities in the amplifier cause distortions in the HF voltage (rectifying) to occur.

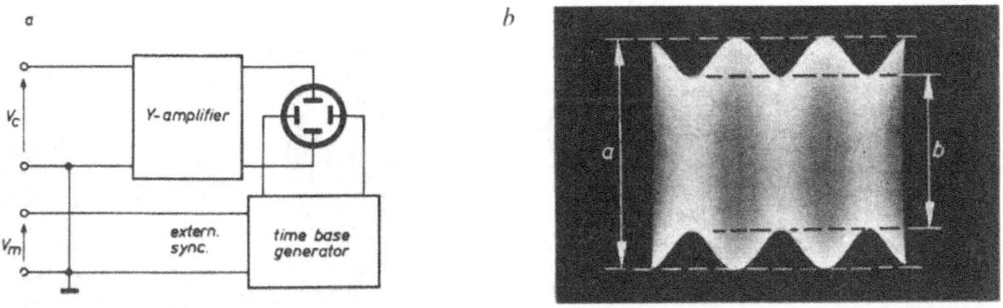

Fig. 23-2 Line deflection linear with time with a fraction (1/3) of the modulation frequency

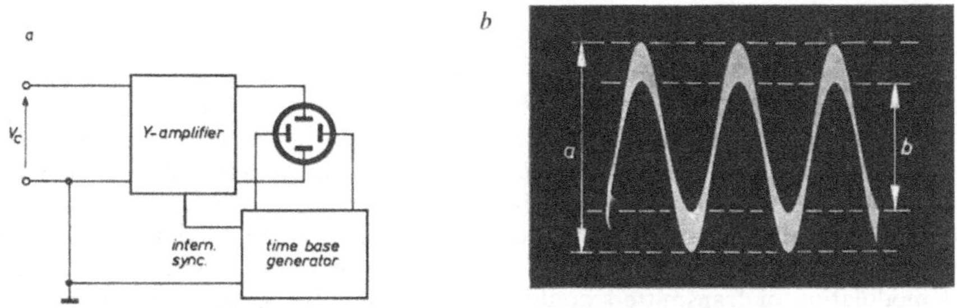

Fig. 23-3 Line deflection linear with time and a fraction (1/3) of carrier frequency

is obtained of the wave form of the HF voltage, in which the amplitudes fluctuate in the low frequency rhythm of the modulation voltage. Over the range of amplitude fluctuation a fluorescent area occurs. The limits *a* and *b* of the change in amplitude for determining the degree of modulation according to Eq. (23.3) can be found from this with ease. This method has the advantage that only the HF carrier frequency has to be fed to the oscilloscope. Its use is limited to those frequencies at which the time base generator of the oscilloscope makes it possible to set a sufficiently small time scale.

It is obvious that for the display of the degree of modulation it is not the time dependence with respect to the modulation frequency or the carrier frequency, but the dependence of the HF amplitude on the amplitude of the modulation voltage which has to be directly observed. The circuit for this is shown in Fig. 23-4*a*. The modulated HF voltage, if necessary, after amplification, is again used for vertical deflection, while the modulating voltage provides the time (horizontal) deflection [108]).

In this way a pattern showing the dependence of the carrier amplitude on the instantaneous values of the modulating voltage is obtained. Provided no non-linearities occur, this pattern is bounded both at the top and the bottom by straight

[108]) Such an arrangement is particularly useful, because in oscilloscopes generally the upper cut-off frequency of the amplifier for the vertical deflection is often higher than the upper cut-off frequency for the horizontal deflection.

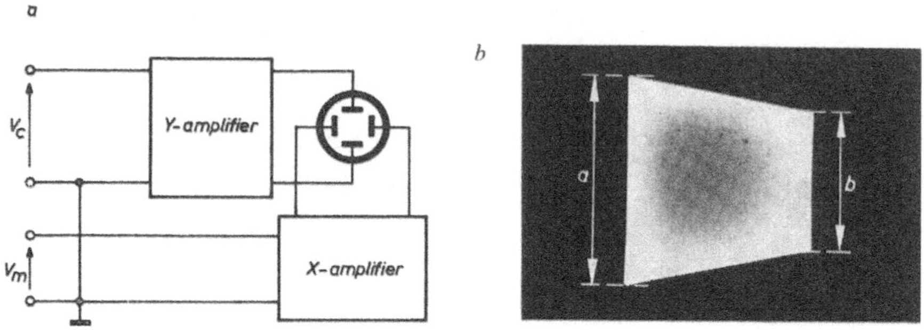

Fig. 23-4 Dependence of the HF amplitude on the amplitude of the modulating voltage

lines. This is known as a "modulation trapezium" because of its characteristic shape. Here too, the limits *a* and *b* necessary for measurement can be obtained, as can be seen from Fig. 23-4*b*. As will be shown by the oscillograms of Fig. 23-8, this circuit is particularly favourable for monitoring the degree of modulation of transmitters with varying modulation (wireless transmitters). When such measurements are carried out on transmitters, both voltages, the carrier frequency HF voltage and the low-frequency modulation voltage, are available. In such cases amplification of the deflection voltages would be unnecessary, so that the degree of modulation of transmitters could also be studied by means of normal oscilloscope tubes operating at frequencies up to about 100...200 Mc/s. In the case of receivers, however, the modulation voltage is obtained from the LF part of the receiver, i.e. after demodulation. Amplification of the signal voltages is then usually necessary, so that the frequency range for such investigations would be limited by the frequency range of the amplifiers of the oscilloscope. If, however, the carrier frequency is higher than the upper cut-off frequency of the vertical amplifier, observation of the degree of modulation can still be carried out on the lower intermediate frequency part of the receiver. Many measuring receivers, the Philips "GM 4010" field-intensity meter, for instance, have special connections for this purpose (Figs. 23-8 and 23-9). As the low-frequency modulation voltage can be taken off in any case (for monitoring and similar tasks), connection for this method of measurement is particularly easy.

The circuit in Fig. 23-5*a* shows yet another way of measuring amplitude modulation, and here too only the HF carrier voltage need be fed in. The HF voltage is connected to both pairs of plates, but with a phase difference, preferably 90°. This phase shift could be obtained, as has often been described in this book, for instance in Figs. 12-46 and 12-56, by means of *RC*- and *CR*-networks. As we are dealing with HF voltages, however, the wiring capacity would adversely affect the steps taken to obtain the desired phase shift. It is therefore advisable to obtain the phase shift by means of two resonant circuits tuned to the carrier frequency and coupled to the carrier voltage.

Fig. 23-5*a* shows such a circuit. The phase shift can be obtained by counter-current detuning of the normal tuner, for instance, by detuning capacitors C_1 and C_2. The circuits should be coupled to the voltage source via as high an impedance as possible (for example, a small coupling capacitor C_{k1}). For a 45° shift in phase

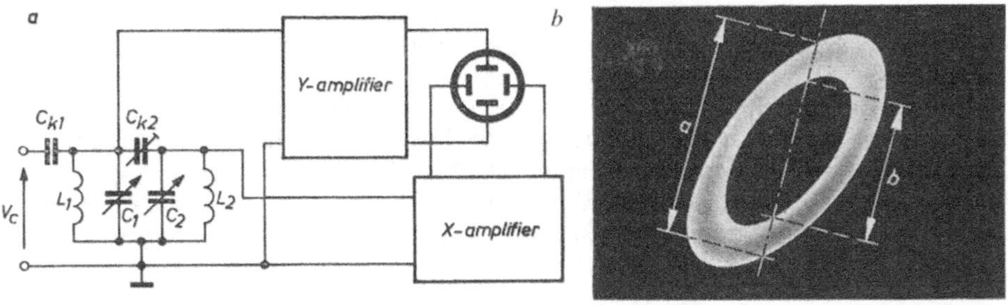

Fig. 23-5 Ring oscillogram by means of phase-shifted carrier frequency voltages on the pairs of deflection plates

the individual circuit must be detuned only to the extent that the voltage across it falls to $1/\sqrt{2} = 0.707$ of the value of the resonance voltage. If one circuit is detuned to lower and the other to higher frequencies, the required phase difference of 90° is obtained. The circuit can be more closely coupled, however, as in bandfilters. At critical coupling ($k = d$) a voltage is obtained on the secondary circuit also, which is shifted in phase by 90° compared with the primary circuit. If the deflection amplitudes in both directions are adjusted so as to be equal, a ring-shaped oscillogram is obtained, the thickness of the ring being dependent on the degree of modulation.

In the oscillogram of Fig. 23-5*b* the phase difference between the deflection voltages was only about 50°. Nevertheless it suffices to determine the degree of modulation in the way indicated. A straight line must be drawn through the centre point of the ellipse, and the distances a and b must again be measured. For the d series of recordings in Fig. 23-6 the circuit was adjusted to produce an approximately circular oscillogram (for $m = 0$). If the high-frequency carrier voltage contains harmonics it is impossible to adjust a completely circular oscillogram as the path of the spot is subject to additional deflection by the harmonics during each cycle of the fundamental wave. This, however, does not influence the measurement of the degree of modulation.

If sufficient voltage is available, the signal voltages can be used without any special amplification for deflection as in the circuit in Fig. 23-4*a*. Since here the capacitances of the deflection plates and of the leads to the oscillating circuit capacitances (C_1 and C_2) are in parallel, the circuits can be tuned with them. They then no longer impose an undesirable capacitive load as in the case of untuned connection to a vertical amplifier. This is a very real advantage in measuring the degree of modulation of carrier voltages of particularly high frequency (up to about 100...200 Mc/s).

Fig. 23-6 shows a series of oscillograms for measuring the degree of modulation according to the four methods described, for $m = 0$, $m =$ about 7%, 20%, 52% and overmodulated. While overmodulation can be readily recognized in the oscillograms corresponding to the methods shown in Figs. 23-2*a* and 23-4*a* (Fig. 23-6*a* and *c*), this is not so with the other two methods. The most suitable method can be chosen according to the task and the equipment available.

In displaying the degree of modulation by means of the modulation trapezium

(Figs. 23-4*a* and *b* and Fig. 23-6*c*) photographs were shown in which there is no phase difference between the time deflection and the waveform of the modulating voltage. This is, however, not always the case, and particularly not for all modulation frequencies. If there is a phase shift, then the upper and lower edges of the trapezium are bounded not by lines but by ellipses, which give some indication of the phase shift (Chapter 11, Fig. 11-12). For the purpose of comparison with the series of recordings in Fg. 23-6*c*, Fig. 23-7 shows a similar series of such oscillograms, this time with a phase difference of about 46°.

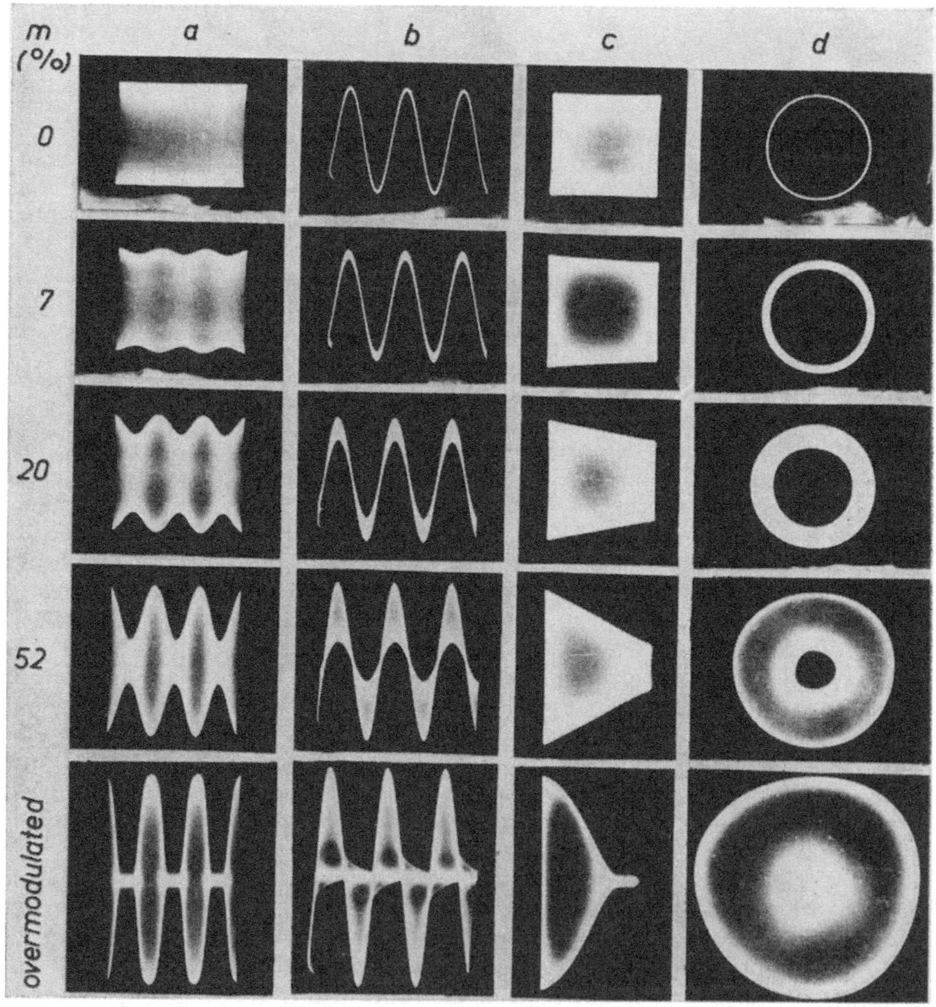

Fig. 23-6 Collection of oscillograms with $m = 0$, 7, 20 and 52% and when overmodulated as shown in the four methods given in Figs. 23-2 to 23-5

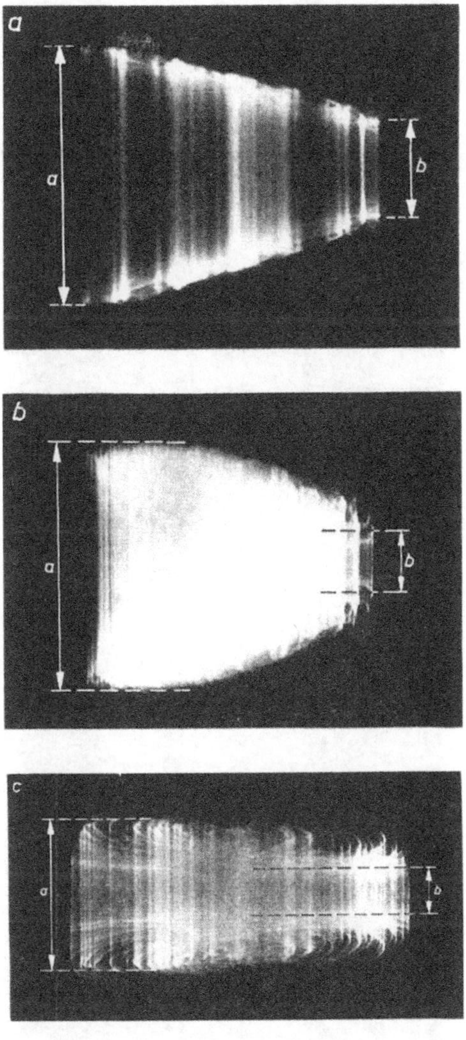

Fig. 23-8 Modulation-trapezium-recordings according to Fig. 23-4 with various wireless transmitters. *a*) Simple trapezoid pattern without phase shift; *b*) bending of upper and lower edges of pattern as a result of non-linear distortion; *c*) varying phase-shifting at individual modulation frequencies

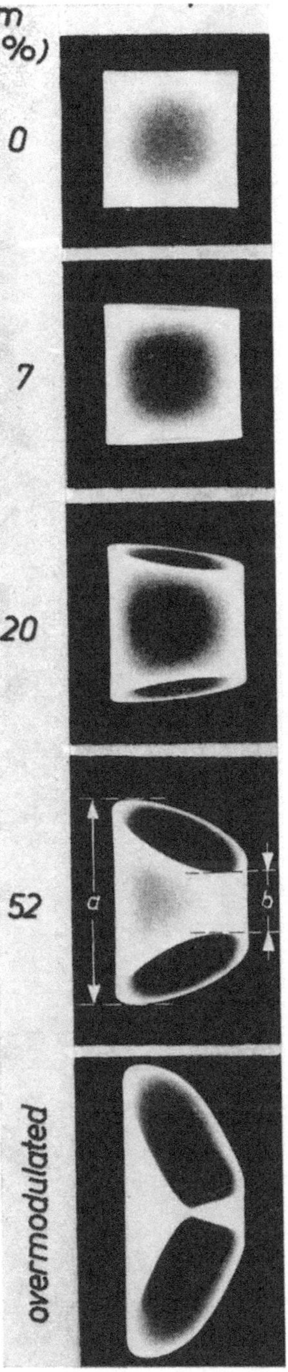

Fig. 23-7 Oscillograms as in row *c* of Fig. 23-6, but with about 46° of phase shift between line deflection with modulation voltage and the modulation waveform on the carrier voltage

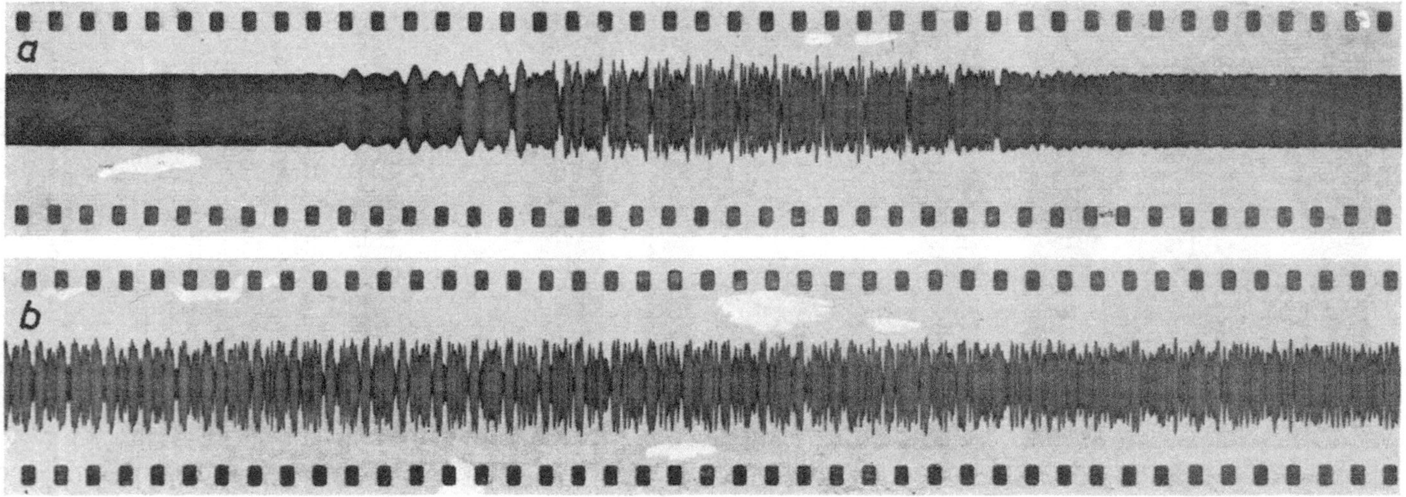

Fig. 23-9 Recordings on moving recording material for portraying the instantaneous values of the degree of modulation in wireless transmissions during a period of time
a) Speech (short word); *b*) music with strong modulation

As already mentioned, the modulation trapezium is particularly suitable for studying a transmitter with a constantly changing degree of modulation. The pattern width varies with the amplitude of the modulation voltage, while the ratio of the vertical sides of the trapezium varies with the degree of modulation. In Fig. 23-8 three other modulated trapezium oscillograms are shown which were recorded with music modulation when receiving programmes from wireless transmitters. In Fig. 23-8*a* an average degree of modulation of 42% was measured. The boundary lines of the trapezium are fairly straight and show only a slight phase shift. Fig. 23-8*b* does not show any considerable shift in phase either, although the limiting line is curved. This means that the modulation (or demodulation) was non-linear, for the boundary line of the trapezium is identical with the modulation characteristic curve (see also Figs. 17-12 and 17-13). The highest degree of modulation shown was found to be 68%.

The oscillogram in Fig. 23-8*c* represents a modulation trapezium, showing that at the different modulation frequencies varying phase shifts occur. The highest degree of modulation in this recording is about 49%.

For the purpose of studying the degree of modulation of a transmitter over a longer period it is equally possible to record a modulation trapezium photographically. Care should be taken to ensure that the photographic material is protected from exposure to excessive light by suitably screening the recording camera. Such an oscillogram shows clearly what degree of modulation has been mainly used.

If it is intended to study the degree of modulation at individual instants over a given period of time, then a moving film record is required (see Part IV, Ch. 32). For this, the horizontal deflection is omitted and the spot is only moved vertically by the modulated voltage. The time resolution is now obtained by the horizontal travel of the photographic material. In the case of a high-frequency carrier modulated with speech, singing or music, the speed of film movement must be about 1 to 3 m/s in order to obtain a sufficiently clear time resolution. The two photographs in Fig. 23-9 are examples of such investigations and were taken at 1 m/s. The photographic strip *a* was taken during a fairly short word, and *b* during the transmission of music. Sections showing great degrees of modulation were chosen from a long strip for these illustrations. For clear observation of the music-modulated voltage a film speed of 2 m/s might seem more favourable. These photographs were made while the oscilloscope was connected to the IF section of a wireless receiver with the usual intermediate frequency of about 470 kc/s, and before demodulation. GEVAERT "electrotype" recording paper was used. A blue-fuorescent tube (DB 10-6) must be fitted in the oscilloscope for this. At such relatively high film speeds brightness must also be adequate. A total acceleration voltage of 3.2 kV (2 kV post-acceleration voltage) was therefore used in the Philips "GM 5654" oscilloscope.

CHAPTER 24

REPRESENTATION OF THE FREQUENCY SPECTRUM OF MODULATED HF VOLTAGES AND OF THE FREQUENCY PANORAMA OF TRANSMITTERS

24. 1 Modulation and frequency spectrum

In amplitude modulation, which alone is considered here, the waveform of the HF voltage is represented by the equation

$$v_t = V_0 \cdot \{ 1 + m \cdot \sin (\omega_m \cdot t) \} \cdot \sin (\omega_c \cdot t) \, ,$$

in which v_t is the instantaneous value of the voltage at time t, V_0 the amplitude of the carrier voltage, $\omega_m = 2 \cdot \pi \cdot f_m$ the angular frequency of the modulation voltage, m the degree of modulation and ω_c the angular frequency of the HF carrier voltage. Eq. (25.1) can also be written trigonometrically as the sum of three frequencies:

$$v_t = V_0 \cdot \sin (2 \cdot \pi \cdot f_c \cdot t) + \frac{1}{2} \cdot m \cdot V_0 \cdot \cos [2 \cdot \pi \, (f_c - f_m)] -$$

$$\frac{1}{2} \cdot m \cdot V_0 \cdot \cos [2 \cdot \pi \, (f_c + f_m)] \, . \qquad (24.2)$$

In this equation $\omega_c = 2 \cdot \pi \cdot f_c$ and $\omega_m = 2 \cdot \pi \cdot f_m$. It shows, among other things, that in modulation, in addition to the carrier frequency f_c, two new frequencies $f_c - f_m$ and $f_c + f_m$ occur.

For the study of modulation processes it is generally desirable to be able to represent in some way the amplitude of the carrier wave and of these new side waves ("side bands") whose frequencies according to Eq. (24.2) differ from the carrier frequency f_c by $\pm f_m$. This is relatively simple with the use of an oscilloscope, as in shown in the following section.

24. 2 Circuits for representing the frequency spectrum of an amplitude-modulated HF carrier

Fig. 24-1 shows the circuit diagram of the apparatus required. The modulation of the HF carrier under investigation is obtained in the normal way by applying a modulation voltage to a modulator stage. A normal HF generator which could be externally modulated was used for the recordings reproduced in this section. The modulated HF voltage is mixed in a frequency modulator (the Philips "GM 2886" unit was used for these pictures) with a voltage derived from an auxiliary oscillator and, controlled by the oscilloscope time base unit, frequency-modulated. Thus the frequency of the input carrier wave as well as the frequencies of its side bands are also frequency-modulated in the rhythm of the time base

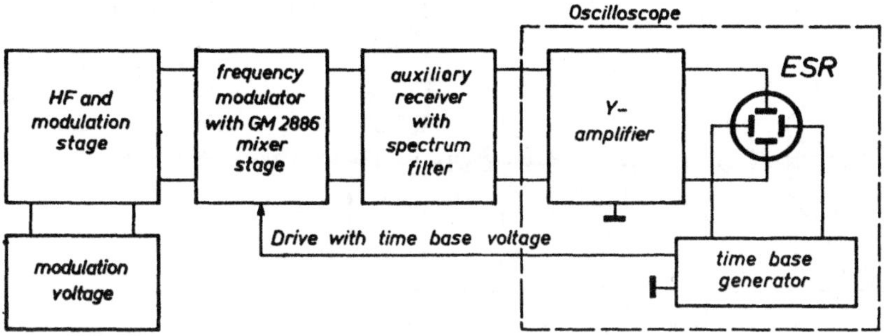

Fig. 24-1 Arrangement for displaying the frequency spectrum of amplitude-modulated high frequency

frequency of the oscilloscope. The frequency swing must be at least as great as the frequency difference between the extreme side bands. The frequency modulator used for Figs. 24-3 and 24-4 could be adjusted to give a maximum frequency swing of about 70 kc/s.

The output voltages of the frequency modulator, whose frequencies progressively increase linearly with time in accordance with the sawtooth amplitude of the time base and then drop back to their original value, are fed to a narrow band filter. The bandwidth of this filter should be kept as small as possible compared with the difference between the two side band frequencies.

In those periods of time in which the output voltage of the frequency modulator traverses the passband range of the filter due to control of its frequency by the time base voltage, the voltage delivered by the Y-amplifier of the oscilloscope produces a vertical deflection of the spot corresponding to the voltage which has passed the narrow band filter. As the horizontal deflection voltage of the oscilloscope also controls the frequency variation, deflections at the relevant frequencies of the spectrum on a horizontal frequency scale are obtained. According to the passband curve of the filter, pictures of narrow resonance curves (with high frequency) occur side by side, similar to those shown and discussed in the treatment of the representation of passband curves of normal resonant circuits in Ch. 18 (Figs. 18-5 and 18-7). As, however, the task is only to determine the frequency and amplitude of the frequency components, only simple vertical lines should be represented. This could be achieved by using several narrow band filters. It should be remembered, however, that the frequency retention time of the voltages passing the filter, should, as far as possible, be twenty times greater than the time obtained from the reciprocal of the half-width [Eq. (18.12].

For very narrow pictures of the individual frequency spectra only very low time base speeds can be employed. But under these conditions a coherent picture cannot be seen, only a wandering spot or a trace of varying length moving to the right. It is therefore necessary to find a compromise between these conflicting demands to suit the requirements of the moment. It is best to use oscilloscope tubes with a P-screen and orange filter (in the case of American tubes the $P7$-screen). In the patterns shown the time base frequency was about 0.6 c/s. The oscilloscope tube was a DP 10-6. A critically coupled radio-bandfilter for 468 kc/s, dedamped by a cathode-follower in the so-called Q-multiplier was used as a filter [1] [2]. The circuit used is shown in detail in Fig. 24-2.

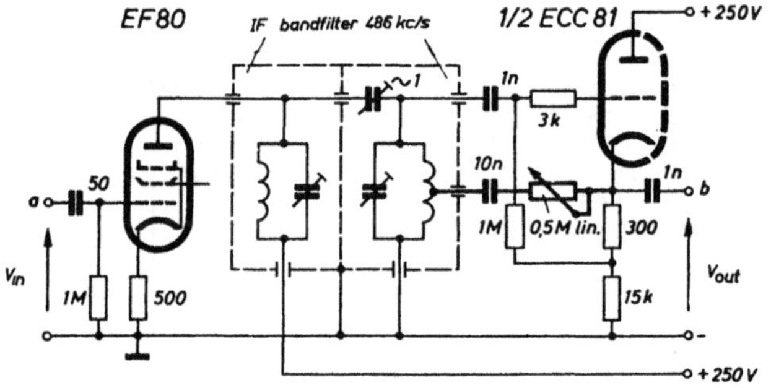

Fig. 24-2 Circuit of narrow-band filter with feed-back coupling over a cathode-follower stage ("Q-amplifier") for the subsequent recordings

An EF 80 pentode connected as a pre-amplifier and separator stage follows the bandfilter the circuits of which are screened from one another and have only a loose capacitive coupling. The secondary circuit of this filter drives a cathode-follower stage for which one system of an ECC 81 valve was used. The output voltage V_{out} is taken as usual from the cathode of this valve and fed to the oscilloscope. For dedamping, a feedback current is led from the cathode via a variable resistor of 0.5 MΩ maximum to a tap on the coil of the second circuit (shown in thick lines in the pattern on the screen). It is thus possible to dedamp the bandfilter so that the bandwidth of the passband curve is only a few hundred c/s.

Feedback via a cathode-follower has the advantage of extreme stability compared with feedback in the usual way by means of a valve connected as an anode amplifier. Whereas in anode amplifiers the feedback and thus the dedamping of the filter circuits can vary considerably due to fluctuations of the supply voltages and other influences, with the cathode-follower it remains substantially constant. It is, of course, a strongly feedback-coupled circuit. Although the gain of the cathode-follower is always less than unity, self-excitation can be achieved or almost achieved with such a circuit if the feedback voltage is stepped up. This is done here by connecting the feedback wiring to a coil tap.

In this way resonance curves of high selectivity, or, as they are sometimes called, high quality-factor Q, can be obtained under conditions of good stability.

24. 3 Recordings of the frequency spectra of modulated voltages

Figs. 24-3 and 24-4 show oscillograms that were obtained in this manner. In Fig. 24-3a-c the carrier wave was modulated to a modulation factor of about 35%. In *a* the modulation frequency was 5 kc/s, in *b* it was 10 kc/s and in *c* it was 20 kc/s. The fact that the side bands are not equidistant on both sides of the pattern of the carrier is because of slight non-linearity of the frequency swing in the frequency modulator resulting from the relatively strong modulations experienced.

In the oscillogram in Fig. 24-3*d* the modulation frequency was 20 kc/s, and the degree of modulation was 100%. As can be seen from Eq. (24.2) the amplitudes of the side bands, if modulation is undistorted (at $m = 1$), are half the carrier frequency amplitude ($^1/_2\ m\ V_0$). In this figure it can be clearly seen at the right flank of the spectrum markings that the variation speed of the frequency was

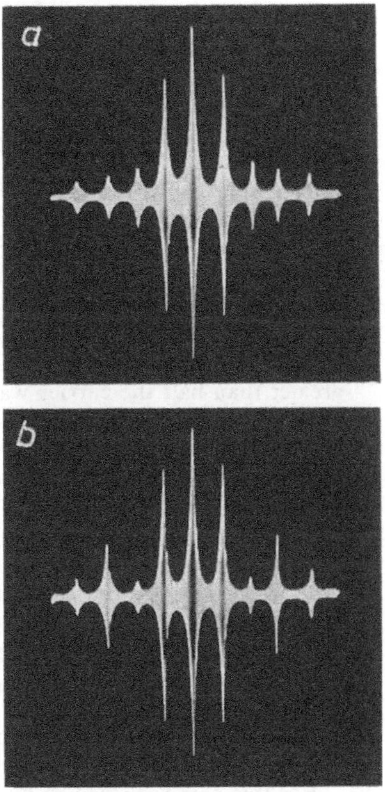

Fig. 24-4 Frequency spectrum in modulation with distorted voltage

a) Moderate overmodulation of the modulator stage of the transmitter

b) greater overmodulation of the modulator stage

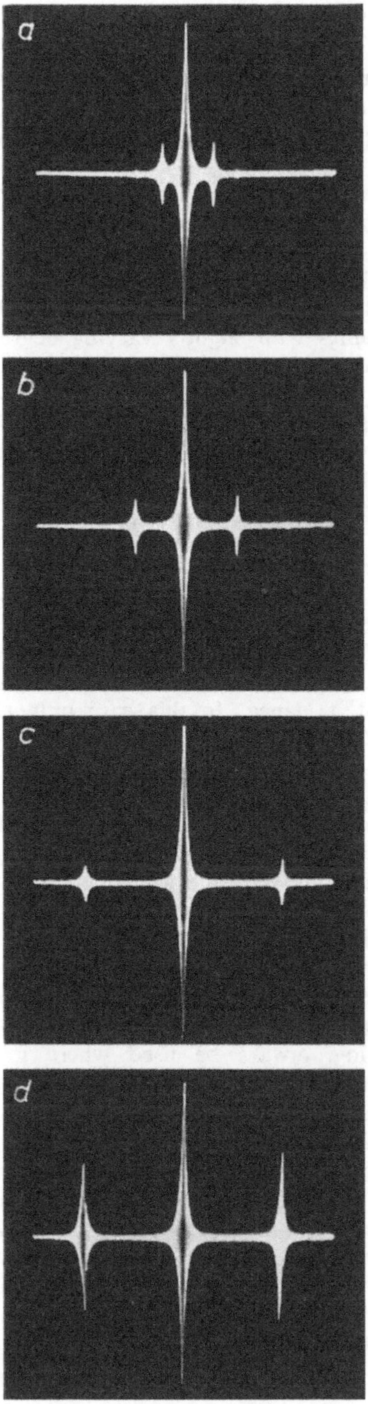

Fig. 24-3 Spectra of a carrier modulated with a single modulation frequency:

a) modulation frequency 5 kc/s

b) modulation frequency 10 kc/s

c) modulation frequency 20 kc/s (degree of modulation for *a*) to *c*) about 35%)

d) modulation frequency 20 kc/s (degree of modulation about 100%)

too high for perfect reproduction of the individual tuning curves (see also Fig. 18-5). This small distortion can be accepted, however, as the pattern of the filter curves is itself not of interest. These curves are only required to indicate the amplitude and frequency of the individual components of the frequency spectrum. The inequality in amplitude of the side bands might be due to certain asymmetries in the dedamped filter or in the modulation [3].

Two side bands for every modulation frequency occur when broadcasting speech or music. Likewise, the distortion of a sinusoidal voltage due to over-modulation in the modulator stage can produce multiples of the fundamental frequency, which in turn also create new side bands when modulating a high frequency carrier. Their distance from the carrier is then equal to the modulation frequency. The oscillograms in Fig. 24-4 show the other recordings with slightly varying degrees of over-modulation. If the side band amplitudes of the modulation wave are greater than half the carrier wave amplitude, it is clear that there is over-modulation. Whereas in Fig. 24-4a the amplitudes of the side bands progressively decrease, in b, in which the distortion was increased, the amplitude of the second harmonic of the modulation has decreased, while the amplitudes of the third harmonic have become greater.

Such investigations are especially interesting when the characteristics of very narrow pulses (radar) are to be studied [4]. WOSCHNI has shown in several publications [5] [6] [7] [8] how this method can be used to display frequency-modulated oscillations and Bessel functions as well as distortions in frequency modulation. It is thus also possible to display the individual frequency components of a low-frequency non-sinusoidal voltage, in other words, to carry out a frequency analysis in which the whole spectrum can be studied at a glance. In this case only the pattern of the frequency components left (or right) of the pattern of the carrier frequency voltage is of interest.

For this a mixer stage in which the spectrum is transferred to an auxiliary carrier is employed. After demodulation, in which the auxiliary carrier is suppressed, the pattern of frequency spectrum of the voltage under investigation [9] [10] is obtained in a way similar to that already shown. Similarly, spectra which do not occur directly as electric voltages, but for instance, as the spectrum of the light emission of a radiant material, can also be studied [11].

This process, by which the frequency spectrum is transferred to the carrier frequency by mixing and then frequency-modulated, after which the individual frequency components can be selected by a narrow-band filter (hence also known as the beat-note process) can, generally speaking, always be used where it is necessary to obtain a survey of phenomena involving a variety of frequencies in a given range. Switchable filters can also be used [12].

In monitoring wireless transmissions it is desirable to have a survey — a "panorama" — of the transmitting stations over a certain range of wavelengths. Radio amateurs are also interested in such a device which will show how the amateur waveband in question is occupied, or where a new transwitter makes its appearance. For these purposes receivers have been developed which, in a way similar to that described, provide a panorama of the wavelength in a given range.

24. 4 Panoramic receivers and panoramic oscillograms

The principal employed in Fig. 24-1 applies fundamentally also to panoramic

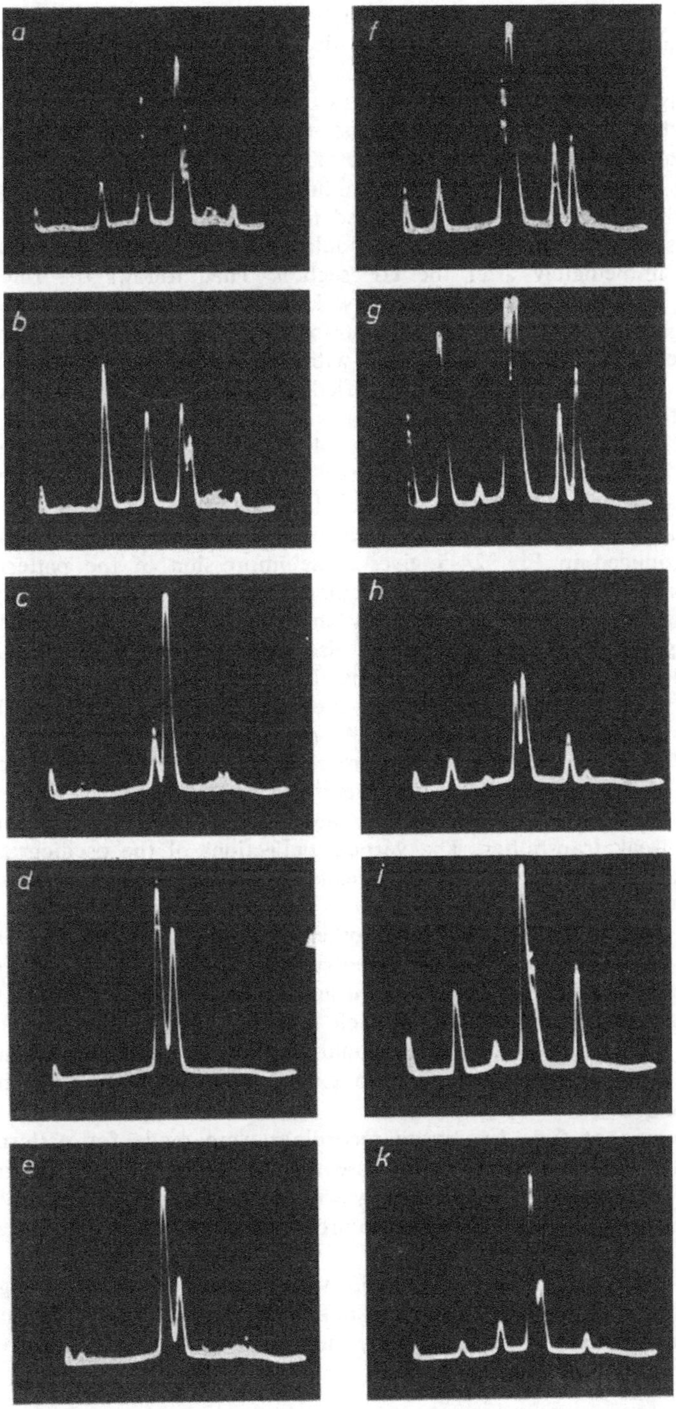

Fig. 24-5 Oscillograms of a frequency panoramic receiver during the observing of short-wave amateur transmitters (discussed in text)

receivers. In this case, however, the antenna signal is first applied to a receiver whose transmission range must be at least as wide as the frequency panorama under investigation (several Mc/s; usually adjustable). The frequency curve must therefore be as flat as possible over this broad transmission range. The response curve of the IF section of such receivers must therefore, as with television receivers, be made equally broad by means of circuits with staggered tuning of over-critically coupled bandfilters (or both). The mixing stage in which the average output frequency of the receiver input is converted to the frequency of the filter, and whose output voltage is also frequency-modulated by the time deflection voltage, is connected immediately after the HF section. Then follows the filter, whose bandwidth can now be much greater (a few kc/s to several hundred kc/s) as in recording the modulation spectrum of a transmitter. As the intermediate frequency in this case is usually high compared with the wobbulating frequency, the output voltage of the filter can also be rectified and smoothed by means of circuit elements of sufficiently large time constants.

For every transmitter DC pulses in the form of the filter tuning curve are obtained, so that an indication of every transmitter working within the adjusted receiving range appears on the oscilloscope screen [13] [14].

The set of ten oscillograms recorded while receiving radio-amateur short wave transmission and reproduced in Fig. 24-5 gives some impression of the patterns thus obtainable. In patterns *a* and *b* looking from left to right are seen firstly, two transmitters whose field strength is constantly changing. During the recording of pattern *a* the voltage of the second transmitter also fluctuated due to amplitude modulation. Further to the right in these patterns there are two transmitters the frequencies of which are so close together that they could not be satisfactorily separated by the filter of the panorama receiver. As the bandwith of the one filter is kept roughly equal to that of a normal amateur receiver set, it could be concluded that useful reception of either of these transmitters was only possible by special means (quartz filter, LF bandwidth compression and suchlike). At the extreme right there is another weak transmitter. The vertical deflections of the oscillogram can easily be provided with the familiar *S*-value scale for the sensitivity ranges of the receiver.

Pictures *c*, *d* and *e* were recorded one after another somewhat later. Two of the transmitters can now be seen to have good field strength. The transmitter on the left of the pattern corresponds to that in patterns *a* and *b*. To the right of it a new one has made its appearance, and it could be well separated out by the receiver filter. Is can therefore be assumed that both could be well received by normal amateur receiving sets. These three pictures also give a clear impression of the fluctuations in field strength.

During a further period of time five more recordings were made for patterns *f*-*k*. Once again two transmitters can be seen close together in the middle of the pattern. The field strength of the one on the right was somewhat stronger in pattern *g* as compared to *f*, and became so weak during the recordings that further reception was impossible.

In a similar manner the various Postal Authorities wish monitor wavelength range with panoramic receivers. If an unknown transmitter is spotted on a certain frequency, its identity can quickly be established by tuning a normal monitor receiver to the frequency.

CHAPTER 25

ADJUSTMENT OF HIGH IMPEDANCE WIDEBAND VOLTAGE DIVIDERS BY SQUARE PULSES OR SYMMETRICAL SQUARE VOLTAGES

25. 1 High impedance wideband voltage dividers

In oscilloscope amplifiers and in amplifiers used in pulse technique and in many similar tasks it is necessary to divide voltages and this division has to remain constant up to high frequencies. If leads are terminated with their surge impedance, which is usually low (50...70Ω), a wideband division of the voltages can be undertaken without the inevitable parallel capacitance through wiring or valves reducing the upper out-off limit to an unacceptable extent. But as a certain voltage value is always required from these low-impedance resistors (e.g. 1 V_{pp}), this means that a by no means negligible amount of power is consumed by a low impedance voltage divider. So long as the voltage on the divider (60 Ω, 1 V_{pp}) is intended to be small, the power required also remains relatively low (about 17 mW). As soon as the divider voltage to be applied is higher, a considerable amount of power has to be provided. The power absorbed by a resistor is proportional to the square of the voltage applied (for 10 V_{pp} at 60 Ω about 1.7 W would be needed). Such loads are not permissible, for instance, at the input of an oscilloscope, they are impermissible in fact in any case as they would have to be provided by the voltage sources under investigation. Therefore steps must be taken to make voltage division at the input of oscilloscope amplifiers and within amplifier circuits as highly resistive as possible. (In special cases low-impedance division is possible by using cathode-follower stages. The cathode-follower stages then supply the power to the divider.)

Fig. 25-1 shows some high-impedance voltage dividers. That in Fig. 25-1a is used, for instance, if the sawtooth voltage generator for the time base is to be connected to the *X*-direction amplifier. If the adjusting capacitor C_v is ignored for the moment, then the *RC*-network consisting of R_v and the valve input capacitance C_g, would act as a low-pass filter (R_g is $> 1/\omega_{cu}\, C_g$ and therefore negligible) and would reduce the upper cut-off frequency of the circuit by several orders of magnitude under certain circumstances. To compensate for the effect

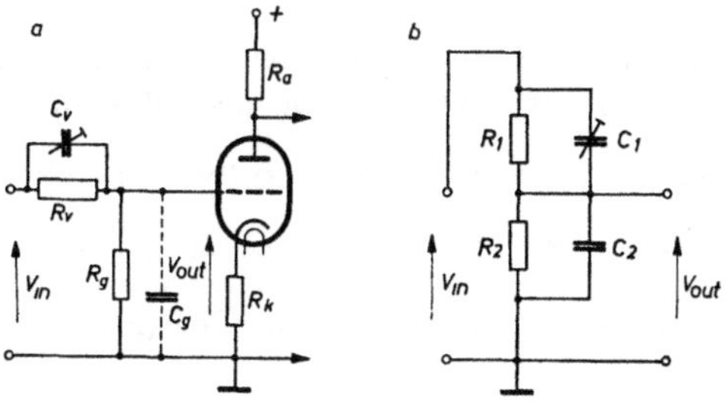

Fig. 25-1 Application of wideband voltage dividers *a*) Voltage division in the grid lead of an amplifier valve *b*) input voltage divider of oscilloscope amplifiers

of valve capacitance C_g, as already described in Part I, Ch. 5.21 "Setting the deflection amplitude", a capacitor C_v is connected in parallel to the series resistor R_v and is of such value that the time constant $R_v \cdot C_v = R_g \cdot C_g$. The conditions are similar in the case of the high resistance input voltage dividers as incorporated in the vertical amplifier input or as provided in the form of a voltage divider probe. The basic circuit diagram of a divider of this kind can be seen in Fig. 25-1*b*. To discover whether the dividing ratio remains the same for all frequencies throughout the wide range required in the case of such "compensated" dividers, it is best to use a generator for square pulses or for symmetrical square voltages and an oscilloscope. To clarify the equations which apply, these relationships will be dealt with in more detail in the following pages.

25. 2 Waveform of the output voltage of a wideband voltage divider at various adjustments

As NEETESON has shown and THIELE has described, the time constant T_T of a divider according to the circuit in Fig. 25-1*b* is given in the following equation:

$$T_T = (C_v + C_g) \cdot \frac{R_v \cdot R_g}{R_v + R_g} \tag{25.1}$$

[1] [2].

For the general case, the output voltage V_{out} after a short-duration voltage surge of the input voltage V_{in} after time t is obtained from the equation

$$V_{\text{out}} = V_{\text{in}} \cdot \frac{R_g}{R_v + R_g} \cdot \left(1 + \frac{R_v \cdot C_v - R_g \cdot C_g}{R_g \cdot (C_v - C_g)} \cdot e^{-\frac{t}{T}}\right). \tag{25.2}$$

At the commencement of the voltage surge the capacitors are discharged and therefore have the effect of short-circuits. The output voltage quickly rises to the value $V_{g(0)}$ with the speed of the input voltage, the value of $V_{g(0)}$ being given by the voltage division at the capacitors according to the equation

$$V_{g\,(0)} = \frac{C_v}{C_v + C_g} \cdot V_{\text{in}}\,. \tag{25.3}$$

The oscillogram in Fig. 25-2*a* shows this in the case in which

$$\frac{C_v}{C_v + C_g} < \frac{R_g}{R_v + R_g}\,.$$

(When R_v, R_g and C_g are given, C_v is thus too small.)

After the voltage value according to Eq. (25.3) has been attained, the further rise of the output voltage to the value determined by the ratio $\frac{R_g}{R_v + R_g}$ takes place much more slowly, in fact at a time constant T_T according to an exponential function [of factor $e^{-\frac{t}{T}}$ in Eq. (25.2)].

If, on the other hand, $\frac{C_v}{C_v + C_g} > \frac{R_g}{R_v + R_g}$ (that is, C_v is too large), then the voltage at first shoots up to a value which in turn is given by the capacitive attenuation, and exceeds the voltage which would correspond to the resistance division ratio

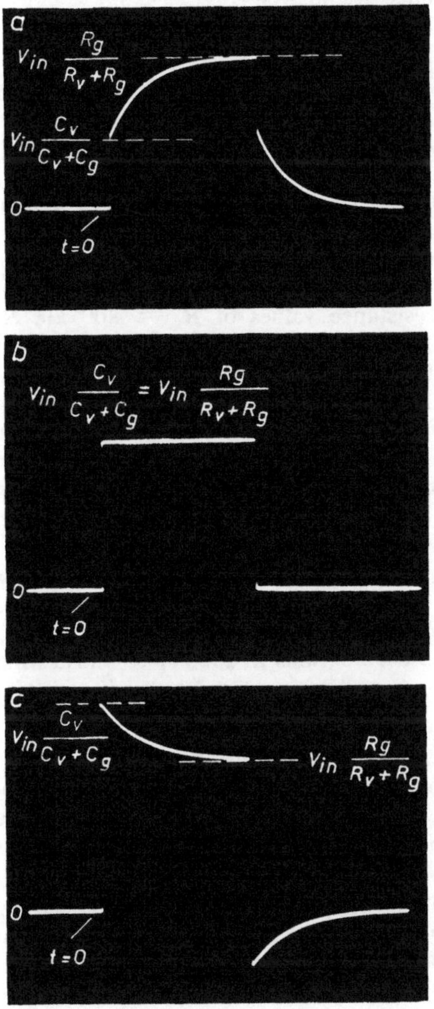

Fig. 25-2 Output voltage of a divider by a 120 μs pulse at various values of C_v. *a)* C_v too small; *b)* C_v correctly adjusted; *c)* C_v too large

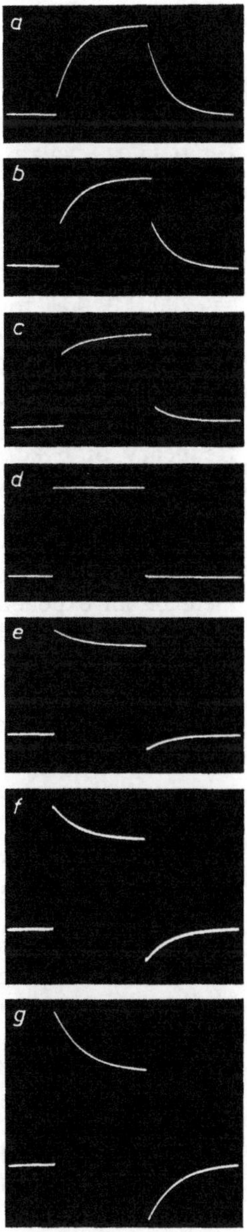

Fig. 25-3 Voltage forms of the output voltage of a divider as in Fig. 25-2, but at seven different settings of the trimming capacitor C_v for greater clarity

(for direct voltage and low frequencies) $\frac{R_g}{R_v + R_g}$. Then — once more with the time constant determined by the factor $e^{-\frac{t}{T}}$ — it again drops according to an exponential function to the value determined by the resistance attenuation (Fig. 25-2*c*).

From these considerations it is clear that the output voltage will only rise immediately to the value determined by the resistance attenuation ratio $\frac{R_g}{R_v + R_g}$ (without overshoot) if

$$R_v \cdot C_v = R_g \cdot C_g \,, \tag{25.4}$$

i.e. if in Eq. (25.2) the second term in the bracket with the exponential function (25-2b) becomes zero.

For various settings of the adjusting capacitor C_v, the series of oscillograms in Fig. 25-3 gives an additional survey of the waveform of the output voltages at various values of C_v for a divider having resistance values of R_v = 800 kΩ, R_g = 200 kΩ, i.e. for a division ratio of 200/(200 + 800) = 1/5. The output capacitance C_g was 60 pF and thus C_v should be 15 pF for the correct adjustment. The signal pulses 120 μs in width were taken from a Philips "GM 2314" pulse generator. To ensure that the rising flank of the output voltage can be clearly seen from the very outset, the signal pulses were delay-triggered with a repetition frequency of 6 kc/s by the method described in Part II, Ch. 13, the time base generator being directly triggered by the AC controlling voltage (circuit as in Fig. 13-2).

It can be seen that, even when C_v has a low value, the output voltage always commences with a rapid rise. Only if this capacitance were wholly absent would there be an exponential waveform for the *RC*-network (formed in this case by R_v and C_g) from the outset, as can be seen in the oscillograms in Fig. 11-34 illustrating such cases.

25. 3 Adjusting compensated voltage dividers with pulse voltages

The oscillograms in Figs. 25-4 and 25-5 indicate that in order to obtain oscillograms which are easy to study, the most favourable ratio of the width of the signal pulses to the time constant of the divider circuit must be chosen. In this example T_T, the time constant according to Eq. (25.1), is taken as = 94 μs.

As the three oscillograms in Fig. 25-4*a* have a pulse duration of 1000 μs, only very slight changes at the commencement of the pulses are seen, indicating that the adjustment is incorrect. For good reading clearer distortions are imperative.

The three oscillograms in Fig. 25-4*b* with a pulse width of 400 μs show the oscillogram distortion essential for judging the pattern and the level of adjustment of the divider much more clearly. In the *c* series of oscillograms the pulse width was only 100 μs; the optimum compensation is now very clearly visible. [In describing these patterns, the terms "overshoot" or "roof slope" are often used, depending upon the ratio of the pulse duration to the time constant of the circuit. These concepts are the same thing, at least as far as this investigation is concerned.]

In the three oscillograms in Fig. 25-4*c* the voltage drop — or rise — can be observed most clearly of all. But, as is clear from the oscillograms in Fig. 25.5, with narrower pulses the clearest changes in the oscillograms are obtained when the pulse width is made roughly equal to the time constant of the circuit. Even

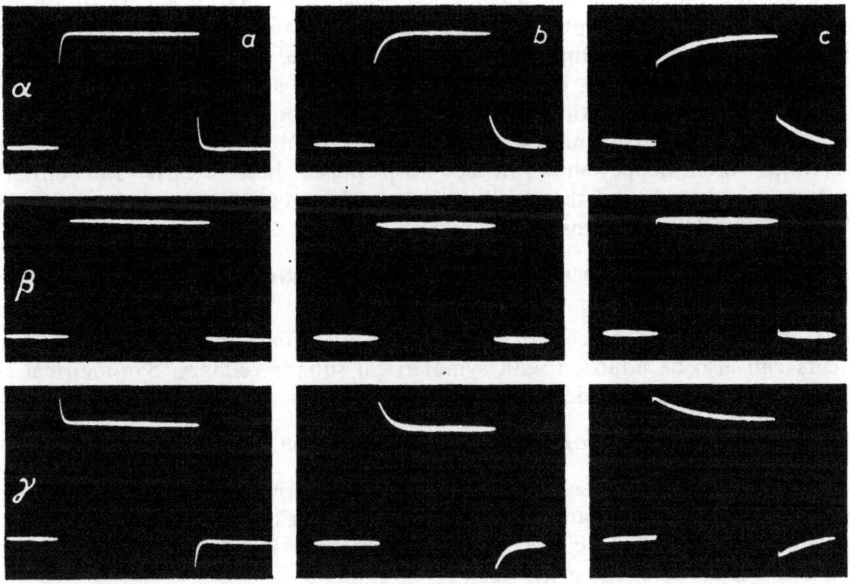

Fig. 25-4 Influence of pulse width on interpretability of the oscillogram *a*) Pulse width 1000 μs; *b*) pulse width 400 μs; *c*) pulse width 100 μs

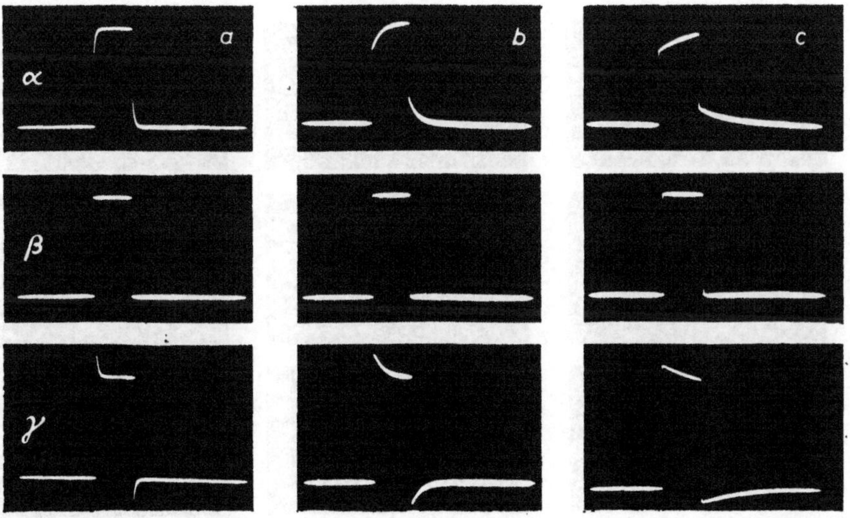

Fig. 25-5 Judging the state of balance of a compensated divider with narrow pulses. *a*) Pulse width 300 μs; *b*) pulse width 60 μs; *c*) pulse width 15 μs

in the case of pulses whose duration is only a fraction of the time constant, however, as in the three oscillograms of Fig. 25-5*c*, the extent to which the adjustment of the divider approximates to the optimum can be clearly judged.

The sloping portion of the oscillogram which only shows a part of the voltage drop — or rise — becomes almost horizontal at optimum adjustment so that its inclination is a standard for judging the accuracy of compensation of the divider. The time coefficients for these oscillograms were adjusted in each case to give clear patterns. The condition must of course be laid down for such pulses, that their rise time is sufficiently short in proportion to their duration and that the observation oscilloscope shows these steep pulses both clearly and unflattened. The input capacitance of the oscilloscope of 20 . . . 60 pF must be taken into consideration in such measurements.

25. 4 Adjustment of compensated wideband voltage dividers with symmetrical square voltages

Similarly, as already described in the case of square pulses, wideband voltage dividers can also be adjusted with symmetrical square voltages. Symmetrical square voltages can also be regarded as a pulse sequence in which the sequence frequency is equal to the reciprocal of twice the pulse duration ($f_F = \frac{1}{2T_d}$). This is of particular importance, because generators for symmetrical square voltages with rise times of only 30...40 ns are relatively easier to make than pulse generators with adjustable sequence frequencies and are therefore inexpensive (e.g., the Philips "GM 2324" square voltage generator).

The oscillograms in Fig. 25-6 show recordings with symmetrical square voltages, made with the divider previously described at various frequencies. Thus the most favourable frequencies for judging such a divider are obviously between 2000 and 5000 c/s.

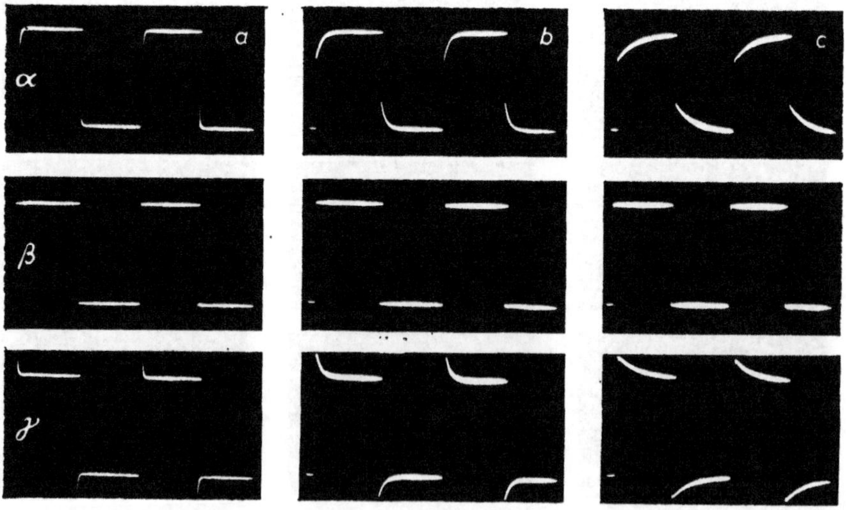

Fig. 25-6 Judging the state of balance of a compensated divider with symmetrical square voltages. *a*) 500 c/s; *b*) 2500 c/s; *c*) 10.000 c/s

25.5 Compensation and adjustment of specially high impedance wideband voltage dividers

In the method described in the previous sections only the effect of the lead capacitance connected to the output of the divider is compensated. This is permissible provided that the "upper" resistance of the divider (R_1 in Fig. 25-1*b*) has no considerable distributed capacitance to the screening; it must therefore be as short as possible.

If particularly large voltage division ratios are necessary, but the output resistance of the divider has to be high, this first resistance (behind the voltage source) is very great. This is sometimes in fact the first requirement, as then the signal source is only slightly loaded by the oscilloscope. The high division ratio must be accepted as a rather undesirable disadvantage. In most cases resistance values of > 10 MΩ are required. Such resistances can usually be built up to have a low inductance and satisfactory stability only by means of connecting several resistors in series. The divider resistor is thus fairly bulky and the individual resistors have capacitances to the casing (a highly resistive divider of this sort has to be well screened), which, however, can be considered as being lumped at the connecting points as indicated in Fig. 25-7*a* (C_{q1}, C_{q2}, C_{q3}). If the attempt is made, as previously described, to compensate such a divider by means of a single capacitance in parallel with the series of resistors ($R_{v1} + R_{v2} + R_{v3}$), oscillograms like the three shown in Fig. 25-8 are obtained. At all adjustments of the compensation capacitor (C_v) there always remains in the pulse roof of the square voltage a characteristic bending which cannot be overcome by adjusting the capacitor.

On examining the circuit of Fig. 25-7*a* more closely the explanation is readily discovered. The same number of *RC*-networks occur as there are shunt capacitances, and the frequency dependence of the networks can only be compensated by

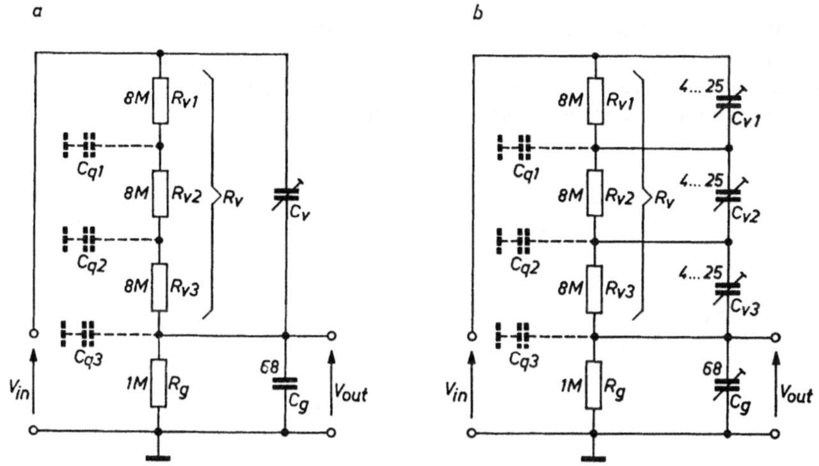

Fig. 25-7 For compensating particularly highly resistive wideband voltage dividers
a) Wrong compensation by a single compensating capacitor over the partial resistors of R_v
b) correct compensation by special capacitors, in each case parallel to the partial resistors of R_v

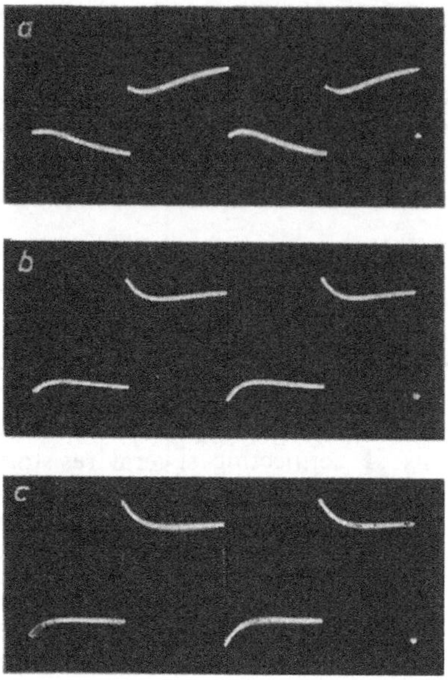

Fig. 25-8 Oscillograms of wrong compensation according to Fig. 25-7*a*. *a*) C_v too small; *b*) C_v set as well as possible; *c*) C_v too large

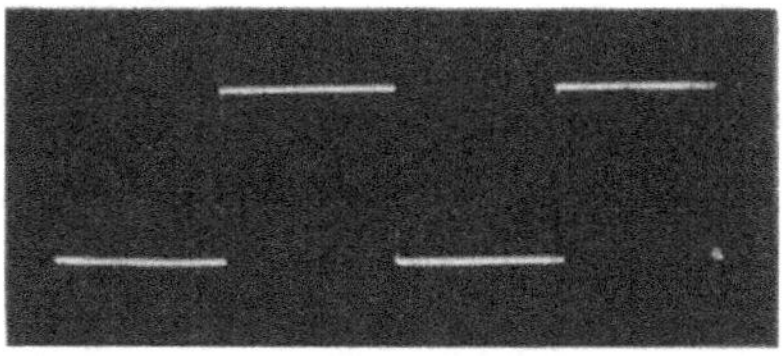

Fig. 25-9 Output voltage of a very highly resistive wideband voltage divider, correctly compensated according to 25-7*b* and well trimmed (frequency of the square wave voltage 4.4 kc/s, rise time about 40 ns)

connecting a separate compensating capacitor in parallel with each series resistor as shown in the circuit of Fig. 25-7*b*. The shunt capacitor C_{q3} of the last partial resistor R_{q3} can be assumed to be in parallel with the output capacitor C_g.

Such an arrangement is best adjusted by connecting the square voltage source to the individual partial resistors in turn, commencing with the divider series resistor (R_q) connected to the output resistor (R_g).

If the dividers are now correctly adjusted step by step by means of the compensating capacitors, a very high impedance divider (up to about 100 MΩ, in this case 25 MΩ with a dividing ratio of 25 : 1) can be safely used, as shown in the oscillogram in Fig. 25-9, to obtain satisfactory wideband voltage division and good transmission of square voltages of steep slope.

CHAPTER 26

RECORDING LOW-FREQUENCY CHARACTERISTIC CURVES

26. 1 Methods of measurement

The frequency dependence of four-terminal networks in low-frequency technology can be displayed by means of a suitably frequency-modulated signal. As the decay time of such circuit elements is considerable, a relatively slow rate of wobbulation must be used. As in HF investigations, each new frequency traverse should not take place until the voltage in the circuit elements has decayed sufficiently. It therefore seems suitable to carry out the frequency change manually and observe the filter characteristic of the object under study on a long-persistence fluorescent screen.

The process about to be described was introduced by TERMAN in 1943. [1]. Here the frequency characteristic is traced directly by the moving spot, using direct voltages for deflection, so that it can only be applied with oscilloscopes having DC vertical amplifiers.

The basic principle of this method is indicated in Fig. 26-1. A low-frequency voltage source must be used in which the complete low-frequency range can be covered by simply turning a knob. The output amplitude must not fluctuate more than ± 3% (e.g., Philips "GM 2308" low-frequency beat oscillator).

The horizontal deflection of the spot is by means of a direct voltage obtained after rectification of the voltage from the signal generator after it has been passed through a frequency-dependent network. By suitably rating the circuit elements of this network, the output voltage and thus the horizontal deflection can be made linear with the frequency, or, as is most usual, proportional to the logarithm of the frequency. By completely rotating the regulating knob for frequency adjustment of the low-frequency generator, the spot moves from left to right on the oscilloscope screen.

The output voltage of the system under investigation can now be used for vertical

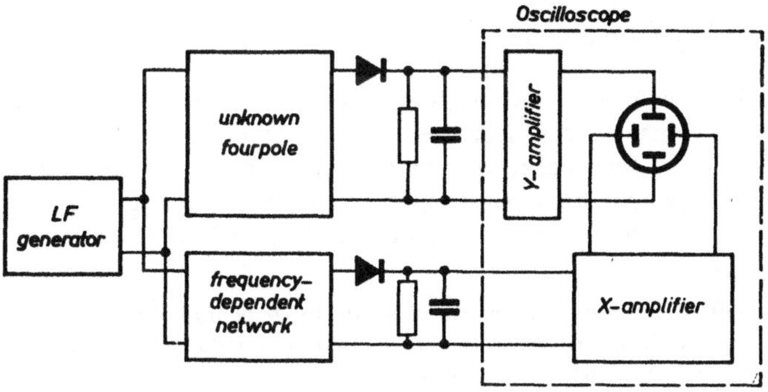

Fig. 26-1 Circuit diagram for recording LF characteristic curves

deflection either directly or after rectification. In the first case the characteristic of frequency dependence is indicated by the "envelope" of the oscillogram, as shown in Fig. 26-3*a*. After rectification, only the boundary appears, as can be seen from Fig. 26-3*c*. If necessary a vertical amplifier, so constructed that the vertical deflection is logarithmically proportional to the amplitude, can also be connected between the output of the system under investigation and the rectifier.

26.1.1 The frequency-dependent network and the rectifier

The circuit in Fig. 26-2*a* shows the network with rectifier and filter for horizontal deflection. It has been found in practice that at least four networks are required to obtain a satisfactory logarithmic frequency scale. The dependence of the output amplitude is achieved in four ranges in which the resistance ratios are as 30:12:6:3 and the capacitance ratios are as 30:10:3:1, and hence the time constant ratios are as 300:120:18:3. The terminating resistance is 20 kΩ. The voltage appearing here is rectified by a germanium diode OC 85, and after smoothing the DC ripple voltage with a *CRC-filter* is led to the input of the amplifier for horizontal deflection [109]).

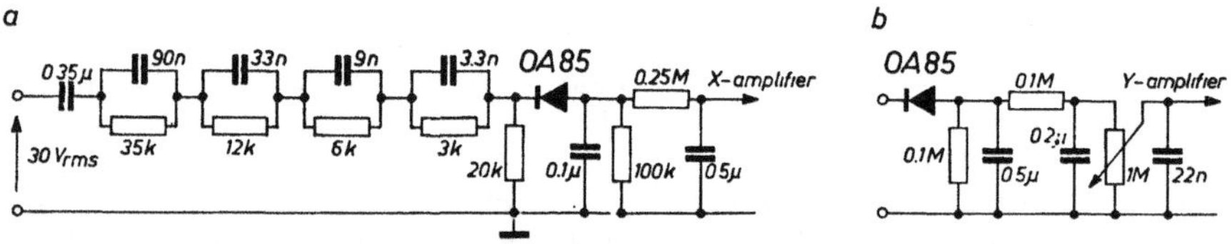

Fig. 26-2 Frequency-dependent network, rectifier and filter
a) Network with rectifier and filter for line deflection
b) rectifier and filter for frame deflection

The circuit of Fig. 26-2*b* shows the rectifier and the filter used in the vertical deflection circuit for the oscillograms reproduced in this section.

As the 100 kΩ load of the input resistor of the oscilloscope amplifier reduced the efficiency of the filter, the oscilloscope input was connected in such a way that the first valve works without an input resistor (Philips "GM 5656" oscilloscope). In this circuit it is replaced by a 1 MΩ variable potentiometer. In order to avoid hum, the slider of this potentiometer is decoupled to earth by a capacitor of 22 nF.

26.1.2 Calibration

The scale for the oscillograms shown here was obtained by photographing the individual traces. For the vertical division the null trace was first photographed with the time base unit working. Then direct voltages of suitable values were applied step-by step via a resistor decade to the vertical amplifier until the upper limit of the desired screen dimensions was reached with the 11th trace.

---

[109]) Germanium diodes are more useful here than electronic valve diodes, as the latter cause a premature beat of the starting current. The zero point is thus shifted by a certain amount, which could be rather awkward in calibration.

The vertical traces were supplied directly by the output voltage of the LF generator, attenuated as required. After adjusting the LF generator to the frequency stage required for the screen, one photographic recording was made in each case. To accentuate the calibration lines for of 100 c/s, 1 kc/s and 10 kc/s, the exposure time ($^1/_2$ s) was five times that of the other traces (1/10 s). As there is still a slight ripple at 50 c/s, a flat ellipse results here in a double trace.

The horizontal deflection voltage must be cut off while photographing the vertical null trace in order to prevent simultaneous horizontal shifting of the spot. The frequency of the voltage for vertical deflection can be chosen as described (several hundred c/s).

This admittedly somewhat complicated process for obtaining a test grid can be simplified by making it only once and matching the negative to the separately photographed negatives of the characteristics and enlarging them together. A transparent negative of the test grid can be made and placed on the enlargement paper when the curves are enlarged.

If an oscilloscope with a DC amplifier having an illuminated graticule is available, it is possible to use a much simpler method. The original scale is then replaced by one having engraved on it divisions as shown in the present oscillograms and corresponding to the frequency response of the auxiliary network. If the grid and spot intensities are correctly adjusted, the grid can be photographed at the same time as the characteristic. Before each measurement the gain in both directions must be so adjusted that the grid is visible from the null point to the end of the frequency range for the horizontal deflection and vertically from zero to the topmost trace at the maximum value of the output voltage.

26. 2 Examples of application

If the low-frequency output voltage of the four-terminal network under investigation is used directly for vertical deflection in a circuit [2] according to Fig. 26-4 [110]), the frequency dependence curve appears as the "envelope" of a fluorescent area, as can be seen in the oscillogram in Fig. 26-3*a*.

Of course, only the upper half of the figure symmetrical about the centre line is sufficient to judge the frequency response. It is therefore best to double the gain and adjust the pattern vertically so that its centre line coincides with the edge of the grid as in the oscillogram in Fig. 26-3*b*.

After rectification the characteristic of the frequency response appears when the LF-oscillator is tuned over in its frequency range, as the path of a moving fluorescent spot, as displayed in the oscillogram of Fig. 26-3*c*. The thickening of the curve at the commencement is due to a low-frequency residual ripple. It cannot be completely avoided, as otherwise the time constants of the filter elements would be too great for the rectified voltages, and the frequency variation would occur extremely slowly. With the values of circuit elements indicated, a range of from 30 c/s to 15 kc/s can be traversed. Because the graduation of the beat frequency oscillator

[110]) The recordings were made with the Voigtländer-Philips "PP 1011" camera on Scopix-G film and the "GM 5656" oscilloscope. Aperture 16 was used for the frequency characteristics. Photographic treatment intensifies the contrast, and the inner part of the fluorescent surface appears darker here than on the screen itself. In the vicinity of the zero crossover point the speed of the spot deflection in the vertical direction is greatest when it corresponds to the waveform of the sine function.

Fig. 26-3 Frequency characteristics of a low-pass filter for modulation amplifiers for amateur transmitters

a) LF voltage direct to the deflection plates

b) as *a*), but amplitude doubled and centre line of the oscillogram shifted to zero line of raster

c) characteristic with rectified voltage

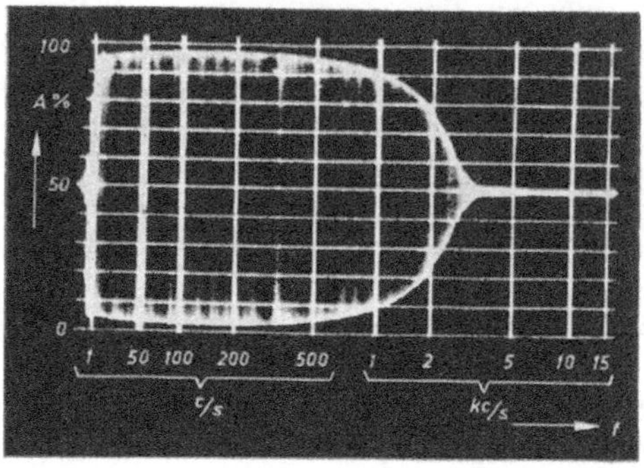

a

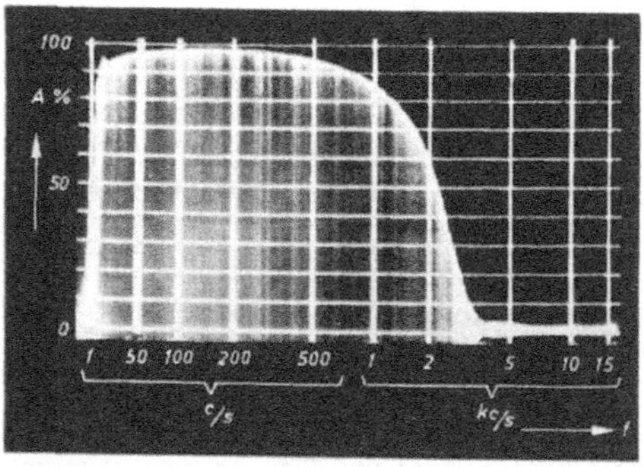

b

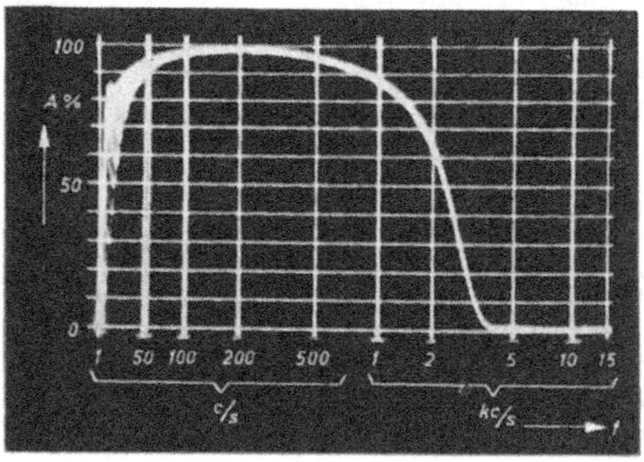

c

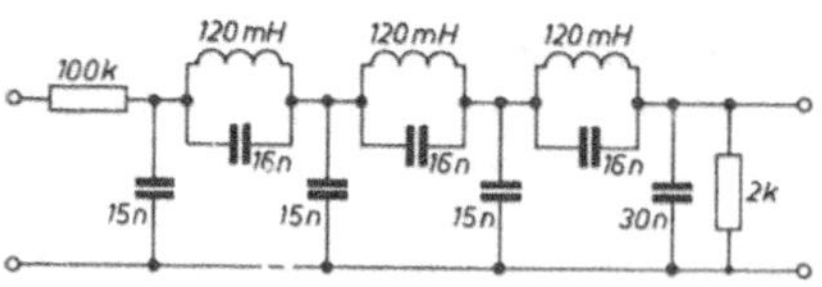

Fig. 26-4 Circuit of the filter with frequency dependence according to oscillograms in Fig. 26-3

frequency scale is logarithmic, the traverse should start slowly and then speed up if the characteristic is to be uniformly strong. After some practice it is not difficult to carry out the frequency change in such a way that satisfactory oscillograms are obtained. Of course, the frequency can also be varied mechanically, but the additional cost would not usually be worth while.

Fig. 26-5 shows the frequency dependence given by this arrangement as a whole. From 30 c/s to 10 kc/s the deviation of the frequency response from the straight lies within the reading error (± 2%).

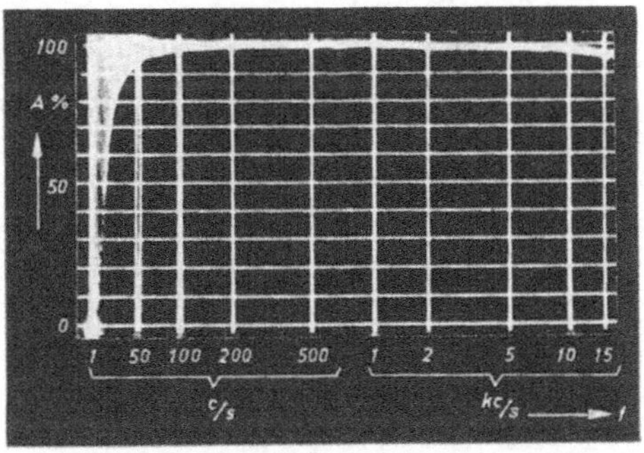

Fig. 26-5 Frequency dependence of whole arrangement without object of study

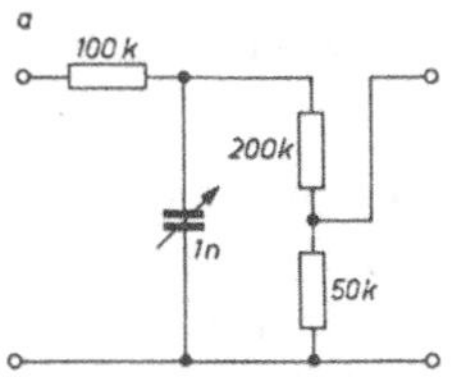

b

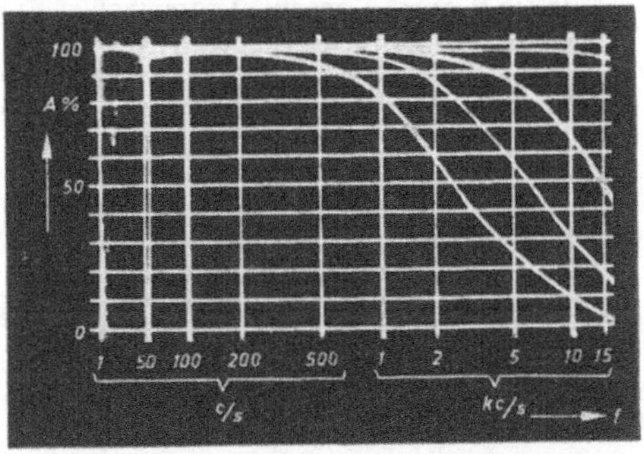

Fig. 26-6 Low-pass filter. *a*) circuit; *b*) characteristics for various adjustments of the capacitor

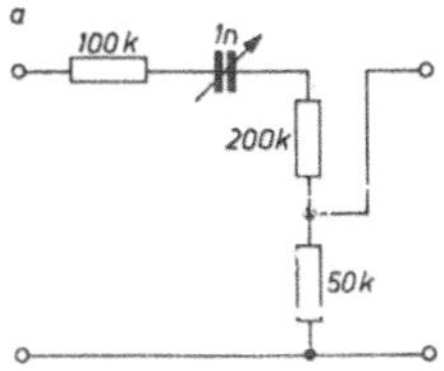

b

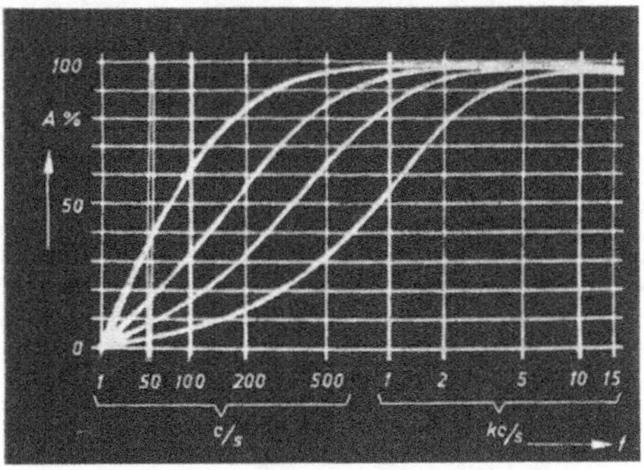

Fig. 26-7 High-pass filter. *a*) circuit; *b*) characteristics for various settings of capacitor

a

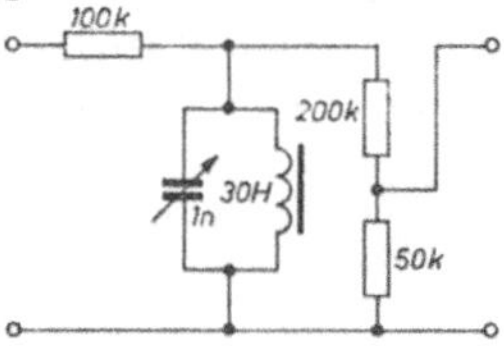

b

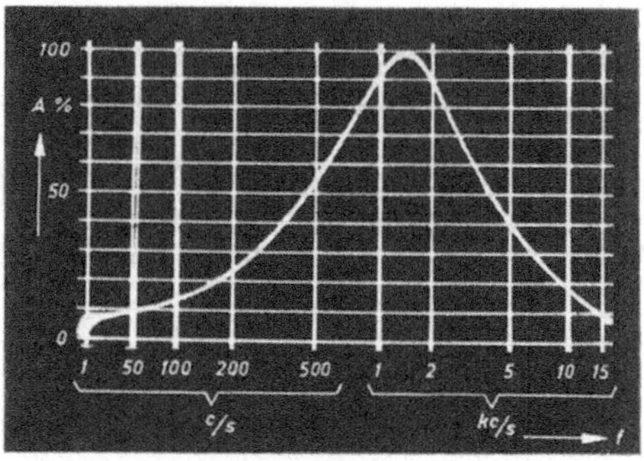

Fig. 26-8 Parallel resonant circuit fourpole. *a*) Circuit; *b*) characteristic with resonance at about 1400 c/s

a

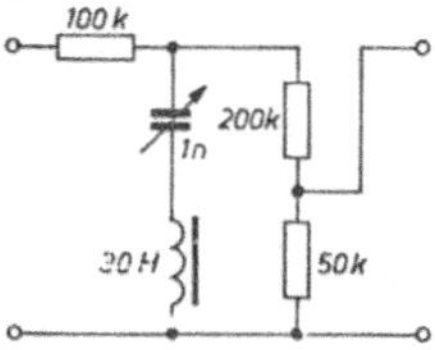

b

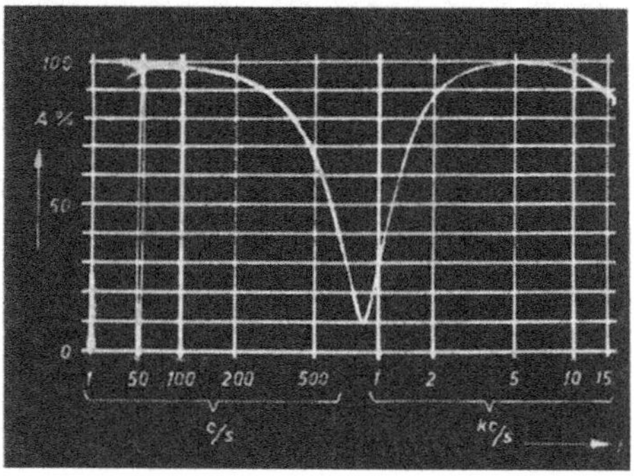

Fig. 26-9 Series resonant fourpole. *a*) Circuit; *b*) characteristic with resonance at about 800 c/s

The oscillograms in Figs. 26-6*b*, 26-7*b*, 26-8*b* and 26-9*b* illustrate the application of this process to the four-terminal networks of Figs. 26-5*a*, 26-7*a*, 26-8*a* and 26-9*a*. They show the frequency response of a circuit which can be seen on the left (as *a*) of the oscillograms in each case. The scale figures were added later.

CHAPTER 27

SOME EXAMPLES FROM ELECTRO-ACOUSTIC PRACTICE

27. 1 Possible applications of the cathode ray oscilloscope to electro-acoustics

In this field too the cathode ray oscilloscope can be used to great advantage in many types of measurement and investigation. For instance, the precise calibration of tuning forks and the tuning of musical instruments, indeed, of whole orchestras, has been carried out by means of oscilloscopes and duly described [1] [2]. TAK has described the measurement of the reverberation time of a room, in which the reverberation of sound pulses is picked up by microphones, the resulting voltage amplified exponentially and displayed with an oscilloscope [3] [4]. The rhythm of the shock waves and the exponential gain increases are controlled by the time base voltage of the oscilloscope. If the time base frequency is adjusted so that the oscillogram has a roughly constant amplitude throughout the whole width of the pattern, then the reverberation time is equal to the duration of one time base cycle. In this way the dependence of the reverberation time upon the frequency of the sound pulses can be investigated.

It is true that in general the transmission characteristics of electro-acoustic equipment and transmission lines is judged by measuring the frequency response with sinusoidal voltage sources or sound sources producing simple harmonic notes (see Ch. 26). However in view of the fact that very many sound generators are excited by pulse-shaped impulses, the study of electro-acoustic equipment (microphones, loudspeakers) by means of pulse voltages is becoming increasingly important [5].

In a large number of telecommunication engineering investigations it is necessary to analyse the frequency components of speech and music. In this connection the reader's attention is drawn to Ch. 24 and the relevant bibliography. By means of the methods there described the frequency spectra of sound frequency phenomena can be represented with clarity on the oscilloscope screen. Fluctuations in the speed of tape recorders can be studied by comparing the phase of a played-back alternating voltage and the voltage phase of tape recorder head [6].

The present chapter will be limited to showing a number of oscillograms of speech oscillations and music.

27. 2 Oscillograms of variations in sound pressure

A microphone is generally used as the transducer. It translates the sound pressure fluctuations into electrical voltage changes, which are fed to the oscilloscope. The type of microphone used must accord with the demands as to fidelity. Test microphones of special design are needed where particularly high standards must be maintained. They usually give low output voltages, but little or no difficulty is experienced in amplifying them sufficiently, if need be, with special pre-amplifiers, so that sufficiently large deflections occur on the oscilloscope screen. The circuit is so simple that a special circuit diagram and detailed description need not be given here.

If an attempt is made to record the curve produced when vowels are pronounced, it is found that with normal photographic recording with the time base repeating, multiple traces occur in the pattern due to small amplitude fluctuations, and these give a blurred impression. It is therefore advisable to ensure either, as in Fig. 27-1, that the shutter opening time is shorter than the forward trace duration of the time deflection (there is then a gap in the pattern) or, better still, the changes in voltage are recorded as a moving film picture.

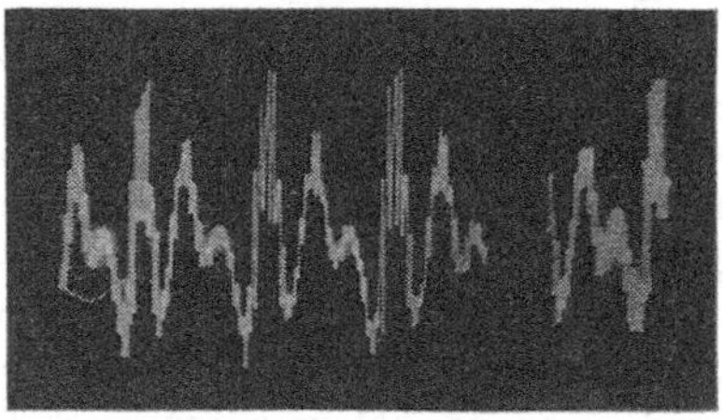

Fig. 27-1 Spoken vowel "a" recorded with time base unit working repeatedly (german pronunciation)

The oscillograms in Fig. 27-2 show the patterns of the vowels "*a*", "*e*", "*i*", "*o*" and "*u*" pronounced by the male voice, the movement of the film being 1 m/s.

Naturally, the waveform of these curves is contingent to a large extent on the articulation and idiosyncracies of the speaker. It is therefore possible to obtain differing sound pictures of the same vowel from one and the same speaker. On closer examination it is found that the fundamental and the arrangement of the harmonics for the individual vowel patterns are characteristic. For the sake of better time resolution, Fig. 27-3 shows once again, with a time-marking of 1000 c/s, the oscillogram of the vowel "*i*", which has a very large number of harmonics, this time recorded at a speed of 3 m/s.

When comparing it with the oscillogram of "*i*" in Fig. 27-2, certain differences in the amplitudes of the harmonics are noticeable. This is due to the fact that they were not recorded simultaneously, but one after the other.

As a final example, Fig. 27-4 shows some moving pictures of speech and music. The time scale of 1000 c/s is shown in (*a*); photograph (*b*) represents the waveform of the German word "Arbeit" (work), containing the wovels *a*, *e* and *i*. By comparing with Fig. 27-2 we can establish a certain connection with regard to pulse sequence and curve waveform. Photograph (*c*) represents the voltage waveform of a piano in its middle register, and (*d*) is that of a mixed orchestra [7] [8] [9].

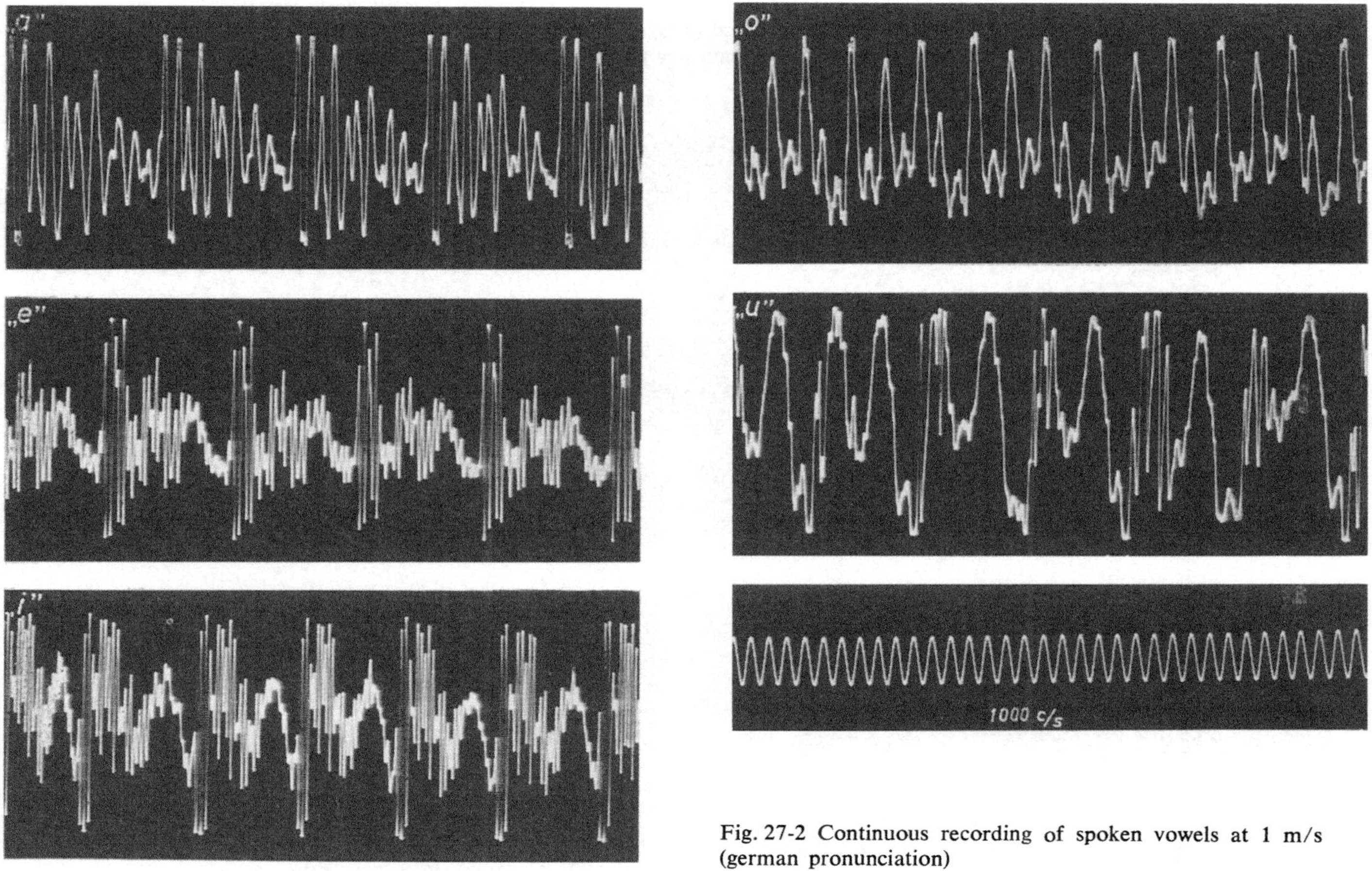

Fig. 27-2 Continuous recording of spoken vowels at 1 m/s (german pronunciation)

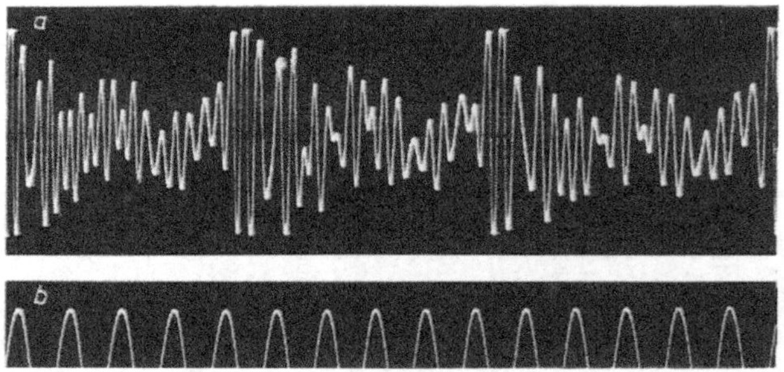

Fig. 27-3 Continuous recording of vowel "e" with 3 m/s. *a*) Vowel "e" (german pronunciation) *b*) 1000 c/s (sinusoidal line cut off to save space)

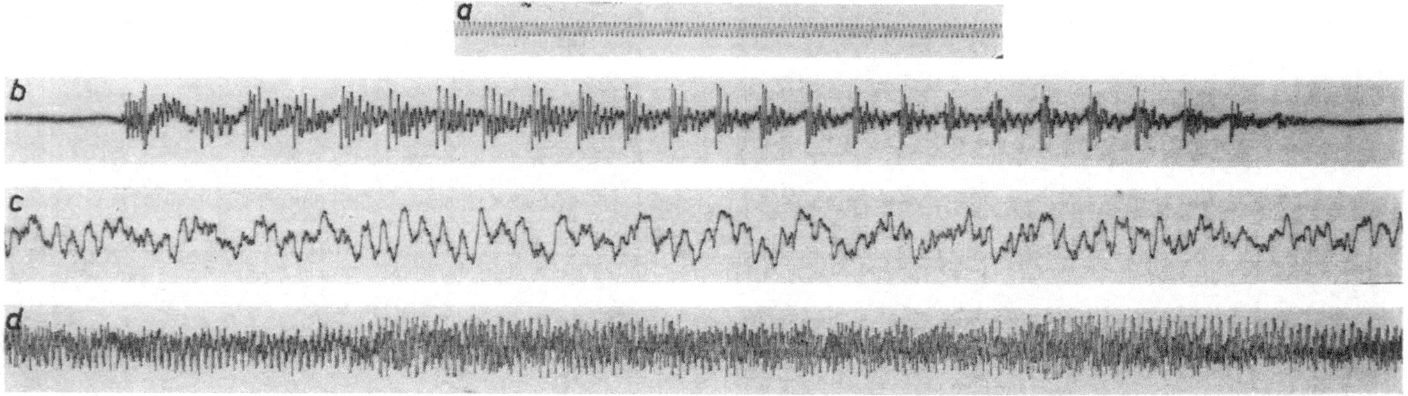

Fig. 27-4 Continuous recordings of speech and music. *a*) Time scale 1000 c/s for *b*) and *c*), 1 m/s; *b*) the word "Arbeit" (work) 1 m/s; *c*) piano in middle register; 1 m/s; *d*) mixed orchestra, 0.5 m/s

CHAPTER 28

MEASURING THE ACTION OF BETWEEN-LENS SHUTTERS OF CAMERAS

The widespread use of synchronized flashlight has made it more important to have exact knowledge of the action of camera shutters. Of the many existing methods of measuring the opening time of camera shutters, the clearest and most accurate results are obtained by using the cathode ray oscilloscope. There are two basic methods. In the one, the shutter to be studied is mounted in a camera with which a photograph is taken of the path traced on the oscilloscope screen by the spot travelling at a known speed. In the other method the shutter is mounted between a light source and a photocell. During the opening of the shutter, light falls on the photocell. The cell then delivers a proportional voltage which serves to deflect the spot vertically. By simultaneously deflecting the spot in the horizontal direction with a voltage linear with time, a complete picture of the shutter action and time function appears on the screen.

28. 1 Measuring the opening time by recording a spot trace of known speed

The longer the line traced on the screen, the more accurate will be the results obtained by the first method. With the means usually available in test laboratories the most convenient way is to produce a sinusoidal trace [1]. This can be obtained quite simply by applying the alternating voltage of a beat-frequency oscillator to the vertical deflection plates. For shutter times of between 1/5 and 1/100 s, the 50 c/s mains voltage will be found to be sufficient. In the basic diagram shown in Fig. 28-1 for such an arrangement the time base generator frequency is adjusted

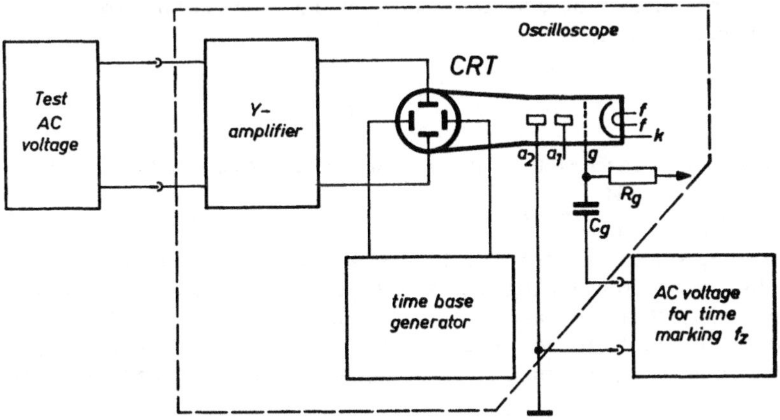

Fig. 28-1 Circuit for recording the oscillograms in Fig. 28-2*a* to *e*

so that from 3 to 6 cycles of the voltage on the Y-plates appear on the screen. The pattern should be made stationary in the way specified in the instructions for use of the oscilloscope.

The frequencies of the vertical and horizontal deflection must be so chosen that the duration of the whole pattern will be greater than the expected opening time of the shutter. (The flyback time must be as short as possible.) There will thus be a gap in the oscillogram, and the position of this during the movement of the spot will indicate the instants at which the shutter opens and closes. From the measuring frequency f_M and the number of cycles photographed N_M, including fractions of such cycles, the shutter opening time is given by:

$$T_o = \frac{N_M}{f_M}\,. \tag{28.1}$$

Fig. 28-2*a* shows such an oscillogram. With a nominal opening time of 1/25 sec and a measuring frequency of 80 c/s, the number of cycles counted is $3^1/_4$, so that according to Eq. (28.1) the actual opening time is 40.7 (40) ms. The nominal opening time is shown in brackets, as it will continue to be throughout this chapter. It is clear from this recording that interpretation will be inaccurate when simply using a sinusoidal measuring voltage without adequate time marking. The speed of the spot is not constant, as it follows the cosine function.

As already seen [111]), it is possible for time marks to control the brightness of the trace by simply modulating the beam with an alternating voltage of known frequency. The trace will then be punctuated and not continuous. The interval between the dots appearing on the screen as a result of the time-marking frequency f_Z corresponds to the time T_Z equal to the reciprocal of this frequency:

$$T_Z = \frac{1}{f_Z}\,. \tag{28.2}$$

In the example in Fig. 28-2*b* the measuring frequency was 50 c/s and the time-marking frequency 1000 c/s. The shutter was set at 1/10 s, corresponding to 100 ms. The number of recorded dots must now be counted to obtain the interpretation. The number of dots D divided by the time-marking frequency f_Z gives the actual open time, thus:

$$T_o = \frac{D}{f_Z}. \tag{28.3}$$

In order to obtain the results in ms, f_Z must be expressed in kc/s. In this example we therefore have $\frac{116}{1} = 116$ ms.

When using sinusoidal voltages a limit is set to the accuracy of the measurement by the dots being crowded together at the peaks of the curves but widely spaced at the null intersecting point and in the rest of the trace.

It would thus be better to use a voltage with a linear rise and fall, that is, one with a triangular waveform. Such voltages can be obtained from generators

[111]) Part II, Ch. 10, Fig. 10-1.

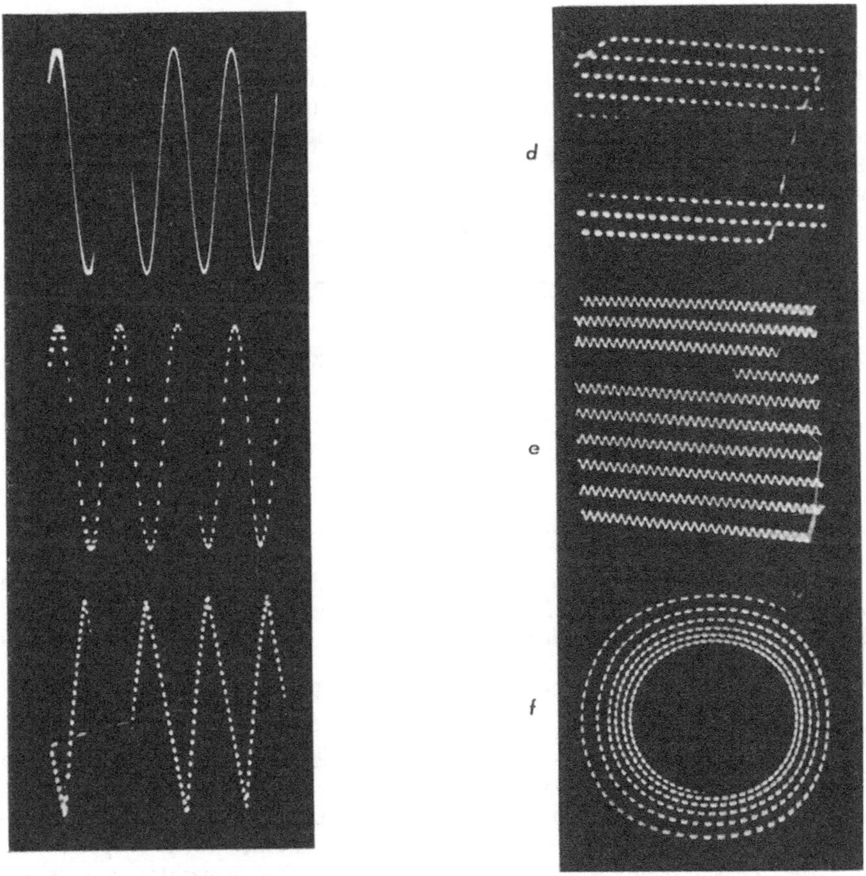

Fig. 28-2 Shutter time recordings as traces of known duration

specially developed for the purpose. Usually, square-wave voltage sources are available, however, such as a square-wave voltage generator (e.g., Philips "GM 2324"). It is then quite simple to integrate a rectangular waveform to obtain a triangular voltage. The simplest way is to use an RC-network as shown in Fig. 28-3. As a guide to the rating of such an integration circuit, it should be pointed out that $R \geqq 5 \cdot \frac{1}{\omega \cdot C}$, in which ω is equal to $2 \cdot \pi \cdot f_M$ (see footnote 78 in Part II, Ch. 11, p. 330.

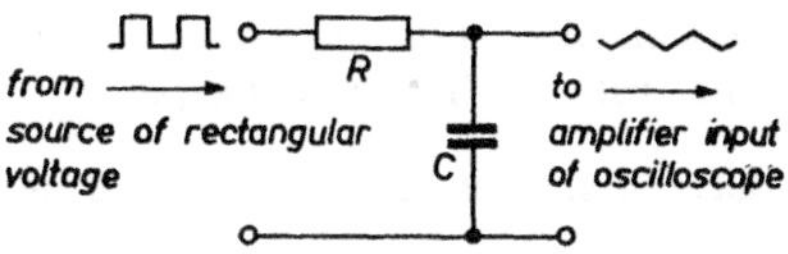

Fig. 28-3 *RC*-network for electrical integration of rectangular to triangular voltages

Table 28-1 gives the appropriate values for R and C for various frequency ranges. They were the values used for the recordings reproduced here. The desired effect can only be obtained if the capacitor has a quality grading of I.

The capacitor and leads to the vertical amplifier usually require electrostatic screening against stray fields from the mains voltage, which are always present. This applies also for the capacitors in the phase-shifting circuits for producing a spiral trace (Fig. 28-5).

TABLE 28-1 RATING THE INTEGRATION CIRCUIT (Fig. 28-3)

Frequency range [c/s]	R MΩ	C
5 50	3.0	0.1 μF
50 250	3.0	25 nF
250 750	1.0	10 nF
750 2000	1.0	2 nF

Fig. 28-2*c* shows an oscillogram of this kind. What is of particular interest in comparison with Fig. 28-2*b* is the spacing of the dots, of which there are 155. The exposure time was adjusted to 1/50 s = 20 ms, and the time-marking frequency was 8.4 kc/s. According to Eq. (28.3) the opening time was thus: $\frac{155}{8.4} = 18.5$ ms. To obtain a long trace it is very easy to produce a line-shaped oscillogram, similar to a television picture. To do this, sawtooth voltages are applied to both pairs of deflection plates in the cathode ray tube. The ordinary time base generator can, of course, be used for the horizontal deflection. A second sawtooth voltage is required for the vertical deflection. If another oscilloscope is available while the measurements are being made, its time base generator can be used to supply the required vertical deflection voltage. Its frequency must be so adjusted that the duration of the pattern is again somewhat longer than the expected open time of the shutter. (The frequencies for the two directions of deflection can also be changed over, so that vertical lines appear instead of horizontal ones.) The time base generator with the lower (frame) frequency must be synchronized with the line frequency, to keep the lines stationary on the screen.

An oscillogram of this kind is shown in Fig. 28-2*d*. The exposure time adjusted for this was 1/100 s = 10 ms. The number of dots counted out was 153. With a time-marking frequency of 12.5 kc/s, the opening time is found to be $\frac{153}{12.5} = 12^1/_3$ ms.

When time-marking by brightness modulation of the trace is impossible, or for any reason undesirable, the time-marking voltage can also be used for additional deflection of the spot. When suitably rated it creates a ripple throughout the trace. The distance between the peaks of this ripple, which is usually a sinusoidal voltage, corresponds once more, according to Eq. (28.2), to time $T_Z = \frac{1}{f_Z}$. The voltages

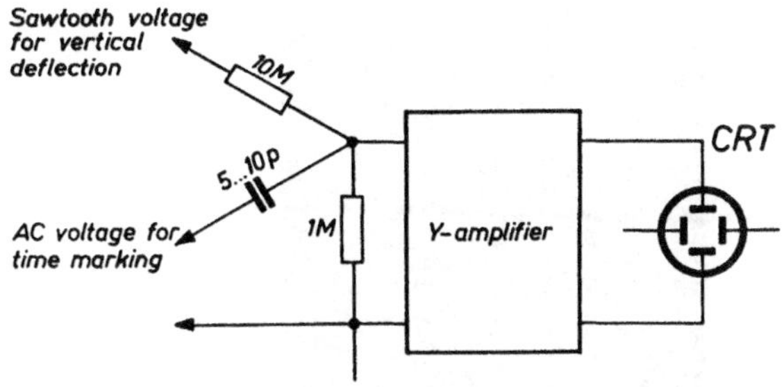

Fig. 28-4 Addition of an AC voltage for vertical deflection as time marking for the oscillograms in Figs. 28-2*e* and 28-6

from time-marker generators can be used in the same way. For the oscillogram of the sort shown in Fig. 28-2*e*, the voltage of a beat-frequency oscillator (Philips "GM 2307") was applied, together with the vertical deflection voltage, to the input of the vertical amplifier. The simple circuit is shown in Fig. 28-4. To avoid reaction, both voltages are applied via high impedances. The small capacitor of about 6 pF for the time-marking voltage can, of course, always be replaced by a high ohmic resistor of appropriate value. The frequency of the alternating voltage used here for the time-marking was 2500 c/s and the adjusted exposure 1/10 s = 100 ms. The count resulted in 276 cycles of the time-marking frequency, and thus the open time T_0 = 110 ms. In patterns with sinusoidal or triangular vertical deflection voltages it is expedient to add the time-marking voltage to the time base voltage for additional horizontal movement of the spot. Connection must then always be made via a low coupling capacitor, to block possible direct voltages at this point of the oscilloscope from the AC voltage source for the time-marking. To obtain the longest possible oscillogram on a circular screen, the obvious way is to produce a spiral trace. It is obtained by applying the decaying voltage of a periodically excited oscillating circuit to both plates, so that the component voltages have a phase difference of 90° [2] [3] (see also Chapter 22). The time base generator in the oscilloscope can be used to excite the oscillating circuit, but its voltage must not reach the deflection plates, which must receive only the phase-shifted voltage from the LC circuit, as shown in the instructions for use (Fig. 28-5). Here, once more, time-marking is introduced by intensity modulation,

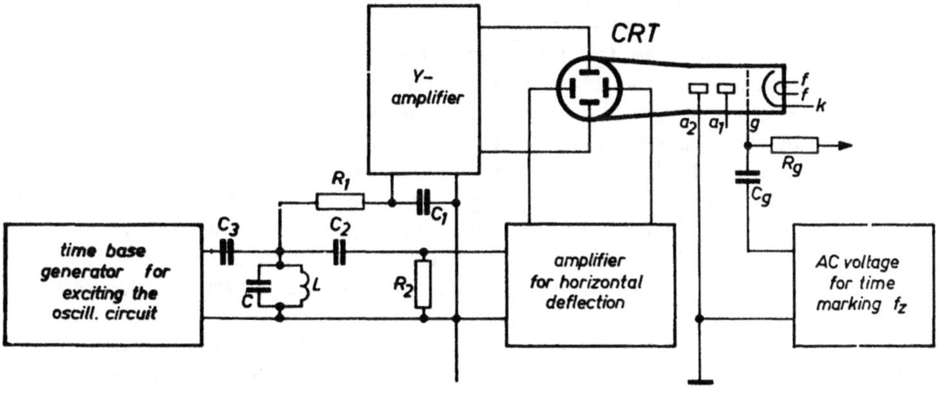

Fig. 28-5 Layout for producing the spiral trace for the oscillogram in Fig. 28-2*f*

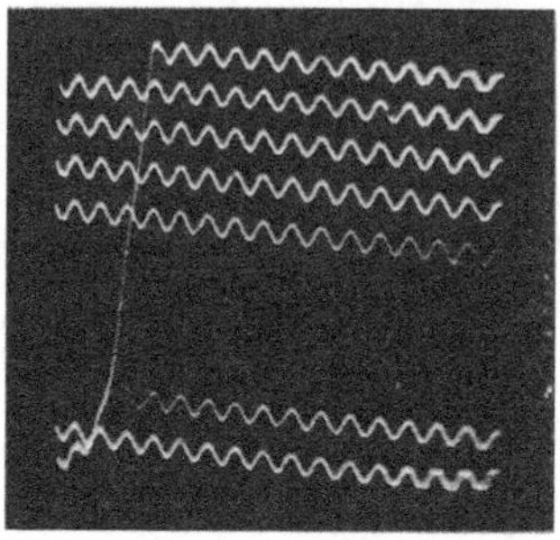

Fig. 28-6 Oscillogram as in Fig. 28-2*e*, but for 1/200 s

giving a pattern on the screen as shown by the oscillogram in Fig. 28-2*f*. The exposure time for this recording was 1/25 s = 40 ms. The dots with a time-marking frequency of 10 kc/s were counted and came to 371. (The trace described on the screen was altogether about 60 cm long.) The opening time was, therefore: $\frac{371}{10} = 37.1$ ms. The oscillogram in Fig. 28-6 was obtained by the method described in Fig. 28-2*e* for an exposure time of 1/200 s and a time-marking frequency of 15 kc/s. It can be seen that it is now fairly difficult to count the open time, as the beginning and end of the oscillogram are poorly defined. This is due to the fact that the times taken for the shutter to open and close are no longer negligibly short in relation to the open time. The trace now appears gradually, depending on the photographic conditions set.

From this oscillogram the shutter time might be estimated at 6.5 ms (5 ms). For the study of short opening times it is, therefore, better to use the other method which gives o clear picture of the process of opening in dependence upon time.

28. 2 Recording the action-time function of the shutter

As shown in Fig. 28-7, the shutter is placed between a light source and a photocell. When the shutter is opened, light falls on the photocell thus causing a proportional voltage drop across the anode resistor. This voltage is then used, after passing through a DC amplifier, for vertical deflection of the spot.

As the recording to be made is of non-recurrent phenomenon, use has to be made — somewhat before the shutter is opened — of a single-stroke time base deflection as described in the instructions for use of the oscilloscope. To avoid premature exposure of the photographic material by the quiescent spot, it must either be deflected beyond the screen area or, as indicated in Fig. 28-7, it must be blocked by the negative voltage of a B_3 dry battery. (In some Philips oscilloscopes no battery is needed). Using a Philips "GM 3156" oscilloscope, the single-stroke time base is triggered by short-circuiting a negative voltage applied to the terminal for external synchronization. In the circuit in Fig. 28-7 this is not done directly, but via a capacitor of 0.1 μF. To enable the capacitor to discharge it is shunted by a resistor of 2 MΩ. (This layout makes it possible to trigger the deflection process repeatedly.)

After the various relay contacts have been properly adjusted, armature *A* is actuated when the cable release is depressed, whereupon the spot is released and the single time base starts. Fig. 28-8 shows an oscillogram, obtained in this way,

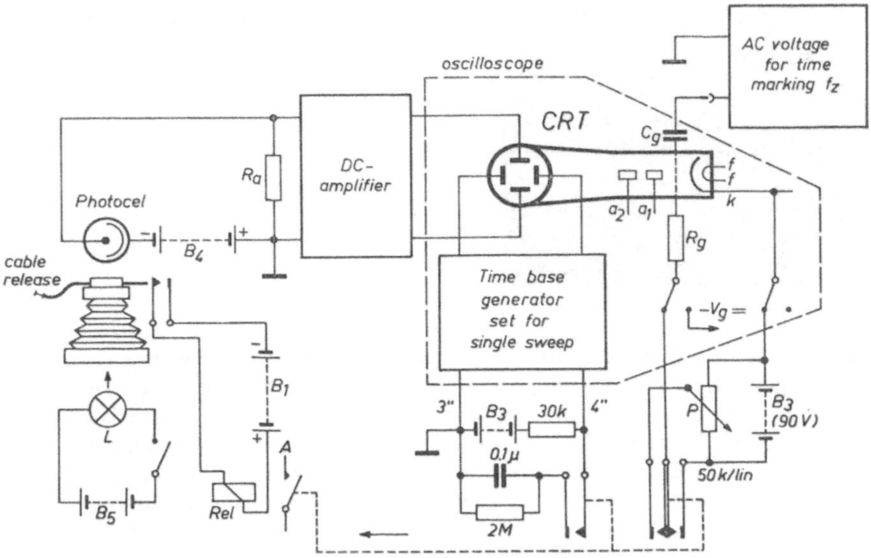

Fig. 28-7 Layout for recording shutter opening time curve for Figs. 28-8 and 28-9

for a between-lens shutter with an opening time of 1/5 s. The time-marking used here is a second oscillogram of a voltage with a frequency of 100 c/s, recorded below the pattern of the opening process. The opening time is seen to be 190 (200) ms.

The use of oscilloscopes the time base of which can be triggered simplifies this process a great deal. The unblanking is then automatically triggered by the time base, which, in turn, is easily initiated by an external control voltage, a contact or even the rising flank of the signal voltage.

Time marking can also be effected with this method by modulating the trace brightness with an alternating voltage of known frequency, as already described (Figs. 28-2*b*, *c*, *d* and *f*). This possibility is also indicated in Fig. 28-7. Fig. 28-9 shows two more oscillograms recorded for a miniature camera with shutter speeds of 1/250 s and 1/500 s. It should be noted that it is always the mean value of the open times which is given. This means that, to estimate the shorter times, the points were counted from half the height of the oscillogram. The value for the open time (shutter fully open) are approximately 3.4 and 2.2 ms. The mean open time is 4.2 (4.0) and 2.85 (2.0) ms respectively.

Fig. 28-8 Oscillogram taken with arrangement as in Fig. 28.7 at 1/5 s with time calibrating frequency 100 c/s

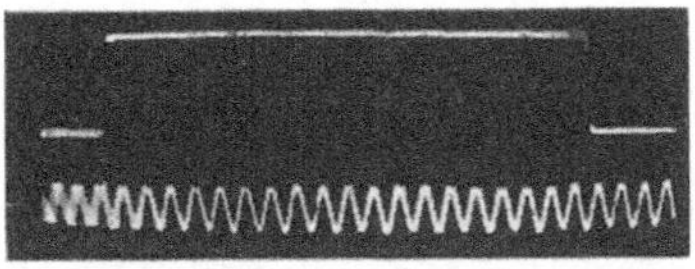

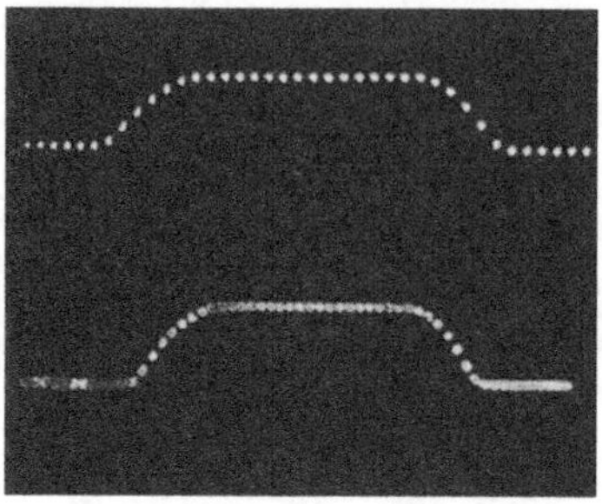

Fig. 28-9 Shutter-opening time curves of a miniature camera. Above: 1/250 s, f_z = 5000 c/s, T_0 = 4.2 ms; below: 1/500 s, f_z = 10,000 c/s, T_0 = 2.8 ms

In the method first described (Figs. 28-2 and 28-6) cathode ray tubes with the shortest possible afterglow (e.g., Philips DB 10-78) are required. In the second method, on the other hand, it is best to use tubes with long-persistence screens (Philips DN 10-78 or DP 10-78), so that the non-recurrent phenomena will remain visible for some time. With the extremely long-persistence tube DP 10-78, the trace remains fluorescent with 1% of the original intensity of luminescence even after 12 s. This is quite long enough to allow of good observation of the waveform. To avoid glare from the bright spot of light during the measurement it is advisable to fit an orange-yellow filter over the screen. By using long-persistence tubes of this kind, measurements of action time can be carried out without the need for photographic recording.

CHAPTER 29

RECORDING THE WAVEFORMS OF THE LUMINOUS FLUX AND IGNITION CURRENT OF FLASH-BULBS AND INVESTIGATING THE WORKING OF SYNCHRONOUS CONTACTS

The eye reacts only to the incidence of luminous radiation, but photographic film reacts to the work performed by the light. It is only luminous intensity which stimulates the eye, but the film integrates the effect of radiant energy over a period of time; the density of the image produced on film therefore corresponds to the product of intensity and time, the work done by the light.

By adjusting the exposure time in daylight and incandescent light, it is possible to control the amount of work which is performed by the light. With flashlight the situation is different. Here the maximum effective exposure time is predetermined by the burning time of the flash material and the variation of the light output during its combustion within the flash-bulb. This applies equally to the "open flash" process and to synchronized flash-light with *X*-contacts and, under certain conditions, with *M*-contacts. Only by using a cathode ray oscilloscope is it possible to obtain a true picture of the behaviour of the luminous flux in flashlight photography and hence of the amount of luminous energy available. By means of the oscilloscope any required information can be obtained and the oscillogram can be provided with time-markings at any characteristic points of the curve.

29. 1 Luminous flux

The required circuit is basically the same as that shown in Fig. 28-7. In place of the shutter, the flash-bulb is now placed near the photocell, preferably at a distance of three feet, to avoid overloading the photocell or the amplifier. A grey filter is found to be very useful as a further precaution. The release contacts are adjusted in such a way that the electron beam is initiated before the flash and then the time base is started. A DC amplifier is again required to amplify the photocell voltage to the value needed for satisfactory deflection of the spot. With this arrangement, the luminous flux versus time curves of Philips PF 14, PF 25, PF 45, PF 56 and PF 110 Photoflux lamps were recorded one above the other on a single photograph (Fig. 29-1). Time-marking was again provided by recording below the oscillogram the pattern of a sinusoidal voltage having a frequency of 200 c/s, with the same time scale, of course. The distance between the peaks corresponds to 5 ms. Two or more lamps can be easily compared by this method, though it should be borne in mind that such oscillograms only reproduce the characteristic of the particular samples used. For a general assessment the usual manufacturing spread must be taken into account. The oscillograms should not be confused with the curves published by the makers, which represent the average values measured on a large number of samples.

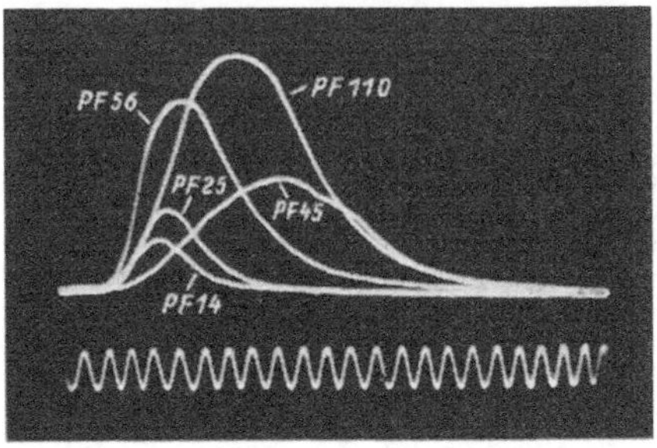

Fig. 29-1 Luminous flux waveforms of different flash-bulbs; time-marking: 200 c/s = 5 ms.

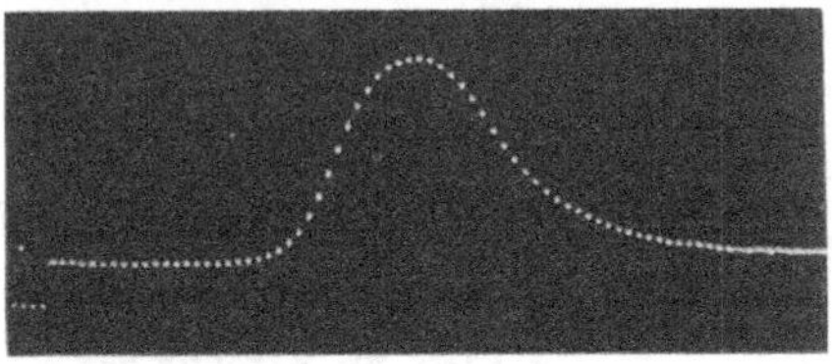

Fig. 29-2 Luminous flux waveform of Photolux lamp PF 3 N. Time-marking by intensity modulation: 2000 c/s

The oscillogram of the small PF 3N flash-bulb is seen in Fig. 29-2; in this case brightness modulation of the trace has been used for time marking. In order to judge the behaviour of the flux curve in flashlight photography it is important to have an exact marking of the instant at which the bulb is ignited. In Fig. 29-2 this instant can be observed at the left-hand side of the oscillogram. It was obtained by taking a portion of the ignition voltage from the ignition battery and feeding it via a high-value resistor to the grid of the input valve grid of the vertical amplifier. The voltage division resulting from this depends on the ratio of the series resistor to the grid resistor and must be such that the oscillogram is displaced vertically sufficiently to allow of clear interpretation. By counting the dots from this "threshold", the significant times can be obtained (f_Z = 2000 c/s). Here, for instance, the ignition time $t_{\text{ign}} = 14^1/_4$ ms, peak time $t_p = 18^1/_2$ ms and light emission time $t_l = 8^3/_4$ ms.

Care must of course be taken during the measurement to prevent the flashlight from falling, directly or indirectly, on the screen of the cathode ray tube. The result would be the complete blanking out of the oscillogram. For continuous recordings the oscilloscope is best installed in an adjoining room.

To obtain correct ignition conditions the value of the ignition current required, or the electrical work in watt-seconds needed for ignition must be known. The oscillogram of the ignition current during the switching-on period provides this information clearly.

29. 2 Ignition current waveform

The basic diagram for this measurement is shown in Fig. 29-3. Reference should also be made to remarks on the layout in Fig. 28-7. Ignition is caused by four nickel-iron accumulator cells supplying a voltage of 3.3 V. A resistor R_e of 0.5 Ω is connected in the ignition circuit. Across this resistor appears a voltage linearly proportional to the current. This voltage is passed via the DC amplifier to the *Y*-plates for vertical deflection. This resistor also increases the internal resistance of the voltage source at the same time, so that ignition conditions resemble those for dry-battery ignition. To assess the current intensity from the vertical deflection a calibration current is provided by changing over switch *S* to position 1 and adjusting ammeter *A*, by means of resistor R_v, to 0.5 A. The displaced

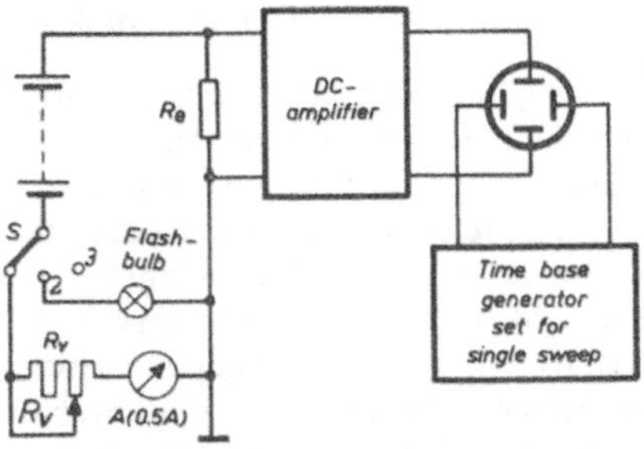

Fig. 29-3 Layout for recording ignition current waveforms

zero line resulting from this is recorded separately and added to the oscillogram. In this way information on the absolute value of the occurring currents is also obtained.

Fig. 29-4 shows a selection from a large series of oscillograms recorded in this manner, depicting the waveforms of Photoflux lamps PF 14, PF 25, PF 45, PF 56 and PF 110 respectively. The horizontal line near the centre of the patterns is the 0.5 A current marking. At the instant of switching on, when the primer pellet is cold and the internal resistance of the cold bulb is only about 1 Ω, a steep current surge of over 1 A appears. This falls very rapidly, however, and remains at a value of roughly 0.5 A for a time which varies according to the size of the different types of bulb.

As these curves have steep portions, time marking by intensity modulation was dispensed with, and instead, a third oscillogram of a 200 c/s alternating voltage was recorded beneath the waveform under investigation. A 5 ms time marking was used. Here too, it should be emphasized that these recordings should not be judged as manufacturer's mean values, but only as results obtained on single samples. The burning times for the series PF 14, PF 25, PF 45, PF 56 and PF 110 are 23, 32.5, 42.5, 80 and 72.5 ms respectively. Assuming that the voltage on the lamp during ignition was roughly 3.0 V, the electrical work required for ignition can be calculated as 38, 32, 36, 144 and 122 mWs respectively.

PF 14

PF 25

PF 45

PF 56

PF 110

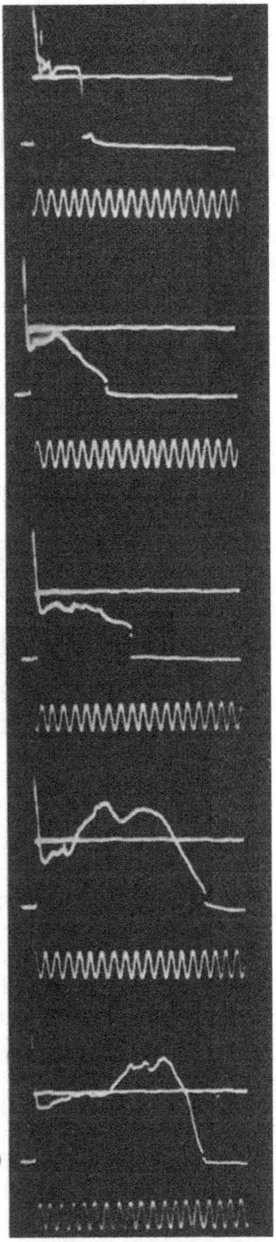

Fig. 29-4 Oscillograms of ignition current of Photoflux lamps; time-marking 200 c/s

29. 3 Shutter action, synchronizers and luminous flux waveform of flash-bulbs

Usually one oscillogram showing the instant during the operation of the shutter at which the flash-light contact is made is sufficient for examining the action of synchronizer contacts. This can be simply done by connecting the input of the vertical amplifier which amplifies the voltage from the photocell to a direct voltage source of little voltage via a small capacitor. If the other pole of the voltage source is connected to the camera casing, the surge produced by the charging current of the capacitor appears on the oscillogram as a peak. In every other respect the circuit is fundamentally the same as in Fig. 28-7. The oscillogram in Fig. 29-5 is the result of testing the *X*-contact of the between-lens shutter of a 6×6 camera, using a time-marking voltage of 500 c/s. The instant at which the *X*-contact closes coincides with the commencement of the peak in the curve of the shutter opening time and is indicated by a downward blip. (The exposure time is 1/25 s.) As expected, the contact closes only after the shutter is fully open.

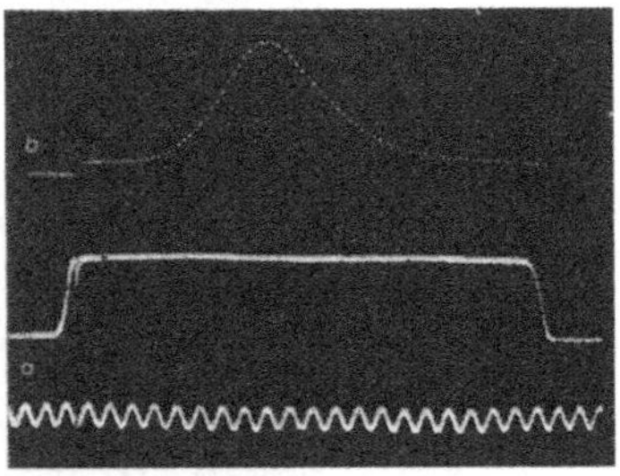

Fig. 29-5 Shutter-time curve for 1/25 s exposure, with blip marking the beginning of ignition with *X*-contact and luminous flux of type PF 14 flash-bulb; time-marking: 500 c/s = 2 ms

Fig. 29-5 shows the enlarged luminous flux versus time curve of a PF 14 Photoflux lamp. The step at the beginning of this curve, the moment at which the synchronizer contact closes and the corresponding blip in the oscillogram of the shutter opening time all occur simultaneously. The average open time of the shutter embraces 21 cycles of 500 c/s, i.e. 2 × 2 = 42 (40) ms. The peak duration of 17 ms and the time of light emission — 11 ms — both lie well within the open time of the shutter. This shows that the total light is used when taking the photograph. This is different at an exposure of 1/50 s. The time-marking chosen in this instance is 1000 c/s. The relevant oscillograms are shown in Fig. 29-6. The mark at the instant at which the synchronizer contact closes after the shutter has been opened is more clearly recognizable. The mean open time is found to be 19 (20) ms.

The luminous flux versus time curve of a PF 25 flash-bulb, also enlarged and with the same time scale, is shown above the curve of the shutter open time in Fig. 29-6. Its peak time of 21 ms and its light emission time of 15.5 ms are set against a shutter open time of only 19 ms. Since the flash-bulb does not reach its peak intensity until 21 ms after switching on, the shutter closes before half this time has elapsed and not even half the available light has been made use of. A badly underexposed negative will be the outcome. For this reason, when using standard flash-bulbs of 16...20 ms peak time with *X*-contacts, it is best never to use shorter times than 1/25 s.

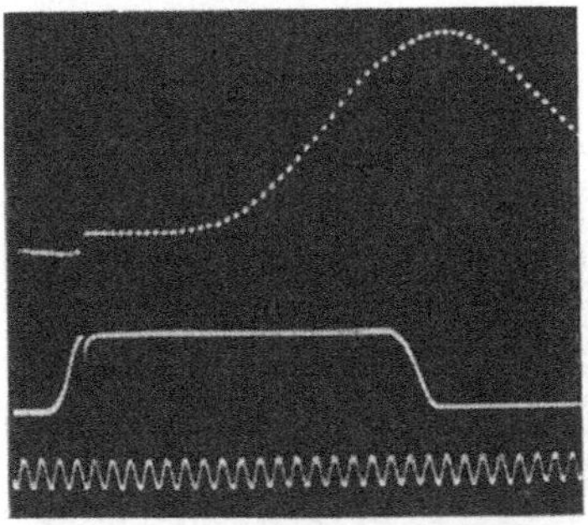

Fig. 29-6 Shutter-time oscillogram for 1/50 s with time marking for beginning of ignition with X-contact and luminous flux of PF 25 flash-bulb; time-marking: 1000 c/s = 1 ms

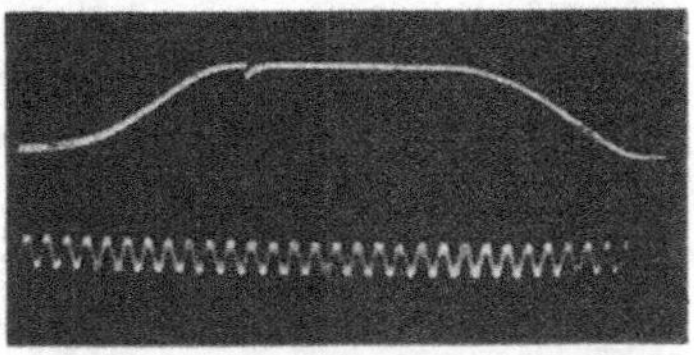

Fig. 29-7 Shutter-time oscillogram for 1/500 s with X-contact synchronization; time-marking: 5000 c/s = 0.2 ms

It is of importance to users of flash-bulb apparatus to know exactly when the contact closes at even shorter exposures. This information is shown in the oscillogram in Fig. 29-7 for an exposure of 1/500 sec with time marking of 5000 c/s corresponding to 0.2 ms. The results obtained are: mean shutter open time 3.7 (2.0) ms, "open time" (during which the shutter is fully open) 2.2 (2.0) ms. After the synchronizer contact closes, the shutter remains open for nearly 2.0 ms. Since the emission time of flash-bulb apparatus is only a few ms or even a fraction of one ms, this contact is quite sufficient. The oscillograms in Fig. 29-8 were taken when testing the between-lens shutter of a miniature camera with an M-contact. They represent exposures of 1/25, 1/50, 1/100, 1/250 and 1/500 s respectively. The time marking voltage is 500 c/s, corresponding to 2 ms. The blip appearing when the M-contact closes now occurs before the shutter opens. The luminous flux versus time curve of a PF 14 flash-bulb has been reproduced above these oscillograms so that the markings for the instants of contact for flash-bulb and camera lie exactly one above the other. In this way it is possible to observe immediately the agreement of the peak time of $18^3/_4$ ms and the emission time of $12^1/_2$ ms with the shutter time of the camera. It is also of interest to examine the total duration of the switching-on process. This can be shown in an oscillogram by applying, instead of a pulse, an alternating voltage of suitably small amplitude and appropriate frequency through the synchronized contact to the input of the vertical amplifier. During the time the contact is closed, a ripple is superimposed on the oscillogram of the shutter time. The peaks of this ripple represent a time marking corresponding to $1/f_Z$. A series of oscillograms for the same exposure times as in Fig. 29-8 are reproduced in Fig. 29-9, but enlarged to produce oscillograms expanded along the time axis. The frequency of the voltage used for marking the contact times has also been adapted to the different time base expansions. It can be seen from these recordings that the X-contact also closes. This has no bearing on the ignition process, however, the flash-bulb having already been ignited by the M-contact. Even at 1/250 sec and 1/500 sec, the corresponding ripple is found at the peaks of the oscillograms, although not very distincly at 1/500 sec. Fig. 29-10 shows a greatly time-expanded oscillogram prepared specially for investigating the X-contact at a time marking

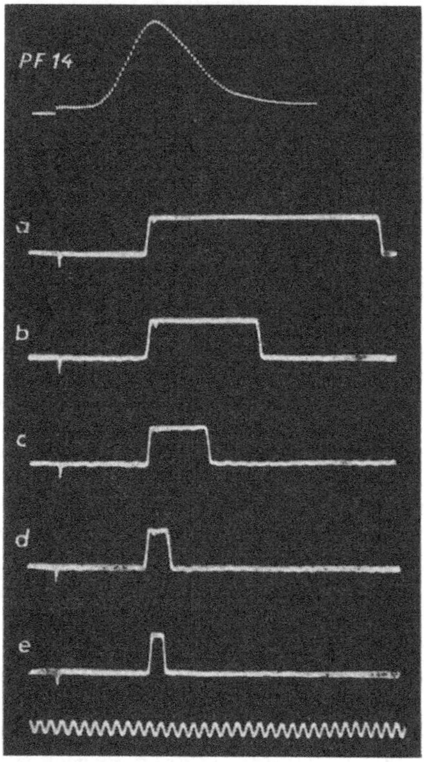

Fig. 29-8 Shutter-time curves with *M*-contact synchronization and luminous flux of a PF 14 flash-bulb

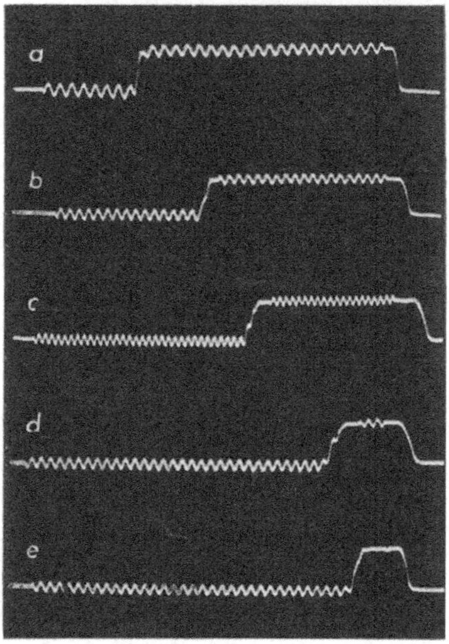

Fig. 29-9 Shutter-time oscillograms with superimposed ripple to mark the closed time of *M*- and *X*-contacts

frequency of 5000 c/s corresponding to 0.2 ms. From this it appears that the contact is closed for almost 1 ms, which is quite sufficient for flash-bulb apparatus. Contact is made in this case for about 0.6 ms after the shutter is fully open.

It is obvious that the time marking for the switching-on process can also be used for interpreting the whole oscillogram of the shutter action. The deflection along the horizontal axis is linear with time, so that the time scale thus obtained can be applied to the whole pattern. The values obtained from the oscillograms in Figs. 29-9 and 29-10 are set out in the following table:

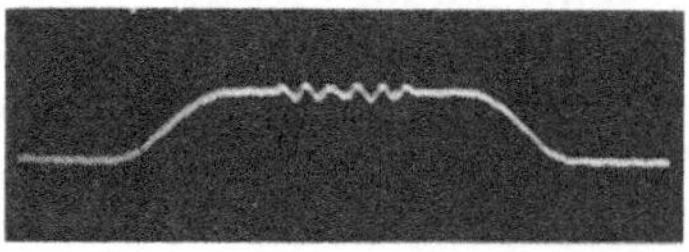

Fig. 29-10 Shutter-time curve for 1/500 s, with ripple to mark closed time of *X*-contact alone

TABLE 29-1

Nominal exposure time	Measured opening time	Time marking f_z
a) 1/25 s (40 ms)	45.2 ms	500 c/s
b) 1/50 s (20 ms)	21.9 ms	1000 c/s
c) 1/100 s (10 ms)	14.2 ms [112])	2000 c/s
d) 1/250 s (4 ms)	4.2 ms	2000 c/s
e) 1/500 s (2 ms)	2.8 ms	2000 c/s
1/500 s (2 ms)	2.85 ms	5000 c/s

Table 29-11 Nominal shutter-times, measured shutter-times and time-marking frequency in the oscillograms of Figs. 29-9 and 29-10

We have here an example of the variety of information that can be gathered from oscillograms of this kind regarding the action of synchronizer contacts.

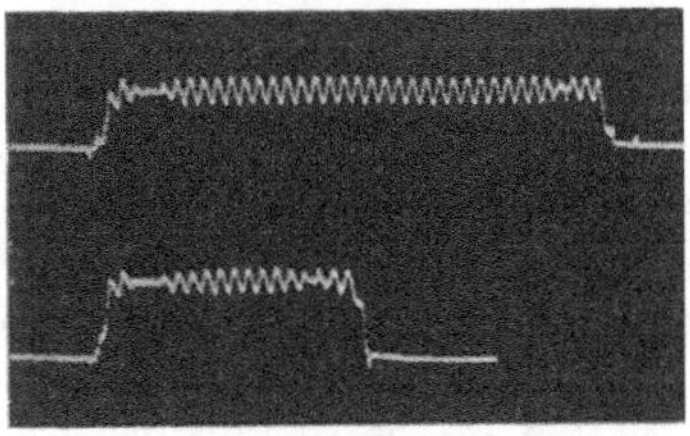

Fig. 29-11 Shutter-time curves with ripple to mark the close time of a X-contact built into a camera. Above: 1/25 s; below: 1/50 s

Finally, Fig. 29-11 gives two oscillograms indicating the contact times of an *X*-contact subsequently built into a camera by hand. The shutter times are 1/25 and 1/50 s respectively. They show clearly that the contact is interrupted shortly after the shutter opens and shortly before it closes. As is confirmed by practical recordings, it can be used for normal *X*-contact work at an exposure time of 1/25 s. It is, however, obvious that the action of this contact cannot be compared with that of manufactured contacts in cameras of repute, as illustrated by the oscillograms in Figs. 29-8 and 29-9.

112) This open-time was excessive. It must be pointed out again that the oscillograms represent individual results obtained by the author using whatever cameras happened to be available.

CHAPTER 30

STUDY OF MECHANICAL VIBRATIONS BY MEANS OF ELECTROMAGNETIC AND ELECTRODYNAMIC PICKUPS

30. 1 Observing non-electrical phenomena in general

For the study of the behaviour of non-electrical phenomena by means of oscilloscopes, these phenomena must first be converted by transducers to proportional electrical voltages. Thus, for instance, in Ch. 14 the light from glow lamps and fluorescent lamps is converted by photocells to proportional voltages and then fed to the oscilloscope for observation. The display of the fluctuations in sound pressure corresponding to speech and music by means of microphones has been dealt with in Ch. 27. In Ch. 28 the movement of between-leens shutters of cameras has been described in detail.

Oscilloscope measuring techniques have attained considerable importance in the study of mechanical changes of state as these techniques often represent the only possibility of improving on the results obtainable by traditional methods, and of increasing sensitivity in measurement. The advantage always offered by electronic measuring is that the voltage supplied by the pickup can be amplified as required by the task in hand. It is therefore possible to make the pickup apparatus small and light, so that they impose little load on the system under investigation. The voltage generated in the pickup by the phenomenon under investigation is then very small and must be very considerably amplified, and this is provided for in cathode ray oscilloscopes.

Although devices working on the principle of capacitance measurement and fed by carrier frequency [1] are used as pickups, in this section and the following one only the investigation of mechanical phenomena by means of pickups working on electromagnetic and electrodynamic principles will be considered.

30. 2 Magnetic vibration pickup

30.2.1 ACTION AND PROPERTIES

In this type of pickup, a basic diagram of which is shown in Fig. 30-1 and a typical construction in Figs. 30-2*a* and *b*, there is a permanent magnet *M* surrounded by a coil *C*, so that the lines of force of the magnet also pass through the coil. The part *B*, whose vibration movements are to be studied, is placed in front of the pole of the magnet at a distance *A* from it. *B* must either be ferromagnetic, or a small plate of ferromagnetic material such as a transformer lamination must be affixed to it.

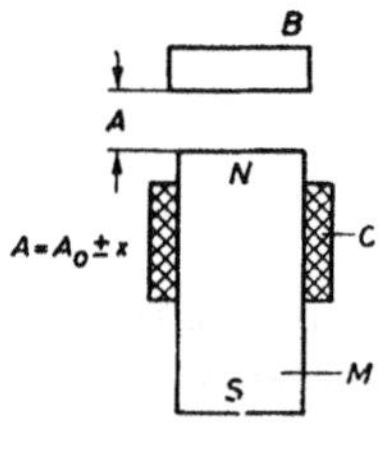

Fig. 30-1 Principle of the electromagnetic pickup

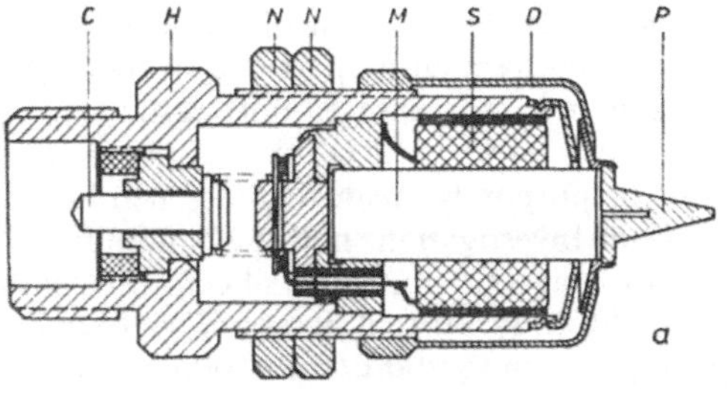

Fig. 30-2 Philips "PR 9262" Electromagnetic Pickup. *a*) Section; *b*) external view with connecting cable. *H*, stainless steel casing; *M*, permanent magnet; *S*, coil; *D*, non-magnetic cover; *P*, removable cap for field concentration; *C*, connecting pin; *N*, threaded fastening rings

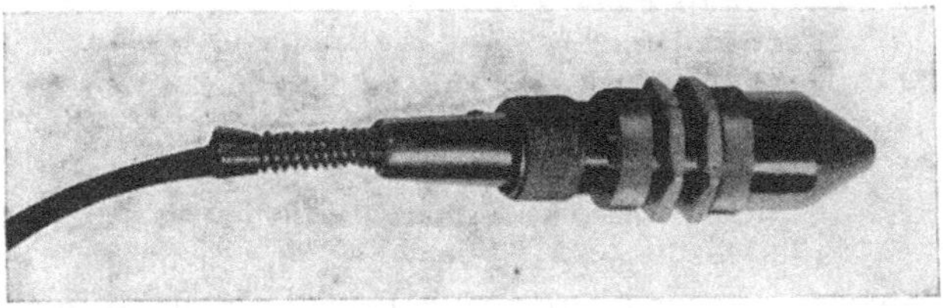

When *B* vibrates in the direction of the magnetic field, the latter undergoes a corresponding change and a voltage is induced in the coil. This voltage is proportional to the change in field $\mathrm{d}\Phi/\mathrm{d}t$ and is thus proportional to the time rate of change of distance *A*, that is to say, it is proportional to $\mathrm{d}A/\mathrm{d}t$.
From this it is clear that:

$$\frac{\mathrm{d}\Phi}{\mathrm{d}t} = \frac{\mathrm{d}\Phi}{\mathrm{d}A} = \frac{\mathrm{d}x}{\mathrm{d}t} \tag{30.1}$$

in which x is the change of position from the position of rest A_0.

Thus the pickup delivers a voltage proportional to the speed of movement. The dependence of sensitivity upon distance is illustrated in Fig. 30-3. These curves apply,

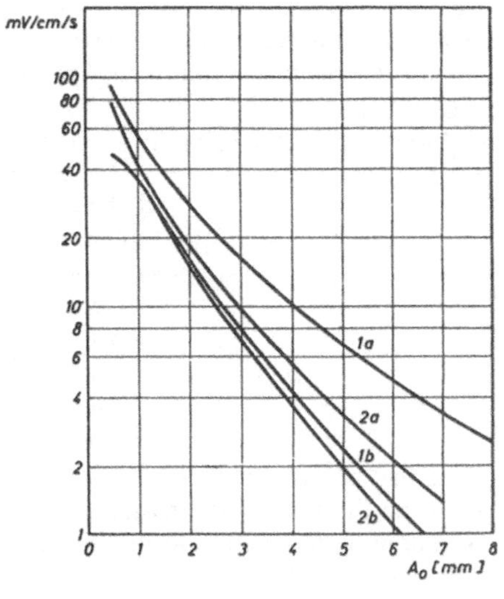

Fig. 30-3 Sensitivity of the "PR 9262" pickup in dependence upon the distance A_0. Curves 1 apply for the pickup without field concentration cap, curves 2 with cap, curves *a* apply for oscillating ferromagnetic parts thicker than 2 mm, while curves *b* refer to non-ferromagnetic oscillating parts on which a silicon platelet was affixed skew to the axis of the pickup

however, only for vibration frequencies up to about 200 c/s. Sensitivity also decreases on account of eddy currents occurring both in the oscillating part and in the pickup. These eddy currents not only attenuate the induced voltage, but also damp the movement of the oscillating part. Eddy currents can be reduced to a minimum if the electrical resistance of the oscillating part is purposely made high. When specific ferromagnetic machine parts are the subject of investigation, nothing can be done to change their electrical conductivity; with non-magnetic parts the silicon-iron lamina attached to the object causes only very slight eddy current losses. The frequency characteristic of sensitivity is then completely flat up to 200 c/s; at 1000 c/s it drops by 15% and at 2500 c/s by about 23%.

In the case of larger and thicker parts the frequency characteristic drops earlier; but as the eddy current losses depend on the electrical properties of the particular parts, no definite figures can be quoted. With the usual types of steel a fairly flat characteristic can be expected.

With this type of pickup there will, theoretically, always be a certain non-linear distortion of the output voltage even when the movement of the vibrating part is sinusoidal. Moreover, a second harmonic is liable to occur. If, however, the distance of the vibrating part is at least ten times the oscillation amplitude and the latter is not more than 0.5 mm, these distortions will not be serious. The polarities of the pickup and the coil connection to contact C are so chosen that a rapid approach of the oscillating part to the pickup causes a positive-going voltage and thus an upwards deflection on the oscilloscope.

30.2.2 REACTION OF THE PICKUP ON THE VIBRATING PART

Between the pickup and the magnetic body whose vibration movements are to be observed there is a magnetic attraction of about 100 to 500 gr at a distance of 1.0 to 0.1 mm. This is equivalent to the effect of a non-linear spring with a negative spring constant mounted between the pickup and the vibrating object. The effect of this cannot be forecast easily in individual cases; it must suffice to say that the natural frequency of the vibrating system is reduced. The size of this "spring constant" again depends on the distance between the magnet and the vibrating part, and the magnetic and electrical properties of the latter. In the case of a large, thick ferromagnetic object the following approximate values apply:

TABLE 30-1

DEPENDENCE OF THE SPRING CONSTANT UPON DISTANCE BY MAGNETIC ATTRACTION

Distance [mm]	Spring constant [kg/cm]
0.2	—8.0
0.5	—3.3
1.0	—1.3
2.0	—0.45

It has already been mentioned that the induction of eddy currents in the object opposite the pole of the magnet produces a damping effect in the system, the amount of damping depending on the given conditions. For example, with a mild steel object at a distance of 0.25 mm from a pickup and oscillating at a frequency of 150 c/s, the damping resistance is stated to be 475 dyn/cm/s. The data sheet supplied by the makers contain more detailed information.

30.2.3 EXAMPLES OF APPLICATION AND TYPICAL OSCILLOGRAMS

Due to the above-mentioned limitations it is clear that this form of pickup will not give results that will satisfy the most stringent requirements. Such pickups are also affected by strong pulsating magnetic fields which generate interference voltages. In spite of this, however, they can be used for a number of applications, particularly if only the frequency (or number of revolutions) has to be observed and it is of minor importance whether the output voltage corresponds to the vibration waveform exactly or not. For exact measurements it is advisable to use the amplitude pickup that will be described in the next section.

This type of pickup is also useful as a revolution counter. All that is then necessary is to mount the pickup in the vicinity of the revolving part near the perimeter, so that the field is amplified — or attenuated — one or more times per revolution. At these instants voltage pulses occur and can be used, according to one of the processes described in Part II, Ch. 12, for measuring the number of revolutions. One advantage of this type of pickup is that it requires no auxiliary voltages but is itself an active two-terminal network. A further advantage is that it is not in physical contact with the object under investigation.

The voltage pulses generated in the pickup can also be used to synchronize an oscillogram displaying conditions which depend on the number of revolutions. The pickup is also useful for synchronizing stroboscopes. If, for instance, screws with protruding heads are driven into the perimeter of the revolving part at an angle of 90° from one another, four pulses per revolution are obtained when these screw-heads are in motion in front of the pickup. It is thus possible to synchronize — or trigger — the oscillogram every quarter revolution.

The voltage delivered is, as has already been said, proportional to the speed of movement of the oscillating object, as indicated in the curves in Fig. 30-3. Often, however, it is the picture of the vibration path — not dx/dt, but x — which is required. As already described in detail in Part II, Ch. 11.12 and 11.9 "Electrical integration" and "Electrical differentiation", it is relatively easy to integrate a voltage waveform electrically by means of an *RC*-network or differentiate it by means of a *CR*-network, as will be shown in the description of an electrodynamic pickup in the next section.

By this means a voltage can be obtained (within the frequency range of the integration device) proportional to the vibration path x, or a voltage can be produced by means of a differentiation circuit corresponding to the first derivative of the speed curve.

The oscillogram in Fig. 30-4 shows the recording of the oscillation amplitude of a rotating object during its passage through the critical revolutions. Using an electronic switch a simultaneous time scale was traced with a 25 c/s alternating voltage. From this the critical speed for 10 c/s is seen to be 600 rpm. Another application can be seen in the sketch in Fig. 30-5. Here, the radial movement (i.e. eccentricity) of a revolving object can be represented on the oscilloscope screen. To do this, the movement is observed by two pickups spaced 90° apart. If the voltages of these pickups, after integration and amplification, are fed to the two pairs of plates of the cathode ray tube, the two voltage components of the pickups influence the cathode

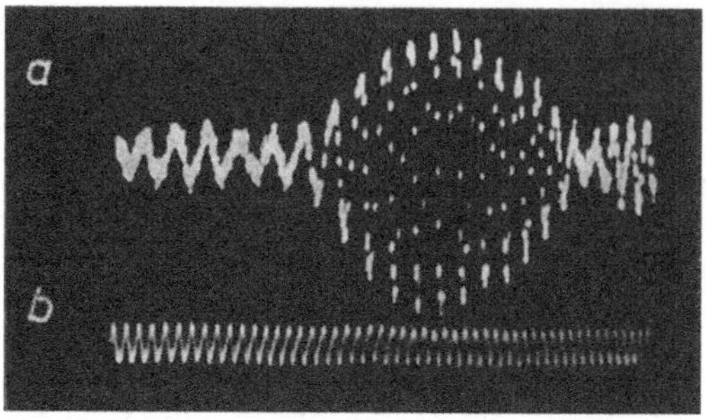

Fig. 30-4 Determining the critical number of revolutions. $D_{cr} = 600$ rpm. *a*) Process; *b*) time calibration 25 c/s

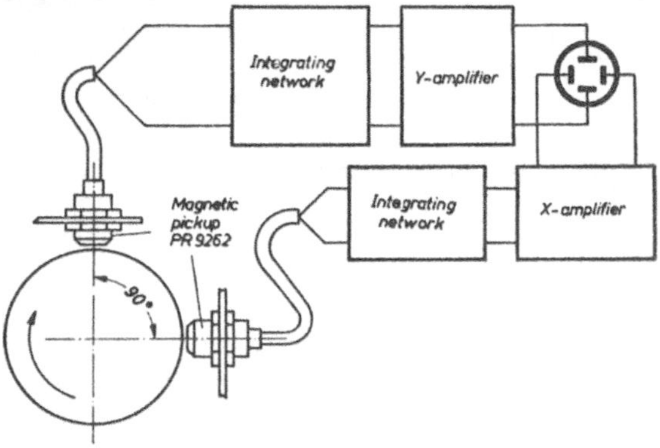

Fig. 30-5 Layout for observing the radial movements of a body rotating symmetrically with respect to its axis

ray in such a way that the spot describes enlarged the same path as the centre of the moving part. This part must be absolutely circular, however. Irregularities in its surface cause additional interference voltages. But, as in these tasks the change in distance due to the displacement of the centre point is usually many times greater than the possible departure from the true circular shape of the revolving object, it causes little disturbance.

In the same way errors in the contour of a nominally circular revolving body can also be studied. The condition must then be made that the amount of play in the bearing must be negligibly small and that there is no shifting of the centre of rotation. The distance between the pickup and the revolving part must also be as small as possible.

As a practical example, a spinning can as used in artificial silk manufacture was studied. A spinning can of this sort, mounted on a vertical, hollow axle, is driven by a vertical electric motor. This axle is free-moving so that the ratio of its number of natural oscillations to the number of working revolutions is low. As this spinning can is made of magnetically indifferent materials, a portion of its perimeter was covered with iron powder at the level at which the pickups were fixed.

Fig. 30-6 shows an oscillogram obtained in this way. For these recordings there was available neither an oscilloscope with horizontal amplifier nor any other auxiliary amplifier for the horizontal deflection voltage, but as an expedient a Philips "GM 4581" electronic switch was used as a horizontal amplifier. The vertical line caused by the second channel had to be put up with (it is only visible in Fig. 30-7 and has been blanked out in Fig. 30-6).

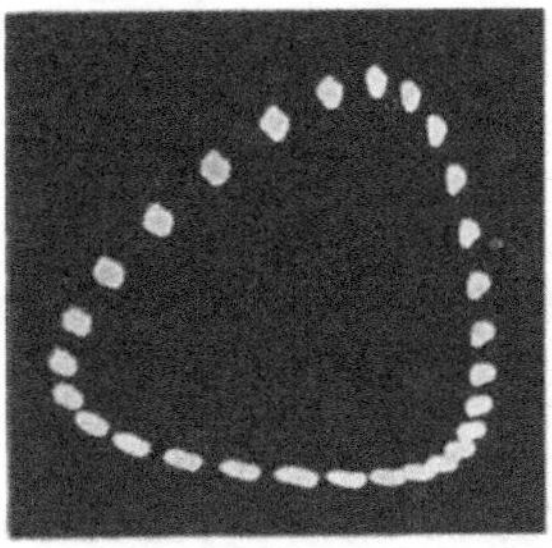

Fig. 30-6 Radial movement of a spinning can at almost maximum number of revolutions. Brightness modulated with 500 c/s to determine the rpm

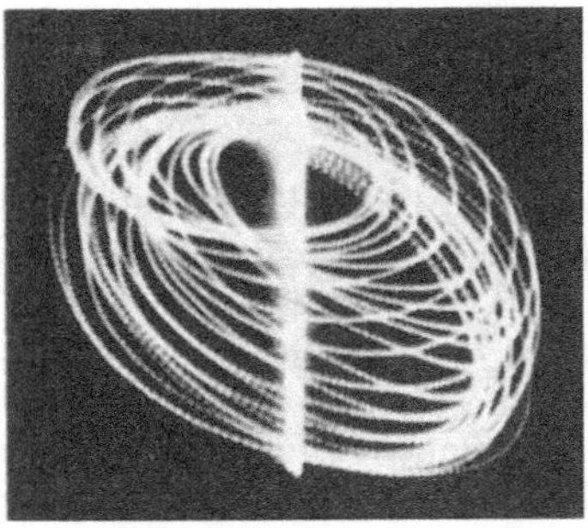

Fig. 30-7 Radial movement of spinning can at maximum rpm during a registering period of 1/5 s

The triangular form of the pattern was first thought to be due to the drive motor having a three-point mounting [2]. It might be suspected in this case, however, that this oscillogram is of the special hypocycloid form (Fig. 12-39*a*), which can occur as a result of precession oscillations. The oscillogram of Fig. 30-7 gives another impression of the radial movements of the can during an exposure time of 1/5 sec. The behaviour of the spinning can at various critical rotational speeds may be gathered from oscillograms *a* and *b* of Fig. 30-8. The line caused by the electronic switch has been shifted electronically in these recordings and does not cause any interference. At the critical speeds the axle of the spinning can oscillate, so that the so-called precession oscillations takes place, and the cycloid form of oscillogram results.

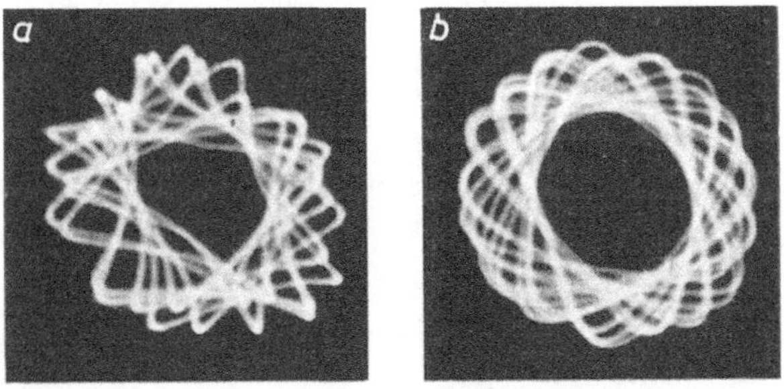

Fig. 30-8 Spinning can movements at critical rpm

As the path of the spot is not due to purely sinusoidal voltages, these do not show as uniform a waveform as in Figs. 12-38*h* and 12-34*a* in Part II, Ch. 12. They are fundamentally comparable however.

If an alternating current is applied to the coil of this pickup, it generates an alternating field at its free end, and with this field ferromagnetic parts can be excited to oscillation. The coil can constantly carry a current of 15 mA (for short periods 25 mA). At distances of 2.0 to 0.25 mm the instantaneous value of the power exerted on the part is thus from 1.2 to 6 g. In spite of this low power, systems capable of oscillation can be successfully excited provided the damping is not too great.

30. 3 Electrodynamic pickup for measuring the absolute value of mechanical oscillations

30.3.1 ACTION AND PROPERTIES

The Philips "PR 9260" pickup will be described as an example and is illustrated in Figs. 30-9 and 30-10. Two coils, S_1 and S_2, are suspended on two resilient diaphragms V in such a way that they form a so-called seismic system [3] [4]. The pickup can be used vertically or horizontally as shown at *a* and *b* in Fig. 30-11. If the resilient system of the pickup is set in motion by a mechanical oscillation in the signal frequency range of the device, it can be considered that the coils are standing still in space and that the magnetic field is moving to and fro over them. The resonance frequency of the flexible system is 12 c/s. In coil S_1 there is induced an alternating voltage similar to that in the magnetic pickup and proportional to the speed of the movement. As the magnetic field in this pickup is very powerful, even oscillations

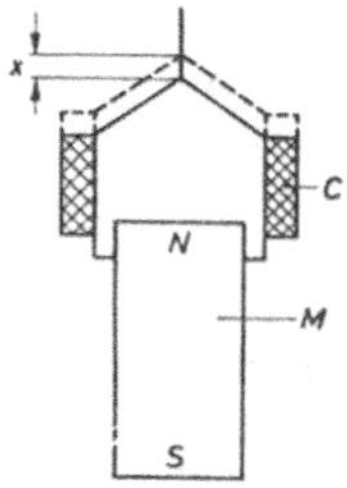

Fig. 30-9 Principle of the inductive pickup

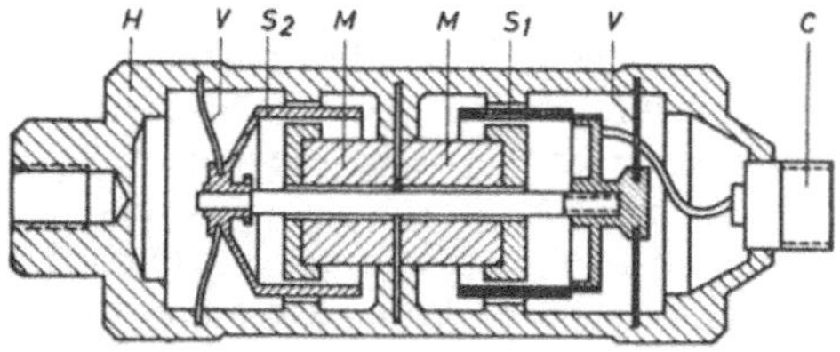

Fig. 30-10 Longitudinal section of the Philips "PR 9260" inductive pickup. M permanent magnets; S_1 pickup coils (c. 2250 Ω); S_2 damping coils (copper ring); V diaphragm springs; H body of the pickup; C cable connection

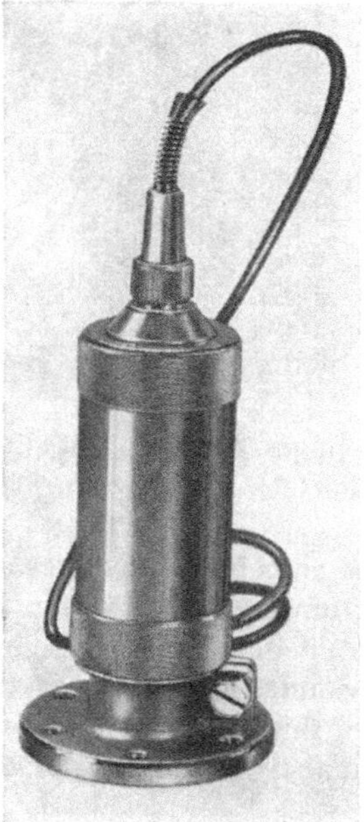

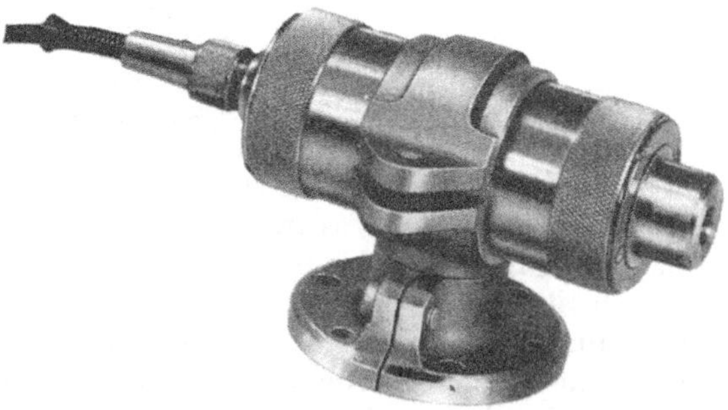

Fig. 30-11 "PR 9260" Absolute pickup. *a*) Fixed for observation of vertical oscillations; *b*) as *a*) but for horizontal oscillations

of very small amplitude give voltages adequate for examination, provided the mass of the oscillating system under investigation is great compared to the 570 g weight of the pickup.

The coil S_2 consists of a copper ring and serves for damping purposes. So-called half-critical damping is chosen. An oscillating system is critically damped when the free decay is just aperiodic. Half-critical damping ($d = 1$) offers the advantage that the movement amplitudes of the system are already equal to the absolute movement at natural frequency, but in the case of surges its natural movement decays relatively quickly. It is only the phase ratio between relative and absolute movement which is less favourable than with critical damping. The output voltage at frequencies above the resonance frequency for a constant oscillation speed is almost constant at 302 mV per cm/s ($\pm$ 1.2%) without load. If the pickup is connected to an auxiliary unit for integration, differentiation and calibration, which will be described presently, the pickup gives an output voltage of $V_{\text{out rms}} = 200$ mV per cm/s.

With measurements in the vicinity of the resonance frequency and below, a correction according to the curves in Fig. 30-12 is required. As a result of the electrodynamic damping, the influence of temperature is relatively small. These correction curves are recorded separately for each pickup, so that at frequencies > 12 c/s a measurement uncertainty $< \pm 2\%$ and below 12 c/s $< \pm 5\%$ can be guaranteed. The correction curves illustrated were recorded for pickups mounted at angles of 20° and 70° to the perpendicular. They are thus applicable for mounting positions from 0° . . . 45° and 45° . . . 90° to the perpendicular.

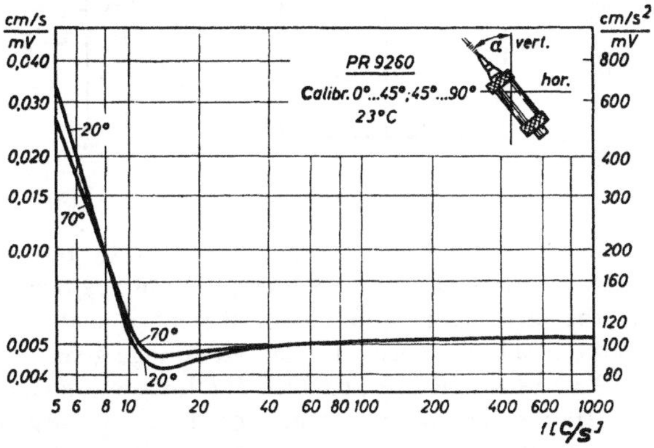

Fig. 30-12 Correction curves for the "PR 9260" Absolute pickup with scales for velocity (cm/s) and acceleration (cm/s²); calibration temperature 23 °C

This pickup is sensitive to direction, so that it can be used to measure the size and position of the resultant of the vibration of an object by recording the vibration amplitudes in terms of space coordinates. The output voltage is proportional to the cosine of the angle which the axis of the pickup makes with the direction of vibration. At right angles to the axis of the pickup it is zero.

30.3.2 INTEGRATION, DIFFERENTIATION AND CALIBRATION OF THE SIGNAL VOLTAGE

Like the other device described in this section, this pickup delivers a voltage proportional to the instantaneous velocity of the vibrating body. To obtain a voltage proportional to the vibrating path, that is, to the instantaneous amplitude of the oscillation, this voltage must be integrated by means of an *RC*-network. Fig. 30-13 shows the basic diagram of the Philips "PR 9250" auxiliary unit, which among other things can be used for integration. By means of switch S_1, the output voltage of the pickup can be fed (suitably attenuated for matching) via a *CR*-network for differentiation or via an *RC*-network integration, or directly to the oscilloscope. For selecting intermediate values of the output voltage a variable potentiometer (R_3) is provided.

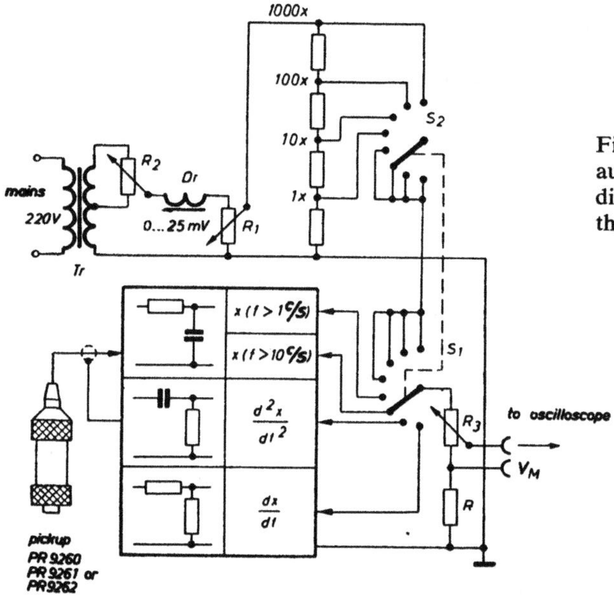

Fig. 30-13 Circuit of the "PR 9250" auxiliary unit for integration, differentiation and calibration of the pickup voltage

If the pickup current is connected to the output potentiometer via the *CR*-network (time constant 33.1 μs), the voltage occurring on the resistor is proportional to the differential quotient of the input voltage. As the input voltage corresponds to the oscillation speed, the output voltage gives a picture of the dynamic acceleration.

Two positions are provided for integration, namely for frequencies > 1 c/s and > 10 c/s. The time constants of these *RC*-networks are 0.302 s and 30.2 ms. With the larger time constant an output voltage of 1 mV_{rms} is obtained for a vibration amplitude of 10 μm, and with the smaller time constant an output voltage of 1 mV for 1 μm. If the frequency is not < 10 c/s, it is advisable to use the last-mentioned integration network. As oscilloscopes used for this purpose usually have deflection coefficients of $DC_{rms} = 1$ mV/cm ($\hat{=} DC_{pp} \approx 3$ mV/cm), 1 cm of deflection is obtained per micron of oscillation amplitude. In the case of smaller vibration amplitudes the sensitivity of indication can be further increased by using a pre-amplifier. Oscillation measurement is often used for balancing and similar tasks in which the phase of the oscillation amplitude is of importance. Fig. 30-14 shows the dependence of the phase angle of the pickup voltage upon the frequency, and Fig. 30-15 the frequency dependence of the amplitude and phase errors of the integration network. As indicated in Fig. 30-14, a movement of the pickup in the

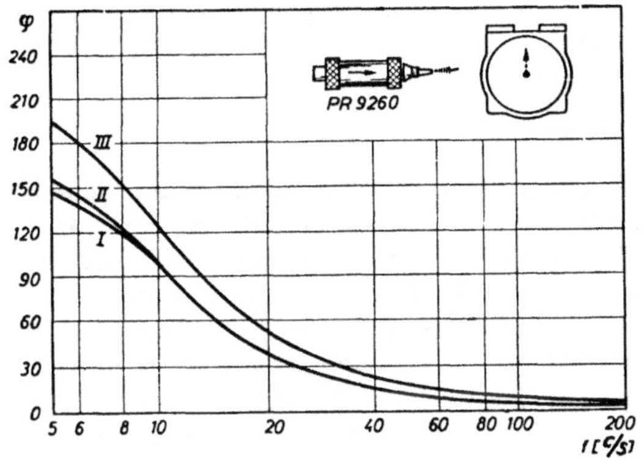

Fig. 30-14 Dependence of the "PR 9260" pickup output voltage phase upon frequency. I pickup unloaded or in dx/dt position of the "PR 9250" auxiliary unit; II position x ($f > 1$ c/s) of the auxiliary unit; III position x ($f > 10$ c/s) of the auxiliary unit.

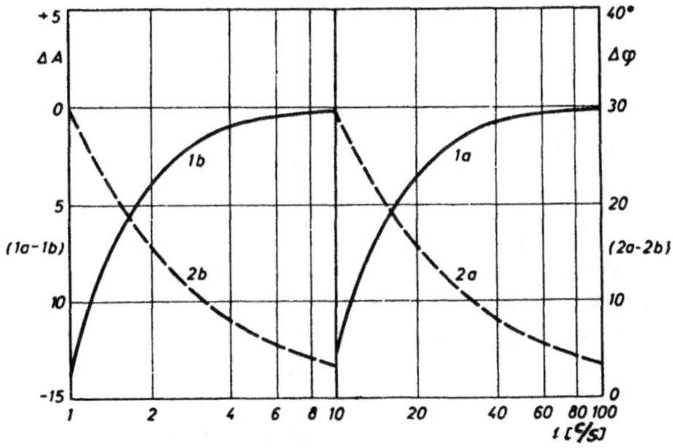

Fig. 30-15 Amplitude error (1) and phase error (2) of the integration elements in the "PR 9250" auxiliary unit; *a*) switching position x ($f > 10$ c/s); *b*) switching position x ($f > 1$ c/s)

direction of the connection contact results in a positive-going output voltage and thus an upward deflection of the spot on the oscilloscope.

In measuring vibration acceleration, errors of amplitudes and phase are small enough to be negligible. A 1 mV output voltage is obtained at an acceleration of 100 cm/s^2.

For determining the output voltage amplitude (or the vibration speed) with a Philips "PR 9250" auxiliary unit, the unit can be connected to a 50 c/s alternating voltage from the lighting mains via switch S_2 (Fig. 30-13), which is coupled to S_1. This voltage can be adjusted to peak values of 1 mV to 25 V (the calibration corresponds to the rms values $\times \sqrt{2}$). The adjustment is carried out in steps by means of switch S_2, and for the intermediate values by means of the potentiometer R_1, the knob of which has a large scale dial. By comparing the deflection with an equal calibration voltage, the voltage delivered by the pickup via the auxiliary unit can be determined immediately and thus also the amplitude of the vibrations.

A marking voltage can be connected to the resistor R to mark a given phase when using balancing phase indicator "PR 9280", or as a time-marking. In this way, for instance, the magnetic pickup "PR 9262" can be used to add a pulse to the signal voltage at the dead centre adjustment of a machine or other point in the revolution. The oscillogram will thus have distinct voltage peaks which indicate these instants.

If the voltage from the "PR 9262" pickup is too small it can be simply amplified by means of one of normal preamplifiers such as the "GM 4574". It is also possible to apply a voltage pulse via a contact actuated by a cam.

30.3.3 EXAMPLES OF APPLICATIONS AND OSCILLOGRAMS

The measuring certainty, especially at frequencies of several hundred c/s, depends on how well the mechanical connection between pickup and oscillating part can be made. If the pickup is merely pressed manually on to a massive steel object, the output voltage at 250 c/s may be about 4% too high, and at 400 c/s about 10% too high. If, on the other hand, the pickup is screwed by its base firmly on the vibrating object, errors of this magnitude result only if the frequency is doubled. These discrepancies are due to the fact that the mechanical junction with the mass of the pickup is more or less elastic and thus forms an oscillating system with its own natural frequency known as the "contact resonance frequency" or "contact resonance" for short. For accurate results, this frequency must be increased to at least five times the highest measuring frequency by making an inelastic contact between the pickup and the oscillating body. If the pickup is firmly fixed, the contact resonance frequency can be raised from 1000 c/s to 3000 c/s. If the measuring frequency is 1/3, 1/4, 1/5 or 1/10 of the contact resonance, the "build-up error" becomes 12.5%, 6.7%, 4.2% or 1%.

Oscillograms *a* and *b* in Fig. 30-16 were made with this pickup to study the vibration of a machine-shop floor when subject to shocks. Apparatus for taking moving pictures was not available so that the various observations had to be recorded singly. They are reproduced here side by side. The waveform of a 50 c/s voltage was recorded simultaneously as a time scale. These patterns show beats of the decaying vibration oscillations, which are due to the oscillations being reflected by the walls and interfering with the "primary" floor oscillations at the measuring point. This method can be used for the dynamic study not only of buildings but also of road surfaces [5] [6].

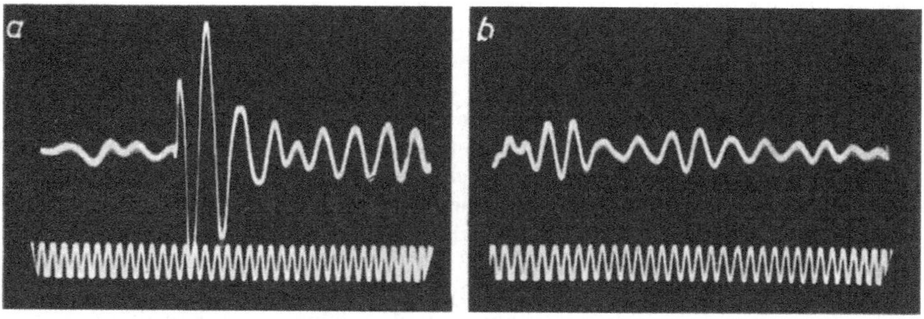

Fig. 30-16 *a*) and *b*) Oscillations of a building floor after a jolt

The moving film records in Fig. 30-17 are a further example and represent the bearing vibrations of a small motor when starting up and stopping. In oscillogram *a*, for which a slow film feed speed was used, both the starting and stopping can be

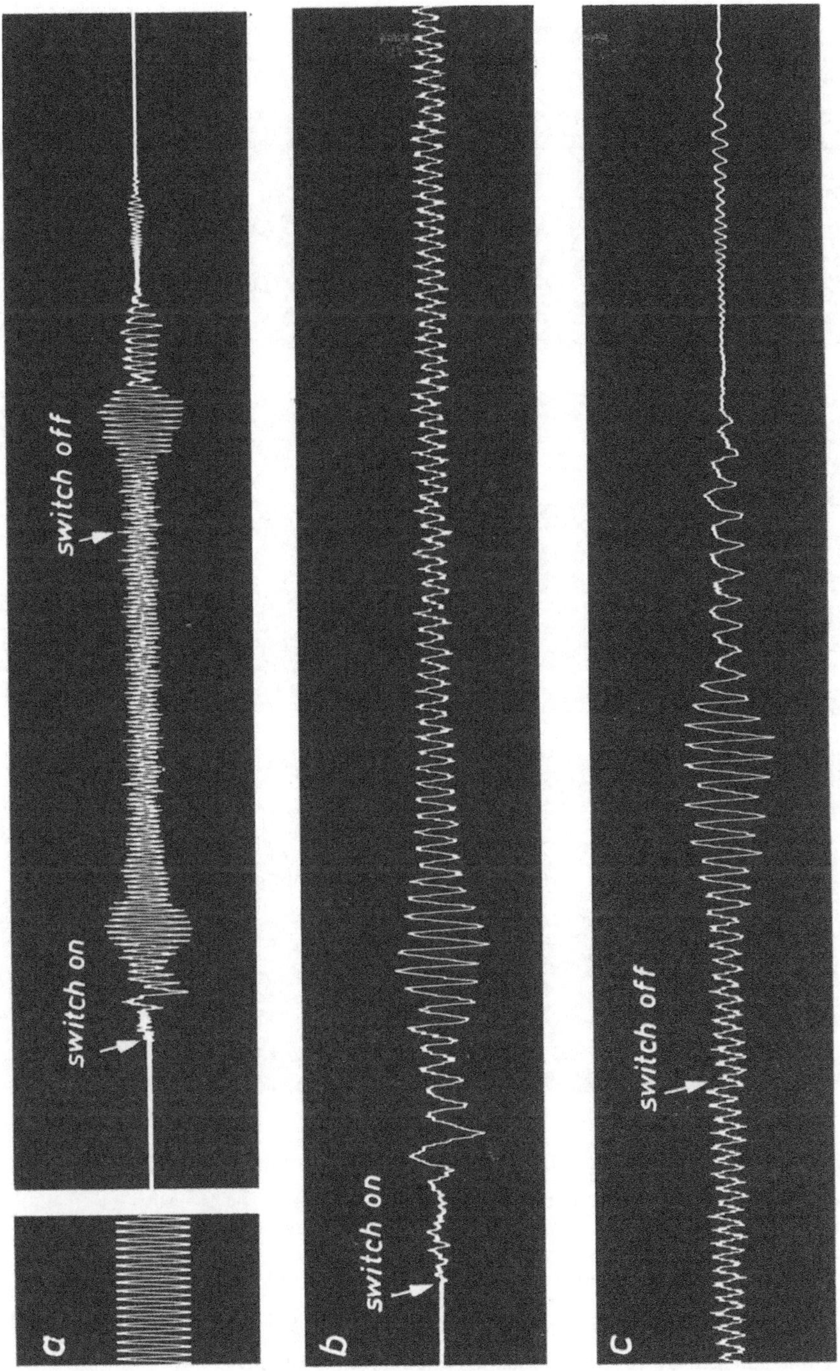

Fig. 30-17 Continuous recording of bearing oscillations of a small motor when starting and slowing down. *a*) Transport speed 2 cm/s; *b*) starting; *c*) slowing down [transport speed for *b*) and *c*) 6 cm/s]; before *a*) at left-hand side, calibration 40 μm

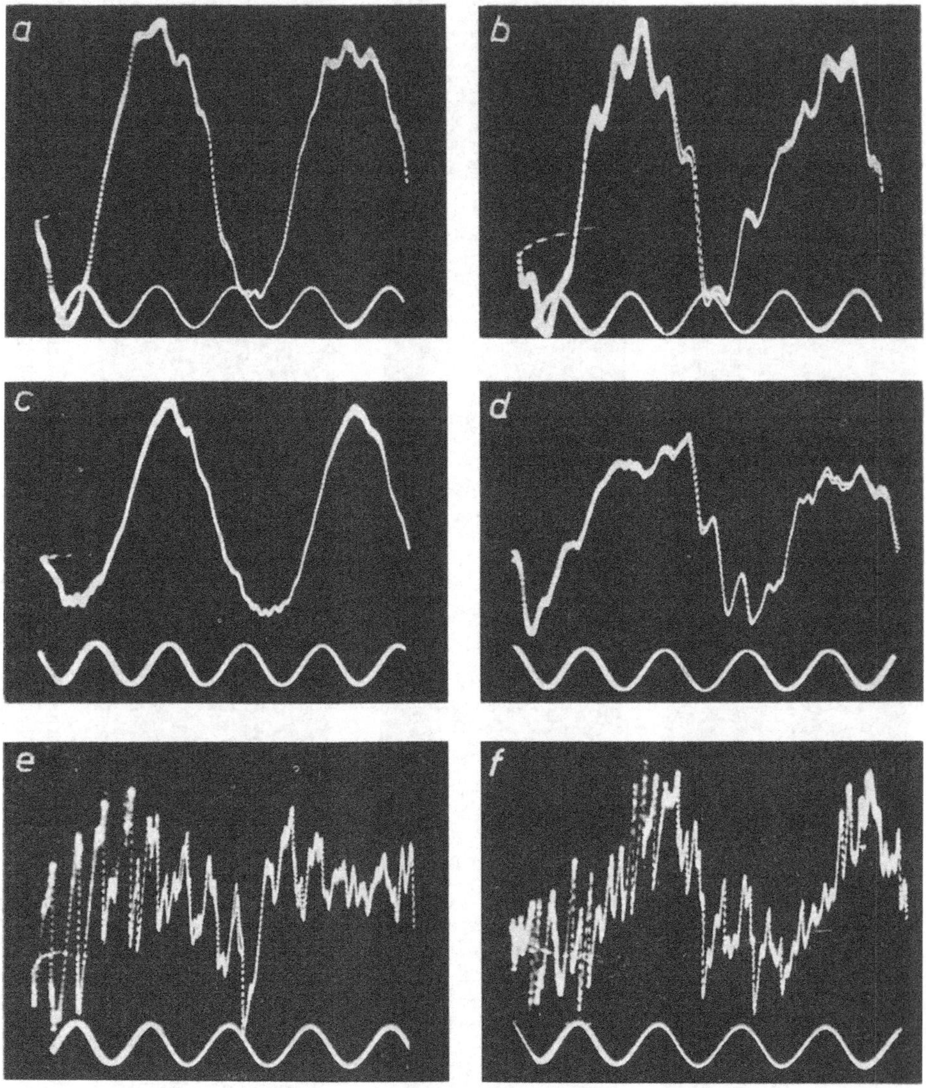

Fig. 30-18 Bearing oscillations in a Diesel installation with compressor.

a) Generator on full load, with compressor, amplitude 190 μm;
b) generator on full load, without compressor, amplitude 63 μm;
c) no-load running with compressor, oscillation amplitude 180 μm;
d) no-load running without compressor, oscillation amplitude 40 μm;
[for *a*) to *d*) $x = f(t)$]; *e*) as *c*); *f*) as *d*) [for *e*) and *f*) $\frac{dx}{dt} = f(t)$]. The most favourable picture height was adjusted by varying the gain. Time marking: 50 c/s.

clearly seen with resonance occurring at the critical speed. It is impossible to determine accurately the critical speed in this oscillogram as the individual oscillation cycles are too close together. In *b* and *c* the speed at which the film was fed was trebled; the critical speed is seen to be 1300 rpm (21 c/s). Oscillogram *a* was used for amplitude calibration and corresponds to 40 μm.

The oscillograms in Fig. 30-18 are of the vibrations of the bearing of a ship's auxiliary diesel generator. A compressor was also coupled to this unit. The unit had no special bearing between diesel and generator, so that the study of the bearing vibrations was of particular interest for investigating balance problems. It can be clearly seen from the recordings that the bearing vibrations increased considerably when the compressor was connected.

30. 4 Electrodynamic pickup for measuring relative vibrations

While the previously described pickup with its seismic coil system is intended for the measurement of the absolute value of mechanical vibrations, the Philips "PR 9261" pickup is mainly intended for the study of the vibrations of a part relative to a reference object. Its construction, shown in section in Fig. 30-19, is similar to the absolute pickup "PR 9260", which is shown in section in Fig. 30-10 and in photographs in Fig. 30-11.

In this pickup, the tracer stylus, carried on loose bearings and connected to the coil system, is pressed forward by diaphragm springs. When measuring a relative oscillation the pickup must be fixed to the larger part and the tracer stylus must be pressed against the oscillating part so that the flexible system is, as far as possible, situated centrally between the two. Depending on the material against which the stylus is pressed, a suitable cap or roller can be placed on its tip so that sensitive surfaces and revolving parts are not scratched, but the resonant contact frequency remains sufficiently high. The tension of the springs is about 850 g, the weight of the flexible system only about 10 g. The mass of the cap must of course by added to this.

The maximum acceleration at which the movement system of the scanned part can follow, is 85 g, i.e. about 1000 cm/s^2. Again, as with the absolute pickup, the voltage generated is proportional to the vibration speed. The auxiliary unit "PR 9250" can also be used to obtain, by integration, a voltage proportional to the instantaneous amplitudes, or, by differentiation, a voltage proportional to the acceleration. The sensitivities are then the same as those given for the "PR 9260" pickup. Here, too, the maximum amplitude is $\pm$ 1 mm.

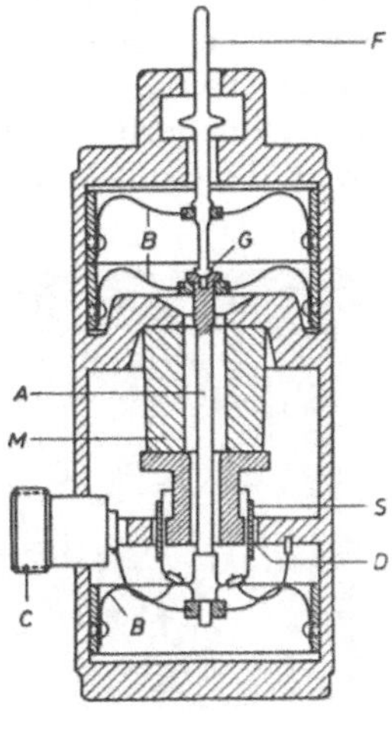

Fig. 30-19 Longitudinal section of "PR 9261" relative pickup. *A* is the carrier pin for moving system; *B*, diaphragm springs; *C*, connection; *D*, air gap; *F*, pickup pin; *G*, ball-bearing connection of pickup pin to system pin; *M*, permanent magnet; *S*, oscillation coil

Fig. 30-20 Recording the oscillations of a lathe with the relative pickup

Every pickup is supplied with detailed instructions for use. An example of a possible application of the pickup can be seen in the arrangement shown in Fig. 30-20, for the study of the vibration of a lathe tool with respect to its clamping. The "PR 9261" is screwed to a steel plate on the lathe bed.

Oscillograms *a* and *b* of Fig. 30-21 show recordings of vibration speed and vibration amplitude respectively. The pattern of a voltage of 250 c/s was traced under this curve as a time-marking, using an electronic switch for the purpose. The trace of a 50 c/s voltage with an amplitude of 5 mV has been added for calibration of the vertical deflection. SALJÉ has published a detailed report on a study of machine tools using these oscillation pickups [7].

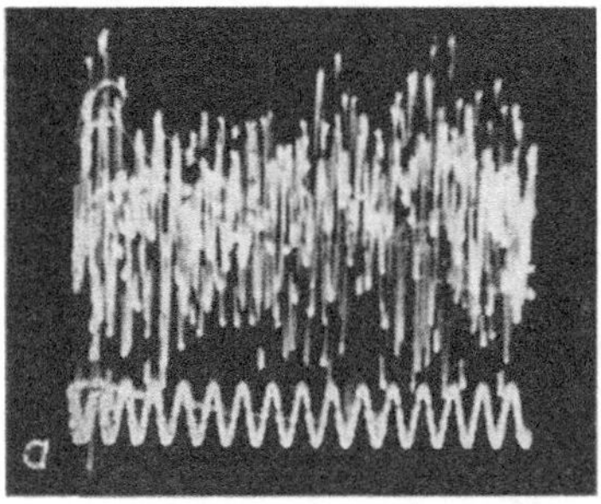

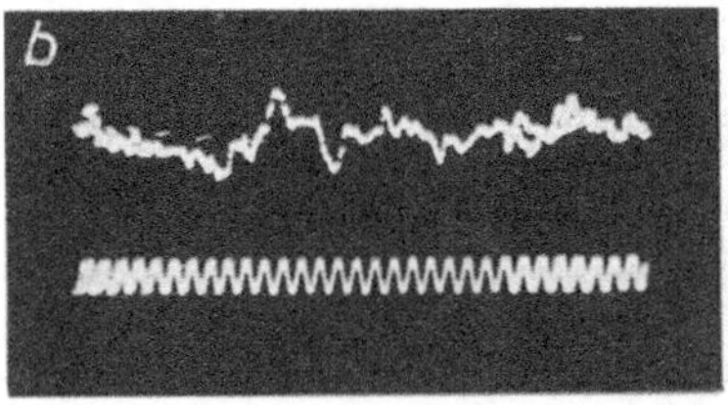

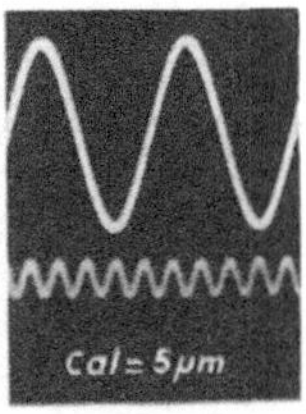

Fig. 30-21 Lathe oscillations. *a*) Speed; *b*) amplitude [for *a*) and *b*) thrust 0.28 mm/s, calibration frequency 250 c/s, amplitude calibration 5 mV]

This pickup can also be used to excite mechanical oscillations. At a maximum sinusoidal current of 20 mA_{rms} a maximum force of 87 g is obtained, which is sufficient power to excite small oscillatory systems at their resonance frequency and, given low damping, larger mechanical oscillation systems also.

The relatively high impedance of the "PR 9261" pickup rises steeply with the frequency, so that — when the supply voltage is constant — the exciting current and the force exerted decrease. For use as an oscillation exciter, a special model, "PR 9271", is made which can supply a current of 60 mA_{rms} max. at an impedance of about 500 Ω. This current can be maintained at frequencies up to 16,000 c/s.

For more stringent requirements there are oscillation exciters available developing forces up to 3 kg ("PR 9270"), and even several hundred kilograms. Thus, even a very large object (an aircraft) can be excited to oscillation and its behaviour studied by means of pickups and oscilloscopes.

CHAPTER 31

STUDY OF DYNAMIC STRAIN PROCESSES AND OBSERVATION OF MECHANICAL VIBRATIONS BY MEANS OF STRAIN GAUGES

31. 1 Strain gauges

Because of their great range of application, strain gauges are being used more and more as pickups. The imperfection of early forms led to their being used at first for orientation testing rather than for measurement proper. Thanks to improvement in manufacture it is now possible to use strain gauges, for example, in load cells for weighing purposes, measuring uncertainty of only a few parts per thousand being achieved.

As far back as 1938/39 Philips supplied a combination of materials known as special material "GM 4470", which could be used to make strain gauges in the form of layer resistors [1]. The actual resistor strip made of aquadag (a colloidal solution of graphite) was "painted" on to a strip of pressboard with a drawing pen. The outer surface of this carrier strip was then metallized. Narrow strips with the aquadag resistor in the middle were cut crosswise from the pressboard, and leads, coated with the same metal — the bronze lacquer used at the time for screening the bulbs of radio valves — were affixed. This strip was glued to the object under study in a way similar to that to which, as will be described in due course, modern strain gauges are mounted. The changes in resistance according to strain and deformation of the object were studied. Of course, only very modest demands as to measuring certainty and reproducibility could be made on a strip of this nature. Its function was, in fact, similar to that of a carbon microphone. Similar strips were supplied by the firm of AEG [2]. Thanks to SIMMONS and RUGE, strain gauges began to be used in the U.S.A. in 1939. The change in the properties of a resistor wire of these strain gauges was used as the basis of the measurement.

The manufacture of such strain gauges was taken up by European firms in 1945, and they were used in this continent on a large scale. The basic construction has been retained to date, although the wire material, measuring element carrier (paper or plastic) and adhesive have all been much improved. A gauge consists fundamentally of a resistor wire glued to a carrier base -usually paper- in loops, zig-zags or spirals. The sketch in Fig. 31-1 shows the most common type with the thin resistor wire (1) stuck on the paper carrier (2) in the form of a zig-zag. The connecting wires (3) are soldered to the ends of the resistor wire and are also anchored with glue to the paper carrier to relieve the strain on the soldered joints (4). The resistance of this gauge is usually 600 Ω, but there are also smaller gauges of 300 and 120 Ω. The dimensions of the paper carrier are usually 29 × 14 and 20 × 8.5 mm, those of the actual gauge surface usually between 8 × 6 and 4 × 3 mm [3].

The gauge is carefully glued with a special adhesive to the surface of the object the mechanical strain in which is to be studied. In order to obtain accurate measure-

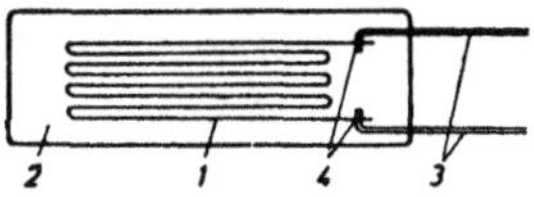

Fig. 31-1 Structure of a strain gauge

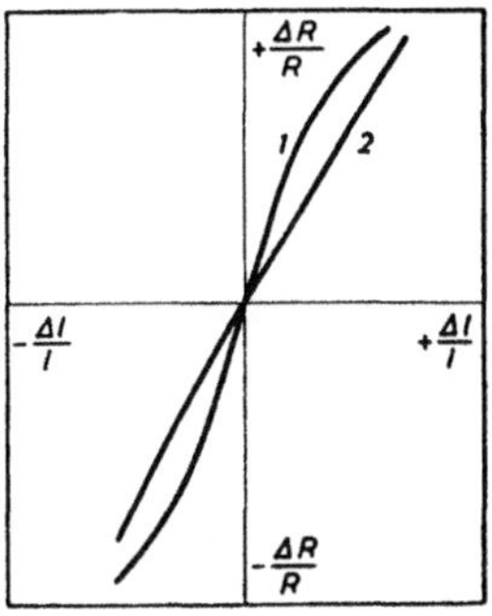

Fig. 31-2 Dependence of the relative change in resistance upon the relative change in length of resistor wires. 1 Chromenickel wire; 2 Konstantan

ments it is essential for the gauge to be completely and homogeneously bonded without air-bubbles to the object under study. The instructions for bonding, which is described in detail in the literature on the subject, should be studied and followed meticulously [4] [5].

These requirements mean that the gauge cannot be removed from its base nor be used for more than one set of observations.

It should be noted that the insulation resistance of the gauge to the base and also between the two connecting wires and the base must be at least 1000 MΩ, to avoid grave errors in measurement due to irregular fluctuations of the insulation resistance. The insulation resistance should be measured with a not too high voltage (about 20 V).

After sufficient drying, all the deformations of the object under test are communicated, via the adhesive layer and the paper base, to the resistor wire, which follows the deformations faithfully. The variation in length of the resistor wire produces a variation in resistance, and within the working range this is exactly proportional to the change in length. It should be noted particularly that these strain gauges follow cyclic loads accurately up to about 50,000 c/s.

The ratio of the resistance change to the corresponding change in length is generally denoted the *k* factor of the strain gauge. It is found from the following equation:

$$k = \frac{\Delta R}{R} : \frac{\Delta l}{l} . \qquad (31.1)$$

Its value is usually about 2. Fig. 31-2 shows the dependence of the change in resistance upon the elongation for two different kinds of wire. The slope of the curve corresponds to the *k* factor. In the case of constantan (2) it is obviously constant over a far greater range than chrome-nickel wire (1). Furthermore, the temperature coefficient of the former is lower than that of the latter. This is important, because every resistance variation in the measuring arrangement is interpreted as a change in length.

Normal strain gauges can be used at temperatures up to 70 °C, but these have to be "baked" on to the object under study with a special adhesive.

31. 2 Relative elongation, material tension and dynamic measurements using strain gauges

The total resistance change as a result of a given elongation can be regarded as a mechanical tension σ occurring during measurement, its value being given by the equation:

$$\sigma\ [\mathrm{kg/mm^2}] = E \cdot \frac{\Delta l}{l} = E \cdot \frac{\Delta R_{cs}}{k \cdot R_{cs}}, \tag{31.2}$$

when E represents the modulus of elasticity. This makes it possible to use strain gauges in every problem of strain measurement [6].

The greatest permissible elongation is about 0.5% or 5.10^{-3}. At elongations of this order the hysteresis can still be kept sufficiently small if the gauge is subjected before the measurement to repeated maximum pre-loading. In the case of smaller elongations, corresponding more closely to practical applications, the hysteresis remains negligibly small [7]. FINK and ROHRBACH have dealt critically with the various aspects of this in a publication [8].

The simplest circuit which can be used to observe the change in resistance of a strain gauge is shown in Fig. 31-3. From a battery of voltage V_B a constant current of 10 mA (max. 25 mA) flows through a series resistor R_v and the strain gauge. A voltage drop occurs across the strain-gauge resistor, its value being:

$$V_s = V_B \cdot \frac{R_s}{R_v + R_s}. \tag{31.3}$$

In the dynamic changes of the length of the object under investigation, the voltage drop in the gauge changes by the slight amount dV_s. By differentiation of Eq. (31.3:)

$$dV_s = V_B \cdot \frac{R_v \cdot R_s}{(R_v + R_s)} \cdot \frac{dR_s}{R_s}. \tag{31.4}$$

The following applies, however, for the relative change in resistance of the strip:

$$\frac{\Delta R_s}{R_s} = k \cdot \frac{\Delta l}{l} = k \cdot \frac{\sigma}{E}. \tag{31.5}$$

The voltage change dV_s (or ΔV_s) is proportional to the mechanical tension occurring in the material, provided this is within the limits of elasticity of the object. The sensitivity of this measuring process, i.e. the change in output voltage per unit of elongation (e.g. 1% of elongation), is determied by the factor k and the maximum permissible voltage for the strain gauge given by the product $R_s \cdot I_{s\max}$.

For example, with a series resistor R_v of 1800 Ω and a gauge resistance of 600 Ω, and using a battery voltage of 45 V, a current of 18.8 mA is obtained, which is permissible for a limited period of time. Thus the sensitivity of the circuit at a k factor of 2 is about 17 mV for 1‰ elongation. As the relative changes in length are usually very small, an elongation $\Delta l/l \cdot 10^{-6}$ is called a micro-strain (μS).
1‰ strain thus corresponds to 1000 units of micro-strain (1‰ = 1000 μS)

A voltage change of 1.7 mV gives a deflection of 3.4 cm on a "PM 3236" oscilloscope when the deflection coefficient is 0.5 mV/cm. As, however, the material tensions are generally much less than 1000 μS, an attempt must be made to make the circuit more sensitive.

The best solution is the bridge circuit, as shown in Fig. 31-4. If only one strip is used as an "active" gauge which experiences the mechanical changes of state, the voltage across the diagonal branch of the bridge is:

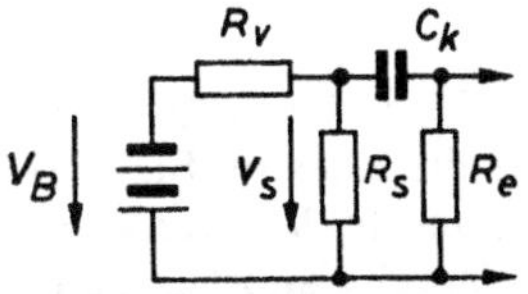

Fig. 31-3 Simple circuit for measuring dynamic strains without static components

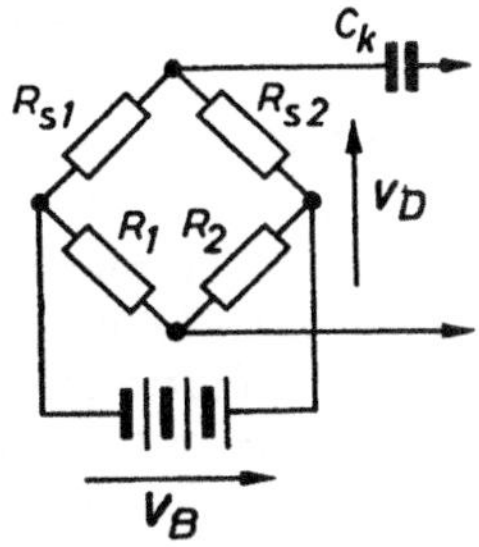

Fig. 31-4 Bridge circuit for measuring dynamic strains with several active strain gauges to increase sensitivity

$$V_D = \frac{V_B}{4} \cdot \frac{\Delta R}{R} . \tag{31.6}$$

If the other bridge resistors are also rated at 600 Ω, and the supply voltage is 20 V, the current through the resistors is about 16 mA, and the sensitivity of the bridge for $1^0/_{00}$ is 10 mV. However, one can usually affix two gauges on opposite sides of parts under such mechanical stress, so that one is elongated and the other compressed. If these gauges, R_{S1} and R_{S2}, are arranged in the bridge circuit, as shown in Fig. 31-4, the effects due to the change in resistance, are additive, i.e. the sensitivity is doubled; in the example quoted it is 20 mV for $1^0/_{00}$. With 100 μS a voltage of 2 mV is obtained, giving a 4 cm deflection on the "PM 3236" oscilloscope.

It is sometimes even possible to use four gauges in such a way that two are elongated and two compressed by the mechanical stress. If they are connected in the bridge circuit so that opposed changes occur each time in the other branches, the bridge sensitivity is then:

$$V_D = V_B \cdot \frac{\Delta R}{R} . \tag{31.7}$$

It is therefore four times greater as compared with that of Eq. (31.6). In the example given, 4 mV per 100 μS was obtained and thus a deflection of 8 cm per 100 μS oscilloscope screen. As the stress on the material is often of the order of several hundred microstrains, an oscilloscope such as the "PM 3236" can well be used alone in this circuit. Such a circuit is usually known as a "full-bridge circuit", while those with two gauges — both or only one active — are known as "half-bridge circuits".

In the circuit shown in Figs. 31-3 and 31-4 a coupling capacitor (C_k) is connected to the output of the signal voltage. An output voltage is thus obtained representing only the *change* of the variable of state under examination. The frequency of this change must be higher than the lower limiting frequency of the *CR*-network formed by C_k and the input resistance of the measuring circuit (R_{in}). If $C_k = 10$ μF and $R_{in} = 1$ MΩ, the cut-off frequency is 0.2 c/s. Frequencies > 1 c/s are thus transmitted relatively well. In the circuit shown in Fig. 31-3 the interposition of a capacitor is certainly needed as the voltage change of 17 mV as against the quiescent voltage drop of 11.3 V is very slight. It would be impossible, for instance, to connect the voltage from the strain gauge direct to the input of an oscilloscope via a DC amplifier, as the shift of the working point of the amplifier by the quiescent voltage would be much too great. This could be avoided only by a compensating circuit, but this would be to all invents and purposes a bridge circuit. This will be dealt with in more detail in a later section.

31. 3 Measurement of dynamic strain without static components

Bridge circuits are usually employed for this purpose. Fig. 31-5 shows the circuit and Fig. 31-6 an exterior view of such a unit constructed on the lines of Fig. 31-4. Resistor R_2 is identical with R_2 in Fig. 31-4, $R_3 + R_4$ correspond to R_1. Resistor R_6 limits the current in the event of sockets 1 and 3 being accidentally short-circuited. If only one active gauge is employed, switch S_1 is turned to position 1. Resistor R_1 is then the fourth bridge resistor. If two strain gauges are available, of which one could be passive or the two used in opposition to each other, switch S_1 should be placed at position 2. To calibrate the circuit, a key is used to connect a resistor R_p in parallel for a short time with the partial resistor R_4 of the one branch of the bridge, so that a reading corresponding to a strain of $1^0/_{00}$ on the active gauge is obtained.

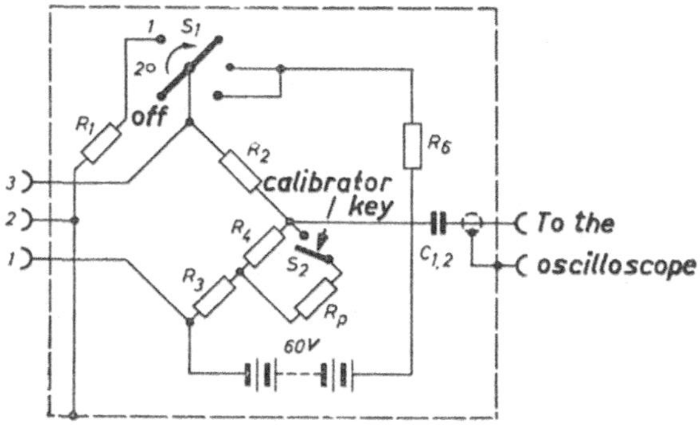

Fig. 31-5 Practical unit circuit according to the principle of Fig. 31-4

Fig. 31-6 Exterior of the Philips unit "PT 1205" according to the layout in Fig. 31-5

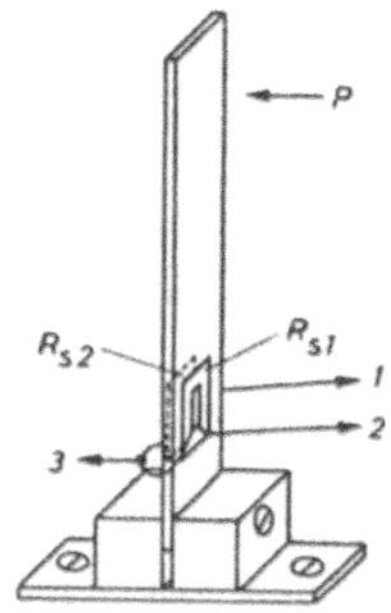

Fig. 31-7 Arrangement for recording the mechanical properties of a sample of band steel

The examination of a piece of steel strip (150 × 15 × 2 mm) as to resonance frequency and damping will be described as an example of such a measurement. The piece of steel was clamped as illustrated in Fig. 31-7. Two strain gauges were affixed at the foot, corresponding to gauges R_{S1} and R_{S2} in Fig. 31-4. These gauges were connected to the "PT 1205" strain measurement unit, and the diagonal voltage of this to a "GM 5666" oscilloscope. The oscilloscope was set for triggering operation, so that no pattern is visible on the screen at first. If the leaf spring is excited by a blow or by plucking, it oscillates at its natural frequency. A voltage change in the form of a decaying oscillation occurs in the bridge circuit. Its first peak triggers the time base and the unblanking of the oscilloscope.

Oscillograms obtained in this way are shown in Fig. 31-8. If the slip is excited by a blow (*a*), the first oscillation cycle is somewhat distorted due to cushioning by the striking tool, but if the strip is pulled back and suddenly released (*b*), the piece of steel oscillates freely. A 10 mV amplitude-calibration voltage of the oscilloscope (a clipped 50 c/s voltage) was used to record a time scale (*c*). The natural frequency of the sample was found to be 54.5 c/s. In a further oscillogram (*d*) the deflection for 1⁰/₀₀ strain was obtained by repeatedly depressing the calibration key (S_2 in Fig. 31-5). As a half-bridge circuit was used with the two gauges in opposition, this calibration corresponds to 0.5⁰/₀₀ strain.

From the patterns of the decaying oscillation according to Ch. 22, 7 cycles can be counted before the amplitude has dropped to half, i.e. the damping of the material is $d = 0.22/7 = 4.1\%$.

The principle of measurement described here can be used for innumerable tasks [9] [10] [11].

31. 4 Measuring dynamic strain with static components

In the various tasks encountered in practice it must often occur that, in contrast to the previously described observations, the static components of the strain must also be shown. This can be done using a unit based on the circuit shown in Fig. 31-5 in conjunction with an oscilloscope having a highly sensitive DC amplifier. The capacitor C_k must be short-circuited. Furthermore, the bridge must be balanced accurately to zero voltage by means of the diagonal supply before the measurement

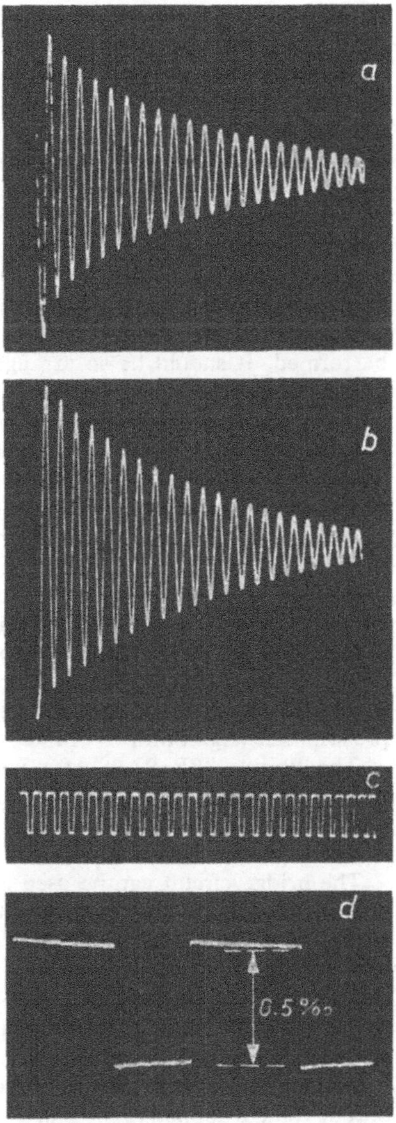

Fig. 31-8 Oscillograms obtained with the arrangement in Fig. 31-7 after the free end of the band steel has been subjected to a sudden jolt.
a) On being struck; *b*) after plucking; *c*) time calibration 50 c/s; *d*) amplitude calibration (0.5⁰/₀₀ strain), resonant frequency 54.5 c/s at dimensions of 150 × 15 × 2 mm

is made. As already described, with two active gauges in a half-bridge circuit at 20 V bridge voltage, a deflection of 1 cm is obtained for 150 micro-strain. It should be pointed out, however, that in observing the static components it is essential that the circuit elements should be of high stability. (The "PT 1205" unit was actually not intended for such measurements.) The strain gauges too, must be affixed so as not to cause interfering resistance deviations due to changes in temperature. The steps which have to be taken are described in the literature on the subject [4] [5] [7]. If only one active gauge can be used, it is advisable to use an identical strip in the opposite branch as a bridge resistor. This must be glued on to the object under study in such a way that it undergoes no mechanical stresses but does undergo the same temperature changes as the active strip.

As already mentioned, a complete bridge consisting of four gauges can also be formed. It should be so arranged that the gauges which are elongated and compressed respectively are included in other branches. As all four are subject to the temperature change simultaneously, the temperature influence cancels itself out. A condition is, of course, that temperature influences do not produce additional mechanical stresses in the object under study, as these would naturally find expression as changes in the resistance of the gauges and result in shifting the zero point.

So long as it is a matter of observing a few hundred micro-strains, a gain of 8000 to 10,000 is sufficient to enable the diagonal voltage of the bridge to produce a useful deflection on the oscilloscope screen. For higher sensitivity requirements the gain would have to be at least 3 to 5 times greater or, better still, one order of magnitude higher. DC amplifiers capable of such high gain and of sufficient stability, will no doubt be produced at some future time. It is therefore necessary either to limit investigations to the observation of the changes of state or to connect a suitable AC preamplifier (e.g., Philips "PM 4574") in series, or to use a carrier frequency amplifier.

The Philips "PR 9300", for instance, is such a bridge with a carrier frequency measuring device, the basic diagram of which can be seen in Fig. 31-9, the supply unit being omitted for the sake of simplicity. The exterior of this strain-measuring bridge is shown in Fig. 31-17, the second unit from the right.

The bridge circuit can be used either for one active gauge or for two (half-bridge circuit). The strain gauges — or the active strips and the temperature compensating strip — are connected to sockets 1′, 2′ and 3′ at A. The circuit is supplied at point *G* with a 4 kc/s alternating voltage via sockets 1, 2 and 3 from a stable valve oscillator (7). The balance of the bridge with the other two branches is not obtained by means of resistors but with a differential variable capacitor (C_1). The other differential variable capacitor (C_2) levels out capacitive asymmetries. For this purpose switch S_3 is set to position *P* when the phase of the amplified voltage is shifted by 90°. The output voltage applied to the phase-sensitive rectifier then corresponds to the capacitive component of the bridge asymmetry.

The diagonal voltage of the bridge is amplified by a particularly stable AC amplifier of high gain (about 40,000 times, valves 1, 2, 3 and 4). Their output voltage is fed via bandpass filter (*BP*) (± 1 kc/s bandwidth) to a phase-sensitive demodulator consisting of a valve (5) and a rectifier circuit (*D*), and also to output sockets ("carrier frequency", 7 and 8). The rectified current flows to the measuring instrument (*M*) and to a low-pass output filter (*LP*), which passes only a voltage whose frequency corresponds to the variable of state, but not the 4 kc/s carrier frequency. From the terminating 1 kΩ resistor a direct voltage is obtained at the "modulation"

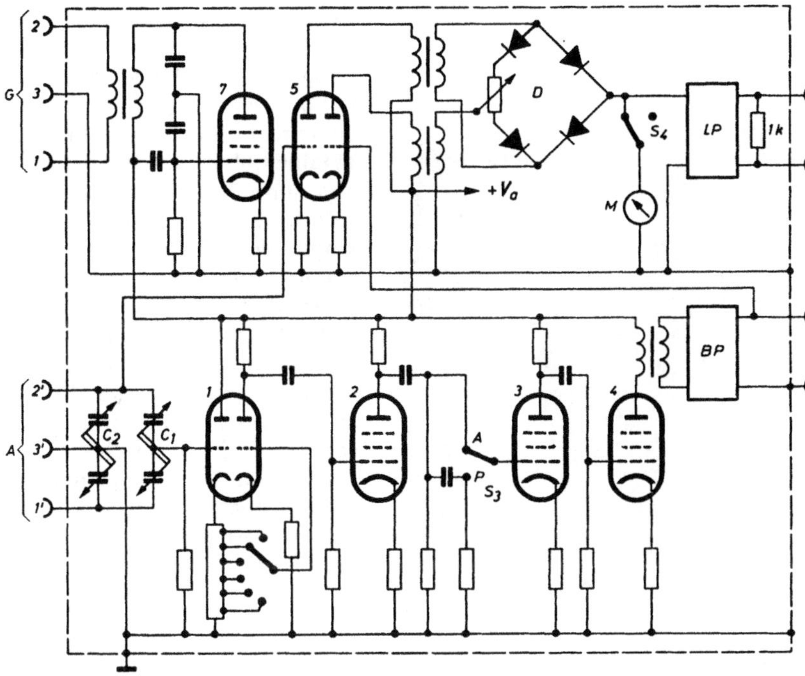

Fig. 31-9 Fundamental circuit diagram of the Philips "PR 9300" carrier frequency measuring bridge

sockets (5 and 6) which is directly proportional to the condition under investigation.

Static and slowly changing magnitudes can be read off the instrument by means of a large dial (Fig. 31-17). The maximum sensitivity of this bridge corresponds to full deflection for 300 micro-strain. One degree of the scale thus corresponds to 10 μS of the active gauge. If a half-bridge is used with two opposed active gauges, the smallest measuring range is 150 units and 1° is equal to 5 μS. When investigating dynamic changes of state the oscilloscope can be connected to the "modulation" sockets (5 and 6) as well as to the "carrier wave" sockets (7 and 8). As will be seen in the oscillograms in Figs. 31-12, 31-13 and 31-16, either a direct picture of the waveform of the change of state is obtained on the oscilloscope screen — an oscilloscope with a DC amplifier is required for this, however — or a pattern of the carrier voltage modulated by the change of state. Any oscilloscope with an AC amplifier whose upper frequency limit lies above 4000 c/s can be used for the latter.

In order to satisfy higher demands as to measuring sensitivity and for more diverse applications, the Philips "PT 1200" measuring bridge was evolved. An exterior view is shown in Fig. 31-10. The sensitivity is three times that of the unit previously described, full deflection corresponding to 100 μS. With a carrier frequency of 6000 c/s it allows of observation of elongation changes at frequencies up to 1200 c/s. In the input circuit which is also suitable for "full bridge circuits" (four active strain gauges),

Fig. 31-10 Philips "PT 1200" carrier frequency measuring bridge

the voltages of the gauge bridge are fed via a special circuit to a bridge circuit of much higher resistance. This not only enables perfect balance of the combined strain gauge bridges to be obtained, but also greatly facilitates and simplifies the process of bridge balancing. The state of balance of amplitude and phase can be observed simultaneously on a "magic eye".

Another development, the Philips "PR 9302" strain gauge measuring bridge, has a maximum sensitivity of measurement of 0.6 μS. It represents at the present time the limit of what can be achieved in this respect. Its application in conjunction with an oscilloscope is on the same lines as already described for the observation of dynamic changes in strain.

31. 5 Examples of oscillograms recorded during the investigation of the movement of a leaf spring and pressure roller on an eccentric cam drum

Among the many possible applications of the measuring processes using strain gauges, the observation of large mechanical displacements has achieved considerable importance [12].

As an example and description of the possibilities for such observations in conjunction with an oscilloscope, the oscillograms in Figs. 31-12, 13 and 14 are given. Here is recorded the movement of a pressure-roller (Fig. 31-11) which is pressed

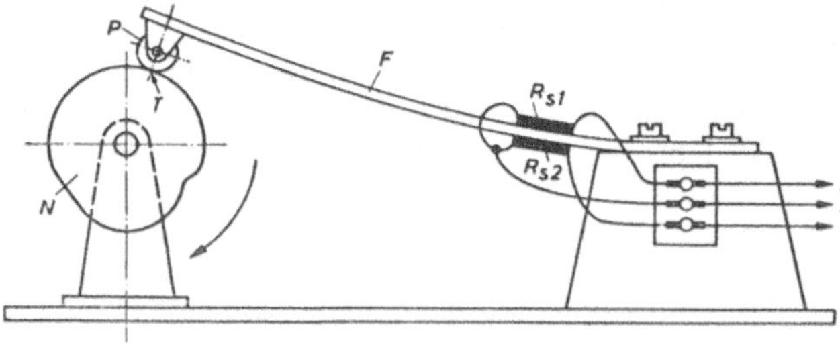

Fig. 31-11 Model for demonstrating the measurement of large mechanical dislocations with strain gauges

against a cam drum by a spring and thus displaced. (This arrangement is also used to demonstrate the use of strain gauges in large displacements.)

On rotation of the cam, which is coupled to the shaft of an electric motor supplied by a variable transformer for maintaining the desired motor speed, the spring F goes through movements corresponding to the changes in distance between the pheriphery of the cam drum and the centre of its spindle. Hence, one of the two strain gauges R_{S1} and R_{S2} is elongated and the other compressed, and the variation of their resistance is a measure of the distance from roller R to the centre of the cam drum spindle.

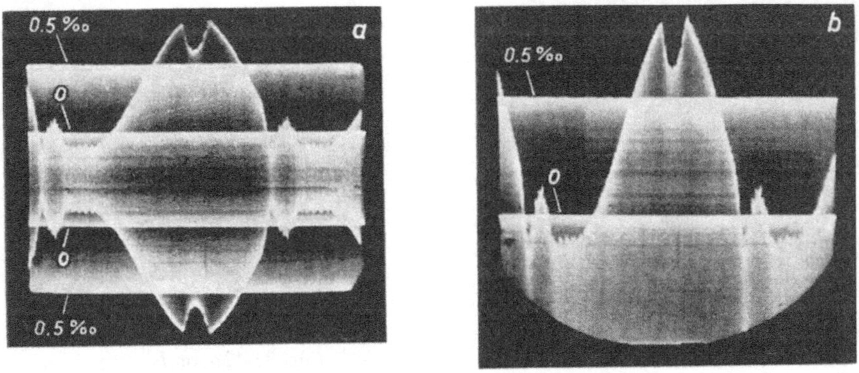

Fig. 31-12 Carrier oscillograms obtained with the arrangement in Fig. 31-11. *a*) Normal adjustment; *b*) vertical deflection doubled as opposed to *a*) and middle shifted to lower edge of the screen

For the oscillograms in Figs. 31-12*a* and *b*, a "GM 5656" oscilloscope was connected to the "carrier voltage" output of a "PT 1200" measuring bridge. This gives the pattern of the carrier voltage modulated by the extension of the spring F. The point in Fig. 31-11 marked T serves as the reference point. If the cam drum is stopped at this point, the reference zero point (0) in the oscillogram is obtained. The roller movements were recorded using the 2‰ range of the carrier frequency bridge. The calibration key was depressed for calibration in the same range. At 1‰ detuning the indication is 50% of full deflection. As a half-bridge circuit and opposed gauges were used, the shift in the level shown in the oscillogram represents 0.5‰ or 500 micro-strain. If the modulated carrier wave is observed in this way, the available area of the screen is not fully exploited since both the positive and the negative side are displayed.

If the oscilloscope amplifier can be sufficiently driven, it is possible to show only one (preferably the positive) side on the screen, as shown in the oscillogram of Fig. 31-12*b*. The vertical resolution is now twice that in Fig. 31-12*a*, so that details, such as the natural oscillations of the leaf-spring in the lower part of the oscillogram become clearer. It should be pointed out that the static component of the change of state is retained in the pattern with the carrier-wave voltage. It is important to bear this in mind when an oscilloscope with only an AC amplifier is available.

In an oscilloscope with a DC amplifier, the "modulation", that is to say, the

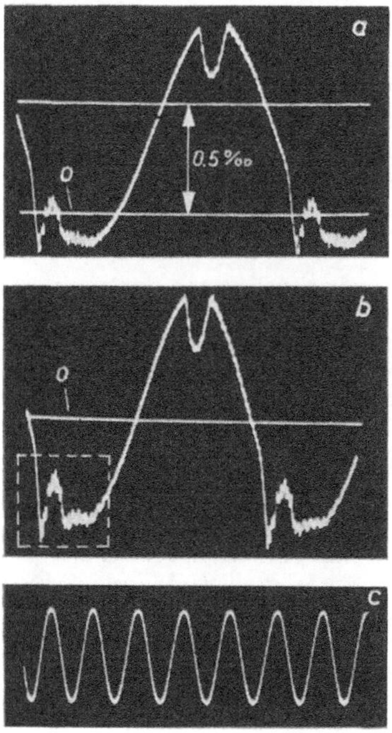

Fig. 31-13 Oscillograms of the "modulation" output. *a*) Oscilloscope with DC amplifier; *b*) over AC amplifier; *c*) calibration of time scale 50 c/s

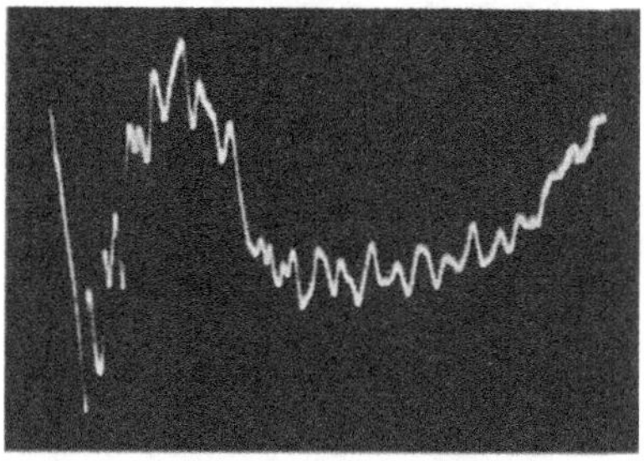

Fig. 31-14 Part of oscillogram from Fig. 31-13*a* or *b*

voltage directly proportional to the changes under investigation, can be represented directly as in the oscillogram in Fig. 31-13*a*. Here the oscillogram, the zero level (0) and the calibrating deflection (0.5‰) are traced as lines, so that all details can now be seen much more clearly than in Fig. 31-12*b*.

If such a voltage was fed to an oscilloscope via an AC amplifier, the static component would be lost. Fig. 31-13*b* shows the oscillogram of 31-13*a* again, but this time using an AC amplifier. The zero level remains but not at the reference line. Furthermore, the parts of the curve above and below the zero line are equal.

The 50 c/s voltage, used with this oscilloscope for time calibrating, has been recorded in *c* to indicate the rpm. The indication for one cycle of the oscillogram in *a* and *b* was found to be 8.7 c/s or 520 rpm.

To obtain a clear display, the lower part of the oscillogram indicated by the square in Fig. 31-13*b* in Fig. 31-14 is reproduced at increased gain and triggering with a smaller time scale (1/4 of the original oscillogram or 1/3 of a cycle). The natural oscillations of the leaf spring are now much more distinct.

31. 6 Measuring the torsional oscillations of shafts

This task can also be carried out successfully with strain gauges [13] [14]. For this purpose, four gauges are affixed askew on the shaft, each at an angle of 45° to the axis of the shaft. A complete bridge circuit of four gauges, as shown in Fig. 13-15, is formed. The supply, and the lead-off of the diagonal voltage must be carried out by means of collector rings. It is essential, for satisfactory and accurate measurement, that the contact resistance between the sliprings and the brushes is small [15] [16].

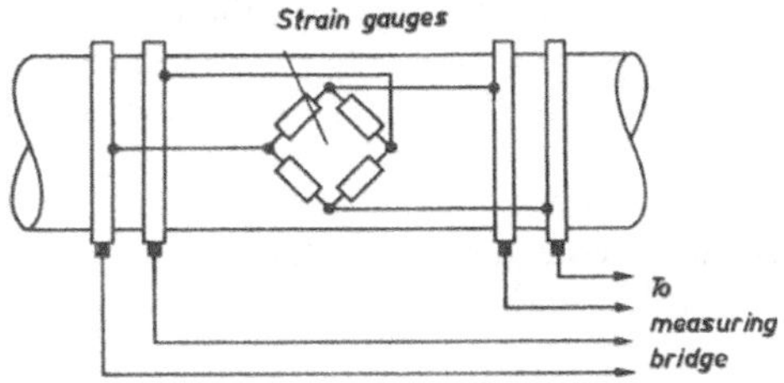

Fig. 31-15 Measuring torsion oscillations with a complete bridge circuit of strain gauges; supply of bridge current and take-off of diagonal voltage by slip rings

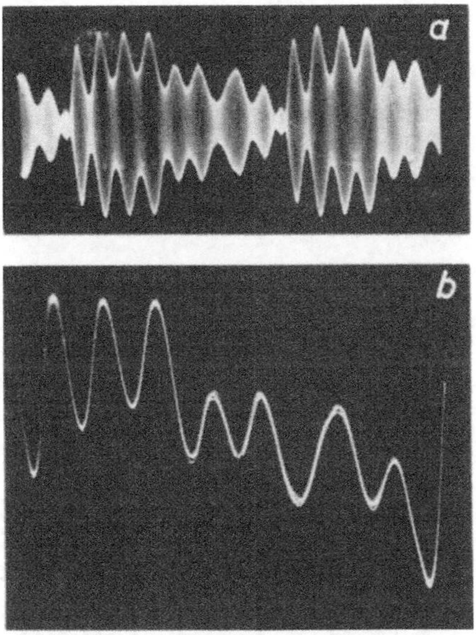

Fig. 31-16 Measuring torsion oscillations. *a*) Carrier pattern; *b*) DC voltage proportional to entropy

BERGHAUS [17] has shown how small torques — only torques not torsional oscillations — can be measured. The tension at the surface of the shaft caused by torsion produces changes in the gauge resistance and this can be observed by the processes described.

Fig. 31-16 shows two oscillograms made on a small diameter shaft by means of the smallest gauges ("PR 9214") and the "PR 9300" measuring bridge.

In 31-16*a* is shown the oscillogram, with the carrier voltage, of two cycles of the change of state and in 31-16*b,* one cycle of the demodulated voltage and indicating to the torsion oscillations. On comparison of these two patterns it should be observed that in the pattern of the carrier frequency a part of the negative peak at the end of every cycle is bent upwards. This is due to the fact that in this case the negative deflection exceeds the null balance of the bridge. In such cases, in contrast to observation of one-side loads, where null balancing is easy, the bridge should always be unbalanced to such an extent that the demodulation of the carrier voltage always occurs in the same region of the sensitivity curve of the bridge circuit. The unbalance must therefore be so great that even the lowest negative amplitude does not drop below the null-balance level (oscillograms in Fig. 32-12).

31. 7 Simultaneous observation of the strain at several points

In studying the mechanical behaviour of large objects (machines, aircraft, bridges and suchlike) it is often necessary to make observations at several or even a considerable number of measuring points, if not simultaneously, at least one after the other in rapid succession [18].

It is possible, of course, as has been described by LEVIS, to provide a number of measuring points with strain gauges, connect them to the same number of oscilloscopes standing in a row and photograph the screens of the oscilloscopes together [19]. This method does in fact give an accurate account of the waveforms of all the changes at the points observed, but, due to the high costs involved, can only be undertaken in isolated cases. It is usually sufficient if simple manual throw-over switching to the measuring point is possible.

The throw-over switch method, in which the gauges are switched over to the measuring points by means of relays, has attracted considerable attention. Fig. 31-17 shows such an arrangement with measuring bridge and oscilloscope [20] [21].

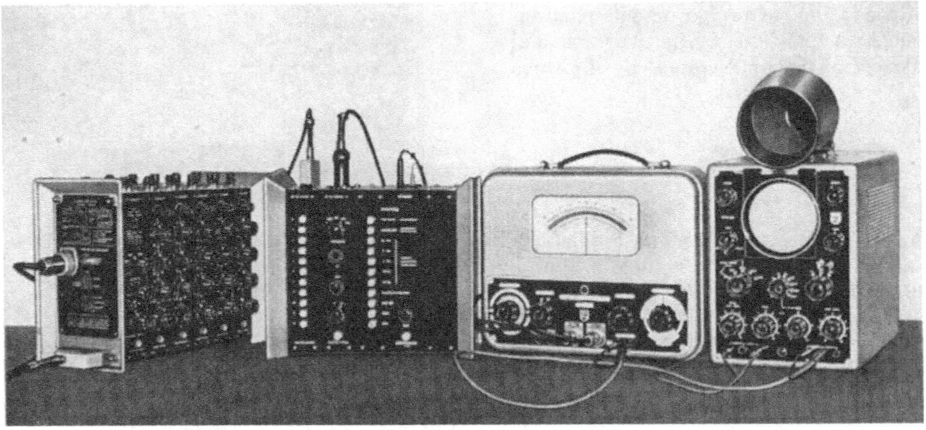

Fig. 31-17 Measuring arrangement for 5 measuring places with balancing equipment, switch-box, strain gauge bridge ("PR 9300") and oscilloscope ("GM 5656")

Since on each strain-measuring bridge only one point can be balanced for amplitude and phase, a number of measuring points presuppose the same number of balancing possibilities. The piece of apparatus on the left of the arrangement seen in Fig. 31-17 serves for this purpose. In it, up to 10 balancing units ("PT 1210") can be built up on the building-brick principle to form a calibration unit ("PT 1211"). It can be used to balance both half bridges and full bridges. Switching over to the individual measuring points is carried out in each case by two special relays controlled by the throw-over switchbox ("PT 1220", left centre in the picture). The measuring points can be selected manually by means of the left-hand vertical row of keys. It can also be done automatically by adding a step-by-step switch relay to the "PT 1221" unit. The switching times for each measuring point can be chosen in five steps from about 1/10 to 10 sec. A clearly defined degree of bridge unbalance to suit the deflection at the measuring points can be adjusted on this calibrating unit. As the calibrating unit is left switched on twice as long as the measuring points, the deflection due to the calibrating units can be clearly distinguished from the deflection due to the measuring points. Using the step-by-step relay it is possible to stop the cycle at any chosen measuring point selected within the set sequence. The throw-over switch unit emits a pulse which can be used, among other things, to trigger off the horizontal deflection of the oscilloscope.

31. 8 Examples of oscillograms when switching over to ten measuring points

Fig. 31-18 shows an example of the observation of mechanical strain tensions of a loaded bar. (This example can be extended to the study of larger objects, e.g., bridges.)

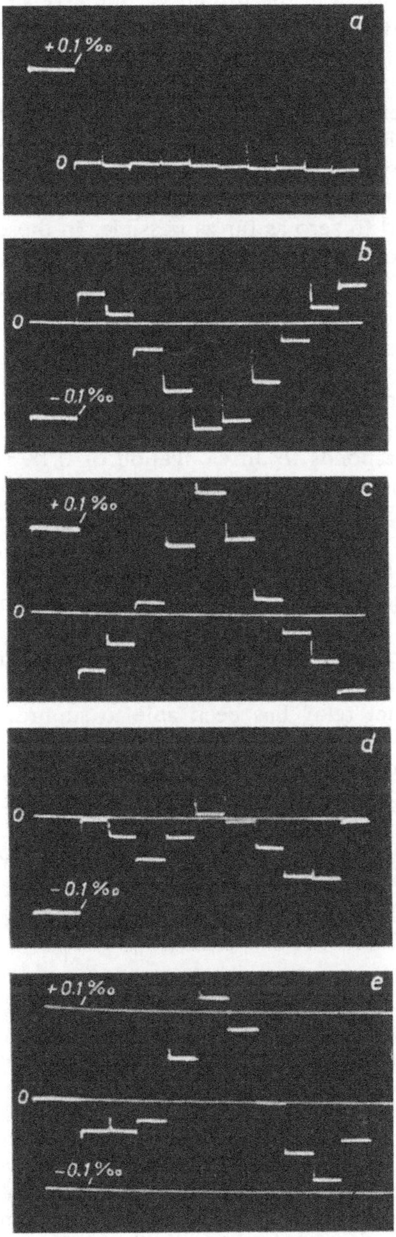

Fig. 31-18 Portraying material tensions on a loaded bar with various types of loads and bearings (details in text). Left, loads on loaded bar; right, corresponding oscillograms

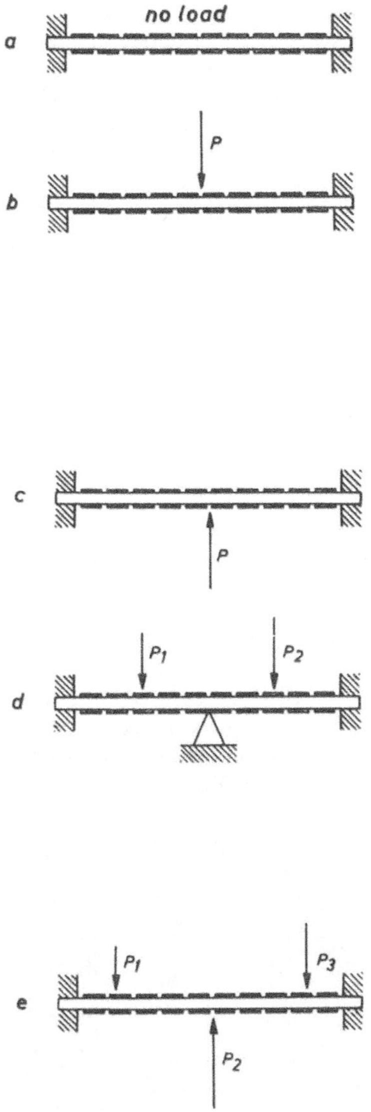

Ten half-bridge gauges have been affixed along the bar as indicated in the sketches in Fig. 31-18 (left-hand side), i.e. one gauge above and one below the bar. As a result of forces exerted on the bar in the directions indicated by arrows, mechanical tensions occur on the surface and can be measured with the individual gauges. This group of gauges was connected to the "PT 1200" measuring bridge via the arrangement shown in Fig. 31-17. By depressing the keys singly in sequence, the desired measuring point in the bridge is switched in. A particularly clear picture is obtained if the switching-over is carried out cyclically by the automatic arrangement and the time base generator of the oscilloscope is triggered off by the pulse at the end of every cycle. As is shown in Fig. 31-18, by oscillograms thus obtained, deflections occur in steps and the deflection for every stage from the null point is the measure of the mechanical tension at the measuring point in question.

In the first picture all the gauges were balanced to zero as far as possible. In those which follow, the effect of various directions of one force and of several forces is illustrated. In oscillogram *d* the tension states are shown for two forces and three bearings. The length of the arrows represents approximately the strength of the forces, which were exerted by weights occupying a small base area.

At first they are seen twice as long as the deflection stage caused by the calibrating unit. The adjustment was to 0.2°/oo, corresponding to 200 micro-strain. As the measuring points in half bridges employ two opposed gauges, the calibration indication must be divided by 2. The calibration thus corresponds to an elongation of 0.1°/oo or 100 μS. If required, the polarity can be changed and matched to the waveform of the pattern produced by measurement. The zero line was photographed on to all the oscillograms as a second recording. In the oscillogram of e another type of calibration was used. It was switched to zero for the oscillogram itself. In three subsequent photographs the zero level, the deflection for +100 μS and the deflection for —100 μS were added.

Behind the individual step-stages, small switch-over blips can still be discerned in these oscillograms. In the final model of this balancing and switch-over unit, the provision of an additional terminating contact relay has been able to suppress these blips.

Of course, similar pictures are obtainable by connecting the carrier frequency voltages, thus producing a fluorescent area outlined in the form of steps. No specific examples of this are offered, but the reader is referred to the oscillograms in Fig. 31-12 and the corresponding parts of the text. What is said there applies similarly to the representation of states of tension at several measuring points on the oscilloscope screen. Fig. 31-19 illustrates yet another type of observation of a variable of state

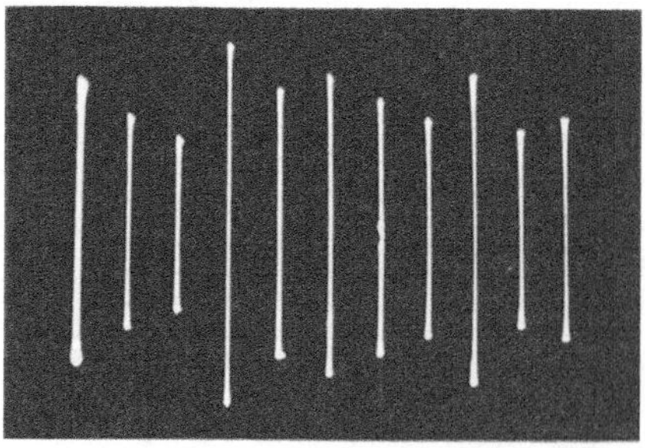

Fig. 31-19 Simultaneous indication of several entropies by means of selector switch and stepped voltage for the line deflection

at several points. The horizontal deflection is, however, not carried out by means of a voltage linear with time, but a stepped voltage. This can be obtained very simply if a switch coupled to the measuring point throw-over unit switches the X-plates over to a voltage divider having equivalent steps. For this purpose the carrier frequency voltage is best used direct for the display.

This oscillogram was made with an experimental model of the measuring point switch-over unit previously described. As the delay time at the calibrating unit was twice as long, the calibrating deflection can be seen very clearly owing to its greater brightness.

Switch-over devices of this sort are also useful for observing resistance thermometers and other resistance or voltage transducers at a number of points simultaneously. Provision has been made for units capable of dealing with up to 100 measuring points. Similar devices are also produced by the firm of Dr. P. E. Klein, Tettnang, Bodensee, Germany, particularly for temperature measurements, and are known under the trade name "Multiscop". With this unit it is possible to show on the oscilloscope screen temperature conditions at up to 50 measuring points simultaneously. All these processes are intended for the observation of relatively slow changes of state. The condition is indicated in each case only for a fraction of the total time of observation in inverse proportion to the number of measuring points. The advantage in using a cathode ray oscilloscope for indication instead of electromechanical indicators is that the scanning is very rapid and the variable of state can be read off on the oscillogram practically simultaneously for all the measuring points (P-type screen) [22].

31.9 The use of strain gauges as a measuring element for special tasks

The oscillograms in Fig. 31-20 are intended as an indication of the wide range of application of the strain gauge. They show the waveform of the pulse in the wrist. For this task an active gauge was pressed against the wrist by means of an elastic bandage. One end of the gauge was clamped. As the elongations are slight, a gain of about 300,000 had to be employed for this oscillogram (Philips "GM 3156"

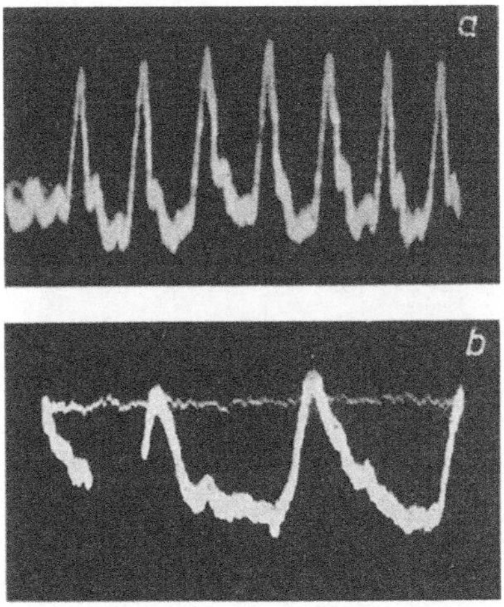

Fig. 31-20 Pressure curves of wrist pulse. *a*) 7 cycles; *b*) overexposed recording of 2 cycles for portraying the small voltage oscillations in the flyback pattern

oscilloscope and "GM 4570" preamplifier). At such high gain (if carrier frequency amplifiers are not available) observation is limited to the dynamic changes of state, and the static component cannot be indicated since such high DC gain could only be made sufficiently stable at prohibitive cost.

Oscillogram b has been included to show, in the flyback pattern, that the thickening of the oscillogram in a caused by rapid voltage fluctuations is probably due to muscular tensions.

Final remarks to chapters 30 and 31

In these sections it has not been possible to do more than discuss the fundamental principles of the almost unlimited range of applications of the cathode ray oscilloscope for the measurement of mechanical phenomena. As has been said in the introduction to Ch. 30 there are usually several methods which can be used with the oscilloscope, and suitable pickups are available for observing practically every sort of physical condition [23] [24] [25].

Part IV

PHOTOGRAPHIC RECORDING AND LARGE PICTURE PROJECTION OF OSCILLOGRAMS

CHAPTER 32

PHOTOGRAPHIC RECORDING OF THE SPOT TRACE

32. 1 Importance of photographic recording and the special conditions governing it

The interpretation of rapid, non-recurrent processes (single transients) is only possible with the aid of photographic recording. But screen patterns of cyclic processes, which can usually be examined when stationary, must also be photographed as single pictures if accurate interpretation is required. By enlarging these photographs, to several times their natural size if necessary, a convenient record of the phenomenon is obtained. Modern tubes with small spot diameter and a correspondingly clear trace make it possible to obtain reading accuracies of 1% with ease, as has been shown in Part II, Ch. 7.5 "Accuracy of reading" (p 267).

If the method of measuring and the results obtainable are to be communicated to others who are interested, then photographic recording is essential. Only the reproduction of the results in the form of original photographs, as has been done in this book, can in fact provide the information required for explaining a new method of measurement.

Photographic recording is always needed if a coherent impression is to be gained of the variations of an unsteady phenomenon, such as the starting and slowing down of a machine or the like over a considerable period of time. In such cases the recording of the deflections of the spot during the observation must be obtained on recording material moving at a constant rate. The speed at which the recording material is fed thus determines the time-resolution or the time scale.

Due to the unusally wide variety of applications of the cathode ray oscilloscope, oscillograms often have to be made in circumstances where, in view of the special conditions obtaining, the investigators have limited previous experience. On the other hand, the proper use of the means made available through photographic techniques makes it possible to use simple photographic equipment and yet achieve results which would generally require considerable outlay.

In this section of the book the applications referred to will be dealt with individually. In view of the great popularity of amateur photography among those engaged in technical work, it is assumed that the reader possesses the required fundamental knowledge of photographic technique. Should this not be the case, the relevant literature on the subject should be consulted [1] [2].

32. 2 Equipment for photographic recording

It is possible in principle to photograph the patterns on the screen in the simplest way possible without special apparatus, by bringing the light-sensitive material into contact with the fluorescent screen. This process has been used and described on various occasions [3], but it requires a completely darkened studio. Moreover, because of the risk of unwanted exposure due to stray light, only medium-sensitive photographic material can be used (usually photo-sensitive paper) and this, in turn, requires inconveniently long exposure times. The method has therefore not become very widespread and is only applied when none of the equipment which are about to be described are available or obtainable.

Use is generally made of a normal camera and an adaptor tube or apparatus specially designed for the recording of oscillograms, not excluding the continuously moving film recorder. Where it is just a case of recording the dependencies displayed on the fluorescent screen in the form of single pictures, normal cameras are generally used.

The two-lens mirror reflex camera "Rolleicord"[113]) has a reputation of many years standing for making single exposures. As an example of this type of photography Fig. 32-1 shows the Philips single-exposure recording equipment "PM 9300" which uses the "Rolleicord Vb", a specially designed model for this purpose by Messrs. Franke u. Heidecke, Brunswick, in conjunction with Philips. This recording device not only permits exposures to be made on conventional photographic material (Negativ-Rollfilm B II 8) but also on "Polaroid" Land-Film [113]). While Fig. 32-1 shows the design for taking pictures on conventional material, Fig. 32-2 represents the same design for recording on "Polaroid" material.

Fig. 32-1 Philips single-picture recording equipment "PM 9300". Camera: "Rolleicord Vb", photo-tube "GM 5687" and connexion-flange "GM 5685". Supplementary for normal material

[113]) "Rolleicord" is the registered trade mark of the "Rollei-Werke Franke und Heidecke" Braunschweig, Germany and "Polaroid" and "Polapan" are registered trade marks of the Polaroid corporation, Cambridge, Mass. USA.

With "Polaroid" material we obtain directly inside the camera a finished positive picture after a developing time of only 10 seconds without any other means.

Focussing and viewing the oscilloscope screen with this camera is carried out by means of a frosted glass screen through a special view-finder lens. In order to adjust the focussing difference in Polaroid exposures the view-finder accessory "Rolleicopi" (Fig. 32-2) is inserted in the view-finder channel between frosted glass screen and the camera body. The scale of reproduction of this design is about 1 : 2.3. The lens-objective is a Schneider-Xenar, a four-lens design with an aperture ratio 1 : 3.5 and focal length of $f = 75$ mm, allowing a diaphragm setting between 3.5 and 22.

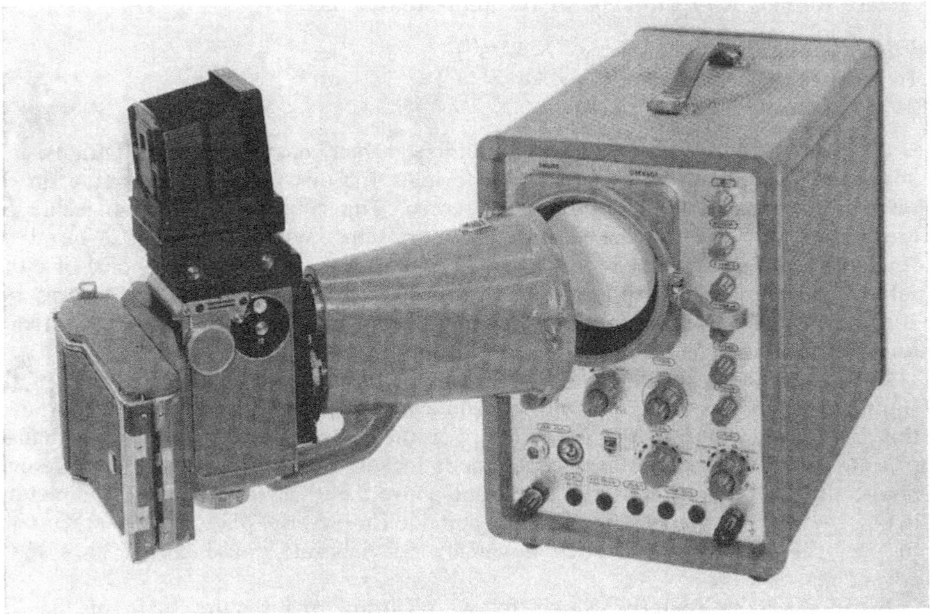

Fig. 32-2 Philips single-picture recording equipment "PM 9300". Camera "Rolleicord Vb". Photo-tube "GM 5687". Connexion-flange "GM 5685". Supplementary for Polaroid material

A synchron-compur shutter for speeds between 1 sec and 1/500 sec is used. A choice of X- or M-contact synchronization is possible. The X-contact is particularly useful for releasing single processes which have to be photographed.

The complete outfit consists of:
"Rolleicord" mirror reflex camera Vb
"Polaroid" cassette
View-finder accessory "Rolleicopi"
Photo-tube GM 5687 with one flange each for Philips oscillographs with valves of 10 cm and 13 cm diameter (GM 5685 and GM 5686)
2 supplementary lenses (Rolleinar 4) for exposures with normal material
2 lens hoods

2 extra counters for 16 and 24 exposure settings
2 view-finder and film masks each for picture formats of 4 × 5.5 cm and 2.8 × 5.5 cm respectively
Synchronizing contact cable
Flexible release
Tripod screw
Case

When using normal film the supplementary lenses Rolleinar 4 must be placed in front of the camera lens and the view-finder. When using the Polaroid cassette correct focussing is done without supplementary lenses.

When using conventional photographic material we can choose three different picture formats with this type of recording device, namely:

12 exposures 6 × 6 cm ($2^1/_4'' \times 2^1/_4''$)
16 exposures 4 × 5.5 cm ($1^5/_8'' \times 2^1/_8''$)
24 exposures 2.8 × 5.5 cm ($1^1/_8'' \times 2^1/_8''$)

As the 6 × 6 format is not wholly utilized when reproducing oscillograms, it is advisable to use one of the other two formats for this purpose, depending on the raster dimensions on the oscillograph screen. The counter is interchangeable for the purpose of selecting the various transport steps. With "Polaroid" material the frame dimensions are 6 × 8 cm ($2^3/_4'' \times 3^1/_2''$). Here we have the choice of either setting for one single exposure or, with narrow oscillograms, two exposures one above the other on one frame. For this purpose the cassette can be adjusted vertically to three positions (centre, top and bottom) by means of a lever.

It is a particular advantage of this camera that the automatic coupling of the shutter winder and film feed mechanism can be easily uncoupled. For instance, this is necessary during the observation of the phase position of various changes of state in which the time base generator is locked in synchronism with a reference phase, and the pictures are to be taken one above the other (for example, the pictures in Ch. 14). As this type of camera is also suitable for pictorial photographs 60 × 60 mm in size, the acquisition of such a camera for laboratory use would have many advantages.

Because of the remarks which follow regarding the picture scale of the line (Fig. 32-7), it is usual to adjust the resulting oscillogram picture on the film to an area with a diameter of $<$ 30 mm. The use of photographic material of cine-film size with a useful width of 24 mm, and hence the use of miniature cameras, is obvious as the next step. A very large selection of emulsions is available. This material is supplied in considerable lengths, and this size is therefore also inexpensive.

Among miniature cameras, the "monocular" reflex cameras are particularly suitable for this purpose (e.g., "Contaflex", "Ultramatic", "Exacta-Varex", "Edixa-reflex", "Practica"). With these cameras it is possible to obtain satisfactory sharpness of focus, completely free from parallax, while changing from one oscilloscope to another. One point must be noted as compared with "binocular" reflex cameras, however. During recording, as well as for a short time before and after, the recorded object is invisible on the view finder ground-glass plate, as the mirror which directs the light to the view finder must be moved upwards out of the way to allow the light to reach the film. As the exposure times for single oscilloscope photographs are known to lie between 1/10 and $^1/_2$ sec, the oscillogram is invisible in the view-finder for this space of time or a little longer (the movement of the mirror requires some additional time). But as it is often necessary to photograph processes in which undesired and spontaneous changes may occur, the recording equipment should be such as will permit the screen to be observed directly during the course of photographing as well. For this purpose the tube for keeping out ambient light during observation through the ground-glass plate can be provided with a visor opening in addition. The picture is focussed according to the pattern on the ground-glass plate, while the photograph is observed through the visor opening in the tube.

With all other types of camera, focussing must be carried out with a ground-glass plate on the plane of photographic film. Should a camera be used with one particular type of oscilloscope only, the focussing can be done once and for all and the camera fixed to the tube at the correct distance. But if there have to be changes from one oscilloscope to another, this type of focussing would be very inconvenient. Focussing must always be done very accurately, since a faulty adjustment, even by as little as a fraction of millimetre may cause a blurred picture in view of the very short distance between the photographic lens and the screen, particularly with the full diaphragm opening so often essential for critical photographs.

When a discontinuous change of state is to be observed over a considerable period of time, the width of the fluorescent screen is insufficient. The details then lie so close together, even when the duration of the time deflection can be extended as required, that the waveform could not be discerned in detail. Apart from special processes by means of which the time can be extended to occupy several lines and hence a sufficiently large period of time shown on a sufficiently small time scale, recording in such cases, even for the sake of observation alone, must be done with photographic material continuously moved at a correspondingly high speed. Horizontal deflection is then not employed, so that the spot is deflected only vertically by the voltage delivered by the vertical amplifier proportional to the change of state. The time expansion is obtained by moving the photographic material at right angles to the deflection due to the signal voltage — in this case horizontally — at a constant speed.

Fig. 32-3, showing the Philips "PP 1014" camera, is an example of such an equipment. The pattern on the screen is copied by the lens (Lytax "Acron" 1 : 3.2, focal length 85 mm) in the plane of the photographic apparatus. The scale of copying can be chosen such that all screen patterns of oscilloscope tubes of a diameter of 7 cm and above fill the photographic plate fully (for single shots 24 × 24 mm).

The camera can also be used for other purposes such as, for example, object movement studies. For this the objects in question are illuminated by a stroboscope in a darkened room while in movement. The continuous photographs thus give a picture of one phase of movement for every flash of the stroboscope. For such purposes it is an advantage for the camera that focussing is possible at a distance of up to 5 m from the objects under study.

In order to obtain a fine focus a deflecting mirror on the camera is swiveled into the light path by a lever (Fig. 32-3), whereupon the oscillogram can be observed on a frosted-glass surface through the fivefold magnifier (mirror-reflex fine focussing). By turning the front part of the lens the correct adjustment is easily found.

In addition to the lens there is a Prontor-Press shutter (self-setting) with shutter speeds of 1, 1/2, 1/4, 1/8, 1/15, 1/30, 1/60 and 1/125 sec, as well as *B* and *T* settings. An iris diaphragm adjustable in steps of 3.2 to 32 makes it possible to adjust the light ratios to the photographic conditions.

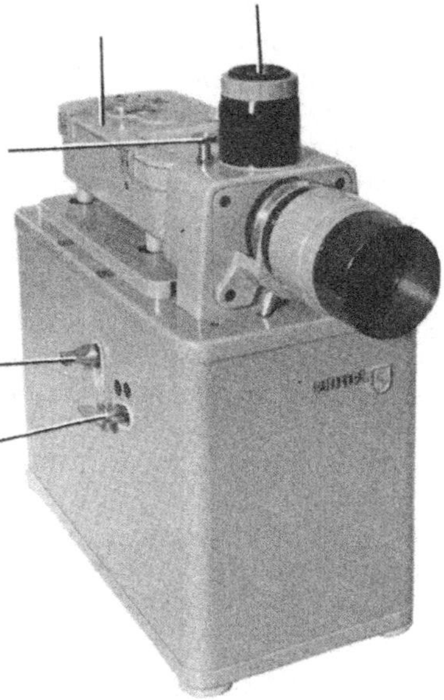

Fig. 32-3 Philips standard recording camera "PP 1014" for motion film and single pictures

The horizontally placed magazine (with film feed device) can be mounted on slide rails on the lens head and, if required, can be replaced by another. Fig. 32-4 shows the inside of this magazine open. The accommodation for and feed path of the film (15 m roll) can be seen so clearly in the illustration that no further explanation is needed. A marking peg is mounted in the magazine lid so as to be light-proof (bottom, left). Small notches can be cut in the edge of the film by pressing down this peg. It is thus possible to mark certain sections of the film so that they can be recognized by touch even in the dark. In long continuous shots or especially when it is essential to work at high feed speeds and the length of the "take" is not accurately known, this method can be used to indicate where individual sections of film should be cut for developing. This prevents the accidental cutting through a part of the actual picture.

The feed mechanism is driven via a set of toothed gearwheels and a countershaft by an external-rotor asynchronous motor (mains-synchronized to 2 m/s) with a mechanical coupling.

By selecting the pairs of geared weels and switching the countershaft, feed speeds can be adjusted in 10 stages (mains-synchronized) from 1 cm/s to 2 m/s. Complete synchronism is not obtainable in the two further stages, namely 3 and 4.75 m/s, but as the coupling slip is relatively constant the feed speeds are known with sufficient accuracy for most applications. If required, brightening of the oscillogram can be traced as a time scaling by an AC or pulse voltage of known frequency, or an electronic switch can trace the pattern of such a voltage. In the ranges up to 2 m/s the margin of error in feed speed readings, relative to mains frequency, is $< 1\%$.

For continuous photographic recording, the shutter should be set to *B*. By pressing down the button for the mechanical coupling (Fig. 32-3) the shutter is also opened

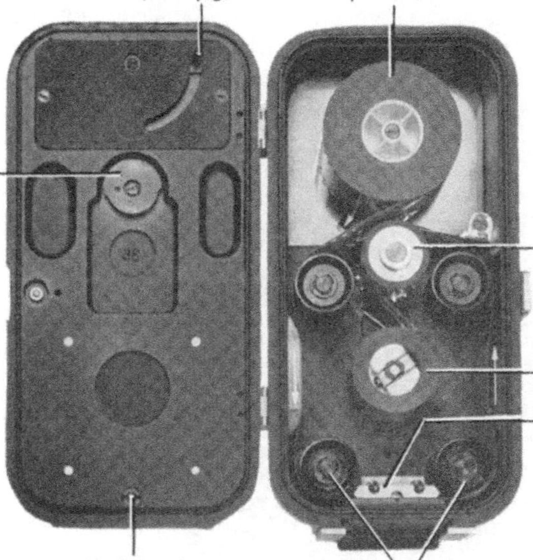

Fig. 32-4 Magazine "PP 1015" of Philips camera "PP 1014"

automatically for as long as the coupling is connected. The camera shutter contains a synchronous X-contact, which closes immediately the shutter opens. By means of this contact it is possible to start the process, if it can be released by triggering, at the same moment at which the shutter is wide open and the film is being fed. It is particularly advantageous to use this contact for unblanking of the trace.

For moving records, the stationary spot must be set to the intensity corresponding to the requisite feed speed. Should it happen that some considerable time elapses after the oscilloscope is set, before the deflection occasioned by the measuring process takes place, there is a danger that the spot will darken the screen by burning. To prevent this from occurring, the following steps should be taken:

Te spot is first adjusted to the required intensity. The spot intensity is then suppressed by switching on a bias voltage which is provided in many types of oscilloscope and is accessible by means of sockets. If the connections of this switch to the lower voltage are short-circuited by the synchronous contact, the brightening will occur only at the instant at which the deflection due to the measuring process is expected to take place. In this way the fluorescent screen can be protected to a great extent. This camera can also be used for single 24 × 24 mm pictures. The film is then fed by means of the hand winder on the magazine. The large store of film (15 m) is then found to be a great advantage.

Although the Philips "PP 1014" camera has a satisfactory performance as a versatile universal camera, the Zeiss "PP 1021" camera offers perhaps the greatest

Fig. 32-5 Zeiss precision recording camera type PP 1021

possible range of applications and the highest standard of performance. The lens is a Zeiss-Sonnar with an aperture ratio of 1:1.5 and focal length of 50 mm. The feed speed can be switched in 15 stages by means of planetary gearings from 0.04 mm/s to 4 m/s. In addition, a drum cassette can be fitted which will enable the film to be fed at speeds up to 51 m/s. At the photograph in Fig. 32-5 shows, the operation of this recording camera is to a large extent automatic. The various possible operations can be initiated from a press-button control panel. For moving film records with times greater than 1 s, an adjustable, electronic preselector giving a range of times from 1 to 16 s is provided. When taking single pictures, film feed may be either by press-button or automatic after each photograph. It is also possible to take a series of photographs one above the other. These two recording cameras are equipped to take 35 mm perforated cine film. The Siemens recording camera is also equipped to take film 100 mm in width [4].

32. 3 "Writing speed" and the influence factors governing it

The photographic performance of such a recording equipment depends upon the "maximum writing speed". By this is meant the speed of movement of the fluorescent spot on the screen which, under the given photographic conditions gives the minimum blackening capable of being copied when using negative materials. When paper is used, a minimum perceptible blackening is obtained. Blackening considered sufficient for copying purposes (denoted by B) is $B = 0.1$ above the basic fog of the emulsion.

For determining the blackening of a negative the light transmitted through the exposed part is compared with the light allowed through the unexposed portion, i.e. unblackened area. This gives the degree of transmission or transparency. The negative logarithm of this quotient is thus the measure of the blackening. Blackening 1 therefore indicates a difference in light of 1 : 10, blackening 2 a difference of 1 : 100, and 0.1 a difference of 1 : 1.26 [1] [5].

As CUSTERS has derived in detail, the greatest attainable writing speed v_{max} with such an equipment for photographing oscillograms can be determined by the equation [6]

$$v_{s\,max} = a \cdot \frac{L^2}{(m+1)^2} \cdot \frac{V_a \cdot I_s}{d} \cdot \eta \cdot \beta \,. \tag{32.1}$$

In this equation

a is the absorption number of the photographic lens
L is the aperture ratio of the lens or "power" (D/f)
m is the image scale (reduction or enlargement)
V_a is the total acceleration voltage of the electron beam
I_s is the beam current
d is the diameter of the spot on the screen
η is the screen efficiency (dependent on V_a, Fig. 33-3)
β is a factor which must be determined experimentally (dependent on the spectral energy distribution of the fluorescent light in proportion to the spectral sensitivity of the photographic emulsion).

The ratio of the aperture D of the lens to the focal length f is the effective opening. It is also known as the "power" (L) of the lens. Thus:

$$L = \frac{D}{f} \,. \tag{32.2}$$

A lens is the more powerful, the greater the aperture at the same focal distance. Fig. 32-6 is intended to explain these relationships in greater detail. While it is sufficient in the case of photographs taken at distances of > 1 m ($f \ll g$) to judge the luminous intensity of the image of the object portrayed on the photographic plane by taking into account the aperture ratio alone, in the case of the relatively short distances g in photographs of oscillograms the copying scale m must be taken into consideration. The applicable relationships can be derived according to the sketch in Fig. 32-6.

The formula:

$$\frac{1}{g} + \frac{1}{b} = \frac{1}{f} \tag{32.3}$$

applies, and f once more represents the focal distance. It can be assumed that the

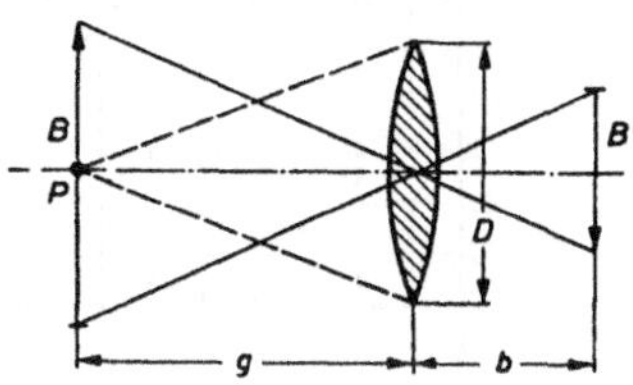

Fig. 32-6 Copying ratio of a lens

light from the oscilloscope screen is emitted diffusely according to the Lambert cosine law, Eq. (33.2). If for instance, a square millimetre at P on the oscillogram radiates a certain amount of light C, the lens receives a part of this light proportional to $(D/g)^2$. This quantity of light is focussed on an area of the photographic plane, which in the ratio $(b/g)^2$ is smaller (or larger) than one square millimetre. The intrinsic brilliance is thus increased (or reduced) by the lens. The brightness H on the photographic plane is then

$$H = C \cdot \left(\frac{D}{g}\right)^2 \cdot \left(\frac{g}{b}\right)^2 = C \cdot \left(\frac{D}{b}\right)^2 . \tag{32.4}$$

From Eq. (32.3) we then obtain

$$f = \frac{b \cdot g}{b + g} = \frac{b}{\frac{b}{g} + 1} , \tag{32.5}$$

so that

$$b = f \cdot \left(\frac{b}{g} + 1\right) . \tag{32.6}$$

Thus, according to Eq. (32.4), the brightness is found to be

$$H = C \cdot \left(\frac{D}{f}\right)^2 \cdot \frac{1}{\left(\frac{b}{g} + 1\right)^2} . \tag{32.7}$$

The ratio b/g, however, represents the image scale m in Eq. (32.1). It also determines the attainable writing speed to a great extent. The relation of the factor $[1/(m + 1)]^2$ and the image scale m is shown in Fig. 32-6. When photographing oscillograms there is a tendency to choose the image scale in such a way that the picture is not too small; an attempt is thus often made to disp!ay in the ratio of 1 : 1 as far as possible. But as can be gathered from the curve in Fig. 32-7, the factor $[1/(m + 1)]^2$ then becomes only 0.25. When reduced to 1 : 2.5 it becomes 0.5, that is to say it is doubled; and at 1 : 6 it becomes 0.75, or three times as great. It can be gathered from this that it is advantageous to display the oscillogram on a somewhat reduced scale if a high

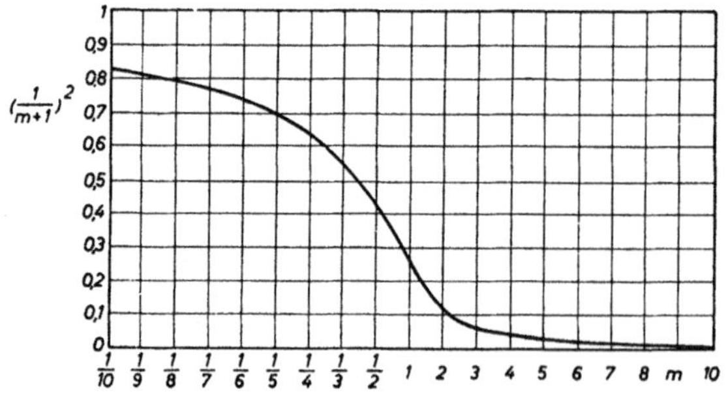

Fig. 32-7 Influence of the copying scale m on the factor $\left(\frac{1}{m+1}\right)^2$ and as a result on the intrinsic brilliance in the photographic plane

recording speed is to be obtained. On the other hand, it can be seen from this curve that the factor $[1/(m + 1)]^2$ increases but little at reductions below 1 : 5. The photographs would, moreover, then be so small that even the grain in the enlargement would cause difficulties and interpretation be made uncertain through other causes also. The image scale is therefore usually chosen between 1 : 2 and 1 : 5. Thus all the factors of Eq. (32.1), except β, can be assumed to be known.

Factor β, however, which practically represents the "sensitivity" of the particular photographic material for a screen pattern with a given spectral energy distribution cannot, unfortunately, be determined by any of the processes for determining photographic sensitivity, such as DIN, ASA, BSI, Scheiner or other similar ones. All these methods are based on the determination of the blackening threshold of the photographic material with approximately white light (for the DIN method using a light corresponding to the colour temperature 5000 °K of a black body), while the light of the oscilloscope tubes varies greatly in spectrum according to the material of which the screen is made. The usual sensitivity data of the photographic materials are therefore invalid for oscilloscope photographs. It is thus far from easy to determine the most favourable conditions for photographing oscillograms with the exposure meters commonly used in photography. It must also be particularly borne in mind that in oscillograms it is the intrinsic brilliance which must be determined, while exposure meters integrate the light from the object observed.

An attempt has been made to measure the intrinsic brilliance of the screen pattern directly, using a photocell (a photomultiplier) illuminated via a small aperture. A condition here is that this aperture must be smaller than the thickness of the spot trace. Such a light meter made by the firm of HILTRON (G. HILLE, Munich 55) is shown in Fig. 32-8. As, however, the spectral sensitivity of the photocell never agrees with the colour sensitivity of the film, even with this aparatus calibration is necessary for the particular photographic conditions obtaining at the time.

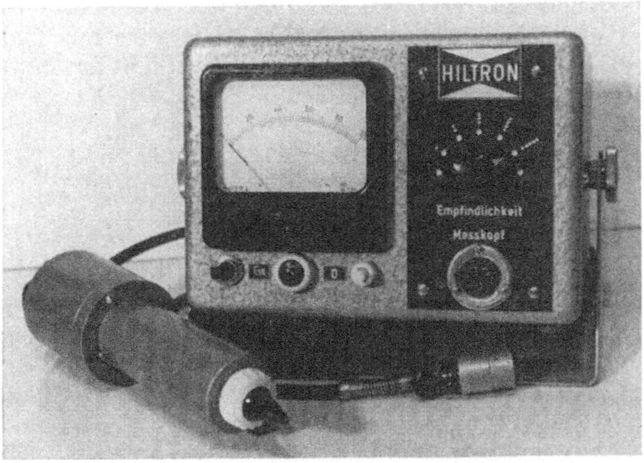

Fig. 32-8 Hiltron light meter

The only thing that can be done therefore, is to determine experimentally for a given composition of the light colour of the screen and the spectral sensitivity of film under otherwise known photographic conditions what the maximum writing speed $v_{s\ max}$ (or the writing speed v_s for a given greater blackening, better than 0.1) should be.

The values for $v_{s\ max}$ so obtained can also be used in the same way for other photographic conditions.

32. 4 Spectral energy distribution of the light emitted by the screen, and properties of the photographic material

As can be seen from the curves in Figs. 2-23 *a, b, c,* the spectral distribution of the light radiated from the different screens varies a great deal according to the screen material; in other words, the light from a particular type of screen includes a greater proportion of a given colour than of others. Only in special cases can it be taken to be white. On the other hand, the photographic recording materials differ in their sensitivity to the various spectral components of white light.

Without special measures, photographic emulsions are actually only sensitive to to blue and violet light. By means of special additives, called sensitizers, it was first possible to make the material sensitive to yellow-green light "orthochromatic" film -and later to the whole spectrum of visible light, including red- the "panchromatic" film.

For highest photographic efficiency the colour-sensitivity of the film should correspond to the spectral distribution of the light of the fluorescent screen. It is for this reason that the best results are generally to be expected with blue fluorescent *B*-type screens. The curves of the spectral energy distribution of the Philips *B, G* and *P*-fluorescent screens are reproduced in Fig. 32-9 together with the colour sensitivity curves of the high-sensitivity orthochromatic "X-ray film", of the "Perpantic" and of the "Peromnia" film made by PERUTZ. It must be pointed out that the blackening curves of the films were not photographed with light of equal spectral energy, but, as is usually done, with incandescent lamp light, corresponding to a black body at 2700 °K [7] [8].

The measurement of the absolute colour sensitivity with an arrangement having a thermocouple as a measuring probe gives much higher sensitivity values for the range of the shorter waves (from about 550 nm onwards) than for the blue part of the spectrum, as is to be seen in Fig. 32-9 [9].

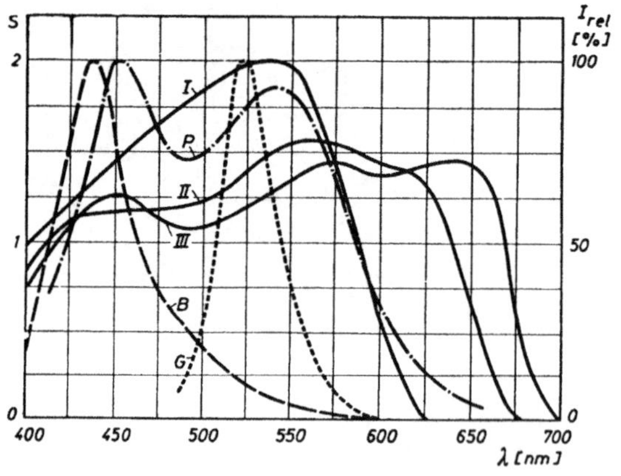

Fig. 32-9 Spectral energy distribution of Philips fluorescent screens and spectral sensitivity of Perutz films. *B* blue-persistence screen; *G* green-persistence screen; *P* (yellow) long-persistence screen; I Perutz X-ray film; II Perutz-Perpantic film (17 or 18 ° DIN); III Perutz-Peromnia film (21 or 25 DIN)

From these curves it is evident that the blue fluorescent screen is suitable for every kind of photographic material. For photographing images on the green-yellow fluorescent *G*-screen, which has the highest efficiency for visual observation, the film material must also be of maximum sensitvity for the green-yellow part of the spectrum. Material like the Agfa "Fluorapid" film, GEVAERT "Scopix-G" or the PERUTZ "X-ray film" is most suitable for the purpose.

The photographic material on the market suitable for producing oscillograms can be divided into the following groups -classified according to general sensitivity:

1. Ultra-sensitive panchromatic of about 25 DIN (250 ASA) to above 30 DIN (800 ASA), for example: Agfa Isopan Record (about 34 DIN corresponding to 2000 ASA), Kodak Tri-X (27 DIN, 400 ASA), Perutz Peromnia (25 DIN, 250 ASA), Ilford HPS (30 DIN, 800 ASA) and HP 3 (27 DIN, 400 ASA), as well a Polapan" 3000 (36 DIN, 3200 ASA) and Polapan 4000 (27 *v* DIN, 400 ASA).
2. High-sensitive panchromatic, for example: Ilford SP 3 (22 DIN, 125 ASA), Agfa-Isopan ISS (21 DIN, 100 ASA), Perutz Peromnia (21 DIN, 100 ASA), Gevapan 33 (23 DIN, 160 ASA).
3. High-sensitive orthochromatic, especially the so-called X-ray emulsions, for example: Gevaert "Scopix-G", Agfa-Fluorapidfilm, Perutz-X-ray film. On account of their specific field of application the customary sensitivity data have been omitted for these materials.
4. Blue-sensitive photographic material only, for example: Gevaert "Scopix-G", as well as all printing papers such as Gevaert "Oscilloscript", Agfa-Registrierpapier, "Typon" printing paper and similar. When using these materials it is absolutely essential to operate with blue-light oscilloscope tubes ("B"-, "P 11"- screen), as otherwise the photographic efficiency is greatly impaired.

For exposures under average conditions panchromatic material with sensitivities of about 21 DIN (100 ASA) generally available should be quite sufficient. With "Polaroid" material the type 32 Polapan 400 (27 DIN, 400 ASA) is very satisfactory for general purposes.

Enlargements of up to about sevenfold can be made of normal film material with 21 DIN without noticing the grain. Since, however, as a rule no larger oscillogram enlargements than 3...5-fold are required, the 25 DIN (250 ASA) materials can be used equally well in practice. The high sensitivity in that case provides a welcome reserve when conditions of exposure become more difficult (trace extension).

Also the "Polaroid" material the type 32, "Polapan" 400, can be used as universal material. The Polaroid materials of higher sensitivity are inclined to emphasize the brightness gradation of the screen picture. This is particularly noticeable when the luminous spot velocities on the screen vary greatly (voltage diagrams with steep slopes and flat tops).

Provided there is sufficient brightness we can naturally also use less sensitive material for oscillograms (14 DIN, 20 ASA to 17 DIN, 40 ASA). It should be noted, however, that in that case these materials work "harder", i.e. they tend to reproduce the brightness gradations particularly strongly.

Some very sensitive panchromatic films (23 DIN and over) are not so suitable as might be expected from the published sensitivity data. This is due to the fact that the increase in sensitivity has been provided mainly in the red portion of the spectrum, with a view to their use for photography with incandescent lamps; emulsions such as these are usually not much more sensitive in the green-yellow range [10].

Under the name of "Scopix-G" and *B*, the film of GEVAERT supplies two X-ray films which are also suitable for photographing oscillograms. The spectrograms of these two films can be seen in Figs. 32-10*a* and *b*.

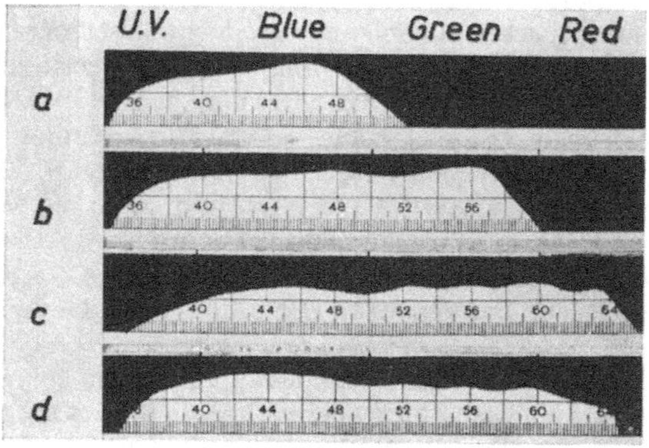

Fig. 32-10 Spectrograms of Gevaert films *a)* Scopix-B; *b)* Scopix-G; *c)* Gevapan (for *a)* to *c)* artificial light 2850 °K); *d)* Gevapan (daylight 5900 °K)

Oscillograms can also be recorded on colour film. As HILLE has described, filters of different colours can be used in multiple photographs to give different colours to the individual voltage curves, so that they can be distinguished from one another even when there is a large number of them, [11]. The condition must be made, of course, that the spectrum of the screen light is sufficiently wide. The white fluorescent *W*-screen is recommended for such purposes. It has however been found that very useful colour photographs can be obtained with *N* or *P*-screens. The efficiency of these screens is much higher than that of the *W*-screen.

If many oscilograms have to be obtained at the lowest cost, it is also possible to use films specially designed for documentation work, such as Agfa "Agepe" and "Agepan". These films give very strong contrast, however, and are more suitable for line than area-shaped oscillograms.

Under the trade name "Polaroid" the Polaroid Corporation, Cambridge, Mass. USA, offers a photographic material which is developed within 10 seconds inside the camera (see also section on "Photographic equipment").

"Polaroid" films consist of a spool of ultra-sensitive negative film and a roll of special paper for reproducing the positive picture [12]. Small capsules containing a developing paste are stuck in front of each frame on the carrier paper-strip for the printing paper. After exposure the carrier paper-strip is pulled out of the camera. The negative and the positive are thus pulled through between two steel rollers as shown in the schematic diagram of such a camera in Fig. 32-11. This squeezes the developing paste out and spreads it evenly between negative and positive. The developing process thus begins immediately and is completed after 10 s. The positive picture can now be torn off. It is then sponged over for fixing with a special sponge supplied, which contains a fixative and a protective glaze.

Type 32, "Polapan" 400 (27 DIN, 400 ASA) and type 36, "Polapan" 3000 (36 DIN, 3200 ASA) are suitable Polaroid materials for this kind of photographic equipment. Type 32 can be recommended as a suitable universal material. "Polaroid"

Fig. 32-11 Polaroid attachment to Rolleicord V b

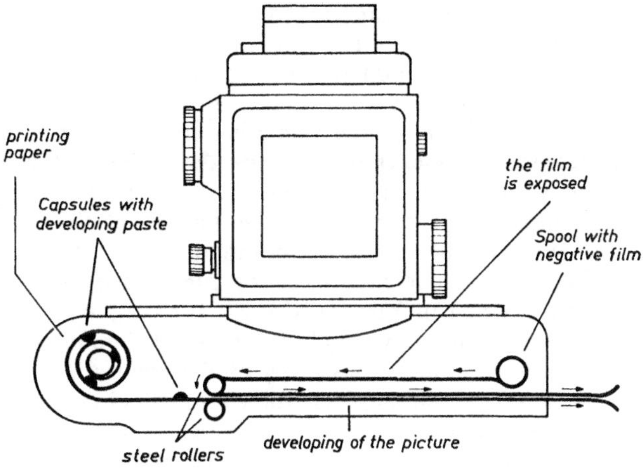

materials with higher sensitivity tend to emphasize the brightness gradations. One should bear this in mind when screen pictures with greatly varying spot velocities have to be taken (voltages with steep slopes and flat tops).

A positive is obtained which can be duplicated by any of the many processes familiar. NEIDAMAST has described [13] how this is best achieved with large numbers of oscillograms.

Another property of the film, the blackening curve or gradation, should be borne in mind, particularly when processing oscillogram photographs.

If a photo-sensitive material is exposed in stages to increasing amounts of light, corresponding increases in blackening are found after development. As can be seen in Fig. 32-12, steeper slopes of the gradation curves are obtained with increasing developing time. The slope of the gradation curve is denoted by γ and called "gamma" for short. The whole surface of the film is fogged if exposed too long, and this is indicated by a rise at the beginning of the blackening curves. From Fig. 32-12 it is clear that it is not advisable to extend the developing period beyond 8 minutes.

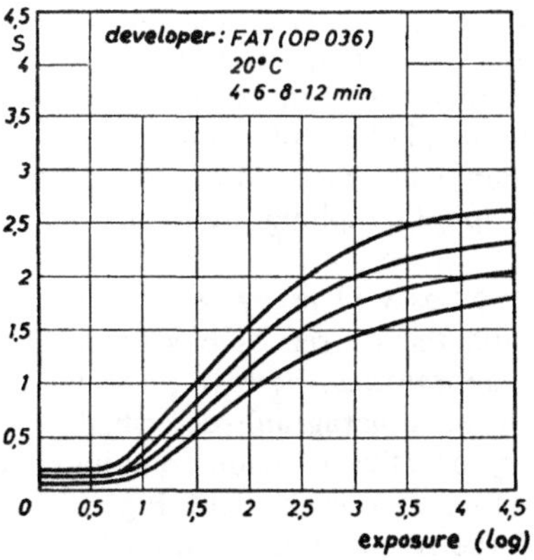

Fig. 32-12 Gradation curves of the Perutz "Peromnia 25" miniature-camera film resulting from varying periods of development in the Perutz fine-grain-equalizing tank developer at 20 °C

32. 5 Measuring the maximum writing speed

The maximum writing speed can be determined with one single stroke photograph if it is ensured that the spot traces a path, as a non-recurrent process, at a constantly increasing or decreasing speed. All that need than be done is to find the place on the photograph where the blackening can just barely be copied.

An elegant solution of this task is to make the fluorescent spot on the screen trace the path of a logarithmic spiral (Figs. 28-2f and 28-5). As the blackening curve rises slowly at the beginning (Fig. 32-12), though that is actually where the difference in blackening is to be determined, it is not a disadvantage if the test oscillogram shows the decrease in blackening in stages. The interesting section is then more easily recognized. The pattern of the dependence upon time of the decaying oscillation of a resonant circuit is very suitable for this purpose.

How a resonant circuit can be triggered by the time base voltage has been discussed in detail in Part III, Ch. 22. It should be noted here that the resonant circuit can be triggered very simply by the — differentiated — flyback voltage of the time base generator in the observing oscilloscope, it being possible to couple this voltage via a small capacitor. If the voltage of the circuit is applied to the vertical amplifier of the oscilloscope, either a non-recurrent or a repeatedly traced pattern of the decaying oscillation of the circuit is obtained, according to the mode of working of the time base generator.

With sinusoidal voltages the highest writing speed occurs when the fluorescent spot passes through the null position. If f_0 is the resonant frequency of the circuit, λ the wavelength of the oscillation pattern and A the oscillation amplitude, then the speed of movement of the fluorescent spot and, when the recording is studied [114]), the writing speed at these places is given as the geometrical sum of horizontal and vertical deflection by the following equation:

$$v_{s,\,\max} = f_0 \cdot \sqrt{\lambda^2 + (2 \cdot \pi \cdot A)^2}\,. \tag{32.8}$$

If the first term of the sum, λ^2, is very small compared with $(2 \cdot \pi \cdot A)^2$, it can be neglected and we obtain

$$v_{s,\,\max} = 2 \cdot \pi \cdot f_0 \cdot A\,. \tag{32.9}$$

For a chosen resonance frequency of the circuit, $2 \cdot \pi \cdot f_0 = \omega_0$ is constant, so that the writing speed can be determined from the oscillation amplitude at which the desired limit of blackening is observed. Once the trace speed in the case of given photographic conditions is known, not only can be photographic conditions for recording non-recurrent processes (moving film shots) be fixed, but also aperture and exposure time can be calculated for correctly photographing stationary single pictures.

Such investigations, described by CUSTERS and BLOK as well as by MOURIC, necessitate the use of a microphotometer for determining the degree of blackening [6] [14] [15]. Such apparatus is not available in all laboratories, however, for which reason a simple experimental method is in the following paragraph described.

To determine the border blackening, an auxiliary negative is made containing in the middle across the pattern a strip with the blackening on a bright field. Fig. 32-13*a* shows a positive of this (the line is now white and the surrounding field dark). It is advisable to make the thickness of this line roughly the same thickness as the oscillogram trace. This negative is then laid on a ground glass plate which is illuminated from below, and on top is placed the oscillogram of Fig. 32-13*b* intended

[114]) It would be very useful if a distinction were always made in publications between the speed of deflection or movement and *writing* speed (to be used only for recording).

to be used to determine the trace speed. For Fig. 32-14, both negatives were placed together and enlarged. In this way it is possible to discover immediately, by means of the measuring line across the oscillogram, the zero crossover at which the desired blackening is to be found. (For the photographs reproduced here a blackening in excess of that required for $v_{s,\max}$ had to be obtained, as it is very difficult to reproduce such slight contrasts successfully on the printed page.)

Measurement of the writing speed is only useful if data on the adjusted brightness of the fluorescent spot can be given as well. For photographs whose absolute maximum writing speed is to be determined, this is not of such great importance; in any case most favourable conditions will be created so that similar results are obtained when the measurements are repeated.

However, whenever oscillograms have to be photographed under altered conditions of brightness, a standard of measure for the light performance is required. Reference has already been made in Fig. 32-8 to the "Hiltron" light meter. As the spectral sensitivity of the photocells is never identical with that of the film, new calibration for every type of film must be carried out by means of test photographs. In the method described by FELDT [16], similar steps have to be taken. As an extension of the measuring method used in the manufacture of oscilloscope tubes, he recommends tracing a square area at the adjusted brightness and measuring its light with a barrier-layer photocell. As the photocell is provided with a filter such that its spectral sensitivity is corrected to that

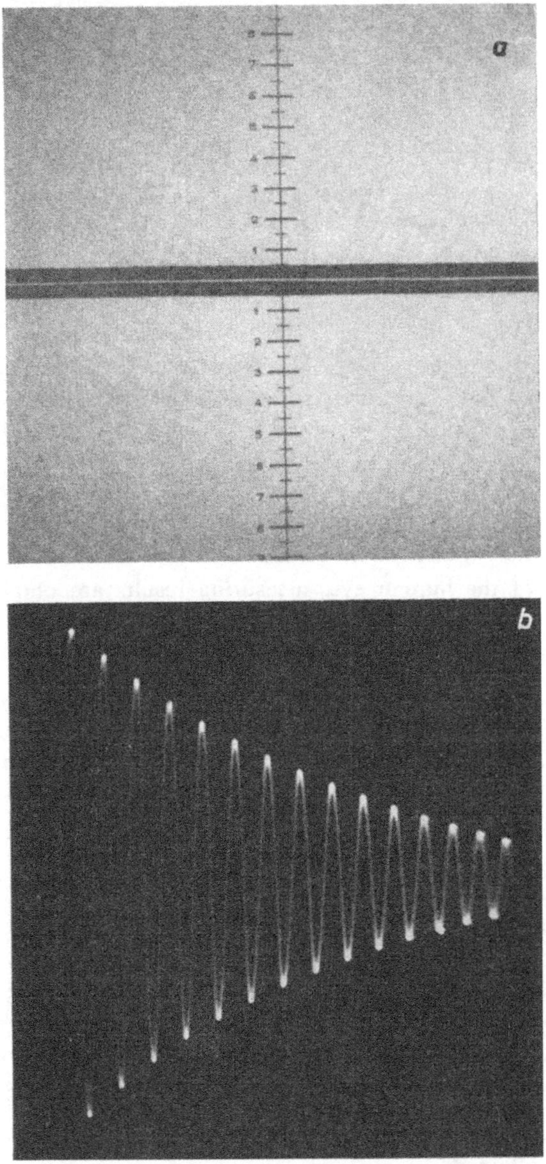

Fig. 32-13 Comparison negative (*a*) and photograph taken from the purpose of determining the recording speed (*b*)

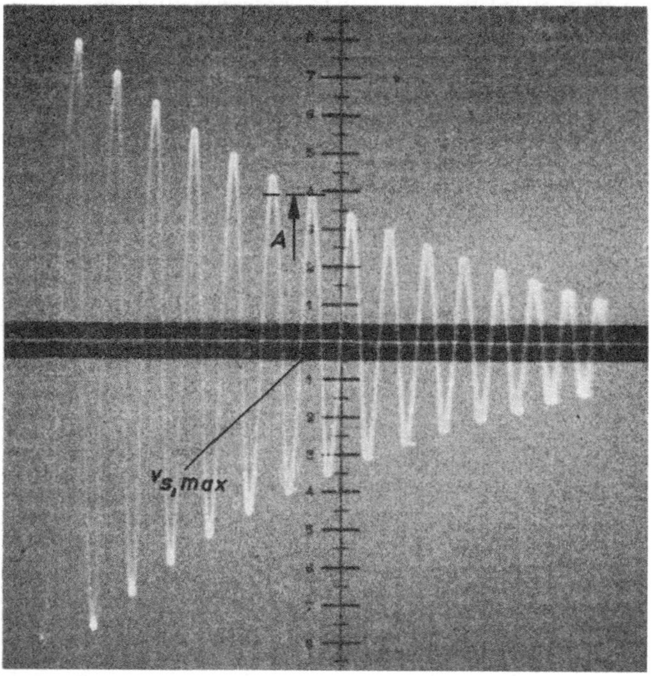

Fig. 32-14 The two photographs in Fig. 32-13 superimposed for the evaluation of Fig. 32-13*b*

of the human eye, measuring results are obtained which correspond to the visual impression. But in this case, however, it is the photographically very efficient blue portion which is measured with considerable imprecision, so that calibration must be carried out here too for the various conditions.

It has also been suggested that the current of the post-acceleration electrode should be measured as an indirect indication of the picture intensity. Apart from the fact that the actual beam current cannot be measured by this means (the screen itself has no connection and the secondary emission is not taken into consideration), the spread of screen efficiency between individual tubes is not taken into account either. However, the measurement of the current to the last acceleration electrode does give a very fair indication for a particular tube. The writing speeds determined by the different methods described change but little during the lifetime of the tube.

The adjuster for the spot intensity must of course be calibrated on the oscilloscope itself. Apart from the fact that these adjustment knobs are not always provided with scales, the scale distribution is to coarse and the change of intensity is usually so abrupt, that turning the knob a few degrees affects the whole adjustment range of the brightness. It cannot, therefore, be defined with sufficient precision. In the case of oscilloscopes in which the unblanking is carried out via a valve circuit (circuits in Figs. 4-63, 4-64 and 4-65), the additional recommendation is made that certain intensity adjustments should not be carried out until thermal equilibrium has been attained.

However, if grid and cathode are directly accessible in the oscilloscope (PHILIPS "GM 3156", "GM 5653" and "GM 5654"), a very finely regulable brightness adjustment can be achieved by an indirect method. First of all, the luminous intensity of the screen pattern is turned down with the brightness-adjusting knob of the oscilloscope, to such an extent that the oscillogram practically disappears. By means of an additional positive voltage applied between grid and cathode, an external potentiometer can be used to adjust the intensity as gradually as is usually desired. (Caution is needed, for this voltage source is usually at a cathode voltage of > — 1 kV to chassis.) The relation between the obtainable writing speed and this additional grid voltage can then be determined. Fig. 32-15 shows such a curve measured for the DN 9-5 tube with a total acceleration voltage of about 1 kV.

This type of indirect measurement of intensity also offers the advantage that it applies to all tubes of the same type. It is known, in fact, that though the cut off points of the characteristic — the cut-off voltage — can vary as between individual tubes, the slope of the characteristic is the same for all within reasonable limits, as it depends on the geometrical structure of the tube electrode system. In general, both this method and the method of measuring the current to the last acceleration electrode make it possible to obtain results with a high degree of reproductibility.

Fig. 32-16 shows two more curves for Agfa -"ISS" 21 DIN and "Isopan"- 17 DIN, also using the DN 9-5 tube, showing the influence of the total acceleration voltage on the writing speed [115]).

For the photographs in Fig. 32-16, most favourable conditions for the highest possible trace speed were provided in each case. The image scale was 3.5 : 1, the lense power 1 : 4. It can be seen that with the "ISS" film the writing speed rises from an initial value of about 0.8 km/s at 1 kV to about 20 km/s at a total acceleration voltage of 5 kV. It should be explained that these films were not quite new and that later individual photographs gave far more favourable results.

A very useful method for measuring the intrinsic luminous intensity of the

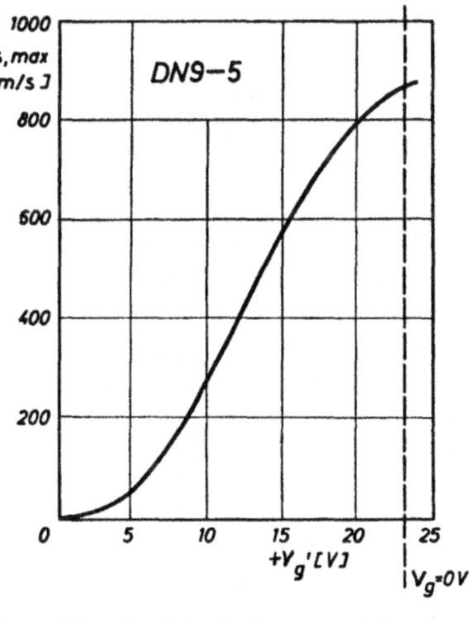

Fig. 32-15 Maximum writing speed $v_{s,max}$ in dependence on an additional, positive grid voltage

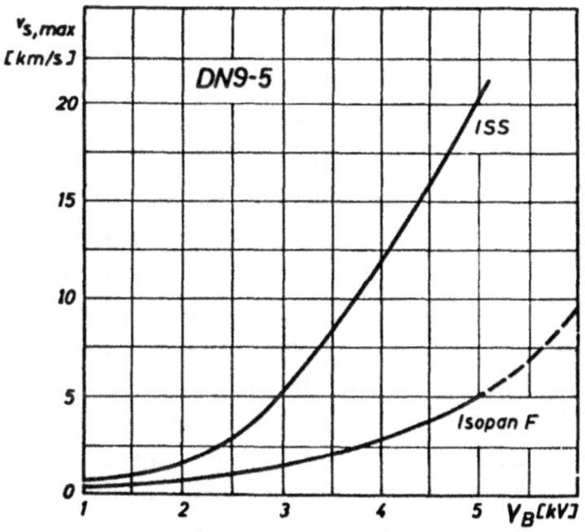

Fig. 32-16 Maximum writing speed in dependence upon the acceleration voltage for two types of Agfa films

[115]) Although these investigations were carried out some considerable time ago on an older type of tube, the results should still be suitable for giving the reader a clear indication of what can be achieved practically. The results can be applied by analogy to modern tubes as well, for instance, to the DH 10-78.

oscillogram is by comparing the screen pattern with a surface or lines illuminated in the same colour. By means of an extremely simple method described by FÄLKER and HÜCKING, the contrast filter plate, usually mounted between the floodlight scale and the oscilloscope screen, is placed in front of the floodlight grid (facing the observer) [17]. The brightness of the raster is then so altered by adjusting the filament current of the glowlamps used for floodlighting the screen, that the raster appears as bright as the oscillogram. The parts of the oscillogram most suitable for this comparison are those close to the raster boundary lines, or where the raster boundary lines cross those parts of the oscillogram where good photographic reproduction is essential. If the adjusting knob for the floodlight raster is provided with a scale, then the dependence of the trace speed on this adjustment can also be determined.

One disadvantage of this method is the fact that light of different spectral distributions, observed through the same filter does not give light of the same colour. Thus, for instance, it is relatively difficult to obtain, by filtering, light from an incandescent lamp which contains but a small blue component, light comparable with that obtained with the blue *B*-tubes, which are the most important for photographic records. It is a well-known fact in photometric practice that the measurement of differences in light intensity becomes more difficult as the sources to be compared differ spectrally.

An improvement can be obtained if the comparison light of the raster is separately filtered. This comparison light can then be better adjusted to the colour of the screen pattern. But this only applies for a certain filament temperature of the bulb on the floodlit scale. If the brightness of the scale is reduced by reducing the temperature of the filament, then the spectrum of the light has a greater component near the red region, while the blue component rapidly decreases. A completely satisfactory solution is only obtainable if the bulbs always burn at the full voltage — their filters are adjusted to this light — and the brightness of the raster is varied by grey filters in front of the bulbs. This could be arranged to take place continuously or in stages. Adjustment in stages generally makes it easier to judge the most favourable adjustment. It should also be borne in mind that for such a comparison the graticule should be as broad as the oscillogram trace if possible.

But even without this perfection there is a very simple way of obtaining a good indication by comparing the raster light with the brightness of the oscillogram. It is best in this case to adjust the brightness of the pattern on the screen to that of the raster with the bulbs operating at full brightness. If the adjustment is made at approximately the same brightness there is the added advantage that the same photographic conditions apply to both the raster and the oscillogram. After some practice, it soon becomes possible to adjust the brightness ratio of oscillogram and raster so that the trace of the oscillogram remains a fine line, as can be expected from the point focus of modern tubes, but the raster remains well and completely exposed without, however, appearing too bright in relation to the oscillogram. Uniform illumination of the graticule is of the greatest importance in this respect.

Finally, the results of a number of measurements of writing speeds is given below. It must be emphatically pointed out that these are the results of individual measurements in each case with a random combination of properties both of the oscilloscope tubes and of the films used. It might very well be the case that, using another tube or a different emulsion number of the same type of film, different results might be obtained. These results cannot be taken therefore in any way as an indication or

WRITING SPEEDS MEASURED UNDER DIFFERENT PHOTOGRAPHIC CONDITIONS

Cathode Ray Oscilloscope	Cathode Ray Tube	Acceleration Voltage	Aperture Ratio of the Lens	Image Scale	Recording Material	Measured Writing Speed	Writing Speed for Lens Aperture 1 : 1.5
Philips GM 5602	DN 10-78	4.1 kV	1 : 4	3 : 1	Agfa-Isopan-Ultra	4 cm/μs (40 km/s)	29 cm/μs (290 km/s)
					Agfa-Fluo-Rapid	3.5 cm/μs (35 km/s)	25 cm/μs (250 km/s)
					Agfa-Isopan-Record	7.7 cm/μs (77 km/s)	55 cm/μs (550 km/s)
Philips GM 5602	DB 10-78	4.1 kV	1 : 4	3 : 1	Gevaert-Scopix G	5.9 cm/μs (59 km/s)	42 cm/μs (420 km/s)
					Agfa-Isopan-Ultra	6 cm/μs (60 km/s)	42.5 cm/μs (420 km/s)
					Agfa-Fluo-Rapid	1.5 cm/μs (15 km/s)	10.7 cm/μs (107 km/s)
					Agfa-Isopan-Record	10 cm/μs (100 km/s)	71 cm/μs (710 km/s)
Philips GM 5603	DN 13-79	10 kV	1 : 1.5	4 : 1	Agfa-Isopan-Record	30 cm/μs (300 km/s)	30 cm/μs (300 km/s)
	DB 13-78	10 kV	1 : 2.8	4 : 1	Agfa-Isopan-Record	20 cm/μs (200 km/s)	80 cm/μs (800 km/s)

guide to quality. For such purpose long series of photographs would have to be taken, which would be a costly proceeding. These data are only intended for use as some indication as to what can be expected under the photographic conditions stated. For these investigations the pictures were processed every time in fresh developer, as the results obtained with the repeated use of one and the same batch of developer rapidly deteriorate [18].

As can also be seen in Table 32-1, the maximum writing speed does increase rapidly with the acceleration voltage (Fig. 32-16 [18]). Oscilloscopes from which particularly high writing speeds are demanded (surge voltage oscilloscopes) are operated at total acceleration voltages of 10 kV and more. Thus writing speeds in excess of 10,000 km/s are possible.

Even with the Philips "GM 5602" oscilloscope, writing speeds of more than 100 km/s (10 cm/μs) have been attained with a total acceleration voltage of 4 kV with a DB 10-78 tube and Agfa "Isopan-Record" film (about 34 DIN).

Under the same photographic conditions writing speeds of about 3000 km/s (300 cm/μs) were attained with the "GM 5603" oscilloscope with a total acceleration voltage of 10 kV with the DB 13-79 tube.

It should be pointed out in addition that, in the case of film for reproduction purposes, as for instance, Agfa "Agepe", which is attractive for use in long series of exposures because of its low price, the trace speed is limited to only about 1/8 to 1/10 of the maximum trace speed of high-sensitivity film.

The velocities are sometimes given in km/s. The values in cm/μs, which are also often used, are calculated simply by dividing the former by 10.

32.6 Exposure

32.6.1 SINGLE PICTURES

The writing speed required for photographing non-recurrent processes can be readily obtained, as can be seen from the previous section.

From a knowledge of the writing speeds available for a non-recurrent process the correct exposure time for a stationary picture of a cyclic process under given photographic conditions can also be calculated. It should be borne in mind, however, that although for photographing short non-recurrent processes a high screen load (beam current) can be permitted (this applies also to the maximum possible writing speed), in the case of stationary pictures the beam current must be kept within reasonable limits to avoid damage to the fluorescent screen. Its permissible value depends on the length of the path travelled by the fluorescent spot, care being taken that the permissible screen load quoted in the tube data is never exceeded. In general it will be 1/10 of the writing speed laid down for non-recurrent processes. If the writing speed v_s thus available is less than the deflection speed v_a in those parts of the oscillogram of importance for the picture, then, in a first approximation, the open time of the camera shutter should be so selected that the complete oscillogram is traced at least $\frac{v_a}{v_s}$ times during this period. To ensure that the whole oscillogram is exposed uniformly, this quotient must be made up to a whole number.

If the time base frequency is f_X, or T_X is the duration of time base cycle, the necessary open time T_V of the shutter is:

$$T_V = \frac{1}{f_X} \cdot \frac{v_a}{v_s} = T_X \cdot \frac{v_a}{v_s} . \tag{32.10}$$

Neglecting the flyback time, it is assumed that the oscillogram is having its effect on the film throughout the whole time deflection cycle. In triggered operation it should be borne in mind that the duration of a time deflection cycle is often considerably longer than the period of time during which the actual pattern is visible.

Particularly when considerable time expansions are employed, the visible portion of the time is only a small fraction of the duration of a time deflection cycle. If, for example, the voltage waveform of a single line of a 625-line television picture is displayed, the time deflection cycles occur at intervals of 20 ms (triggered by the frame frequency of 50 c/s), while the time period displayed is only about 80 μs. But the pattern on the screen is only unblanked during this short period of time, so that conditions for actual photography only obtain during a period represented by the ratio of the time duration of the forward trace to the duration of a time deflection cycle, i.e. T_H/T_X, which in this example is 80/20,000, that is, 1/250, and must be made more favourable to correspond to Eq. (32.10). Inserting instead of time T_X the usually directly known trigger frequency as $1/f_X$ and the forward trace time T_H, we obtain from Eq. (32.10):

$$T_V = \frac{1}{f^2{}_X \cdot T_H} \cdot \frac{v_a}{v_s}. \tag{32.11}$$

In this example $v_a = 5$ km/s, corresponding to a slope steepness of 2 μs/cm, and $v_s = 80$ km/s, so that $T_V = 1/3$ s. As v_s was determined with an aperture of 1 : 2.6, but the oscillogram in Fig. 19-34b, for instance, was photographed with aperture 2.0, the exposure time of 1/10 s is approximately the correct one.

In Eqs. (32.10) and (32.11) it was assumed that the same aperture ratio and image scale are used as for the determination of the trace speed v_s. Taking different diaphragm ratios and other image scales into consideration, we obtain by using the deflection duration T_X, the generally valid equation

$$T_V = T_X \cdot \frac{v_a}{v_s} \cdot \left(\frac{L_s}{L_a}\right)^2 \cdot \frac{\left(\dfrac{1}{m_s + 1}\right)^2}{\left(\dfrac{1}{m_a + 1}\right)^2}. \tag{32.12}$$

In this equation L_s is the aperture and m_s the diaphragm ratio for determining the writing speed; L_a is the luminous intensity and m_a the diaphragm ratio of the photograph. When triggering is employed, this equation must also be multiplied by the factor $\dfrac{1}{f^2{}_X \cdot T_H}$. The values for $\left(\dfrac{1}{m + 1}\right)^2$ can be read from the curve in Fig. 32-7 for the various values of m. In general, the exposure time which has been calculated should, as is usual in photography, be doubled at least as a measure of safety.

It has been assumed up till now that the writing speed necessary for the photograph (as a measure for the necessary light output) is greater than that determined for the writing speed under these conditions. If, however, as often happens, the deflection speeds are less than the writing speed corresponding to this brightness adjustment, the diaphragm must be stepped down (i.e. set for a lower luminous intensity-larger stop number) to prevent overexposure. The shutter should remain open at least for a whole time deflection cycle, so that a complete, gapless photograph of the pattern on the

Fig. 32-17 Oscillogram obtained with open-time of camera shutter shorter than the cycle-duration of a picture

screen and not one like that in Fig. 32-17, for instance, or partial double exposure is obtained. As it is impossible to set the open time of the shutter exactly to the duration of the picture, it is advisable to make the open time equal to 5 deflection cycles at least. In the case of a change of state at 50 c/s for instance this means that for the display of a cycle the shutter must be set to 5 × 20 ms = 1/10 s. To prevent excessive exposure of the film, the diaphragm and thus the amount of light falling on the film must be made correspondingly smaller. In the case of such a photograph the exposure time is laid down and the diaphragm aperture must be determined. From Eq. (32.12) we obtain (for $m_a = m_s$):

$$L_a = L_s \cdot \sqrt{\frac{T_X \cdot v_a}{T_V \cdot v_s}} . \tag{32.13}$$

If it is assumed, for instance, that $L_s = 1 : 2.6$, $T_X = 40$ ms (display of two cycles of a 50 c/s process), $v_a = 6.28$ m/s (amplitude 2 cm), $T_V = 1/5$ s (5 traces and $v_s = 50$ m/s), we obtain $L_a = 1/16$; thus the exposure should be 1/5 s at step 16.

When photographing laminar oscillograms (pictures of modulated H.F.) it must be understood that the spot instead of making a simple line trace must now pass over this whole area. Experimental investigations have shown that, compared with a line oscillogram for a picture of equal intensity, the lighting conditions must be 25 times better.

This means that if an oscillogram of a line picture of three cycles made, for instance, with high-sensitivity film with diaphragm 8 requires an exposure time of 1/25, an exposure of 1 s will be required when photographing the corresponding illuminated area. If the deflection duration permits this (it must be several times less than 1/10 s), the same result can also be obtained with an exposure time of 1/10 s, but using diaphragm 2.6. For laminar oscillograms the brightness will usually be intensified just before the exposure. This is permissible since the specific screen load is now slight. More favourable photographic conditions are then obtained.

It is especially important in oscillogram photographs that a note is made of the measuring data and photographic conditions for every picture. Otherwise it may occur that even a very successful photograph is worthless if the relevant data are unknown or incomplete. It is advisable when taking single photographs, to number them provisionally in a notebook. After they have been developed the useful photographs are given a new serial number to correspond to the record system. Counting devices are often used with equipment for photographing oscillograms. These expose an illuminated number beside the oscillogram after each photograph. This has the disadvantage that ervery photograph has a serial number, even the less successful ones. The present writer scratches the record number on the useful photographs with a sharp slate pencil after they have been developed (Fig. 32-22*b*).

32.6.2 MOVING FILM RECORDING

The correct conditions for moving film recording can be determined directly from what has been said about the writing speed. The exposure conditions are of course the same for both non-recurrent phenomena and moving film recording. Moving film recording is also used for observing and recording non-recurrent phenomenon, particularly when the process to be studied varies relatively quickly and erratically over a considerable period of time. The advantages and working conditions offered by the various recording cameras have been discussed in detail in Ch. 32.2 "Equipment for photographic recording". For this purpose it is best for the oscilloscope tube to have a short-persistence blue-type screen, as otherwise the persistent trace is included in the photograph and the oscillogram appears as blurred. Fig. 32-18, which was photographed with a *G* (*P*1)-screen, shows this effect very clearly. As can be gathered from the curves in Fig. 2-23*a*, the light emission of the *G*-tube decays relatively slowly immediately after excitation, and this rate of decay does not increase until after about 20 ms have elapsed. After 20 ms it still retains about 15% of the excitation light. During this time it can -in the case of well adjusted photographic conditions- still be able to blacken the film. As against this, it is clear from the curve of the so-called persistent screens, that their luminous intensity drops relatively rapidly at first and only decays slowly after it has dropped to about 1%. This light is still satisfactory for visual observation, but is insufficient for photographic recording since the intensity is not strong enough. Useful moving film recording can also be made with the *P*-screen. The feed speed of the film must not exceed 50 cm/s, however. Here it is possible to turn to good account the fact that the excitation light of these tubes lies more in the short wave range (blue), while persistence takes place in the long wave range (yellowish orange). If only blue-sensitive material, such as recording paper or film like Gevaert "Scopix-B" is used, then only the excitation light is recorded and the blurring effect does not take place. If, however, material of wider spectral sensitivity must be used, the same results can be obtained by mounting a blue filter in front of the oscilloscope tube. Excellent recordings at up to 3 to 5 m/s are also obtainable with screens whose persistence time is not so extremely long (Philips *N*-screen or the American phosphor *P*2) and require no blue filter. If a recording apparatus with a drum cassette is not available, the greatest time resolution of the change of state under study is limited to 4 to 5 m/s by the maximum feed speed. However, one other possibility should be indicated which serves not only to increase the time expansion by several times, but also offers a number of other advantages.

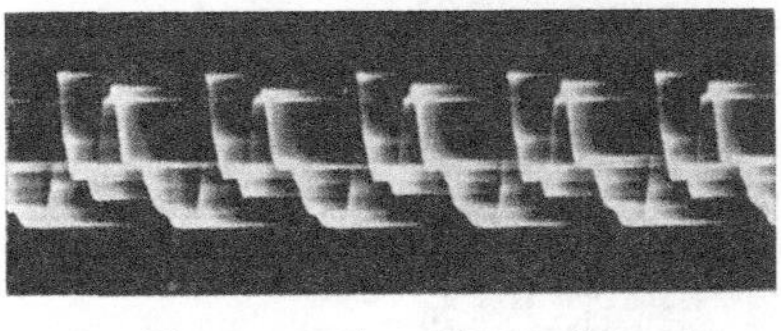

Fig. 32-18 Blurring effect due to too intensive persistence of the *G*-screen

Fig. 32-19 shows first of all a normal moving oscillogram for the sake of explanation. It portrays the waveform of the current pulses in the cathode lead of a directly heated full-wave rectifier valve, when the associated apparatus is switched on. So long as the voltage on the charging capacitor in the circuit of the following filter is still small, current flows during each half-cycle at a value permitted by the cathode temperature at the moment. If the capacitor is only charged with rectified current a charging current can only flow so long as the capacitor voltage is lower than that occurring across the rectifier valve. Current pulses are thus obtained (see also Fig. 3-3*d*). This cannot be very clearly seen, however, in the photograph in Fig. 32-19, as these pulses are too short in proportion to the observation time of the whole process. In order to show the waveform of these pulses and the changes in their duration with some degree of clarity, the speed at which the film is fed would have to be several times greater. But then the whole recording would be very much longer.

It is, however, possible both to have a clear view over the whole of a fairly considerable period of time and to recognize the details of the change, if — as has been shown by ŠILINK — a moving film oscillogram was recorded in lines [19]. For this purpose the time base generator must be so used and the direction of deflection changed round, that the time base voltage now serves for vertical deflection, while the fluorescent spot is deflected in the horizontal direction by what would normally be the vertical voltage. If the photographic material in the camera moves constantly in a horizontal direction, we obtain a photograph which, as can be seen from Fig. 32-20, shows the change of state in lines, time-expanded $3^1/_2$ times in this case. The moving speed of the photographic material was the same, 9.5 cm/s for both photographs in Fig. 32-20. The time resolution in Fig. 32-20*b* now corresponds to about 33 cm/s. As can be clearly seen, only a small fraction of half a cycle is taken up by the flyback between the individual lines, which causes practically no error.

Line recording brings with it the advantage that a satisfactory time resolution is obtained in a restricted area of film. This is also particularly useful when a spontaneous change of state is expected at some unknown instant within a considerable period of time, but has to be suitably time-expanded. (The lines can then be recorded very close together.) If it is desired to obtain this time resolution at a correspondingly high feed speed, it would involve an unnecessarily large consumption of material during the waiting times.

The one blemish that might be found in the picture in Fig. 32-20*b* is the extreme slope of the line trace. On the screen, as is shown in Fig. 32-20*a*, the oscillogram is displayed traced straight, as is usual, in the direction of deflection. This pattern is,

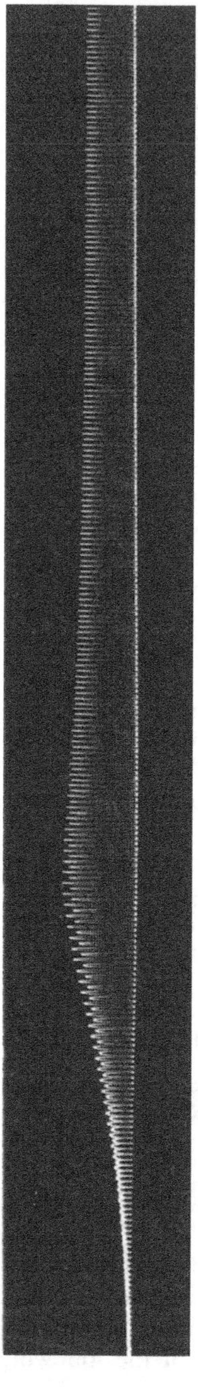

Fig. 32-19 Moving film oscillogram of the current pulses through a directly heated two-way rectifier valve to the charging capacitor of the filter. Film transport speed is 9.5 cm/s

however, obliquely distorted by the feed of the photographic material. This small imperfection can be easily overcome by tracing the screen pattern with a corresponding predistortion, as shown by the oscillogram in Fig. 32-21*a*. The moving film record in Fig. 32-21*b* thus gives approximately horizontal lines. (The correction was a little exaggerated in this oscillogram, with the result that the lines now lean a little to the left.)

This correction can be obtained by applying a current from the sawtooth generator to one pole of the vertical amplifier output. To obtain the photograph in Fig. 32-21*b* the output of the time base generator was connected, via the series connection of a resistor and a capacitor, directly to the low-resistance output of the vertical amplifier (see also circuit in Fig. 11-4). The visible, yet slight non-linearity of the time deflection can be explained by the additional load imposed in this way. It is therefore advisable to interpose an impedance matching device in such a circuit, for instance, a cathode follower stage.

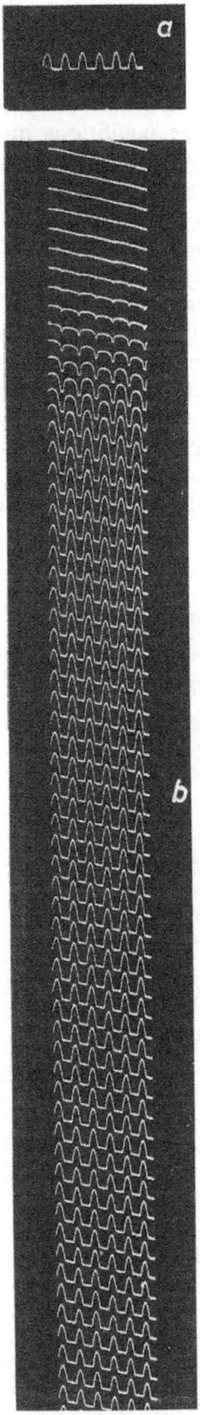

Fig. 13-20

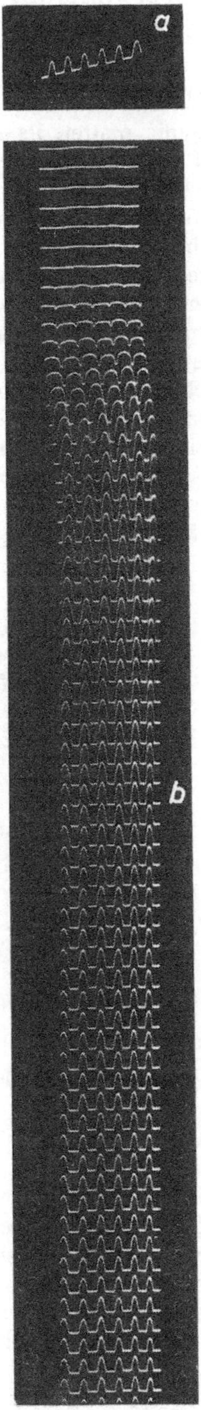

Fig. 13-21

Fig. 32-20 Process in Fig. 32-19 time-expanded $3^1/_2$ times by means of line recording. *a*) Single picture on the oscilloscope screen; *b*) moving oscillogram

Fig. 32-21 As 32-19, but with horizontal lines due to pre-distorted screen pattern. *a*) screen pattern; *b*) moving oscillogram

32. 7 Processing the photographs

32.7.1 HANDLING THE PHOTOGRAPHIC MATERIAL

As the photographic material contains organic matter, its storage conditions must not be too damp, too dry or too warm (if possible the temperature should be less than 20 °C). In particular, if stored for long periods in laboratories or factory shops which are usually very dry and often warmer than 20 °C, as in the case of central heating with no special humidification of the air, the film can easily become brittle.

Most photographic emulsions are covered with a thin protective layer intended to guard against slight scratches and, by means of anti-static additives against lightinlike exposure. In spite of this it sometimes happens when particularly dry materials are wound on to the basket spools of the developing boxes or even during movement when taking moving film records, that such lightinlike exposures occur. This takes place when the film carrier, which is a good insulator when dry, charges up through friction against pieces of cloth or by rubbing against dry fingers, and its potential becomes so great that it sets up a brush discharge on the surface of the film. These sparks fog the photo-sensitive layer. Photographs can be rendered useless in this way. Fig. 32-22 *a* and *b* show the results of two such flashes.

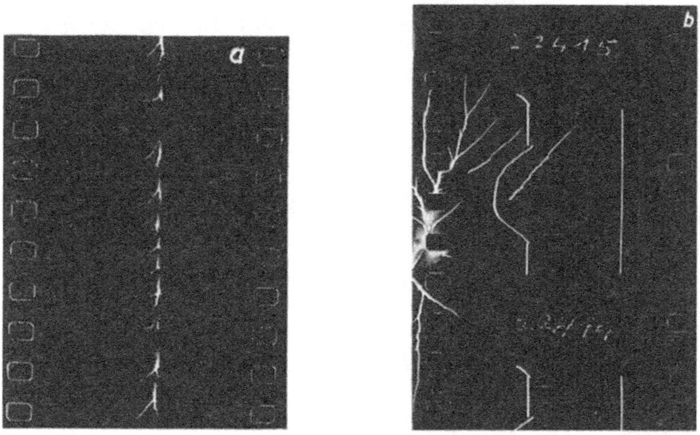

Fig. 32-22 Lightinlike exposure on sections of the record. *a*) Moving oscillogram; *b*) individual strips of film on which the lightinlike exposure is probably due to friction during loading of the film

32.7.2 DEVELOPING

The expenditure on purchasing developing equipment depends on the amount of photographic work it is intended to undertake. For occasional laboratory photographs it will suffice if the exposed material is developed in the familiar developing cans. Usually these boxes, in which the film is placed between the so-called Correx bands in the dark, are most suitable. These boxes can be quickly dried out and then immediately used afresh. It is, in practice, often necessary to cut off and develop a piece of film perhaps containing test photographs. In the case of material sensitive only to blue light (recording paper) a dark red light can be allowed, but is is advisable to avoid every kind of dark-room lighting when using highly sensitive photographic material. One can quickly accustom oneself to performing the few operations required in complete darkness.

For the simplest requirements a light-proof room at least one meter square is sufficient. Here the film is taken out of the camera and placed in the developing can. Everything else can be done in daylight. A properly equipped dark room is of course required for making enlargements.

For developing oscillograms on 35 mm cinefilms, the "Jobo" five-stage tank has proved very successful. It can hold five 1.6 m lengths of photographic material and can be used to develop, fix and wash up to 8 m (cut in four pieces, of course). Fig. 32-23 shows a picture of this tank with its loading basket and Correx bands.

Fig. 32-23 "Jobo" five-level tank with loading basket

If photographic film is used, it is even possible to charge each level with the double amount of film. Great care must be taken, however, that the double lengths of film, that is, a maximum of about 3.2 m in each stage, are folded so that the light-sensitive layer is facing outwards. If they are accidentally so placed that they face inward, they will become stuck together and the whole film will be spoiled. The side with the sensitized layer can be recognized even in the dark, because the film kinks to that side. In case of doubt the film can be moistened at the edge of the side in which the sensitive layer is thought to be, and then scratched with the finger-nail. The layer becomes soft and flakes off. With double-loaded film, for instance, the total charge of the magazine of the "PP 1014" Philips moving film recording camera can be developed in one batch.

When a considerable number of photographs have to be dealt with, developing cans are no longer sufficient and equipment for carrying out the development in tanks or in developing machines must be acquired. The photographic industry can supply equipment to meet any requirement. Even the equipment and instructions necessary for setting up a dark room are obtainable from various firms. The brochure on "the smallest laboratory for black and white photography" issued by Agfa is particularly recommended.

Generally speaking, oscillograms should be developed "hard", that is to say with developers the action of which is rather vigorous. The highest writing speeds can then be employed. Those developers known as *X*-ray developers are most suitable of all, as they are intended to give negatives with good contrast and ensure the best use of the exposure. They have the additional advantage of a relatively short developing time (3-6 min.). Their somewhat coarser grain is no serious drawback in the case of oscillogram photographs, as no particularly big enlargements are required.

But when oscillograms covering an illuminated area have to be photographed (for instance, the illustrations in section 23), particularly if they are not traced by voltages linear with time and are hence not uniformly bright, developing must be "softer" to prevent the contrast from being undesirably great on the photograph. In that case either the *X*-ray developer is used for a very short time only (2 min) or a normal equalizing developer of the "Rodinal", "Perinal" or some similar type is used [20].

An absolutely fine grain developer should not be used, as it would then reduce the writing speed to half or less. The same applies to photographs with varying deflection speed, such as pulse voltages with steep slopes.

While developing should usually be done at 20 °C, the temperature can be raised to 22 °C where it is desired to exploit sensitivity to the full. At 22 °C it was often possible to reach writing speeds of twice the value of these employed at 20 °C.

32.7.3 ENLARGING

As explained in detail under "Writing speed" and the influences governing it, it is an advantage to photograph the oscillogram reduced in size. Usually, therefore, an enlargement is required to enable the oscillogram to be interpreted. The bibliography already referred to gives more detailed information as to the many publications intended for the amateur photographer. In view of this it is not necessary to consider this aspect further. It should only be pointed out that, contrary to pictorial photography, the prime requirement is good "contrast". The aim must be to reproduce the oscillogram as nearly as possible pure white on a completely black background. Hard or extra hard enlargement paper is therefore generally used. For fast working, sensitive silver bromide paper is commonly used. In the case of particularly faint photographs, in which even the use of extra hard silver bromide paper does not produce sufficient contrast, it is possible, by using sufficiently powerful enlargers with projection lamps, to use the less sensitive "extra hard" silver chloride paper.

However, when enlarging oscillograms of illuminated surfaces, as shown in Part III, Ch. 23, the contrast must not be pushed too far during enlargement, to avoid its becoming excessive over the luminous surface. Care must therefore be taken that normal grade paper is used to give at most a moderately "hard" enlargement. It should be pointed out that not all firms use the same description for the various grades [21]. When using products of different makes it is advisable to make test exposures with these materials in stages, doubling the duration of the exposure for successive sections for purposes of comparison.

The change of blackness over the area of the photograph is graduated and shows no sharp edges. The intrinsic brilliance of the spot on the oscilloscope screen is not constant, of course, since it varies with the intensity of the electron beam and therefore decreases from the centre outwards. This fact should be taken into consideration when the trace in the enlargement is intended to be as thin as possible. A broader or a thinner line is obtained according to the variation of density of the trace, which depends on the duration of the exposure. The much enlarged section of a part of an oscillogram which is shown in Fig. 32-24 serves to demonstrate this. From step to step the exposure time was doubled in successive sections of the trace. In the last part even the dense centre of the trace appears gray. This exposure would therefore be too long. The most favourable exposure is the one which precedes it, and on it the trace is narrow but still pure white. In general, the blackening of the oscillogram varies over the complete trace, and particularly where the line is the thinnest. The dimensions of the enlargements can be made to conform to requirements and can be of any size desired. Usually enlargements are either 6 × 9 cm or 6 × 6 cm, according to the aspect ratio of the oscillogram. Most of the oscillograms shown in this book have been enlarged to these dimensions, but in special cases have been enlarged to even greater size.

Fig. 32-24 Influence of duration of exposure on the line thickness of the oscillogram trace at enlarging

As HOFFMANN has shown, oscillograms can also be enlarged on blue print paper by using special apparatus [22]. This can be of particular interest if it is frequently required to enlarge considerable numbers of oscillograms. As an example may be quoted the growing practice of submitting oscillograms along with measurement record sheets, and/or descriptions and delivery documents as incontestable evidence.

32.7.4 RETOUCHING

An oscillogram can be expected to display trend of a charging quantity faithfully and without distortion. This assumption would discountenance any kind of touching up whatsoever, as, even the less experienced person will readily agree. In principle, therefore, nothing should ever be changed on the trace itself. Parts of the voltage waveform which cannot be reproduced satisfactorily by electrical or photographic means, should be omitted and their absence accepted. Only in the case of particularly faint negatives should it be permitted to touch up the background of the positive, without, however, correcting the oscillogram itself.

This principle was consistently adhered to with all the oscillograms dealt with in this book. In all the screen patterns the traces of the spot have been shown unchanged and without additions.

CHAPTER 33

LARGE PICTURE PROJECTION OF OSCILLOGRAMS

33. 1 Need for enlarged reproduction of oscillograms

The surface area of the fluorescent screen of the normal 10 cm or 13 cm diameter cathode ray tubes, is adequate for the saticfactory display of all ordinary oscillograms; indeed it is for this reason that the 10 cm and 13 cm dimensions have been selected and more or less standardized. Even fine details can be observed at normal reading distance without difficulty. In special cases it may be advantageous to use magnifying glasses when making critical examinations, especially in the case of tubes with a sharp spot focus.

The size of these oscillograms also satisfies all the requirements of photographic recording on negative material, which permits of a very wide range in enlargement. The case is different when the oscillogram has to be traced on transparent paper to provide a rapid and simple record for later interpretation [1]. For such purposes a fair-sized enlargement is desirable. If the oscillograms have to be shown to a large audience in schools and lecture halls as a direct illustration of the subject under discussion, it is necessary to enlarge them to between 5 and 30 times.

Provided the oscillogram is sufficiently bright it can be shown enlarged on the projection screen by means of a suitable lens mounted in front of the fluorescent screen. Most oscilloscopes of good quality are provided with a suitable device for mounting such an optical system.

33. 2 Optical system for projection

The most suitable optical system and type of projection screen obtainable to suit requirements as to size and brightness of the projected picture should be selected. As a considerable distance is required between the oscilloscope and the projection screen, the focal length of the optical system must be at least 150 mm. If the brightness of the oscillogram is considerable and particularly large projected pictures are not required, an aperture ratio of 1 : 4, such as in the Philips "GM 8024" optical system is sufficient. On a good projection screen (Ch. 33.4 "Projection screen") pictures covering an area of $^1/_4$ m^2 can be shown at satisfactory brightness. For pictures of greater dimensions, optical systems with an aperture ratio of 1 : 2.8 or better are required. A system of this sort, for instance, the Astro-Pantachar 1 : 1.8, is supplied by Philips as the "PP 1042" projection optical system. Fig. 33-1 shows this lens and its mounting. Its optical qualities satisfy stringent requirements, and the screen patterns even on tubes of optimum spot focus are reproduced without introducing any perceptible error. Its focal length is 150 mm. The structure of the system and aberration curves can be seen in Figs. 33-2*a* to *d*. It comprises a compensated anastigmat with four single lenses of slight curvature and flat image field. In view of its use as a lens for the projection of slides, a very considerable degree of colour correction over the essential part of the visible spectrum from 400 to 600 nm

Fig. 33-1 Philips "PP 1042" projection lens mounted on an oscilloscope head

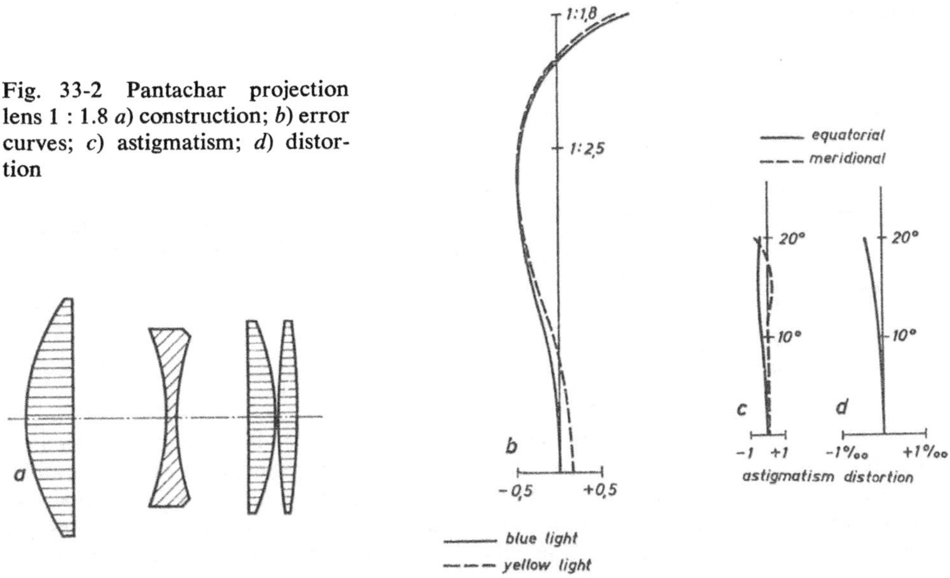

Fig. 33-2 Pantachar projection lens 1 : 1.8 *a*) construction; *b*) error curves; *c*) astigmatism; *d*) distortion

has been introduced, so that it is now suitable for the projection of light of practically every visible colour.

33. 3 Cathode ray tubes

Even although when such powerful lenses are used oscillograms can be reasonably enlarged at normal working voltages without special measures being taken, it will readily be understood that an increase in the intrinsic brilliance of the light spot is very desirable in certain circumstances. It must always be remembered that, just as in the television picture, the intrinsic brilliance of the spot on the screen must be sufficient to enable its luminous flux, distributed over the total projected area, to give an impression of sufficient brightness. When it is realized that the spot diameter of good cathode ray tubes is only a fraction of a millimetre, it becomes clear that this demands extremely high intrinsic brilliance. The simple expedient of increasing the beam current of the cathode ray tube is only a partial and limited solution, as without increasing the beam velocity a stronger electron current produces a defocusing effect and thus reduces the increase of the intrinsic brilliance. Th energy contained in the electron beam is given by the following Eq.:

$$N_E = I_E \cdot V_A \,. \tag{33.1}$$

It is therefore obvious that it should be raised by increasing the acceleration voltage V_A.

Moreover, it is of particular importance that the degree of screen efficiency [cand/W] shown in Fig. 33-3 should, in conformity with the statements in Part I, Ch. 2, increase with the acceleration voltage V_A. To prevent the deflection sensitivity from decreasing in inverse proportion to the acceleration voltage, it is advisable to obtain the total acceleration voltage as far as possible by means of a high post-acceleration voltage (Part I, Ch. 2.10 "Post-acceleration").

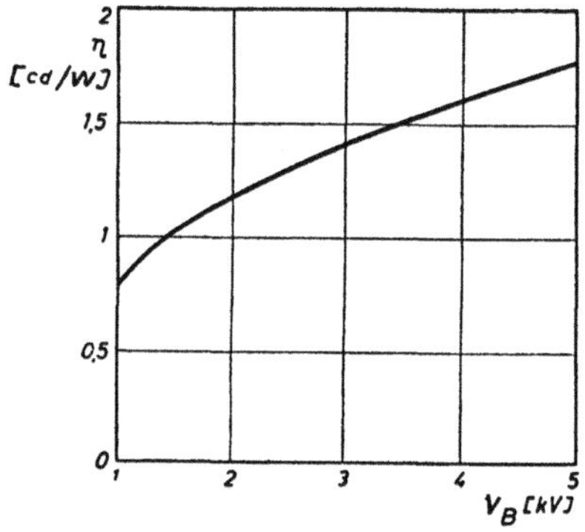

Fig. 33-3 Degree of efficiency of the fluorescent screen of an oscilloscope in dependence upon acceleration voltage

There need be no serious apprehension on the score of blackening the screen as a result of the high load. From experience it has been confirmed, time after time, that it is possible to increase the luminous intensity to the maximum value — at high acceleration voltages —for the time required to display experiments, without the slightest risk. In any case most fluorescent screens recover to a greater or lesser degree fairly rapidly after a short period of overload if a pattern containing luminous areas is formed on the screen afterwards for a short time. Of course, the maximum load should only be used so long as it is really required. Moreover, when good lenses are used, screen brightnesses at which no blackening of the screen occurs are quite sufficient, particularly in the case of oscillograms which are not too small in area. For patterns which produce luminous areas (for instance, oscillograms of modulated voltages) it is in any case always possible to adjust for maximum brightness, as then the load occasioned by the small fluorescent spot is spread over a large surface. Particular care must be taken that the deflection voltage is not accidentally cut off, as then the fluorescent spot remains stationary and will certainly burn the screen at that spot. For general large projection, cathode ray tubes with a *H*-screen (*P* 31) are best used as their greenish-yellow light coincides very closely with the maximum sensitivity of the human eye.

If it is required, however, to display very slow processes such as, for instance, heart muscle tensions, slowly changing mechanical conditions and the like, to a fair-sized audience, it is more advisable to use particularly long-persistence tubes of type *P* (*P*7) in the oscilloscope. To prevent the bright fluorescent spot of the excitation light from dazzling the eye, it is recommended that, just as in the case of direct observation of such screens, an orange filter should be employed for projection. The yellow, long-persistent light, particularly after the observer has become somewhat adapted to the dark, is then more readily perceptible.

It should also be pointed out that for such tasks a very satisfactory projection can be obtained when the fluorescent spot path of a blue-persistent tube is projected on to a screen covered with a phosphorescent material. Then, of course, a filter is not required, either for the tube or for the screen.

33. 4 Projection screen

It follows from the preceding explanations that the luminous flux as it passes from the oscilloscope tube to the screen and finally to the eye of the observer, must be used as economically as possible. This means that not only the proportion of the flux reflected from the projection screen, but also the distribution of its luminous intensity, is of fundamental importance in determining the luminous intensity of the projected picture as observed by the spectator. In the language of the lighting engineer it is known as the "characteristic of light reflection".

Screens such as are often used for projecting slides during lectures, and consisting of linen stretched on a collapsible frame, are quite unsuitable for these purposes because of the extraordinarily low index of reflection (the light transmitted is about equal to the reflected light). "White" projection screens, which are available in the majority of lecture rooms, although having a higher coefficient of reflection, have a reflection characteristic as shown in Fig. 33-4, curve IV. They are almost completely diffusely reflecting surfaces with the same intrinsic brilliance in all directions. The Lambert cosine law applies to them. The luminous intensity I_a of a completely diffusely reflecting surface with the reflection coefficient ρ and receiving luminous flux at an angle α to the normal is given by the following equation:

$$I_a = \frac{\rho \cdot \Phi}{\pi} \cdot \cos \alpha \,. \tag{33.2}$$

If such a surface is illuminated perpendicularly, the intrinsic brilliance of the visible surface section appears equally bright from all observation points. The intrinsic brilliance index is therefore a segment of a sphere [2].

For the present purpose the screens which are desirable not only have a high coefficient of reflection, but above all emit as far as possible the whole of the luminous flux in the direction of the observer; in other words their reflection properties are directional. Such projection screens are known as selective screens; the best known type being the so-called pearl screen or beaded screen, consisting of a white linen base on which small beads of glass are glued. In order to gain some insight into practical results, the characteristic of light reflection of four samples of common types of projection screens [116]) were recorded by means of a miniature camera projector and a luxmeter. The luxmeter was provided with a tube socket 10 cm long with an opening of about 2 cm^2 for the purpose of measuring the intrinsic brilliance.

[116]) Mechanische Weberei Bad Lippspringe (Teutoburger Wald) and Schumann, Hamburg.

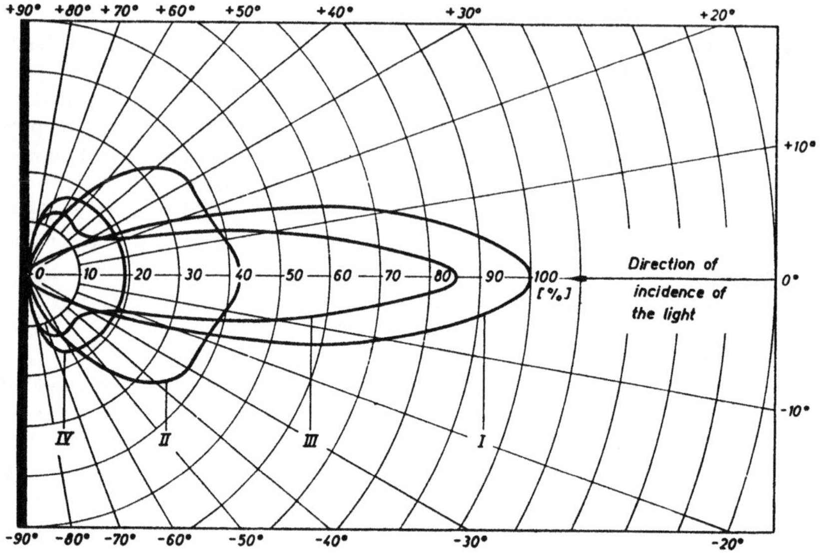

Fig. 33-4 Reflection curves of various types of projection screens. I Admira-Extra (smooth silver screen); II Schumann-fluted screen (silver screen with vertical, cylindrical section grooves); III Atlanta-Extra (pearl screen); IV Blankana-Extra (white, lacquered linen)

Fig. 33-4 shows the results in a relative, linear scale. The luminous intensity of the silver screen (curve I) in the direction of projection was taken as 100%. From this it can be clearly seen that, although the dense type IV (Blankana) matt-white varnished screen is not particularly suitable for the purpose, it is very considerably superior to the makeshift screens previously mentioned.

To obtain directional reflection one is often inclined to use the pearl screen which is such a favourite for home-cinema and colour-slide projection. This screen does in fact give a maximum of reflected light several times greater as can be gathered from curve III in Fig. 33-4, yet the angle is particularly narrow, with the result that only a few observers in the vicinity of the direction of maximum reflection enjoy a satisfactory picture. When used in the home it is much easier than in a hall filled with spectators to ensure that the small number of those taking part are so placed as to be seated in the favourable area.

The pearl screen has also another disadvantage which must be taken into consideration, and that is its unusual behaviour if the incident light is at some other angle to the screen than a right angle. Fig. 33-5 shows four curves of the screens under discussion with the light striking the screen at an angle of 30° to the vertical.

The white screen (IV), as expected, shows a curve similar to that obtained with vertical projection. The reflection coefficient is now somewhat less. While the reflection curves of the silver screens (I and II) distort in accordance with the rule that the angle of reflection of light is equal to the angle of incidence, the reflection of the pearl screen is seen to be in the direction of the light source. This is due to the fact that the light is reflected from the back of the individual glass beads, which have the effect of small concave reflectors. But this means that the pearl screen cannot, as is the case with "real" reflecting surfaces, cause light to be distributed. The greater portion of the reflected light always returns towards the projecting oscilloscope and its immediate environment [3].

As can be seen from Figs. 33-8*a* and *b*, it is usually necessary to direct the light from the projecting apparatus upwards to the projection screen, and, by suitably inclining the screen above the apparatus and the projectionist, so direct the light towards the auditorium, that the greatest possible number of the audience can see a bright projected picture. But according to the curves in Fig. 33-5, this is only possible with the so-called silver screens. In general, the most favourable would be a screen with a type I reflection curve. Type II screens are only required when projection has to be from the front in a wide auditorium. The maximum light impression is then less than with type I.

The reflection curves discussed hitherto should be regarded practically as rotation-symmetrical bodies. They have the same form both sideways and vertically. This is an advantage when the seating in the auditorium extends to a considerable height, as is indicated, for instance, in the sketch in Fig. 33-8*b*, for it ensures that vertically too, the majority of the observers are placed in the most favourable range of the reflected light.

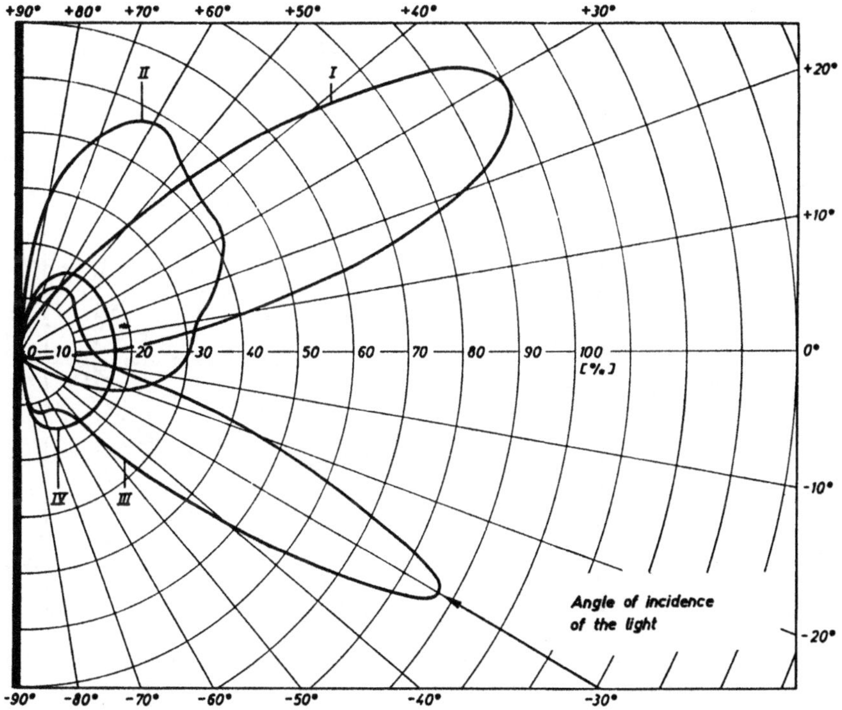

Fig. 33-5 Reflection curves of various screens as in Fig. 33-4, but with steeply incident light

The most suitable curve is that of the so-called "silver screen" — type I — (layer of sprayed metal). It has a favourable form but deteriorates rapidly at the sides. The half-power angle is usually given as a measure of the properties of the projection screens in this respect. This is the angle of reflected light at which the reflected luminous intensity has fallen to half the maximum value. Table 33-1 shows the collated results of these measurements. As the screen must nearly always be inclined for this purpose, projection surfaces mounted on card or similar material are better than screens which can be rolled up.

The only disadvantage of the silver screen is that it is easily scratched and damaged. If it has to be transported a good deal it should be protected with a cover of soft material.

TABLE 33-1 TABLE OF SCREENS FOR THE CURVES IN FIGS. 33-4, 33-5 AND 33-6.

Projection screen	Curve	Half-power angle
"Admira Extra" (silver screen)	I	30°
Schumann silver screen	II	100° [117]
"Atlanta Extra" (pearl screen)	III	24°
"Blankana Extra" (white linen with matt lacquer)	IV	160°
Back-projection screen (horizontal)	I	45°
(vertical)	II	20°

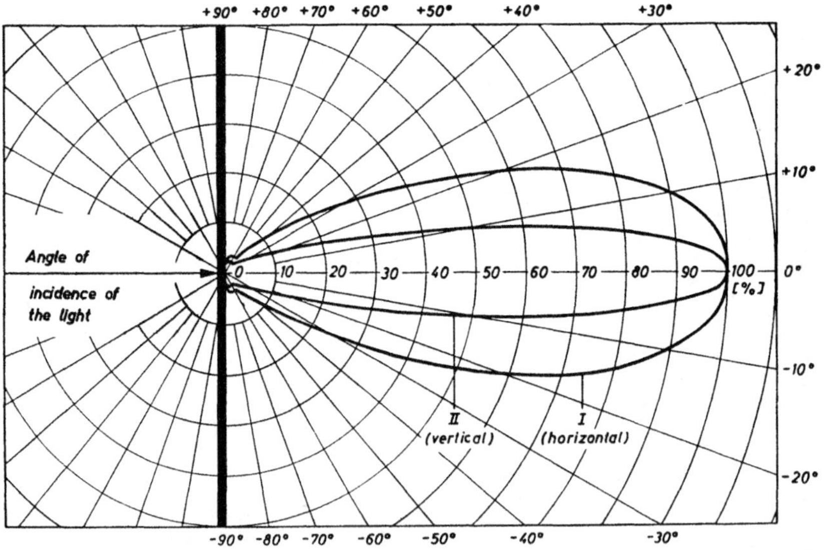

Fig. 33-6 Curves of the selective transparency of a back-projection screen of a Philips home-projection receiver. I in the horizontal direction; II in the vertical direction

[117]) Only 32° in the vertical direction.

If the observers are arranged on one level, a screen which concentrates the light on that level is very favourable. The projection screens of transparent plastic material used in the Philips "TD 2312A" home-projection television receiver have such properties [4] [5]. The screen is 45 × 35 cm in size and has been developed for back-projection. The light source, seen from the position of the viewer, is behind the screen. This screen has a so-called Fresnel lens pressed into it at the side facing the lens, and at the front a pattern consisting of small hyperbolic lines overlaid by a sinusoidal wave, thus forming small lenses.

By this means a favourable concentration of the transmitted light is obtained and at the same time there is the certainty that viewers situated at the sides see a uniformly bright picture [4].

Fig. 33-6 shows the curves of light transmission in the horizontal direction (I) and the vertical direction (II) on the same scale. The limits at half transmission are thus about 45° and 20°. The maximum illumination is about 20% higher than for the silver screen (I in Fig. 33-4). The use of such a screen for display of oscillograms can be seen in the photograph in Fig. 33-7. Philips projection screens produced for their home-projection television receivers, which have become unserviceable due to slight damage, scratches etc. can be used to advantage for this purpose.

The most favourable alignment with respect to the viewers is achieved by tilting the screen downwards and correspondingly raising the oscilloscope and lens. If the oscillogram occupies the whole area of the screen it is possible to obtain a picture in which an audience of up to 300 persons can still make out details.

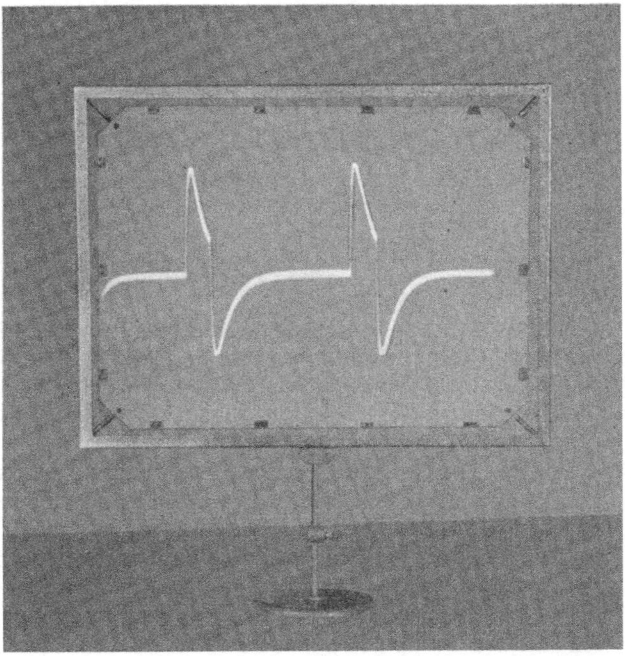

Fig. 33-7 Embodiment of a back-projection screen according to Fig. 33-6

33. 5 Placing the oscilloscope, projection screen and viewers

Because of the large proportion of directed reflection, the placing of the oscilloscope and of the projection screens as well as the seating arrangements for the viewers must be so selected that the reflected light (or, in the case of a back-projection screen, the diffused light) is directed towards the middle of the group of viewers. On the basis of a half-power angle of 30° for the silver screen, arrangements like those shown in Figs. 33-8*a* and *b* can be used in a lecture-room.

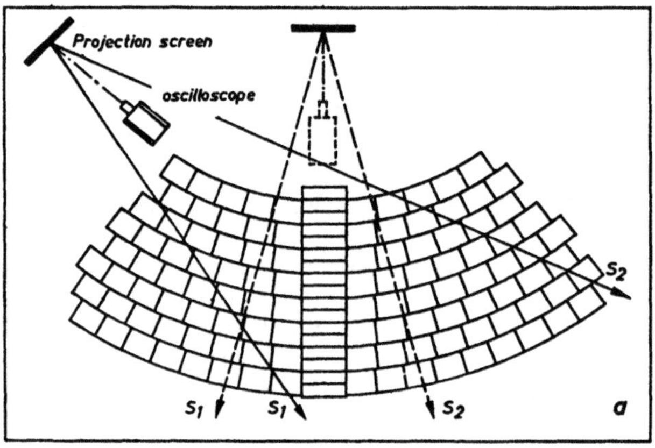

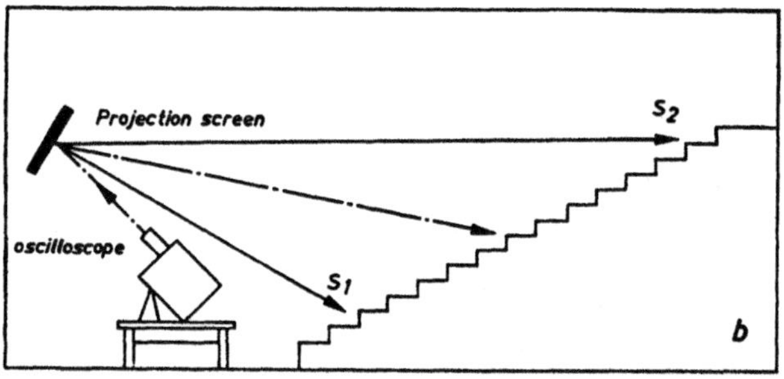

Fig. 33-8 Suitable arrangement of the projection equipment in a lecture room; *a*) in the horizontal direction, *b*) in the vertical direction
a. Projection screen
b. Projection screen oscilloscope

From Fig. 33-8 it is clear that the half-power angle of 30° is sufficient in a vertical direction, and that even in rather large lecture-rooms all those present within the area covered see a picture of adequate intensity. It is rather more difficult to find a satisfactory solution in the horizontal direction, as in most lecture halls the seating is arranged in breadth. If the central arrangement of the projection equipment shown in dotted lines, which is always attempted when lecture-rooms are being equipped, is adopted, a considerable number of the viewers would find themselves outside the range of the half-power angle and in fact in the area over which there is a rapid drop in the light intensity. The arrangement in the left (or right) corner sketched in full lines, in which the viewers sit at different distances from the screen, is much more favourable, as the luminous intensity of the picture remains well within acceptable limits [118]).

It is advisable to arrange the projection oscilloscope table so that the auxiliary equipment usually required can also be set up on it [6].

The distance from the oscilloscope to the projection screen is a multiple of the focal length, corresponding to the required enlargement. With a focal length of 185 mm ("GM 8024" projection-optical system) an enlargement of about 3.5 times is obtained at a distance of 1 m, and of about 15 times at a distance of three metres. With a focal length of 150 mm ("PP 1042" projection-optical system) the distances are 0.75 to 5 m and the enlargement ranges from 3 to 30 times.

The picture distortion occasioned by the slight inclination of the lens axis to the perpendicular on the screen has been found in practice to be barely noticeable. Of course, the top and the sides are seen the wrong way round. The latter is desirable in the case of upwards projection (from the front), as the direction of view is also reversed. The top-to-bottom reversal will not usually be important. If it is objectional it would be necessary for the connections of the pairs of plates to be interchanged, or an auxiliary amplifier for phase reversal of the oscilloscope output would have to be used. As no great gain is required from this amplifier, and the amplitudes can be small, a wide frequency band can be obtained without any deterioration of the performance of the *Y*-amplifier.

In the case of back projection, the sides have to be reproduced in the correct position, so that the direction of deflection must be reversed in both coordinates. Fig. 33-9 shows the practical setting up of apparatus for upwards projection together with an original oscillogram, the time sequence of the decaying oscillations of a resonant circuit being displayed. On a "silver" projection screen one metre square

Fig. 33-9 Projection equipment with screen pattern

118) To a certain extent in directed reflection the luminous intensity indicator is also subject to these laws [2].

in area. Projection equipment and oscillogram were photographed separately. For the photograph of the room the screen was covered with black paper, and for the photograph of the oscillogram the room was darkened.

33. 6 Enlarged reproduction of oscillograms by means of industrial television apparatus

Although no direct large-picture projection takes place in the sense used in this chapter, the use of television apparatus will be dealt with for the sake of completeness. It is to be expected that in the future well equipped lecture-rooms will have an industrial television installation with several direct view receivers at their disposal. This would enable oscillograms to be shown to a larger number of viewers. The oscillogram, like any other object to be televised, is scanned by a television camera and transmitted via the system to the receivers or to a large-picture projector (for instance, Philips "VE 2601" or "EL 5750").

A very important advantage of this method is of course the possibility of using suitable photographic optical systems (frontlens attachments) to show particularly small oscillograms as a large picture. Brightness and contrast can be amplified by the television installation. Fig. 33-10 shows an example of an oscillogram photographed on a normal table receiver after being transmitted over a rather old Philips industrial television system. It shows a part of the T.V. vertical synchronous pulses, time-expanded. From this it can be seen that the transmission of such oscillograms is satisfactory in spite of the poor brightness of time-expanded screen patterns. In this case an iconoscope was used in the camera. Using a vidicon as in the Philips "GM 4930" camera, screen patterns of even poor brightness can be transmitted. A blue-persistant (DB10-6) tube operated at 3 kV total acceleration voltage matching the colour-sensitivity of the iconoscope in the oscilloscope was used for the photograph in Fig. 33-10.

Because of the different spectral sensitivity of the vidicon (Resistron), greenish-yellow phosphorescent oscilloscope tubes (*G* - *P*1 - or *N* - *P*2 types) are most suited when these camera tubes are used.

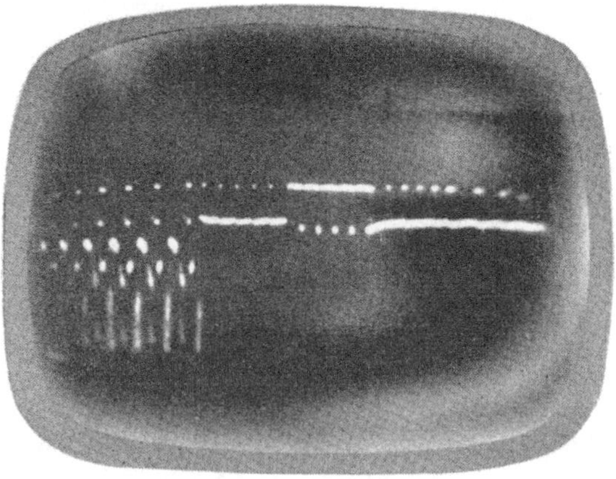

Fig. 33-10 Oscillogram on the screen of a television receiver, transmitted over industrial television circuit

CONCLUSION

One of the prime objectives of this book has been not only to serve as a complete and detailed introduction to the technique used in the use of traditional types of cathode ray oscilloscopes and to give examples of such applications, but also to point out what further possibilities of application are offered by the modern high-performance oscilloscope.

However, inexpensive apparatus has also been recently introduced, the correct use of which makes it possible to obtain results approaching those obtained with high-performance oscilloscopes. An attempt has therefore been made in this book to show, by means of clear examples, how optimum results can be obtained with these oscilloscopes also. Indeed, these hints should be of particular value in those cases in which only limited funds are available.

Developments which may be expected in the near future have been discussed in detail as far as was possible at the time of completion of this manuscript for the English edition in winter 1962/63. The broad fundamental treatment of the functions of the oscilloscope ensures that readers will be able to understand without difficulty and apply successfully new or special methods which may become available later.

Although the number of practical examples quoted is far greater than in the preceding book "The Cathode Ray Oscilloscope", only a small selection from the constantly growing number of possible applications could be included. Nevertheless it is hoped that a comprehensive picture has been presented and that in particular those readers who are interested in the measurement of non-electrical magnitudes, will have been given a clear explanation by examples and discussions of the general fundamentals and of the ways in which the cathode ray oscilloscope can best be used.

BIBLIOGRAPHY

Chapter 1 Construction of the cathode ray oscilloscopes

[1] ITTMANN, G. P., *An oscillograph apparatus,* Philips Tech. Rev., May 1936, Vol. **1** pp 147-151, 7 illustrations.

Chapter 2 The cathode ray tube

[1] VON ARDENNE, M., *Über eine indirekt geheizte Kathode für Braunsche Röhren,* Fernsehen u. Tonfilm, Vol. **4** No. 6 (1953), pp 67-69, 6 illustrations.

[2] ITTMANN, G. P., *Control of the beam intensity in cathode ray tubes,* Philips Tech. Rev., Vol. **1** No. 3 (1936), pp 91-94, 6 illustrations.

[3] DARBYSHIRE, J. A., *The electron gun of the cathode-ray tube.* Electronic Engineering, December 1955, pp 523-528, 2 illustrations.

[4] THEILE, R., a. WEYRES, TH., *Grundlagen der Kathodenstrahlröhren,* 2 Aufl. 1950, Berlin, Technischer Verlag Herbert Cram, 145 pp, 172 illustrations.

[5] SAY, M. G., *Cathode-ray tubes.* London, George Newnes Ltd., 1954, 213 pp, 120 illustrations.

[6] WEINRIB, E. A., a. MILJUTIN, W. I., *Elektronenoptik,* Berlin, VEB Verlag Technik 1964, 182 pp, 155 illustrations.

[7] FEINBERG, R., *A high speed oscillograph cathode-ray tube for the direct recording of high current transients,* Electronic Engineering, December 1956, pp 540-541, 6 illustrations.

[8] PIEPLOW, H., *Entwicklung und Aufgaben der Elektronenstrahl-Oszillografischen Messtechnik,* Jahrbuch der AEG-Forschung, Vol. **8**. December 1941, pp 161-174, 18 illustrations.

[9] PIEPLOW, H., *Messgenauigkeit und Messgrenzen technischer Elektronenstrahloszillographen,* ATM. Vol. **I**, May 1949, No. 8340-5 and **II** Jan. 1950, No. 8340-6, 3 illustrations.

[10] GUNDERT, E., *Dimensionierung von Kathodenstrahlröhren.* In: Elektronenröhren-Physik, Hrsg. von E. Rothe, München, Franzis-Verlag 1953, pp 70-79, 2 illustrations.

[11] BOUCKE, H., *Vervollkommnungen auf dem Gebiet der Elektrokardiographie,* Funktechnische Monatshefte, 1937, Heft 10, p 319-323, 16 illustrations.

[12] *Die Philips Kathodenstrahlröhre DG 9-3. Eine Kathodenstrahlröhre mit Korrektur zur nicht-symmetrischen Ankopplung der Ablenkspannung an einem Plattenpaar.* Philips Monatsheft für Apparatefabrikanten Nr. **52** (1937), pp 93-98, 10 illustrations.

[13] PARKER, P., *The defects of electrostatic focussing and deflecting systems,* Electronics, London, Edward Arnold & Co. 1950, pp 75-80, 5 illustrations.

[14] SCHWALGIN, K., *Eine Kathodenstrahlröhre für symmetrische und asymmetrische Ablenkung,* Elektronik, Nr. **4** (1955), pp 82-84, 8 illustrations.

[15] GUNDERT, E., *Der Trapezfehler bei Oszillographenröhren,* Telefunken-Zeitung, Vol. **28**, No. 99 (1953) pp 89-94, 8 illustrations.

[16] SCHWALGIN, K., *Ablenksysteme von Oszillografenröhren für Messzwecke,* Radio Mentor Vol. **3** (1957), pp 139-142, 16 illustrations.

[17] HOLLMANN, H. E., *Die Braunsche Röhre bei sehr hohen Frequenzen,* Funktechnische Monatshefte, No. **3** (1932), pp 284-293, 8 illustrations.

[18] LEWIS, J. A., a. WELLS, F. H., *Millimicrosecond pulse techniques,* Pergamon Press Ltd., London, 1954. p. 184.

[19] PIERCE, J. R., *Travelling-wave oscilloscope,* Electronics, November 1949, pp 97-99, 5 illustrations.

[20] Germershausen, K. J., Goldberg, S., a. McDonald, D. F., *A high-sensibility cathode-ray tube for millimicrosecond transients,* IRE Transactions on electron devices, April 1957, p. 152-158, 8 illustrations.

[21] Garlick, G. F. J., *The physics of cathode ray tube screens,* Electronic Engineering, Aug. 1949, pp 287-291, 9 illustrations.

[22] Rottgart, K. H. J., *Elektronenstrahlanregung von Leuchtstoffen,* Funk und Ton, Berlin-Borsigwalde, Verlag für Radio-Foto-Kinotechnik GmbH. Vol **7** (1953) No. 8, pp 385-397, 10 illustrations.

[23] de Boer, F., a. Emmens, H., *Sedimentation of fluorescent screens in cathode ray tubes,* Philips Tech. Rev. Vol. **16** No. 8 (1955), pp 232-236, 5 illustrations.

[24] Kröger, F. A., *Application of luminescent substances,* Philips Tech. Rev., Vol. **9** (1947) No. 7 pp 217-224, 4 illustrations.

[25] Verwey, E. J. W., a. Kröger, F. A., *New views on oxidic semiconductors and zinc-sulphide phosphors.* Philips Tech. Rev., Vol. **13** No. 4 (1951) pp 90-96, 3 illustrations.

[26] Kröger, F. A., a. de Groot, W., *The influence of temperature on the fluorescence,* Philips Tech. Rev., Vol. **12** No. 1 pp 6-14, 10 illustrations.

[27] Kröger, F. A., a. Bril, A., *Saturation of fluorescence in television tubes,* Philips Tech. Rev., Vol. **12** No. 4 pp 120-128, 7 illustrations.

[28] Bril, A., a. Klasens, H. A., *The efficiency of fluorescence in cathode-ray tubes,* Philips Tech. Rev., Vol. **15** No. 2 (1953) pp 63-72, 10 illustrations.

[29] Bornemann, I., a. Thurley, I., *Die Bestimmung der Emissionsfarbe von Bildschirmen,* Nachrichtentechnik, Vol. **6** No. 1 (1956) pp 23-26, 4 illustrations.

[30] Dirbach, W., *Zur Leuchtfarbe des Bildröhrenschirms,* Telefunken-Zeitung, Vol. **29** No. 112 (1956) pp 105-108, 6 illustrations.

[31] Veegens, J. D., *A cathode ray oscillograph,* Philips Tech. Rev. Vol. **4** No. 4 (1939) pp 198-204, 13 illustrations.

[32] Custers, J. F. H., *The recording of rapidly occurring electric phenomena with the aid of the cathode ray tube and the camera,* Philips Tech. Rev., Vol. **2** No. 5 (1937) pp 148-155, 7 illustrations.

[33] Blok, L., *An apparatus for the measurement of scanning speeds of cathode ray tubes,* Philips Tech. Rev., Vol. **3** No. 7 (1938), pp 216-219, 5 illustrations.

[34] Czech, J., *Kamera-Aufnahmen von Elektronenstrahloszillogrammen.* Zeitschrift für angewandte Photographie, Vol. **III** No. 5 pp 65-71, 9 illustrations.

[35] de Gier, J., *A cathode ray tube with post-acceleration,* Philips Tech. Rev., Vol **5** Sept. 1940, pp 253-260, 14 illustrations.

[36] White, W. G., *Cathode-ray tubes with post-deflection acceleration,* Electronic Engineering, March 1949, pp 75-79.

[37] Allard, L. S., *An "ideal" post deflexion accelerator* C.R.T. Electronic Engineering, Nov. 1950, pp 461-463, 6 illustrations.

[38] Schlesinger, K., *Progress in the development of post-acceleration and electrostatic deflection,* Proc. of the I.R.E., May 1956, pp 659-667, 13 illustrations.

[39] Czech, J., *Großprojektion von Oszillogrammen,* Funk und Ton, Vol. **6**, 1952, pp 363-368 and p. 492, 8 illustrations.

[40] de Gier, J., a. van Rooy, A. P., *Improvements in the constructions of cathode ray tubes,* Philips Tech. Rev., Vol. **9** No. 6 (1947) pp 181-195, 6 illustrations.

[41] Puckle, O. S., *Time bases,* Section *Cross-talk,* London: Chapman & Hall Ltd., p 241.

[42] *Die Bildgüte einer neuzeitlichen Oszillographenröhre,* Telefunken-Zeitung, Vol. **29** No. 112, June 1956, pp 124-125, 4 illustrations.

[43] Schwarz, E., *Über Nachbeschleunigung bei Braunschen Röhren.* Fernsehen und Tonfilm, Vol. **6** June 1935, pp 37-40 and July 1935, pp 47-49, 9 illustrations.

[44] Rider, J. F., *Encyclopaedia on cathode-ray oscilloscopes,* New York, Publisher J. F. Rider Publisher Inc. 1950, Multigun Tubes, pp 72-75.

[45] BULLOCK, T. M., *Double-beam C-R tube in biological research.* Electronics, July 1946, pp 103-105, 8 illustrations.

[46] VON FERRONI, E., *Neuerungen auf dem Gebiet der Elektronenstrahl-Oszillographen.* Enlarged special reprint from: Siemens-Zeitschrift, Vol. **25** No. 5 (1951), 20 illustrations.

[47] KATZ, H., a. WESTENDORF, E., *Erreichung hoher Schreibgeschwindigkeiten mit einer abgeschmolzenen, rein elektrostatisch arbeitenden Braunschen Röhre,* Zeitschrift für technische Physik, Vol. **20** No. 7 (1939) pp 209-212, 5 illustrations.

[48] VON BORRIES, B., *Die Kathodenstrahl-Oszillograph. Entwicklungsstand, Anwendung und Vergleich.* VDI-Z., Vol. **80** (1936), No. 37, pp 1135-1141, 18 illustrations.

[49] VON BORRIES, B., a. RUSKA, E., *Über die Beurteilung und den objektiven Vergleich der Messleistung von Kathodenstrahloszillographen.* Archiv für Elektrotechnik. Vol. **XXXIV** (1949) No. 3 pp 161-166, 3 illustrations.

[50] SCHWALGIN, K., *Kathodenstrahlröhren für Stosspannungsprüfung,* Elektronische Rundschau, Vol. *9* No. 11 (1955), pp 396-397, 4 illustrations.

[51] BUCH, W., *Die neue Stosspannungsanlage für die Untersuchung und Prüfung von Überspannungsschutzgeräten,* Siemens-Zeitschrift, No. **7** (1955), pp 253-259, 12 illustrations.

[52] VON BORRIES, B., a. RUSKA, E., *Hochleistungsoszillographen mit abgeschmolzener Braunscher Röhre,* Archiv für Elektrotechnik, Vol. **XXXIV** (1940) No. 2 pp 106-114, 12 illustrations.

[53] ROCHELLE, R. W., *Cathode-ray-tube beam intensifier,* Electronics, Oct. 1952 pp 151-153, 3 illustrations.

[54] WILSON, W., *Transient recorder,* pp 61-63, 2 illustrations. *Recurrent surge oscillograph.* pp 70-75 and pp 124-142, 6 illustrations. From: *The cathode ray oscillograph in industry,* London: Chapman and Hall Ltd.

[55] HERRNKIND, O. P., *Oszillographenröhren,* Elektronik, No. **8** (1956) pp 213-218, 4 tables.

[56] SOLLER, TH., STARR, M. A., a. VALLEY jr., G. E., *Cathode ray tube displays. The dark-trace cathode-ray-tube screen and its development,* New York, Toronto, London: McGraw-Hill Book Comp. Inc., p 664.

[57] ROTTGART, K. H. J., a. BERGTOLD, W., *Bildröhren für Industrie-Fernsehen,* Elektronische Rundschau, Vol. **10** (1956) No. 2 p 47-49, 8 illustrations.

[58] DIETRICH, W., *Der Blauschreiber, ein neues Gerät zum Aufzeichnen nichtperiodischer Vorgänge,* NTZ. No **11** (1956) pp 504-507, 3 illustrations.

[59] WINKLER, ST., a. NOZIK, S., *Operation of CRT storage devices,* Electronics, Oct. 1954 pp 185-187, 4 illustrations.

[60] NOZIK, S., BURTON, N. H., a. NEWMAN, S., *Dark-trace display tube has high writing speed,* Electronics Dec. 1954 pp 155-156, 5 illustrations.

[61] HUBBY, A. G., a. WATSON, R. B., *Scriptoscope shows messages on C-R tube,* Electronics July 1952, pp 144-145, 2 illustrations.

[62] KNOLL, M. a. KAZAN, B., *Storage tubes and their basic principles,* John Wiley and Sons, Inc., New York, Chapman and Hall, Ltd., London. 1952, 143 pages, 34 illustrations.

[63] SMITH, S. T. a. BROWN, H. E., *Direct viewing memory tube,* Proceedings of the I.R.E., Sept. 1953, pp 1167-1171, 4 illustrations.

[64] KATES, J., *A method for improving the read-around-ratio in cathode-ray storage tubes,* Proceedings of the I.R.E., Aug. 1953, pp 1017-1023, 20 illustrations.

[65] WINKLER, ST., a. NOZICK, S., *Operation of CRT storage devices,* Electronics October 1954, pp 184-187, 4 illustrations.

[66] KRAMER, A. S., *Cathode-ray storage tubes,* Electronics 1959, January 23, pp 40-41.

[67] *Elektronenstrahl-Speicherröhren für direkte Beobachtung,* Elektronik 1959, No. **11** pp 345-348, 4 tables.

[68] KRAMER, A. S., *Cathode-ray storage tubes for special purposes.* Electronics June 30, 1959.

[69] KNOLL, M. a. HARTH, W., *Niveauverschiebungsdiagramme zur Beschreibung der La-*

dungsprozesse in Speicherröhren, ETZ-A Vol. **78**, No. 15, 1.8.1957, pp 543-548, 7 illustrations.

[70] JENSEN, A. S., SMITH, J. P., MESNER, M. H. a. FLORY, L. E., *Barrier grid storage tube and its operation,* RCA Review Vol. **9** (1948), pp 112-135, 19 illustrations.

[71] GIBBONS, D. J., *The barrier grid storage tube,* Electronic Engineering October 1961, pp 630-636, 10 illustrations.

[72] LEHRER, N. H., *Selective erasure and nonstorage writing in direct-view halftone storage tubes.* Proc. of the Inst. of Radio Eg., Vol. **49** (1961) No. 3, pp 567-573, 9 illustrations.

[73] CAWKELL, A. E. a. REEVES, R., *Transient storage oscilloscope.* Electronic Technology, February 1960, pp 50-59, 13 illustrations.

Chapter 3 Power supply unit

[1] VALETON, J. J. P., *High-tension generators for large-picture projection television,* Philips Tech. Rev., Vol. **14** No. 1 (1952) pp 21-32, 15 illustrations.

[2] KELLER, H., *Ein Netzanschlussgerät für die Braunsche Röhre,* Funktechnische Monatshefte No. **2** (1933) pp 67-68, 4 illustrations.

[3] DAMMERS, B. C., HAANTJES, J., a. VAN SUCHTELEN, H., *Applications of the electronic valve in radio receivers and amplifiers,* Series Electronic valves, book V, chapter VIII. D, Design calculation of high-tension rectifiers, Philips Technical Library, 1951.

[4] KAMMERLOHER, J., *Gleichrichter,* Hochfrequenztechnik, Leipzig: Verlag Wintersche Verlagshandlung, Part **III** (1942) pp 189-236.

[5] KÜHN, R., *Die Bemessung von Netzgleichrichter-Transformatoren,* Elektronische Rundschau Vol. **9** (1955) Nos. 7-10 pp 266-268, 296-300,332-334 and 405-406, 19 illustrations.

[6] JURRIAANSE, T., *A voltage stabilizing tube for very constant voltage,* Philips Tech. Rev., Vol. **8** No. 9 (1946) pp 272-277, 6 illustrations.

[7] CUBASCH, F., *Spezialröhren. Eigenschaften und Anwendungen,* Berlin-Borsigwalde, Verlag für Radio-Foto-Kinotechnik, GmbH.

[8] LINDENHOVIUS, H. J., a. RINIA, H., *A direct current supply apparatus with stabilized voltage,* Philips Tech. Rev., Vol. **6** No. 2 (1941) pp 54-61, 11 illustrations.

[9] ELMORE, W. G., a. SANDS, M., *Power supplies and control circuits,* Electronics, Experimental Techniques, New York, London: McGraw-Hill Book Comp. Inc., 1949. Chap. 7 pp 363-410, 21 illustrations.

[10] GÜNTHER, H., *Stabilisierung von Gleichspannungen,* Funk und Ton, Vol. **5** (1951) No. 3, pp 124-132, 4 illustrations.

[11] WILMORE, A. P., *The cathode follower as a voltage regulator,* Electronic Engineering, Sept. 1950 pp. 399-400, 7 illustrations.

[12] PERRY, B. J., *Cathode-follower type power supplies,* Electronic Engineering, Dec. 1956 pp 517-520, 9 illustrations.

[13] KRÖNER, K., *Dimensionierung und Berechnung von elektronisch stabilisierten Gleichspannungsquellen,* Elektronik, Nos. **2-5** (1957) pp 43-48, 107-112, 139-140, 43 illustrations.

[14] WOUK, V., *High voltage supply uses electronic filter,* Electronics, Aug. 1955 pp 154-155, 3 illustrations.

[15] SOLLER, TH., STARR, M. A., a. VALLEY jr., G. E., *Cathode ray tube displays,* High Voltage power supply for cathode-ray-tubes, New York, London: McGraw-Hill Book Comp. Inc., 1948 pp. 163-183, 20 illustrations.

[16] BIGALKE, A.: *Elektronenstrahl-Oszillograph, Netzgeräte,* ATM, Sept. 1950 J 8344-4. 2 pages, 4 illustrations.

[17] SULZER, P. G., *Series-resonant high-voltage supply,* Electronics, Sept. 1952 pp 156-157, 3 illustrations.

[18] LOWE, A. E., *A stabilized radio-frequency. E.H.T. supply,* Electronic Engineering, Feb. 1955 pp 85-86, 3 illustrations.

[19] Stephens, R. B., *Mains-voltage stabilizers,* Philips Tech. Rev., Vol. **17** No. 1 (1955) pp 1-9, 16 ullistrations.
[20] Benson, F. A., a. Seamon, M. S., *An A.C. voltage stabilizer,* Electronic Engineering, June 1956 pp 260-265, 9 illustrations.
[21] Nentwig, K., *Einiges über den Betrieb von Kathodenstrahlröhren,* Funktechnische Monatshefte, No. **10** (1933), pp 391-393, 4 illustrations.
[22] See [15] pp 555-558. Magnetic shielding.
[23] *Shielding CRT's from magnetic fields,* Electronic Industries, Dec 1960, p. 112-116, 3 illustrations.

Chapter 4 Time base unit

[1] Keller, H., *Ein Kippgerät mit Synchronisation zur Zeitablenkung bei der Braunschen Röhre,* Funktechnische Monatshefte, Oct. 1933 pp 412-414, 8 illustrations.
[2] Richter, H., *Elektrische Kippschwingungen,* Leipzig: Verlag S. Hirzel 1940.
[3] Puckle, O. S., *Time Bases (Scanning Generators),* London: Chapman & Hall Ltd., 2nd edn. 1952.
[4] Jager, J., *Comments on circuits for generation of time-base voltages,* Philips' Electronic Application Bulletin, Vol. **X** No. 1 (1948), pp 15-29, 25 illustrations.
[5] Kinne, E., *Kippgerät für Fernsehempfänger,* Funk und Ton, Vol. **6** (1961) No. 8 pp 429-434, 11 illustrations.
[6] *Ein Kippspannungsgerät für Kathodenstrahloszillographen,* Philips' Technische Monatshefte No. **33** (1936) pp 45-50, 5 illustrations.
[7] *A versatile oscilloscope,* Philips' Electronic Application Bulletin, Vol. **17** (1956) No. 1 pp 1-10, 13 illustrations.
[8] Aschen, R., a. Chambas, R., *The transitron effect in H.F. pentode type EF* **42**, Philips' Electronic Application Bulletin, Vol. **X** No. 10 (1949) pp 221-226, 13 illustrations.
[9] Briggs, H. B., *The miller-integrator,* Electronic Engineering, Aug., Sept. and Oct. 1948 pp 243-247, 279-284 resp. 325-329, 17 illustrations.
[10] Brunetti, C., *The transitron oscillator,* Proc. of the I.R.E., Vol. **27** (1939) No. 2, p. 88.
[11] Cocking, W. T., *Linear saw-tooth oscillator,* Wireless World, No. **6** (1946) p. 176.
[12] Kerkhof, F., a. Werner, W., Television, vol I, Philips Technical Library, 1952, pp 154-185 (new edition in active preparation).
[13] den Hartong, H., a. Muller, E. A., *Oscilloscope time-base circuits,* Wireless Engineer, Oct. 1947 pp 287-292, 8 illustrations.
[14] Kuehn, R. L., *Signal triggered sweep magnifies pulse widths,* Electronics, April 1956 pp 146-147, 3 illustrations.
[15] Fleming, L., *Trigger adapter for transient oscillograms,* Electronics, April 1955 pp 159-161, 2 illustrations.
[16] Thackeray, D. P. C., *Triggered microsecond sweep generators,* Electronic Engineering, Sept. 1955 pp 317-401, 11 illustrations.
[17] Rao, V. N., a. Sankarasubramanyan, V.: *An expanded time-base using a miller-integrator,* Electronic Engineering, June 1955 pp 273-274, 2 illustrations.
[18] Bigalke, A., *Die selbsttätige Aufnahme einmaliger Vorgänge mit dem Elektronenstrahl-Oszillographen,* ETZ. V. **59** No. 15 (1938) pp 105-107, 6 illustrations.
[19] Attree, V., *A single sweep time-base,* Electronic Engineering, May 1948 pp 160-161, 2 illustrations.
[20] Pfeiffer, H., *Die Erzeugung von Rechteckimpulsen mit Multivibratoren,* Nachrichtentechnik, Vol. **6** No. 4 (1956) pp 166-170, 15 illustrations.
[21] Renwick, W., a. Pfister, M. A., *A design method for direct-coupled flip-flops,* Electronic Engineering, June 1955 pp 246-250, 6 illustrations.
[22] Armstrong, H. L., *Bistabile circuits using triode-pentodes,* Electronics, July 1956 pp 210, 212, 214 and 216, 3 illustrations.

[23] THIELE, G., *Berechnungsanleitung für Flip-Flop Schaltungen,* Elektronische Rundschau, Vol. **11** (1957) Nos. 7, 8, 9. pp 212-215, 250-252 and 274-276, 15 illustrations.

[24] FERGUSON, A. E., *Feedback in time-base circuits,* Electronic Engineering, June 1952 pp 280-281, 2 illustrations.

[25] TUCKER, E. C., *Delayed sweep and expanded sweeps. Principles of radar,* New York, London: McGraw-Hill Book Comp. Inc. 2nd edn. 1946 pp 27 and 28 of section 3.

[26] FARLEY, F. J. M., *Elements of pulse circuits,* London, Menthuen & Co. Ltd. 138 pages (Fig. 49, p 81). 74 illustrations.

[27] SCHMITT, O. H., *Thermionic trigger,* Journ. Sci. Instr. 1938 Vol. **15** p. 24, 2 illustrations.

[28] ELMORE, W. C., a. SANDS, M., *Electronics. Experimental techniques,* New York, London: McGraw-Hill Book Comp. Inc. 1947 Chap. 2: Trigger circuits. pp 76 and 100, 5 illustrations.

[29] BUYER, E. M., *Line selector checks television waveforms,* Electronics, Sep. 1953 pp 153-155, 5 illustrations.

[30] FISHER, J., *Television picture line selector,* Electronics, Mar. 1952, pp 140-143, 8 illustrations.

[31] MACEK, O., *Ein Zeilenwähler mit Eigensynchronisierung für die Fernsehtechnik,* Frequenz **10** (1956) No. 6 pp 193-197, 11 illustrations.

[32] MOTHERSOLE, P. L., *A television line selector unit,* Electronic Engineering, Dec. 1956 pp 520-523, 6 illustrations.

[33] CZECH, J., *Besondere Zeitdehnungsverfahren bei Elektronenstrahl-Oszillographen,* FTZ. No. **8** (1954) pp 425-430, 8 illustrations.

[34] DAMMERS, B. G., *A time-base generator for frequencies of 1 to 100 Mc/s.,* Philips Electronic Application Bulletin, Vol. **X** No. 7 (1949) pp 157-166, 8 illustrations.

[35] HARDY, D. R., JACKSON, B., a. FEINBERY, R., *The recording of high-speed single stroke electrical transients,* Electronic Engineering, Jan. and Feb. 1956 pp 8-12 and 75-79, 1 illustration

[36] WHITEWAY, F. E., *The recording of high-speed single-shot phenomena,* Proc. Inst. Electr. Eng. **107** (1960), 36, pp 614-623, 8 illustrations.

[37] JANSSEN, J. M. L., *An experimental "stroboscopic" oscilloscope for frequencies up to about 50 Mc/s, I. Fundamentals,* Philips Techn. Rev. **12** (1950), 2, pp 52-59, 7 illustrations and Michels, A. J., II. *Electrical build-up,* **12** (1950), 3, pp 73-82, 12 illustrations.

[38] REEVERS, R. J. D., *The Recording and Collocation of Waveforms,* Electr. Engg., March 1959, pp 130-137, and April 1959, pp 204-212, 25 illustrations.

[39] QUEEN, J. G., *The Monitoring of High-Speed Waveforms,* Electronic Engg., October 1952, pp 436-441, 10 illustrations.

[40] *Simple Stroboscopic Oscilloscope for Displaying Pulses with Short Rise Times and High Repetition Frequencies,* Philips Electronic Application Bulletin, Vol. **19**, Nr. 3, pp 115-120, 4 illustrations.

[41] LOUIS, H. P., *Messung von Signalen im Zeitbereich von Nanosekunden mittels Abtastoszillografen,* Elektronische Rundschau Nr. 4/1960, pp 137-144, 9 illustrations.

[42] AMODAI, J. J., *Converting Oscilloscopes For Fast Rise Time Sampling,* Electronics, June 1960, pp 96-99, 5 illustrations.

[43] FARBER, A. S., *Sampling Oscilloscope for Millimicrosecond Pulses at a 30-Mc Repetition Rate,* The Rev. of Scientific Inst., **31**, Nr. 1. January 1960, pp 15-17, 6 illustrations.

[44] BUSHOR, W. E., *Sample Method Displays Millimicrosecond Pulses,* Electronics, July 31, 1959, pp 69-71, 7 illustrations.

[45] CARLSON, R., KRAKAUER, S., MAGLEBY, K., MONNIER, R., VAN DUZER, V. and WOODBURY, R., *Sampling Oscillography,* Hewlett-Packard-Application Note 36, Nov. 1959, 7 pages, 2 illustrations.

[46] *A Versatile New DC-500 MC Oscilloscope with High Sensitivity and Dual Channel*

Display, Hewlett-Packard Journal, Vol. 11, No. 5-7, Jan.-March 1960, 8 pages, 19 illustrations.
[47] *C.R. Abtastoszillograf für 500 MHz,* Radio Mentor, September 1960, Nr. 9, pp 720-721, 2 illustrations.
[48] SUGERMAN, R., *Sampling Oscilloscope for Statistically Varying Pulses,* The Rev. of Scientific Inst., **28**, Nr. 11 (1957), pp 933-937, 10 illustrations.
[49] MCAUSLAN, J. H. L., *A dekatron C.R.O. time marker,* Electronic Engineering, Dec. 1952 pp 567-568, 6 illustrations.
[50] STEINBERG, P., *Gated time markers for C.R.O. display,* Electronics, Mar. 1954 pp 150-151, 5 illustrations.
[51] GREGSON, M., *Calibration bars for single sweep time-bases,* Electronic Engineering, May 1952 pp 239-240, 3 illustrations.

Chapter 5 Deflection amplifiers

[1] SCHUBERT, J., *Das Brummen indirekt geheizter Verstärkerröhren,* from: Elektronenröhren-Physik, Edn. H. Rothe, Munich: Franzis Verlag. Vol. **1** pp 202-215 and Vol. **2** pp 368-373, 10 illustrations.
[2] ROTHE, H., a. KLEEN, W., *Elektronenröhren als Anfangsstufenverstärker,* Bücherei der Hochfrequenztechnik, Leipzig: Akademische Verlagsgesellschaft Geest & Portig KG. Vol. **3**. Section: Das Klingen von Verstärkerröhren, pp 374-385.
[3] LEVIS, J. A. D., a. WELLS, F. H., *Millimicrosecond pulse techniques.* London: Pergamon Press 1954.
[4] KLEIN, P. E., *Ein neuer Trägerfrequenz-Messverstärker für Elektronen- und Lichtstrahl-Oszillographen,* Radio Mentor, No. **7** (1954) pp 374-377, 9 illustrations.
[5] BIERMASZ, A. L., a. HOEKSTRA, H., *The measurement of changes in length with the aid of strain gauges,* Philips Tech. Rev. Vol. **11** No. 1 (1949) pp 23-31, 12 illustrations and Philips: Instructions for use of strain-measuring bridges **PR** 9300 and **PT** 1200.
[6] MEYER-BRÖTZ, G., a. FELLE, K., *Die Dimensionierung von Transistor-Breitbandverstärkern,* NTZ. No. **6** (1956) pp 498-503, 9 illustrations.
[7] STRUTT, M. J. O., *Verstärker und Empfänger,* Berlin: Springer-Verlag 1943, Section **5**: Spontane Spannungs- und Stromschwankungen (Rauschen) pp 36-49.
[8] ROTHE, H., a. KLEEN, W., *Elektronenröhren als Anfangsstufen-Verstärker,* Leipzig: Akademische Verlagsgesellschaft Geest & Portig KG 1948, **III**: Störerscheinungen in Elektronenröhren. Das Rauschen der Elektronenröhren, pp 263-373.
[9] KOSMAHL, H., *Rauschen und Grenzempfindlichkeit gittergesteuerter Röhren,* Elektronische Rundschau, Vol. **9** (1955) No. 3 pp 103-108, 6 illustrations.
[10] *Rauschen,* Nachrichtentechnische Fachberichte, No. **2** (1955) p 82.
[11] GILLESPIE, A. B., *Signal, noise and resolution in nuclear counter amplifiers,* London: Pergamon Press Ltd. 1953 and [8] pp 315-321.
[12] ROTHE, H., a. KLEEN, W., *Elektronenröhren als Anfangsstufen-Verstärker,* Leipzig: Akademische Verlagsgesellschaft Geest & Portig KG 1948 pp 315-321.
[13] FELDTKELLER, R., *Einführung in die Theorie der Rundfunksiebschaltungen,* Leipzig: Verlag S. Hirzel 1945.
[14] DE BRUIN, S. L., a. DORSMAN, C., *A cathode ray oscilloscope for use in tool making,* Philips Tech. Rev., Vol. **5** Oct. 1940 pp 277-285.
[15] ROTHE, H., a. KLEEN, W., *Elektronenröhren als Anfangsstufen-Verstärker.* Bücherei der Hochfrequenztechnik. Vol. **3**. Leipzig Akademische Verlagsgesellschaft Geest & Portig KG., 2nd Edn. 1948. Chapter: Breitbandverstärker.
[16] KERKHOF, F., a. WERNER, W., Television, Philips Technical Library (1952), chapter Wide-band amplifiers, pp 205-314 (new edition in preparation).
[17] STRACK, H., a. WALTER, S., *Die Erweiterung des unteren Übertragungsbereiches von Breitbandverstärkern durch Anwendung von Kompensationsschaltungen,* Nachrichtentechnik, Vol. **6** No. 5 (1956), pp 208-212, 8 illustrations.

[18] STRUTT, M. J. C., *Verstärker und Empfänger,* Berlin, Göttingen, Heidelberg: Springer-Verlag 1943, Section: Vorverstärkerstufen. pp 104-109.

[19] BARTELS, H., *Grundlagen der Verstärkertechnik,* Leipzig: Verlag S. Hirzel. 2nd edn. 1944. Chapter V: Vorverstärker. Section 5: Verstärker für breite Frequenzbänder. pp 158-163.

[20] HAANTJES, J., *Judging an amplifier by means of the transient characteristic,* Philips Tech. Rev., Vol. **6** No. 7 (1941) pp 193-201, 10 illustrations.

[21] BEDFORT, A. V., a. FREDENDALL, G. L., *Analysis, synthesis and evaluation of the transient response of television apparatus,* Proc. of the I.R.E., Oct. 1942 pp 440-457, 23 illustrations.

[22] VAN SLOOTEN, J., *Experimental testing of electrical networks by the means of the unit function responce.* Philips Tech. Rev., Vol. **12** No. 8 (1951), pp 233-239, 11 illustrations.

[23] BRÜCK, L., *Gegenkopplungsschaltungen,* Telefunken-Zeitschrift, No. **77** (1937) pp 9-23, 18 illustrations.

[24] TELLEGEN, B. D. H., *Inverse feed-back,* Philips Tech. Rev., Vol. **2** No. 10 (1937), pp 289-294, 7 illustrations.

[25] DEKETH, E. T. H. J., *Fundamentals of radio-valve technique,* Series on electronic valves, vol I, Audio-frequency negative feedback, pp 377-408, Philips Technical Library, 1949.

[26] BRÜCK, L., *Frequenzgang und Schwingneigung gegengekoppelter Verstärker,* Telefunkenröhre, No. **14** (1938) pp 237-253, 18 illustrations.

[27] KLÜNNER, W., *Begrenzung der Gegenkopplung durch Phasendrehung,* Elektronik, No. **2** (1954) pp 11-13, 6 illustrations.

[28] BODE, H. W., *Network analysis and feedback amplifier design,* Toronto New York, London: Verlag D. van Nostrand Co. Inc. 10th impr., Jan. 1955.

[29] PETERS, J., *Einschwingvorgänge, Gegenkopplung, Stabilität,* Berlin, Göttingen, Heidelberg: Springer-Verlag.

[30] ROTHE, H., *Elektronenröhren-Physik,* München: Franzis-Verlag 1953. Die Grenzempfindlichkeit gittergesteuerter Röhren. Part II: Der Einfluss von Rückkopplungen auf die Geräuschzahl von Trioden. pp 246-254, 4 illustrations.

[31] STRUTT, M. J. O., *Verfahren zur Verringerung der Störungen im Vergleich zum Signal am Ausgang von Verstärkern,* ETZ, Edn. A. Vol. **73** (1952) No. 20 pp 649-653, 5 illustrations.

[32] NOTTEBOHM, H., *Das Rauschen einer Verstärkerstufe mit Spannungsgegenkopplung,* Elektronische Rundschau, Vol. **10** (1956) No. 3 pp 57-62, 9 illustrations.

[33] MILLMANN, J., a. TAUB, H., *Cathode compensation linearizes video stage,* Electronics, Nov. 1955 pp 156-157, 2 illustrations.

[34] DILLENBURGER, W., *Über die Linearisierung des Frequenzganges der Verstärkung in Breitbandverstärkern durch Gegenkopplung,* Frequenz, Vol. **5** (1954), No. 1 pp 1-5, 5 illustrations.

[35] MÜLLER, J., *Die Übertragung der Sprungfunktion durch den gegengekoppelten Verstärker,* FTZ., No. **12** (1951), pp 547-551, 11 illustrations.

[36] SOWERBY, J. M. C., *Electronic circuitry,* Wireless World, Sep. 1948 p 321 and *Der Ausgangswiderstand des Katodenverstärkers.* Funk und Ton (now: Elektronische Rundschau), Vol. **2** (1948) No. 12, pp 657-658, 3 illustrations.

[37] GEYGER, W., *Der Katodenverstärker.* Funk und Ton (now: Elektronische Rundschau), Vol. **2** (1948) No. 3 pp 119-124, 2 illustrations.

[38] ROTHE, H., a. KLEEN, W., *Elektronenröhren als Anfangsstufenverstärker,* Leipzig: Akademische Verlagsgesellschaft Geest & Portig KG 1948. Vol. **3**. pp 198-201.

[39] AMOS, S. W., *Valves with resistive loads,* Wireless Engineer, April 1949 pp 119-123, 11 illustrations.

[40] ROSS, S. F. G., *Design of cathode-coupled amplifiers,* Wireless Engineer, July 1950 pp 212-215, 5 illustrations.

[41] DILLENBURGER, W., *Der Katodenverstärker,* Funk und Ton (now: Elektronische Rundschau), Vol. **5** (1951) No. 4, pp 190-193, 6 illustrations.

[42] FLOOD, J. E., *Cathode-follower input impedance,* Wireless Engineer, Vol. **28** (1951) No. 335, p 231 and Funk und Ton (now: Elektronische Rundschau), Vol. **6** (1952) No. 3 pp 158-161, 5 illustrations.

[43] DAMMERS, B., HAANTJES, J., OTTE, J. a. VAN SUCHTELEN, H., Application of the electronic valve in radio receivers and amplifiers, Series on electronic valves, vol V, chap VI. B, Phase inverters and splitters, pp 18-31, Philips Technical Library, 1951.

[44]) FUNK, S., *Kathodengekoppelte Röhrenstufen,* Hochfrequenztechnik und Elektroakustik, Vol. **63** (1953) No. 2 pp 40-46, 4 illustrations.

[45] BRIGGS, B. H., *The anode follower,* R.S.G.B. Bulletin, Vol. **22** (1947) pp 138-143, 7 illustrations.

[46] KLEIN, P. E., *Über die Regelung des Verstärkungsgrades bei Breitband-Messverstärkern ohne Verschlechterung des Frequenzganges,* Funktechnische Monatshefte No. **6** (1940) pp 81-84, 11 illustrations.

[47] BATCHER, E. E., a. MOULIC, W., *The electronic engineering handbook,* New York: Electronic Development Assiciates. Part **III**, section 11: Electron Tube Circuit Applications. Using a Tube as a Series Resistance, pp 325-332, 11 illustrations.

[48] MÜLLER-LÜBECK, K., *Der Kathodenverstärker in der elektronischen Messtechnik,* Berlin, Göttingen, Heidelberg: Springer-Verlag 1956.

[49] LEVY, L., *Higher pentode gain.* Electronics, July 1956 pp 190, 192, 194 and 196, 2 illustrations.

[50] FARLEY, F. J. M., *Elements of pulse circuits,* London: Menthuen Co. Ltd. 1956, 138 pages, 74 illustrations.

[51] GODDARD, M. J., *The development and design of direct-coupled cathode-ray oscilloscopes for industry and research,* Journal of the British Institution of Radio Engineers, April 1955 pp 179-197, 20 illustrations.

[52] CROWTHER, G. O., *A very high input impedance oscilloscope probe unit,* Electronic Engineering, June 1955. pp 242-245, 3 illustrations.

[53] SCHLEGEL, H. R., a. NOWAK, A., *Impulstechnik,* Hannover: Fachbuchverlag Siegfried Schütz 1955, 600 pages.

[54] TUCKER, E. C., *Principles of radar.* New York, London: McGraw-Hill Book Comp. Inc. 2nd edn. 1946, 2/21 Traveling Waves on Artificial Transmission Lines, pp 288-290 and Chapter VIII: Radio Frequency Lines, pp 8-1 - 8-84.

[55] SCHWEITZER, H., *Dezimeterwellen-Praxis,* Berlin-Borsigwalde: Verlag für Radio-Foto-Kinotechnik GmbH 1956, **III**. Leitungstechnik. pp 42-67.

[56] SODARO, J. F., *Delay-line design,* Electronics, June 1955 176 pages, 2 illustrations.

[57] CARLEY, W. S., a. SEYMOUR, E. F., *High-impedance artificial delay line,* Electronics, April 1953 pp 188-194, 8 illustrations.

[58] GINZTON, E. L., HEWLETT, W. R., ASBERG, J. H., a. NOE, J. D., *Distributed amplification,* Proc. of the I.R.E., Aug. 1948 No. **8** pp 956-969, 22 illustrations.

[59] KELLEY, G. G., *A high speed synchroscope,* The Review of Scientific Instruments, Vol. **21** (1950) No. 1 pp 71-76, 11 illustrations.

[60] YU, Y. P., KALLMANN, H. E., a. CHRISTALDI, P. S., *Millimicrosecond oscillography,* Electronics, July 1951, pp 106-111, 10 illustrations.

[61] SCHARFMAN, H., *Distributed amplifier covers* **10** *to* **360** *Mc/s,* Electronics. July 1952, pp 113-115, 5 illustrations.

[62] KALLMANN, H. E., *Electromagnetic delay lines,* From: Components Handbook, Edn. of J. F. Blackburn. London: McGraw-Hill Book Comp. Inc., 1949 pp 191-217, 23 illustrations.

[63] PITSCH, H., *Lehrbuch der Funkempfangstechnik,* Leipzig: Akademische Verlagsgesellschaft Geest & Portig KG 1948, Par. 273: Der direkt gekoppelte Verstärker, pp 426-428, 4 illustrations.

[64] KESSLER, G., *Der Gleichspannungsverstärker I,* ATM. Z 634-8. July 1952, pp 163-166 and II. Z 634-9. Sept. 1952, pp 211-214, 28 illustrations.

[65] STOCKHUSEN, W., *Niederfrequenzverstärker mit neuer Kopplung,* Funktechnische Monatshefte, No. **11** (1933), pp 443-446, 6 illustrations.

[66] ROTHE, H., a. KLEEN, W., *Elektronenröhren als Anfangsstufen-Verstärker,* Leipzig: Akademische Verlagsgesellschaft Geest & Portig KG 1948, Chap. 7: Der Gleichstrom- und Gleichspannungsverstärker. pp 92-102, 9 illustrations.

[67] FUNK, S., *Kathodengekoppelte Röhrenstufen,* Hochfrequenztechnik und Elektroakustik, Vol. **63** (1953), No. 2 pp 40-46, 4 illustrations.

[68] ROSS, S. G. F., *Design of cathode-coupled amplifiers,* Wireless Engineers, July 1950 pp 212-215, 5 illustrations.

[69] ASHER, H., *A D.C.-A.C. amplifier for use in physiology,* Electronic Engineering, May 1951 pp 170-172, 4 illustrations.

[70] STROJNIK, A., *Zum Entwurf des netzbetriebenen Elektrokardiographenverstärkers,* Frequenz, Vol. **6** (1952) No. 10 pp 305-312, 12 illustrations.

[71] SCHWARZER, F., *Elektro-encephalographie,* ATM. V 664-1, Feb. 1957, pp 25-28, 2 illustrations.

[72] RICHARDS, J. C. S., *An improved type of differential amplifier,* Electronic Engineering, July 1956, pp 302-305, 6 illustrations.

[73] RICHARDS, J. C. S., *A wide band differential amplifier of unity gain,* Electronic Engineering, Nov. 1956, pp 499-501, 4 illustrations.

[74] YARWOOD, J., a. LE CROINETTE, D. H., *D.C. amplifiers,* Electronic Engineering, Jan., Feb. and Mar. 1954, pp 14-18, 64-70 and 114-117, 23 illustrations.

[75] BARNETTE, W. E., a. GIACOLETTO, L. J., *Differential amplifier for null detection,* Electronics, Aug. 1955, pp 149-151, 6 illustrations.

[76] GINZTON, E. L., *D.C. amplifier design techniques,* From: Electronics for Engineering. Edn. from Markus and Zeluff. New York, London: McGraw-Hill Book Comp. Inc., 1945, pp 384-388, 9 illustrations.

[77] REYNER, J. H., *Direct-coupled oscilloscope,* Electronics, July 1948, pp 102-106, 5 illustrations.

[78] RUPPEL, W., *Über Röhrenverstärker für extrem niedrige Frequenzen.* NTZ. No. **11** (1955), pp 595-602, 15 illustrations.

[79] GOLDBERG, L., *Universal direct-coupled differential amplifier,* Electronics, Oct. 1951, pp 128-131, 7 illustrations.

[80] MILLER, ST. E., *Sensitive D.C. amplifier with A.C. operation,* Electronics, Nov. 1941, pp 27-31 and 105-109, 12 illustrations.

[81] EARNSHAW, J. B., *Heater voltage compensation for D.C. amplifiers,* Electronic Engineering, Jan. 1957, pp 31-35, 15 illustrations.

[82] HORN, L., *Gleichspannungsverstärker mit Subminiaturröhren,* Funkschau, No. **7** (1957), pp 179-181, 5 illustrations.

[83] ROBERTS, H. L., *Low capacitance probe for video testing,* Electronics, April 1956, pp 196, 198, 200, 202 and 204, 3 illustrations.

[84] CROWTHER, G. O., *A very high input oscilloscope probe unit,* Electronic Engineering, June 1955, pp 242-245, 3 illustrations.

[85] SHIMMINS, A. J., *Cathode-follower performance,* Wireless Engineer, Dec. 1950, pp 289-293, 11 illustrations.

[86] HARRIS, E. J., *An electrometer impedance convertor,* Electronic Engineering, Mar. 1951, pp 109-110, 2 illustrations.

[87] BROWN, D. E., a. KANDINAH, K., *A very high impedance valve voltmeter,* Electronic Engineering, July 1952, pp 320-321, 2 illustrations.

[88] THOMASSON, J. G., *Multi-valve cathode follower,* Wireless World., July and Aug. 1957, pp 310-313 and 373-377, 13 illustrations.

Chapter 7 Amplitude measurements

[1] PIEPLOW, H., *Messgenauigkeit und Messgrenzen technischer Elektronenstrahloszillografen,* ATM. May 1949, J 8340-5, 3 illustrations and Jan. 1950. J. 8340-6, 3 illustrations.
[2] ROBINSON, G., a. VAN ALLEN, R., *Precision measurements with a cathode-ray oscilloscope,* The Review of Scientific Instruments, Vol. 23 (1952) No. 12, pp 701-706, 12 illustrations.
[3] v. ARDENNE, M.,*Ein Präzisions-Elektronenstrahloszillograph mit wenigen μ Schreibfleckdurchmesser,* Nachrichtentechnik, Vol. **5** (1955), pp 481-489, 16 illustrations.
[4] GARMERSHAUSEN, K. H., GOLDBERG, S., a. MCDONALD, D. F., *A high sensibility cathode-ray tube for millimicrosecond transients,* IRE Transactions on Electron Devices, April 1957, pp 152-158, 8 illustrations.
[5] CZECH, J., *„FTO 2" Oszillograf für hohe Ansprüche,* Der Elektronenstrahl-Oszillograf, Berlin-Borsigwalde: Verlag für Radio-Foto-Kino-Technik GmbH 1955, 21 illustrations.
[6] SCHROEDER, O., *Hilfsgerät für Impulsspitzenmessungen,* Funk-Technik, Berlin-Borsigwalde: Verlag für Radio-Foto-Kino-Technik GmbH. Vol. **11** (1956) No. 14, 417 pages, 6 illustrations.

Chapter 8 Null indication in A.C. bridge circuits

[1] BRAILSFORD, H. D., *Measuring coil characteristics without an impedance bridge,* Electronics Manual for Radio Engineers, Vol. **20**, Feb. 1949, pp 239-242.
[2] OETKER, R., *Das Braunsche Rohr als Indikator für Wechselstrombrücken.* Frequenz, No. **2** (1951), pp 33-38, 8 illustrations.
[3] PESCHKE, H., *Prüfung und Abstimmung von Sendern,* Funktechnische Monatshefte, No. **12** (1940), pp 177-181, 13 illustrations.
[4] HOLLEUFEL, W., *Oszillograf als einfacher Nullindikator in Wechselstrombrücken,* Die Elektro-Post, No. **10** (1952), pp 191-192.
[5] COLE, R. H., *Use of the cathode-ray tube for comparison of capacities,* The Review of Scientific Instruments, June 1941, pp 298-300.
[6] FROMMER, I. C., *A.C. null indicator,* Electronics, Oct. 1951, pp 136, 138, 140, 160 and 164.
[7] VAN SUCHTELEN, H., *Comments on magnetic sorting bridges,* Philips' Electronic Application Bulletin, No. **12** (1949), pp 261-273 and Wilson, W.: The Cathode Ray Oscillograph in Industry, London: Chapman & Hall Ltd., 1948, pp 155-157.
[8] CROWHURST, N. H., *Transformer iron losses,* Electronic Engineering, Oct. 1951, pp 396-403.

Chapter 9 The electronic switch

[1] HEINS VAN DER VEEN, A. J., *Testing amplifier output valves by means of the cathode ray tube,* Philips Tech. Rev., Vol. **5** No. 3 (1940), pp 62-68, 17 illustrations.
[2] TE GUDE, H., *Elektronische Schalter und Wähler,* ETZ. No. **20** (1954), pp 680-686, 24 illustrations.
[3] CARPENTIER, E. E., *An electronic switch with variable commutating frequency,* Philips Tech. Rev., Vol. **9** No. 77 (1947/1948), pp 340-346.
[4] *Gleichzeitige Wiedergabe mehrerer Vorgänge auf dem Leuchtschirm eines Oszillographen,* Philips' Elektronisch Messen, Vol. **2** No. 4 (1948), pp. 2-16, 18 illustrations.
[5] BUDLONG, W. A., a. LUTZ, B. C., *An automatically synchronized electronic switch,* The Review of Scientific Instruments, Vol. **21** (1950) No. 2, pp. 167-168, 3 illustrations.
[6] JOHANNSEN, K., *Ein synchronisierbarer Elektronenschalter mit 20-30.000 Hz Schaltfrequenz,* Radio Mentor, No. **1** (1954), pp 33-38, 22 illustrations.
[7] JOCHEMS, P. J. W., a. STIELTJEN, F. M., *Apparatus for testing transistors,* Philips Tech. Rev., Vol. **13** No. 9 (1952), pp 254-265, 18 illustrations.

[8] Donaldson, P. E. K., *A multi-channel oscilloscope for electrophysiology*, Electronic Engineering, Feb. 1957, pp 78-83, 10 illustrations.
[9] Ludwig, H., a. Hind, J. E., *Electronic switch eliminates transients*, Electronics, Oct. 1955, pp 163-165, 4 illustrations.
[10] Spooner, A. M., *Electronic switch for television*, Electronic Engineering, May 7956, pp 196-199, 5 illustrations.
[11] Groenendyke, G. M., a. Lorper, G. B., *Cathode-ray display of saismic recording*, Electronics, May 1955, pp 160-165, 7 illustrations.
[12] Cordes, H. B., *Low-frequency switch for recording transients*, Electronics, May 1954, pp 168-169, 3 illustrations.
[13] Wagner, H. M., a. Herrink, J. F., *Self-switching R-F-amplifier*, Electronics, June 1947, pp 128-131, 8 illustrations.
[14] Dorsman, C., a. de Bruin, S. L., *An electron switch*, Philips Tech. Rev., Vol. **4**, Sep. 1950, pp 267-271, 13 illustrations.
[15] Classen, R., Gundlach, F. W., a. Lentze, F., *Ein Bildrasterverfahren zur gleichzeitigen Abbildung mehrerer Vorgänge mit dem Elektronen-Einstrahl-Oszillographen*, ATM. Sept. 1951, J 8344-5. T 106. 5 illustrations.
[16] Wood, K. E., a. Keenan, T. C., *A multi-trace cathode-ray tube display*, Electronic Engineering, Mar. 1956, pp 105-107, 5 illustrations.

Chapter 10 Uses of intensity modulation

[1] Czech, J., *Genaue Messung des Öffnungsverlaufes von Zentralverschlüssen photographische Kameras mit Elektronenstrahloszillographen*, Philips' Elektronisch Messen, Vol. **3** No. 8 (1953), pp 3-18, 14 illustrations.
[2] Tucker, M. J., *A note on electronic analogue integration and differentiation*, Electronic Engineering, Jan. 1953, pages: 35.
[3] Klein, P. E., *Differenzierung, Integrierung*. Elektronenstrahloszillographen, Vol. **1**, Berlin: Weidmannsche Verlagsbuchhandlung 1948, pp 102-103.
[4] Janssen, I. M. L., a. Ensing, L., *The electron-analogus, anapparatus for studying regulating mechanisms*, Philips Tech. Rev., Vol. **12** No. 9 and 11 (1951), pp 257-271 and 319-335, in particular pp 329-331.
[5] Wilson, W., *The cathode ray oscillograph in industry, automatic brilliancy control*, London: Chapman & Hall Ltd., 1948, 46 pages.
[6] Rochelle, R. W., *Cathode-ray-tube-beam intensifier*, Electronics, Oct. 1952, pp 151-153.
[7] de Klerk, I., *Automatic C.R.T. trace brightening for varying amplitude R.F. signals*, Electronic Engineering, Sept. 1953, pp 388-389, 7 illustrations.
[8] Elmore, W. C., a. Sands, M., *Electronics experimental techniques. A trace-brightening circuit*, New York, Toronto, London: McGraw-Hill Book Comp. Inc., 1949, 307 pages.
[9] Niesten, I. G., a. Wijnterp, W., *Het registreren van polaire diagrammen van enkele electrische grootheden bij constante frequentie met behulp van electronenstraaloscillograaf en camera*. De Ingenieur. Vol. **58** No. 32 (1946), pp 33-38, 6 illustrations, and *Berekening en registratie van enkele polaire stroomfiguren aan een autotransformator met variabele overzetting, welke met een constante weerstand is belast*. De Ingenieur. Vol. **59**. June 1947, pp 23-27, 6 illustrations.
[10] Sulzer, P. G., *Vector voltage indicator*, Electronics, June 1949, pp 107-109.
[11] McKay, D. M., *Projektive three-dimensional displays*, Electronic Engineering, July and August 1949, pp 249-254 and 281-286.
[12] Admiral, D. J. H., *Tracing vector locus diagrams by means of a cathode-ray oscilloscope*, Philips' Electronic Application Bulletin, No. **1** (1950), pp 2-14.
[13] Schäfer, O., *Ortskurvenschreiber für den Tonfrequenzbereich*, ATM. June 1952, I 36-8. pp 137 and 138.

[14] GENÇ, SELÂHATTIN: *Ein Kathodenstrahloszillograph zur Untersuchung von Funktionen zweier Variablen,* Technische Mitteilungen PTT. No. **9** (1950), pp 342-348, 18 illustrations.

[15] GREVEL, R., GUNDLACH, F. W., a. HERKLOTZ, H., *Die Erzeugung von dreidimensionalen Messdiagrammen mit dem Kathodenstrahloszillographen,* ATM. Nov. 1951, J 8344-6.

[16] JANSSEN, J. M. L., a. MICHELS, A. J., *An experimental "stroboscopic" oscilloscope for frequencies up to about 50 Mc/s,* Philips Tech. Rev., Vol. **12** Nos. 2 and 3 (1950), pp 52-60 and pp 73-83.

[17] FOSTER, B. C., *A simple valve comparator,* Electronic Engineering, May 1952, pp 220-223.

[18] CLASSEN, R., GUNDLACH, F. W., a. LENTZE, F., *Ein Bildrasterverfahren zur gleichzeitigen Abbildung mehrerer Vorgänge mit dem Elektronen-Einstrahl-Oszillographen,* ATM. Sept. 1951, J 8344-5.

Chapter 11 Phase measurements

[1] BENSON, F. A., a. CARTER, A. O., *Phase-angle measurements,* Electronic Engineering, June 1950, pp 238-242.

[2] SABAROFF, S.,*Technique for distortion analysis,* Electronics, June 1948, pp 114-117.

[3] NIJENHUIS, W., *Measurements of phase angles with the help of the cathode ray tube,* Philips Tech. Rev., Vol. **5** No. 7 (1940), pp 208-214.

[4] PATCHETT, G. N., *A versatile phase-angle meter,* Electronic Engineering, May 1952, pp 224-229.

[5] BARTELS, H., *Messung des Phasenmasses,* From: Grundlagen der Verstärkertechnik, Leipzig: Verlag S. Hirzel 1949, pp 24-27.

[6] THIEDE, H., *Die Umwandlung zweier phasenverschobener Spannungen in zwei phasengleiche Spannungen mit einem durch die Phase bestimmten Spannungsverhältnis.* Funk und Ton, Vol. **2** (1948) No. 3, pp 111-118.

[7] GRÜBEL, G., *Phasenmesser für Labor und Prüffeld,* Funk und Ton, Vol. **3** (1949) No. 6, pp 315-319.

[8] GRÜBEL, G., *Zwei einfache Summe-Differenzschaltungen.* Funk und Ton (now: Elektronische Rundschau), Vol. **3** (1949) No. 11/12, pp 591-593.

[9] RUHRMANN, A., *Hochfrequenz-Phasenmessung mit direkter Anzeige,* ATM. May 1950, V 3631-3, T 52/53.

[10] WITTKE, H., *Fehlerfreie Differentiation,* Elektronische Rundschau, Vol. **11** (1957) No. 1, 7 pages, 1 illustration.

[11] PITSCH, H., *Stromquellen-Ersatzschaltung.* Lehrbuch der Funkempfangstechnik, Leipzig: Akademische Verlagsgesellschaft Geest & Portig KG 1948, pp 6-7.

[12] GÜNTHER, H., *Tafeln zur Umwandlung von Reihenschaltungen komplexer Widerstände in äquivalente Parallelschaltungen,* Funk-Technik, Vol. **6** (1961), no. 2. pp 52-53.

[13] WITTKE, H., *Elektrische Integrationsverfahren,* Frequenz. Vol. **9** (1955) No. 2, pp 49-57, 12 illustrations.

[14] WITTKE, H., *Fehlerfreie elektronische Integration,* Elektronische Rundschau, Vol. **11** (1957), No. 3 pp 73-74, 2 illustrations.

[15] *Eine neue Messmethode zur Fehlerortung in Kabeln und Freileitungen,* Frequenz, Vol. **6** (1952) No. 7, pp 213-215.

[16] REVEREY, G., *Die zerstörungsfreie Prüfung von Isolatoren,* ETZ. Edn. A. Vol. **73** (1952) No. 14, pp 451-455.

[17] TERMAN, F. E., *Square-wave testing,* From: Radio Engineers' Handbook, New York, London: McGraw-Hill Book Comp. Inc., 1943, pp 968-971.

[18] MOSS, H., *Cathode ray tube traces,* Electronic Engineering, Sep. 1949, pp 52-55.

[19] SCHIFFEL, R., *Empfängerprüfung mit dem Multivibrator,* Das Radio Magazin, Vol. **9** (1949), pp 254-256.

[20] KÖHLER, A., *Verstärkerprüfung mit Rechteckschwingungen,* Das Radio Magazin, Vol. **13** (1949), pp 379-382.

[21] MEYER-EPPLER, W., *Die Messung der Frequenzcharakteristik linearer Systeme durch einmalige oder wiederholte Schaltvorgänge,* FTZ. No. **4** (1951), pp 174-182.

[22] MÜLLER, J., *Die Bestimmung des Amplituden- und Phasenganges von linearen Übertragungssystemen mit Hilfe von Rechteckwellen.* FTZ. No. **5** (1951), pp 211-220.

[23] SCHLEGEL, H., *Elemente der Impulstechnik,* Radio Mentor. I. Rechnerische Behandlung der Impulse und ihrer Verzerrungen durch einfache Vierpole. No. **5** (1951. pp 183-189. II. Der Impulsübertrager. No. **7** (1951). pp 336-230. III. Impulsverstärker. No. **11** (1951). pp 554-558.

[24] ZIMMERMANN, H., *Der Anteil des ZF-Verstärkers am Einschwingvorgang eines Fernsehempfängers,* FTZ. No. **12** (1951). pp 537-542. 9 illustrations.

[25] BRYAN, H. E., *Square wave testing simplified.* Audio Engineering. Oct. 1951. pp 14-15.

[26] HERSHLER, A., a. SEIDMAN, A. H., *General purpose short-pulse generator,* Electronics, Aug. 1953, pp 182-183.

[27] MÜLLER, J., *Die Prüfung von Fernsehübertragungssystemen mit Hilfe von Rechteckwellen,* Funk und Ton (jetzt Elektronische Rundschau), Vol. **6** (1952), No. 12 pp 617-631, 21 illustrations.

Chapter 12 Frequency measurements

[1] VAN SUCHTELEN, H., *Measuring the rate of watches with a cathode-ray oscillograph,* Philips Tech. Rev., Vol. 9 No. 10 (1947/48). pp 316-319.

[2] BRIGGS, B. H., *The miller-integrator.* Electronic Engineering, Aug. Sep. and Oct. 1948, pp 243-247, 279-284 and 325-329, and Funk und Ton, Vol. **2** (1948) No. 12, pp 653-655.

[3] BARKHAUSEN, H., *Einführung in die Schwingungslehre,* Leipzig: Verlag S. Hirzel 1940, pp 53-59.

[4] MÖLLER, W., *Die Braunsche Röhre,* Berlin-Tempelhof: Jakob Schneider Verlag, 4th Edn. 1949, pp 176-194.

[5] FELDTKELLER, R., *Einführung in die Theorie der Rundfunk-Siebschaltungen,* Leipzig: Verlag S. Hirzel, 3. Aufl. 1945, pp 54-61, Illustrations 20a, b, c, d and 31.

[6] KANBERG, H., *Die Breite des Mitnahmebereichs bei der Steuerung eines selbsterregten Röhrengenerators durch ganze Vielfache seiner Eigenfrequenz,* Funk und Ton (jetzt Elektronische Rundschau), Vol. **3** (1949) No. 9, pp 497-505.

[7] LAPORTE, H., *Die Messung von elektrischen Schwingungen aller Art nach Frequenz und Amplitude,* Halle: Verlag Knapp 1949.

[8] KLEIN, P. E., *Zeit- und Kurzzeitmessungen mit Elektronenstrahloszillographen,* Berlin: Weidmannsche Verlagsbuchhandlung, Edn. 1949.

[9] BADER, W., *Frequenzvergleich durch Zykloiden,* Archiv für Elektrotechnik. Vol. **XXXIV** (1948), No. 2/3, pp 115-124.

[10] CZECH, J., a. RODRIAN, G., *Darstellung von Vorgängen der analytische Mechanik mit dem Elektronenstrahloszillografen,* Funk und Ton, Vol. **4** (1950) No. 5, pp 239-249.

[11] WILSON, W., *Directional synchroscope. The cathode ray oscillograph in industry;* London: Chapman & Hall Ltd., 1948, p. 90, Fig. 67.

[12] LEWER, S. K., *Comparison of frequencies,* from: The Cathode-Ray Tube Handbook, London: Sir Isaac Pitman & Sons Ltd., 1947, p. 86. Fig. 32.

[13] REICH, H. J., *Circuits for oscillographic frequency comparison,* Review of Scientific Instruments, Vol. **8** Sep. 1937, p. 348.

[14] RANGACHARI, T. S., *The harmonic comparison of radio frequencies by the cathode-ray oscillograph,* Exper. Wireless and Wireless Eng., Vol. **5** May 1928, p. 264.

[15] RANGACHARI, T. S., *The super-position of circuit motions.* Exper. Wireless and Wireless Eng. Vol. **6** April 1929, p. 184.

[16] DORF, H., *Audio patents.* Audio Engineering. May 1951, pp 2-4, 4 illustrations, and JAYNES, E. T., USA-Patent No. 2,541,067.

Chapter 13 Representation of the rising flank of pulsed-shaped signal voltages by oscilloscopes without delaying elements in the vertical amplifier

[1] LIMANN, O., *Phasenschieber für oszillographische Zwecke,* Funkschau, No. **21** (1955) pp 471-472, 6 illustrations.

Chapter 14 Recording the waveforms of the luminous flux, current and voltage of fluorescent lamps

[1] BLOK, L., *High-frequency oscillations in sodium lamps,* Philips Tech. Rev., Vol **1** Mar. 1936, pp 87-90, 6 illustrations.
[2] RUEGG, W., *Heutiger Stand der Radioentstörung von Leuchtstoffröhren,* Bull. Schweiz. elektro-techn. Ver., Vol. **44** (1953), No. 18 pp 804-807, 3 illustrations.
[3] ORANJE, P. J., *Gas discharge lamps,* Philips Technical Library, 1951.
[4] ENDLER, H., *Die Betriebstechnik der Leuchtstofflampe,* Elektrotechnik, Vol. **34** No. 45 (1952).
[5] KRETZMANN, R., *Industrial electronics handbook,* Philips Technical Library, 3rd edition, 1959.
[6] ROHLOFF, E., *Das Verhalten der Leuchtstofflampen bei höheren Frequenzen.* ETZ. Edn. B.H. 1 (1958), pp 1-6, 10 illustrations.

Chapter 16 The display of hysteresis loops

[1] GRIESSEN, B., a. ZWAAG, H., *Aufnahme der Wechselstrommagnetisierungskurven mit Hilfe elektronischer Messgeräte.* Industrie-Elektronik. Vol. **6** (1955). pp 12-15. 11 illustrations.
[2] JONKER, G. H., a. VAN SANTEN, J. H., *The ferro electricity of titanates,* Philips Tech. Rev., Vol. **11** No. 6 (1949), pp 183-192.
[3] ETTINGER, G. M., *Butterfly curve tracer for magnetic materials.* Electronics, Mar. 1953, pp 119-121.

Chapter 17 Recording the characteristic curves of crystal diodes, transistors and electron tubes

[1] HEINS VAN DER VEEN, A. J., *Testing amplifier output valves by means of the cathode ray tube,* Philips Tech. Rev., Vol. **5** No. 3 (1940), pp 61-68, 17 illustrations.
[2] FOSTER, B. C., *A simple valve comparator,* Electronic Engineering, May 1952, pp 220-223, 9 illustrations.
[3] KAMMERLOHER, J., a. KREBS, H., *Entwicklung eines Kennlinienschreibers für Senderöhren,* Funk und Ton, Vol. **8** (1954) No. 9, pp 453-470, 34 illustrations.
[4] JOCHEMS, R. J. W., a. STIELTJES, F. H., *Apparatus for testing transistors,* Philips Tech. Rev., Vol. **13** No. 7 (1952), pp 254-265, 18 illustrations.
[5] KURSHAW, J., LOHMAN, R. D., a. HERZOG, G. B., *Cathode-ray tube plots transistor-curves,* Electronics, Feb. 1953, pp 122-127, 8 illustrations.
[6] O'NEILL, B. J., a. GUTTERMANN, A., *Methods and equipment for transistor testing,* Electronics, July 1953, pp 172-175, 8 illustrations.
[7] PANKOVE, J. I., *Pulsed curve tracer for semiconductor testing,* Electronics, Sep. 1954 pp 172-173, 4 illustrations.
[8] HUNTER, L. P., a. BROWN, R. E., *Production testor for transistors,* Electronics, Oct. 1950, pp 96-99, 6 illustrations.
[9] DAMMERS, B. C., HAANTJES, J., OTTE, J., a. VAN SUCHTELEN, H., *Applications of electronic valves in radio receivers and amplifiers,* Series on electronic valves vol IV, chapter IV. A, Parasitic effects and distorsion due to curvature of valve characteristics, R.F. and I.F. amplifying valves, pp 308-328, Philips Technical Library, 1950.

[10] ROTHE, H., a. KLEEN, W., *Messung der nichtlinearen Verzerrungen. Elektronenröhren als Anfangsstufenverstärker,* Leipzig: Akademische Verlagsgesellschaft Geest & Portig KG 1948, pp 135-136, 2 illustrations.

[11] BARTELS, H., *Bestimmung des Modulationsfaktors,* From: Grundlagen der Verstärkertechnik, Leipzig: Hirzel-Verlag 1949, p. 30, 1 illustrations.

Chapter 18 Recording the bandpass curves of HF-circuits, radio and television receivers

[1] SCHADOW, R., *Die Darstellung von Resonanzkurven durch den Kathodenstrahl-Oszillographen,* Funktechnische Monatshefte. No. **3** (1934), pp 105-109, 12 illustrations.

[2] PITSCH, H., *Lehrbuch der Funkempfangstechnik,* Leipzig: Akademische Verlagsgesellschaft Geest & Portig KG 1949, Par. 409: Die Nachstimmung mittels Impedanzröhre, pp 651-654, 5 illustrations.

[3] KLEIN, P. E., *Elektronenstrahl-Sichtgeräte in Medizin und Technik,* Berlin: Weidmannsche Verlagsbuchhandlung 1952, Section 304: Frequenzgang, pp 102-127, 49 illustrations.

[4] MEINKE, H., a. GUNDLACH, F. W., *Taschenbuch der Hochfrequenztechnik,* Berlin, Göttingen, Heidelberg: Springer-Verlag 1956, Section 57: Blindröhren bei kleiner Gitterspannung, pp 1129-1133, 4 illustrations.

[5] CORMACK, A., *Wide-range variable-frequency oscillator,* Wireless Engineer, Vol. **28** (1951), No. 336, pp 266-270, 10 illustrations.

[6] VON DUHN, H., *Apparat zur Ausmessung von Resonanzkurzen von einfachen Schwingungskreisen und Bandfiltern,* Funktechnische Monatshefte, No. **6** (1940), pp 84-86, 3 illustrations.

[7] SCROGGIE, M. G., *Design for a wobbulator,* Wireless World, Oct. 1950, pp 369-372, 8 illustrations.

[8] VILBIG, F., *Lehrbuch der Hochfrequenztechnik,* Leipzig: Akademische Verlagsgesellschaft Geest & Portig KG 1953, Vol. I: Verhalten des Serienresonanzkreises bei Anlegen einer veränderlichen Frequenz. pp 241-244, 5 illustrations.

[9] JUST, G., a. SIEWERT, W., *Ein Impedanzkurvenschreiber für Quarze,* Radio Mentor, No. **8** (1956), pp 504-509, 10 illustrations.

[10] KÜPFMÜLLER, K., *Die Systemtheorie der elektrischen Nachrichtenübertragung,* Zürich: Hirzel-Verlag, 1952.

[11] TURBOWITSCH, I. T., *Über die Messfehler bei der Bestimmung von Frequenzkurven mit Hilfe der Wobbelmethode,* Nachrichtentechnik, Vol. **4** No. 12 (1954), pp 523-524.

[12] RÖSLER, G., *Dynamische Untersuchungen an Ratiodetektoren,* Funk-Technik, Vol. **12** (1957), No. 3 pp 68-70, 7 illustrations.

[13] GRIESE, H. J., *Verfahren zur Messung der Selektionseigenschaften von Fernsehempfängern,* Archiv für elektrische Übertragung, Vol. **9** (1955), No. 4 pp 167-170, 7 illustrations.

[14] DE BOER, H. J., a. VAN WEEL, A., *An instrument for measuring group delay,* Philips Tech. Rev., Vol. **15** No. 11 (1954). pp 307-316, 12 illustrations.

[15] HUEVELMANN, C. J., a. VAN WEEL, A., *Group delay measurements,* Wireless Engineer, May 1956, pp 107-113, 8 illustrations.

[16] VAN WEEL, A., *Phase linearity of television receivers,* Philips Tech. Rev., Vol. **18** No. 2 (1956), pp 33-51, 30 illustrations.

[17] SCHÖNFELDER, H., *Ein Gerät zur statischen Messung und oszillographischen Darstellung der Phasen- oder Phasenlaufzeitkurve,* Frequenz, Vol. **10** (1956), No. 10 pp 309-318, 9 illustrations.

[18] KROEBEL, W., a. WEGNER, L. A., *Ein Gerät zur oszillographischen Darstellung des Phasen- und Amplitudenganges von video- und zwischenfrequenten Netzwerken,* Rundfunktechnische Mitteilungen, Vol. No. **2** (1957), pp 37-44, 22 illustrations.

Chapter 19 Investigations in television engineering

[1] DEMUS, E., *Untersuchungen an Fernseh-Studioeinrichtungen mit Hilfe von Synchronisierimpulsen veränderlicher Phasenlage,* FTZ, No. **5** (1953), pp 208-213, 14 illustrations.

[2] MÜLLER, J., *Stand der Normung von Prüf- und Messverfahren für die Fernseh-Übertragungstechnik,* Techn. Hausmitteilungen des NWDR, Vol. **7** (1955), No. 11/12, pp 209-216, 15 illustrations.

[3] MÜLLER, J., *Die Eigenschaften von Fernsehleitungen und deren Messung,* Der Fernmeldeingenieur, Vol. **10**, No. 9, Sept. 1956, 30 illustrations.

[4] MÜLLER, J., a. DEMUS, E., *Betriebsprüfverfahren für internationale Übertragungslinien,* FTZ. No. *3* (1955), 7 illustrations.

[5] *IRE standards on television:* Methods of Testing Monochrome Television Broadcast Receivers, 1960. Proceedings of the IRE. June 1960, pp 1124-1154, 12 illustrations.

[6] *Messtechnische Richtlinien für die Fernsehübertragungstechnik.* Fernmeldetechnisches Zentralamt Darmstadt VEa. Fernsehausschuss der FuBK, Arbeitsausschuss für Prüf- und Messverfahren.

[7] WEAVER, L. E., *Die exakte Messung von Videopegeln,* Rundfunktechnische Mitteilungen, Hamburg. Vol. **V** 1961, No. 6, pp 261-263, 7 illustrations.

[8] MÜLLER, J., *Anforderungen und Messungen an Fernsehweitverbindungen,* NTZ, 1960, No. 7, 22 illustrations.

[9] LEGLER, E., *Messgeräte der Fernsehtechnik,* NWDR, Techn. Hausmitt. Vol. **7** (1955), pp 71-76, 6 illustrations.

[10] DEMUS, E., *Die betrieblichen Messverfahren und Messeinrichtungen im Fernseh-Übertragungsdienst,* Der Fernmeldeingenieur, Vol. **10**, No. 12, Dec. 1956, 22 illustrations.

[11] DILLENBURGER, W., *Fernseh-Messtechnik,* Fachverlag Schiele und Schön, GmbH., Berlin (1960), 373 pages, 352 illustrations.

[12] MACEK, O., *Fernsehempfänger-Messtechnik,* ATM, Part I, General: February 1953, V 373-5, pp 41-42, 6 illustrations.
Part II, Messung der Übertragungseigenschaften von Fernsehempfängern: February 1954, V 373-6, pp 29-30, 3 illustrations.
Part III, Verfahren zur Messung der Übertragungseigenschaften im Video-Frequenz-Teil: May 1954, V. 373-7, pp 101-104, 8 illustrations.
Part IV, Verfahren zur Messung der Übertragungseigenschaften des Zwischenfrequenzverstärkers und Hochfrequenzteils, June 1954, V 373-8, pp 125-128, 6 illustrations.
Part V, Messungen im Ablenkteil des Empfängers, Sept. 1954, V 373-9, pp 205-208, 3 illustrations.
Part VI, Linearitäts-, Empfindlichkeits- und Trennschärfe-Messungen im Bild- und Tonteil, October 1954, V 373-10, pp 235-238, 6 illustrations.

[13] MACEK, O., *Fernsehsender-Messtechnik,* ATM (Archiv für technisches Messen),
Part I: March 1958, V 373-17, pp 45-48, 5 illustrations.
Part II: 1958, V 373-18, pp 93-96, 16 illustrations.
Part III: August 1958, V 373-19, pp 165-166.

[14] HAMILTON, G. E., a. ILOWITE, R., *Grey-scale generator,* Electronics, November 1952, pp 143-145, 4 illustrations.

[15] WEBB, R. C., *Synthetic-pattern generator for the solution of certain instrumentation problems in television,* RCA-Review, June 1954, pp 187-202, 12 illustrations.

[16] FRÖLING, H. E., *Prüfzeile und elektronisches Testbild,* Radio Mentor, 1959 No. 4, pp 244-247, 5 illustrations.

[17] KIVER, M. S., *Television and F-M receiver servicing,* D. van Nostrand Company, Inc., New York-London.

[18] GROB, B., *Basic television - Principles and serving,* McGraw-Hill Book Company, Inc., 1949.

[19] SWALUW, H. L. a. VAN DER WOERD, J., *Introduction into T.V. service,* Philips Technical Library, 1963.

[20] RIDER, J. I., *Picture book of TV troubles,* Vol. **3,** Video I-F and Video Amplifier Circuits.

[21] KERKHOF, F., a. WERNER, W., *Television,* section 4.5.1. Separation of the field-synchronizing signal by means of an RC integrator, pp 101-103, Philips Technical Library, 1952 (New edition in active preparation).

[22] SCHRÖDER, W., *Messung von Röhren-Betriebswerten bei der Horizontalablenkung,* Funk-Technik, Vol. **11** (1956), No. 12, pp 391-393, 12 illustrations.

[23] FISHER, J., *Television picture line selector,* Electronics, March 1952, pp 140-143, 9 illustrations.

[24] BRUYER, E. M., *Line selector checks television wareforms,* Electronics, Sept. 1953, pp 153-155, 5 illustrations.

[25] MACEK, O., *Ein Zeilenwähler mit Eigensynchronisierung für die Fernsehtechnik,* Frequenz, Vol. **10** (1956), No. 6, pp 193-197, 11 illustrations.

[26] MOTHERSOLE, P. L., *A television line selector unit,* Electronic Engineering, Dec. 1956, pp 520-523, 6 illustrations.

[27] WOLF, H., *Ein Fernseh-Zeilenwähler,* NTZ, 1959, No. 5, pp 239-242, 9 illustrations.

[28] BÖDEKER, H., *Verfahren zur Beurteilung und Kennzeichnung des Sägezahn-Testsignals bei der Prüfung von Fernsehleitungen,* FTZ, No. 6 (1955), 4 illustrations.

[29] *IRE standards on television:* Measurement of Differential Gain and Differential Phase, 1960. February 1960, pp 201-208, 10 illustrations.

[30] FRÖLING, J., *Das Prüfzeilenverfahren beim Fernsehen.* NWDR-Hausmitteilungen, Vol. **7** (1955), No. 7/8, pp 129-138, 7 illustrations.

[31] FRÖLING, J., *Die internationale Entwicklung des Prüfzeilenverfahrens beim Fernsehen.* Frequenz, January 1959. No. 1, pp 1-10, 5 illustrations. 1959, No. 5, pp 147-155, and 1959, No. 6, pp 175-183.

[32] POUYFERRIE, A. a. FRACHET, G., *Die Prüfzeilen des französischen Fernsehens.* Rundfunktechnische Mitteilungen Hamburg, Vol. **4** (1960), pp 153-157, 7 illustrations.

[33] WOBST, J., *Die Messung der Phasenlaufzeit von Videoverstärkern und -netzwerken.* Technische Mitteilungen des BRV, No. 2, May 1958, pp 37-40, 12 illustrations.

[34] *IRE standards on video techniques:* Definitions of Terms Relating to Television, 1961. Proceedings of the IRE, July 1961, pp 1193-1195.

[35] *Normblatt* DIN 450 60, Fernsehen (Begriffe).

Chapter 20 Investigating matching conditions and measuring impedances in the ultra-short wave band by means of a long transmission line

[1] VAN HOFWEEGEN, J. M., *The measurement of impedances, particularly on decimetre waves,* Philips Tech. Rev., Vol. **8** No. 1 (1946), pp 16-24, 8 illustrations.

[2] VAN HOFWEEGEN, J. M., *Impedance measurement with a non tuned lecher system,* Philips Tech. Rev., Vol. **8** No. 9 (1946), pp 280-288, 8 illustrations.

[3] SCHWEITZER, H., *Dezimeterwellen-Praxis,* 126 pages, 141 illustrations.

[4] KERKHOF, F., a. WERNER, W., *Television,* chapter 9, The transmission line or feeder, pp 315-334, Philips Technical Library, 1952 (New edition in preparation).

[5] GUNDLACH, F. W., *Grundlagen der Höchstfrequenztechnik,* Berlin Göttingen, Heidelberg: Springer-Verlag 1950, 500 pages, Part C: Die elementaren Wellen auf Doppelleitungen, pp 205-266.

[6] MEGLA, G., *Dezimeterwellentechnik,* Leipzig: Fachbuchverlag, 3rd. Edn. 1954, pp 29-120, 352 pages.

[7] VILBIG, F., *Lehrbuch der Hochfrequenztechnik,* Leipzig: Akademische Verlagsgesellschaft Geest & Portig KG 1953, Elektrische Vorgänge auf Leitungen, pp 319-408, 110 illustrations.

[8] HOLM, W. A., How television works, section 6. I: Fundamentals of wireless transmission, Philips Technical Library, second edition 1963.
[9] MEINKE, H., a. GUNDLACH, F. W., *Taschenbuch der Hochfrequenztechnik.* Berlin, Göttingen, Heidelberg: Springer-Verlag 1956, pp 185-203.
[10] ZINKE, O., *Hochfrequenz-Messtechnik,* Leipzig: Verlag S. Hirzel. 2nd edn. 1947. Section H: Scheinwiderstands- und Dämpfungsmessungen. p. 161.
[11] MEINKE, H. H., *Theorie der Hochfrequenzschaltungen,* München: R. Oldenbourg Verlag 1951, Par. 28: Die Leitung als Messgerät. pp 193-201. 8 illustrations.
[12] TERMAN, F. E., *Radio engineers handbook.* New York, London: McGraw-Hill Book Comp. Inc. 1943. Section: Transmission Lines. pp 172-197, 20 illustrations.
[13] WITTING, R., *Kreisbogenförmige Messleitung für Höchstfrequenzen mit oszillografischer Aufzeichnung der Spannungsverteilung.* Nachrichtentechnik, Vol. **4** Sep. 1954, pp 378-380, 3 illustrations.
[14] BAUER, J., *Special applications of ultra-high-frequency wideband sweep generators,* R.C.A. Review, Vol. **8** (1947), pp 564-575, 7 illustrations.
[15] VAN DEN HOOGENBAND, J. C., a. STOLK, J., *Reflection and impedance measurements by means of a long transmission line,* Philips Tech. Rev., Vol. **16** No. 11 (1955). pp 309-320, 18 illustrations.
[16] KRAUSSE, W., *Die Ermittlung von Amplitudenverhältnissen im Höchstfrequenzgebiet mit Hilfe einer Messleitung bei unbekannter Indikatorkennlinie,* Nachrichtentechnik, Vol. **6** No. 2 (1956), pp 83-86, 7 illustrations.
[17] BERGTOLD, F., *Einführung in die Arbeit mit dem Smith-Diagramm,* Funk-Technik, Vol. **12** (1957), No. 16, pp 565-566, 9 illustrations.
[18] SMITH, P. H., *Transmission-line calculator,* From: Electronics for Engineers. Edn. von J. Marcus and V. Zeluff. New York, London: McGraw-Hill Book Comp. Inc., 1st Edn. 5th impression, 1945, pp 326-331, 4 illustrations.
[19] FÖRSTER, G., *Fernseh-Bildstörungen durch Fehlanpassung und Verstimmung des Tuners,* Funkschau, Nos. 3 and 5, (1957), pp 67-70 and 123-125, 27 illustrations.
[20] MENZEL, H., *Einfache Fehlerortsbestimmung bei schadhaften HF-Leitungen,* Radio und Fernsehen, No. 5 (1957), pp 148-149, 3 illustrations.

Chapter 21 Measuring transit time and investigating matching conditions in cables by means of pulse voltages

[1] VAN VELTHOVEN, TH. M. W., *The oscilloscope, type "GM* **5660**", Communication News, Vol. **VIII**. June 1953, pp 139-146, 11 illustrations.
[2] SCHLEGEL, H. R., a. NOWAK, A., *Impulstechnik,* Hannover: Fachbuchverlag Siegfried Schütz 1955.
[3] TUCKER, E. C., *Principles of radar.* New York, London: McGraw-Hill Book Comp. Inc., 1946.
[4] RÖSCHLAU, H., *Die Anwendung der Impulstechnik zur Prüfung von Fernseh-Übertragungseinrichtungen,* NWDR-Hausmitteilungen, Vol. **5** (1953), No. 9/10, pp 187-190, 4 illustrations.
[5] KADEN, H., *Über das Verhalten von Kabeln mit Wellenwiderstandsschwankungen bei Fernseh- und Messimpulsen,* Archiv der elektrischen Übertragung, Vol. **7** (1953), pp 157-162 and 191-198.
[6] KRAUTKRÄMER, J., a. KRAUTKRÄMER, H., *Werkstoffprüfung mit Ultraschall nach dem Echolotverfahren,* Elektronische Rundschau, Vol. **9** (1955), No. 6 pp 238-241, 12 illustrations.

Chapter 22 Determining the characteristic qualities of resonant circuits and bandpassfilters from the pattern of the decaying oscillation after shock excitation

[1] BLOK, L., *An apparatus for the measurement of scanning speeds of cathode ray tubes,* Philips Tech. Rev., Vol. **3** No. 7 (1938), pp 216-219, 5 illustrations.

[2] CZECH, J., *Darstellung abklingender Schwingungen als stehendes Bild auf der Kathodenstrahlröhre,* VDI-Z, Vol. **84** (1940), No. 5, pp 83-85, 6 illustrations.

[3] FRICKE, H., *Die oszillografische Dämpfungsmessung an Resonanzkreisen,* Funktechnischer Vorwärts, Vol. **13** No. 7 (1943), pp 81-84, 9 illustrations, and *Probleme um die Trennschärfe.* No. 7 (1943), pp 90-94, 14 illustrations.

[4] NENTWIG, K., *Dämpfungsmessungen,* Radio Mentor, No. **6** (1948), pp 251-253, 3 illustrations.

[5] MEYER-EPPLER, W., *Die Messung der Frequenzcharakteristiek linearer Systeme durch einmalige oder wiederholte Schaltvorgänge,* FTZ, No. 4 (1951), pp 174-181, 18 illustrations.

[6] VAN SLOOTEN, J., *Experimental testing of electrical networks by the means of the unit function response,* Philips Tech. Rev., Vol. **12** No. 8 (1951), pp 233-239, 11 illustrations.

[7] OLSSON, C. O., a. ORLIK-RÜCKEMANN, K., *The Dampometer,* Electronic Engineering, Oct. 1954, pp 420-428, 11 illustrations.

[8] PUNGS, L., *Grundzüge der Hochfrequenztechnik,* Bücher der Technik, Wolfenbütteler Verlagsanstalt GmbH 1948, 106 pages, 80 illustrations.

[9] VILBIG, F., *Lehrbuch der Hochfrequenztechnik,* Vol. I: Elektrische Vorgänge in quasistationären Kreisen, pp 205-265, 52 illustrations.

[10] PITSCH, H., *Lehrbuch der Funkempfangs-Technik,* Leipzig: Akademische Verlagsgesellschaft Geest & Portig KG. 855 pages, 894 illustrations.

[11] FELDTKELLER, R., *Einführung in die Theorie der Hochfrequenz-Bandfilter,* Stuttgart: Verlag S. Hirzel 1953, 193 pages, 110 illustrations.

[12] TERMAN, F. E., *Radio engineers' handbook.* New York, London: McGraw-Hill Book Comp. Inc., 1ste edn. 8th impr. 1943, Part 3: Circuit Theory, pp 135-172, 37 illustrations.

Chapter 23 Some methods for measuring the amplitude modulation of high-frequency voltages

[1] HEYBOER, J. P., a ZIJLSTRA, P., *Transmitting valves,* Series on electronic valves vol. 7, chapter 5: Modulation of an R.F. amplifier, pp 90-137, Philips Technical Library, 1951.

[2] VILBIG, F., *Lehrbuch der Hochfrequenztechnik,* Vol. *II.* Akademische Verlagsgesellschaft Becker & Faber KG 1942, 14. Chap. 1: Die Modulation von Sendern, pp 218-267, 60 illustrations.

[3] TERMAN, F. E., *Radio engineers' handbook,* New York, London: McGraw-Hill Book Comp. Inc., 1943, Part **7**: Modulation and Demodulation. pp 531-588, 53 illustrations.

[4] MEINKE, H., a. GUNDLACH, F. W., *Taschenbuch der Hochfrequenztechnik,* Berlin, Göttingen, Heidelberg: Springer-Verlag 1956, Abschnitt U: Modulation, pp 1048-1149, 141 illustrations.

Chapter 24 Represention of the frequency spectrum of modulated HF voltages and the frequency panorama of transmitters

[1] HARRIS, H. E., *Simplified Q-multiplier,* Vol. **25** (1951), No. 5. pp 130-134, 3 illustrations.

[2] NONNENMACHER, W., *Ein selektiver Verstärker mit* **1** *Hz Bandbreite,* Elektronische Rundschau, No. **5** (1956), pp 125-128, 7 illustrations.

[3] FINKBEIN, U., *Modulationsverzerrungen durch Bandfilter,* Funk und Ton, Vol. **4** (1950), No. 1, pp 11-15, 4 illustrations.

[4] KROEBEL, W., *Anzeige- und Messgeräte mit gleitender Oszillator- und Generatorfrequenz,* FTZ, No. **2** and **3** (1951), pp 70-76 and 106-110, 14 illustrations.

[5] WOSCHNI, E. G., *Zwei Verfahren zur Untersuchung der Modulationsverzerrungen bei Frequenzmodulation,* Nachrichtentechnik, Vol. **2**. No. 6 (1952), pp 170-177, 22 illustrations.

[6] WOSCHNI, E. G., *Aufnahme von Spektren frequenzmodulierter Schwingungen sowie von Bessel-Funktionen auf dem Braunschen Rohr.* Hochfrequenztechnik und Elektroakustik, Vol. **63** (1953), No. 2, pp 46-49, 8 illustrations.

[7] WOSCHNI, E. G., *Messanordnung zur Messung des Frequenz- und Phasenhubes sowie der quadratischen Modulationsverzerrungen bei FM- und PM-Sendern auf dem Braunschen Rohr,* Nachrichtentechnik, Vol. **4** No. 5 (1954), pp 200-203, 10 illustrations.

[8] WOSCHNI, E. G., *Verzerrungsmessungen bei Frequenzmodulation.* Nachrichtentechnik, Vol. **5** No. 2 (1955), pp 52-55, 8 illustrations.

[9] STEINBACH, S., *Ein schnell arbeitender Suchton-Spektrograf für den Sprachfrequenzbereich,* Nachrichtentechnik, Vol. **6** Nos. 8 and 9 (1956), pp 365-368 and 396-400, 43 illustrations.

[10] SOANES, S. V., *Audio frequency spectro analysis,* Electronic Engineering, June and July 1952, pp 268-270 and 312-318, 17 illustrations.

[11] HANSEN, N. R., *A simple pulse spectrograph,* Journal of Scientific Instruments, Vol. **34**. Oct. 1957, pp 402-404, 4 illustrations.

[12] KAULE W., a. JOHNE, A., *Tonfrequenz-Spektrometer,* Nachrichtentechnik, Vol. **6** No. 1 (1956), pp 35-39, 20 illustrations.

[13] DE LORENZI, J., *FS-Frequenz-Panoramaempfänger,* Funk-Technik, Vol. **11** (1956), No. 4, pp 92-93, 4 illustrations.

[14] THOMASSON, D. W., *Panoramic display,* Electronic Engineering, July 1949, pp 259-261, 4 illustrations.

Chapter 25 Adjustment of high impedance wideband voltage dividers by means of square pulses or symmetrical square voltages

[1] NEETESON, P. A., *Electronic valves in pulse circuits,* Philips Technical Library, second edition 1959.

[2] THIELE, G., *Berechnungsanleitung für Flip-Flop-Schaltungen.* Elektronische Rundschau, Vol. **11** (1957), Nos. 7, 8 and 9, pp 212-215, pp 250-252 and pp 274-276, 29 illustrations.

Chapter 26 Recording low-frequency characteristic curves

[1] TERMAN, F. E., *Curve tracing systems for observing and recording amplification characteristics.* Radio Engineers' Handbook, New York, London: McGraw-Hill Book Comp. Inc., 1943, pp 966-968.

[2] MÖLLER, C., *Zum Entwurf von Tiefpassfiltern für Modulationsverstärker,* Funk-Technik, Vol. **5** (1950), No. 2, pp 48-49, 5 illustrations.

Chapter 27 Some examples from electro-acoustic practice

[1] VAN DER POL, B., a. ADDINK, C. C. J., *The pitch of musical instruments and orchestras,* Philips Tech. Rev., Vol. **4** No. 7 (1939), pp 205-210, 6 illustrations.

[2] ADDINK, C. C. J., *Precise calibration of tuning forks,* Philips Tech. Rev., Vol. **12** No. 8 (1950), pp 228-232, 3 illustrations.

[3] TAK, W., *Measurement of reverberation,* Philips Tech. Rev., Vol. **8**, No. 3 (1946), pp 82-88, 12 illustrations.

[4] TAK, W., *Measuring reverberation time by the method of exponentially increasing amplification,* Philips Tech. Rev., Vol. *9*, No. 12, (1947/48), pp 370-378, 14 illustrations.

[5] MAYO, C. G., BEADLE, D. G., a. WHARTON, W., *Equipment for acoustic measurements,* Electronic Engineering, Sep., Oct., Nov. 1951, pp 326-331, 8 illustrations. pp 368-373, 10 illustrations resp. pp 424-428, 9 illustrations.
[6] WICKER, R. G., *Simplified "wow" and "flutter" measurement,* Wireless World, Feb. 1956, pp 97-98, 3 illustrations.
[7] *Klangstruktur der Musik,* Ed by Fritz Winckel, Berlin-Borsigwalde: Verlag für Radio-Foto-Kinotechnik GmbH 1955, 224 pages, 140 illustrations.
[8] REICHHARDT, W., *Grundlagen der Elektroakustik,* Leipzig: Akademische Verlagsgesellschaft Geest & Portig KG 1952, 445 pages.
[9] ENDRES, W., *Methoden der Sprachanalyse und Sprachsynthese und ihre Bedeutung für die Nachrichtentechnik,* Der Fernmeldeingenieur, Vol. **11**. No. 11 (1957), pp 1-40, 25 illustrations.

Chapter 28 Measuring the action of between-lens shutters of cameras

[1] WILSON, W., *The cathode-ray-oscillograph in industry,* London: Chapman & Hall Ltd. 1948, "Timepieces", pp 112-113, 2 illustrations.
[2] BLOK, L., *An apparatus for the measurement of scanning speeds of cathode ray tubes,* Philips Tech. Rev., Vol. **3** No. 7, July 1938, pp 221-224, 5 illustrations.
[3] CZECH, J., *Darstellung ablingender Schwingungen als stehendes Bild auf der Kathodenstralröhre,* VDI-Z., Vol. **84** (1940), No. 5, pp 83-85, 6 illustrations.

Chapter 30 Study of mechanical vibrations by means of electromagnetic and electrodynamic pickups

[1] HAGENDOORN, P. J., a. REYNST, M. F., *Anelectrical pressure indicator for internal combustion engines,* Philips Tech. Rev., Vol. **5**, Dec. 1940, pp 348-356, 12 illustrations.
[2] WORMSBECHER, H., *Anwendung elektronischer Messgeräte in der Kunstseide-Industrie,* Philips Elektronisch Messen, Vol. **3** No. 7, pp 2-12, 16 illustrations.
[3] SEVERS, J., *An electrodynamic pickup for the investigation of mechanical vibrations,* Philips Tech. Rev., Vol. **5**, Aug. 1940, pp 230-237, 10 illustrations.
[4] WINSLADE, R., *Vibration measurements,* Electronic Engineering, Dec. 1952, pp 553-557, 9 illustrations.
[5] VAN SANTEN, G., *Mechanical vibrations,* Philips Techn. Library, third enlarged edition, 1961.
[6] V. D. POEL, C., *Dynamische Prüfung von Strassendecken,* Industrie-Elektronik, Vol. **3**, No. 1 (1955), pp 9-12, 10 illustrations.
[7] SALJÉ, E., *Die Werkzeugmaschine unter dynamischer Belastung.* Industrie-Anzeiger. Nos. 27 and 36 (1955), pp 355-365 and pp 501-507, 84 illustrations.

Chapter 31 Study of dynamic strain processes and observation of mechanical vibrations by means of strain gauges

[1] DE BRUIN, S. L., *The investigation of rapidly changing mechanical stresses with the cathode ray oscillograph,* Philips Tech. Rev., Vol. **5** No. 1 (1940), pp 26-28, 8 illustrations.
[2] HERRMANN, P. K., *Neuzeitliche Messverfahren nichtelektrischer Antriebsgrössen,* ETZ, Vol. **64**, No. 25/26 (1943), pp 349-353, 9 illustrations.
[3] *Dehnungsmesstreifen-Messtechnik. Theorie und Praxis,* Hamburg: Elektro Spezial GmbH., 32 pages, 57 illustrations.
[4] KOCH, J. J., BOITEN, R. G., and others, *Strain gauges, Theory and applications,* Philips Technical Library 1951.
[5] FINK, K., *Grundlagen und Anwendungen des Dehnungsmesstreifens,* Düsseldorf: Verlag Stahleisen GmbH, 1952, 219 pages, 191 illustrations.

[6] RÖTSCHER, F., a. JASCHKE, R., *Dehnungsmessungen und ihre Auswertung,* Berlin, Göttingen, Heidelberg: Springer-Verlag 1939, 120 pages, 191 illustrations.

[7] ROHRBACH, CH., *Das Dehnungsmesstreifen-Verfahren,* ATM. I. Allgemeines über Dehnungsmesstreifen und Zubehör. J 135-4. July 1953. pp 139-142. 5 illustrations. II. Eigenschaften technischer Dehnungsmesstreifen. J 135-5. Oct. 1953. pp 237-240. 10 illustrations. III. Handhabung technischer Dehnungsmesstreifen. J 135-6. July 1954. pp 135-138. 3 illustrations. IV. Elektrische Messtechnik. J 135-7. Feb. 1955. pp 41-44. 5 illustrations. V. Elektrische Messgeräte. J 135-8. July 1955. pp 133-136. 11 illustrations. VI. Verschiedenes. J 135-9. Oct. 1955. pp 231-234. 3 illustrations.

[8] FINK, K., a. ROHRBACH, Ch., *Eigenschaften und Handhabung technischer Dehnungs*messtreifen, Archiv für das Eisenhüttenwesen, Vol. **23** No. 1/2 (1952), pp 75-81, 14 illustrations.

[9] BIEBER, E., *Schlagkraftmessungen an Dauerschlagwerken mittels Dehnungsmesstreifen,* Industrie-Elektronik, Vol. **5** No. 1 (1957), pp 20-24, 6 illustrations.

[10] EBERT, W., *Messungen an Druckmaschinen mit Dehnungsmesstreifen,* Industrie- Elektronik, Vol. **3** No. 3/4 (1955), pp 8-11, 7 illustrations.

[11] MÄKELT, H., *Der Tiefziehvorgang auf der Kurbelpresse.* Industrie-Anzeiger, No. III (1956), pp 298-302, 8 illustrations.

[12] SCHUH, W., *Messen grosser mechanischer Verlagerungen mittels Dehnungsmesstreifen in Anwendung bei Schwingungsversuchen,* Industrie Elektronik, No. **6** (1955), pp 3-7, 9 illustrations.

[13] FRIESE, TH., *Schraubenschub- und Drehmomentmessungen auf Schiffen,* Industrie-Elektronik, Vol. **4** No. 5/6 (1956), pp 3-7, 8 illustrations.

[14] DITTRICH, O., *Drehmomentmessung mit Dehnungsmesstreifen,* Elektronik No. 2/3 (1957), pp 37-40, 6 illustrations.

[15] CURTIN, W. F., *Measurement of stress in rotating shafts,* Electronics, July 1945, pp 114-122, 11 illustrations.

[16] WACHTER, J., *Dehnungsmessung an umlaufenden Maschinenteilen,* VDI-Z. Vol. **98** (1956), No. 3, pp 93-97, 14 illustrations.

[17] BERGHAUS, H. J., *Statische und dynamische Messung kleiner Drehmomente mittels Dehnungsmesstreifen,* Industrie-Elektronik, Vol. **4** No. 5/6, (1956), pp 10-11, 4 illustrations.

[18] LÖFFLER, K., *Schwingungsmessungen an Flugmotoren-Kurbelwellen und Luftschrauben,* Elektronik, No. 2 (1956), pp 35-43, 24 illustrations.

[19] LEWIS, A. D. M., *Strain-testing railroad bridges,* Electronics, Sep. 1951, pp 117-119, 5 illustrations.

[20] BRANDT, W., *Fortschritte in der Vielstellenmesstechnik,* Industrie-Elektronik, No. 6 (1955), 5 illustrations.

[21] BRANDT, W., *Neue Möglichkeiten in der Umschaltautomatik bei Dehnungsmesstreifen,* Industrie-Elektronik, No. 2 (1957), pp 10-16, 13 illustrations.

[22] KLEIN, P. E., *Neue Wege zur Überwachung von automatischen Prozessen mit elektronischer Vielfachanzeige,* Elektronik, 1959, Nr. 3, p 79-86, 21 illustrations.

[23] VON BASEL, C., *Mechanisch-elektrische Umwandler zur Messung mechanische Grössen,* ATM., J 86-3, June 1954, pp 139-142, 18 illustrations.

[24] KLEIN, P. E., *Geber mit Widerstands-Systemen zur elektrischen und elektronischen Messung nichtelektrischer Grössen,* Elektronik, No. **11** (1955), pp 272-278, 24 illustrations.

[25] PFIER, P. M., *Elektrische Messung mechanischer Grössen,* Berlin, Göttingen, Heidelberg: Springer-Verlag 1948, 307 illustrations.

Chapter 32 Photographic recording of the path of the spot trace

[1] MUTTER, E., *Kompendium der Photographie,* I. Bd. Die Grundlagen der Photographie, Berlin-Borsigwalde: Verlag für Radio-Foto-Kinotechnik GmbH 1957, 360 pages, 156 illustrations.

[2] FINK, L., *Photographische Messtechnik,* München, Berlin: R. Oldenbourg Verlag 1940, 222 pages, 174 illustrations.

[3] *Über die Aufnahme von Oszillogrammen auf photographisches Papier.* Philips Miniwatt-Monats-Heft, No. 87 (1941), pp 74-76, 4 illustrations.

[4] SÖRENSEN, CH., a. KÜBLER, A., *Eine neue Registrierkamera für Elektronenstrahl- Oszillographen,* Siemens-Zeitschrift, No. 4 (1956), pp 205-206, 2 illustrations.

[5] v. ANGERER, E., *Wissenschaftliche Photographie,* Leipzig: Akademische Verlagsgesellschaft Becker & Erler KG. 3rd Edn, 1943.

[6] CUSTERS, J. F. H., *The recording of rapidly occuring electric phenomena with the aid of the cathode tube and the camera,* Philips Tech. Rev., Vol. **2** No. 5 (1937), pp 148-155, 9 illustrations.

[7] SCHILLING, A., *Fortschritte auf dem Gebiet der Kinefilm-Emulsionen für Aufnahmezwecke,* Agfa-Veröffentlichungen, Vol. **VI** (1956), pp 65-75, 8 illustrations.

[8] KLÖTZER, F., *Röntgen-Emulsionen,* Phototechnik und -Wirtschaft, Berlin-Borsigwalde: Verlag für Radio-Foto-Kinotechnik GmbH, No. **9** (1954), pp 437-438, 1 illustration.

[9] BILTZ, M., *Absolute Farbenempfindlichkeit photographischer Schichten,* Agfa-Veröffentlichungen, Vol. *IV* (1934), pp 26-34, 2 illustrations.

[10] HILLE, G. M., *Fotografische Aufnahme- und Registriermaterialien für technische Zwecke,* Elektronik, No. **9** (1956), pp 236-241, 25 illustrations.

[11] HILLE, G. H., *Der Farbfilm in der Oszillografentechnik,* Photographie und Wissenschaft, Vol. **6** No. 1 (1957), pp 17-19, 9 illustrations (in eight colours).

[12] *Die "Land"-Fotografie in der Industrie,* Industriekurier, Vol. **5** No. 105 (1952), 5 illustrations.

[13] NAIDAMAST, D., *Quantity reproduction of photo-oscillograms,* Electronics, Sep. 1954, pp 264, 266 and 268, 2 illustrations.

[14] BLOK, L., *An apparatus for the measurement of scanning speeds of cathode ray tubes,* Philips Techn. Rev., Vol. **3** No. 7 (1938), pp 221-224, 5 illustrations.

[15] MOURIC, K., *O zapisovací rychlosti elektronového paprsku u oscillografu,* Elektrotechnicky Obzor, Vol. **32** (1940), No. 5, pp 1-5 (special impression).

[16] FELDT, R., *Photographing patterns on cathode-ray tubes,* Electronics for Engineers, New York, London: McGraw-Hill Book Comp. Inc., 1945, 1st Edn. pp 93-98, 4 illustrations.

[17] FÄLKER, R., a. HÜCKING, E. E., *Zur Schirmbild-Fotografie,* Elektronische Rundschau, Vol. **11** (1957), No. 11, pp 332-335, 13 illustrations.

[18] *Anwendung der Kathodenstrahlröhre DN 9-5,* Philips Miniwatt-Monatsheft, No. 81 (1940), pp 117-126, 11 illustrations.

[19] SILINK, K., *Das Zeilenregistrierverfahren bei der Kathodenstrahloszillographie,* Funktechnische Monatshefte, No. 1 (1941), pp 7-10, 11 illustrations.

[20] KOEPPE, H. H., *Gamma-Zeitkurven für Perutz-Perinal und Perutz-Film,* Phototechnik und -Wirtschaft, No. 9 (1954), page 28, 1 illustration.

[21] MUTTER, E., *Ein Hilfsmittel zur Kennzeichnung der Papiergradation: Das "Gradameter",* Phototechnik und -Wirtschaft, No. **9** (1954), pp 25-26, 1 illustration.

[22] HOFFMANN, K., *Vergrösserung oszillographischer Kleinbildaufnahmen auf Lichtpauspapier,* ETZ., Edn. A. Feb. 1953, page 82, 2 illustrations.

Chapter 33 Large picture projection of oscillograms

[1] MAROUN, W., *Auswerten von Leuchtschirmbildern mit einem Nachzeichengerät,* Elektronik, No. **7** (1956), 178 pages, 2 illustrations.

[2] KEITZ, H. A. E., *Light calculations and measurements,* Philips Technical Library, 1955 (new edition in preparation).
[3] JENSEN, H., *Bildwandprobleme,* Philips-Kinotechnik, No. **8** (1954), pp 3-7, 2 illustrations.
[4] VAN ALPHEN, P. M., a. RINIA, H., *Projection-television receiver,* Philips Tech. Rev., Vol. **10** No. 3 (1948), pp 69-78, 11 illustrations.
[5] HILKE, O., *Heimprojektionsempfänger,* From: Fernsehen, Edition from G. LEITHÄUSER and F. WINCKEL. Berlin, Göttingen, Heidelberg: Springer-Verlag 1953, pp 259-301, 50 illustrations.
[6] BONUS, E., *Projektionsoszillograph für Vorlesungen,* Industrie- Elektronik, No. 5/6 (1956), 8 pages, 2 illustrations.

INDEX

GPSR Compliance

The European Union's (EU) General Product Safety Regulation (GPSR) is a set of rules that requires consumer products to be safe and our obligations to ensure this.

If you have any concerns about our products, you can contact us on ProductSafety@springernature.com

In case Publisher is established outside the EU, the EU authorized representative is:

Springer Nature Customer Service Center GmbH
Europaplatz 3
69115 Heidelberg, Germany

Batch number: 10837688

Printed by Printforce, the Netherlands